HEAT TRANSFER

HEAT TRANSFER, VOLUME I
by Max Jakob

HEAT TRANSFER, VOLUME II
by Max Jakob
In preparation

ELEMENTS OF HEAT TRANSFER AND INSULATION
by Max Jakob and G. A. Hawkins

HEAT TRANSFER

BY

Max Jakob

RESEARCH PROFESSOR OF MECHANICAL ENGINEERING
ILLINOIS INSTITUTE OF TECHNOLOGY

CONSULTANT IN HEAT RESEARCH
ARMOUR RESEARCH FOUNDATION

NON-RESIDENT RESEARCH PROFESSOR
OF HEAT TRANSFER, PURDUE UNIVERSITY

Volume I

New York · JOHN WILEY & SONS, *Inc.*

London CHAPMAN & HALL, *Limited*

PRINTED IN THE UNITED STATES OF AMERICA

PREFACE

For many years the author's main activity consisted in following up and analyzing the literature and doing research in heat transfer. During the past twelve years he has also had the opportunity to give elementary, intermediate, and graduate courses in heat transfer at Illinois Institute of Technology. The present book is a comprehensive product of such work and may, therefore, serve both as a treatise and as a textbook.

Its main purpose is to show the development of the ideas which led to the present state of our knowledge in the branch of science and engineering called heat transfer. This has required a somewhat historical treatment of the material.

The student of the progress in this field will find that the foundation of the theory of thermal conduction was laid in France more than 125 years ago. The basic equations of flow and similarity which play such a dominant role in heat convection were derived by English scientists of the nineteenth century. Their application to heat convection, however, was first demonstrated in Germany in the first decade of the twentieth century, following the discovery of the laws of heat radiation in the second half of the last century.

Starting early in our century, British and American engineers and physicists contributed, at an ever-increasing rate, to the collection of experimental data and to the application and improvement of the principles and theories which, as in some other fields, had first been suggested in continental Europe. Today, most work in the field of heat transfer is carried out and the results are published in the English-speaking countries, with the gravitational center of such work and literature in the United States.

Because of these facts the author believed that only a rather international survey would cover the high lights in the path of development. In particular, it seemed to be indicated to allow ample space for the German literature of the 25 years before Hitler. Since, obviously, German science has doomed itself for a long time to come, knowledge of the German language among students will decrease accordingly, and the earlier basic literature will not be accessible. The main contents of this literature, including the most important derivations, have been presented in such detail by the author that study of the original papers will be necessary only in exceptional cases.

Somewhat differently from other books on heat transfer, this book does not attempt to deal, in three separate parts, with conduction, convection, and radiation, because the peculiarities and interrelation of these three modes become more evident in parallel treatment. Therefore the first part of the book contains the basic equations, mostly differential equations, of all three branches of heat transfer; the second part deals with the most characteristic properties occurring in each of these branches; and only then are the solutions of the equations, with all their consequences, shown in the conventional separation into special parts for each of the three kinds of heat flow. Each of these parts contains straight solutions, similarity treatments, graphical and experimental methods, etc. Finally the combination of the different modes will be shown for selected topics of practical application, such as thermometry and heat exchangers. This part and the preceding special part on radiation will be presented in a second volume in order to keep the first volume at a feasible size.

Concerning the method of presentation of the material, the fact had to be considered that the book, like the courses of the author, should serve the needs of mechanical, chemical, electrical, and civil engineers and physicists alike. Therefore a middle road was chosen between the highly mathematical procedure in some books and the presentation of empirical facts in others.

A rather modest knowledge of mathematics will be needed for the understanding of this book. Therefore it may be used not only for graduate but also for intermediate courses and for the private study of those interested in the field. The non-mathematician working on a problem of heat transfer should be able to set up the pertinent differential equations with suitable boundary conditions and to decide what features are essential and which details may be neglected to keep or make the equation solvable or at least usable without solution. Such decision is his main job and cannot be left to the mathematician, who will generally not be prepared to judge the physical side of the problem. The solution of a difficult differential equation, on the other hand, can be left to the mathematician if necessary. For this reason derivations of solutions are only exceptionally shown in this book. In general it is considered sufficient to present an available solution and to show by differentiation that the solution is in agreement with the general differential equation under consideration. For the benefit of readers with little mathematical experience, the mathematical procedure is explained in considerable detail in the first chapters and parts so that these users of the book may become acquainted with such typical features as finding and substituting suitable boundary conditions. Special items, such as the methods of finite differences, conformal mapping, fictitious heat sources, and operational calculus, are explained in an elementary manner.

As in the use of mathematics, some restriction was observed in the presentation of empirical material. As coworker of Landolt-Boernstein's tables of physical and chemical properties, the author collected abundant data on the thermal conductivity of all kinds of substances. However, it was believed that the present book should not be overloaded with this or any other empirical material, but that such data should be used only for illustration, for verification, or for showing the deficiencies of the theoretical analysis. For this reason the empirical material has been distributed over the text and presented wherever the analysis required it.

On the other hand, a considerable part of the book is devoted to the understanding of such physical peculiarities of matter as play a role in heat transfer. This is a topic neglected in most books on heat flow. Tabulation of the numerical values of these properties, though useful for application, does not help the student to comprehend the enormous differences occurring. In this book attention is directed to the influences of chemical composition, state of phase, physical structure, quality of surface, temperature, etc., which are reflected in the faculty of matter to promote or reduce the flow of heat.

More space than in most other books on heat transfer has also been devoted to experimental methods and techniques in order to give at least an idea of some of the various procedures of determining thermal conductivities, characteristics of convection, temperature distributions, and paths of heat flow.

The formal organization of the contents of the book is self-explanatory. Sections, equations, tables, figures, and problems are identified by numbers running through each chapter and added to the chapter number. By strictly logical disposition and fine subdivision of the contents into sections, the table of contents expedites the location of any major subject. In order to facilitate the finding of numerical data, the tables and those figures which contain them are indicated in two lists following the table of contents.

Problems to be solved by the student are not given at the end of each chapter but are assembled in Appendix A.

Extreme care has been taken in the choice and consistent use of symbols, particularly of subscripts and special signs. A list of the symbols most often needed, including some important explanations and recommendations, is given as Appendix B. Almost without exception the symbols adopted by the American Standards Association have been used.

The physical dimensions are generally expressed in units of the British Gravitational System. Wherever other systems are employed this fact is clearly indicated. Use of different systems is facilitated by a brief table of conversion factors in Appendix C.

A remark may be in order regarding the symbols for the basic units of heat energy and temperature which occur so frequently in the book. The author uses the single letter B for the British thermal unit and F for the Fahrenheit degree which is identical with the Rankine degree when employed as a unit. Similarly, C represents the centigrade and Kelvin degree alike, since these are also identical units. The signs °F, °R, °C, and °K are reserved for indicating points on the Fahrenheit, Rankine, centigrade, and Kelvin scales, the same temperature having four different values in these temperature scales. During a number of years the author's experience has been that distinguishing between symbols for temperature units and positions on temperature scales is in the interest of clarity and simplification.

For citation of literature, the name of the author and the year of publication (with the addition of *a*, *b*, etc., for more than one publication by the same author in the same year) are indicated in the text as references to the bibliography (Appendix D), and, vice versa, the pages of the text where the references occur are indicated in each entry of the bibliography. Here the author wishes to apologize for having cited so many of his own publications, being well aware of the fact that their number does not mean much about their significance. However, in a treatise, as well as in a textbook, the interest of the student is animated if he sees that the author deals with fields of his own activities, with which he supposedly is most familiar. It is further understandable that those subjects which an author has personally treated will seem to him the ones best fit for discussion in detail and, therefore, will preferably be cited; otherwise he would probably not have chosen them for his research. More important publications of other research workers may have been omitted or not dealt with in detail. No attempt was made to approach completeness. However, the author seriously endeavoured to give credit to all pertinent publications used and, in particular, to the first communication about any subject under consideration. Moreover, the names of coauthors of any publication are also separately cited, so that the bibliography includes a complete alphabetical list of the authors mentioned in the text. This facilitates locating a publication in case a reader remembers a coauthor's name only.

Most topics can easily be located by means of the detailed lists of contents, tables, and illustrations preceding the text. Therefore, the subject index (Appendix E) was kept relatively brief; special rather than more general items should be looked up in this index.

Any suggestions regarding subject matter and literature references, the addition of which seems to be desirable, will be highly appreciated. It may be possible to incorporate some of them in the second volume of this book.

Originals of photos and charts have been placed at the author's disposal by several colleagues in his field of work. Grateful acknowledgment of such cooperation is made at the places where this material is used.

Finally, the author wishes to express his thanks to Dr. Madan L. Ghai and Mr. Stothe P. Kezios for their valuable assistance in the preparation of Sects. 17–5, 6, and 7.

MAX JAKOB

Chicago, Illinois
February 1949

CONTENTS

PART B. PROPERTIES OF MATTER SIGNIFICANT FOR HEAT TRANSFER

PART C. HEAT CONDUCTION IN SIMPLE BODIES

LIST OF TABLES

SELECTIVE LIST OF FIGURES

NOTE. Only figures containing quantitative information about a topic are listed.

CHAPTER 1

Introduction

1-1. The Three Kinds of Heat Transfer

Heat energy originates from other kinds of energy according to the first law of thermodynamics; transmission of heat energy is due to the peculiar property of matter called temperature, and to the second law of thermodynamics, which claims that free migration of heat energy is always in the direction from a body at higher temperature to a body at lower temperature.

Because sources of energy exist all over the universe, heat energy is produced all over the world, and the exchange of heat energy is a phenomenon almost as universal as gravity.

But whereas the main laws of thermodynamics are of similar simplicity as the laws of mechanics, transmission of heat is much more complicated than the mechanical transmission of matter, which is governed by the laws of mechanics only. Properly, the science of heat transfer cannot be considered a single branch of science, but may be compared with a system of branches, belonging to trees of different classes, entangled with each other and even grown together.

As usual, we shall distinguish three modes of heat transmission, known as heat conduction, heat convection, and heat radiation. The storage of heat energy also plays an important role. Before entering the specific discussion of the three kinds of heat exchange, their peculiarities may be characterized in a general way.

Heat conduction, considered from an atomistic point of view, is due to the elastic impacts of molecules in gases, to longitudinal oscillations in solid non-conductors of electricity, and to the motion of electrons in metals. From a phenomenological point of view, it means the exchange of heat between contiguous bodies or parts of a body which are at different temperatures. The basic law of heat conduction is in complete analogy to the law of electrical conduction.

Heat convection is the transportation and exchange of heat, due to the mixing motion of different parts of a fluid. It is governed by the laws of fluid dynamics in combination with the law of heat conduction.

Heat radiation is identical with light radiation and belongs therefore to optics, in its geometry as well as in its dynamics. Speaking of "heat" radiation means only that the main interest is focused on the influence of temperature upon that kind of energy transportation through a transparent medium or the empty space which is called radiation.

Storage of heat energy is the change of enthalpy (heat contents) of a medium in the path of heat transmission. It occurs when the temperatures involved are changing with time, and it depends on the heat capacity of the medium.

Heat transfer in its threefold manifestation, and in its combination with the storage of heat energy, deals with almost all branches of physics: mechanics, heat, acoustics, optics, and electricity. The development and application of the science of heat transfer are further based on mathematical analysis, and it may be added that both physics and mathematics have not only served in, but have been greatly enriched by, the studies of heat transfer.

Although generally the three kinds of heat transmission occur together, entangled as mentioned before, fortunately one or the other often prevails in practical cases. Therefore, separate laws, governing each kind of heat transfer, have been developed and may be used in such cases. Superposition of these laws is also often possible and used.

The basic law * of heat conduction is

$$q = k \frac{A}{L} \Delta t \tag{1-1}$$

It originates from Biot [1804, 1816] but is generally called Fourier's equation, because Fourier [1822] used it as a fundamental equation in his analytic theory of heat.†

For heat convection the following equation was first recommended by Newton [1701]:

$$q = hA \cdot \Delta t \tag{1-2}$$

It is often called "Newton's cooling law." However, it is rather a definition of h than a law as will be seen later.

* Explanation of the symbols is given after Eq. 1-3.

† Fourier's papers on the subject go back to 1811, but publication was delayed for many years.

For the total radiation, equations similar to

$$q = \sigma A T^4 \tag{1–3}$$

have been used, since Stefan [1879] found this relation and Boltzmann [1884] proved it theoretically for a perfectly black surface.

In all these equations q denotes the time rate of heat flow, i.e., the heat energy flowing through a constant area A in unit time, and q is assumed to be unvariable in time.‡

Equation 1–1 relates to a plane plate of the thickness L with a perfectly insulated edge, the two free surfaces being held at the temperature difference Δt. The factor k, which for the time being may be considered a constant, is called thermal conductivity.

Equation 1–2 relates to the heat transfer between a surface and a fluid in contact with it, their temperature difference being Δt. The factor h is called surface coefficient of heat transfer or film coefficient of heat transfer or, briefly, coefficient of heat transfer.§

Equation 1–3 relates to the emission of radiation from a surface at the absolute temperature T. The factor σ is a natural constant and is known as the constant of Stefan-Boltzmann's law, or the constant of total black radiation, as the total radiation of a perfectly black surface is often called. For surfaces not absolutely black σ is to be replaced by another factor.

The basic equation of heat storage is

$$q = \rho c_p V \frac{\Delta t}{\Delta \tau} \tag{1–4}$$

where q is the heat energy stored in unit time, in the volume V of a medium of density ρ and specific heat c_p when the temperature increases by Δt in a time interval $\Delta \tau$.

1–2. The Place of Heat Transfer in the System of Sciences

The particular place of heat transfer in the system of sciences can be fixed by considering the physical units occurring in some branches of science.

Proceeding from the less to the more complicated cases, one may begin with that most simple science, which operates without dimensions, or with lengths only, *mathematics*. In this science, *algebra* considers

‡ Often the time rates of heat flow are simply called heat conduction, convection, and radiation. Therefore these words, ending with the syllable "tion" have a double meaning, either the one just mentioned or that of a particular sort of heat exchange.

§ Some writers call it unit conductance.

relations between numbers only, using no dimensions at all, whereas *geometry* deals with relations between lines and angles, thus using length dimensions only. *Kinematics*, dealing with relations between lines, angles, and time, and so operating with length and time units, has its place between mathematics and *mechanics*. The latter is split into two branches: *statics*, dealing with relations between geometrical configurations and forces, thus using length and force units, and *dynamics*, devoted to the relations between geometrical configurations, time, and forces (or masses), with the dimensions of length, time, and force (or mass).

In the science of *heat*, a new category appears which has not been recognized as a particular dimension by some classical physicists and actually is an outsider in the system of dimensions, the temperature. By proper definition even the electrical and magnetical quantities could be expressed in length-mass-time units. Thus, for instance, 1 ohm = 10^9 cm/sec. The representation of temperature in units of length, time, and force (or mass) was possible only in the so-called *kinetics of heat*. This really is a dynamics of molecules and atoms, which adds the atomistic explanation of temperature to that of pressure, so that no new dimension for the definition of temperature is needed. In those branches of science, however, in which the phenomenological point of view prevails, namely, in *thermodynamics* and in *heat transfer*, temperature could not be superseded by the three basic units. The concept of heat energy, on the other hand, did not present particular difficulties, because according to the first law of thermodynamics heat energy has the same dimension as mechanical energy.

Thermodynamics considers the temperature phenomenon as it appears, i.e., phenomenologically, and connects it with the phenomena of volume, pressure, and mass. The dimensions used are length, mass (or force), and temperature. It is remarkable that the time unit is not used. At least it does not occur in the three basic laws of thermodynamics. This seems strange at the first glance, because thermodynamics deals with devices, such as engines, where time apparently plays an important role. But in those devices time occurs only as far as mechanical dynamics and friction or heat storage and exchange are considered. From the purely thermodynamical viewpoint it is irrelevant whether a process or cycle is performed fast or slowly.

In *heat transfer* and *chemical kinetics*, finally, the tendency of different bodies to come to the same temperature is essential, and because each case of heat exchange is a process of equalization against resistances, time steps in. This is the item which links Eqs. 1–1, 2, 3, and 4, and at the same time distinguishes them from thermodynamical equations. It would be possible, however, to include the processes of equalization into thermodynamics. Then the science of heat transfer would be a branch of non-equilibrium thermodynamics. In any case the dimensions needed are length, mass (or force), time, and temperature.

Thus, of all branches of science quoted in this short systematology, heat transfer and chemical kinetics are the only ones which need as many as four basic units. Connecting this fact with the one explained in Sect. 1–1, that these four dimensions can be combined in three modes of heat transfer, which are different in principle, it can be foreseen that heat transfer is a rather complex section of science.

PART A. BASIC EQUATIONS OF HEAT TRANSFER

CHAPTER 2

The Basic Equation of Heat Conduction

2-1. Separation of Heat Conduction from Convection and Radiation

As mentioned before, one or the other of the three kinds of heat transmission often prevails, so that it is possible to deal with each of them separately. Beginning with heat conduction, it may first be discussed how far this actually can be separated from convection and radiation.

Heat conduction is the exchange of heat energy between warmer and colder bodies which are in direct contact. Theoretically this exchange is considered as a transfer of kinetic energy of the molecules of the warmer body to those of the colder body, but its mechanism is well known only for gases.

In solids intermolecular radiation may play a part. Furthermore, it is difficult or impossible to separate from conduction the radiation through the pores of a solid body. Therefore, the internal radiation is generally included in the definition of thermal conductivity. Later on it will be seen more in detail that the radiation through small pores is almost proportional to the temperature drop, exactly as the heat flow by conduction.

Dealing with fluids, it must further be considered that heat convection often interferes with heat conduction. Heat convection without conduction does not occur, but conduction without convection may. For convection brings warmer parts of a fluid into contact with colder ones, causing conduction. Conduction without convection, on the other hand, can occur when warmer and lighter portions of a fluid are placed above cooler and heavier ones. Convection is also avoided sometimes by reducing the pressure of a gas so much that the buoyancy can no longer overcome the gas friction. This will also be dealt with more in detail later on. For the time being it may have been made sufficiently evident that conduction can virtually be isolated from radiation and convection.

2–2. The Physical Dimension of Thermal Conductivity

Introducing a new physical quantity necessitates first of all finding its physical dimension, expressed in terms of a consistent system of units. This will now be done with the thermal conductivity k, which has been defined by Eq. 1–1. The simplest method of determining the dimension of one quantity in an equation from the known dimensions of the other quantities occurring in the equation is to assume that each quantity has the numerical value 1, so that only the product of its units remains. This may be expressed by writing the units in brackets.

Thus, using the physical system of units, with the gram-calorie (cal) as heat unit, Eq. 1–1 becomes *

$$\frac{[\mathrm{cal}]}{[\mathrm{sec}]} = [k][\mathrm{cm}^2]\frac{[\mathrm{C}]}{[\mathrm{cm}]}$$

where $[k]$ denotes the dimension of k. Hence

$$[k] = [\mathrm{cal\ sec^{-1}\ cm^{-1}\ C^{-1}}] \tag{2–1}$$

In British technical units analogously

$$\frac{[\mathrm{B}]}{[\mathrm{hr}]} = [k][\mathrm{ft}^2]\frac{[\mathrm{F}]}{[\mathrm{ft}]}$$

and

$$[k] = [\mathrm{B\ hr^{-1}\ ft^{-1}\ F^{-1}}] \tag{2–2}$$

In metric technical units

$$\frac{[\mathrm{kcal}]}{[\mathrm{hr}]} = [k][\mathrm{m}^2]\frac{[\mathrm{C}]}{[\mathrm{m}]}$$

where kcal denotes kilocalorie, which is equal to 1000 cal. Thus

$$[k] = [\mathrm{kcal\ hr^{-1}\ m^{-1}\ C^{-1}}] \tag{2–3}$$

Finally in the physical system with the joule (equal to the watt-second) as heat unit

$$\frac{[\mathrm{joule}]}{[\mathrm{sec}]} = [\mathrm{watt}] = [k][\mathrm{cm}^2]\frac{[\mathrm{C}]}{[\mathrm{cm}]}$$

or

$$[k] = [\mathrm{joule\ sec^{-1}\ cm^{-1}\ C^{-1}}] = [\mathrm{watt\ cm^{-1}\ C^{-1}}] \tag{2–4}$$

Some factors for conversion of numerical values from one system of units to another are given in Appendix C.

* Concerning the symbols for temperature degrees, see Appendix B, Note 6.

2-3. Analogy between Thermal and Electrical Resistances

As mentioned already, a perfect analogy exists between heat conduction and electrical conduction. This will be dealt with in several sections of this book. In particular, a close relationship between the two kinds of conductivity will be demonstrated. At present only a formal relation will be used which exists between Ohm's law

$$i = k_e \frac{A}{L} \Delta e = \frac{\Delta e}{R_e} \tag{2-5}$$

and the law of heat conduction

$$q = k \frac{A}{L} \Delta t = \frac{\Delta t}{R} \tag{2-6}$$

The symbol i in Eq. 2-5 denotes the electrical current, k_e the electrical conductivity, Δe the electrical potential difference (electromotive force), and R_e the electrical resistance. Equation 2-6 defines a thermal resistance

$$R = \frac{L}{kA} \tag{2-7}$$

corresponding to the electrical resistance. With $[k]$ according to Eq. 2-4, q appears in watts, whereas i is given in amperes, if the usual electrical units are employed.

Using the resistance instead of its reciprocal value, the conductance, presents the same advantage for bodies connected in series as in electrical circuits: the resistances can be added to each other. Considering, for instance, a wall composed of two layers, 12 and 23, having the same area A (see Fig. 2-1), the sum of the resistances is

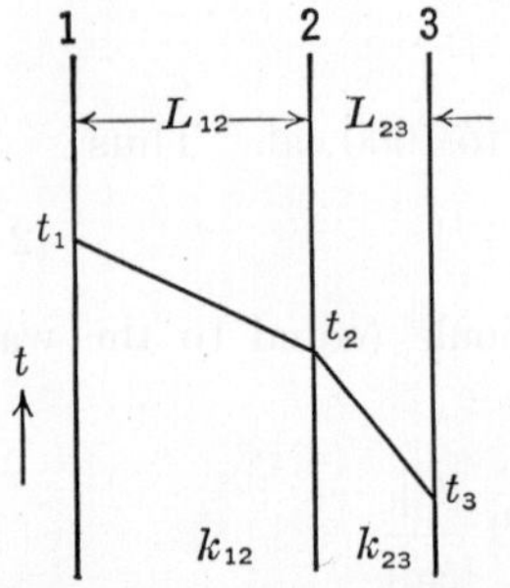

k_{12} and k_{23} are thermal conductivities

FIG. 2-1. Temperature distribution in a composite wall.

$$R_{13} = R_{12} + R_{23} = \frac{1}{A}\left(\frac{L_{12}}{k_{12}} + \frac{L_{23}}{k_{23}}\right)$$

and the rate of heat flow becomes

$$q = \frac{t_1 - t_3}{R_{13}}$$

The temperature does not jump at the plane 2, but has a bend there. The slopes of the lines $\overline{t_1t_2}$ and $\overline{t_2t_3}$ are reciprocal to the conductivities k_{12} and k_{23}.

Likewise in analogy to the laws of electric resistance, a system consisting of two thermal resistances R_a and R_b in parallel has the resultant resistance

$$R = \frac{R_a R_b}{R_a + R_b}$$

2–4. Derivation of the Differential Equation of Heat Conduction in a Substance at Rest

Equation 1–1 can also be written in the differential form

$$dQ = -kA \frac{dt}{dx} d\tau \tag{2-8}$$

where dQ is the heat conducted in the direction x during the time interval $d\tau$.

The minus sign follows from the second law of thermodynamics, according to which heat automatically always flows from points of higher temperature to points of lower temperature. Thus, if dx is taken positive in the direction of a positive heat flow, the temperature difference dt must be negative, and these conditions are consistent with each other only if the minus sign is used.

In the following derivation of the differential equation of heat conduction in a non-steady state, the existence of heat sources or sinks is provided, but movements of parts of the body, as, for instance, occur in the flow of a fluid, will not be considered yet. The thermal conductivity will first be assumed to be constant.

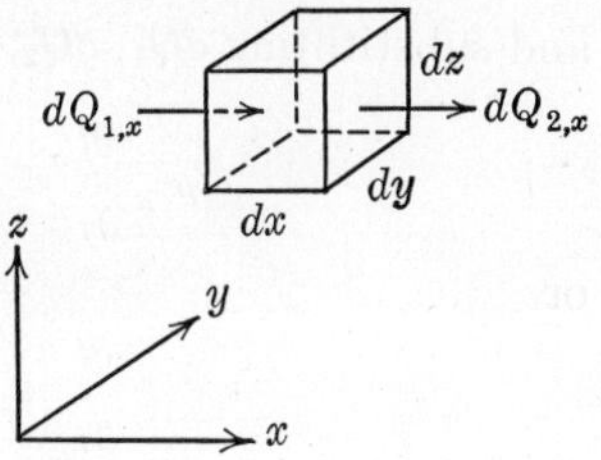

FIG. 2–2. Heat conduction through a parallelepiped.

For a parallelepipedon of differential size (Fig. 2–2) a heat balance will be set up which is valid in the differential of time $d\tau$. The heat entering from the left will be

$$dQ_{1,x} = -k(dy \cdot dz) \frac{\partial t}{\partial x} d\tau$$

and the heat leaving at the right side will be

$$\begin{aligned} dQ_{2,x} &= -k(dy \cdot dz) \frac{\partial}{\partial x}\left(t + \frac{\partial t}{\partial x} dx\right) d\tau \\ &= -k(dy \cdot dz)\left(\frac{\partial t}{\partial x} + \frac{\partial^2 t}{\partial x^2} dx\right) d\tau \end{aligned}$$

Similar equations exist for the heat quantities $dQ_{1,y}$, $dQ_{2,y}$, $dQ_{1,z}$, $dQ_{2,z}$, which are conducted in the directions y and z.

The total heat energy entering the small body in the time $d\tau$ may be

$$dQ_1 = dQ_{1,x} + dQ_{1,y} + dQ_{1,z}$$

and the heat energy going out may be

$$dQ_2 = dQ_{2,x} + dQ_{2,y} + dQ_{2,z}$$

The heat energy stored in the differential body will be

$$dQ_3 = \rho c_p(dx \cdot dy \cdot dz) \frac{\partial t}{\partial \tau} d\tau$$

where ρ is the density and c_p is the specific heat of the substance. Finally, heat energy may be developed in the differential body in the amount of

$$dQ_4 = q'''(dx \cdot dy \cdot dz)\, d\tau$$

where q''' is the heat energy developed in unit volume and time.

Using now the law of the conservation of energy,

$$dQ_1 + dQ_4 = dQ_2 + dQ_3$$

and substituting dQ_1, dQ_2, dQ_3, and dQ_4 from the equations above,

$$\rho c_p \frac{\partial t}{\partial \tau} = k\left(\frac{\partial^2 t}{\partial x^2} + \frac{\partial^2 t}{\partial y^2} + \frac{\partial^2 t}{\partial z^2}\right) + q''' \tag{2-9}$$

or

$$\frac{\partial t}{\partial \tau} = \alpha\left(\frac{\partial^2 t}{\partial x^2} + \frac{\partial^2 t}{\partial y^2} + \frac{\partial^2 t}{\partial z^2}\right) + \frac{q'''}{\rho c_p} \tag{2-10}$$

where a new quantity has been introduced, the "thermal diffusivity" †

$$\alpha = \frac{k}{\rho c_p} \tag{2-11}$$

According to its derivation, Eq. 2-9 expresses the fact that the sum of the heat energies entering a body and developed in it is equal to the sum of the heat energies stored in and leaving the body. Considering its mathematical form, the equation links the change in time of the heat contents with the change in space of the heat flow.

The thermal diffusivity α is a property of the substance in which heat is transferred and stored, as is the thermal conductivity k. Since, according to Eq. 2-11, α is the ratio of a quantity which is proportional

† This name was suggested by Sir William Thomson (Lord Kelvin) [1880, 1884].

to the heat conduction (numerator) to a quantity which is proportional to the heat storage (denominator), heat units are cancelling in the dimension of α, which becomes [cm^2/sec] in the physical system, [ft^2/hr] in the British technical system, and [m^2/hr] in the metric system of units.

If the medium through which the heat is conducted is inhomogeneous, then the thermal conductivity may have different values k_x, k_y, and k_z in the directions x, y, and z. When the derivation which has led to Eq. 2–9 is modified, the following more general equation is obtained:

$$\rho c_p \frac{\partial t}{\partial \tau} = k_x \frac{\partial^2 t}{\partial x^2} + k_y \frac{\partial^2 t}{\partial y^2} + k_z \frac{\partial^2 t}{\partial z^2} + q''' \qquad (2\text{–}12)$$

If the thermal conductivity is a function of the temperature, the first term on the right becomes $\frac{\partial}{\partial x}\left(k_x \frac{\partial t}{\partial x}\right)$, and the two following terms take an analogous form. In this way Eq. 2–12 generalizes to

$$\rho c_p \frac{\partial t}{\partial \tau} = \frac{\partial}{\partial x}\left(k_x \frac{\partial t}{\partial x}\right) + \frac{\partial}{\partial y}\left(k_y \frac{\partial t}{\partial y}\right) + \frac{\partial}{\partial z}\left(k_z \frac{\partial t}{\partial z}\right) + q''' \qquad (2\text{–}13)$$

Solutions of the partial differential equation of heat conduction will be dealt with in Part C (Chapters 8 and 10 to 15).

CHAPTER 3

The Basic Equations of Heat Convection

3–1. A Difficulty in Principle concerning the Coefficient of Heat Transfer

The fundamental equation for the heat transfer between a surface at the temperature t_s and a fluid in contact with it at the temperature t is

$$q = hA \cdot \Delta t = hA(t_s - t) \tag{3–1}$$

which is identical with Eq. 1–2. The coefficient of heat transfer h, defined by this equation, has the dimension [B hr^{-1} ft^{-2} F^{-1}] in British technical units, and [kcal hr^{-1} m^{-2} C^{-1}] in metric technical units.

Equation 3–1 seems magically simple, but this apparent simplicity was a grave handicap to progress in the field of heat convection. For a long time it has been assumed that the factor h is a constant. This is not true in any way.

The complication begins with a difficulty in principle in the definition of the temperature t to be used in Eq. 3–1. The temperature of a fluid in contact with a surface at a different temperature changes its value with the distance from the wall. What temperature then is to be put into the equation? For an extended fluid like air in contact with a vertical wall it seems best to take the temperature far from the wall, theoretically at infinite distance. For a defined cross section, like that of a tube through which a fluid passes, the temperature in the axis or the mean temperature could be taken. But what is the mean temperature?

One might average the temperatures found by a traverse through the cross section of the fluid. Another procedure would be to integrate the amounts of heat energy carried through each element of the cross-sectional area and to divide the integral through the rate of total mass flow. The two methods can give quite different results, as will be explained in detail later on. This uncertainty may have first revealed the fact that heat transfer by convection actually is not an independent phenomenon that needs a special and new formula, defining it like Eq. 1–2, but that it is a case of heat conduction, complicated by the flow of the fluid and governed by the laws belonging to both processes. The factor h, then, is not a constant, but a function of many variables which are specific for the flow and heat conduction of a fluid. Theoretically it must be possible to avoid Eq. 1–2 or 3–1 entirely, and to develop any problem of heat transfer without using the coefficient of heat transfer h. This could be done by bringing Eq. 3–1 into the form

$$q = \frac{k}{L} A(t_s - t) \tag{3-2}$$

where k is the thermal conductivity of the fluid, and L the thickness of a layer of it in contact with the wall in which the temperature would change linearly with the distance from the surface. The transformation consists in replacing h by k/L. But this procedure fails practically because the temperature does not change as in Fig. 3–1, but as in Fig. 3–2, i.e., mainly in a thin layer close to the wall, where the fluid is at rest or moves only slightly. Hence the thickness L is not well defined.

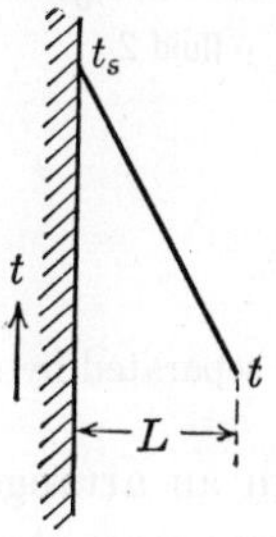

FIG. 3–1. Incorrect sketch of temperature drop on a vertical surface.

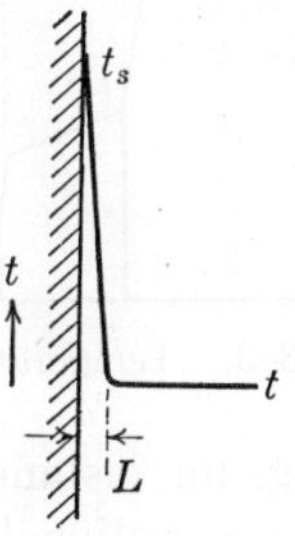

FIG. 3–2. Correct sketch of temperature drop on a vertical surface.

However, even if an approximative value for L could be introduced in the equations, it would be practically impossible to measure t in the distance L from the surface as given in Fig. 3–2, because this distance actually is much too small. Thus, although Eq. 3–2 will be used frequently, the main question in the theory and practice of heat transfer by convection is to determine the function h.

3–2. The Heat Resistance at a Surface and the Overall Resistance of Solids and Fluids in Series

Equation 3–2 can be used for applying the definition of a heat resistance (Sect. 2–3) to a fluid in contact with a solid. According to Eqs.

2-6 and 7, the heat resistance will be

$$R = \frac{t_s - t}{q} = \frac{L}{kA} = \frac{1}{hA} \tag{3-3}$$

This, like Eq. 3-2, means that the resistance against heat transfer by convection can be represented by the heat resistance of a layer of the fluid having the thickness L.

Operating with heat resistances, the heat transmission through solid bodies and fluids in series can be dealt with in a very simple manner.

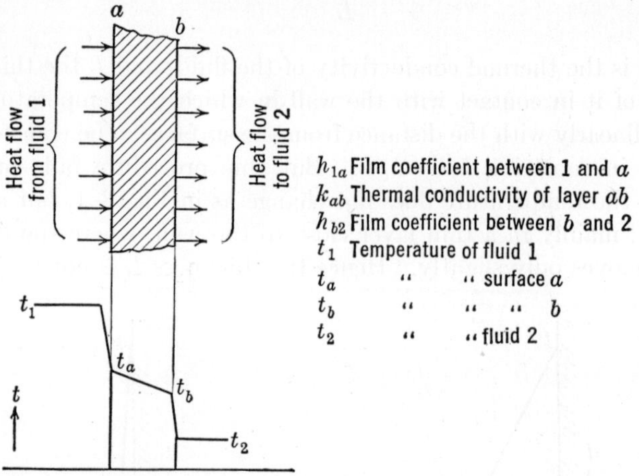

FIG. 3-3. Temperature drop between two fluids separated by a wall.

Considering, for instance, the heat flow through an arrangement with constant cross-sectional area A (see Fig. 3-3) one may start from the equation

$$q = UA(t_1 - t_2) \tag{3-4}$$

by which an "overall coefficient of heat transfer" U is defined. Correspondingly, the overall resistance would be

$$R = \frac{t_1 - t_2}{q} = \frac{1}{UA} \tag{3-5}$$

It will be proved that the overall resistance of the arrangement can be expressed by

$$R = R_{1a} + R_{ab} + R_{b2} \tag{3-6}$$

where the double subscripts relate to the layers in Fig. 3-3 with the same letters.

Substituting the thermal conductivity and coefficients of heat transfer, according to Eqs. 2–7 and 3–3,

$$\frac{1}{UA} = \frac{1}{h_{1a}A} + \frac{L}{k_{ab}A} + \frac{1}{h_{b2}A}$$

or

$$\frac{1}{U} = \frac{1}{h_{1a}} + \frac{L}{k_{ab}} + \frac{1}{h_{b2}} \tag{3–7}$$

or, according to Eqs. 2–6, 3–3, and 3–5,

$$\frac{t_1 - t_2}{q} = \frac{t_1 - t_a}{q} + \frac{t_a - t_b}{q} + \frac{t_b - t_2}{q} \tag{3–8}$$

Since Eq. 3–8 is algebraically right, Eq. 3–6 must also be right.

Generalizing this result to any series Σ of resistances R_h between plane walls and fluids, and R_k of the walls,

$$R = \Sigma R_h + \Sigma R_k \tag{3–9}$$

and

$$\frac{1}{U} = \Sigma\frac{1}{h} + \Sigma\frac{L}{k} \tag{3–10}$$

This includes plates of different thermal conductivities connected in series, i.e., composed walls.

3–3. Interdependence of Heat Convection and Conduction

These equations were almost the only scientific tools which, until the beginning of this century, were used by engineers for the investigation of heat convection. The result was a rough and inconsistent knowledge of the heat exchange between solid walls and fluids; that is, one recognized sometimes *what* happened, but never *why* it happened. This was neither satisfactory nor sufficient.

The first step to a more intimate knowledge of the process of heat transfer by convection consisted in recognizing the interdependence of heat convection and conduction.

Between two solid bodies in perfect contact, heat is transmitted without a discontinuity of temperature. If temperature jumps are observed, they are due to imperfect contact. In this case the heat flow must cross a gas layer which, though very thin, may have a considerable heat resistance owing to the extremely small thermal conductivity of gases.

In fast-flowing liquids and gases the velocities and temperatures are equalized by turbulent motion. Therefore only small differences of velocity and temperature are observed, if one proceeds perpendicularly to the direction of the main flow. If the gradients remained small up to the wall, a jump of the velocity value down to zero and a jump of the temperature value to that of the surface would occur. But by exact measurements no such unsteadiness was found. Today it is known that in a very thin boundary layer the velocity decreases steeply but steadily from that of the bulk of the fluid to zero at the wall. Likewise the temperature gradient on the wall is large, but not infinite. Only at very small gas pressure do jumps in velocity and temperature occur on the surface, owing to sliding of the fluid, as will be shown later on.

The continuity described indicates that the heat transfer between solid walls and fluid (except that by radiation) is governed only by the laws of the flow of fluids and heat conduction. Properly, heat convection is a combination of mechanical and thermal flow, namely, the transportation of heat energy by relatively large masses of the fluid, in which the molecular transportation called heat conduction is included. This means that the fluid carries a product of specific weight, specific heat, and temperature like a load to a place of lower temperature, and from there it returns with a lighter load; but at the same time another sort of heat exchange exists, dependent on thermal conductivity instead of heat capacity, and independent of the mass flow.

The difficulties of all problems of heat transfer by convection are due to this combination of mass flow and heat conduction, both being governed by rather complicated laws, which will be dealt with in the following sections. Whereas for heat conduction, a single, relatively simple partial differential equation has been found, the following analysis will lead to a system of several differential equations. As usual and practical, first the flow of a fluid and then the heat conduction in a flowing fluid will be considered. The equations of both these processes will only partly be derived. They will, however, be presented in their most general form.

3-4. The Differential Equations of the Flow of Fluids

The flow of fluids, like any dynamical process, is governed by Newton's fundamental equation:

$$\text{Force} = \text{Mass} \times \text{Acceleration}$$

In order to obtain an expression for the acceleration the change of the velocity v_x of a fluid in a path element dx and in the time interval $d\tau$ will be considered. Mathematically this change can be represented by the total differential

$$dv_x = \frac{\partial v_x}{\partial \tau} d\tau + \frac{\partial v_x}{\partial x} dx \tag{3-11}$$

The first term on the right represents the change of velocity at a fixed place x in the time interval $d\tau$, and the second term gives the difference of velocity at the time τ between two points in the fluid, distant by dx.

Dividing by $d\tau$ yields the acceleration

$$\frac{dv_x}{d\tau} = \frac{\partial v_x}{\partial \tau} + \frac{\partial v_x}{\partial x} \cdot \frac{dx}{d\tau} = \frac{\partial v_x}{\partial \tau} + v_x \frac{\partial v_x}{\partial x} \tag{3–12}$$

In this equation, $\partial v_x/\partial \tau$ originates from a change of the state of flow at the place x, and is called local acceleration. The term $v_x \dfrac{\partial v_x}{\partial x}$ represents a part of the acceleration which is due to the fact that the particle under consideration comes from a neighboring place, where it had a different velocity. This is called convective acceleration. If velocity components v_y and v_z in the directions y and z also exist, local accelerations in these directions must be considered in the same way, and Eq. 3–12 generalizes to

$$\begin{aligned}\frac{dv_x}{d\tau} &= \frac{\partial v_x}{\partial \tau} + \frac{\partial v_x}{\partial x} \cdot \frac{dx}{d\tau} + \frac{\partial v_x}{\partial y} \cdot \frac{dy}{d\tau} + \frac{\partial v_x}{\partial z} \cdot \frac{dz}{d\tau} \\ &= \frac{\partial v_x}{\partial \tau} + v_x \frac{\partial v_x}{\partial x} + v_y \frac{\partial v_x}{\partial y} + v_z \frac{\partial v_x}{\partial z}\end{aligned} \tag{3–13}$$

Similar equations exist for $dv_y/d\tau$ and $dv_z/d\tau$. They follow from Eq. 3–13 by cyclic variation of x, y, z, and v_x, v_y, v_z.

The force occurring in Newton's fundamental equation is composed of three different types of force acting on a unit volume of the fluid:

1. Inertia forces G_x, G_y, G_z, as gravity.
2. Dynamical forces P_x, P_y, P_z, as the pressure drop.
3. Frictional forces V_x, V_y, V_z, caused by viscosity and wall friction.

Finally, the mass, related to unit volume, is equal to the density ρ of the fluid. Thus Newton's equation, applied to the motion of unit volume of the fluid in the direction of the three coordinates, respectively, becomes

$$\left.\begin{aligned}\rho \frac{dv_x}{d\tau} &= \rho\left(\frac{\partial v_x}{\partial \tau} + v_x \frac{\partial v_x}{\partial x} + v_y \frac{\partial v_x}{\partial y} + v_z \frac{\partial v_x}{\partial z}\right) = G_x + P_x + V_x \\ \rho \frac{dv_y}{d\tau} &= \rho\left(\frac{\partial v_y}{\partial \tau} + v_x \frac{\partial v_y}{\partial x} + v_y \frac{\partial v_y}{\partial y} + v_z \frac{\partial v_y}{\partial z}\right) = G_y + P_y + V_y \\ \rho \frac{dv_z}{d\tau} &= \rho\left(\frac{\partial v_z}{\partial \tau} + v_x \frac{\partial v_z}{\partial x} + v_y \frac{\partial v_z}{\partial y} + v_z \frac{\partial v_z}{\partial z}\right) = G_z + P_z + V_z\end{aligned}\right\} \tag{3–14}$$

There is no difficulty in determining the forces G and P. In general, the only inertia force which has to be considered is gravity, which may act in the direction z. Consequently G_x and G_y would be zero.

G_z generally occurs as buoyancy, according to Archimedes' law, which states that the unit volume of a fluid of density ρ in a medium of the density ρ_0 is rising under a force

$$G_z = g\rho_0 - g\rho = g\rho\left(\frac{\rho_0}{\rho} - 1\right) \tag{3-15}$$

The difference $\rho_0 - \rho$ in a homogeneous fluid is due to a corresponding temperature difference $T - T_0$.

For any fluids

$$\frac{1}{\rho} = \frac{1}{\rho_0}[1 + \beta(T - T_0)] \tag{3-16}$$

where β is a coefficient of thermal volume expansion defined by this equation. Hence

$$G_z = g\rho\beta(T - T_0) \tag{3-17}$$

For gases, as far as deviations from the ideal state are negligible,

$$\frac{\rho_0}{\rho} = \frac{T}{T_0}$$

Substituting this in Eq. 3-16,

$$\beta = \frac{1}{T_0}$$

so that Eq. 3-17 becomes

$$G_z = g\rho\frac{T - T_0}{T_0} \tag{3-18}$$

The force caused by the pressure drop of a fluid which flows in the direction x apparently is $-(\partial p/\partial x)\,dx$, related to the length dx and the unit of the cross-sectional area perpendicular to x. For unit volume it becomes

$$P_x = -\frac{\partial p}{\partial x} \tag{3-19}$$

It is more difficult to find the frictional forces. Stokes [1845] derived the equation

$$V_x = \mu\left(\frac{\partial^2 v_x}{\partial x^2} + \frac{\partial^2 v_x}{\partial y^2} + \frac{\partial^2 v_x}{\partial z^2}\right) + \frac{\mu}{3} \cdot \frac{\partial}{\partial x}\left(\frac{\partial v_x}{\partial x} + \frac{\partial v_y}{\partial y} + \frac{\partial v_z}{\partial z}\right) \tag{3-20}$$

in which μ is a property of the flowing substance.* Two similar equations for V_y and V_z follow by cyclic variation of x, y, z and of v_x, v_y, v_z.

Equations 3–14, 15, 19, and 20, employed for the three coordinate directions, form the system of Stokes's equations.

For incompressible fluids (liquids) the second term on the right side of Eq. 3–20 becomes zero, as will be proved in Sect. 3–6. For this case the equations had been derived before Stokes and are called Navier's [1823] equations.

If the forces V are zero, the equations turn into the system of Euler's [1754] equations for frictionless flow of fluids.

3–5. Dynamic and Kinematic Viscosity

The property μ, occurring in Eq. 3–20, is defined by the equation

$$S_s = \mu \frac{dv}{dy} \tag{3–21}$$

where v = the velocity of the fluid,
y = a direction perpendicular to that of the flow,
S_s = the shearing stress in a plane perpendicular to y.

This equation has been set up by Newton as a hypothesis. Its concept of μ found experimental evidence when the French physician, Poiseuille, used it in his law of the flow of fluids through capillary tubes. Finally it has been proved in the kinetic theory of gases.

According to Eq. 3–21, μ has the dimensions [dyne cm^{-2} sec] or [g cm^{-1} sec^{-1}] in the physical system. Because the "dynamical" units of force or mass occur in the dimension of μ, Jakob [1928] has suggested calling it dynamic viscosity. The unit of dynamic viscosity in the physical system is called poise, in honor of Poiseuille; one hundredth of this unit, called centipoise, is used most frequently for liquids, because the dynamic viscosity of water at 20°C is almost equal to 1 centipoise.

In the British gravitational system of units the dimension of μ is [lb ft^{-2} hr], the pound being the unit of force; in the metric technical system the dimension of μ is [kg m^{-2} hr] with the kilogram [kg] as a force. For conversion factors see Appendix C. Conversion factors of μ for almost all imaginable combinations have been published by Hawkins, Solberg, and Sibbitt [1941].

Dividing both sides of Eq. 3–14 by ρ, the viscosity terms on the right

* Poisson [1831] had come close to this equation before Stokes. But instead of the specific factors μ and $\mu/3$, his equation contained two unknown constants. For a relatively simple derivation of Eq. 3–20 see Rouse's [1938] *Fluid Mechanics*, Chapter VII.

side (see Eq. 3–20) assume a factor $\nu = \mu/\rho$ which is called the kinematic viscosity. In the physical system ν has the dimension [cm^2 sec^{-1}], in the British gravitational system [ft^2 hr^{-1}], in the metric system [m^2 hr^{-1}]. The units of force or mass do not appear in these dimensions, but only the units used in kinematics (see Sect. 1–2). Thus the name kinematic viscosity is chosen very well.† For the unit of ν in the physical system Jakob [1928] has suggested the name stokes. See conversion factors in Appendix C.

3–6. The Differential Equation of Continuity of Flow

In addition to Stokes's differential equations of flow, the condition of continuity must be satisfied.

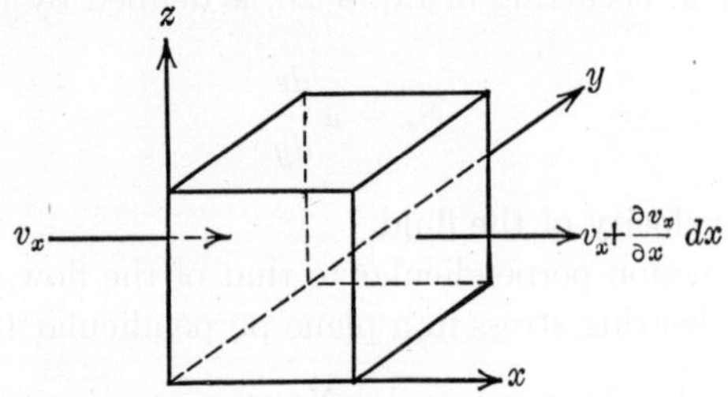

FIG. 3–4. Fluid flow through a parallelepidon.

Again a differential parallelepiped, with edges in the coordinate directions, will be considered (Fig. 3–4). The mass of the fluid entering at the left side in a time interval $d\tau$ is

$$dm_{1,x} = \rho v_x (dy \cdot dz)\, d\tau$$

and the mass leaving at the right is

$$dm_{2,x} = \left(\rho v_x + \rho \frac{\partial v_x}{\partial x}\,dx + v_x \frac{\partial \rho}{\partial x}\,dx\right)(dy \cdot dz)\, d\tau$$

Similar equations exist for the masses $dm_{1,y}$, $dm_{2,y}$, $dm_{1,z}$, $dm_{2,z}$, carried in the directions y and z.

Thus the total mass entering in the time interval $d\tau$ is

$$dm_1 = dm_{1,x} + dm_{1,y} + dm_{1,z}$$

and the mass leaving is

$$dm_2 = dm_{2,x} + dm_{2,y} + dm_{2,z}$$

† Instead of the name dynamic viscosity for μ, the denotation absolute viscosity is often used. This name is not to be recommended because μ is not more absolute than ν is.

Further, because the density in the differential body may vary with time, there will be a storage of mass

$$dm_3 = \frac{\partial \rho}{\partial \tau} (dx \cdot dy \cdot dz)\, d\tau$$

According to the law of conservation of mass, one obtains

$$dm_1 - dm_2 = dm_3$$

and by substitution of the above values

$$-\rho\left(\frac{\partial v_x}{\partial x} + \frac{\partial v_y}{\partial y} + \frac{\partial v_z}{\partial z}\right)(dx \cdot dy \cdot dz)\, d\tau$$

$$-\left(v_x \frac{\partial \rho}{\partial x} + v_y \frac{\partial \rho}{\partial y} + v_z \frac{\partial \rho}{\partial z}\right)(dx \cdot dy \cdot dz)\, d\tau = \frac{\partial \rho}{\partial \tau}(dx \cdot dy \cdot dz)\, d\tau$$

or

$$\rho\left(\frac{\partial v_x}{\partial x} + \frac{\partial v_y}{\partial y} + \frac{\partial v_z}{\partial z}\right) + \left(v_x \frac{\partial \rho}{\partial x} + v_y \frac{\partial \rho}{\partial y} + v_z \frac{\partial \rho}{\partial z}\right) + \frac{\partial \rho}{\partial \tau} = 0 \tag{3-22}$$

For incompressible fluids (liquids) the second and third term disappear, and Eq. 3-22 simplifies to

$$\frac{\partial v_x}{\partial x} + \frac{\partial v_y}{\partial y} + \frac{\partial v_z}{\partial z} = 0 \tag{3-23}$$

3-7. The Differential Equation of Heat Conduction in a Moving Substance

Now the differential equation of heat conduction, derived in Sect. 2-4, must be extended to the case of a moving fluid.

In this case the temperature increment of a particle of the fluid in the time interval $d\tau$ becomes

$$dt = \frac{\partial t}{\partial \tau} d\tau + \frac{\partial t}{\partial x} dx + \frac{\partial t}{\partial y} dy + \frac{\partial t}{\partial z} dz$$

just as the velocity suffers changes in space and time (see Sect. 3-4). Dividing by $d\tau$,

$$\frac{dt}{d\tau} = \frac{\partial t}{\partial \tau} + v_x \frac{\partial t}{\partial x} + v_y \frac{\partial t}{\partial y} + v_z \frac{\partial t}{\partial z}$$

Thus dQ_3, as given in Sect. 2–4, is to be replaced by

$$dQ_3 = \rho c_p(dx \cdot dy \cdot dz)\, dt = \rho c_p(dx \cdot dy \cdot dz)\left(\frac{\partial t}{\partial \tau} + v_x \frac{\partial t}{\partial x} + v_y \frac{\partial t}{\partial y} + v_z \frac{\partial t}{\partial z}\right) d\tau$$

so that Eq. 2–9 turns into

$$\frac{dt}{d\tau} = \frac{\partial t}{\partial \tau} + v_x \frac{\partial t}{\partial x} + v_y \frac{\partial t}{\partial y} + v_z \frac{\partial t}{\partial z} = \alpha\left(\frac{\partial^2 t}{\partial x^2} + \frac{\partial^2 t}{\partial y^2} + \frac{\partial^2 t}{\partial z^2}\right) + \frac{q'''}{\rho c_p} \quad (3\text{–}24)$$

In this equation the terms $v_x(\partial t/\partial x)$, $v_y(\partial t/\partial y)$, and $v_z(\partial t/\partial z)$ are convection terms, analogous to the terms $v_x(\partial v_x/\partial x)$, etc., of the flow Eq. 3–13.

3–8. Boundary Condition for the Surface of a System with Heat Convection

Finally a boundary condition for the confining surface of a system in which heat convection occurs must be considered. The general equation for the heat exchange by conduction and convection is

$$dq = -k \frac{\partial t}{\partial n}\, dA \quad (3\text{–}25)$$

where n is the direction of the heat flow normal to the surface element dA, and k is the thermal conductivity of the fluid close to the surface.

Equation 3–25 may readily be transformed into an equation in which $-k(\partial t/\partial n)$ is replaced by $h \cdot \Delta t$ (*cf.* Eqs. 3–1 and 2). Only by this transformation does the practical coefficient of heat transfer h enter the system of basic equations. If radiation at the surface must be considered, this often can be done by using an "equivalent" value of h which includes the effect of radiation.

Some straight solutions of the differential equations of heat convection presented in this chapter will be dealt with in Chapter 22.

CHAPTER 4

The Basic Equations of Heat Radiation

4–1. Stefan-Boltzmann's Law

In Sect. 1–1 only the most popular law of heat radiation was mentioned, namely, Stefan-Boltzmann's law, according to which the total emission of radiant energy from a perfect black surface of area A_1 in unit of time is

$$\Phi_b = \sigma A_1 T^4 \tag{4–1}$$

This equation for the radiant flux * Φ_b plays in radiation a role similar to that of Newton's Eq. 1–2 in the heat transmission by convection. Both equations are integral relations, involving complicated phenomena without giving an insight into their mechanism. However, whereas the factor h in Newton's equation actually is a function, more obscuring than elucidating the apparent difficulties, Stefan-Boltzmann's factor σ is a true natural constant, and whereas Newton's equation has never found any scientific derivation or explanation and is actually only a very rough practical rule or at the best an equation of definition, Stefan's equation happens to be a basic law of nature and has been derived thermodynamically by Boltzmann. A somewhat modified proof will be shown in Sect. 4–6, but first some other fundamental relations about heat radiation will have to be presented. For the time being, only the physical dimension of the constant σ of Stefan-Boltzmann's law will be given in different systems of units.

In the physical system with the calorie as heat unit

$$[\sigma] = [\text{cal sec}^{-1}\ \text{cm}^{-2}\ \text{C}^{-4}] \tag{4–2}$$

In the British technical system

$$[\sigma] = [\text{B hr}^{-1}\ \text{ft}^{-2}\ \text{F}^{-4}] \tag{4–3}$$

* Different symbols are used for the emitted flux Φ_b and the net heat exchange by radiation q_b, the subscript b standing for "perfect black."

In the metric system

$$[\sigma] = [\text{kcal hr}^{-1}\ \text{m}^{-2}\ \text{C}^{-4}] \tag{4-4}$$

In the physical system with the joule as heat unit

$$[\sigma] = [\text{joule sec}^{-1}\ \text{cm}^{-2}\ \text{C}^{-4}] = [\text{watt cm}^{-2}\ \text{C}^{-4}] \tag{4-5}$$

In these equations the units of temperature difference, centigrade [C] and [F] are used which are identical with the temperature-difference units in the Kelvin and Rankine absolute-temperature scales, whereas the temperatures themselves in these different scales are related to each other by the equations

$$T\ °\text{K} = t\ °\text{C} + 273.2 \tag{4-6}$$

and

$$T\ °\text{R} = t\ °\text{F} + 459.7 \tag{4-7}$$

As mentioned in Appendix B, the signs °K, °C, °R, and °F will be used only to indicate the place of the temperature in the Kelvin, Centigrade, Rankine, and Fahrenheit scales, whereas [C] ≡ [K] and [F] ≡ [R] stand for the units in these scales.

The most reliable numerical values for σ are 0.1742 (10^{-8}) in the British and 4.96 (10^{-8}) in the metric system. Conversion factors are given in Appendix C.

4-2. Kirchhoff's Law

Basic for all kinds of radiation is Kirchhoff's [1859, 1860] law. It maintains that the better absorbing of two surfaces is also the one which emits more radiation or, more exactly: If a surface at any given temperature absorbs n times more of an arriving radiation than another surface of equal area at the same temperature, then at the same temperature the first surface emits n times more of the same kind of radiation than does the second surface.

In order to prove this law, some definitions are first needed. When radiation hits the surface of a body, a certain fraction of it, α, is absorbed, another fraction, τ, is transmitted through the body, and a third part, ρ, is reflected. Then apparently

$$\alpha + \tau + \rho = 1 \tag{4-8}$$

Reflection may be specular or diffuse. A beam hitting a mirror leaves it again as a defined and limited beam under an angle, against the normal to the surface, which is equal to the angle of incidence. A diffusely reflecting surface, on the other hand, converts the beam into a bundle of beams equally distributed over the space. A surface which neither reflects or allows to pass any radiation would have the absorptivity $\alpha = 1$. It is called a perfect black surface or, briefly, a black surface. Such surfaces do not exist in nature. Soot, for instance, which appears entirely black to our eyes, absorbs only about 95% of the radiation.

According to a suggestion of Kirchhoff, however, a perfect black surface can be produced by using a hollow enclosure provided with only a very small hole. The area of the opening can be considered a surface having the absorptivity $\alpha = 1$.

This can easily be understood. The hollow space may be surrounded by walls, having everywhere the same and constant temperature. Any radiation which enters the small hole will suffer many reflections on the inside surfaces and will be weakened at each reflection, so that virtually nothing of the original radiation will get out again. Thus the area of that hole behaves like a very black surface absorbing almost all the arriving radiation. The trick of this arrangement consists in providing many subsequent absorptions by surfaces with an absorptivity α much smaller than 1, instead of just one absorption by a surface of high absorptivity. Hence the radiation which leaves the hole is not reflected, but is emitted by the inner surface of the wall. It can be considered the emission of an entirely black surface at the place of the small opening.

To prove this more strongly, the concept of the mechanical equilibrium of radiation will be used. According to the electromagnetic theory, radiation hitting a surface exerts a pressure upon it. If a thin plate were placed inside a hollow enclosure, the pressure of radiation on both sides of the plate would be the same. Otherwise the plate would be moved and external work would be done in contradiction to the second law of thermodynamics, according to which heat energy cannot be transformed into external work unless differences of temperature exist. Because the plate can be placed anywhere in the hollow space and in any direction, it follows that every radiant beam must correspond to a beam of the same intensity running the opposite way. It can be proved that this holds also for rays of a fixed wavelength and for polarized rays, i.e., rays formed by electromagnetic oscillations in one plane.† Now a beam may be considered that comes out of a small hole in the enclosure; it could be compensated for by a beam thrown into the hollow space from an outside surface equal to the inside surface in structure and temperature. The entering beam, however, would be almost entirely absorbed inside the hollow space, independently of the quality of the inside sur-

† Kirchhoff [1860] in his original paper claims only the proof for monochromatic and for polarized radiation as his contribution to the problem. But the value of the law consists just in this generalization.

face. Therefore also the whole small bundle of rays which leaves through the opening and is balanced by the incoming radiation must be independent of the properties of the inside surface. Therefore the intensity of each of its rays will depend only on the temperature. To understand this more easily, the hole can be supposed to be closed, and the incoming radiation replaced by that issuing from the closing surface element.

Thus, in a hollow enclosure having an inner surface of uniform and constant temperature, an equilibrium of radiation exists which depends only upon the temperature, and particularly not upon the kind of the inside surface.

Since this law can be applied to any frequency or wavelength, the emitted radiation must have a fixed spectral composition which also depends only upon the temperature. Hence not only any radiation of different intensity, but also any radiation of a different spectral composition would mean a disturbance of the equilibrium of the hollow-enclosure radiation, and the equilibrium would recover automatically.

Now the radiation of an arbitrary surface may be considered, which reflects diffusively and is in equilibrium with some radiation arriving from all sides of the hemispherical space above the surface, i.e., radiation from a source at the same temperature as the surface considered. The symbols W and W_b may denote the emitted and arriving radiant energy per unit area of the surface and unit time; these quantities are called radiant flux densities. Let α be the absorptivity and ρ the reflectivity of the surface, whereas the transmissivity τ may be zero.‡ Then W_b can be represented in two ways, by the equations

$$W_b = \rho W_b + W \tag{4-9}$$

and

$$W_b = \rho W_b + \alpha W_b \tag{4-10}$$

Equation 4-9 indicates that the rates of arriving and leaving energy are equal, and Eq. 4-10 indicates that the arriving energy is partly reflected and partly absorbed. From these two equations follows Kirchhoff's law in the form

$$\frac{W}{W_b} = \alpha \tag{4-11}$$

If $\alpha = 1$, then $W = W_b$; hence the flux density of radiation in equilibrium with that of a diffuse radiating surface is always the same, namely,

‡ Heat radiation incident upon metals is entirely absorbed within a depth smaller than the wavelength. Experiments by Eckert [1938] showed that electric nonconductors, whether optically transparent or opaque, absorb heat radiation within 1 mm depth, except a few minerals like salt (NaCl) which can be used as windows in measurements on heat radiation.

equal to the flux density W_b of a black body of the same temperature. Since generally $\alpha < 1$, it follows that the flux density of emission, W, of a real surface is always smaller than that of a black body. The ratio W/W_b is often called emissivity and denoted by ϵ. Thus absorptivity and emissivity of a surface at any fixed temperature are identical by definition. For two surfaces with the absorptivities α_1 and $\alpha_2 = \alpha_1/n$, it follows from Eq. 4–11 that

$$\frac{W_1}{W_2} = \frac{\alpha_1}{\alpha_2} = n$$

as asserted in the first paragraph of this section. Because this holds for radiation of any wavelength, it holds also for the total radiation.

4–3. Calculation of the Total Radiation by Means of Lambert's Law

The expression diffuse radiation has already been used without exact definition; it means equal distribution of radiant flux density to all directions of the space.

In Fig. 4–1 an element of a diffusely radiating surface is marked by dA_1. The rate of energy radiated by dA_1 in the direction OM is proportional to the cosine of the angle ϕ between OM and ON, the normal to

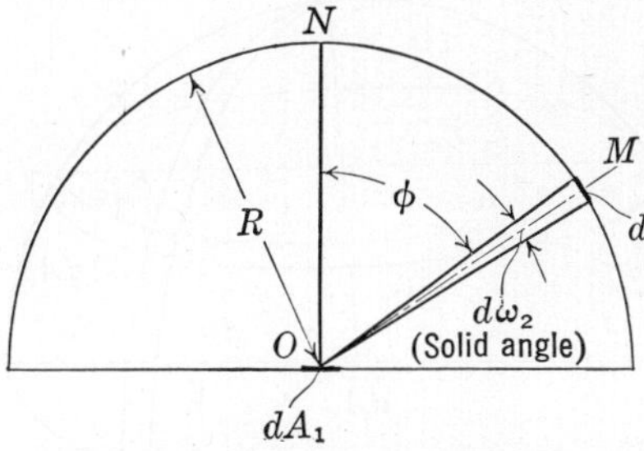

FIG. 4–1. Geometrical data for the derivation of Lambert's law.

dA_1, because, from M as eyepoint, a surface $dA_1 \cdot \cos \phi$ is seen which, at equal space distribution of radiation, appears as bright as the area dA_1 seen from N. This is known as the cosine law of Lambert [1760], who claimed that diffusely reflecting surfaces radiate with this distribution. In Sects. 4–9 and 10 deviations of the radiation of different surfaces from this law will be dealt with.

For black radiation, Lambert's law holds exactly. Thus, for a black surface the total radiation can be calculated by assuming the emission to be distributed according to this law. To apply it, some geometry of radiation will be needed.

Referring to Fig. 4–1, the differential of radiation which comes from the black surface dA_1 and hits the surface dA_2 may be expressed by

$$d^3\Phi_b = N_b(dA_1 \cos \phi)\, d\omega_2 \tag{4–12}$$

According to this equation, $d^3\Phi_b$ is proportional to an emissivity factor N_b (called areal radiant intensity) which is proportional to the density of total emission W_b of the black surface dA_1. The third-order differential $d^3\Phi_b$ is used because later three differentials will appear as factors on the right side.

The differential $d\omega_2$ is the solid angle under which dA_2 is seen from O. It may be recalled that, in analogy to the definition of an ordinary angle ϕ by the quotient L/R, where L is an arc subtended by ϕ and R is the radius of a circle, a solid angle is defined by $\omega = A/R^2$, where A is an area on a sphere of radius R, which is cut out by a cone issuing from the center of the sphere. Corresponding differential angles are $d\phi = dL/R$ and $d\omega = dA/R^2$.

If $R = 1$, then $d\phi = dL$ and $d\omega = dA$, and by integrating over the whole perimeter of the circle and the surface of the sphere with unit radius, $\phi = 2\pi$ and $\omega = 4\pi$, respectively.

Returning to Eq. 4–12, the solid angle ω_2 is to be expressed in coordinates of spherical trigonometry R, ϕ, ψ, as defined by Fig. 4–2. Herein

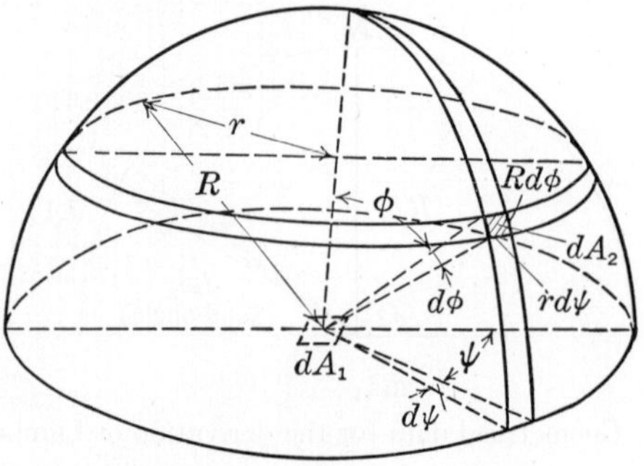

FIG. 4–2. Geometrical data referring to diffuse radiation into the hemispherical space.

the pole distance is ϕ, the spheric latitude is $(\pi/2) - \phi$, and the spheric longitude is ψ. From the figure it follows that

$$dA_2 = (R \cdot d\phi)(r \cdot d\psi) = (R \cdot d\phi)(R \sin \phi \cdot d\psi)$$

$$d\omega_2 = \frac{dA_2}{R^2} = \sin \phi \cdot d\phi \cdot d\psi$$

and by substitution in Eq. 4–12

$$d^3\Phi_b = N_b \cdot dA_1 \cdot d\psi \cdot \sin \phi \cdot \cos \phi \cdot d\phi \tag{4–13}$$

In order to obtain the radiation from dA_1 into the whole hemispheric space above the plane of dA_1, Eq. 4–13 must be integrated from $\phi = 0$ to $\phi = \pi/2$ and from $\psi = 0$ to $\psi = 2\pi$. This yields

$$d\Phi_b = N_b\,dA_1 \int_{\psi=0}^{\psi=2\pi} d\psi \int_{\phi=0}^{\phi=\pi/2} \sin\phi \cdot \cos\phi \cdot d\phi = \pi N_b\,dA_1 \quad (4\text{–}14)$$

and for a finite area A_1

$$\Phi_b = \pi N_b A_1 = W_b A_1 \quad (4\text{–}15)$$

4–4. The Space Density of Radiation

Whatever radiant energy may be, it certainly does not travel with infinite velocity, although it travels enormously fast. Therefore in a finite portion of the space, crossed by radiation, a finite amount of energy must exist at any time. Dividing this energy by the volume yields the so-called space density of radiation.

In Fig. 4–3 a small volume V of any shape is indicated around the center O of a sphere whose surface A_1 is supposed to be black and whose

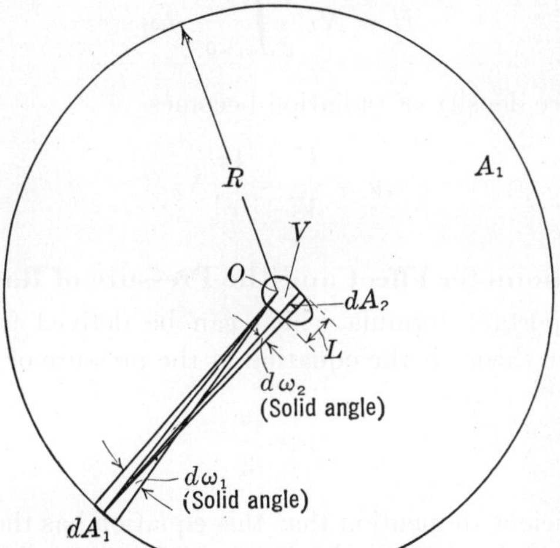

FIG. 4–3. Geometrical data for the calculation of the space density of radiation.

volume may be very large compared with V. Every beam arriving in V comes from a spot of this sphere, e.g., from dA_1. A bundle of rays emitted from dA_1 may fill the solid angle $d\omega_2 = dA_2/R^2$. The radiation needs the time τ for crossing volume V along a line of the length L. Thus $\tau = L/c$, where c is the light velocity. In this time all energy d^2U in the almost cylindrical volume $L \cdot dA_2$ leaves the space V and is replaced by the same amount of energy entering it. According to Eq. 4-12, with $d\omega_2 = dA_2/R^2$ and $\phi = 0$, one obtains §

$$d^2U = \tau \cdot d^3\Phi_b = \tau N_b \frac{dA_2}{R^2} dA_1 = \frac{L}{c} N_b \frac{dA_2}{R^2} dA_1$$

and by integration over the whole volume V

$$dU = \frac{1}{c} N_b \frac{dA_1}{R^2} \int_V L \cdot dA_2 = \frac{N_b}{c} \cdot \frac{dA_1}{R^2} V \tag{4-16}$$

This must be integrated over the whole surface A_1 of the sphere. To that purpose A_1 is looked at from the center O. Then the element dA_1 is cut out by the solid angle $d\omega_1 = dA_1/R^2$. Substituting this in Eq. 4-16,

$$dU = N_b \frac{V}{c} d\omega_1$$

and by integrating over the whole sphere,

$$U = N_b \frac{V}{c} \int_{\omega_1=0}^{\omega_1=4\pi} d\omega_1$$

Hence the space density of radiation becomes

$$u = \frac{U}{V} = \frac{4\pi}{c} N_b \tag{4-17}$$

4-5. The Radiometer Effect and the Pressure of Radiation

Another important formula which can be derived from Maxwell's electromagnetic theory is the equation of the pressure of radiation

$$p = \frac{u}{3} \tag{4-18}$$

It may be sufficient to mention that this equation has the same form as that derived for the pressure in the kinetic theory of gases, where u is the kinetic energy of all molecules in unit volume.

On surfaces hit by radiation two different mechanical effects have been observed which work in opposite directions, the radiometer effect and the effect of the pressure of radiation.

A light disc with one side blackened, the other side bright, may be suspended in an evacuated vessel, and both surfaces may be exposed to

§ Since the order of any differential can be chosen arbitrarily, d^2U is used here, considering that only two differentials appear in the calculation.

radiation of the same intensity. Then the disc is moved in the direction of that radiation which hits the black surface. William Crookes, in 1874, constructed the first radiometers, so-called light-mills, driven by this effect. The black surface, because of its larger absorptivity, becomes warmer than the bright one. Therefore, the gas molecules hitting the black side are rejected with a greater velocity than their speed of arrival; the recoil (repulsion) of these molecules is considered as the force which moves the disc. This phenomenon has nothing to do with the pressure of radiation, but only with the heat exchange between the disc and the gas in the vessel. Radiometers are used for the measurement of the radiation. Two wings, consisting of tinfoil, mica, or goldfoil, and arranged symmetrically in an evacuated glass balloon of about 4 in. diameter, are suspended on a quartz string. One side of each wing is blackened by soot, and the source of radiation is located so that its beams hit the blackened sides only. The resulting torsion is observed by mirror and scale. The effect begins at a gas pressure of some millimeters of mercury, reaches its maximum at about 0.02 mm (dependent on the kind of gas), and decreases with further decreasing pressure. The deflection is proportional to the radiation.

The pressure of radiation, on the other hand, originates from the close analogy between radiation and the state of an ideal gas. At normal incidence a radiation of areal radiant intensity N_b erg sec^{-1} cm^{-2} exerts a pressure $p = 4\pi N_b/3c = u/3$ upon a perfect mirror. For a black body the pressure would be only half of this; however, the same amount would be emitted in equilibrium and the recoil due to emission would also be the same, so that again $p = u/3$. More generally, if ρ is the reflectivity, then according to Maxwell-Bartoli's fundamental law of the electromagnetic theory of radiation, the pressure becomes $2\pi N_b(1 + \rho)/3c$ and the recoil amounts to $2\pi N_b(1 - \rho)/3c$. This means that the radiation carries an impulse, mc, like a gas. The pressure of radiation on a mirror-like surface ($\rho = 1$) corresponds formally to the law of elastic impact, according to which under perfect reflection $2mc$ is exchanged.

Qualitatively the existence of the pressure of radiation was stated first by Lebedew in 1901 and by Nichols and Hull. Precise measurements of Gerlach and Golson in the year 1923 showed an agreement with the theory within 2%.

Assuming the so-called solar constant ‖ as about $\frac{1}{30}$ cal sec^{-1} cm^{-2} = 442 B hr^{-1} ft^{-2}, one has calculated that the pressure of the radiation of the sun upon the earth is about 0.4 mg/sq m or $16.2(10^6)$ kg = $36(10^6)$ lb upon a disc of the diameter of the earth.

‖ The value of 400 B hr^{-1} ft^{-2} is often used in air-conditioning calculations.

4–6. Derivation of Stefan-Boltzmann's Law

Now we are prepared for the thermodynamical derivation of Stefan-Boltzmann's law [Boltzmann 1884], which is based on the assumption that an ideal-gas state exists in a space "filled" with black-body radiation.

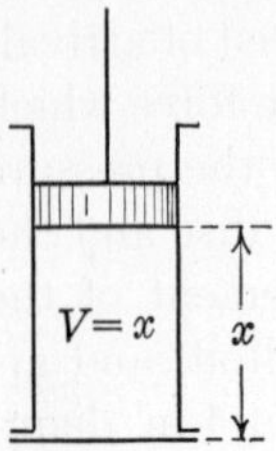

FIG. 4–4. Cylinder with piston of unit cross-sectional area.

Consider the radiation inside a cylinder with a piston of unit cross-sectional area (Fig. 4–4). The volume below the piston is

$$V = x \tag{4–19}$$

The internal energy of the radiation inside this volume is

$$U = Vu = xu \tag{4–20}$$

From this by differentiation

$$dU = x \cdot du + u \cdot dx \tag{4–21}$$

Now substitute p, V, and dU from Eqs. 4–18, 19, and 21 in the first law of thermodynamics:

$$dQ = dU + p \cdot dV \tag{4–22}$$

This yields

$$dQ = x \cdot du + u \cdot dx + \frac{u}{3}\,dx = x \cdot du + \frac{4}{3}\,u \cdot dx \tag{4–23}$$

Using the definition of entropy,

$$dS = \frac{dQ}{T} = \frac{x}{T}\,du + \frac{4}{3} \cdot \frac{u}{T}\,dx \tag{4–24}$$

However, dS must be a perfect differential of two variables. Since, for a black hollow space, u is a function of temperature only

$$du = \frac{du}{dT}\,dT$$

and by substitution in Eq. 4–24

$$dS = \frac{x}{T} \cdot \frac{du}{dT}\,dT + \frac{4}{3} \cdot \frac{u}{T}\,dx \tag{4–25}$$

where T and x are the only independent variables.

Now, using the equation which defines dS as a perfect differential,

$$dS = \frac{\partial S}{\partial T}\,dT + \frac{\partial S}{\partial x}\,dx \tag{4–26}$$

and equating the corresponding terms in Eqs. 4–25 and 26,

$$\frac{\partial S}{\partial T} = \frac{x}{T} \cdot \frac{du}{dT} \tag{4-27}$$

and

$$\frac{\partial S}{\partial x} = \frac{4}{3} \cdot \frac{u}{T} \tag{4-28}$$

Differentiation of Eq. 4–27 with respect to x and of Eq. 4–28 with respect to T yields the equity

$$\frac{1}{T} \cdot \frac{du}{dT} = \frac{4}{3} \cdot \frac{T\dfrac{du}{dT} - u}{T^2} = \frac{4}{3} \cdot \frac{1}{T} \cdot \frac{du}{dT} - \frac{4}{3} \cdot \frac{u}{T^2}$$

or

$$\frac{du}{dT} = 4\frac{u}{T} \tag{4-29}$$

Herefrom by integration

$$\ln u = 4 \ln T + \ln \frac{4\sigma}{c}$$

or

$$u = \frac{4}{c}\sigma T^4 \tag{4-30}$$

where $\ln (4\sigma/c)$ is the constant of integration, c is the light velocity, and σ is an arbitrary constant.

Now, according to Lambert's law

$$\Phi_b = \pi N_b A_1 \tag{4-15}$$

and further

$$u = \frac{4\pi}{c} N_b \tag{4-17}$$

Substituting N_b from Eq. 4–15 in Eq. 4–17,

$$u = \frac{4}{c} \cdot \frac{\Phi_b}{A_1} \tag{4-31}$$

By equating Eqs. 4–30 and 31, Stefan-Boltzmann's law is obtained in its usual form:

$$\Phi_b = \sigma A_1 T^4 \tag{4-1}$$

Comparison of Eqs. 4–1 and 15 shows that the emissivity factor

$$N_b = \frac{\sigma}{\pi} T^4 \tag{4-32}$$

Stefan-Boltzmann's law holds only for black surfaces. To understand the reason for this fact and to learn what properties of radiating surfaces are responsible for the particular kind of their radiation, the distribution of radiation over the range of wavelengths must now be considered. So far we have learned only that a fixed spectral distribution belongs to black radiation. But what distribution?

4-7. Wien's Displacement Law

The first step in answering the question of spectral distribution of radiation was taken by Wien [1893] in the development of the so-called displacement law. This will not be derived in all details; however, the main features of the derivation will be given.

Again radiant energy in a space is considered as an ideal gas. The displacement law claims that in an adiabatic change of the radiation-gas from a state 1 to a state 2 the wavelength of any portion of it changes inversely to the absolute temperature, i.e.,

$$\lambda_1 T_1 = \lambda_2 T_2 \tag{4-33}$$

In the derivation of Stefan-Boltzmann's law the two main laws of thermodynamics were applied to a gas-filled rigid cylinder with heat-absorbing walls and a movable piston. Wien in his derivation dealt with a spherical shell whose specular-reflecting wall can be expanded by pressure like that of a rubber ball. He further assumed a very small specular-reflecting sphere in the center, so that energy is radiated to and fro in radial direction only. The space density u_r of this radiation will obviously decrease with increasing radius r of the spherical shell, and it can be proved that for any value of r

$$u_r = \frac{2N_b}{cr^2}\, dA_1 \tag{4-34}$$

where dA_1 is the area of the small inner reflecting sphere. Referring to N_b, Wien had to assume an element dA_1 which first would radiate like a black body and later was replaced by a perfect mirror, whereupon reflection between the inner and outer spheres would set in and continue infinitely.

It may further be mentioned that for a hollow sphere with the radius r the local space density at this radius becomes

$$u_r = \frac{u}{3} = p \tag{4-35}$$

where u is the average space density in the sphere, and p the pressure of radiation as defined formerly.

With these definitions a procedure exactly analogous to that of Boltzmann, given in Sect. 4–6, leads to

$$\frac{du_r}{dT} = 4\frac{u_r}{T} \tag{4–36}$$

By substitution of u_r from Eq. 4–35 into 36, Stefan-Boltzmann's law in its differential form (Eq. 4–29) is obtained.

Considering now an adiabatic change of state by expanding the outer reflecting sphere, proceeding similarly as in Eqs. 4–26 to 28, and taking $dS = 0$,

$$4u_r\,dr + r\frac{du_r}{dT}\,dT = 0$$

and by integration between $T = T_1$ and T_2 and between $u_r = u_1$ and u_2

$$\frac{u_1}{u_2} = \left(\frac{r_2}{r_1}\right)^4 \tag{4–37}$$

This is not in contradiction to Eq. 4–34, which refers to a system at rest, whereas Eq. 4–37 deals with the change of state of a moving system.

Further, according to Stefan-Boltzmann's law,

$$\frac{u_1}{u_2} = \frac{T_1^{\,4}}{T_2^{\,4}} \tag{4–38}$$

and from Eqs. 4–37 and 38

$$\frac{T_1}{T_2} = \frac{r_2}{r_1} \tag{4–39}$$

The rays, travelling with light velocity c towards the surface of the spherical envelope, which may expand with the radial velocity v, suffer a change of wavelength according to Doppler's principle. At each reflection on the envelope surface the wavelength λ, existing with walls at rest, is increased by $2\lambda v/c$, and for the limit $v \to 0$ the calculation leads to

$$d\lambda = \lambda\frac{dr}{r}$$

or

$$\frac{\lambda_1}{\lambda_2} = \frac{r_1}{r_2} \tag{4–40}$$

Comparison of Eqs. 4–39 and 40 yields the displacement law, Eq. 4–33.

By a sagacious combination of the law of total radiation and the displacement law, Wien eventually came to the expression

$$N_{b,\lambda} = 2\frac{c^4}{\lambda^5}\,\Phi\left(\frac{\lambda T}{c}\right) \tag{4–41}$$

where Φ is a function, unknown for the time being, and $N_{b,\lambda}$ is the monochromatic areal radiant intensity of unpolarized black radiation at wavelength λ and temperature T. This intensity is defined by

$$d^3\Phi_{b,\lambda} = N_{b,\lambda}\, d\lambda (dA_1 \cos\phi)\, d\omega_2 \tag{4–42}$$

which corresponds to Eq. 4–12 and gives the radiation of a bundle of unpolarized rays in the wavelength range from $\lambda - (d\lambda/2)$ to $\lambda + (d\lambda/2)$, emitted from the surface dA_1 under the angle ϕ into the solid angle ω_2. The polarized radiation would be half this amount. The intensity $N_{b,\lambda}$ has the dimension [cal sec^{-1} cm^{-3}] in the physical system.

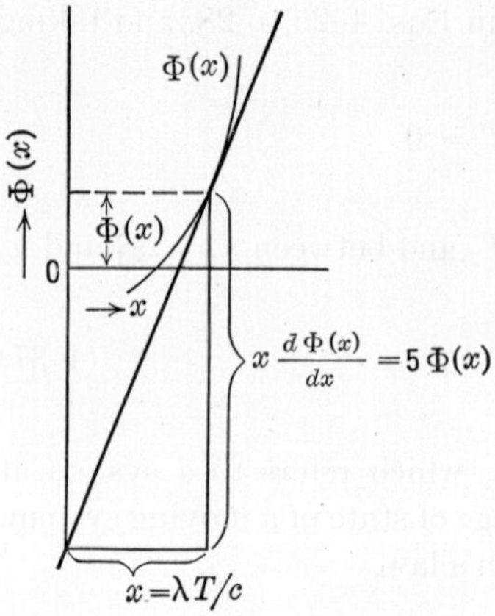

FIG. 4–5. Construction to find the maximum of the monochromatic areal radiant intensity, $N_{b,\lambda}$.

It can be demonstrated that $N_{b,\lambda}$ has a maximum, with respect to λ. Because for constant λ and increasing T the function Φ is increasing, it must increase also for constant T and increasing λ.

With a new variable, $x = \lambda T/c$, Eq. 4–41 yields the following condition for a maximum or minimum:

$$\frac{dN_{b,\lambda}}{d\lambda} = -2c^4\left[\frac{5}{\lambda^6}\,\Phi(x) - \frac{1}{\lambda^5}\cdot\frac{T}{c}\cdot\frac{d\Phi(x)}{dx}\right] = 0$$

or

$$x\frac{d\Phi(x)}{dx} = 5\Phi(x) \tag{4–43}$$

According to experience, a minimum of intensity does not exist; so a maximum must occur. If, for different values of x, the function $\Phi(x)$ is given, the maximum can be found by a simple trial-and-error construction (Fig. 4–5), which corresponds to Eq. 4–43 and does not need any further explanation.

Denoting λ and T at the maximum point by a star, it follows that

$$\lambda^* T^* = B \tag{4–44}$$

where B is a constant. Hence the maximum of radiation will be displaced to smaller wavelengths if the temperature is raised. Experiments have led to $B = 0.288$ cm C.

Since the intensity of sun radiation has a maximum at about 0.5 μ, Eq. 4–44 yields as temperature of the sun surface

$$T_{\text{sun}} \approx \frac{2880}{0.5} = 5760°\text{K}$$

in satisfactory agreement with other calculations and observations.

As another consequence of Wien's displacement law Eqs. 4–41 and 44 lead to the following expression for the maximum of intensity:

$$N_{b,\lambda}{}^* = CT^5$$

where C is a constant.

Most interesting in this work of Wien's is the fact that he was able to draw some important conclusions concerning the radiation at distinct wavelengths, without knowing much about the distribution of radiation over the wavelengths. But with this performance, the possibilities of classical thermodynamics were exhausted, as will be seen now.

4–8. Planck's Law of Radiation

Whereas the three laws of radiation, which had been derived by means of classical thermodynamics, namely, Kirchhoff's law, Stefan-Boltzmann's law, and Wien's displacement law, were exactly confirmed by experience, the question of how at a certain temperature the intensity of radiation is distributed to wavelengths in the spectrum of a radiating black surface could not be answered by thermodynamics. For this purpose hypotheses about the very mechanism of the process of radiation had to be used. The situation was similar to that occurring in the theory of gases, where the mechanism of atomic motion in the molecule had to be considered in order to find numerical values of the specific heat.

In both cases statistical methods were successfully employed. The first who tried to develop the general law of radiation by means of statistics was Wien [1896]. Using Maxwell's law of the distribution of velocity, he derived the equation

$$N_{b,\lambda} = 2c_1 \frac{\lambda^{-5}}{e^{c_2/\lambda T}} \tag{4–45}$$

where c_1 and c_2 are empirical constants.

Lord Rayleigh [1900] and Jeans [1905], on the other hand, starting from the law of equipartition of energy, i.e., distribution to the so-called degrees of freedom of energy, obtained

$$N_{b,\lambda} = 2c_3\lambda^{-4}T \tag{4-46}$$

where c_3 is a constant.

Both equations are in agreement with Eq. 4-41. But Wien's equation apparently fails at high values of λT; at $T = \infty$ it would lead to a finite intensity $N_{b,\lambda} = 2c_1\lambda^{-5}$, which cannot be. Rayleigh's equation, on the other hand, fails at small wavelengths; at $\lambda = 0$ it yields $N_{b,\lambda} = \infty$ in contradiction to experience.

From these discrepancies Planck [1900, 1901] concluded that there must be a lack in classical statistics on which both equations were based. Therefore, he developed a new kind of statistics, known as quantum theory. According to Planck, radiation consists of a multitude of single, finite parts $h\nu$, where ν is the frequency of an oscillator producing the radiation, and h is a new, universal constant, peculiar to the quantum theory. The elements of radiation or quantum elements $h\nu$ are distributed to the vibrations of the oscillators inside the molecules, according to the laws of probability. One of these is a theorem of Boltzmann, which states that the entropy is proportional to the natural logarithm of the probability of state; the factor of proportionality k is called Boltzmann's constant. This is identical with the gas constant when the law of ideal gases is written for the mass of one single molecule.

From these elements Planck has built up theoretically the equation for the intensity of monochromatic, unpolarized radiation, normal to the radiating surface:

$$N_{b,\lambda} = 2c^2h\,\frac{\lambda^{-5}}{e^{ch/k\lambda T} - 1} = 2c_1\,\frac{\lambda^{-5}}{e^{c_2/\lambda T} - 1} \tag{4-47}$$

in which $c = 29.985(10^9)$ cm/sec = the light velocity,
$h = 6.55(10^{-27})$ erg sec = Planck's quantum constant,
$k = 137.2(10^{-18})$ erg/C = Boltzmann's constant,
$c_1 = c^2h = 5.889(10^{-6})$ erg cm^2 sec^{-1} = the first constant of Planck's law,
$c_2 = ch/k = 1.435$ cm C = the second constant of Planck's law.¶

The relationship between $N_{b,\lambda}$ and N_b is expressed by the equation

$$N_b = \int_{\lambda=0}^{\lambda=\infty} N_{b,\lambda}\cdot d\lambda \tag{4-48}$$

¶ It may be noted that the constant of Rayleigh-Jeans's Eq. 4-46 was found to be $c_3 = \pi ck$.

in which $N_{b,\lambda}$ may be substituted from Eq. 4–47 and $x = c_2/\lambda T$ may be introduced for convenience. Comparing, then, Eqs. 4–32 and 48 yields the constant σ of Stefan-Boltzmann's law as the following function of the universal constants c, h, and k:

$$\sigma = \frac{2\pi k^4}{c^2h^3}\int_0^\infty \frac{x^3\,dx}{e^x - 1} = \frac{2\pi^5 k^4}{15c^2h^3} = 40.803\,\frac{k^4}{c^2h^3} \tag{4–49}$$

Using the numerical values of c, h, and k as given above

$$\begin{aligned}\sigma &= 5.75(10^{-12})\text{ internat. watt cm}^{-2}\text{ C}^{-4}\\ &= 1.374(10^{-12})\text{ cal sec}^{-1}\text{ cm}^{-2}\text{ C}^{-4}\\ &= 0.1737(10^{-8})\text{ B hr}^{-1}\text{ ft}^{-2}\text{ F}^{-4}\end{aligned}$$

in close agreement with the best direct measurements.

Applying Wien's maximum condition, $dN_{b,\lambda}/d\lambda = 0$ (see Eq. 4–43), to Planck's Eq. 4–47,

$$e^{-c_2/\lambda^*T^*} + \frac{c_2}{5\lambda^*T^*} - 1 = 0$$

Herefrom:

$$\lambda^*T^* = \frac{c_2}{4.965} = \frac{1.435}{4.965} = 0.2883\text{ cm C} \tag{4–50}$$

corresponding to the experimental results (see Sect. 4–7).

The intensity of linear polarized radiation normal to the radiating surface is $N_{b,\lambda}/2$, that into the spherical space above the radiating surface element is $\pi N_{b,\lambda}$.

Comparison of Eqs. 4–45 and 47 reveals how close Wien had come to the true solution. Only the term -1 in the denominator of Eq. 4–47 was missing in Wien's equation. This term, however, has the effect of bringing the intensity to an infinite value for infinitely high temperature, as obviously must be the case.

Expanding e^x in Planck's equation and omitting all terms with a higher than first power of x, i.e., neglecting the influence of small values of λT, yield the Rayleigh-Jeans equation.

The distribution of the intensity to the wavelengths, as following from Planck's law, is represented by Fig. 4–6. The ordinates are values of $W_{b,\lambda} = \pi N_{b,\lambda}$, belonging to the ordinary (non-polarized) radiation into a hemispherical space. The wavelengths are given in microns.

Heat radiation occurs in the range * of 0.8 to 400 μ; the range of visible light is 0.4 to 0.8 μ, that of ultraviolet radiation is 0.02 to 0.4 μ,

* 1 μ = 1 micron = 0.001 mm.

that of X-rays is 10^{-6} to 0.02 μ. Electrical wireless radiation is partly identical with long-wave heat radiation. the wavelength range being 100 μ to some miles.

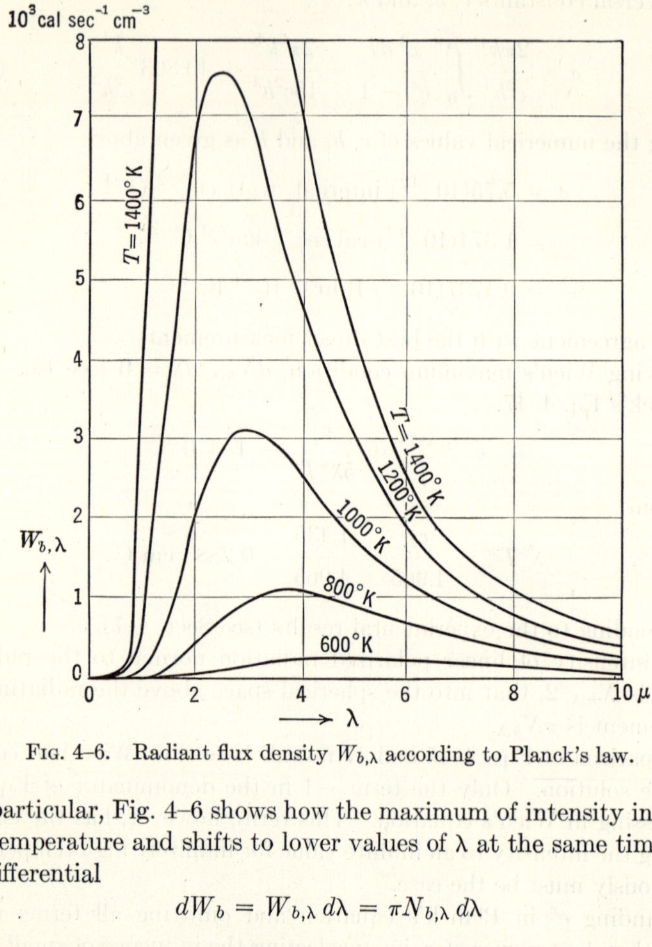

FIG. 4-6. Radiant flux density $W_{b,\lambda}$ according to Planck's law.

In particular, Fig. 4-6 shows how the maximum of intensity increases with temperature and shifts to lower values of λ at the same time.

A differential

$$dW_b = W_{b,\lambda}\, d\lambda = \pi N_{b,\lambda}\, d\lambda$$

of the area below any of the isothermals T = constant, given in Fig. 4-6, can be considered as the radiant flux density of the radiation of a black surface for a wavelength range $d\lambda$ at the wavelength λ. The total area below the isothermal line gives the radiant flux density of total radiation

$$W_b = \pi \int_{\lambda=0}^{\lambda=\infty} N_{b,\lambda}\, d\lambda \tag{4-51}$$

This is only another expression for Eq. 4-48.

4-9. The Radiation of Electrical Insulators

Planck's law has been derived for a black radiating surface only. Fortunately there exist surfaces which radiate at almost the same distribution, though with smaller absolute values of the intensity. This means that for each wavelength the intensity of radiation is smaller by a fixed ratio than that for a black surface.

Such a surface is called a "gray" radiator. Its total radiant flux is given by the equation

$$d\Phi_{gr} = \epsilon_{gr}\sigma T^4 \, dA_1 \tag{4-52}$$

where ϵ_{gr} is the emissivity of the gray surface.† Most of the electrical non-conductors and semi-conductors practically can be considered gray radiators.

More thorough information about radiation of electrical insulators has been obtained from the electromagnetic theory, according to which for radiation normal to the surface the reflectivity is

$$\rho_{i,0} = 1 - \epsilon_{i,0} = \left(\frac{n-1}{n+1}\right)^2 \tag{4-53}$$

where ϵ = the emissivity,†

$$n = \frac{\sin \phi}{\sin \chi} \tag{4-54}$$

is the coefficient of refraction (see Fig. 4-7), and the subscripts i and 0 indicate "insulator" and "$\phi = 0$."

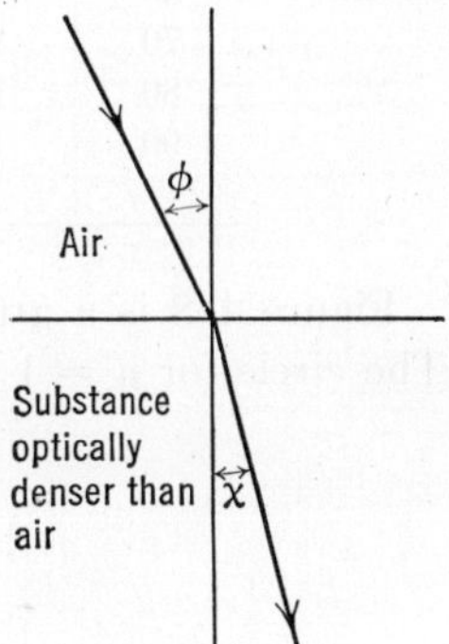

Fig. 4-7. Refraction of a ray.

In the range of long-wave radiation n is known for a few substances only; it varies with the wavelength to some extent. For an angle of incidence or emission ϕ and an angle of refraction χ the reflectivity of an insulator is given by Fresnel's formula:

$$\rho_{i,\phi} = 1 - \epsilon_{i,\phi} = \frac{1}{2} \cdot \frac{\sin^2(\phi - \chi)}{\sin^2(\phi + \chi)} \left[1 + \frac{\cos^2(\phi + \chi)}{\cos^2(\phi - \chi)}\right] \tag{4-55}$$

By combination of Eqs. 4-53, 54, and 55 a law of distribution of radiation over the hemispheric space was found [see E. Schmidt 1927, 1934; E. Schmidt and Eckert 1935] which can be compared with Lambert's

† According to the definition given immediately after Eq. 4-11.

law. The results of the rather complicated calculation are given in Table 4–1, in which n is supposed to be independent of ϕ.

TABLE 4–1

EMISSIVITY $\epsilon_{i,\phi}$ OF ELECTRICAL INSULATORS

$\phi°$	n			
	1.0	1.41	2	3
0	1.000	0.970	0.889	0.750
10	1.000	0.970	0.889	0.750
20	1.000	0.970	0.889	0.750
30	1.000	0.969	0.888	0.749
40	1.000	0.966	0.881	0.746
50	1.000	0.955	0.869	0.742
60	1.000	0.925	0.839	0.728
70	1.000	0.846	0.763	0.690
80	1.000	0.628	0.573	0.567
90	1.000	0.000	0.000	0.000

Figure 4–8 is a graphical representation of the values of Table 4–1. The circle for $n = 1$ represents the equal distribution of the density of

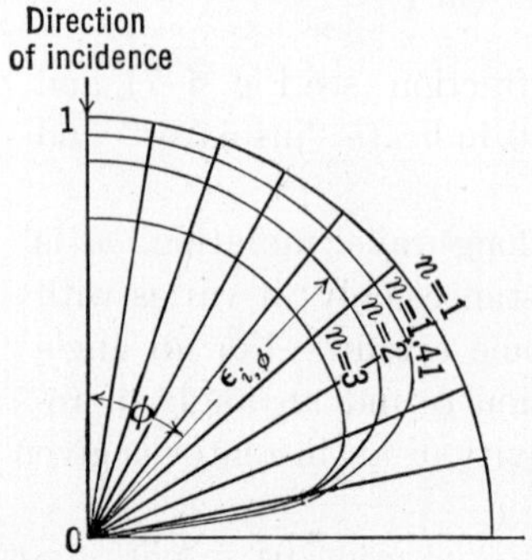

FIG. 4–8. Distribution of emissivity of electric insulators.

radiation $N_{b,\phi} = N_b$ of a black body. For higher values of n, the curves approach the shape of a circle up to about 70°. Above that angle the emissivity $\epsilon_{i,\phi}$ decreases considerably, and at $\phi = 90°$ it becomes zero. Therefore Lambert's law holds only approximately but is sufficient for most practical purposes.

In the same way as, for a black surface, Eq. 4–14 was derived from Eq. 4–12, the following formula for total radiation of an electrical insulator is obtained:

$$d\Phi_i = dA_1 \int_{\psi=0}^{\psi=2\pi} d\psi \int_{\phi=0}^{\phi=\pi/2} N_{i,\phi} \sin\phi \cdot \cos\phi \cdot d\phi$$

$$= \pi \cdot dA_1 \int_0^{\pi/2} N_{i,\phi} d \sin^2 \phi \qquad (4\text{–}56)$$

where

$$N_{i,\phi} = \epsilon_{i,\phi} N_{b,\phi} = \epsilon_{i,\phi} N_b$$

Substituting this in Eq. 4–56 and taking $\epsilon_{i,\phi}$ from Eq. 4–55, with χ from Eq. 4–54, the last integral in Eq. 4–56 can be computed, the simplest way being graphical integration, using the values of Table 4–1. In this way the values of the total emissivity $\epsilon_i = d\Phi_i/d\Phi_b$ of Table 4–2 have been calculated. This table shows that, according to the theory, all electrical non-conductors with a refractive index $n < 3$ emit 75% or more of the radiation of a black body. Because generally $n < 2$, the emissivity of many insulators exceeds even 90% of the black radiation. Thus very small emissivities do not occur with non-conductors. At higher temperatures, however, things become more complicated, as will be seen later on.

TABLE 4–2

Total Emissivity of Insulators

n	1	1.41	2	3	4	5
ϵ_i	1.00	0.91	0.83	0.72	0.63	0.56

An interesting relation exists between the total emissivity ϵ_i and the emissivity $\epsilon_{i,0}$ (normal incidence). Values calculated from Eqs. 4–55 and 56 are given in Table 4–3. It is very strange and remarkable that values of $\epsilon_i/\epsilon_{i\,0} > 1$ occur. We shall come back to this later on.

TABLE 4–3

Ratio of Total and Normal Emissivity of Insulators

$\epsilon_{i,0}$	0.45	0.5	0.55	0.6	0.65	0.7
$\epsilon_i/\epsilon_{i,0}$	1.04	1.025	1.005	0.995	0.98	0.97
$\epsilon_{i,0}$	0.75	0.8	0.85	0.9	0.95	1.0
$\epsilon_i/\epsilon_{i,0}$	0.96	0.95	0.94	0.935	0.935	1.0

4–10. The Radiation of Electrical Conductors

The heat radiation of electrical conductors can also be found theoretically from the electromagnetic theory, provided the radiating surfaces are smooth and the wavelength of radiation is not too small. [See E. Schmidt 1927, 1934; E. Schmidt and Eckert 1935.]

In this case the coefficient of refraction for waves entering a metal can be expressed by

$$n = c\sqrt{k_e \mu_m \tau_c} \tag{4–57}$$

where c = 30(10^9) cm/sec = the velocity of light,
k_e = the electric conductivity in electromagnetic absolute units [sec/cm^2],
μ_m = the magnetic permeability [dimensionless],
τ_c = the period of oscillation of the electromagnetic wave in seconds.

The reflectivity and emissivity for normal incidence of radiation ($\phi = 0$) can be represented by the equations

$$\left.\begin{aligned} \rho_{m,\lambda,0} &= 1 - \frac{2}{n} \\ \text{and} \qquad \epsilon_{m,\lambda,0} &= \frac{2}{n} \end{aligned}\right\} \tag{4–58}$$

where the subscripts m and 0 relate to metallic surface, and $\phi = 0$. Because of the high frequencies of the radiated waves, it can be assumed that $\mu_m = 1$ for ferromagnetic metals as for air and vacuum. Substituting n from Eq. 4–57 in 4–58, and introducing, further, instead of k_e its converse value, the electric resistivity r_e in ohm centimeters, and λ/c instead of τ with the wavelength λ in centimeters, the following equation is obtained:‡

$$\rho_{m,\lambda,0} = 1 - \epsilon_{m,\lambda,0} = 1 - \frac{2}{\sqrt{30}}\sqrt{\frac{r_e}{\lambda}} = 1 - 0.365\sqrt{\frac{r_e}{\lambda}} \tag{4–59}$$

According to Kirchhoff's law, the emissivity $\epsilon_{m,\lambda,0}$ perpendicular to a surface can be calculated from

$$\epsilon_{m,\lambda,0} = \frac{N_{m,\lambda,0}}{N_{b,\lambda}} \tag{4–60}$$

‡ Since 10^{-9} ohm corresponds to 1 cm/sec in electromagnetic absolute units, the numerical value of r_e in Eq. 4–59 is 10^9 times smaller than it would be without the indicated change of units. Therefore a factor $\sqrt{10^{-9}}$ had also to be added in the denominator, yielding 30 instead of 30(10^9). It will be noticed that the factor 0.365 has the dimension ohm$^{-0.5}$.

where $N_{b,\lambda}$, as given by Planck's law (Eq. 4–47), is the intensity of an ordinary (non-polarized) bundle of rays, normally emitted by a black surface, and $N_{m,\lambda,0}$ is the corresponding surface density of radiation from a metallic surface.

By comparison of Eqs. 4–47, 59, and 60, Aschkinass [1905] derived the equation

$$N_{m,\lambda,0} = 0.365\sqrt{\frac{r_e}{\lambda}}\,N_{b,\lambda} = 0.73\sqrt{r_e c_1}\,\frac{\lambda^{-5.5}}{e^{c_2/\lambda T} - 1} \qquad (4\text{–}61)$$

in analogy to Planck's law.

From this he concluded that the location of the maximum of radiation can be found from

$$\lambda^* T^* = D \qquad (4\text{–}62)$$

corresponding to Wien's displacement law (Eq. 4–44) for black radiation, but with a constant $D = 0.2660$ cm C, whereas in Eq. 4–44 the constant was $B = 0.2880$ cm C.

By substituting $N_{m,\lambda,0}$ from Eq. 4–61 in

$$N_{m,0} = \int_{\lambda=0}^{\lambda=\infty} N_{m,\lambda,0}\,d\lambda \qquad (4\text{–}63)$$

Aschkinass arrived at

$$N_{m,0} = 0.73 c_1 \sqrt{r_e}\left(\frac{T}{c_2}\right)^{4.5} \int_0^{\infty} \frac{x^{3.5}\,dx}{e^x - 1}$$

where $x = c_2/\lambda T$. From this, by means of the theory of Γ functions

$$N_{m,0} = 0.73 c_1 \sqrt{r_e}\left(\frac{T}{c_2}\right)^{4.5} 12.27 \qquad (4\text{–}64)$$

Now for pure metals

$$r_e \approx r_{e,273}\,\frac{T}{273} \qquad (4\text{–}65)$$

where $r_{e,273}$ is the electrical resistivity at 0°C = 273°K, and T is the temperature in °K.

Introducing, further, $c_2 = 1.430$ cm C (instead of 1.470, as did Aschkinass),

$$N_{m,0} = 1.084(10^9) c_1 \sqrt{r_{e,273}}\left(\frac{T}{100}\right)^5 \qquad (4\text{–}66)$$

Thus the total radiation of pure metals increases with the 5th power of the absolute temperature, whereas for the black body the 4th power occurs.

The validity of Aschkinass' equations is limited by the following considerations. Maxwell's electromagnetic theory is based on the assumption of absolutely homogeneous bodies; as soon as the distance of the molecules plays a part, i.e., at wavelengths below about 4 μ, a body cannot be considered anymore as homogeneous with respect to radiation. Since wavelengths above 4 μ are prevailing only below the temperature of about 400°C, it was questionable whether Aschkinass' equations would hold above 400°C. Experiments of Lummer and Kurlbaum [1898] and Lummer and Pringsheim [1899], however, showed that for platinum $N_{m,0}$ is proportional to T^5 in the range of 700 to 1800°K. Thus Aschkinass' equation seems to be valid up to much higher temperatures than had been expected. For alloys with a very small temperature coefficient of r_e, however, Aschkinass presumes deviations from his equations to occur at relatively low temperature.

By comparison of Eqs. 4–47 and 61 and substitution of r_e from Eq. 4–65

$$\frac{N_{m,\lambda,0}}{N_{b,\lambda}} = 0.0221 \sqrt{\frac{r_{e,273}T}{\lambda}} \tag{4–67}$$

According to Kirchhoff's law, the absorptivity of the total radiation of a metallic surface in normal direction becomes

$$\alpha_{m,0} = \frac{N_{m,0}}{N_b}$$

The numerator is given by Eq. 4–66; the denominator by Eq. 4–32. Using again Eq. 4–65,

$$\alpha_{m,0} = \frac{\sqrt{r_{e,273}}\,T}{28.50}$$

The distribution of metallic heat radiation with respect to the angle of incidence is well known since Davisson and Weeks [1924] derived the relation between the electric conductivity and the emissivity of a bright metal from Fresnel's equation of optical reflection. First, however, the more simple, though less exact, derivation by E. Schmidt [1927] will be presented.

Reference is made to the simple Eq. 4–58 for the reflection of normal radiation ($\phi = 0°$) by a metallic mirror. For oblique incidence ($\phi > 0°$) the theory of electromagnetism shows that the reflectivity depends on the angle of incidence and on the polarization of the ray.

If the electric field of a wave oscillates normal to the plane of incidence, which is formed by the directions of the ray and of a line perpendicular to the mirror,§ the reflectivity is approximately given by

$$1 - \rho_{m,\phi,n} = \frac{2 \cos \phi}{n} \tag{4–68}$$

The reflection of this radiation increases with ϕ. For a wave with vibration of the electrical field parallel to the plane of incidence the reflectivity is given by

$$1 - \rho_{m,\phi,p} = \frac{2}{n \cos \phi} \tag{4–69}$$

Here, the reflection decreases with increasing ϕ. For normal incidence both equations become identical with Eq. 4–58. These relations hold only for values of n which are large compared to 1.

For frequencies such as occur with wavelengths of about 4 μ, the refractive index, according to Eq. 4–57, is $n = 87$ for the well-conducting copper, $n = 12$ for the poorer-conducting mercury, i.e., n is large compared to $n = 1$. With this assumption Eqs. 4–68 and 69 are valid up to rather large angles ϕ. Now, again by means of Kirchhoff's law, the reflectivities or the emissivities for both kinds of rays can be calculated:

$$N_{m,\lambda,\phi,n} = (1 - \rho_{m,\phi,n})N_{b,\lambda,\phi,n} = \frac{2 \cos \phi}{n} N_{b,\lambda,\phi,n} \tag{4–70}$$

and

$$N_{m,\lambda,\phi,p} = (1 - \rho_{m,\phi,p})N_{b,\lambda,\phi,p} = \frac{2}{n \cos \phi} N_{b,\lambda,\phi,p} \tag{4–71}$$

where the subscripts ϕ, n, or p mean: incidence angle ϕ, and oscillation normal or parallel to incidence plane, respectively. Considering that each of the two intensities $N_{b,\lambda,\phi,n}$ and $N_{b,\lambda,\phi,p}$ amounts to half that of the unpolarized black radiation, $N_{b,\lambda,\phi}$, and that $N_{b,\lambda,\phi} = N_{b,\lambda}$, addition of Eqs. 4–70 and 71 yields

$$N_{m,\lambda,\phi} = N_{m,\lambda,\phi,n} + N_{m,\lambda,\phi,p} = \frac{1}{n}\left(\cos \phi + \frac{1}{\cos \phi}\right) N_{b,\lambda} \tag{4–72}$$

In this equation n depends on the period of oscillation and on the electric conductivity of the metal. The sum, given by Eq. 4–72, is graphically represented in Fig. 4–9. The ratio $N_{m,\lambda,\phi,n}/N_{b,\lambda}$ is drawn as a vector issuing from the point O and describes a sphere with the diameter $1/n$. The ratio $N_{m,\lambda,\phi,p}/N_{b,\lambda}$ is also drawn as a vector, the locus of its end being a plane, parallel to the radiating element dA_1 at the distance $1/n$. Addition of the vectors $\overline{ON}$ and $\overline{OP}$ gives the vector $\overline{OS}$, instead

§ Defined in this way, the plane of incidence is the so-called plane of polarization.

of $\overline{OL}$, which would be the vector if Lambert's law were also valid for metals. In contrast to this law, the radiation has a minimum for $\phi = 0°$ and increases with ϕ, so that at $\phi = 75°$ it is already twice as

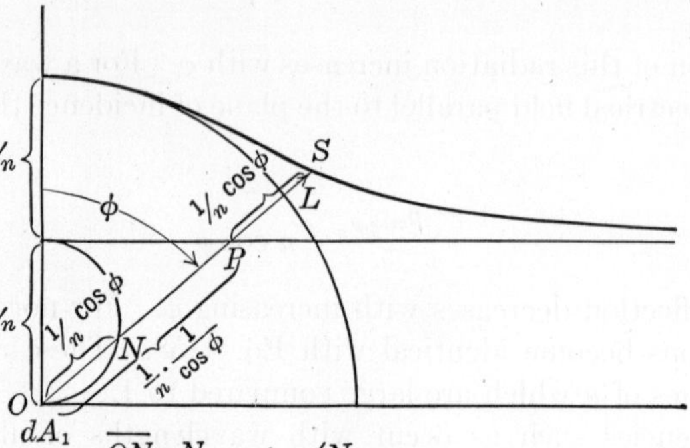

FIG. 4–9. Distribution of emissivity of electric conductors.

large as for $\phi = 0°$. For angles close to 90°, however, the intensity cannot increase to ∞, since $N_{m,\lambda,\phi,p}$ can never surpass the value $N_{b,\lambda,\phi,p} = \frac{1}{2}N_{b,\lambda}$, and thus

$$\overline{OP} = \frac{N_{m,\lambda,\phi,p}}{N_{b,\lambda}} \leq \frac{1}{2}$$

If, for instance, $n = 20$, then the radius $\overline{OL} = 2/n = 1/10$. Therefore $\overline{OP}$, which cannot exceed $\frac{1}{2}$, can only attain five times the length of the radius $2/n$ as a maximum.

Therefore the approximative Eqs. 4–68 and 69 will fail close to $\phi = 90°$. However, using them through the whole quadrant seems to influence only moderately the amount of the total radiation.

Hence the ratio of the total to the normal emission can be calculated as follows:

By integration of $N_{m,\lambda,\phi}$, as given in Eq. 4–72, over the hemispherical space, in analogy to the procedure in Eqs. 4–12 to 15, the total radiation for any wavelength becomes

$$\begin{aligned} W_{m,\lambda} &= \int_{\omega=0}^{\omega=2\pi} N_{m,\lambda,\phi} \cos\phi \cdot d\omega \\ &= \int_{\phi=0}^{\phi=\pi/2} N_{b,\lambda} \frac{1}{n}\left(\cos\phi + \frac{1}{\cos\phi}\right) \cos\phi \sin\phi \cdot d\phi \int_{\psi=0}^{\psi=2\pi} d\psi \\ &= \frac{8}{3} \cdot \frac{\pi}{n} N_{b,\lambda} \end{aligned} \tag{4–73}$$

But, according to Eq. 4–72, for $\phi = 0$

$$N_{m,\lambda,0} = \frac{2}{n} N_{b,\lambda} \tag{4-74}$$

By comparison of Eqs. 4–73 and 74

$$W_{m,\lambda} = \tfrac{4}{3}\pi N_{m,\lambda,0} \tag{4-75}$$

Integration with respect to λ from $\lambda = 0$ to $\lambda = \infty$ leads to the surface density of metallic total radiation

$$W_m = \tfrac{4}{3}\pi N_{m,0} \tag{4-76}$$

This is to be compared with

$$W_b = \pi N_b$$

for black radiation (see Eq. 4–15).

H. Schmidt and Furthmann [1928] uttered doubts as to this result; they demonstrated that Eq. 4–69 holds only if $\cos \phi$ is large compared with $1/n$, and for this reason the integration which leads to Eq. 4–73 is not exact. Therefore in a more recent paper E. Schmidt and Eckert [1935] used Davisson and Week's equation without simplifications.

First, Eq. 4–58 had to be replaced by

$$\rho_{m,\lambda,0} = \frac{2n^2 - 2n + 1}{2n^2 + 2n + 1}$$

$$= 1 - \frac{2}{n} + \frac{2}{n^2} - \frac{1}{n^3} + \frac{1}{2n^5} - \frac{1}{2n^6} \cdots \tag{4-77}$$

so that, instead of Eq. 4–59,

$$\rho_{m,\lambda,0} = \frac{60\lambda/r_e - 2\sqrt{30\lambda/r_e} + 1}{60\lambda/r_e + 2\sqrt{30\lambda/r_e} + 1}$$

$$= 1 - \frac{2}{\sqrt{30}}\sqrt{\frac{r_e}{\lambda}} + \frac{2}{30} \cdot \frac{r_e}{\lambda} + \cdots \tag{4-78}$$

Aschkinass, in deriving Eq. 4–61, had used the two first terms of this series only, Foote [1915] used the three terms of the right side of Eq. 4–78, but Schmidt and Eckert considered also the following terms in their approximative formula:

$$\rho_{m,\lambda,0} = 1 - \epsilon_{m,\lambda,0} = 1 - 0.365\sqrt{\frac{r_e}{\lambda}} + 0.0464\frac{r_e}{\lambda} \tag{4-79}$$

By integration

$$\rho_{m,0} = 1 - \epsilon_{m,0} = 1 - 0.576\sqrt{r_e T} + 0.124 r_e T \tag{4-80}$$

Also Eqs. 4–68 and 69 for oblique radiation had to be replaced by the more complicated equations of the exact electromagnetic theory of light:

$$\rho_{m,\lambda,\phi,n} = \frac{60\lambda/r_e - 2\sqrt{30\lambda/r_e}\cos\phi + \cos^2\phi}{60\lambda/r_e + 2\sqrt{30\lambda/r_e}\cos\phi + \cos^2\phi} \tag{4–81}$$

$$\rho_{m,\lambda,\phi,p} = \frac{60(\lambda/r_e)\cos^2\phi - 2\sqrt{30\lambda/r_e}\cos\phi + 1}{60(\lambda/r_e)\cos^2\phi + 2\sqrt{30\lambda/r_e}\cos\phi + 1} \tag{4–82}$$

For unpolarized radiation $\rho_{m,\lambda,\phi}$ can be found by representing the intensity of unpolarized radiation as the sum

$$N_{m,\lambda,\phi} = N_{m,\lambda,\phi,n} + N_{m,\lambda,\phi,p}$$

Because for black radiation

$$N_{b,\lambda,\phi} = 2N_{b,\lambda,\phi,n} = 2N_{b,\lambda,\phi,p}$$

Kirchhoff's law yields

$$\epsilon_{m,\lambda,\phi} = \frac{\epsilon_{m,\lambda,\phi,n} + \epsilon_{m,\lambda,\phi,p}}{2}$$

or

$$1 - \rho_{m,\lambda,\phi} = \frac{(1 - \rho_{m,\lambda,\phi,n}) + (1 - \rho_{m,\lambda,\phi,p})}{2}$$

or

$$\rho_{m,\lambda,\phi} = \frac{\rho_{m,\lambda,\phi,n} + \rho_{m,\lambda,\phi,p}}{2} \tag{4–83}$$

Schmidt and Eckert substituted the reflectivities from Eqs. 4–81 and 82 in Eq. 4–83, graphically integrated this equation over the hemispheric space, and represented their result by empirical equations for two ranges, namely:

For $0 < r_e/\lambda < 0.5$,

$$\epsilon_{m,\lambda} = 0.476\sqrt{\frac{r_e}{\lambda}} - 0.148\,\frac{r_e}{\lambda} \tag{4–84}$$

For $0.5 < r_e/\lambda < 2.5$,

$$\epsilon_{m,\lambda} = 0.442\sqrt{\frac{r_e}{\lambda}} - 0.0995\,\frac{r_e}{\lambda} \tag{4–85}$$

Further, by integration from $\lambda = 0$ to $\lambda = \infty$:
For $0 < r_eT < 0.2$,

$$\epsilon_m = 0.751\sqrt{r_eT} - 0.396r_eT \tag{4–86}$$

For $0.2 < r_eT < 0.5$,

$$\epsilon_m = 0.698\sqrt{r_eT} - 0.266r_eT \tag{4–87}$$

Finally, they considered that, according to Kirchhoff's law and Stefan-Boltzmann's law for a temperature T,

$$W_m = \epsilon_m W_b = \epsilon_m \int_0^\infty W_{b,\lambda}\, d\lambda = \epsilon_m \sigma T^4 \tag{4–88}$$

Substituting ϵ_m from Eqs. 4–86 and 87,
For $0 < r_eT < 0.2$,

$$W_m = \sigma(0.751\sqrt{r_e}T^{4.5} - 0.396r_eT^5) \tag{4–89}$$

For $0.2 < r_eT < 0.5$,

$$W_m = \sigma(0.698\sqrt{r_e}T^{4.5} - 0.266r_eT^5) \tag{4–90}$$

These equations replace Aschkinass' Eq. 4–66 for the total radiation of metallic surfaces.

If $\epsilon_{m,0}$ has been found by an experiment, r_eT may be expressed as a function of $\epsilon_{m,0}$ by means of Eq. 4–80. Substituting, then, r_eT in Eq. 4–86 or 87, values of $\epsilon_m/\epsilon_{m,0}$ can be found, corresponding to those for non-metallic surfaces (see Table 4–3).

Some values, calculated in this way, are given in Table 4–4. The table indicates that the factor $\frac{4}{3}$ of Eqs. 4–75 and 76 is only a limit for

TABLE 4–4

RATIO OF HEMISPHERIC TO NORMAL EMISSIVITY OF A METALLIC SURFACE

$\epsilon_{m,0}$	0	0.05	0.1	0.15	0.2	0.25	0.3	0.35
$\epsilon_m/\epsilon_{m,0}$	1.33	1.27	1.225	1.185	1.145	1.105	1.075	1.055

very small values of $\epsilon_{m,0}$, and that for metals the total emissivity becomes 1.33 to 1.05 times the normal emissivity, whereas for electrical insulators the corresponding ratio has been found to be between 1.05 to 0.95. (See Table 4–3.)

Figure 4–10 shows the theoretical values, according to Tables 4–3 and 4.|| They will later be compared with experimental values. (See Tables 7–3 and 7.)

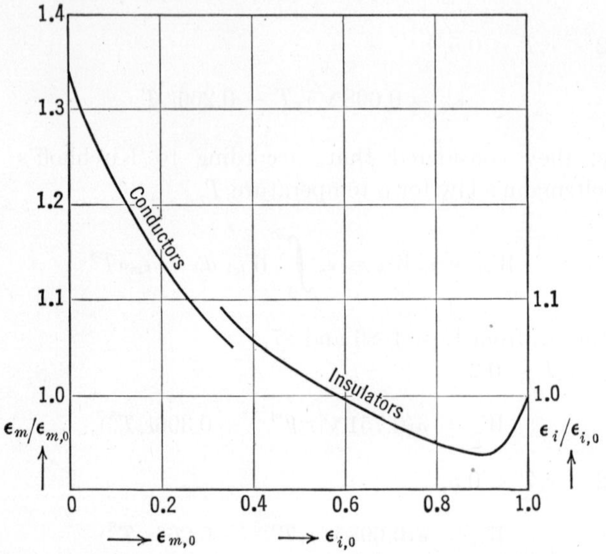

FIG. 4–10. Ratio of hemispheric to normal emissivity of radiating surfaces.

4–11. Gas Radiation

Because gases in the range of visible light do not appreciably absorb radiation, it was assumed for quite some time that they do not absorb and emit radiation at all. Extension of the spectral investigation to the infrared range, however, revealed that, owing to changes in the energy level of interatomic oscillation and of molecular rotation, some gases absorb radiation and therefore, according to Kirchhoff's law, must also emit rays of the same wavelengths. For many years this seemed to be of interest to physicists only. Schack [1924] was possibly the first to recognize the practical importance of gas radiation, particularly in furnaces.

Whereas solid bodies show a continuous spectrum, selective radiation is characteristic for gases. This means that they emit radiant energy

|| The curve for insulators is considerably lower than the curve of E. Schmidt and Eckert [1935] who made mistakes in the numerical calculations. The present new calculation was undertaken when a student in my class, Mr. J. Kordig, found that the value of $\epsilon_{i,\phi}$ for $n = 3$, $\phi = 80°$ in Table 4–1 was 0.567, whereas E. Schmidt [1927] had found 0.540. Our calculation was extended up to $n = 10$. The gap between the curves for insulators and conductors at this point may be caused by the approximations used in the derivation of Eqs. 4–80, 86, and 87.

only in some narrow regions of wavelength which are called bands. For most gases the width of the bands is so small that no considerable amounts of energy are radiated; but radiation has to be considered in quite a few gases, particularly some which take a part in combustion processes, as carbon dioxide, water vapor, carbon monoxide, methane, sulfur dioxide.

For solid bodies it was found sufficient to use the concept of "absorptivity" without considering the mechanism of absorption which, of course, is hidden to rough inspection from outside. The absorption of a gas, however, can be analyzed more easily. The law of absorption is as general as, and perhaps even more general than, Kirchhoff's law. It can be represented by the simple statement that in a fixed differential path element the fraction of radiation absorbed is the same for any radiant density, or that absorption of energy is proportional to the local intensity and to the differential distance travelled by the radiation. Let $N_{\lambda,0}$ be the initial intensity of radiation,¶ and $N_{\lambda,x}$ the intensity when the rays have travelled from $x = 0$ to $x = x$.* Then the law of absorption may be expressed by

$$dN_{\lambda,x} = -m_\lambda N_{\lambda,x}\,dx \tag{4–91}$$

where m_λ is a factor of proportionality. The subscript is necessary because m_λ depends upon the wavelength λ, which, however, is no variable in Eq. 4–91. Hence integration of the equation yields

$$\ln N_{\lambda,x} - \ln C = -m_\lambda x$$

where C is a constant. For $x = 0$

$$\ln N_{\lambda,0} - \ln C = 0$$

so that

$$\ln \frac{N_{\lambda,x}}{N_{\lambda,0}} = -m_\lambda x$$

or

$$N_{\lambda,x} = N_{\lambda,0} e^{-m_\lambda x} \tag{4–92}$$

Here m_λ plays the same part as the so-called logarithmic decrement does in the equation of a damped mechanical oscillation. Therefore it may be called the logarithmic decrement of radiation. It has the dimension

¶ So far, N has been defined as the intensity for unit solid angle. Speaking of parallel rays in the following deduction means that bundles of differential thickness will be considered, corresponding to differential solid angles, $d\omega$.

* Intensity of radiation is proportional to the number of electrons in a portion of space. Equal numbers of these suffer equal absorption. This is the reason why $dN_{\lambda,x} \sim N_{\lambda,x}$.

[cm^{-1}] in the physical, and [ft^{-1}] in the British technical system of units.

The decrement m_λ is a very peculiar property of a gas. Apparently it depends upon the type and average state of oscillation and upon the number of the molecules hit by the radiation. Thus, it will depend on the kind of gas, its temperature, and its state of incitement, in other words, on the location and strength of the bands and also on the pressure, because the number of molecules is proportional to the pressure. Considering the last item, one can introduce

$$m_{\lambda,p} = m_{\lambda,a}\frac{p}{p_a} \tag{4–93}$$

where $m_{\lambda,a}$ is the decrement at a standard pressure p_a, for instance, atmospheric pressure. For gas mixtures p is the partial pressure, but it will be shown later that for some of them a more complicated relation exists.

According to Eq. 4–92, the absorbed amount of energy will be

$$N_{\lambda,0} - N_{\lambda,0}e^{-m_\lambda x} = N_{\lambda,0}(1 - e^{-m_\lambda x})$$

Using this, an "apparent absorptivity" of a gas layer of thickness x for radiation in the direction x may be defined by

$$\alpha_{\lambda,x} = \frac{N_{\lambda,0} - N_{\lambda,0}e^{-m_\lambda x}}{N_{\lambda,0}} = 1 - e^{-m_\lambda x} \tag{4–94}$$

In contrast to the absorptivity α_λ of an opaque surface, the apparent absorptivity $\alpha_{\lambda,x}$ is not a property of the absorbing substance alone, but depends also on a length x.

For $x = \infty$ the absorptivity becomes $\alpha_{\lambda,\infty} = 1$. Thus, a gas layer of infinite thickness can be considered a black body, as far as radiation in the wavelength range of a band is concerned.

Let this range be $\Delta\lambda = \lambda_2 - \lambda_1$. Then by means of Planck's law for black radiation

$$\Delta N_b = \int_{\lambda_1}^{\lambda_2} N_{b,\lambda}\, d\lambda \tag{4–95}$$

where ΔN_b is represented by the area shaded in Fig. 4–11, and may be called the areal intensity of black gas radiation of the band.

By means of Eq. 4–94 a more general apparent absorptivity $\alpha_{\lambda,A,V}$ of an arbitrary gas volume V can be found through which rays of the wavelength λ are transmitted, coming from or going to a solid surface A which envelopes the gas body.

The surface A may radiate diffusely to all directions. Therefore the integration of the absorbed radiation must be extended to all directions. Thus $\alpha_{\lambda,A,V}$ will depend on m_λ and the shape and size of the gas body. For simplification the subscript λ will now be omitted in deriving a general relation for the apparent absorptivity of a gas body.

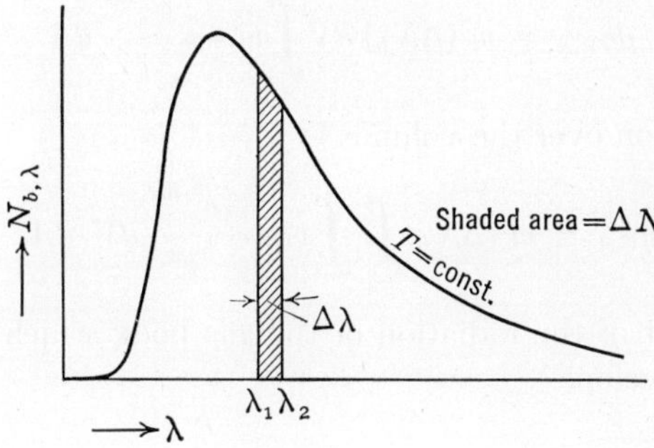

FIG. 4-11. Areal intensity ΔN_b of black gas radiation.

As has been found by Koenigsberger [1903] † the total radiation of a volume element dV of a gas is

$$dq_V = 4\pi m(\Delta N_b)\, dV \tag{4-96}$$

so that the radiation into a solid angle $d\omega$ will be

$$d^2q_V = 4\pi m(\Delta N_b)\, dV \frac{d\omega}{4\pi} = m\,(\Delta N_b)\, dV \cdot d\omega \tag{4-97}$$

Starting from this equation, Nusselt [1923], whose derivation we are following, assumed that m is independent of the temperature, and that the gas volume and its envelope are at the same temperature. Figure 4-12 represents a gas volume limited by a solid surface with the area A. The solid angle $d\omega_A$ in which a differential volume dV is radiating to the surface element dA at the distance r is given by

$$d\omega_A = \cos\phi \frac{dA}{r^2}$$

Substituting this in Eq. 4-97, one obtains

$$d^2q_V = m\,(\Delta N_b)\, dV \cdot \cos\phi \frac{dA}{r^2}$$

† A simpler and also general proof of Eq. 4-96 has recently been published by Jakob [1947].

Only a part of this will hit the surface element dA, namely,

$$d^2q_{V,A} = d^2q_V e^{-mr} = m\,(\Delta N_b)\,dV \cdot \cos\phi \,\frac{e^{-mr}}{r^2}\,dA$$

By integration over the total surface area A

$$dq_{V,A} = m\,(\Delta N_b)\,dV \int_A \cos\phi \,\frac{e^{-mr}}{r^2}\,dA$$

and by integration over the volume V

$$q_{V,A} = m\,(\Delta N_b) \int_V \int_A \cos\phi \,\frac{e^{-mr}}{r^2}\,dV \cdot dA \tag{4-98}$$

This is that part of the radiation of the gas body which hits the inner surface of its envelope.

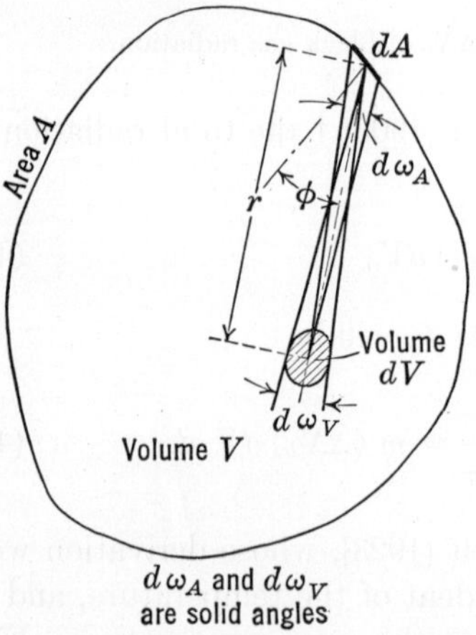

FIG. 4-12. Geometrical data for the calculation of the apparent absorptivity for radiation of a gas body.

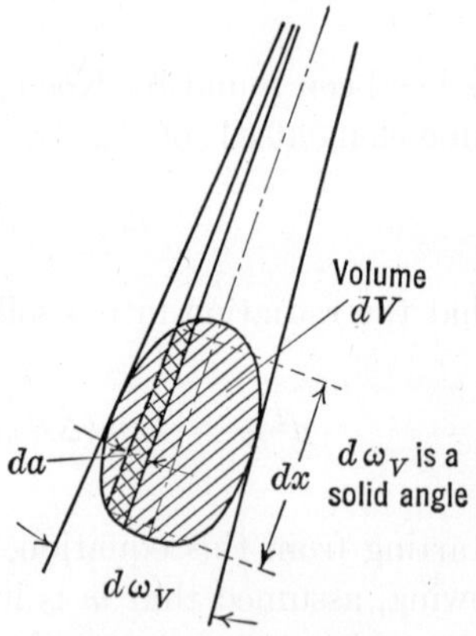

FIG. 4-13. A detail of Fig. 4-12.

Now the emission of this surface was defined as a gray radiator. Therefore, in analogy to Eq. 4-12 and with the denotations of Fig. 4-13, the emission of a surface element dA into the space angle $d\omega_V$ subtended by dV is

$$d^2q_A = N_{gr} \cos\phi \cdot d\omega_V\,dA \tag{4-99}$$

where N_{gr} is an emissivity factor for gray radiation normal to the surface. Again, only a part of this emission hits the volume element dV, namely,

$$d^2q_{A,V} = N_{gr} \cos\phi \cdot e^{-mr}\,d\omega_V\,dA$$

To find the fraction of the arriving energy which is absorbed in the volume element dV, it must be considered that the cross-sectional area of the bundle of rays coming from dA is $r^2\, d\omega_V$. Hence the density of radiation at dV is

$$(dq_{A,V})'' = \frac{d^2q_{A,V}}{r^2\, d\omega_V} = N_{gr} \frac{\cos\phi \cdot e^{-mr}\, d\omega_V\, dA}{r^2\, d\omega_V} = N_{gr} \cos\phi \cdot \frac{e^{-mr}}{r^2}\, dA \tag{4–100}$$

Now the bundle of rays, which crosses dV, may be replaced by still thinner prismatic bundles, and dx may be the length, and da the cross-sectional area of one of these elementary prisms. The radiation crossing it is $(dq_{A,V})''\, da$ and the amount $m(dq_{A,V})''\, da \cdot dx$ will be absorbed in passing through. Thus for the volume element dV the absorbed portion of radiation will be

$$d^2q_{A,V,\alpha} = m(dq_{A,V})'' \int_{dV} (da \cdot dx) = m(dq_{A,V})''\, dV$$

and by substitution of $(dq_{A,V})''$ from Eq. 4–100

$$d^2q_{A,V,\alpha} = mN_{gr} \cos\phi \cdot \frac{e^{-mr}}{r^2}\, dA \cdot dV$$

By integration over V and A

$$q_{A,V,\alpha} = mN_{gr} \int_V \int_A \cos\phi \cdot \frac{e^{-mr}}{r^2}\, dA \cdot dV \tag{4–101}$$

This is the part of the radiation of the surface A which is absorbed by the total gas volume V.

From Eqs. 4–98 and 101

$$q_{A,V,\alpha} = \frac{N_{gr}}{\Delta N_b} q_{V,A}$$

On the other hand, by analogy to Eq. 4–15, the total emission of the surface A is

$$q_A = \pi N_{gr} A \tag{4–102}$$

and therefore the apparent absorptivity $\alpha_{A,V}$ of the gas body becomes

$$\alpha_{A,V} = \frac{q_{A,V,\alpha}}{q_A} = \frac{q_{V,A}}{\pi(\Delta N_b)A} \tag{4–103}$$

so that the radiation of a gas body of the volume V and the surface area A is

$$q_{V,A} = \alpha_{A,V}\pi(\Delta N_b)A \tag{4–104}$$

Finally, by substitution of $q_{V,A}$ from Eq. 4–98 in Eq. 4–103,

$$\alpha_{A,V} = \frac{m}{\pi A} \int_V \int_A \cos\phi \cdot \frac{e^{-mr}}{r^2}\, dA \cdot dV \qquad (4\text{–}105)$$

Herewith the apparent absorptivity of an arbitrary gas body for radiation of wavelength λ is given as a function of the dimensions of that body and of the decrement of radiation m of the gas. Some integrations of this general expression will be given in Chapter 32 (Vol. II).

The equations of this section relate only to gases without suspensions of solid particles. The radiation of flames, however, i.e., of gases which by the radiation of tiny particles of solid carbon compounds become luminous, can be calculated in a similar way, as far as appropriate numerical values of the decrement of radiation m are known. Properly spoken, the radiation of luminous flames is a complex of surface and gas radiation, the surfaces being formed by the solid particles which burn in the gas.

PART B. PROPERTIES OF MATTER SIGNIFICANT FOR HEAT TRANSFER

CHAPTER 5

Viscosity

5–1. The Role of Viscosity

The fundamental equations for heat conduction, convection, and radiation contain some properties of matter which are defined by those equations and can be determined by experiments which are based on them. On the other hand, these properties must be known if one wants to apply the equations for other purposes. In this part of the book we shall not deal with the methods used for determining numerical values of those properties, but only with the results, i.e., what values seem to be the most reliable, how their order of magnitude and change with physical conditions, as temperature and pressure, can be explained, and whether and how far some of the properties are interdependent.

It will not be necessary to deal with well-known properties, as density, thermal expansibility, and specific heat, which also occur in the equations, but mainly with viscosity, thermal conductivity, and absorptivity for radiation. Although thermal conductivity is possibly the most important and popular of these properties, we shall start with viscosity, because learning about its physical signification helps in understanding that of the thermal conductivity of fluids and because according to Stokes's equations viscosity controls all forms of heat convection.

As explained in Sect. 3–5, there are two kinds of viscosity, the dynamic and the kinematic viscosity. The latter plays a particular role in the theory and application of similarity, and will be found to be analogous to the thermal diffusivity.

5–2. Viscosity of Liquids

Whereas a tremendous amount of experimental data on the viscosity of liquids has been accumulated, the theory of liquids is still in its beginning. Therefore only empirical facts can be given.

The most striking and generally known phenomenon is the dependence of viscosity upon temperature. For water, for instance, the dynamic viscosity μ decreases with increasing temperature as shown in Table 5-1.

TABLE 5-1

DYNAMIC VISCOSITY μ OF WATER

t [°C]	0	20	40	60	80	100
μ [cp]	1.789	1.005	0.653	0.470	0.356	0.282
t [°C]	150	200	250	300	350	374
μ [cp]	0.185	0.136	0.112	0.095	0.070	0.0375

At 20°C the viscosity of water is close to 1 cp (centipoise), as mentioned already in Sect. 3-5. The values above 100°C relate to the saturated state of water; 374°C is very near to the critical point where water and steam become identical. Referring to absolute temperatures T, an increase from $T = 273$ to $T = 647$°K, i.e., at the ratio 1:2.4, causes a decrease of μ from 1.8 to 0.0375 cp, i.e., almost at the ratio 48:1.

For other liquids empirical equations with three constants have been found suitable. Reference may only be made to the investigation of 87 liquids by Thorpe and Rodger [1894, 1897], and for illustration a few numerical values are given in Table 5-2.

TABLE 5-2

DYNAMIC VISCOSITY μ OF SOME LIQUIDS IN CENTIPOISE (CP)

Liquid	Chemical Formula	Temperature [°C]			
		0	50	100	150
Hexane	C_6H_{14}	0.402	0.247	.....	.....
Toluene	C_7H_8	0.765	0.422	0.272	0.185
Benzene	C_6H_6	0.900	0.433	0.263	0.170
Ethyl alcohol	C_2H_6O	1.77	0.702	0.325	0.166
Water	H_2O	1.79	0.550	0.282	0.184
Aniline	C_6H_7N	10.2	1.85	.....	.....
Sulfuric acid	H_2SO_4	54	10.3	.....	.....
Glycerine	$C_3H_8O_3$	5400	175	.....	.....
Castor oil	....	7200	128	.....	.....

The influence of pressure is not considerable in the practical range of heat-transfer work. The viscosity of water at 100°C, for instance, increases by about 20% only if the pressure is increased from 1 to 300 atmospheres.

In considering the influence of chemical composition and constitution the choice of an appropriate basis of comparison is important. It is known, for instance, that the viscosity of organic liquids belonging to the same homologous series is approximately the same at the boiling point.

With mixtures of liquids the ordinary mixing rule is almost never applicable. At 0°C, for instance, for water $\mu = 1.79$ cp, and for ethyl alcohol $\mu = 1.77$, but a mixture of 64% water and 36% alcohol has a viscosity of 7.34 cp. Therefore it is difficult to estimate the viscosity of a mixture even if the viscosity of its liquid components is known. Fortunately plenty of experimental data are given in physical-chemical tables.

Finally it may be mentioned that, because the density of most liquids is of the order of magnitude of 1 g/cm^3, the numerical value of the kinematic viscosity in physical units does not differ much from that of the dynamic viscosity in poise.

5-3. Theoretical Relations about the Viscosity of Gases

Some strange relations concerning the viscosity μ and the thermal conductivity k of gases can be understood only by means of the kinetic (statistical) theory of gases; as a matter of fact, the possibly most peculiar phenomenon, the independence of μ and k of the pressure in a rather wide range, has been discovered theoretically first, and only afterwards checked by experiments. Furthermore, the so-called Prandtl number, a dimensionless group of properties, which is of influence in most processes of heat convection, also plays a role in the statistical theory. For these reasons some simple derivations of the gas theory will be presented.

A gas between two horizontal plates will be considered, the lower at rest, the upper moved horizontally with the constant velocity u_L in the direction x (see Fig. 5-1). Then, according to experience, the gas is carried along with linearly increasing velocity, which at the height z above the lower plate will be

$$u = \frac{u_L}{L} z$$

The molecules at this height will move in all directions of the space with very different velocities, but their average apparent velocity will be in the direction x and of the magnitude u. The average free path λ of the molecules is defined as the average length of their way between two subsequent impacts.

By the random movements of the molecules mechanical momentum will be transported from places above z to places below z. It is required to find the momentum carried through a horizontal area of 1 cm^2 at the height z. As an approximation it will be assumed that the molecules travel only parallel to the three coordinate axes x, y, and z, one third of them parallel to each axis, and therefore one sixth in the positive and negative directions. Let n be the number of molecules in 1 cm^3 and w

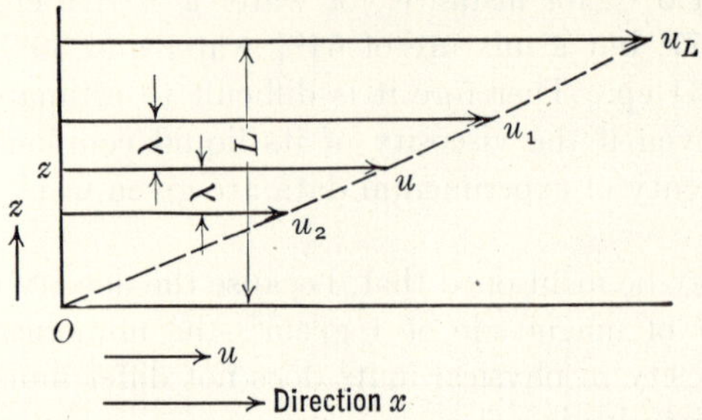

FIG. 5-1. Velocity distribution between a fixed and a moving plate.

their average velocity in the z direction.* Then, in 1 sec $nw/6$ molecules will hit the considered area from above. The last time they suffered impacts from other molecules was at a distance λ above z, i.e., at $z + \lambda$, and at that time they assumed the average velocity u_1, which the molecules at that height had in the x direction. Thus they arrive at the height z with the velocity

$$u_1 = u + \lambda \frac{\partial u}{\partial z}$$

For the same reasons, the molecules coming from below arrive at z with the velocity

$$u_2 = u - \lambda \frac{\partial u}{\partial z}$$

Those coming from above in 1 sec carry the momentum $(nw/6)(mu_1)$ downwards, those from below carry $(nw/6)(mu_2)$ upwards, m being the mass of one molecule. Adding these effects, we find the momentum $(nw/6)m(u_1 - u_2) = (nw/3)m\lambda \cdot \partial u/\partial z$ carried downwards per second. Now, according to the law of impulse, this change of momentum in

* The random velocities of the molecules in the x and y directions will also be equal to w.

unit time corresponds to a force of the same amount acting on an area of 1 cm^2, at the height z, i.e., the shearing stress S_s, so that

$$S_s = \frac{nw}{3} m\lambda \frac{\partial u}{\partial z} \tag{5–1}$$

On the other hand, according to the definition of dynamic viscosity,

$$S_s = \mu \frac{\partial u}{\partial z} \tag{5–2}$$

Consider further that the gas density is

$$\rho = nm \tag{5–3}$$

From Eqs. 5–1, 2, and 3

$$\mu = \frac{nw}{3} m\lambda = \frac{1}{3}\lambda w\rho \tag{5–4}$$

Although this equation is based on the rough assumption that molecules travel in the three directions of coordinates only, it comes very close to the actual behavior. More exact derivations lead to

$$\mu = \eta\lambda w\rho \tag{5–5}$$

where η is a constant higher than the value $\frac{1}{3}$ which originates from Maxwell (0.35 according to Boltzmann).

Because the average molecular velocity w is constant for a given temperature, and the free path length λ is reciprocal to the number of molecules or to the density ρ, Eq. 5–4 shows that μ is independent of the density or pressure. This surprising result was first derived from the theory by Maxwell [1860]; when confirmed by experiments, it became one of the strongest supports of the kinetic theory.

However, it holds only for pressures at which the laws of ideal gases can be assumed. If the pressure is so high that the gas approaches the behavior of a saturated vapor, μ varies considerably with pressure, as we shall see.

At very small pressure, on the other hand, the average path length of molecules is not small compared with the dimensions of the vessels or tubes which contain the gas. Then the irregular molecular motion cannot be replaced by the flow of parallel layers, as was done in connection with Fig. 5–1, and therefore the derivation which led to Eq. 5–1 is not valid any more. In this derivation only mutual collisions of the molecules and no collisions with the walls were considered. But at pressures below, say, 0.1 mm Hg, molecules may fly from one spot of the wall of a narrow vessel or tube to another spot of the wall without hitting any other molecule on the way. For air at 0°C and atmospheric pressure, for instance, the average free path length is about 10^{-5} cm. At 0.1 mm Hg the free path would be only 0.1 cm. In a tube of 1 cm diameter this would already cause sensible deviations from the relation derived above; in a capillary tube of 1 mm diameter this relation would fail entirely. Therefore gases at a very good vacuum act as though they were less viscous and as though, for instance in the arrangement of Fig. 5-1, an unsteadiness of velocity occurred at the wall. The gas seems to slide at the wall.

The influence of temperature can be derived from the statement of Maxwell that from a kinetic viewpoint the absolute temperature T of a gas is equivalent to the average of the kinetic energy $mw^2/2$ of the molecules. Thus, if the free path length λ were independent of temperature, Eq. 5-4 or 5 would require that μ be proportional to $\sqrt{T}$. Measurements of μ, however, have shown a stronger increase of μ than proportional to $\sqrt{T}$. This is considered in Sutherland's [1893] formula for the average free path of molecules

$$\lambda_S = \lambda_M \frac{1}{1 + (C/T)} \tag{5-6}$$

where the subscripts S and M relate to Sutherland and Maxwell, respectively, and C is a constant value. The term C/T is due to the mutual attraction of the molecules, which reduces the free path length.

This equation is in satisfactory agreement with the experiments and is often used. At high temperature the influence of the kinetic energy, which is taken care of by λ_M, dominates compared with the "potential" correction term C/T. For $T = \infty$, C/T disappears entirely.

Substituting λ_S from Eq. 5-6 in Eq. 5-5,

$$\mu = \eta \lambda_M w \rho \frac{1}{1 + (C/T)} \tag{5-7}$$

and indicating a standard state by the subscript 0,

$$\mu_0 = \eta (\lambda_M)_0 w_0 \rho_0 \frac{1}{1 + (C/T_0)} \tag{5-8}$$

Because

$$\lambda_M \rho = (\lambda_M)_0 \rho_0$$

it follows from Eqs. 5-7 and 8 that

$$\mu = \mu_0 \frac{w}{w_0} \cdot \frac{1 + (C/T_0)}{1 + (C/T)} = \mu_0 \sqrt{\frac{T}{T_0}} \cdot \frac{1 + (C/T_0)}{1 + (C/T)} = \frac{B\sqrt{T}}{1 + (C/T)} \tag{5-9}$$

where

$$B = \mu_0 \frac{1 + (C/T_0)}{\sqrt{T_0}}$$

If μ is measured at two temperatures, the two Sutherland constants B and C can be obtained from Eq. 5–9. For very low temperatures this equation is not sufficient, and an expression with three constants has been developed by Chapman [1912].

Considering that for ideal gases $\rho = \rho_0 T_0/T$, Eq. 5–9 yields the following formula for the kinematic viscosity:

$$\nu = \frac{\mu}{\rho} = \nu_0 \sqrt{\frac{T^3}{T_0{}^3}} \cdot \frac{1 + (C/T_0)}{1 + (C/T)} = \frac{E\sqrt{T^3}}{1 + (C/T)} \qquad (5\text{–}10)$$

where

$$E = \nu_0 \frac{1 + (C/T_0)}{\sqrt{T_0{}^3}}$$

Referring to the pressure influence, $\nu = \mu/\rho$ must be reciprocal to the pressure because in the ordinary range μ is independent of, and ρ proportional to, the pressure. To understand this, the fundamental equation of dynamics, Force = Mass × Acceleration, may be applied to a flowing gas. The force is negative and proportional to μ, and the mass is proportional to ρ, so that $\nu = \mu/\rho$ is proportional to a negative acceleration, i.e., a measure for the retardation, and this, of course, becomes smaller if a larger mass is to be retarded.

Finally it may be mentioned that for gas mixtures the theory leads to very complex equations, owing to the fact that molecules of any constituent collide with each other and with those of the other constituents.

5-4. Numerical Values of the Viscosity of Gases and Vapors

To give an idea of the order of magnitude of the viscosity of gases, a few representative numerical values of μ for atmospheric or smaller pressure are given in Table 5-3.

For air at 0°C $\mu = 0.017$ cp, whereas for water $\mu = 1.79$ cp, i.e., about 100 times larger.

It is remarkable that, in contrast to the behavior of liquids, the viscosity of gases increases with temperature, as follows likewise from Sutherland's Eq. 5–9.

TABLE 5–3

DYNAMIC VISCOSITY $10^6\,\mu$ OF SOME GASES IN POISE

Gas	Chemical Formula	Temperature [°C]					
		−200	−100	0	100	200	300
Benzene	C_6H_6			71	95	120	
Hydrogen	H_2	33	61	85	103	120	138
Carbon tetrachloride	CCl_4			93	122	151	181
Ammonia	NH_3			93	130		
Carbon dioxide	CO_2			139	188		
Nitrogen	N_2		114	165	208	246	
Air	....		120	171	218	259	296
Helium	He	80	140	190	234		
Oxygen	O_2		132.5	193	247	295	

Table 5–4 shows some values of μ for steam, according to the most recent experiments of Hawkins, Solberg, and Potter [1940].

TABLE 5–4

DYNAMIC VISCOSITY $10^6\,\mu$ OF STEAM IN POISE

t [°F]	p [lb/sq in.]						
	100	200	400	600	800	1000	1500
400	185	202					
500	200	212	232	251			
600	216	226	246	259	274	288	323
800	249	257	270	283	295	307	334
1000	283	288	298	309	318	328	

The values in Table 5–4 are considerably higher than those given by Timrot [1940]. A review of all available data (also for water) has been published by Hawkins, Sibbitt, and Solberg [1948].

Table 5–5 contains a few values of the kinematic viscosity of gases.

TABLE 5-5

KINEMATIC VISCOSITY ν OF SOME GASES AT 760 MM HG IN CM^2/SEC

Gas	Chemical Formula	Temperature [°C]					
		−200	−100	0	100	200	300
Hydrogen	H_2	0.10	0.43	0.95	1.6	2.3	3.2
Ammonia	NH_3			0.12	0.24		
Carbon dioxide	CO_2			0.07	0.13		
Air	....		0.059	0.13	0.23	0.35	0.48

As in Table 5-3, the pressure is so low that the gases can be considered to be nearly ideal. At 0°C and saturation conditions, on the other hand, the values of ν would be 0.013 and 0.028 for NH_3 and CO_2, respectively.

Again, it must be mentioned, that the ordinary mixing rule cannot be applied for finding the viscosity of gas mixtures.

5-5. Summary about Viscosity

The main results of the foregoing sections can be summarized as follows:

1. Both the dynamic viscosity μ and the kinematic viscosity ν increase with rising temperature for gases and decrease with rising temperature for liquids.

2. For gases, μ depends only slightly on pressure. Because, on the other hand, ρ is almost proportional to the pressure, $\nu = \mu/\rho$ is reciprocal to the pressure.

3. Whereas μ is much smaller for gases than for liquids, the opposite is true for ν. Thus the heaviest liquid, mercury, has the smallest kinematic viscosity, namely, about 0.0012 cm^2/sec at 20°C, whereas for hydrogen at about 0.01 atmosphere and 0°C, $\nu = 94$ cm^2/sec, so that the practical range of kinematic viscosities is given by the ratio 1/80,000.

4. Extreme values for μ are 0.0075 cp for propane gas, and, almost one million times larger, 7200 cp for castor oil, both at 0°C.

CHAPTER 6

Thermal Conductivity

6–1. Numerical Values of the Thermal Conductivity of Gases and Vapors

Thermal conductivities k will be dealt with in a sequence of increasing numerical values, i.e., beginning with the poorly conducting gases and ending with metals.

The values of k, given in literature, are mostly for "low pressure." This not very exact notation is sufficient because the thermal conductivity, like the dynamic viscosity, of gases is virtually independent of the pressure, provided this is not too far from atmospheric conditions. Before deriving this from the kinetic theory of gases, some numerical values, collected in Tables 6–1, 2, and 3, may give an idea of the order of

TABLE 6–1

THERMAL CONDUCTIVITY OF THE LIGHTEST GASES

10^3k [B hr^{-1} ft^{-1} F^{-1}]

Gas	Chemical Formula	Molecular Weight	Temperature [°C]			
			−200	−100	0	100
Hydrogen	H_2	2	29	67	99	124
Helium	He	4	31	58	82	97

magnitude of k for some representative gases and show some peculiarities which will also be explained later by means of the kinetic theory. The values in these tables belong to such pressures at which the pressure influence is negligible. For instance, the values given for steam in Table 6–2 relate to the superheated state at low pressure.

Tables 6–1 to 3 show that k increases with temperature and that, in general, heavier gases conduct the heat not so well as lighter ones.

TABLE 6–2

THERMAL CONDUCTIVITY OF SOME INORGANIC GASES

$10^3 k$ [B hr^{-1} ft^{-1} F^{-1}]

Gas	Chemical Formula	Molecular Weight	Temperature [°C]				
			−200	−100	0	100	200
Ammonia	NH_3	17	...	...	12.4	17.1	....
Steam	H_2O	18	...	...	.....	14	19
Carbon monoxide	CO	28	3.4	8.1	12.6	.....	....
Nitrogen	N_2	28	4.1	9.4	14.1	17.5	....
Air	....	29	...	9.4	14.1	17.8	21.2
Oxygen	O_2	32	3.6	9.4	14.1	18.1	....
Carbon dioxide	CO_2	44	...	...	8.1	12.1	15.5
Xenon	X	131	...	...	3.32 [1]	....	....

[1] Calculated from Eq. 6–1.

Deviations from the latter rule, for instance, with steam, may partly be due to association of molecules. Also the number of atoms in the molecule plays a role.

TABLE 6–3

THERMAL CONDUCTIVITY OF SOME ORGANIC GASES

$10^3 k$ [B hr^{-1} ft^{-1} F^{-1}]

Gas	Chemical Formula	Molecular Weight	Temperature [°C]			
			−100	0	100	200
Methane	CH_4	16	10.4	17.5	....	....
Ethylene	C_2H_4	28	....	9.7	15.4	....
Ethane	C_2H_6	30	....	10.4	18.5	....
Ethyl alcohol	C_2H_6O	46	....	8.1	12.1	....
Ethyl ether	$C_4H_{10}O$	74	....	7.4	12.8	19.5
Benzene	C_6H_6	78	....	5.0	10.1	16.1
Carbon tetrachloride	CCl_4	154	....	....	5.0	6.4

The thermal conductivity of mixtures, like their viscosity, cannot be calculated by the ordinary arithmetic mixing rule. Wassiljewa [1917] seems to have first observed this. She found that mixtures of oxygen and hydrogen had a much smaller conductivity than the simple mixing rule would give. This has been confirmed by other researchers for mixtures of relatively heavy gases with light ones (hydrogen, helium). For a 50% O_2-H_2 mixture at 0°C, for instance, k is 20% smaller than it would be according to the rule mentioned. Gruess and Schmick [1928] and Schmick [1928] found the opposite behavior with binary mixtures of gases whose constituents have not the same kind of atomic symmetry and not too small Sutherland constants. A mixture of 60% air and 40% ammonia at 80°C, for instance, has an 8% larger conductivity than it would according to the mixing rule. Graphs for a number of binary mixtures are given in an article of Gruess [1933].

As an example of the influence of higher pressure and temperature upon the thermal conductivity, Fig. 6–1 represents the experimental

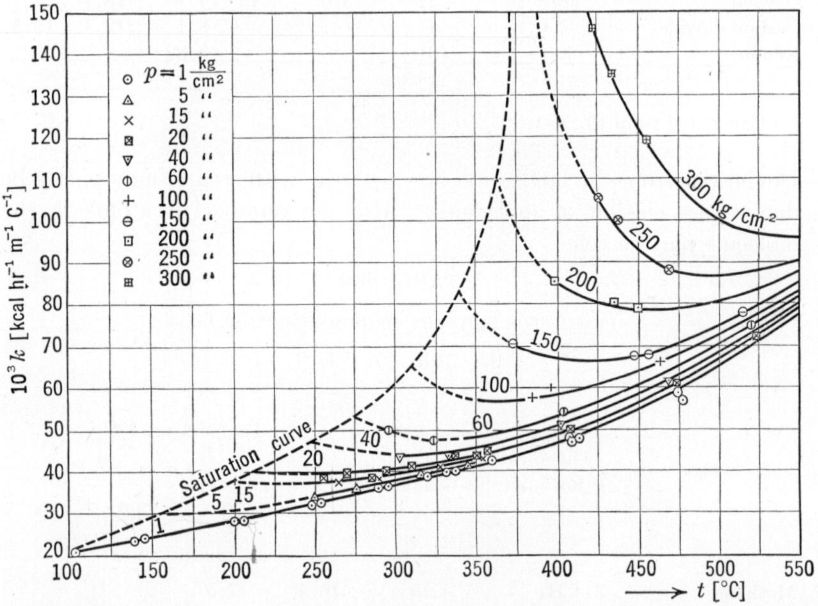

FIG. 6–1. Thermal conductivity of steam according to Timrot and Vargaftik.

results for steam obtained by Timrot and Vargaftik [1940]. The isobaric lines and the saturation curve of this diagram have the same character as the well-known curves of specific heat at constant pressure which were first observed by Knoblauch and Jakob [1906, 1906(*a*)]. In Sect. 6–2 the relationship between thermal conductivity and specific heat will be explained from a theoretical viewpoint. (See also footnote, page 117.)

6–2. Theory of the Thermal Conductivity of Gases

The thermal conductivity of gases has been found to be virtually independent of the pressure in a range which is not too far from atmospheric conditions. This is due to the fact that heat conduction in a gas is a process of diffusion, characterized by the wandering of molecules from warmer to colder zones, and vice versa, and by the exchange of kinetic energy in the collisions of the molecules. It can also be expressed as follows: Heat conduction in a gas is a diffusion in which the gases which diffuse into each other are identical, but are at different temperatures. Thus the diffusing molecules are at different average velocities and carry different amounts of kinetic energy, and diffusion is restricted to their exchange of location and energy.

Considering again that the number of traveling molecules increases in proportion to the pressure, but that their average path length decreases at the same ratio, it follows that thermal conductivity must be independent of pressure, exactly as is dynamic viscosity.

In this simple reasoning a great number of molecules in unit volume is supposed. Therefore it fails at low pressure; in this region,* according to Smoluchowski, the thermal conductivity k is proportional to the pressure p; it approaches zero with the pressure because eventually no molecules are available for the transportation of heat energy.

For heavy gases, k can be considered independent of p above 1 mm Hg; for hydrogen and helium above 20 mm Hg. The upper limit of such dependency may be at one or some atmospheres.

A basic relationship between thermal conductivity k, dynamic viscosity μ, and specific heat at constant volume c_v has been found by the following consideration. Gas friction is due to the retardation of molecules which come from regions of higher mean velocity and transfer their excess of impulse to slower ones. Heat conduction is a quite similar process, derived from the impulse transportation by multiplying the molecular mass m by c_v. Thus, considering Eq. 5–4 and referring to unit temperature difference,

$$k = \epsilon c_v \mu \tag{6–1}$$

where ϵ is a factor, for which values between 2.5 and 1.0 have been found.

This factor will later be dealt with, but first the meaning of c_v from a statistical viewpoint may be recalled. As in Sect. 5–3, it will be as-

* Approximately below 0.05 mm Hg for heavy gases, 0.1 mm Hg for helium, and 0.2 mm Hg for hydrogen.

sumed that the gas molecules move in the direction of the three coordinates only, so that again $nw/6$ molecules per second hit each surface of a cube of 1 cm side length. We may start considering monatomic molecules and assuming them to be elastic spheres. Each molecule receives the momentum $-2mw$ when hitting the wall in a direction normal to its surface, namely, a momentum $-mw$ which stops the movement, and another amount $-mv$ because of the conversion of energy of elasticity into movement in opposite direction. Thus the total change of momentum per second at the wall can be expressed by

$$\frac{d}{d\tau}\Sigma(mw) = \frac{nw}{6}\cdot 2mw = \frac{1}{3}nmw^2$$

where Σ is a symbol for the summation over the volume. According to the basic law of dynamics, this change of momentum is equal to the force exerted upon 1 cm^2 of the wall, i.e., the gas pressure

$$p = \tfrac{1}{3}nmw^2 \qquad (6\text{–}2)$$

Multiplying both sides of the equation by the molecular weight M and the specific volume v of the gas, we obtain

$$pMv = \tfrac{1}{3}(nMv)mw^2$$

or

$$pV_M = \tfrac{1}{3}Nmw^2 \qquad (6\text{–}3)$$

where $V_M = Mv$ is the mole volume, and $N = nMv$ is the number of molecules in a mole, which is called Avogadro's number.

Combining Eq. 6–3 with the characteristic law of ideal gases,

$$\left.\begin{aligned} pV_M &= R_0T = \tfrac{1}{3}Nmw^2 \\ \text{or}\quad T &= \frac{1}{3}\cdot\frac{N}{R_0}mw^2 \end{aligned}\right\} \qquad (6\text{–}4)$$

where R_0 is the universal gas constant.

The internal energy of the mole volume V_M is

$$U = \tfrac{1}{2}Nmw^2 \qquad (6\text{–}5)$$

From Eqs. 6–4 and 5

$$U = \tfrac{3}{2}R_0T \qquad (6\text{–}6)$$

From this, according to the definition of the mole heat at constant volume,

$$C_v = \frac{dU}{dT} = \frac{3}{2}R_0 \qquad (6\text{–}7)$$

a value which has been verified by many experiments.

This holds for a monatomic gas, the molecules of which are said to have three degrees of freedom of motion, corresponding to the three coordinates of the space. With respect to Eq. 6–7, a part of the specific heat, $R_0/2$, can be coordinated to each degree of freedom.

In addition to these degrees of freedom, which are for the three possible directions of translation, three more degrees of freedom are used in dynamics which correspond to the possible rotations about three axes perpendicular to each other.

For monatomic elastic molecules, however, energies of rotation are out of the question, because the atoms have negligibly small momentums of inertia and their impact is considered to be frictionless or, to express it more exactly, because the motion of these molecules is reversed already in the fields of forces surrounding them.

It is different with molecules consisting of more than one atom. A diatomic molecule can be considered a sort of dumbbell. Here, in addition to the three degrees of freedom of translation, rotations may occur around two axes which are perpendicular to the line connecting the two globes of the dumbbell and to each other. The rotation around the connecting line does not yield an appreciable momentum, because the globe radius is too small. Hence, a diatomic molecule has five degrees of freedom.

Triatomic gas molecules have six degrees of freedom because they can rotate around three axes which are perpendicular to each other.

Now, according to the so-called theorem of equipartition, the daring assumption has been made that the same fraction of the specific heat is connected with each degree of freedom. Then the mole heats at constant volume C_v are as the degrees of freedom, each contributing the energy $R_0/2$ per degree temperature increase to the internal energy of a mole. Thus, according to the theory:

For monatomic gases,

$$C_v = \tfrac{3}{2}R_0$$

For diatomic gases,

$$C_v = \tfrac{5}{2}R_0$$

For tri- and more-atomic gases,

$$C_v = \tfrac{6}{2}R_0$$

In Eq. 6–1 ordinary specific heats c_v are used. This is considered in the choice of the factor ϵ.

For monatomic, spherical molecules with any central forces between them, an exact calculation of Chapman [1912] led to $\epsilon = 2.5$.

According to Eucken [1913], a magnifying factor ϵ is needed in Eq.

6–1 because those molecules which are charged with more energy carry this energy faster than the other molecules carry their smaller load, for the following reason: The heat load is proportional to $c_v T$. But T is proportional to w^2. Thus the arithmetic mean velocity cannot be used, as it could in calculating the exchange of momentum. It would lead to a too small average heat exchange because it neglects the fact that the faster molecules are charged more heavily. A similar difficulty occurs in determining the average temperature of a flowing fluid in so far as the temperature of each part of the fluid must be multiplied by its individual velocity. (See Sect. 21–2).

As mentioned already, $\epsilon = 2.5$ if only the translatory degrees of freedom must be considered. If other degrees of freedom, as rotation or intramolecular oscillation, take part in the process, the difficulty is to find whether these energies, in an impact, will be reduced by the same amount as the energy of translation. This problem of thermal equilibrium has been considered thoroughly by Eucken.

In handling Eq. 6–1 numerically, the following average factor can be used:

$$\epsilon = \frac{\epsilon_t c_{v,t} + \epsilon_r c_{v,r} + \epsilon_o c_{v,o}}{c_{v,t} + c_{v,r} + c_{v,o}} = \frac{\epsilon_t C_{v,t} + \epsilon_r C_{v,r} + \epsilon_o C_{v,o}}{C_{v,t} + C_{v,r} + C_{v,o}} \tag{6–8}$$

where the subscripts t, r, and o mean translation, rotation, and oscillation; $c_{v,t} + c_{v,r} + c_{v,o} = c_v$, and $C_{v,t} + C_{v,r} + C_{v,o} = C_v$.

For molecules which carry also energies of rotation and oscillation Eucken [1940] supposed that these forms of energy are independent of the energy of translation, i.e., a fast molecule may sometimes contain only a small amount of the other energies, sometimes relatively much of them. If this is the case, then the energies of rotation and oscillation will be transported in exactly the same way as mechanical momentum, i.e., with $\epsilon_r = 1$ and $\epsilon_o = 1$. Since $\epsilon_t = 2.5$ and $C_{v,t} = 3R_0/2$, the following equation is obtained:

$$\epsilon = \frac{2.5(\frac{3}{2}R_0)}{C_v} + \frac{C_{v,r}}{C_v} + \frac{C_{v,o}}{C_v} = \frac{1.5(\frac{3}{2}R_0)}{C_v} + 1$$

or, if C_v is expressed in cal (gram-mole)$^{-1}$ C^{-1},

$$\epsilon = \frac{4.47}{C_v} + 1 \tag{6–9}$$

From the simple Eqs. 6–1 and 9 the thermal conductivity of gases at atmospheric pressure can be calculated, since the viscosity and specific heat are known for many gases, and $R_0 = 1.986$ cal C^{-1}.

Thus, for diatomic gases with negligible energy of oscillation, as N_2, O_2, and air at room temperature,

$$C_v = C_{v,t} + C_{v,r} = \tfrac{3}{2}R_0 + \tfrac{2}{2}R_0 = \tfrac{5}{2}R_0$$

and $\epsilon = 1.9$.

For diatomic gases with appreciable energy of oscillation, as CO, NO, and Cl_2, adding one degree of freedom for the oscillation along a line which connects the two atoms,

$$C_v = C_{v,t} + C_{v,r} + C_{v,o} = \tfrac{3}{2}R_0 + \tfrac{2}{2}R_0 + \tfrac{1}{2}R_0 = 3R_0$$

and $\epsilon = 1.75$.

These values have to be compared with the following ones found by measurement of μ and c_v at 0°C:

$\epsilon = 1.95$, 1.915, and 1.95 for N_2, O_2, and air, respectively,

and

$\epsilon = 1.865$, 1.83, and 1.803 for CO, NO, and Cl_2, respectively.

However, for H_2, $\epsilon = 2.03$ has been observed at 0°C, whereas at −182.6°C both theory and experiment led to $\epsilon = 2.35$.

Also for some tri- and more-atomic gases the agreement between theory and experiment is perfect, e.g., $\epsilon = 1.67$ for CO_2 and $\epsilon = 1.57$ for C_2H_4 at 0°C, from both theory and experiment. At other temperatures the agreement is less good. Some gases show appreciable deviations, the largest ones, noticed by Eucken, being the following:

$\epsilon =$ 1.89	(theoretical)	and	2.19	(experimental)	for	H_2	at	400°C
1.73	"	"	1.41	"	"	H_2O	"	100°C
1.65	"	"	1.425	"	"	H_2O	"	400°C
1.61	"	"	1.485	"	"	SO_2	"	0°C
1.70	"	"	1.54	"	"	NH_3	"	0°C
1.46	"	"	1.57	"	"	C_2H_6	"	0°C
1.34	"	"	1.48	"	"	$CHCl_3$	"	0°C
1.26	"	"	1.385	"	"	CCl_4	"	0°C

Too small theoretical values, as those for H_2 or C_2H_6, partly are due to the oversimplifying assumption of the factor ⅓ in Eq. 6–2, which influences the last term (1) in Eq. 6–9, and partly indicate a certain coupling of oscillatory and translatory energy, so that the faster molecules also carry more energy of rotation.

Too large theoretical values seem to occur in molecules which possess an electrical dipole momentum, the only exception being found with chloroform gas ($CHCl_3$). For H_2O, usually one of the most unwieldy substances, the difference between the theoretical and experimental value at 100°C exceeds 20% of the experimental value; for all other temperatures and substances it is much smaller. Although a substantiated theory of this behavior is missing, it looks as though rotation and oscillation do not take part in the heat exchange with sufficient speed. For helium this may even occur with the translatory energy ($\epsilon = 2.50$ by theory and 2.43 by experiment).

In general, however, Eq. 6–9 represents the influences on ϵ in a satisfactory manner. With increasing number of atoms, ϵ should approach the value 1. So far, the smallest value observed seems to be $\epsilon = 1.19$ for the fourteen-atomic ethyl acetate ($CH_3 \cdot COO \cdot C_2O_5$). It should be kept in mind that the decrease of ϵ with increasing number of atoms is particularly due to the increase of the oscillatory degrees of freedom and hence of c_v with the number of atoms in the molecule. For the same number of atoms, on the other hand, c_v is almost reciprocal to the molecular weight.

The monatomic helium ($M = 4$) for instance, has $c_v = 0.745$, whereas for the monatomic argon ($M = 40$) $c_v = 0.074$. The diatomic hydrogen ($M = 2$) has $c_v = 2.42$, and the diatomic oxygen ($M = 32$) has $c_v = 0.156$. Compared to such decreases of c_v with increasing molecular weight, the increase of the factor μ in Eq. 6–1 is not very important. At 0°C, for instance, $10^6\ \mu$ for helium is equal to 189 poise, and for argon 210; for hydrogen it is 84, and for oxygen 192.

Therefore the fact, found empirically in Sect. 6–1, that k is smaller for most of the heavier gases, is now understood theoretically. (Compare, for instance, k of C_2H_6O and C_6H_6 in Table 6–3.) Moreover, the exception from this rule in the case of gases with many light atoms is explained. (Compare, for instance k of C_2H_6O and $C_4H_{10}O$ in Table 6–3.)

Eucken [1940] recommends the following procedure to find the thermal conductivity of gases if reliable direct observations are missing: First it must be checked whether the gas molecule has a dipole momentum. (For data see Landolt-Boernstein [1935], page 117 ff.) If no dipole momentum exists, the values of ϵ calculated from Eq. 6–9 should be increased by 2% at low temperatures and by 3% at high temperatures. If a momentum exists, the calculated value of ϵ is to be reduced by 1 to 10%, the latter correction holding for the largest momentums of about $2(10^{-18})$ in absolute units. Except for water vapor this yields thermal conductivities with an accuracy of 1 to 5% for di- and triatomic gases, and of about 5% for gases with more atoms.†

† Greater deviations may occur occasionally. The anomaly mentioned for chloroform is possibly due to experimental errors.

The temperature influence is mainly due to the change of μ with temperature according to Sutherland's equation:

$$\mu = \frac{B\sqrt{T}}{1 + (C/T)} \tag{5–9}$$

Substituting this in Eq. 6–1, we obtain

$$k = \epsilon c_v \frac{B\sqrt{T}}{1 + (C/T)} \tag{6–10}$$

Since in practical ranges, with good approximation,

$$\epsilon c_v = F(1 + GT)$$

where F and G are constants, it follows that

$$k = \frac{BF(1 + GT)\sqrt{T}}{1 + (C/T)} = \frac{K(1 + GT)\sqrt{T}}{1 + (C/T)} \tag{6–11}$$

where $K = BF$.

This equation, with appropriate individual values of C, F, G, K, represents the known experimental data with an accuracy of 0 to 6%, or about 3% as an average, in the range of −180°C to +200°C.

In this range, according to Eq. 6–11, k increases a little less than linearly with the temperature for inorganic gases, as hydrogen and nitrogen, and a little more than linearly for organic gases, as methane, ethane, and benzene. (Compare Tables 6–1 to 3.)

The situation concerning gas mixtures is similar to that of viscosity, and for the same reason. (See Sects. 5–3 and 6–1.)

6–3. Numerical Values of the Thermal Conductivity of Liquids

Again experimental data will be presented first, and then an attempt will be made to make some of them comprehensible by theoretical consideration.

The best-conducting non-metallic liquid is water. Values of k at atmospheric pressure up to 100°C and at saturation pressure from 100 to 260°C are given in Table 6–4.

The conductivity k first increases with temperature up to a maximum at about 120°C and then decreases again. At 100°C the thermal conductivity of water is almost thirty times that of saturated steam.

Aqueous solutions conduct the heat less well than water, and their conductivity decreases with increasing concentration. At room temperature, for a 20% aqueous solution of sulfuric acid $k = 0.32$, and for an 80% solution $k = 0.22$ B hr^{-1} ft^{-1} F^{-1}. Alcohol solutions of these two concentrations have $k = 0.28$ and 0.125, respectively. An increase of k with the temperature has been observed.

TABLE 6–4

THERMAL CONDUCTIVITY OF WATER

$10^3 k$ [B hr^{-1} ft^{-1} F^{-1}]

t [°C]	0	20	40	60	80	100	120
$10^3 k$	322	343	362	377	390	396	397
t [°C]	140	160	180	200	220	240	260
$10^3 k$	396	393	389	383	376	366	353

Other non-metallic liquids also conduct less heat than water does. See, for instance, the following equations by which Kardos [1934] expressed his experimental results, the dimensions being converted into British technical units:

Sulfur dioxide (SO_2) has $k = 0.1036[1 - 0.00264(t - 32)]$

Methyl chloride (CH_3Cl) has $k = 0.0910[1 - 0.00059(t - 32)]$

Methylene chloride (CH_2Cl_2) has $k = 0.222[1 - 0.00164(t - 32)]$

Some other examples, taken from the literature, are:

Glycerine ($C_3H_8O_3$), with $k = 0.16$ at 32°F, and $k = 0.165$ at 132°F

Aniline (C_6H_7N), with $k = 0.10$ at 32°F, and $k = 0.10$ at 132°F

Olive oil, with $k = 0.10$ at 32°F, and $k = 0.095$ at 132°F

Toluene (C_7H_8), with $k = 0.09$ at 32°F, and $k = 0.085$ at 132°F

Experiments on normal primary saturated alcohols by Daniloff [1932] led to the values of Table 6–5. With increasing number of carbon atoms, k first decreases and then increases. Substantially the same decrease of the absolute values of k with increasing temperature has been observed for all alcohols.

Thermal conductivities of seventy-five liquids, most of them at various temperatures and many based on his own experiments, are given in a paper of J. F. D. Smith [1936]. He further gave the following empirical formula of conductivities at 30°C:

$$k = 0.000011 + \frac{(c_p - 0.45)^3}{155} + \frac{\sqrt[3]{\rho/M}}{800} + \frac{\nu^{1/6}}{6000} \qquad (6\text{–}12)$$

in physical units (cm, sec, deg C, and cal). The kinematic viscosity ν, in particular, is expressed in stokes,‡ i.e., in square centimeters per second.

Smith checked this equation with experimental results for forty-six liquids and found a maximum error of 25% for glycerine, the error for all other substances being within 16%, and an average error of only 6.7%.

TABLE 6-5

THERMAL CONDUCTIVITY OF NORMAL PRIMARY SATURATED ALCOHOLS AT 30°C

Kind of Alcohol	Chemical Formula	Number of C Atoms	Molecular Weight	10^3k [B hr^{-1} ft^{-1} F^{-1}]
Methyl	CH_4O	1	32	122
Ethyl	C_2H_6O	2	46	105
n-Propyl	C_3H_8O	3	60	99
n-Butyl	$C_4H_{10}O$	4	74	95
n-Amyl	$C_5H_{12}O$	5	88	94
n-Hexyl	$C_6H_{14}O$	6	102	93
n-Heptyl	$C_7H_{16}O$	7	116	94
n-Octyl	$C_8H_{18}O$	8	130	96
n-Nonyl	$C_9H_{20}O$	9	144	97

One liquid, carbon dioxide, has been investigated up to and above the critical point. However, the results and conclusions of the two observers diverge widely. Whereas Kardos [1934] maintains that k takes very high values in the critical region, as c_p is known to do, Sellschopp [1934], in an elaborate experimental paper, assumes that the increase of k in the neighborhood of the critical point observed by him, as by Kardos, is not real but is due to convective currents. According to Sellschopp, $k \approx 0.035$ B hr^{-1} ft^{-1} F^{-1} at the critical point, whereas Kardos assumes that k approaches infinity. Also, at 60 atmospheres and 10°C, the two observers found very different values, namely, 0.12 (Kardos) and 0.065 (Sellschopp). (See also footnote, page 117.)

6-4. Theoretical Relations about the Thermal Conductivity of Liquids

For a long time only negative statements could be made concerning the theory of liquids. For example, it was known that dissolving glue

‡ In the original paper of Smith ν is expressed in centistokes, and therefore the denominator of the last term of the equation is 10,000 in the original.

in water raises the viscosity greatly, but influences the thermal conductivity only slightly, and it was concluded that Eq. 6–1 cannot hold for liquids; or, it was seen from Table 6–5 that the thermal conductivity of alcohols does not possess the property of additivity.

In liquids, as in gases, heat conduction is due to the movement of the molecules and atoms. However, the random translatory motion of liquids is small and so is the translation of energy due to this motion. The most promising theories of liquid conduction suppose a heat transfer by longitudinal vibrations, similar to the propagation of sound.

The first to base a theory of heat conduction in liquids on this assumption was Paschki [1915]. He developed an equation according to which the thermal conductivity of liquids would be proportional to the sound velocity and to the $4/3$ power of the density.

Independently of this paper, Bridgman [1923] derived the following simple equation:

$$k = 3 \frac{R_0}{N} v D^{-2} \tag{6–13}$$

where R_0 = the universal gas constant,
N = the number of molecules in one mole (Avogadro's number),
v = the velocity of sound in the liquid, and
D = the mean distance of the centers of adjacent molecules, under the assumption that these are arranged in simple cubical array.

The ratio R_0/N is identical with the universal Boltzmann constant, and $\frac{3}{2} \cdot \frac{R_o}{N}$ is the kinetic energy received by an atom of a monatomic substance when heated by 1 C. Hence $3R_0/N$ is the energy taken up by a solid body (half potential and half kinetic). Because in general C_v does not change much in the transition from the solid to the liquid state (except for water, where it jumps to the double amount), $3R_0/N$ may also be used for liquids.

Under this assumption, Bridgman derived Eq. 6–13 as follows: Let there be in the liquid a temperature gradient dT/dx, and a molecular energy $3R_0T/N$, with T as absolute temperature. The difference of energy between neighboring molecules in the direction of the temperature gradient is $\frac{3R_0}{N} \cdot \frac{dT}{dx} D$. This energy difference is to be conceived as being handed down a row of molecules with the velocity of sound. The total energy passing a fixed point of any row of molecules in unit time is the product of the energy difference and the number of such energy steps contained in a row of molecules v cm long, i.e., $\frac{3R_0}{N} \cdot \frac{dT}{dx} D \frac{v}{D}$. The total transfer across unit cross section is the product of the transfer across a single row and the number of rows in unit cross section, or $\frac{3R_0}{N} vD^{-2} \frac{dT}{dx}$. But from the definition of thermal conductivity the transfer is also $k \cdot dT/dx$. Hence, identifying the two last expressions, Eq. 6–13 results.

From this equation Bridgman, using the values of Columns 4 and 5 of Table 6–6, arrived at values k_B, and he compared them with directly observed values k_o. The values of Columns 6 and 8 of Table 6–6 differ

TABLE 6–6

COMPARISON OF COMPUTED AND OBSERVED THERMAL CONDUCTIVITIES OF LIQUIDS AT 30°C

(k_B, k_K, and k_o in erg sec^{-1} cm^{-1} C^{-1})

Liquid	Chemical Formula	Number of C Atoms	$10^{-3}v$ [cm/sec]	$10^9 D$ [cm]	$10^{-3}k_B$	$10^{-3}k_K$	$10^{-3}k_o$
Water	H_2O	0	150	31.0	64.1	59.6	61.0
Methyl alcohol	CH_4O	1	113	40.8	27.9	20.2	21.1
Ethyl alcohol	C_2H_6O	2	114	46.0	22.2	19.8	18.2
n-Propyl alcohol	C_3H_8O	3	124	50.4	20.1	20.7	17.1
Isobutyl alcohol	$C_4H_{10}O$	4	105	53.6	15.2	....	....
n-Butyl alcohol	$C_4H_{10}O$	4	...	....	....	19.8	16.7
Isoamyl alcohol	$C_5H_{12}O$	5	124	56.5	16.0	22.1	14.8
Carbon bisulfide	CS_2	1	118	46.6	22.3	14.7	15.9
Acetone	C_3H_6O	3	114	50.0	18.8	18.2	17.9
Ether	$C_4H_{10}O$	4	92	55.9	12.1	14.0	13.7
Ethyl bromide	C_2H_5Br	2	90	50.2	14.8	11.3	12.0
Ethyl iodide	C_2H_5I	2	78	51.2	12.3	10.3	11.1

slightly from Bridgman's original values, because not exactly the same numerical values for Boltzmann's constant and for k_o of some liquids have been used.

Whereas this table shows some considerable differences between k_B and k_o, due to inaccuracies of the theoretical assumptions and the experimental data, the general agreement is striking in view of the simplicity of the formula. In particular, the theory agrees with the observation that the conductivity of water is 2½ to 5 times that of the organic liquids. Herefrom Bridgman concludes that the high conductivity of water is not due to the molecular peculiarities of water (two or more species of molecules), but mainly to its low compressibility (high value of v) and the relatively small distance D of the centers of the molecules.

The formula gives, further, the right sign for the temperature coefficient of conductivity at atmospheric pressure, both for the organic liquids and water. For the former, v decreases and D increases with rising temperature, so that k would be expected to decrease with rising temperature, as it actually does, whereas for water v increases with rising temperature at a rate more than sufficient to compensate for the increase of D, so that k should increase, as it actually does except at high temperatures (see Table 6-4).

Another interesting performance of Eq. 6-13 is that it fairly reproduces the decrease of k of alcohols with increasing number of C atoms (see Table 6-5), and possibly even the minimum of k.

Kardos [1934(a)] modified the theory by introducing the distance L between the surfaces of adjacent molecules instead of the distance D of their centers. He considers the molecule a unit possessing a certain energy, and the drop of energy and temperature as restricted to the distances between molecules. Attributing the molecular reference energy Q_0 to a plane in the middle between two molecules, he uses the energy drop, $\Delta Q = -L(dQ/dx)$, between the neighboring molecule surfaces which are supposed to be parallel to the middle plane surface. Assuming that the heat energy is transmitted through the substance with the sound velocity v and reducing the energy gradient to that of a quasi-homogeneous body by multiplying with L/D, the following equation for the heat flow per unit area is obtained:

$$q'' = -Lv\left(\frac{1}{D^3}\cdot\frac{dQ}{dt}\right)\frac{L}{D}\left(\frac{dt}{dx}\right)_L$$

where the subscript L indicates that the derivative is taken in the range of L only.

But, according to Fourier's basic equation,

$$q'' = -k\left(\frac{dt}{dx}\right)_D = -k\frac{L}{D}\left(\frac{dt}{dx}\right)_L$$

where the subscript D indicates that the derivative refers to a quasi-homogeneous substance over the distance D. Further, for cubical array of the molecules,

$$\frac{1}{D^3}\cdot\frac{dQ}{dt} = \rho c_p$$

where ρ is the density of the substance. From these three equations

$$k = \rho c_p v L \tag{6–14}$$

It is questionable whether c_p should not be replaced by c_v. However, Kardos considers the use of c_p more justified. He further recommends assuming $L = 9.5(10^{-9})$ cm. The values k_K in Table 6–6 are calculated in this way. Some of them are in better agreement with the observed values than those of k_B, calculated by Bridgman, in particular the value for CS_2; for other ones the contrary is true, particularly for $C_5H_{12}O$. The greatest discrepancy occurs with glycerine, $C_3H_8O_3$, where $57.5(10^3)$ erg sec^{-1} cm^{-1} C^{-1} was calculated, whereas the experiment gave $28.1(10^3)$.

This discrepancy is ascribed to the small compressibility, κ, of glycerine. Kardos claims that the ratio $M/(\rho\kappa)$ is essential for the magnitude of L, where M is the molecular weight. This ratio is 4.10 for water and 36.5 for glycerine. For such high values of the ratio it cannot be assumed any more that $L = 9.5(10^{-9})$. Considering the theory of Van der Waals' molecular forces, Kardos calculated individual values of L for the group of alcohols. For $C_5H_{12}O$ and $C_3H_8O_3$ the values $6.8(10^{-9})$ and $4.8(10^{-9})$, respectively, were obtained. Using the individual values of L, the calculated values of the thermal conductivity of all alcohols approach the observed values with an accuracy of 5% or better. J. F. D. Smith [1936], however, showed that, because of serious errors in the reasoning, Kardos' calculation of L can be considered only semi-empirical.

6–5. Thermal Conductivity of Porous and Loose Insulating Materials

The solid materials are so numerous and different in their behavior that it seems best to divide them into groups, as insulating, building, refractory, and metallic substances.

The insulating and building materials can be considered as mixtures of solid bodies and gases, or as hollow solid bodies, filled with gases.

At this point some general remarks about loose and porous insulating materials seem to be proper. Although gases are the poorest heat conductors, they are not always the best heat insulators. If the spaces filled with gas are so large that heat transfer by convection is considerable, or if the temperature is so high that heat exchange by radiation plays a remarkable part, it may occur that solid bodies present a better protection against heat exchange than gases do, but only if they contain gases, i.e., if they are loosely composed of small particles, like powders, grains, and fibers, or if they are porous. In practice, therefore, such bodies are used as heat insulators with air filling the interstices and pores.

But sometimes a liquid, mostly water, penetrates into the pores, or the pores contain water in their original state, as do all kinds of wood. Then the insulating effect apparently is deteriorated, because water conducts heat almost twenty five times better than air does. Thus, the thermal conductivity of gases and liquids plays an important role in the conduction of composed or porous solid bodies, and particularly of insulating materials.

The following remarks about pores hold also for the interstices of loosely composed bodies, like powders, because both behave very similarly in regard to heat conduction. The porosity greatly influences the weight of a body. For this reason the "apparent" density ρ_a of a solid body must be known in addition to its "true" density ρ_s. The former is defined by the equation

$$\rho_a = \frac{W}{V} \tag{6-15}$$

where W is the weight of the body, and
V is the volume of the body.

Let W_s be the weight of the solid substance, and V_s its volume,
W_g be the weight of the gas in the pores, and V_g its volume,

so that

$$V = V_s + V_g$$

$$W = W_s + W_g$$

Then the porosity P (expressed in percentages of volume) will be

$$P = 100 \frac{V - V_s}{V} = 100 \frac{\dfrac{W_s + W_g}{\rho_a} - \dfrac{W_s}{\rho_s}}{\dfrac{W_s + W_g}{\rho_a}}$$

Because $W_g \ll W_s$, it follows that

$$P = 100 \frac{\rho_s - \rho_a}{\rho_s} \tag{6-16}$$

Thus, for ρ_s = constant, increasing porosity P or gas content means decreasing apparent density ρ_a, and because, on the other hand, increasing gas content means lower thermal conductivity k, decreasing ρ_a is generally connected with decreasing k.

The influence of the pores can be seen from a formula for the apparent conductivity, k_a, of a body which consists of a contingent substance having the conductivity k_s and small particles of conductivity k_p distributed in that substance. This formula, derived by Maxwell and used for the present purpose by Eucken [1940], reads §

$$k_a = k_s \frac{1 - (1 - ak_p/k_s)b}{1 + (a - 1)b} \tag{6–17}$$

where $a = 3k_s/(2k_s + k_p)$,
$b = V_p/(V_s + V_p)$,
V_s = total volume of the contingent substance, and
V_p = total volume of the distributed particles.

The equation is derived for small values of b only; however, Eucken assumes that it may be valid up to $b \approx 0.5$ with good approximation.

Considering the particles as pores, k_p is an apparent conductivity itself, including the influence of radiation through the pores, which, however, can be estimated for a given porous substance and temperature difference. In most cases k_p will be negligibly small compared to k_s. Then Eq. 6–17 simplifies to

$$k_a = k_s \frac{1 - b}{1 + b/2} \tag{6–18}$$

For elliptic or tubular pores, the following formula may be used:

$$k_a = k_s \frac{1 - b}{1 + b/3} \tag{6–19}$$

For extended, more or less plane gaps, assuming that one third of all fissures extend perpendicular to the heat-flow direction, Eucken arrives at the following equation:

$$\frac{1}{k_a} = \frac{1}{k_s} + \frac{b}{3k_p} \tag{6–20}$$

Nusselt [1913], however, obtained a slightly different solution by the following reasoning:

A body with fissures perpendicular to the heat flow may be pictured (Fig. 6–2) as a system of parallel solid plates and air layers of the thicknesses L_{12} and L_{23}, respectively. This will have a better insulating effect than a body with extended gaps and bridges of solid substance crossing the air layers. Assuming that $L_{12} + L_{23}$ is small compared to

§ For convenience $b = P/100$ is introduced in this formula.

dx, the composed system can be handled like a homogeneous body. The thermal conductivities may be k_s for the solid substance and $k_{g,r}$ for the gas including the effect of radiation, and k_a may again be the equivalent or apparent conductivity of the whole body.

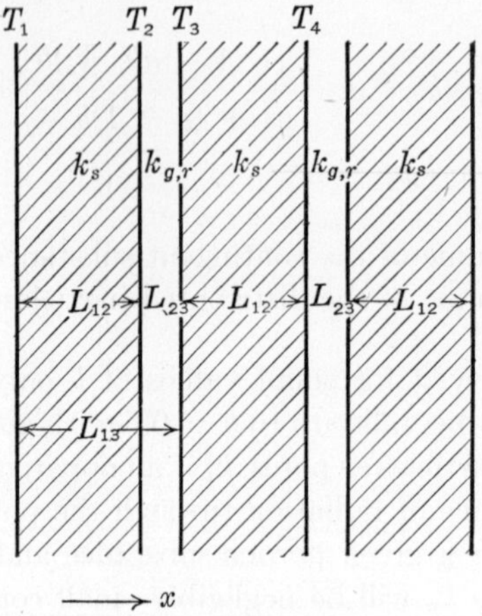

FIG. 6–2. Body with parallel fissures.

Using Ohm's law for the heat flow, the sum of the thermal resistances will be

$$R_{13} = R_{12} + R_{23}$$

or

$$\frac{L_{13}}{k_a} = \frac{L_{12}}{k_s} + \frac{L_{23}}{k_{g,r}}$$

so that

$$k_a = \frac{L_{13}}{\dfrac{L_{12}}{k_s} + \dfrac{L_{23}}{k_{g,r}}} \tag{6–21}$$

Introducing b as defined previously and $k_p = k_{g,r}$ in Eq. 6–21

$$\frac{1}{k_a} = \frac{L_{12}}{L_{13}k_s} + \frac{b}{k_p} \tag{6–22}$$

Comparing this with Eq. 6–20, it is seen that k_p instead of $3k_p$ occurs because a gap all through the body and not covering only a third of the cross section has been assumed. The factor L_{12}/L_{13}, however, is missing in Eucken's equation and should be added there.

The quantity $k_{g,r}$ in Eq. 6–21 can be found by superposition of heat conduction and radiation. First an equivalent thermal conductivity, k_r, that takes care of the heat radiation is derived as follows:

$$\sigma(T_2{}^4 - T_3{}^4) = k_r \frac{T_2 - T_3}{L_{23}}, \quad \text{or} \quad k_r = \sigma \frac{T_2{}^4 - T_3{}^4}{T_2 - T_3} L_{23}$$

Introducing differentials instead of the differences in the last equation,

$$k_r = \sigma \frac{dT^4}{dT} L_{23} = 4\sigma T^3 L_{23}$$

The thermal conductivity of the gas, on the other hand, may be k_g. By addition of the conductances in the gas layer

$$\frac{k_{g,r}}{L_{23}} = \frac{k_g}{L_{23}} + \frac{k_r}{L_{23}} = \frac{k_g}{L_{23}} + 4\sigma T^3$$

or

$$\frac{L_{23}}{k_{g,r}} = \frac{1}{\dfrac{k_g}{L_{23}} + 4\sigma T^3}$$

and by substitution in Eq. 6–21

$$k_a = \frac{L_{13}}{\dfrac{L_{12}}{k_s} + \dfrac{1}{\dfrac{k_g}{L_{23}} + 4\sigma T^3}} \tag{6–23}$$

For very high temperatures the second term in the denominator becomes zero, and the following upper limit for the thermal conductivity of porous bodies is obtained:

$$k_a = \frac{L_{13}}{L_{12}} k_s \tag{6–23a}$$

This, apparently, means that at very high temperature the pores themselves present no resistance at all, but the porous body conducts as though the pores were filled with the solid substance.

If, on the other hand, heat radiation can be neglected at all, as will be the case at very low temperature, then the assumption $k_g \ll k_s$ leads to a lower limit for the heat conductivity

$$k_a = \frac{L_{13}}{L_{23}} k_g \tag{6–23b}$$

as can be derived easily.

At not extremely high temperatures, k_a, calculated from Eq. 6–23, will be considerably smaller than according to Eq. 6–23*a*. On the other hand, the bridges of solid substance across the gas layers increase the heat conduction. In loose materials, like powders or fibers, these bridges are replaced by the many spots where the loose particles are in contact with one another. For these reasons, according to the size and arrangement of the pores or interstices, either a loose or a porous body may be the better insulator. The thermal conductivity of both of them increases more then linearly with the apparent density, beginning with almost the low conductivity value of air at the smallest density. The best insulators are loose cotton, wool, silk, and cork grain with $k < 0.03$ B hr^{-1} ft^{-1} F^{-1} at $\rho_a \approx 0.1$ g/cm^3. Possibly the very best is "expanded" cork. In manufacturing this, calcium carbide is added to fine cork grains; the pressure of acetylene gas, developed inside the cork, expands it, and a very small apparent density results. For such loose cork material, with grains of 1 to 2 mm final thickness, and $\rho_a = 0.05$ g/cm^3, the conductivity $k \approx 0.02$ B hr^{-1} ft^{-1} F^{-1} has been observed.

Similarly low conductivities have been obtained by expanding polystrene, a high-molecular-weight polymer, into a plastic foam. This material, called Styrofoam, was recently described by McIntire and Kennedy [1948]. The smallest conductivity, $k = 0.023$ B hr^{-1} ft^{-1} F^{-1} at 70°F, was obtained with samples having cell sizes between 1/64 and 1/32 in. and the apparent density $\rho_a = 0.027$ g/cm^3. Experiments are being made to reduce the reflectivity of the cell walls by addition of a suitable filler, in the hope of obtaining a conductivity as low as $k = 0.017$.

At the density $\rho_a = 0.5$ cork plates and loose asbestos fabrics have $k \approx 0.05$, and 0.08, respectively. For fine powders, like coal dust and quartz sand, lower values of k at higher ρ_a have been observed, possibly for a reason which will be explained in Sect. 6–6.

With increasing temperature the apparent thermal conductivity of loose and porous insulating materials becomes greater, because the conductivities of the solid substance and of the gas in the pores, and the inner radiation increase. In the range of $\rho_a = 0.05$ to 0.5 g/cm^3, and $t = 0$ to 100°C, the increase will be about 0.1 to 0.5% per degree centigrade temperature rise.

6–6. Thermal Conductivity of Fine Powders

It is well known that generally heat energy is transferred from a wall to a gas without any discontinuity of temperature. There exists, however, an exception to this rule. To understand this, the flow of a gas along a wall may be considered. Close to the wall, the flow velocity is zero, and in a thin boundary layer the velocity is increasing rapidly.

With sinking pressure, the average molecular free path increases, and in narrow vessels or tubes increasingly more molecules will run between the walls without hitting other molecules. Now, according to the kinetic interpretation, gas friction is due to the impulse transportation between the gas molecules only, and not to the elastic impact of walls. Thus walls of narrow tubes and vessels behave as though in their neighborhood the pressure were considerably smaller. If opposite spots of the walls were distant by the mean free path or less, and if all molecules were hitting the walls at right angles and had a free path equal to the mean free path, then not a single molecule would hit another, and, as in a large vessel at absolute vacuum, almost no friction at all would exist. Therefore in a gas flowing along a wall a discontinuity of velocity occurs, which may become considerable at low pressure. The gas slides on the wall, as was observed first by Kundt and Warburg [1875]. They also predicted from the theory that a corresponding temperature discontinuity would be found. In fact, 23 years later Smoluchowski [1898] observed discontinuities up to 7 C in powders the grains of which were separated by gases at very low pressures.

This discontinuity in temperature, which may be designated by Θ, can be replaced by a fictitious continuous change of the gas temperature t by means of the equation

$$k\frac{dt}{dn} = h\Theta, \quad \text{or} \quad \Theta = \xi\frac{dt}{dn} \tag{6–24}$$

where k is the thermal conductivity of the gas,

n is the direction normal to the wall,

h is a coefficient of heat transfer between wall and medium, and

$\xi = k/h$ is a coefficient of discontinuity, defined by Eq. 6–24.

An analogous coefficient ζ for the discontinuity Φ in the velocity v of a gas flowing along a wall has been defined by the equation

$$\Phi = \zeta\frac{dv}{dn} \tag{6–25}$$

where a fictitious continuous change of the gas velocity v is replacing Φ.

By direct measurements it has been found that

$$\left.\begin{aligned} \xi &= 18.4(10^{-6})\frac{760}{p} \quad \text{for air} \\ \xi &= 105.7(10^{-6})\frac{760}{p} \quad \text{for hydrogen} \end{aligned}\right\} \tag{6–26}$$

where p is the gas pressure in millimeters of Hg, and ξ is expressed in centimeters.

According to Eq. 6–24, a continuous temperature change as great as the actual discontinuous change would take place if the wall were shifted in the direction n by a distance ξ almost equal to the molecular free path. Considering two parallel walls, distant by L and different in temperature by Δt, the rate of heat conduction would be

$$q = k \frac{A}{L} \Delta t \tag{6–27}$$

If now between these walls a perfectly conducting plate of the thickness s were inserted, the total distance of the two gas layers would be $L - s$. But instead of this, the effective length would be $L - s + 2\xi$, and by insertion of n plates, Eq. 6–27 would have to be replaced by

$$q = k \frac{A}{L - ns + 2n\xi} \Delta t \tag{6–28}$$

If the gas pressure is reduced, 2ξ increases according to Eq. 6–26, and eventually becomes greater than s. Below this pressure limit the apparent heat conduction of the gas layer becomes smaller, the more sheets are inserted, as Eq. 6–28 indicates.

Smoluchowski [1910] recognized how this peculiarity can be used to produce insulators better than air without extreme reduction of the pressure. He suggested dividing air spaces into tiny elements by filling them with fine powders. Then, even at a rather poor vacuum, the dimensions of the gas cells become of the order of magnitude of the average free path of molecules. Figure 6–3 is a graph with some of Smoluchowski's experimental results.

Curves I, II, and III were obtained with granular powders, curves IV and V with spongy powders. The latter are more effective, but easily agglomerate, thereby reducing the desired effect. ||

Below 1 mm Hg the apparent thermal conductivity of all these powders becomes smaller than that of air; for some of them, including metallic ones, this happens at higher pressure, exceptionally even at atmospheric pressure. Thermal conductivities 3 to 10 times smaller than that of air have been obtained with powders Nos. II to V at 1 mm Hg.

|| A silica aerogel, consisting of a fine powder (6% silica, 94% air; $\rho_a = 0.012\ \text{g/cm}^3$) is used in the material, called "Santocel." By adding an opacifying agent exhibiting a metallic luster of the cell surfaces, a thermal conductivity smaller than that of air seems to have been obtained at atmospheric pressure.

Because of the effect of agglomeration and the difficulty in filling the spaces homogeneously, the practical use of this method of insulation is rather restricted; from a scientific viewpoint, however, it means a triumph of the kinetic theory of the heat conduction of gases.

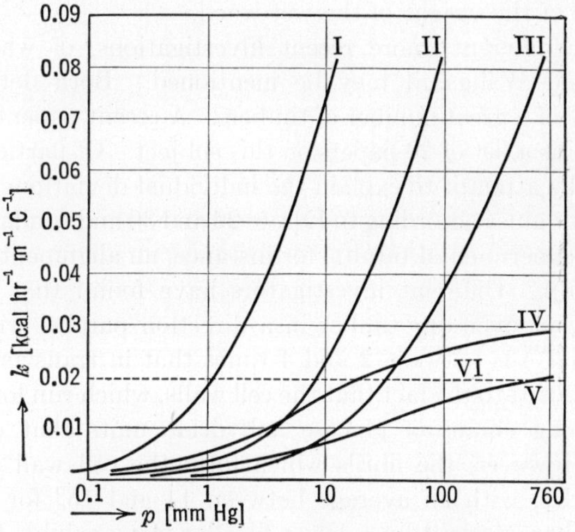

I. Quartz sand, grains 0.26 mm
II. Zinc dust, " 0.028 mm
III. " " , " 0.0062 mm
(I–III: Granular powders)
IV. Diatomaceous earth
V. Lamp soot
(IV–V: Spongy powders)
VI. Air

FIG. 6-3. Apparent thermal conductivity of powders. From Erk [1933].

6-7. Thermal Conductivity of Building Materials

As for insulating materials, the thermal conductivity of building materials increases with the apparent density.

In addition to this, the direction of the fibers and the water content are of influence for all kinds of wood. The very light balsa wood ($\rho_a = 0.1$ to 0.35 g/cm^3) has the lowest conductivity ($k = 0.04$ to 0.05 B hr^{-1} ft^{-1} F^{-1}), whereas oak wood ($\rho_a = 0.65$ to 0.85) has the highest ($k = 0.08$ to 0.12); these figures hold for 15% water content and heat flow across the fibers. In the direction of the fibers the conductivity is almost twice as large. The increase of k with temperature amounts about 0.1% per degree Fahrenheit.

Rowley [1933] investigated 22 different kinds of wood, having apparent densities $\rho_a = 0.4$ to 0.75 g/cm. His results can be roughly represented by the formulas

$$k = 0.053 + 0.1036(\rho_a - 0.400) \quad (6\text{–}29)$$

$$k = 0.060 + 0.1094(\rho_a - 0.400) \quad (6\text{–}30)$$

the former belonging to perfectly dry wood, the latter to a water content of 12% of the weight of the wet wood.

There are different, more recent investigations, of which those of MacLean and Wangaard may be mentioned. Both determined the conductivity of a great number of timbers. A recent paper of Wangaard [1943] contains a list of 20 papers on this subject. Of particular interest is Wangaard's attempt to explain the individual deviations of measured conductivity values according to Eqs. 6–29 and 30 and similar representations of the experimental results, for instance, an alignment chart which he has set up. Different investigators have found that the thermal conductivity of oven-dry timber in a direction parallel with the grain ("longitudinal") is between 2 and 4 times that in transverse direction. This is attributed to the fact that the cell walls, which run longitudinally, consist of long chains of glucose anhydride units that compose the cellulose. However, the fibrils which form the cell wall are tilted at varying angles, with an average between 14 and 15° for Douglas fir. Specimens characterized by a lower fibrillar slope exhibit thermal conductivities as much as 15% below that of normal wood, and those with a greater fibrillar slope, in contrast, show positive conductivity deviations of a similar amount. From this relationship, according to the author's graphs, the following two simple equations for Douglas fir have been derived:

$$\frac{\Delta k_t}{k_t} = 0.535 \sin \phi - 0.135 \quad (6\text{–}30a)$$

and

$$\frac{\Delta k_t}{k_t} = 0.565 - 0.18 \frac{k_p}{k_t} \quad (6\text{–}30b)$$

where ϕ = the "fibril angle,"
k_p = the conductivity in parallel (longitudinal) direction,
k_t = the conductivity in transverse direction, and
Δk_t = the deviation of the transverse conductivity from the "normal" value of k_t.

The author suggests the following interesting application of this experience. The so-called compression wood has long been recognized as a type of defective wood occurring irregularly in the timber of coniferous species. It is abnormally weak, shrinks longitudinally, and is difficult to work with tools. Extreme types of compression wood are readily recognized at sight. The less pronounced forms used to be identified by the high fibrillar slope. Microscopic or X-ray examination of specimens for the detection of compression wood is tedious. The close correlation of thermal conductivity and fibril declination may present a more practical means for classification of timbers, particularly if the simple method of de Sénarmont (see Sect. 9–5) is used, as Wangaard has done in his experiments.

MacLean [1941] published experimental data obtained on 20 hardwoods and 25 softwoods. His results are expressed in two empirical formulas, which can be reduced to the following form:

$$k = \frac{W_s/V}{W/V}\left(0.1158 + m\frac{W_W}{W_s}\right) + 0.01375 \qquad (6\text{–}30c)$$

where V = the volume of the moist sample,
W_s = the weight of the solid substance in the sample,
W_W = the weight of water in the sample,
W = the weight of a volume V of water, and
m = a moisture factor which, according to the original formula, equals 0.233 for $W_W/W_s < 0.40$ and 0.317 for $W_W/W_s \geqq 0.40$.

It is seen that at this ratio the values of k jump by about 30%. However, moisture contents of 40% and more are of no importance. W_s/W varied from 0.29 (for white cedar) to 0.64 (for hickory, mockernut) in MacLean's tests.

There are many building materials on the market made from plants as raw products, such as the one with the trade name Celotex. They cannot be dealt with here. Only a few data about rubber will be mentioned.

For pure plantation rubber the value $k = 0.08$ is given in the literature. For vulcanized rubber, as on the market, k increases from 0.1 at 90% rubber content to 0.16 at 40% rubber content.

The thermal conductivity of such organic building materials as pasteboard and linoleum is of the order of $k = 0.09$.

The behavior of the mineral building materials is similar to that of the insulating substances; the differences are only quantitative. Table 6–7 gives average values of k for bone-dry building materials as dependent on ρ_a. Because of the multiplicity of these substances and the inexactitude of some measurements, individual values may deviate by ±15% from the values of the table, which is given for general information only. The increase of k with ρ_a is more than linear; because the thermal conductivity of the solid constituents of most inorganic building materials is nearly the same, the increase of k with ρ_a is mainly due to the decrease of the volume of the pores.

TABLE 6–7

THERMAL CONDUCTIVITY k OF BONE-DRY BUILDING MATERIALS

$\rho_a[g/cm^3]$	0.2	0.4	0.6	0.8	1.0	1.2
$1000k$ [B hr^{-1} ft^{-1} F^{-1}]	35	45	65	90	120	155
$\rho_a[g/cm^3]$	1.4	1.6	1.8	2.0	2.2	2.4
$1000k$ [B hr^{-1} ft^{-1} F^{-1}]	195	240	300	400	540	765

For illustration a few approximate values of ρ_a will be given: 0.3 for pumice stone, 1.0 for highly porous brick and gypsum board, 1.6 for sand, 1.7 for brick and plaster, 2.3 for sandstone. The density of massive rocks, like gneiss, granite, marble, lies between 2.4 and 3.1 g/cm^3, and their conductivity between 1.2 and 2.4 B hr^{-1} ft^{-1} F^{-1}.

The influence of water content in the pores can be considered by multiplying the values of k from Table 6–7 by a factor f, as given in Table 6–7*a*.

TABLE 6–7*a*

FACTOR f TO CORRECT VALUES OF k IN TABLE 6–7 FOR WATER CONTENT w IN VOL.-%

w	1	2.5	5	10	15	20	25
f	1.30	1.55	1.75	2.10	2.35	2.55	2.75

The considerable influence of humidity will be noticed. It may be mentioned that $w \approx 5$ is the water content of inner walls, and $w \approx 10$ is that of outer walls of brick buildings for moderate climates and heating systems. Great deviations of these averages are not unusual.

For some purposes, like laying of ducts for warm fluids or calculation of heat loss through the foundations of buildings, the thermal conductivity of soil must be known. It varies much with the kind of solid substance, the apparent density, and humidity. Entirely dry soil of $\rho_a \approx 1.65$ has $k \approx 0.2$ and wet soil at $w = 15$ may have $k \approx 0.8$, and at $w = 30$, $k \approx 1.1$.

6-8. Thermal Conductivity of Amorphous and Crystalline Nonmetallic Substances

Comparison of heat conduction by the so-called amorphous or glassy and the crystalline substances is of particular interest because of their entirely opposite behavior. As Eucken [1911] seems to have stated first, the thermal conductivity of amorphous bodies is small at low temperature and increases with the temperature, whereas that of crystals is high and decreases with increasing temperature.

The behavior of the first group can be seen from the values in Table 6-8. The table shows an increase of k with temperature, steep at the beginning, flattening, and becoming almost linear at high temperature.

TABLE 6-8

THERMAL CONDUCTIVITY OF GLASSES AND GLASSY SUBSTANCES

k [B hr^{-1} ft^{-1} F^{-1}]

Temperature [°C]	Heavy Optical Flint Glass	Fire Clay Brick with 58% SiO_2 and 40% Al_2O_3	Silica Brick with 95% SiO_2, ρ_a = 1.81, Burned at 1415°C	Pyrex Glass	Soiled Quartz Glass	Optical Borosilicate Glass
−200	0.175	.....	.....	.....	0.27	0.27
−100	0.335	.....	.....	.....	0.52	0.56
0	0.410	0.51	.....	0.605	0.65	0.69
100	0.435	0.565	0.645	0.67	0.72	0.78
200		0.62	0.665	0.76	0.76	....
400	.....	0.685	0.75	0.93	0.87	....
600	.....	0.75	0.82	1.10	1.00	....
800	.....	0.85	0.90	.....	1.15	....
1000	.....	0.92	0.98	.....	1.26	....

This behavior may be compared with that of crystals. In Table 6-9 the thermal conductivity of the same substance (SiO_2) in amorphous (glassy) and crystalline phase is shown, the latter for conduction normal and parallel to the main crystallographic axis. The figures of this table are striking. At −200°C the crystal conducts the heat 40 to 100 times better than the glass, and at +100°C still 3 to 5 times better, depending upon the orientation of the heat flow with respect to the crystal axis. In this range of temperature the conductivity of the crystal decreases to ⅙ or ⅛ of its value at −200°C, whereas the conductivity of the quartz glass increases to 2½ times the initial value.

TABLE 6–9

THERMAL CONDUCTIVITY OF PURE QUARTZ GLASS AND QUARTZ CRYSTAL

Temperature [°C]	k [B hr^{-1} ft^{-1} F^{-1}]		
	Quartz Glass (Fused Silica)	Quartz Crystal	
		Normal to Axis	Parallel to Axis
−255	....	..	400
−200	0.54	23	58
−100	0.9	10	20
0	1.1	6	11
100	1.2	4	7
200	1.3	..	...

Thorough measurements of the thermal conductivity of several varieties of mica by Powell and Griffiths [1937] and X-ray examinations by Wood [1937] led to a more detailed knowledge of the influence of the structure upon the conductivity. Whereas Indian muscovite mica showed but slight variation of k with temperature over the range from 100 to 600°C (with an average of 0.38 B hr^{-1} ft^{-1} F^{-1}), the conductivity of phlogopite micas (from Canada and Madagascar) decreased, in some cases to one-third of the initial value, if the temperature was increased from 150 to 250°C, and remained constant or increased slightly when the temperature was further raised from 250 to 600°C. The drop in k was accompanied by a more or less abrupt deterioration of the sharp spectral lines, characteristic for a multiple crystal, into irregular bands, thereby pointing to a marked degradation of the mica structure into a mass of crystallites tilted from the original positions by as much as five degrees. The theoretical aspects of this behavior will be discussed in Sect. 6–9. Details about recovery (partial reversibility of the process) may be looked up in the original papers. The absolute values of k are very different for different samples of phlogopite mica (0.26 to 0.38 B hr^{-1} ft^{-1} F^{-1} at 100°C and 0.1 to 0.3 above 250°C).

As another example, the behavior of sulfur will be considered. At −200, −100, and 0°C the thermal conductivities k of rhombically crystallized sulfur are 0.38, 0.24, and 0.175; those of amorphous sulfur (containing also a little crystallic sulfur) are given as 0.09, 0.10, and 0.115, respectively.

It may be mentioned that at the melting point a discontinuous drop of k occurs. This is most obvious with ice and water. Jakob and Erk [1929] in their measurements on ice obtained the values of Table 6–10.

TABLE 6–10

THERMAL CONDUCTIVITY OF ICE

t [°C]	−125	−100	−75	−50	−25	0
k [B hr^{-1} ft^{-1} F^{-1}]	2.27	2.01	1.79	1.60	1.44	1.29

These values decrease with rising temperature because of the crystalline nature of ice. But, according to Table 6–4, $k = 0.322$ for water at 0°C, i.e., exactly ¼ of the value for ice.

Further, the thermal conductivities of different modifications of carbon may be compared. ¶ For natural bituminous coals at 30°C with apparent densities ρ_a between 1.25 and 1.45 g/cm^3, values of $k = 0.115$ to 0.16 have been observed; for anthracite coal, in the same density limits, $k = 0.125$ to 0.20. These values are for conduction across the veins, those parallel to the veins being larger by 3 to 4%.

For artificially treated carbon much higher values are reported; for instance, $k = 1.12$ for retort coal at 30°C, or $k = 1.16$ for a graphitic coal with the electric resistivity $8.9(10^{-3})$ ohm cm. But also these values are small compared with those for crystalline carbon. For an Acheson graphite, for instance, the values given in Table 6–11 were found.

TABLE 6–11

THERMAL CONDUCTIVITY OF A SORT OF GRAPHITE

t [°C]	−150	0	+150	300	500	700
k [B hr^{-1} ft^{-1} F^{-1}]	103	101	97	89	73	54

The thermal conductivity of this sort of graphite at 0°C corresponds to that of a metal, for example, zinc (see Table 6–18).

Incidentally, the thermal conductivity of diamond, another carbon modification, is about 130 B hr^{-1} ft^{-1} F^{-1} at the temperature of liquid

¶ Comprehensive papers about thermal conductivity and diffusivity of coals and cokes have been published by Fritz and Moser [1940] and Fritz [1943].

hydrogen, i.e., about half the conductivity of the best conducting metal, silver, at 0°C.

The values for Acheson graphite, given above, prove that the decrease of k with increasing temperature is not restricted to low or medium temperatures. Table 6–12 shows this behavior up to 1000°C for some other crystals which, like quartz, are main crystalline constituents of refractory materials (see Sect. 6–10).

TABLE 6–12

THERMAL CONDUCTIVITY OF SOME CRYSTALS

Temperature [°C]	k [B hr^{-1} ft^{-1} F^{-1}]			
	Mullite ($3Al_2O_3 \cdot 2SiO$)	Corundum (Al_2O_3)	Periclase (MgO)	Carborundum (SiC)
0	3.4	6	24	41
100	3.0	5	18.5	33.5
200	2.7	4.0	14.5	28
400	2.0	3.4	9.5	21
600	1.7	3.0	6.5	17
800	1.7	2.7	5	13.5
1000	1.3	2.3	3.7	11.5

Though the influence of temperature upon the conductivity of non-metallic crystals is the same, the absolute values of k differ widely. At 0°C the values of k for rhombic sulfur, ice, quartz crystal, and carborundum are as 1:7:50:235.

6–9. Theoretical Relations about the Thermal Conductivity of Non-metallic Crystals at Very Low Temperature

From a theoretical standpoint, the investigation of k at very low temperatures is of special interest because there, according to the trend of the specific heat close to the absolute zero point, almost none of the atoms have a sensible thermal motion. For this reason it had been expected that no heat conduction would be possible. But Eucken's [1911(a)] experiments on quartz and sylvine crystals showed exactly the contrary; he could represent them with ±5 to 10% accuracy in the range of T = 83 to 373°K by the simple empirical equation

$$\frac{k}{k_0} = \frac{T_0}{T} \tag{6–31}$$

where k and k_0 are the thermal conductivities at the absolute temperatures T and T_0 (of melting ice), respectively.

Physical interpretation of this equation met with great difficulties. A very simplified assumption about the transmission of heat in a solid body is that the flow of energy between neighboring places will be proportional to the gradient of the internal energy, because the energy difference is the incentive for equalization. Thus the heat flow through the unit of area in the direction x should be

$$q'' = -B\frac{du}{dx} = -B\frac{du}{dT}\cdot\frac{dT}{dx} = -Bc_v\frac{dT}{dx} \tag{6-32}$$

where u is the internal energy,
c_v is the specific heat at constant volume, and
B is a constant.

On the other hand, by definition

$$q'' = -k\frac{dT}{dx} \tag{6-33}$$

From Eqs. 6-32 and 33 it would follow that c_v/k = constant. With quartz glass, this condition is very roughly satisfied, but with quartz crystal not at all. Therefore, a second effect must exist, opposite to the specific-heat effect and surpassing it extremely at low temperature.

Debye [1914, pp. 46 ff.] succeeded in overcoming this difficulty to a certain extent. He stated that heat energy is not transmitted like a mass from one atom to another, but by thermoelastic waves, which represent the heat motion in a solid body. Bragg [1931] compares a crystal with a three-dimensional spring mattress, the atoms being linked together by springs. Heat causes shivers backward and forward through the solid, displacing the atoms from their normal positions. Thermal conductivity, in those solids which are not electrical conductors, depends upon a transference of these atomic vibrations from hot to cold regions. If the heat-transmitting waves traversed the body quite independently of each other, any agitation of one end of it would immediately rush to the other end at the speed of elastic waves in the body, which for quartz would be 550,000 to 770,000 cm/sec. Hence thermal conductivity would be enormously great. Actually Debye has found, as an essential item, that there are disturbances, and that one of the most important is the heat motion of the crystal lattice itself. This may control the dependence of heat conduction upon temperature. According to Debye's theory, many simultaneous wave trains cross and superimpose upon one another. This would not matter if the waves were harmonic, but they are shown to be unharmonic, and for this reason to give rise to disturbances which increase greatly with the temperature.

Bragg demonstrated how the various wave trains scatter each other, and, most illuminatingly, says that the advancing region of wave agitation has to fight its way towards the cold end of the body through a fog of waves which it itself produces. A. Meissner [1935] uses another simile. He compares the crystal with a chain of many slightly damped electrical oscillators, which are tuned to the same frequency and coupled very loosely. He further assumes that the heat waves are transmitted in a narrow frequency band only. In a quartz crystal, for example, the positive Si atom performs about $30(10^{12})$ oscillations per second against the two negative O atoms, and in the direction of the optical axis there exist about $0.55(10^9)$ such elementary oscillators per centimeter length. Experiments have shown that the velocity of the propagation of heat in quartz is about 0.02 cm/sec, i.e., about 30 or 40 million times smaller than the value mentioned above for the ordinary acoustic wave. This is caused by the great number of elementary oscillators, which form the links of the chain and must be incited one after another. Because $11(10^6)$ oscillators exist on a length of 0.02 cm, only $90(10^{-9})$ sec is available for the swinging up of each oscillator and the transmitting of the heat vibration to the next one. But because the frequency is of the order of 10^{13} per second, the oscillator must perform about one million oscillations until it has swung up to its maximum amplitude. Therefore the oscillators are very little damped and very loosely coupled.

It is obvious that these waves can be formed in crystals only. In a glassy substance with the molecules in complete disorder they cannot exist at all. Concerning the disturbances occurring in crystals, Debye came to a more quantitative solution in the following way:

A single wave is not able to pass straight through the solid body, but a diffusion occurs similar to that of light in a dull medium, or that of heat radiation in a gas.

As in the latter case, the differential equation of the wave intensity is

$$dN_x = -m_s N_x \, dx \tag{6-34}$$

the subscript s standing for "solid," and by integration in the limits $x = 0$ and $x = x$ it follows that

$$N_x = N_0 e^{-m_s x} \tag{6-35}$$

(*cf.* Eqs. 4-91 and 92).

If $m_s x = 1$ or $x = 1/m_s$, the intensity has decreased to N_0/e. Denoting this length x by λ_s, we have what Debye calls the average distance of effect of the thermoelastic wave. λ_s is said to control the heat conduction. If λ_s is large (and m_s small), the heat conduction by the crystal is good; if λ_s is small (and m_s large), the conduction is poor. Eventually Debye arrived at the following formula:

$$k_s = \tfrac{1}{4}\lambda_s w_s \rho_s c_{v,s} \qquad (6\text{–}36)$$

where w_s is the average velocity of propagation of the elastic waves,
ρ_s is the density of the solid, and
$c_{v,s}$ is its specific heat.

This may be compared with the formula for the thermal conductivity of a gas:*

$$k_g = \tfrac{1}{3}\lambda_g w_g \rho_g c_{v,g} \qquad (6\text{–}37)$$

where λ_g is the average free path of the molecules,
w_g is the average molecular velocity,
ρ_g is the gas density, and
$c_{v,g}$ is the specific heat of the gas.

This analogy is of particular interest because in 1923 de Broglie in his quantum mechanics more generally stated that certain processes can be described by means of wave energy as well as by corpuscular motion.

Debye calculated λ_s numerically by expressing the mutual influence of the waves by the coefficients of thermal expansion and of mechanical compressibility, and he showed that λ_s is almost proportional to the inverse value of T, as found experimentally by Eucken. This is due to the fact that the thermal oscillations and therefore the disturbances of the wave trains decrease with the temperature [Eucken 1940].

Table 6–13 contains some of Eucken's experimental results, showing that k_sT does not change much with temperature in a range down to −190°C.

TABLE 6–13

k_sT [CAL SEC^{-1} CM^{-1}] FOR SOME CRYSTALS

t [°C]	NaCl	KCl	$CaCO_3$ (Normal to Main Axis)
−190	4.2	5.3	3.1
− 78	4.9	4.8	2.7
0	4.5	4.5	2.8
100	4.3	4.4	3.2

* Equation 6–37 is obtained by combining Eqs. 5–5 and 6–1.

At still lower temperature, however, Eucken's empirical formula, as well as Debye's and more recent theories [Peierls 1929, Papapetru 1934], fail, as follows from experiments of de Haas and Biermasz [1935] on quartz crystals, extended down to $T = 2.77°\text{K} = -270.4°\text{C}$. One of the surprising results of these experiments is that k of quartz has a maximum at about 10°K. According to the observations, the thermal conductivity k_p (parallel to the crystallographic axis) of one sort of quartz crystal can be exactly represented, down to $24°\text{K} = -249°\text{C}$, by the empirical equation

$$\frac{1}{k} = -0.00676 + 0.000238T \tag{6–38}$$

with T in degrees Rankine (°R) and k in B hr^{-1} ft^{-1} F^{-1}.

Below 24°K the thermal conductivity more slowly increases up to a maximum of about 100 B hr^{-1} ft^{-1} F^{-1} at about 10°K, as was found by interpolation between 15 and 5°K, a region in which observations are missing.

According to the observations, the thermal conductivity at 2.77°K is equal to 51.6 B hr^{-1} ft^{-1} F^{-1}, the same value as observed at about 61°K, and the trend of the curve is in the direction towards zero at the absolute zero point, as had been expected originally (see above). The authors suppose that this behavior is due to the ideal lattice, whereas in the higher temperature range of liquid hydrogen impurities and failures in the lattice seem to have a strong influence.

It is remarkable how in this case theory and experience worked together. First, diminishing of thermal conductivity at very low temperature had been expected, owing to the slowing down of the atomic thermal motion. When just the opposite was observed in the liquid-hydrogen temperature range, this was explained by an elasto-acoustic theory. Again deviations from the theory were observed, and it seemed that they could be handled by modifications of the theory, but eventually the trend to very small conductivity values was observed at extremely low temperature, as had been expected initially. It is seen that experience had the last word, as it always has in science.

6–10. Thermal Conductivity of Refractory Materials

In 1923 the author had to deal with strange deviations of the thermal conductivity of various sorts of magnesite bricks, of about the same density and moderately varying content of MgO (86 to 92%). Three different groups of investigators, for instance, had observed the following values of k in B hr^{-1} ft^{-1} F^{-1}:

At 20°C	At 1000°C	Observer
0.27	0.39	Heyn, Bauer, and Wetzel
2.90	1.94	Goerens
5.32	2.18	Dougall, Hodsman, and Cobb

Whereas some experts attributed these differences to heavy mistakes in one or the other observation, the peculiar fact that in the case of small values k increased with temperature, whereas in the case of large values k decreased, led to the assumption that the bricks of the first group contained mainly crystalline constituents, and those of the two other groups preferably glassy substances. With this assumption it was possible to explain various other peculiarities which had not been understood before; for instance, the observation of Griffiths that by heating white magnesia above 350°C, k suddenly increases by 100%, or that k of chromite bricks decreases after heating above 500°C, as found by Tadokoro. From a paper of Jakob [1923] on these influences of crystalline and amorphous constituents some lines may be quoted: "If my concept of the decisive influence of the composition of refractory bricks upon the heat conductivity is right, it must be possible to influence strongly the conductivity of ceramic and similar materials and its temperature coefficient by suitable mixture and heat treatment. . . . This seems to be a field for systematic research and for improvement of ceramic products uncultivated as yet." Also the behavior of graphite was mentioned in the same paper, and it was suggested to increase k of refractory products by adding graphite.

The suggestion to study the influence of crystalline and glassy structure upon the apparent heat conductivity of refractory materials led Eucken [1932] to some interesting calculations, which were based on the following approximative equations:

For uniform crystals,

$$\frac{1}{k_{c,u}} = AT \tag{6-39}$$

For mixed crystals,

$$\frac{1}{k_{c,m}} = AT + B \tag{6-40}$$

For glass,

$$\frac{1}{k_g} = C + \frac{D}{T} \tag{6-41}$$

where k_c and k_g are the thermal conductivities of crystal and glass, respectively, and A, B, C, and D are constants.

Supposing further that the material can be considered an aggregate of heat resistors of crystalline and amorphous kinds, connected in series, the following equation was obtained:

$$k = \frac{1}{aT + b + c/T} \tag{6-42}$$

where a, b, and c are constants.

This equation generally leads to an increase of k with increasing temperature if T is small, and to a decrease of k with increasing temperature if T is high. With silicon carbide bricks, for example, this behavior has also been found by experiments. The existence of a maximum of k can be understood by the following reasoning. At low temperature the heat resistance of glass is large compared with that of crystal, so that in this range Eq. 6-41 prevails. With increasing temperature the thermal resistance of glass decreases, and that of crystal increases, and eventually Eq. 6-39 becomes dominant. If the material contains only a little glass, the maximum occurs at low temperature; if it contains not much crystalline substance, the maximum occurs at high temperature, becomes less distinct, and possibly disappears entirely.

Eucken further performed some calculations with the assumption that the material is composed of small and irregularly distributed particles of crystal, glass, and air, and that the conductivity of air can be neglected against the conductivities of the basic substances. Using a formula developed by Maxwell for the calculation of the electric resistance of a mixed body, one obtains simple mixing formulas. Even so, application becomes very complicated, as can be seen from Fig. 6-4, a simplified representation of an original graph of Eucken, which refers to a material composed of a sort of glass, a kind of crystal twenty times better conducting, and of 25% pores. By this method Eucken succeeded in representing observed values of k and their change with temperature for various refractory bricks, as far as the pores could be considered spheric and hollow cylindric. With fissure-like pores, however, k actually was much smaller than calculated, particularly at low and medium temperatures; the actual increase of k with temperature was unusually great in this case.

The burning of a refractory material influences k considerably, because of chemical and crystallographic conversions of its constituents, as well as by the reduction of porosity which is caused by sintering. For example, there has been produced a substance called sintered corundum, which is said to contain 99.7% Al_2O_3 and to be formed by a homogeneous net of α-corundum crystals, grown together to a granite-like, extremely tough structure without any glass and pores. The melting point of this material is 2050°C, and the thermal conductivity at 0°C is about 100%

larger, and at 800°C about 30% larger, than that of ordinary corundum, as quoted in Table 6–12.

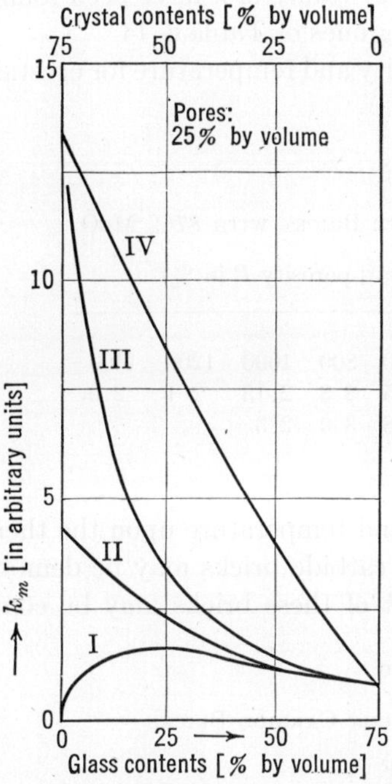

I, II and III	Crystal grains and pores enveloped by an uninterrupted mass of glass.
I	Pores small compared with the crystal grains.
II	Pores of the same size as the crystal grains.
III	Pores large compared with the crystal grains.
IV	Glass particles and pores enveloped by an uninterrupted mass of crystal.

Fig. 6–4. Apparent thermal conductivity of a porous substance composed of glass and crystal.

Because of the influences and variations mentioned it is almost impossible to give values of the thermal conductivity of refractory materials which can be generally used. In the following tables it is intended only to show the magnitude of k for the most common refractory substances.

Table 6–14 contains average values for fire-clay bricks with 54 to 95% SiO_2. Actual values may deviate from these as much as ±30% at 200°C, and ±40% at 1000°C.

TABLE 6–14

Thermal Conductivity of Fire-Clay Bricks with 54 to 95% SiO_2

(k in B hr^{-1} ft^{-1} F^{-1}; apparent density ρ_a in g/cm³)

ρ_a	1.0	1.2	1.4	1.6	1.8	2.0	2.2
k at 200°C	0.28	0.31	0.35	0.42	0.52	0.64	0.84
k at 1000°C	0.40	0.46	0.52	0.60	0.73	0.90	1.20

Bricks with fused Al_2O_3 as the prevailing constituent conduct the heat less well than those with SiO_2 as the main constituent. For alundum bricks (almost pure Al_2O_3, ρ_a = 1.7 to 2.2) values of k have been found which are about half of the corresponding ones of Table 6–14.

Table 6–15 shows the influence of porosity and temperature for crystalline magnesite bricks with 87% MgO.

TABLE 6–15

THERMAL CONDUCTIVITY OF MAGNESITE BRICKS WITH 87% MgO

(k in B hr^{-1} ft^{-1} F^{-1}; ρ_a in g/cm^3; porosity P in %)

ρ_a	P	t [°C] =	200	400	600	800	1000	1200	1400
2.4	36	k =	3.4	2.8	2.5	2.3	2.15	2.1	2.0
2.9	16	k =	...	5.0	4.8	3.0	2.3	....	....

Finally, the influence of composition and temperature upon the thermal conductivity of five kinds of silicon carbide bricks may be demonstrated by Table 6–16. The values of k of these bricks may be compared with those of the crystals of SiC, SiO_2, and Al_2O_3 (Tables 6–9 and 12). The percentage of SiC controls k, the conductivity of the two

TABLE 6–16

THERMAL CONDUCTIVITY OF SILICON CARBIDE BRICKS

k [B hr^{-1} ft^{-1} F^{-1}]

(t_b = burning temperature; ρ_a = apparent density; P = porosity)

Brick No.	Chemical Analysis			t_b [°C]	ρ_a [g/cm^3]	P [%]	Temperature [°C]					
	SiC [%]	SiO_2 [%]	Al_2O_3 [%]				0	200	400	600	1000	1400
I	..	..	..	....	2.06	35	....	....	...	10.6	8.1	6.2
II	90	7	3	1435	2.32	26	14.1	11.5	9.4	8.1	6.1	...
III	77	10	8	1310	2.36	21	....	....	7.0	6.1	4.8	...
IV	57	21	16	1300	2.24	24	....	....	3.4	3.3	2.7	...
V	49	26	20	1310	2.19	22	....	....	2.7	2.3	2.2	...

other constituents being much smaller. The porosity obviously reduces the apparent conductivity considerably.

6–11. Thermal Conductivity of Electrical Insulating Materials

In the design of electrical machines and devices, an important limit of the admissible load is imposed by the behavior of the insulating materials. Most of them do not stand temperatures above 100°C in permanent operation, and there is little hope that in the near future machines may be designed which can be operated at 300°C or more.

It is particularly unfortunate that all organic insulating materials which are used in electrical engineering are poor heat conductors, so that the Joulean heat which is produced in the metallic conductors is stored in just those materials which cannot stand high temperatures.

Here again the suggestion of mixing crystalline and amorphous substances (see Sect. 6–10) was used. A. Meissner [1935] reported that by adding quartz grains to the basic organic substance the thermal conductivity of electric insulating materials could be increased five- to eightfold. As basic plastic material the electric insulating material was used which is known as "compound" and which contains about 20 to 40% bitumen. By adding quartz sand, the dielectric strength of 200 to 400 kv was not changed sensibly.

Broken quartz, as used in porcelain factories, is not recommended as additional substance; it has sharp edges which reduce the heat transmission, possibly by retaining gases. The best material seems to be quartz sand, made up of pure, piezoelectric quartz crystals which are smoothed and mixed properly by nature.

The experimental data for most of the electric insulating materials show great deviations, owing to the difficulties of defining the material in its original state or after special treatment, and of measuring k of thin layers, for example, of paper.

In Table 6–17 values of the heat resistivity $1/k$ of some representative insulating materials are given in watt^{-1} cm C, as usual in electrical engineering (see Sect. 2–2); it must be kept in mind that the values are roughly estimated averages.

The quartz sand with grain thickness $\delta < 0.04$ cm, for which data are given in the table, is mixed from grains of 0.015 to 0.04 cm, mainly those of 0.03 cm. The percentages of quartz given in the table are related to the weight of the basic substance.

By adding quartz to the basic insulating materials, it was possible to reduce the size and copper weight of loud speakers and of magnetic chucks by about one half. Similar improvements were made with magnetic contactors of electric railways. For a magnetic track brake it is reported that, by bedding the magnet coil into the compound-quartz mixture, the temperature excess fell from 95°C to 30°C; for transformers of less than 10 kw, oil or compound can be replaced by the compound-quartz mixture. On a 0.5-kw transformer with compound-quartz mixture, oil, and ordinary compound, temperature excesses as

TABLE 6–17

THERMAL RESISTIVITY $1/k$ OF ELECTRIC INSULATING MATERIALS

Substance	$1/k$ [watt^{-1} cm C]
Air	3900
Silk	2500 to 2000
Silk, varnish treated	700
Cotton	1500 to 1200
Cotton, varnish treated	700
Paper	1000
Paper, soaked with lacquer	700
Micanite	1000 to 600
Compound	1000
Compound + 75% broken quartz, $\delta < 0.01$	500
Compound + 75% quartz sand, $\delta < 0.01$	300
Compound + 75% quartz sand, $\delta < 0.04$	200
Compound + 90% quartz sand, $\delta < 0.04$	150
Transformer oil	750
Transformer oil + 80%(?) quartz sand, $\delta < 0.04$	200
Ebonite	600 to 700
Ebonite + 50% quartz sand, $\delta < 0.04$	450
Mica	200 to 400
Glass } Porcelain }	100 to 200

1:1.2:2.4, respectively, are reported. The advantage, compared with oil cooling, is caused by the homogeneous heat flow, whereas even with well-circulating oil the thermal surface resistance is considerable.

6–12. Thermal Conductivity of Metals

We begin with Table 6–18, which contains numerical values of k for some very pure metals in a wide temperature range. The table, first of all, shows how different the metals are in thermal conductivity, the ratio of k for copper and mercury being about 11:1 at −100°C. Silver, incidentally, is a still better conductor than copper ($k = 242$ at 18°C), and for the purest electrolytic iron $k = 54$ has been observed at 0°C.

TABLE 6–18

THERMAL CONDUCTIVITY OF PURE METALS

k [B hr^{-1} ft^{-1} F^{-1}]

(Values in parentheses are for the liquid state)

Metal:	Copper	Aluminum	Zinc	Nickel	Iron	Platinum	Lead	Mercury
Impurities [%] $\leqq$	0.1	0.3	0.2	0.1	0.1	0.1	....	....
Temperature [°C]								
−200	310	150	70	67	58	45	25	27
−100	235	135	67	....	43.5	42.5	23.5	21.5
0	225	130	65	48.5	41.5	40.5	21	(6.0)
+100	220	130	63	47.5	39.5	41	19.5	(8.1)
200	215	130	61	42.5	35.5	42.5	18	....
400	210	130	54	34	28	44.5	(8.7)	....
600	205	...	(33.5)	37.5	23	47	(8.1)	....
800	...	(51)	....	....	17.5	49.5	....	....

The table further shows that the thermal conductivity of metals, like that of non-metallic crystalline substances, increases with decreasing temperature.

At low temperatures Eucken's equation

$$\frac{k}{k_0} = \frac{T_0}{T} \tag{6–31}$$

can be applied, the subscript 0 relating to 0°C. By differentiation it follows that

$$\frac{dk}{dT} = -\frac{k_0 T_0}{T^2} \tag{6–43}$$

The decrease of k is reciprocal to the square of the absolute temperature and can, therefore, be demonstrated best by experiments at low temperatures. It is further proportional to the thermal conductivity at 0°C. Indeed, as an example, the increases of k for copper and platinum between 0°C and −100°C, according to Table 6–18, are almost exactly in the same ratio as the respective values of k_0.

At higher temperatures the impurities of the metals may cause k to pass through a minimum with increasing temperature. This, for example, occurs with 99.9% pure nickel and platinum, and is observed with most of the less pure metals on the market. With some non-metallic solid bodies, like carborundum bricks, on the other hand, a maximum of k has been observed. Both kinds of behavior can be explained by means of the equation

$$k = \frac{1}{aT + b + (c/T)} \tag{6–42}$$

The constants of this equation depend on the mixture of crystals and amorphous constituents and on the absolute values of their conductivities. Differentiating the equation with respect to T,

$$\frac{dk}{dT} = \frac{c - aT^2}{(aT^2 + bT + c)^2}$$

Herewith the common condition for a maximum or minimum becomes

$$\left(\frac{dk}{dT}\right)_m = 0 = c - aT^2 \tag{6–44}$$

According to this equation, the maximum or minimum occurs at the temperature

$$T_m = \sqrt{\frac{c}{a}} \tag{6–45}$$

and from Eqs. 6–42 and 45 the following equation of the corresponding thermal conductivity is obtained:

$$k_m = \frac{1}{2aT_m + b} = \frac{T_m}{2c + bT} \tag{6–46}$$

From Eq. 6–44 it follows that a maximum or minimum occurs only if both a and c are positive or negative. A further discussion † yields the conditions for a maximum or minimum as shown in Table 6–19.

† I am indebted for most of this discussion to Mr. S. D. Black, who analyzed the maximum and minimum conditions as homework in my class. Recently Mr. Cl. G. v. Fredersdorff directed my attention to the fact that the maximum or minimum condition can be brought into the simple form $(d^2k/dT^2)_m = -2ak^2/T$, showing that a maximum is possible if $a > 0$ and a minimum if $a < 0$, in agreement with Table 6–19.

TABLE 6–19

MAXIMUM AND MINIMUM CONDITIONS FOR k ACCORDING TO EQ. 6–42

(Both a and c must be positive or negative)

Case	Sign of a b c	Sign of $(aT^2 + bT + c)$	Maximum Occurs	Minimum Occurs	Remarks
I	+ + +	+	Always	Never	
IIa	+ − +	+	If $\lvert 2c \rvert > \lvert bT \rvert$	Never	
IIb	+ − +	−	If $\lvert 2c \rvert < \lvert bT \rvert$, i.e., never	Never	If $\lvert 2c \rvert < \lvert bT \rvert$, k would be < 0
IIIa	− + −	−	Never	If $\lvert 2c \rvert > \lvert bT \rvert$, i.e., never	If $\lvert 2c \rvert > \lvert bT \rvert$, k would be < 0
IIIb	− + −	+	Never	If $\lvert 2c \rvert < \lvert bT \rvert$, i.e., always	
IV	− − −	−	Never	If $k < 0$, i.e., never	If $a < 0, b < 0, c < 0$, k would be < 0

The condition $\lvert 2c \rvert > \lvert bT \rvert$, occurring in this table, is equivalent to $\lvert 2aT \rvert > \lvert b \rvert$, as easily can be seen. The table shows that a maximum occurs in cases I and IIa, and a minimum in case IIIb only.

The constants a, b, c can be found from measurements of k at three temperatures. Having experimental values of k at temperatures as low as -200°C facilitates the development of an empirical equation which also holds at high temperatures.

Impurities of metals must be considered as disturbances of the lattice structure. This is the reason why at higher temperatures, where the decrease of k with increasing temperature is small for the crystalline metal, eventually an increase of k takes place as for an amorphous substance, and the absolute value of k becomes smaller at the same time. At 0°C, for example, an impurity of 0.1% causes k of copper, aluminum, nickel, and iron to decrease by about 2 to 3%.

Metallic single crystals, on the other hand, have much higher conductivities than ordinary crystalline metals and a much steeper increase of k with decreasing temperature. The highest thermal conductivity which seems to have been observed at any time is that of a natural copper crystal investigated by Schott [see Landolt-Boernstein 1923, p. 1291] namely, $k = 7050$ B hr^{-1} ft^{-1} F^{-1} at -252.2°C. This is more than 30 times the value given for copper at 0°C in Table 6–18. With other copper crystals Grueneisen and Goens [1927] have found $k = 5320$ at the same temperature, i.e., values of the same order of magnitude. Finally it may be mentioned that, according to Table 6–18, a sudden drop of k occurs when the metal is melting, a behavior which corresponds to that of melting non-metallic crystalline substances.

6–13. The Relation between Thermal and Electric Conductivity of Metals and Alloys

As will be shown later, the measurement of the thermal conductivity of metals is not easy and is always a troublesome job. Fortunately there exists a simple relation between the thermal and the easily measurable electric conductivity. This relation, which has been studied from the theoretical and the practical viewpoint by many workers, is known as the Wiedemann-Franz-Lorenz equation:

$$L = \frac{k}{k_e T} = \text{constant} \tag{6–47}$$

where k = the thermal conductivity,
k_e = the electric conductivity,
T = the absolute temperature, and
L = the so-called Lorenz number.

Whereas Wiedemann and Franz [1853] had assumed k/k_e to be constant, L. Lorenz [1872] derived the foregoing equation. With k in watt cm^{-1} C^{-1}, and k_e in ohm^{-1} cm^{-1}, L appears in $volt^2$ C^{-2}.

A good derivation of Eq. 6–47 originates from Drude [1900]. He assumed that both the thermal and the electric conductivities in metals are due to free electrons inside the metal which behave like gas molecules ("electron gas") and move freely across the lattice of ions which form the skeleton of a metal. Proceeding in a method very similar to the one which we used for developing the equation for the thermal conductivity of a gas, an expression for the thermal conductivity of the electron gas is found, and an expression for the electric conductivity is known from the ordinary theory of electrons. Both conductivities are found to be proportional to the free path of the electrons, and the comparison leads to

$$L = \frac{k}{k_e T} = 3\left(\frac{\kappa}{\epsilon}\right)^2 = \text{constant} \tag{6–48}$$

where κ = the Boltzmann constant,‡ and
ϵ = the elementary charge of an electron.

Refined theories led to the factor 3.3 instead of 3, whereas Lorenz originally had arrived at the factor 2 in this equation. If subscript 0 relates to 0°C, it would follow that $10^9 L_0 = 22.4$ to 24.4 $volt^2/C^2$, or 5.35 to 5.83 cal sec^{-1} C^{-2} ohm.

‡ The usual symbol k for Boltzmann's constant cannot be employed here.

Table 6-20 contains values of the Lorenz number as measured for some pure metals and for some alloys, at temperatures not much different from 0°C.

TABLE 6-20

LORENZ NUMBER (L) FOR PURE METALS AND ALLOYS

Metal	10^9L [volt^2 C^{-2}]	Alloy	10^9L [volt^2 C^{-2}]
Al	22.3	92% Al + 8% Cu	16.5
Cu	22.3	77% Mg + 20% Cu + 3% Si	19.0
Na	22.3	82% Cu + 18% Zn	26.5
Ni	23.3	85% Mg + 15% Cu	27.4
		Steel with 1% C	27.7
Au	23.5	86% Al + 14% Mg	28.3
Zn	24.1	95% Al + 4% Cu + 0.5% Mg	34.7
Mg	24.7	88% Mg + 12% Al	35.8
Pb	24.7	60% Cu + 40% Ni	40.0
Pt	25.1	70% Cu + 30% Mn	44.0
Hg (solid)	28.0		
W	28.8		

Whereas the deviations of L are not great for pure metals, they are for alloys. The deviations are explained by the fact that in addition to the proper "metallic" thermal conductivity, which may be proportional to the electric conductivity, a "non-metallic" thermal conductivity of the lattice must also be considered, such as occurs in electric non-conductors. Its influence will be greater the smaller the electric conductivity of a material is. Thus for very poor electric conductors, L should be very great. Indeed, for an alloy of 48.3% Cd and 51.7% Sb, $10^9L_0 = 2430$ has been observed; for Si, even 170,000.

The influence of temperature upon L is not considerable. Between 0 and 900°C, increments of 15%, 12%, and 26% have been observed with Swedish iron, steel (with 1% C), and nickel, respectively. For tungsten L increases by about 20% between 0 and 1900°C.

Finally it may be remarked that the so-called super-conductivity, i.e., the property of metals to lose any resistance close to the absolute zero point of temperature, holds only for the electric, and not for the thermal, conductivity.

This may be illustrated by the behavior of solid mercury, for which Kamerlingh Onnes and Holst [1914] observed $10^9L = 16$ at -268.4°C and $<16(10^{-7})$ at -269.4°C, the jumping point to super-conductivity being at $-269°C$. It seems to follow that, in the direction towards the absolute zero point of temperature, L rapidly approaches the value zero, although k at the jumping point to electric supraconductivity is still increasing (from 65.5 at $-268.4°C$ to 96.5 at $-269.4°C$). The two latter values are to be compared with those in Table 6–18 for higher temperatures.

A modified equation for the Lorenz number of alloys will be indicated in Sect. 6–14.

6–14. Thermal Conductivity of Alloys

Between the thermal conductivities of alloys and those of pure metals there are differences, as between those of amorphous and crystalline electrical non-conductors. As we have learned, even small additions of other substances have a considerable influence. Let us begin with the technically most important alloys, those with iron as the main constituent. Only very exceptionally is iron used in pure state. Speaking of iron, one generally means an alloy of different metallographic constituents of very different thermal conductivities, namely, ferrite (Fe) with $k = 42$ B hr^{-1} ft^{-1} F^{-1}, cementite, i.e., iron carbide (Fe_3C), with $k = 6$, and pearlite, the eutectic of ferrite and cementite layers, including 0.9% C, with $k = 30$; furthermore, Si, Mn, Ni, Cr, W, Mo, etc., may be present. The different kinds of iron and steel are distinguished by the amounts of these constituents. In addition, the heat treatment changes the thermal conductivity. No wonder that it is very difficult to predict k for any kind of iron or steel, even if the chemical analysis is known, and although thousands of experimental data are available.

At 100°C, for example, malleable cast iron with about $2\frac{1}{2}$% C, has $k \approx 37$ if 82% of the iron is ferrite, but $k \approx 27$ if 91% is pearlite. Cast iron with 3% C has $k \approx 37$ if 2 to 3% of C is graphite, and $k \approx 27$ if no graphite is present. For a non-treated cast iron with 4.5% C, none of which was graphite, $k \approx 8$ was measured; by annealing it 2 hours at 1000°C, so that eventually 4% graphite was present, the conductivity increased to $k \approx 32$. An addition of 1% Si and 1% Ni to cast iron with 3.5% C, 2.4% of which was graphite, reduced k from 35 to 27.5, but an addition of 2 or 4% Ni resulted in $k \approx 24$ B hr^{-1} ft^{-1} F^{-1}.

For wrought steel with 0.15 to 1.5% C, Simidu [1917] has given three empirical formulas, each relating to 30°C:

$$k_u = 22.3 - 4.68\ \mathrm{C} - 7.81\ \mathrm{Mn} - 15.62\ \mathrm{Si} \tag{6–49}$$

$$k_a = 29.5 - 4.68\ \mathrm{C} - 4.81\ \mathrm{Mn} - 9.63\ \mathrm{Si} \tag{6–50}$$

$$k_q = 31.0 - 7.45\ \mathrm{C} - 7.50\ \mathrm{Mn} - 14.98\ \mathrm{Si} \tag{6–51}$$

C, Mn, Si, in the equations, mean the contents of these elements in percentages. The subscripts u, a, q stand for "untreated," "annealed under vacuum at 900°C for one hour," "quenched afterwards."

The influence of the temperature on k_u is given by the equation

$$k_{u,t} = k_{u,30}[1 + \beta(t - 30)] \tag{6–52}$$

where the subscripts t and 30 indicate the temperatures t and 30°C, respectively. Values of the temperature coefficient β, according to measurements of Honda and Simidu [1917], are given in Table 6–21.

TABLE 6–21

TEMPERATURE COEFFICIENT β OF THE THERMAL CONDUCTIVITY OF CARBON STEEL

t [°C]	100	200	300	400	500	600	700	800	900
$-10^6\beta$	0	50	100	150	250	300	350	350	300

Concerning proper iron alloys, it will be mentioned only that thermal conductivities down to 4.5 B hr^{-1} ft^{-1} F^{-1} (for 29% Ni-steel) have been observed, which must be compared with the value at 0°C, given in Table 6–18, or with $k = 54$, which has been observed with very pure electrolytic iron.

It may be noticed that the value of k, given above for nickel-steel, is about 10 times smaller than that for each of the two components Fe and Ni of the alloy.

This behavior is still more evident for some non-iron alloys. At 0°C, for instance, Cd and Sb have $k = 59$ and 9.2, respectively, but an alloy of 33.3% Cd and 66.7% Sb has $k = 0.72$ B hr^{-1} ft^{-1} F^{-1}.

For the time being, one cannot rely on any theoretical conceptions about alloys, but individual experimental results must be used. For some important non-iron alloys values of k are given in Table 6–22.

As with pure metals the thermal conductivities are quite different, but all of them increase with temperature. It may be mentioned only that many aluminum and magnesium alloys of small weight and relatively high thermal conductivity are available.

Finally, with Al-Cu alloys as an example, it is shown how heat treatments may change the thermal conductivity (Table 6–23).

It is understandable that the normalizing effect of annealing is favorable for heat conduction, as it is in the case of carbon steel, according to Eq. 6–50, and that the disorder produced in the molecular structure by quenching has the opposite effect. For the same reason the thermal conductivity abruptly drops when an alloy melts, as is numerically demonstrated for the eutectic tin-zinc alloy in Table 6–22. The same behavior was noticed in dealing with non-metallic crystals and pure metals.

TABLE 6–22

THERMAL CONDUCTIVITY k OF SOME NON-IRON ALLOYS

k [B hr^{-1} ft^{-1} F^{-1}]

(Values in parentheses for liquid alloys)

Temperture [°C]	Monel (67% Ni, 29% Cu)	Eutectic Tin-zinc Alloy (92% Sn, 8% Zn)	Gunbronze (86% Cu, 9% Sn, 4% Zn)	Brass with 30% Zn	Duralumin (Compressed, Heat Treated) (4% Cu, 0.5% Mg)
−200	..	....	..	40	...
−100	..	.....	..	50	76
0	..	35	..	57	93
+100	17	41.5	38	64	105
200	18	(13.5)	40	74	112
400	20	(20)	44	77	..

TABLE 6–23

THERMAL CONDUCTIVITY OF AL-CU ALLOYS (INFLUENCE OF HEAT TREATMENT)

k [B hr^{-1} ft^{-1} F^{-1}]

Cu Contents, %	Cast Non-Treated	Annealed at 450°C	Afterwards Quenched from 500°C in Water and Aged for 4½ Hours	Afterwards Aged for 2 More Weeks
2	86	95	82.5	78
7	83.5	97	79	76
12	72	86	....	..

Empirical modifications of Lorenz' Eq. 6–47 have been suggested by Grueneisen and Goens [1927] and by C. S. A. Smith and Palmer [1935] in order to make it possible to predict k from known values of k_e for certain groups of alloys.

Smith and Palmer's equation can be written as follows:

$$L = \frac{k}{k_e T} = L_a + \frac{a}{k_e T}$$

or

$$k = L_a k_e T + a \tag{6-53}$$

where L_a and a are individual constants for a group of alloys.

For copper alloys Smith and Palmer found that

$$L_a = 5.71(10^{-9}) \text{ cal sec}^{-1} \text{ C}^{-2} \text{ ohm}$$

and

$$a = 0.018 \text{ cal sec}^{-1} \text{ cm}^{-1} \text{ C}^{-1}$$

More recently Powell [1939] obtained the following values:

For iron alloys,

$$L_a = 6.25(10^{-9}) \quad \text{and} \quad a = 0.006$$

For aluminum alloys,

$$L_a = 5.02(10^{-9}) \quad \text{and} \quad a = 0.003$$

For magnesium alloys,

$$L_a = 52.26(10^{-9}) \quad \text{and} \quad a = 0.026$$

The two last pairs of constants do not hold for cast aluminum or magnesium alloys [Eucken 1940].

NOTE (REFERRING TO THE LAST PARAGRAPH OF SECTS. 6-1 AND 3). Only when the printing of this book was almost finished did the author learn of a graphical representation of k for saturated liquid and vapor by which B. Koch and Fritz [1940] correlated experimental data of different workers for water and carbon dioxide. Considering that k for water at high temperatures decreases and for saturated steam increases with increasing temperature, the authors concluded that at the critical point k must be finite and $dk/dT = \pm\infty$. Adjusting and extrapolating the values of Timrot and Vargaftik [1940], they obtained $k = 0.121$ B hr^{-1} ft^{-1}·F^{-1} at the critical point of water and, in a similar way, $k \approx 0.034$ for carbon dioxide. In the same paper values of k in the superheated state are given up to 225 and 90 kg/cm^2 for steam and carbon dioxide, respectively.

CHAPTER 7

Emissivity (Absorptivity)

7–1. Caution in the Definitions

As thermal conductivity and viscosity are the pertinent properties of matter in heat conduction and convection, so emissivity and absorptivity are controlling in any heat exchange by radiation. They are defined by Kirchhoff's law (see Sect. 4–2), and it has been shown that for a body at the same temperature they have the same numerical value.

In order to avoid errors, it must be kept in mind that absorptivity is defined as that part of the arriving radiation of a black surface at a temperature t which is absorbed by the surface considered at the same temperature. Thus, for instance, from the radiation of the sun to a white surface on the earth and the heat energy absorbed by this surface, the absorptivity cannot be found directly because (1) the sun is not a black radiator and (2) the sun temperature is much higher than that of the white surface. But from an experiment in which the emission of the white surface is measured, the emissivity is found directly by division through the emission of a black surface of the same area and temperature (which is known from exact basic measurements), and this emissivity is numerically equal to the absorptivity of the surface under consideration, for the same temperature.

There is another difficulty in the definitions, particularly concerning metallic surfaces, which sometimes has led to errors. This difficulty originates from the fact that no unique law of total radiation holds for different kinds of surfaces.

For the total emission of any solid or liquid body (taken for a surface A, and over the half-spheric space $\omega = 2\pi$) the following empirical equation may be used:

$$\Phi = CA\left(\frac{T}{100}\right)^m \tag{7–1}$$

The divisor 100 is added for practical reasons in order to have only small numerical values in the parentheses. Equation 7–1 is a generalization of the equation for the total radiation of a gray radiator:

$$\Phi_{gr} = C_{gr}A\left(\frac{T}{100}\right)^4 \tag{7–2}$$

which itself is a generalization of Stefan-Boltzmann's equation:

$$\Phi_b = C_bA\left(\frac{T}{100}\right)^4 \tag{7–3}$$

In these equations:

C = a coefficient for the total radiation of any surface,
C_{gr} = " " " " " " " a gray surface,
C_b = " " " " " " " a black surface.

By comparison with Eqs. 4–1 and 52 it is seen that $C_b = 10^8\sigma$ and $C_{gr} = 10^8\epsilon_{gr}\sigma$.

If different values of the exponent m are used for different substances, a value C can be found for most of them which does not vary much with temperature. For metals $m \approx 5.0$ can be taken; for oxides 4.0 to 4.6 will do; and so on.

A second possibility, however, is to admit C to be a function of temperature, but to retain the exponent $m = 4$ as found for the black body. Using now Kirchhoff's law, the absorptivity for total radiation is defined by

$$\alpha = \frac{\Phi}{\Phi_b} \tag{7–4}$$

and therefore

$$\alpha = \frac{C}{C_b} \tag{7–5}$$

Furthermore, the deviations from Lambert's law, dealt with in Sects. 4–9 and 10, will have some effect. Because for gray surfaces Lambert's law is supposed to hold (by definition), as for black bodies:

$$\frac{\Phi_{gr,0}}{\Phi_{gr}} = \frac{\Phi_{b,0}}{\Phi_b}$$

or

$$\frac{\Phi_{gr,0}}{\Phi_{b,0}} = \frac{\Phi_{gr}}{\Phi_b} = \frac{C_{gr}}{C_b} = \alpha_{gr} \tag{7–6}$$

where the subscript 0 relates to the direction normal to the surface ($\phi = 0$).

This means that the absorptivity α_{gr} for gray radiation is the same whether one compares the normal or the semi-spherical intensities of heat flow. This does not hold exactly for electrical insulators, because there is some deviation from Lambert's law (see Sect. 4-9), and it does not hold at all for smooth metallic surfaces, where oblique radiation is much stronger than the normal one (see Fig. 4-9). Therefore for real surfaces

$$\frac{\Phi_0}{\Phi_{b,0}} \gtrless \frac{\Phi}{\Phi_b} \quad \text{and} \quad \frac{C_0}{C_b} \gtrless \frac{C}{C_b}$$

or

$$\alpha = \frac{C}{C_b} = f\frac{C_0}{C_b} = f\alpha_0 \tag{7-7}$$

where f is a factor which previously has been found generally to be smaller than 1 for insulators ($f_i < 1$) and greater than 1 for metals ($f_m > 1$).

In the following tables either values of $\alpha = C/C_b$ or values of $\alpha_0 = C_0/C_b$ will be given, according to which of these quantities has generally been observed for the respective group of bodies. Concerning the conversion from α into α_0 or vice versa, see Tables 4-3 and 4. By adding the factor 100, the absorptivity appears in percentages of

$$C_b = 137.5(10^{-6}) \text{ cal sec}^{-1} \text{ cm}^{-2} \text{ C}^{-4} = 0.1740 \text{ B hr}^{-1} \text{ ft}^{-2} \text{ F}^{-4}$$

In general, the values of absorptivity given in the following sections are "equilibrium" values, i.e., the radiation to be absorbed is supposed to come from a black body at the same temperature. According to Kirchhoff's law, these values are identical with the values of emissivity at that temperature. The total absorptivity for radiation from a body, which is at different temperature or not perfectly black, will depend on the temperature and spectrum of the radiator, and, particularly, upon the location of the intensity maximum of the radiator with respect to the absorptivity maxima and minima of the receiver.

This may be illustrated by a rough comparison, anticipating some data from the following sections. Most polished metals have equilibrium absorptivities of the order of 5% at 100°F, but may absorb 20 to 40% of the sun radiation. In contrast to this, the equilibrium absorptivity of white metal oxides is of the order of 95% at 100°F, but only about 15% of the solar radiation will be absorbed by them.

Though emissivity is a true property of a surface, whereas absorptivity also depends upon the incoming radiation, the term absorptivity is used in giving numerical data in the following two sections. This is done because absorption seems to be more commonly considered than emission in practice and, particularly, in radiation. Unless the contrary is indicated, however, the term absorptivity stands for "equilibrium absorptivity," which is identical with "emissivity."

7–2. Equilibrium Absorptivity of Electrical Insulators

In Table 7–1 absorptivities of different non-metallic solid materials at ordinary temperatures are given. All of them are between 85 and 100%, and there are no great differences due to the roughness or color of the surface. For instance, light gray, polished marble does not absorb less than rough red brick or smooth glass. The most surprising items are that transparent ice reflects as little as the best lampblack coating, namely, only $1 - \alpha = 3.5\%$ of an incoming radiation, and hoarfrost, although entirely white to the eye, has $\alpha = 98.5\%$, and therefore comes closest to the absolute black body of all substances investigated as yet.

TABLE 7–1

ABSORPTIVITY OF SOME NON-METALLIC SURFACES AT ORDINARY TEMPERATURES

Material	State of Surface	100α
Hoarfrost, white	......	98.5
Ice at −9.6°C	Transparent	96.5
Lampblack-waterglass (heavy coat)	......	96.5
Asbestos slate	Rough	96
Lampblack-waterglass (thin coating)	......	94.5
Rubber, hard black	Glossy	94.5
Paper, thin, on black lacquered iron	......	94
Glass	Smooth	93.5
Marble, light gray	Polished	93
Brick, red	Rough, but without great irregularities	93
Quartz, fused	Rough	93
Porcelain	Glazed	92.5
Lacquer, white, on copper, heavy coat	......	92.5
Paper, thin, on tinned iron sheet	......	92
Enamel lacquer, snow white	......	91
Roofing paper	......	91
Gypsum	......	90
Oak wood	Planed	89.5
Rubber, soft, gray	......	86
Lacquer, black or white on copper (thin coat)	......	85

This, of course, originates from Kirchhoff's definition of absorptivity (see Sect. 7-1). If radiation at high temperature, particularly sun radiation, is considered instead of radiation at ordinary temperature, as is done in Table 7-1, things become very different. This has been mentioned in Sect. 7-1 and will be seen more in detail when dealing with applications and also from some of the following examples about the influence of the surface temperature of non-metallic bodies.

For lampblack-waterglass, a very slight decrease of α with increasing temperature has been observed. For coal, $\alpha = 80.5\%$ has been measured at 150°C, and $\alpha = 78\%$ at 600°C. Heilman [1934] investigated the influence of temperature upon the absorptivity of extremely thin paints on silver and obtained the following results:

For black paint $\alpha = 87.5\%$ at 205°C and $\alpha = 96\%$ at 595°C, but for white paint $\alpha = 94\%$ at 185°C and $\alpha = 52.5\%$ at 775°C. It is seen that at temperatures of 600 to 800°C the difference in absorptivity of black and white surfaces becomes very appreciable.* Similarly for a chromium brick $\alpha = 95\%$ at 593°C and $\alpha = 98\%$ at 1095°C, and for a light buff brick $\alpha = 79\%$ at 540°C and $\alpha = 53.5\%$ at 1095°C.

For silica bricks, with 87 to 97% SiO_2, the following values of 100 α have been reported by Wetzler [1927]:

Surface:	Bright, Reflecting	New, Non-reflecting	Used, Rough
At 600°C:	67	75	82
At 1000°C:	75	80	87

It is a strange phenomenon, which may be caused by the other constituents, that a brightly reflecting silica brick, when heated to 1000°C, comes closer to blackness. (See the following remarks about Hild's results.) Similar results have been found with alumina bricks.

Pirani [1939], like Heilman, found a decrease of blackness with increasing temperature for different glassy substances, for instance:

$100\alpha_0$ for:	Quartz Glass	Opaque Quartz (Vitreosil)	Zirconium Silicate	Pyrex Glass
At 300°C:	87.5	90	88	94
At 500°C:	71	82	77.5	84
At 800°C:	46	68	53.5	..

* Special "high-temperature paints" were used, and the result on white paint was checked by a test with a refractory plate as base for the coating because the radiation of silver through the paint might have influenced the result.

According to the same investigator, the following figures hold for some oxides:

$100\alpha_0$ FOR:	ALUMINUM OXIDE	MAGNESIUM OXIDE	THORIUM OXIDE
At 300°C:	59.5	52.5	54
At 800°C:	26	20	21

At still higher temperatures Hild [1932] has measured α_0 of oxides, and for several of them he also determined the influence of the size of the grain. Larger sizes yield higher absorptivity. With very small grains, α_0 does not depend on temperature; with large grains it increases with temperature.

Some results are represented in Table 7–2. In general, the absorptivity of the colored oxides of the heavy metals Zn, Fe, Cr is much larger than that of the white oxides of the light metals Ca, Mg, Al, Si.

TABLE 7–2

TOTAL-RADIATION ABSORPTIVITY OF PURE OXIDES

Substance	Grain Size [microns]	$100\alpha_0$ at	
		850°C	1275°C
CaO	3 to 5	27	27.5
MgO (Produced by heating Mg salt)	1 to 3	29	29
(Electromagnesia)	90 to 120	37.5	49.5
Al_2O_3 (Electrocorund)	2 to 25	35	32.5
	90 to 120	42.5	54
SiO_2 (Quartz crystal powder)	3 to 12	37.5	43
ZnO	<0.5	29.5	65
Fe_2O_3	Very small	57	78
Cr_2O_3	0.5 to 1.5	73	73
	1.5 to 8	87	95

The experiments showed that, up to 1350°C, the absorptivity of the main constituents of refractory bricks, Al_2O_3 and SiO_2, remains below 55% and 45%, respectively. Hild further found that, in mixing white oxides, the ordinary mixing rule holds roughly for α_0, but that small additions of a colored oxide increase the absorptivity considerably. This is particularly important for the effect of refractory bricks, where usually 1 to 3% of iron oxide is present. But adding, for example, 2% Fe_2O_3 to Al_2O_3 increases the absorptivity by more than 75% at 1250°C.

Thus the addition of Fe_2O_3, while not desirable because of deterioration of the ceramic properties, causes an improvement of the heat exchange inside furnaces.

In Table 4-3 theoretical values for the ratio $\epsilon/\epsilon_0 = \alpha/\alpha_0$ of electrical non-conductors were given. It is now possible to compare these with the values found by experiments.

From Table 7-3 it may be seen that the observed and theoretical values of this ratio are in good agreement, and particularly that all of them are below 1, the average of the observed values being 0.96, and that of the theoretical values 0.94.

TABLE 7-3

RATIO OF ABSORPTIVITIES FOR HEMISPHERICAL AND NORMAL RADIATION OF NON-METALLIC SURFACES

Substance	Surface	Temperature [°C]	$100\alpha_0$ Observed	α/α_0	
				Observed	Theoretical
Copper oxide	Smooth, blue black	133	76	0.955	0.96
Corundum abrasive	Grain size 60 to 88 μ, impregnated in bakelite lacquer	84	85.5	0.985	0.94
Fire clay, burnt	Smooth, white	72	91.5	0.945	0.935
Paper	White	95	92	0.965	0.935
Plywood	Not very smooth	70	93.5	0.97	0.935
Glass		91	94	0.93	0.935
Ice	Smooth, transparent	0	96.5	0.95	0.955

Finally, some values for non-metallic liquids may be mentioned. By covering a metallic sheet ($\alpha = 5.8\%$) with an extremely thin oil layer, $\alpha = 6.3\%$ was obtained; but with a $\frac{1}{1000}$-in. layer $\alpha = 66\%$, with a $\frac{1}{200}$-in. layer $\alpha = 75\%$, and with rather thick layers $\alpha = 82\%$ were observed. By wetting an iron sheet ($\alpha = 77\%$) so slightly that the surface still looked dull, the absorptivity increased to $\alpha = 96\%$; when it looked bright, $\alpha = 97\%$ was observed. The water layers used were thinner than $\frac{1}{250}$ in. Wet ice surfaces at 0°C had $\alpha = 96.6\%$; a water jet at 10°C gave $\alpha = 96.5\%$. When a black radiator at 35°C was located behind the jet, the water layer being about $\frac{1}{10}$ in., not the slightest amount of radiation was transmitted.

7-3. Equilibrium Absorptivity of Metals

Pure metals are good reflectors of heat radiation, as of visible light. This and the influence of temperature in a moderate range can be seen from Table 7-4.

TABLE 7-4

ABSORPTIVITY $100\alpha_0$ OF SOME PURE METALS

t [°C]	Ag	Al	Pt	Ni
25	...	4	..	4.5
200	...	...	..	...
300	2	4.5	7	8
400	2.5	5	8	...
500	3	5.5	9	...
600	3	...	10	...

For a high temperature range the following values for tungsten have been found:

At	25°C	$100\alpha_0 =$	3
"	1000°C	"	≈ 16
"	2000°C	"	≈ 30
"	3300°C	"	≈ 40

In Fig. 7-1 the observed values are compared with those calculated from the previously derived theoretical equation:

$$\alpha_0 = 0.576\sqrt{r_e T} - 0.124 r_e T \tag{4-80}$$

The agreement of theory and experiment is satisfactory.

The influence of the state of the surface (polished or rough, oxidized or machined, etc.) can be seen from Table 7-5. Whereas a trace of oxide (tarnishing) does not change the absorptivity, thick oxide layers raise it greatly. Heavily oxidized and very rough surfaces approach the behavior of the black body. The values of the table may hold up to 200°C. For higher temperatures, too, they will not change much.

Some data from the literature are the following:

For an electrolytical iron sheet, with a dense layer of oxide, $\alpha_0 = 78\%$ and 82%, at 127°C and 527°C, respectively.

For machined cast iron, $\alpha_0 = 60\%$ and 70%, at 880°C and 985°C, respectively.

For ground iron sheet or steel cast, $\alpha_0 = 55\%$ and 60%, at 900°C and 1100°C, respectively.

For non-treated wrought iron, $\alpha_0 = 85\%$ and 95%, at 900°C and 1100°C, respectively.

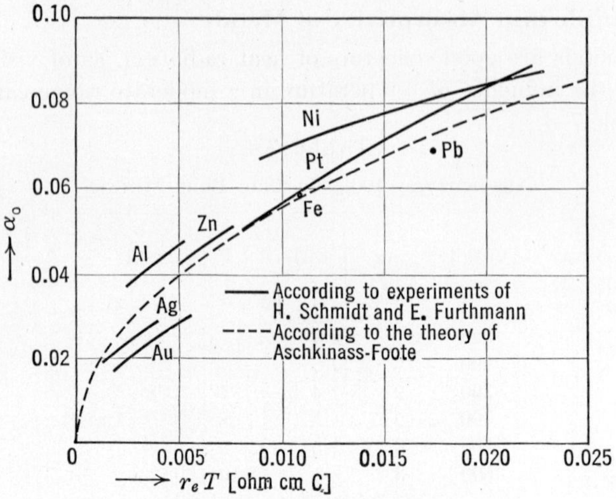

FIG. 7–1. Absorptivity of metals at normal incidence (observed and theoretical values).

From H. Schmidt and Furthmann [1928].

TABLE 7–5

INFLUENCE OF THE STATE OF SURFACE OF METALS UPON THE ABSORPTIVITY AT 25°C

Substance	State of the Surface	$100\alpha_0$
Copper	Polished	3
	Polished, slightly tarnished	3.5
	Shaved	7
	Oxidized, black	78
Brass	Polished	4
	Polished, slightly tarnished	4.5
	After rolling	6
	Fresh rubbed with abrasive	20.5
Tin	Iron sheet, tinned	5.5
	Iron sheet, nickel plated, polished	4.5
	Iron sheet, nickel plated, dull	11
Zinc	Iron sheet, zinc plated, rather bright	24
	Iron sheet, zinc plated, gray	27.5
Iron	Sheet, newly treated with abrasive	24
	Cast, newly machined	43.5
	Sheet, stained	65
	Steel sheet, with skin due to rolling	66
	Steel sheet, with rough or brilliant oxide layer	81
	Cast, with smooth or rough cast skin	81

Some values for aluminum and an aluminum alloy are collected in Table 7–6.

TABLE 7–6

ABSORPTIVITY $100\alpha_0$ OF ALUMINUM AND SILUMIN

Substance	Surface	t [°C]		
		100	300	500
Aluminum, pure	Rolling skin, heavily oxidized	87	63.5	49
	Dull etched, black burnt	51	51	....
	Ordinarily rolled	6	7.5	8.5
	Rolled with polished steel roll	3.5	4.5	5
Silumin cast (Al alloy with 12.5% Si)	Skin from sand casting	32.5	32	31
	Skin from ingot casting	16	19	22
	Highly polished	12.5	20	....

A careful experimental study of the influence of thin oxide coats and lacquer films on aluminum, as well as of slight roughness of the surface, was made by C. S. Taylor and Edwards [1939]. They found an almost linear increase of the absorptivity with the thickness of an oxide coat up to about 2 μ, where $\alpha \approx 63\%$ was obtained. With further increase of thickness, α rose only slightly (73% at 4 μ, 79% at 8 μ thickness). Since 9.3 μ is the wavelength at which the radiation of a black body at 100°F has its maximum, an oxide-coated surface behaves like the massive oxide only if the coating thickness is of that order of magnitude.

For lacquer coatings on aluminum, 2, 4, and 8 μ thick, absorptivities of 28, 44, and 57%, respectively, were observed. Coloring the lacquer had no influence. This, however, holds only for infrared radiation; the reflectivity for light depends considerably upon the color. For instance, lacquer films on aluminum, either clear or with white or black pigment or with yellow or blue dye, all had about 68% absorptivity in the infrared, but 18, 33, 95, 28, and 94% absorptivity, respectively, for light. Another experiment consisted of roughening an aluminum surface with an abrasive cloth so that the specular-reflecting characteristics were completely destroyed; the reflectivity for infrared heat radiation, however, was changed by only a few percent. Finally, the authors remark that the oxide film formed on aluminum in the atmosphere is quite impervious, and this limits its growth and thickness so that its reflectivity in the infrared is only little changed by atmospheric corrosion.

For nickel oxide the following values are given: $\alpha_0 = 40\%$, 60%, 75%, at 300°C, 700°C, 1100°C, respectively, and for copper oxide $\alpha_0 = 55$ to 70% in the same temperature range.

As in Table 7-3 for non-metals, in Table 7-7 the observed values of ratio α/α_0 for metallic surfaces are compared with those found theoretically (*cf.* Table 4-4). Again, the agreement between experiment and theory is satisfactory. Particularly, both show the decrease of α/α_0 with increasing α_0, which has been discussed in Sect. 4-10.

TABLE 7-7

Ratio of Absorptivities for Hemispherical and Normal Radiation of Metallic Surfaces

Substance	Surface	Temperature [°C]	$100\alpha_0$ Observed	α/α_0 Observed	α/α_0 Theoretical
Aluminum	Bright	171	3.9	1.26	1.28
Nickel	Brass, nickel plated, dull	100	4.1	1.14	1.28
Nickel	Brass, nickel plated, polished	100	4.5	1.18	1.28
Manganine	Rolled sheet, blank	119	4.8	1.18	1.27
Chromium	Copper, chromium plated, polished	153	5.8	1.21	1.26
Iron	Etched, dim	147	12.8	1.23	1.20
Aluminum lacquer (aluminum-bronze)		113	25.7	1.11	1.10
Bismuth, cast	Shaved, smoothed	81	34.0	1.08	1.07

Concerning liquid metallic surfaces, for iron at 1300 to 1400°C, $\alpha_0 = 29\%$, for liquid copper $\alpha_0 = 14\%$, have been observed. For mercury $\alpha_0 = 9$ to 12% has been calculated theoretically.

7-4. Absorptivity of Gases

The apparent absorptivity of a gas body, as defined by Eqs. 4-103 and 105, is not a physical property of the gas alone, but simultaneously depends upon the shape and size of the gas volume. The property of matter involved in Eq. 4-105 is the logarithmic decrement of radiation

m_λ, which is defined by Eq. 4-91. This is not an absorptivity, but rather a "weakening factor" of the gas radiation, as Koenigsberger has called it. There are different reliable measurements of m_λ as a function of the wavelength λ, old and recent ones, but it is difficult to form mean values for the width of the bands and to use these for calculating α. This will be dealt with in Chapter 32 (Vol. II). For an infinitely thick gas layer the absorptivity assumes a definite value, which can be calculated from the equation.

$$\alpha_\infty = \frac{\Sigma(\Delta N_b)}{N_b} \tag{7-8}$$

where N_b is the areal intensity of total black radiation, defined by Eqs. 4-12 and 48,

ΔN_b is the areal intensity of the "black gas radiation" of a band, defined by Eq. 4-95, and

$\Sigma(\Delta N_b)$ is a sum taken over all bands of the gas.

The absorptivity α_∞ can be considered a property of the gas, and if the width $\Delta\lambda$ of each of the bands and the distribution of the intensity in this range are known, α_∞ can be calculated by Eq. 7-8. This, however, involves a difficulty in principle. For steam, for instance, besides the main three bands there exist at least two more, which are so weak that their influence may be negligible for thicknesses of gas bodies as occurring in engineering. For an infinitely thick body, however, they may be a considerable percentage of N_b. Therefore for practical purposes it is not very sensible to calculate α_∞, or, as sometimes has been done, to calculate "reduced" values α_∞, not considering bands of minor importance.

So far, the most reliable values have been found by measuring $\Sigma(\Delta N_b)$ as a whole, from the radiation of the gas through layers of different thickness. In this way absorptivities α of gas layers of finite thickness have been found which are convenient for practical purposes. The extrapolation to α_∞, however, is so questionable for the reason indicated above that the use of α_∞ cannot be recommended for the time being, but the use of $\alpha_{A,V}$, which depends on the geometrical dimensions, is more suitable. How to find numerical values of $\alpha_{A,V}$ will be dealt with in Chapter 32 (Vol. II).

PART C. HEAT CONDUCTION IN SIMPLE BODIES

CHAPTER 8

Heat Conduction in Steady State without Heat Gains or Losses

8–1. Simplest Cases of Heat Conduction

In Chapter 2 the basic differential equation of heat conduction was derived. Numerical values of thermal conductivity, which is the main property of matter in this equation, were discussed in Chapter 6. Now we proceed to solutions of the equation.

First, cases at steady state, i.e., without any changes in time, will be dealt with. This condition is almost never strictly, but very often is nearly, satisfied. In terms of the differential equation

$$\frac{\partial t}{\partial \tau} = \frac{k}{\rho c_p}\left(\frac{\partial^2 t}{\partial x^2} + \frac{\partial^2 t}{\partial y^2} + \frac{\partial^2 t}{\partial z^2}\right) + \frac{q'''}{\rho c_p} \qquad (2\text{–}9)$$

it means that $\partial t/\partial \tau = 0$.

For further simplification it will be assumed that no internal heat sources or sinks exist in the body, so that $q'''/\rho c_p = 0$. The temperature differences occurring in the bodies under consideration, however, must originate from heat sources and sinks. These are assumed to be arranged anywhere at the boundaries of the body.

For example, a solid hollow sphere or cylinder with an electrical heater inside and a cooler outside may be considered, no heat sources or sinks being present in the solid mass itself. Also other incidental heat gains or losses may be excluded. Considering, for instance, the axial heat conduction in a metal rod, it is supposed that no appreciable amount of heat is exchanged with the environment, or that such an exchange will be avoided by artificial means. Cases without these restrictions will be dealt with in Chapters 10 and 11.

Exact solutions are possible only for simple bodies, because only these yield simple boundary conditions for the determination of the arbitrary constants of integration. Fortunately, most engineering objects approach one of the following simple shapes: plane plate, cylinder, and sphere. Hence the solutions of the equations for these three shapes are needed in the first place. However, the solution for a cube will also be dealt with. It will be seen how much more difficult the solution becomes for this still simple and regular body.

8–2. Heat Conduction through an Infinitely Wide Plane Plate at Uniform Conductivity

Equation 2–9 has already been simplified by the assumptions $\partial t/\partial \tau = 0$ and $q''' = 0$, mentioned in Sect. 8–1. In addition to this it will be assumed that heat flows only in the x direction; i.e., $\partial^2 t/\partial y^2 = 0$, and $\partial^2 t/\partial z^2 = 0$. Thus, Eq. 2–9 eventually reduces to

$$\frac{d^2 t}{dx^2} = 0 \tag{8–1}$$

This is the differential equation of the temperature distribution across an infinitely extended plane plate with heat flowing in one direction (x) only. If the edge of a plate of finite extension is perfectly insulated, the heat flow across this plate is identical with that through a piece of the same size, cut out from an infinite plate.

By integration of Eq. 8–1

$$t = Mx + N \tag{8–2}$$

where x is the distance from an arbitrary plane, $x = 0$. The constants of integration M and N can be calculated from the temperatures

$$t = t_1 \quad \text{at} \quad x = x_1$$

and

$$t = t_2 \quad \text{at} \quad x = x_2$$

Substituting these boundary conditions in Eq. 8–2,

$$M = \frac{t_1 - t_2}{x_1 - x_2}, \quad N = \frac{t_2 x_1 - t_1 x_2}{x_1 - x_2}$$

and

$$t = \frac{t_1 - t_2}{x_1 - x_2} x + \frac{t_2 x_1 - t_1 x_2}{x_1 - x_2} \tag{8–3}$$

In addition to the temperature distribution along x, given by Eq. 8–3, the heat flow, $q = dQ/d\tau$, is wanted. According to Eq. 2–8,

$$q = -kA \frac{dt}{dx} \tag{8–4}$$

By substituting t from Eq. 8–3 in Eq. 8–4 and performing the differentiation

$$q = -kA \frac{t_1 - t_2}{x_1 - x_2} \tag{8–5}$$

8–3. Radial Heat Conduction in an Infinitely Long Cylinder at Uniform Conductivity

Next to the plane plate, the hollow cylinder is the most important heat-conducting body because of the abundant use of cylindrical tubes

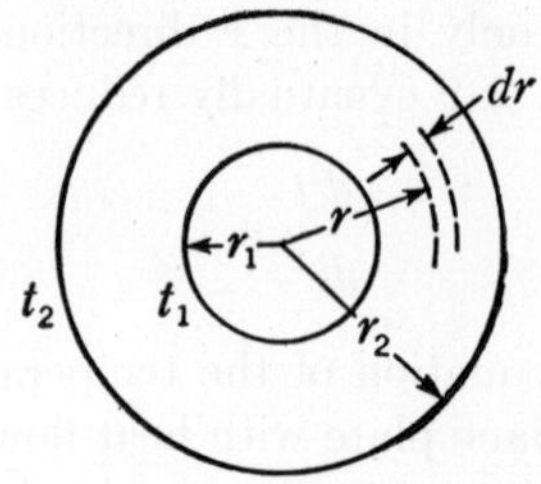

FIG. 8–1. Geometrical data for derivation of radial heat conduction in a cylinder.

in engineering. In an infinitely long cylinder, t_1, t, and t_2 may be the temperatures at the radial distances r_1, r, and r_2, respectively (Fig. 8–1.) Then, for the tube length L, by means of Eq. 8–4,

$$q = -kL \cdot 2\pi r \frac{dt}{dr} \tag{8–6}$$

By integration within the limits r_1 and r_2

$$q = -kL \cdot 2\pi \frac{t_1 - t_2}{\ln r_1 - \ln r_2} \tag{8–7}$$

and by integration within the limits r and r_2

$$q = -kL \cdot 2\pi \frac{t - t_2}{\ln r - \ln r_2} \tag{8–8}$$

From Eqs. 8–7 and 8

$$t = \frac{t_1 - t_2}{\ln r_1 - \ln r_2} \ln r + \frac{t_2 \ln r_1 - t_1 \ln r_2}{\ln r_1 - \ln r_2} \tag{8–9}$$

Though this is the most simple derivation of Eqs. 8–7 and 9, another one may also be given, starting from the differential equation in polar coordinates, which is generally used in dealing with cylinders.

In analogy to the derivation in Cartesian coordinates, presented in Sect. 2–4, a differential cylinder ring of length L, inner radius r, and outer radius $r + dr$ will be considered (see Fig. 8–1).

The heat flow into the cylinder ring is

$$q_i = -kL \cdot 2\pi r \frac{dt}{dr}$$

The heat flow out of the ring is

$$q_o = -kL \cdot 2\pi(r + dr)\left(\frac{dt}{dr} + \frac{d^2t}{dr^2}\,dr\right)$$

Since $q_i = q_o$ for steady heat conduction without heat sources or sinks

$$0 = -kL \cdot 2\pi r \frac{d^2t}{dr^2}\,dr - kL \cdot 2\pi \cdot dr \frac{dt}{dr} - kL \cdot 2\pi \cdot dr \frac{d^2t}{dr^2}\,dr$$

The last term is infinitely small of a higher order than the two first ones because $(dr)^2 \ll dr$. Omitting it leads to the differential equation in polar coordinates:

$$\frac{d^2t}{dr^2} + \frac{1}{r} \cdot \frac{dt}{dr} = 0 \tag{8–10}$$

The general solution of this equation is

$$t = M \ln r + N \tag{8–11}$$

as can be derived easily by introducing a new variable $u = dt/dr$ in Eq. 8–10 or may simply be checked by substitution of t from Eq. 8–11 in Eq. 8–10.

The constants of integration M and N can be calculated from the temperatures

$$t = t_1 \quad \text{at} \quad r = r_1$$

and

$$t = t_2 \quad \text{at} \quad r = r_2$$

Substituting these in Eq. 8–11 yields

$$M = \frac{t_1 - t_2}{\ln r_1 - \ln r_2}, \quad N = \frac{t_2 \ln r_1 - t_1 \ln r_2}{\ln r_1 - \ln r_2}, \quad \text{and Eq. 8–9}$$

Differentiating Eq. 8–9 with respect to r and substituting dt/dr in Eq. 8-6 yield Eq. 8–7.

Equation 8–7, derived for an infinitely long cylinder, can be used for a cylinder of the finite length L if its two ends are perfectly insulated.

For the practical use of Eq. 8–7 Jakob [1926] suggested replacing it by

$$q = \frac{1}{\phi} kA_m \frac{t_1 - t_2}{r_2 - r_1} \tag{8-12}$$

where $A_m = L\pi(r_1 + r_2)$ is a mean area, namely, the area of the cylinder in half the depth of the wall, and

ϕ = a shape factor.

By comparison of Eqs. 8–7 and 12:

$$\phi = \frac{1}{2} \cdot \frac{\dfrac{r_2}{r_1} + 1}{\dfrac{r_2}{r_1} - 1} \ln \frac{r_2}{r_1} \tag{8-13}$$

Table 8–1 contains values of ϕ, calculated by Eq. 8–13.

TABLE 8–1

SHAPE FACTOR ϕ FOR CALCULATION OF HEAT CONDUCTION IN CYLINDERS FROM EQ. 8–12

r_2/r_1	1.0	1.1	1.2	1.3	1.4	1.5	1.6	1.7	1.8
ϕ	1.000	1.001	1.003	1.006	1.010	1.014	1.018	1.023	1.029

r_2/r_1	1.9	2.0	2.2	2.4	2.6	2.8	3.0	3.5	4.0
ϕ	1.034	1.040	1.051	1.063	1.075	1.087	1.099	1.128	1.155

It is seen that for $r_2/r_1 \leqq 1.4$ the radial heat flow in tube walls can be calculated as for a plane plate with a maximum error of 1%.

8–4. Radial Heat Conduction in a Sphere at Uniform Conductivity

Proceeding as for a hollow cylinder, and with the symbols of Fig. 8–1, application of Eq. 8–4 to a hollow sphere yields

$$q = -k \cdot 4\pi r^2 \frac{dt}{dr} \tag{8-14}$$

By integration within the limits r_1 and r_2

$$q = k \cdot 4\pi \frac{t_1 - t_2}{\dfrac{1}{r_1} - \dfrac{1}{r_2}} \tag{8-15}$$

and for the limits r and r_2:

$$q = k \cdot 4\pi \frac{t - t_2}{\dfrac{1}{r} - \dfrac{1}{r_2}} \tag{8-16}$$

Equating the two last expressions of q,

$$t = \frac{t_1 - t_2}{\dfrac{1}{r_1} - \dfrac{1}{r_2}} \cdot \frac{1}{r} + \frac{t_2 \dfrac{1}{r_1} - t_1 \dfrac{1}{r_2}}{\dfrac{1}{r_1} - \dfrac{1}{r_2}} \tag{8-17}$$

Again a second derivation, using the differential equation in polar coordinates, may be given.

In an elementary spherical shell of inner radius r and outer radius $r + dr$ (Fig. 8-1) the incoming heat flow

$$q_i = -k \cdot 4\pi r^2 \frac{dt}{dr}$$

and the outgoing heat flow

$$q_o = -k \cdot 4\pi (r + dr)^2 \left(\frac{dt}{dr} + \frac{d^2t}{dr^2} dr \right)$$

Equating both,

$$0 = -k \cdot 4\pi r^2 \frac{d^2t}{dr^2} dr - k \cdot 8\pi r\, dr \frac{dt}{dr} - k \cdot 8\pi r\, dr \frac{d^2t}{dr^2} dr - k \cdot 4\pi (dr)^2 \frac{dt}{dr} - k \cdot 4\pi (dr)^2 \frac{d^2t}{dr^2} dr$$

The three last terms with $(dr)^2$ and $(dr)^3$ are small of higher order than the two first terms and can therefore be omitted. Thus, the differential equation of the sphere becomes

$$\frac{d^2t}{dr^2} + \frac{2}{r} \cdot \frac{dt}{dr} = 0 \tag{8-18}$$

Its general integral is

$$t = -\frac{M}{r} + N \tag{8-19}$$

as easily can be demonstrated * or checked by substitution in Eq. 8–18.

Again using the boundary conditions,

$$t = t_1 \quad \text{for} \quad r = r_1$$

and

$$t = t_2 \quad \text{for} \quad r = r_2$$

Eq. 8–19 yields

$$M = \frac{t_1 - t_2}{-\dfrac{1}{r_1} + \dfrac{1}{r_2}}, \quad N = \frac{t_2 \dfrac{1}{r_1} - t_1 \dfrac{1}{r_2}}{\dfrac{1}{r_1} - \dfrac{1}{r_2}}, \quad \text{and Eq. 8–17.}$$

The simplification by Eq. 8–12 can also be used for a hollow sphere. In this case

$$A = 4\pi \left(\frac{r_1 + r_2}{2}\right)^2$$

and

$$\phi = \frac{1}{2} + \frac{1}{4} \cdot \frac{r_1}{r_2} + \frac{1}{4} \cdot \frac{r_2}{r_1} \tag{8-20}$$

For r_2/r_1	= 1.0	1.1	1.2	2.0
ϕ	= 1.000	1.002	1.008	1.125

Calculating as for a plane plate causes an error of less than 1% if $r_2/r_1 \leqq 1.2$.

8–5. Heat Conduction through an Infinitely Wide Plane Plate, Considering Variation of Conductivity with Temperature

Until now it has been assumed that k was uniform, i.e., that it did not vary with temperature. According to the discussion of actual values of k in Chapter 6, this is not true in general. However, k can be considered a linear function of temperature in most walls used in engineering. Under this assumption the differential equation of heat conduction can be solved as simply as for constant k.

* By introducing a new variable, $u = dt/dr$, in Eq. 8–18.

Figure 8–2 represents the distribution of t in a plane plate through which heat is conducted.

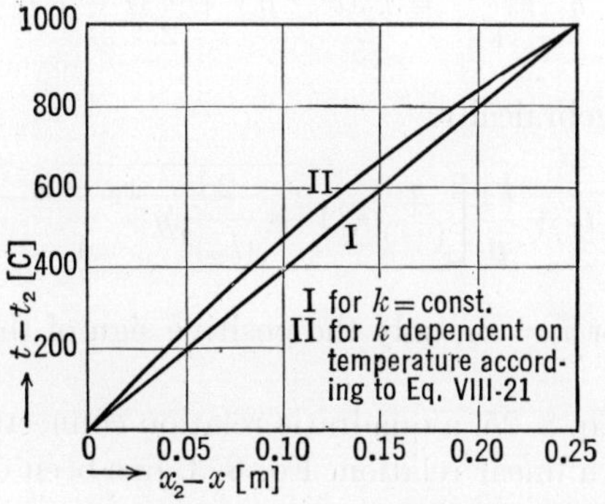

FIG. 8–2. Temperature distribution in plane wall, considering a linear change of conductivity with temperature.

Let

$$k = k_2 + \beta(t - t_2) \tag{8–21}$$

$$k_1 = k_2 + \beta(t_1 - t_2)$$

and

$$k_m = \frac{k_1 + k_2}{2} = k_2 + \frac{\beta}{2}(t_1 - t_2) \tag{8–22}$$

Substitution of k from Eq. 8–21 in the equation

$$q = -kA\frac{dt}{dx} \tag{8–4}$$

yields

$$q = -[k_2 + \beta(t - t_2)]A\frac{dt}{dx} \tag{8–23}$$

By integration within the limits x_1 and x_2 and by introduction of k_m from Eq. 8–22

$$q = k_m A\frac{t_1 - t_2}{x_2 - x_1} \tag{8–24}$$

This indicates that the heat flow can be calculated by means of Eq. 8–5, if only k is replaced by $k_m = (k_1 + k_2)/2$.

Integrating, further, Eq. 8–23 within the limits x and x_2,

$$q \frac{x_2 - x}{A} = k_2(t - t_2) + \frac{\beta}{2}(t - t_2)^2$$

and here from algebraically:

$$t = t_2 + \frac{k_2}{\beta}\left[\binom{+}{(-)}\sqrt{1 + \frac{2\beta}{(k_2)^2} q \frac{x_2 - x}{A}} - 1\right] \qquad (8\text{–}25)$$

Because $t = t_2$ for $x = x_2$, only the positive sign of the square root can be used.

According to Eq. 8–25, a quadratic relation connects t and x, whereas for k = constant a linear relation, Eq. 8–3, has been obtained.

8–6. Radial Heat Conduction in an Infinitely Long Cylinder, Considering Variation of Conductivity with Temperature

The same assumptions concerning thermal conductivity may be made as previously (Eqs. 8–21 and 22), and k will be substituted from Eq. 8–21 in

$$q = -kL \cdot 2\pi r \frac{dt}{dr} \qquad (8\text{–}6)$$

This yields

$$q = -[k_2 + \beta(t - t_2)]L \cdot 2\pi r \frac{dt}{dr} \qquad (8\text{–}26)$$

Integrating between r_1 and r_2 and using Eq. 8–22,

$$q = -k_m L \cdot 2\pi \frac{t_1 - t_2}{\ln r_1 - \ln r_2} \qquad (8\text{–}27)$$

Hence the simple Eq. 8–7 can be used if again k is replaced by k_m.

Integrating, further, Eq. 8–26 within the limits r and r_2,

$$q \frac{\ln r_2 - \ln r}{2\pi L} = k_2(t - t_2) - \frac{\beta}{2}(t - t_2)^2$$

or

$$t = t_2 + \frac{k_2}{\beta}\left[\binom{+}{(-)}\sqrt{1 + \frac{2\beta}{(k_2)^2} q \frac{\ln r_2 - \ln r}{2\pi L}} - 1\right] \qquad (8\text{–}28)$$

Because $t = t_2$ for $r = r_2$, again only the positive sign of the square root can be used.

The distribution of t as a function of r may be best shown on semi-logarithmic graph paper. Plotting r in the logarithmic scale, or, what means the same, $x_2 - x = \frac{\ln r_2 - \ln r}{2\pi L/A}$ in ordinary scale, exactly the same curve of second order is obtained as by representing t according to Eq. 8–25.

It will be seen that in general the graphical treatment of conduction problems for cylinders is simplified by using $\log r$ as one coordinate.

8–7. Radial Heat Conduction in a Sphere, Considering Variation of Conductivity with Temperature

The procedure in the case of a sphere is entirely analogous to that for the cylinder.

Starting from

$$q = -k \cdot 4\pi r^2 \frac{dt}{dr} \tag{8–14}$$

and using again Eqs. 8–21 and 22, integration between r_1 and r_2 yields

$$q = k_m \cdot 4\pi \frac{t_1 - t_2}{\dfrac{1}{r_1} - \dfrac{1}{r_2}} \tag{8–29}$$

Hence again, for k varying linearly with t, Eq. 8–16 holds when k_m instead of k is used.

Integration within the limits r and r_2, on the other hand, leads to

$$q \frac{1}{4\pi}\left(\frac{1}{r} - \frac{1}{r_2}\right) = k_2(t - t_2) + \frac{\beta}{2}(t - t_2)^2$$

or

$$t = t_2 + \frac{k_2}{\beta}\left[\binom{+}{(-)}\sqrt{1 + \frac{2\beta}{(k_2)^2}\, q \frac{1}{4\pi}\left(\frac{1}{r} - \frac{1}{r_2}\right)} - 1\right] \tag{8–30}$$

where again, because $t = t_2$ at $r = r_2$, only the positive sign can be used.

Taking $-1/r$ as abscissa,† as generally is advantageous in problems with spheres, the same temperature curve of second order is obtained as with ordinary coordinates for plane plates according to Eq. 8–25.

† More exactly, $x_2 - x_1 = \frac{1}{4\pi/A}\left(\frac{1}{r} - \frac{1}{r_2}\right)$.

8-8. Generalization of the Equations for Plane Plate, Cylinder, and Sphere

The twelve equations — 8-3, 5, 7, 9, 15, 17, 24, 25, 27, 28, 29, and 30 — can be replaced by two general equations, namely,

$$q = -k_m A \frac{t_1 - t_2}{X_1 - X_2} \tag{8-31}$$

and

$$t = t_2 + \frac{k_2}{\beta}\left[\sqrt{1 + \frac{2\beta}{(k_2)^2} q \frac{X_2 - X}{A}} - 1\right] \tag{8-32}$$

where $X = x$ for the plate,
$X = \ln x$ for the cylinder,
$X = -1/x$ for the sphere,
$x =$ the distance of a point: from the median plane of the plate, from the center line of the cylinder, or from the center of the sphere, respectively,
$A =$ the area, crossed by the heat flow, at the distance $x = 1$,
$k_m = \dfrac{1}{t_1 - t_2}\displaystyle\int_{t=t_2}^{t=t_1} k\, dt,$
$k = k_2 + \beta(t - t_2)$, and
$\beta =$ a positive or negative constant value, so that
$k_m = \dfrac{k_1 + k_2}{2}.$

If $\beta = 0$, so that $k = k_1 = k_2 = k_m =$ constant, Eq. 8-32 simplifies to

$$t = \frac{t_1 - t_2}{X_1 - X_2} X + \frac{t_2 X_1 - t_1 X_2}{X_1 - X_2} \tag{8-33}$$

This will be recognized by inspection of Eqs. 8-3, 9, and 17 or can be checked as follows:

By substitution of q from Eq. 8-31 in Eq. 8-32

$$(t - t_2) + \frac{k_2}{\beta} = \sqrt{\frac{(k_2)^2}{\beta^2} + \frac{2k_m}{\beta} \cdot \frac{X_2 - X}{X_2 - X_1}(t_1 - t_2)}$$

or

$$\beta(t - t_2)^2 + 2k_2(t - t_2) = 2k_m \frac{X_2 - X}{X_2 - X_1}(t_1 - t_2)$$

With $\beta = 0$ and $k_2 = k_m$

$$t - t_2 = \frac{X_2 - X}{X_2 - X_1}(t_1 - t_2)$$

from which Eq. 8-33 is obtained by a simple algebraic conversion.

It may be repeated that by plotting t as a function of x, $\ln x$, or $(-1/x)$, respectively, with equal scale units for Δx, $\Delta(\ln x)$ and $\Delta(-1/x)$, the temperature curves for plate, cylinder, and sphere become identical.

8-9. Heat Conduction in a Cube

The solution of the differential equation for a cube is owed to Nusselt [1909]. Even under the assumption of a uniform thermal conductivity k, it requires a complicated procedure.

It is supposed that a heat source at the center produces the heat flow q, and that the surface is kept at constant temperature. The side length of the cube may be $2a$, and Cartesian coordinates, parallel to the edges, may be used with the zero point in the center of the cube.

The distribution of the temperature in the cube depends upon the following conditions:

1. At any point x, y, z the differential equation of heat conduction (Eq. 2-9) must hold. Considering the simplifying condition of Sect. 8-1, the equation may be written in the form

$$\nabla^2\theta = \frac{\partial^2\theta}{\partial x^2} + \frac{\partial^2\theta}{\partial y^2} + \frac{\partial^2\theta}{\partial z^2} = 0 \tag{8-34}$$

where θ = the temperature excess above surface temperature, and ∇^2 = the usual symbol for the Laplace operator. Equation 8-34 is the well-known partial differential equation of Laplace.

2. At the surface

$$\left.\begin{aligned} x &= \pm a \\ y &= y \\ z &= z \end{aligned}\right\}\theta = 0 \qquad \left.\begin{aligned} x &= x \\ y &= \pm a \\ z &= z \end{aligned}\right\}\theta = 0 \qquad \left.\begin{aligned} x &= x \\ y &= y \\ z &= \pm a \end{aligned}\right\}\theta = 0$$

3. At the center

$$\theta = \infty$$

as can easily be understood and will be dealt with more generally in Sect. 17-4.

4. At any point x, y, z

$$\theta \sim \frac{q}{k}$$

This follows from the fact that in a homogeneous body of given configuration, obviously any temperature difference is doubled if q or $1/k$ is doubled.

The temperature excess θ may be expressed as a sum:

$$\theta = \theta_1 + \theta_2 \tag{8-35}$$

where

$$\theta_1 = \frac{C}{r} = \frac{q}{k \cdot 4\pi r} \tag{8-36}$$

According to Eq. 8–16, θ_1 is the temperature excess in a very large sphere at the radial distance r from the center, above the temperature at a very large radial distance. Therefore θ_2 is the deviation from the temperature which would exist if the body were an infinitely large sphere. According to Eq. 8–36, θ_1 satisfies Eq. 8–34, as can be checked, replacing r by $\sqrt{x^2 + y^2 + z^2}$. Hence, because of Eq. 8–35, θ_2 must also satisfy it, so that

$$\nabla^2\theta_2 = 0 \tag{8-37}$$

Further, because at the cube surface $\theta = 0$, it follows that on the same surface

$$\theta_2 = -\frac{C}{r} \tag{8-38}$$

Following a procedure, which seems to have been suggested by Stokes, θ_2 may be considered to be composed of three components θ', θ'', θ''', which from reasons of symmetry must satisfy the following conditions:

1.	$\nabla^2\theta' = 0$	$\nabla^2\theta'' = 0$	$\nabla^2\theta''' = 0$
2. $\left.\begin{array}{l} x = \pm a \\ y = y \\ z = z \end{array}\right\}$	$\theta' = -\dfrac{C}{r}$	$\theta'' = 0$	$\theta''' = 0$
3. $\left.\begin{array}{l} x = x \\ y = \pm a \\ z = z \end{array}\right\}$	$\theta' = 0$	$\theta'' = -\dfrac{C}{r}$	$\theta''' = 0$
4. $\left.\begin{array}{l} x = x \\ y = y \\ z = \pm a \end{array}\right\}$	$\theta' = 0$	$\theta'' = 0$	$\theta''' = -\dfrac{C}{r}$

By these assumptions the surface condition, $\theta = 0$, is satisfied, except at the twelve cube edges, where jumps of finite size occur. But the temperature at the edges does not influence the temperature field in the cube. To prove that, assume a cylinder around an edge as center line, with the temperatures θ_0 and 0 at the radial distances r_0 and 1, respectively, from the edge. Then the heat flow from the edge is $q_0 = \dfrac{2\pi k\theta_0}{\ln 1/r_0}$, and this becomes zero if r_0 approaches zero, for any finite value of θ_0. Therefore no finite rate of heat will penetrate from the edge into the cube, if the edge is the only part of the surface at finite temperature. This will be dealt with in Sect. 17–4, from a more general viewpoint.

Equation 8–37 can be integrated by the method of particular integrals. The solution for each of the three components θ', θ'', and θ''' is an infinite series, the constants of which must be determined from boundary conditions. As usual in dealing with Laplace's Eq. 8–37, a particular integral can be found as a product of e functions and trigonometric functions, both of which have the peculiarity that double integration leads to the same form of function.

In this way the first of the components becomes ‡

$$\theta' = \frac{C}{r_r} \cdot \sum_{\mu=0}^{\mu=\infty} \sum_{\nu=0}^{\nu=\infty} A_{\mu,\nu} \cosh \lambda\pi x \cdot \cos \frac{2\mu+1}{2a}\pi y \cdot \cos \frac{2\nu+1}{2a}\pi z \quad (8\text{–}39)$$

Conditions 3 and 4 are generally satisfied by Eq. 8–39, and differentiation of this equation shows that condition 1 is satisfied if

$$\lambda^2 = \left(\frac{2\mu+1}{2a}\right)^2 + \left(\frac{2\nu+1}{2a}\right)^2 \quad (8\text{–}40)$$

The factors $A_{\mu,\nu}$ can be determined from condition 2. For $x = a$, in which case $r = \sqrt{a^2 + y^2 + z^2}$, Eq. 8–39 becomes

$$-\frac{C}{\sqrt{a^2+y^2+z^2}} = \frac{C}{r_r} \cdot \sum_{\mu=0}^{\mu=\infty} \sum_{\nu=0}^{\nu=\infty} A_{\mu,\nu} \cosh \lambda\pi a \cdot \cos \frac{2\mu+1}{2a}\pi y \cdot \cos \frac{2\nu+1}{2a}\pi z \quad (8\text{–}41)$$

As usual in the theory of Fourier series, $A_{\mu,\nu}$ is determined by multiplying both sides of Eq. 8–41 by $\cos \dfrac{2\mu_1+1}{2a}\pi y \cdot \cos \dfrac{2\nu_1+1}{2a}\pi z$, and integrating from $y = -a$ to $y = +a$ and from $z = -a$ to $z = +a$.

‡ The constant C, defined by Eq. 8–36, and a constant reference radius r_r are added to make $A_{\mu,\nu}$ dimensionless, in slight modification of Nusselt's derivation.

In this way any term of the right side of the equation becomes

$$\frac{C}{r_r} A_{\mu,\nu} \cosh \lambda\pi a \int_{-a}^{+a} \int_{-a}^{+a} \cos \frac{2\mu + 1}{2a} \pi y \cdot \cos \frac{2\mu_1 + 1}{2a} \pi y \cdot \cos \frac{2\nu + 1}{2a} \pi z \cdot \cos \frac{2\nu_1 + 1}{2a} \pi z \, dy \, dz$$

For all values $\mu \neq \mu_1$ and $\nu \neq \nu_1$ the double integral is zero. For $\mu = \mu_1$ and $\nu = \nu_1$, it becomes equal to a^2. Thus, by the multiplication and integration mentioned, all terms of Eq. 8–41 vanish, except those for $\mu = \mu_1$ and $\nu = \nu_1$, and the result is an equation which contains only μ_1 and ν_1. Omitting now, for simplicity, the subscript 1:

$$\frac{A_{\mu,\nu}}{r_r} = \frac{-1}{a^2 \cosh \lambda\pi a} \int_{-a}^{+a} \int_{-a}^{+a} \frac{\cos \frac{2\mu + 1}{2a} \pi y \cdot \cos \frac{2\nu + 1}{2a} \pi z}{\sqrt{a^2 + y^2 + z^2}} \, dy \, dz \qquad (8\text{–}42)$$

The integration can be performed, using the cosine series. Hence, the summation required by Eq. 8–39 can also be done. The components θ'' and θ''' are found in the same way as θ' by cyclic change of the coordinates.

At an infinitesimal distance from the center the temperature distribution is the same as in a sphere, as easily will be understood. Therefore, according to Eq. 8–36,

$$C = \frac{q}{4\pi k}$$

Hence, the temperature distribution in the cube is given by

$$\theta = \frac{q}{4\pi k r_r} \left(\frac{r_r}{r} + \sum_{\mu=0}^{\mu=\infty} \sum_{\nu=0}^{\nu=\infty} A_{\mu,\nu} \cosh \lambda\pi x \cdot \cos \frac{2\mu + 1}{2a} \pi y \cdot \cos \frac{2\nu + 1}{2a} \pi z \right.$$
$$+ \sum_{\mu=0}^{\mu=\infty} \sum_{\nu=0}^{\nu=\infty} A_{\mu,\nu} \cos \frac{2\mu + 1}{2a} \pi x \cdot \cosh \lambda\pi y \cdot \cos \frac{2\nu + 1}{2a} \pi z$$
$$\left. + \sum_{\mu=0}^{\mu=\infty} \sum_{\nu=0}^{\nu=\infty} A_{\mu,\nu} \cos \frac{2\mu + 1}{2a} \pi x \cdot \cos \frac{2\nu + 1}{2a} \pi y \cdot \cosh \lambda\pi z \right)$$
$$= \frac{q}{4\pi k r_r} B \qquad (8\text{–}43)$$

where B represents the expression in parentheses.

Using Nusselt's numerical calculation, θ is represented in Fig. 8–3 as a function of the radial distance r for $C = 100$ cm C (θ in degrees centigrade, r in centimeters). Curves I, III, and IV belong to a cube of side length $2a = 60$ cm, curve II to a sphere having a diameter $2s = 66\frac{2}{3}$ cm, and the same surface temperature as the cube.

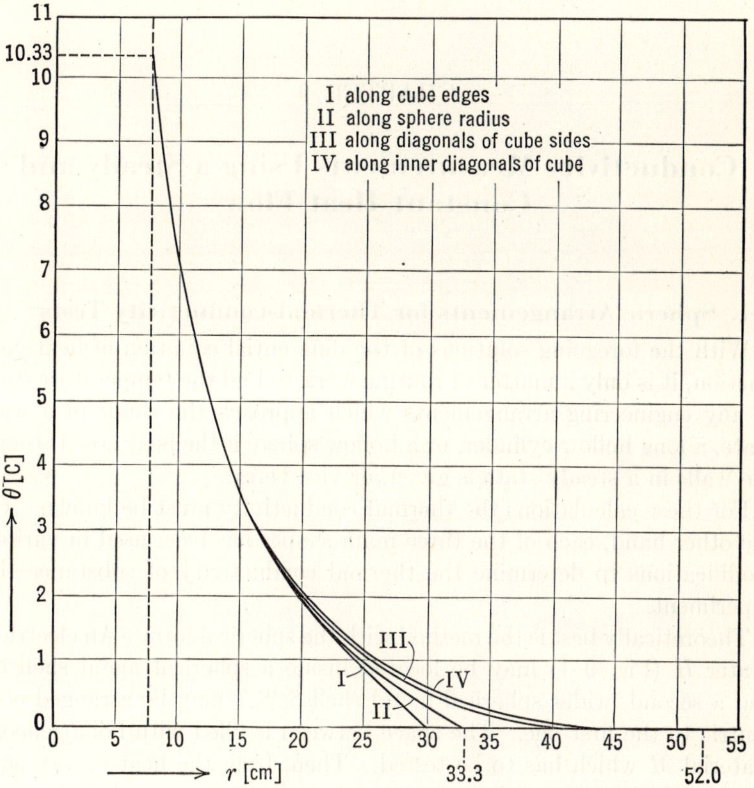

FIG. 8-3. Temperature distribution in cube and sphere.

The temperature distribution in the sphere would be

$$\theta = \frac{100}{r} - 3$$

which at $r = s = 33\frac{1}{3}$ becomes zero, and at $r = 0$ becomes ∞.

Curve I gives the temperature in the directions of the edges x, y, and z, curve III that in the direction of the diagonals of the cube sides (square diagonals), and curve IV that in the direction between opposite corners of the cube (cube diagonals). Curve II is between I and III. At $r = 10$ cm the temperature difference between IV and I is only 0.0014θ, at $r = 15$ cm it is 0.008θ. Up to this radius the temperature distribution in a cube is nearly the same as in a sphere.

CHAPTER 9

Conductivity Measurements Using a Steady and Constant Heat Flow

9–1. Spheric Arrangements for Thermal-conductivity Tests

With the foregoing solutions of the differential equation of heat conduction, it is only a matter of routine work to find the temperature drop in any engineering arrangements which approach the shape of a wide plate, a long hollow cylinder, or a hollow sphere if the heat flow through the walls in a steady state is given, or vice versa.

For these calculations the thermal conductivity must be known. On the other hand, each of the three main shapes has been used in various modifications to determine the thermal conductivity of substances by experiment.

Theoretically best is the method with the spherical form. An electrical heater H (Fig. 9–1) may be located inside a spherical metal shell S_i, and a second, wider spherical metal shell $S_o'S_o''$ may be arranged concentric to the first one. The space between is filled with homogeneous material M which has to be tested. Then, from the heat energy supplied by the heater and the temperatures t_1 and t_2 at two radial distances r_1 and r_2, the thermal conductivity k of the filling material is found by means of Eq. 8–15.

This method has two special advantages: (1) the heat energy supplied by the heater passes the material to be tested, in the required direction (radially), without any losses; (2) the thermal conductivity at different temperatures can be found by a single experiment, if thermocouples are arranged at more than two radial distances.

The sphere method was first used by Péclet [1860] with warm water inside as heater; Péclet, however, did not take care of the condition of steady state. Lees [1898] developed the theory of the method without this mistake but did not apply it because of practical difficulties. Nusselt [1909] succeeded in using the method for testing insulating and building material. The inner spherical shell S_i (Fig. 9–1) of copper, $\frac{1}{20}$ in.

thick and 4 in. in diameter, was provided with a nichrome heater (up to 500°C) or a platinum coil (above 500°C). The outer spherical shell of zinc, 1/25 in. thick and 24 or 28 in. in diameter, consisted of two hemispherical sections, S_o' and S_o'', connected by means of light flanges. When the lower half S_o' was filled with the material to be tested, and the thermocouples were fixed in the plane of the flange F, then the second

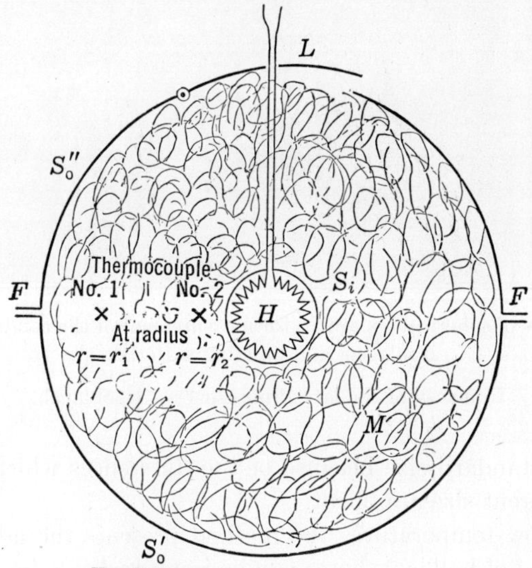

FIG. 9-1. Sketch of spherical apparatus for measurement of the conductivity of loose insulating and building materials.

hemisphere S_o'' was put on and filled by an opening 4 in. wide, which afterwards was closed by a lid L. A photograph showing three spherical devices is reproduced as Fig. 9-1*a*.

It is difficult to fill the hollow sphere homogeneously with loose material, particularly if a definite apparent density is to be obtained, and sagging of the material after filling cannot be prevented entirely.

Whereas this is the greatest difficulty for loose materials, the main difficulty for solid pieces is that special samples must be made. If such samples are to be burnt, as is necessary with refractory materials, it is difficult to get the pieces to fit without gaps when finished, and it is questionable whether the samples have the same apparent conductivity

FIG. 9–1*a*. View of spherical apparatus for measurement of the conductivity of loose insulating and building materials.

Courtesy of the late Professor Osc. Knoblauch.

as bricks of standard size because of the differences which occur when pieces of different size are burnt.

For very low temperatures the method presents the advantage that a small system of hollow spheres can be immersed in a low-temperature bath, for instance, liquefied air, as first was done by Groeber [1908, 1910] (see Fig. 9–2). In this construction the outer metal shell practically assumes the temperature of the liquid, which is easy to determine, and in addition to that, only the temperature of the inner shell must be measured. The figure shows the arrangement in a Dewar vessel, with a tube through which the terminals and the thermocouple wires are led to the outside.

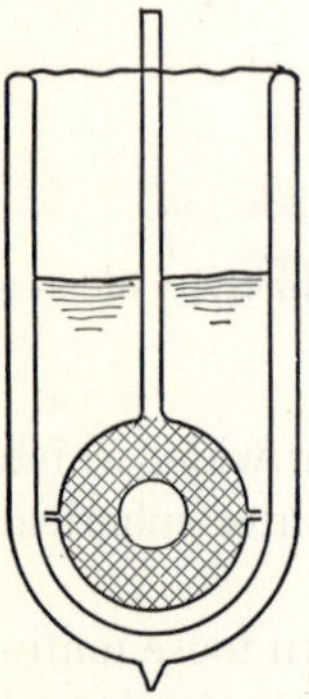

FIG. 9–2. Spherical apparatus for measurement of the conductivity of insulating materials at very low temperatures.

For high temperatures, on the other hand, it is difficult to secure a sufficiently high and equal surface temperature of the outer shell. For this purpose thick layers of insulating material sometimes have been placed around the outer shell, but the insulation increases the heat capacity of the whole arrangement and, consequently, the time necessary for heating up and for adjusting a steady state of heat flow. Time can be saved according to the following reasoning: With a definite rate of heating, the temperatures eventually still slowly increase, but the temperature differences do not change any more. This means that one part of the heat input is used for raising the temperature of the whole arrangement, whereas another part maintains a steady heat flow. That first amount of energy may be calculated as product of weight, specific heat, and observed temperature increase in unit time. By subtracting it from the rate of heat input, the steady-state heat flow is obtained. This procedure was suggested by Jakob [1919(*a*)] in the discussion of a test of van Rinsum [1920], who used a spherical shell composed of eight sections of the refractory material, 9 in. inner diameter, 23¾ in. outer diameter, surrounded by a layer of diatomaceous earth of 17¾ in. thickness. Permanently heating this shell from the inside with 476 B/hr brought the surface of the inner spherical heater to 1060°C in 13 days, which still was 65 C below the steady-state temperature. The temperature difference between the inside and outside surfaces of the spherical shell, however, was already virtually constant at that time.

Of course, the heating-up time could be reduced by reducing the mass of the outside insulation. The most efficient procedure to this end would be to arrange an additional heater outside, say in a third shell. But this would complicate the set-up considerably. In general, the sphere method cannot be recommended for high temperatures.

9-2. Cylindric Arrangements for Thermal-conductivity Tests

The cylinder method of measuring the thermal conductivity of solid substances seems to have been used first by Niven [1905] and Clement and Egy [1909]. The substance to be investigated may be held between two concentric tubes, and heat energy, produced by electric heaters, is conducted in radial-outward direction through the substance. The tubes bear thermocouples for the measurement of the radial temperature difference. The method is particularly appropriate for building, insulating, and refractory materials in shell shape, or for such insulating substances as are applied to steam pipes in loose form, bound by water, and then dried on the pipes. The method is further used for insulating materials which are on the market in the form of bandages, as asbestos ribbons and cords; these are wound around the inner tube in different layers, so that they eventually form a hollow cylinder.

Whereas in a hollow-sphere arrangement no heat losses occur, except by electric-current conducts and thermocouple leads, cylindric devices will lose heat at their ends. Unfortunately a metal tube must be used as the bearer of the inner electrical heating coil in any practical testing device. It is even recommended to cover the heating coil by another metal tube which equalizes the temperature over the heater surface.

Thus a metallic core is formed which conducts heat to the ends, and the effect of axial heat losses must be carefully considered.

There are different means to compensate for this effect. One is to make the tubes as long as possible and to wire the electric heater homogenously all over the length, but to use only a relatively short middle part of the arrangement for the measurement.

If the design is restricted to rather small length, some thermocouples may be fixed on the tubes at different distances from the ends to determine the end losses from the temperature drop along the tube and the thermal conductivity of the tube material. Another procedure is to wire the ends of the heater more narrowly, so that a region with a negligible axial temperature drop may be obtained. A third method is to place "guard" coils * on the ends of the cylindric arrangement. These may be separated from the main tube by insulating air layers or solid discs. If thermocouples are fixed on both sides of these layers, any heat loss can be prevented by heating the guard coils until the temperature difference between the thermocouples on each layer vanishes.

The cylinder method is also used to determine the thermal conductivity of liquids and gases. This application originates from Schleiermacher [1888] and is therefore known as Schleiermacher's method.

It is based on the electric heating of a fluid by a thin wire, which is drawn in the axis of a capillary tube containing the test fluid, for instance, a liquid. The heating wire, usually a platinum wire, serves as an electric resistance thermometer at the same time. The tube, open at the two ends, may be located in the liquid. If it is made of a well-conducting metal, the wall temperature can be assumed to be equal to the liquid temperature outside the tube, which can easily be measured. From the heat energy delivered by the platinum wire and the temperature of the wire, both measured electrically, and from the dimensions of the tube, the thermal conductivity will be found by means of Eq. 8–7. In some investigations the measurements were made while the temperature of the heating wire still increased rather quickly. This led to erroneous results, for only when the temperatures of tube and liquid have risen so much that the heat loss of the whole system is equal to the heating energy, is there no longer any temperature increase, which, according to our remark about the heating up of the sphere insulation (Sect. 9–1), would influence the result and would require proper correction. Recently, however, the method has been improved so that now it can be considered a method of precision.

One of the improvements, due to Eucken [1911(*a*)], is the compensation of the convection on the ends of the tube. A short section of an

* Namely, auxiliary coils guarding the main cylindric arrangement against heat losses.

identical tube, likewise provided with a central platinum wire and open on the two ends, is arranged close to the main tube. If short enough, this piece can be considered a pair of disturbing ends. Connecting its wire, in a Wheatstone-bridge arrangement, as a part of the resistor compensating the resistance of the main wire, only the middle portion of the main wire is used for the measurements, the effect of the two ends being compensated by the auxiliary wire.

This method has not only been employed for the determination of the thermal conductivity of gases and liquids, but has also been developed into a most important means of analysis and control of gases in plants of chemical and mechanical industries.

In Schleiermacher's method, generally $r_2/r_1 \gg 1$. If, instead, $r_2/r_1 \approx 1$, then Eq. 8–7 can be replaced by the formula for plane plates, the curvature being negligible against the radius. This, for example, was done in the determination of the thermal conductivity of water by E. Schmidt and Sellschopp [1932], in which the water was filled into the space between a copper cylinder, 3.0 cm in diameter, which contained an electrical heater, and a 3.1 cm wide copper tube. The temperature drop across the water layer was measured by thermocouples which were located in the two copper bodies. An annulus of only 0.025 cm thickness was used by Sellschopp [1934] for the determination of the thermal conductivity of gaseous and liquid carbon dioxide.

Finally, the cylinder method has also been successfully employed for metals at high temperatures. If the longitudinal heat flow in metal rods is used for the conductivity measurement, relatively small temperature differences between the rod surface and any guard tubes make considerable corrections necessary, owing to the rapid increase of the radiation effect with temperature. For this reason, Powell [1939] in his measurements on steel up to 1000°C employed a pile of concentric flat rings of 3 in. diameter and restricted the measurement to one ring in the middle of the pile. Guard heaters were provided at the top and bottom of the pile so that the longitudinal temperature drop was small. Therefore thermocouples in longitudinal holes could be used. They could be shifted and readily interchanged or renewed. Another major advantage of the radial method is that much smaller temperature differences may be used than with the longitudinal method. This renders the method particularly well suited for the study of metals whose thermal conductivity may vary non-linearly with temperature.

9–3. Twin-plate Systems for Thermal-conductivity Tests

The plane-plate method is the simplest and therefore most frequently applied method; however, it meets with some remarkable difficulties. One of them is the loss of heat at the edges of the plate, which of course is relatively larger than the loss at the ends of cylindric devices. The edge losses, however, can be eliminated by guard-ring heaters, as probably first applied by Berget [1887, 1888] in experiments on mercury. A cylindric tube, 8 in. high and ½ in. wide, was filled with mercury and surrounded by mercury in a cylindric ring space, 2¼ in. wide, which was fairly well insulated against radial heat losses. The upper surfaces of the mercury cylinder and ring were heated by flowing steam. An iron base plate was held at 0°C. By heating from above, convection was avoided. The inner tube, filled with mercury, was led through the base plate into the ice compartment of a Bunsen ice calorimeter, and the heat flow was determined from the mass of melted ice and its latent heat.

For measurements on non-metallic substances, use of the guard ring has been adopted by Poensgen [1912] and now is missing on scarcely any modern device. He designed and employed a twin-plate device (Fig. 9-3) with a thin electric heating plate H_p between two equal plates P_1 and P_2 of the material to be tested, and two cooling plates C_1 and C_2 which sandwich the three plates mentioned. With this arrangement the heat delivered by the electric heater is divided into two virtually equal amounts. These are conducted across the two testing samples, and from these to cooling water or oil which flows through the hollow plates C_1 and C_2. The guard-ring heater H_r is of the same kind and thickness as the main heater H_p. Generally C_1 and C_2 are so wide that their outer parts act as cooling rings. Also the main heater is sometimes made so wide that the outer part of it can be used as the ring heater. This, however, requires a very regular distribution of heating over the whole area and the use of large test plates, extending over the whole surface of the heating and cooling plates. If, on the other hand, a separate ring heater is provided, its heating energy can be regulated independently of the main heating. Therefore it is not indispensable that the test substance be used also between the ring heater and cooler. This space should rather be filled with the same insulating material as the container C, or, preferably, with special pieces having the same thickness as the plates P_1 and P_2 and separated from them by a narrow gap.

A second difficulty occurs in the measurement of the temperature difference between the two faces of a test plate. If thermocouples were attached on these faces, an air gap of variable width would be formed between the test plate and the heating plate or cooling plate, causing large temperature differences over the plate area. This difficulty can be overcome by providing grooves at the surface of the test plate in which the wires of the thermocouples are laid; this, however, is not only inconvenient in practice, but also is not quite reliable, because of irregularities caused by the inevitable air space between the plates. Therefore it is better to provide the grooves in the surfaces of the heating and the cooling plates. In these the temperatures are equalized by metallic conduction, so that the measured temperatures will be fair averages over larger parts of the area than with location of the thermocouples in grooves in the test plate. At all events a good contact of

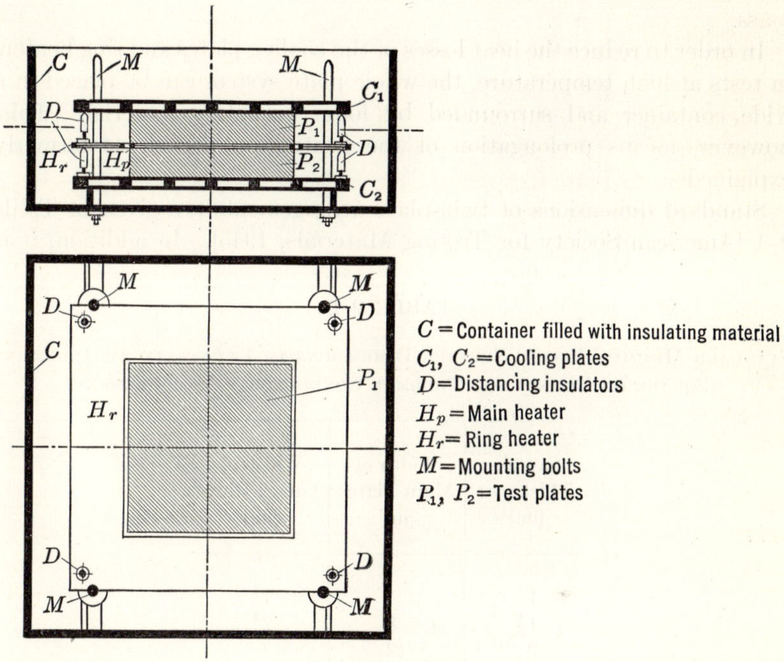

FIG. 9-3. Twin-plate device for measurement of the conductivity of insulating and building materials.

the two surfaces is required, which can be obtained by pressing together the whole system of plates by means of screws or weights. In handling elastic plates the compression must be considered.

Sometimes powder is spread between the plates to fill gaps; this does not help much, owing to the small thermal conductivity of any kind of powder. The best method is to have the plates made as plane as possible.

Of course, a very equal distribution of heating and cooling energy over the heating and cooling plates is important. This is not difficult, as far as the heater is concerned, and requires only regular winding, no irregular air spaces, heating wires or ribbons with small temperature coefficient of electric resistance, and metal plates, equalizing the temperature sufficiently. It is not quite so simple with the cooling plates, because the liquid flowing through these necessarily has different temperatures at the entrance and exit. To compensate for this, the liquid is carried in convenient directions; preferably in such a way that some counterflow takes place and the liquid leaves close to the entrance opening. The mass velocity of the flow should be chosen as great as practically convenient, because the increase in temperature is inverse to the flowing mass.

In order to reduce the heat losses of the cooling plates and ring heaters in tests at high temperature, the whole plate system can be placed in a wide container and surrounded by loose insulating material. This, however, means prolongation of the heating-up period, as formerly explained.

Standard dimensions of twin-plate arrangements are given in Table 9-1 [American Society for Testing Materials, 1945]. In addition, it is

TABLE 9-1

STANDARD MINIMUM LINEAR SURFACE DIMENSIONS IN TWIN-PLATE ARRANGEMENT (SQUARE OR ROUND) FOR THERMAL CONDUCTIVITY MEASUREMENTS

Sample Thickness [in.]	Width of Main Plates [in.]	Width of Guard Ring [in.]
1	4	$1\frac{1}{2}$
$1\frac{1}{2}$	8	$2\frac{1}{4}$
2	12	3
4	12	6

required that the distance between main and guard heater does not exceed $\frac{1}{8}$ in.,† and further information about design and procedure is given in the publication just mentioned.

Twin-plate arrangements for non-metallic substances have been used from the ice-melting point up to 1000°C. At the highest temperatures the system was placed in nitrogen.

For metals the twin-plate method seems to have been used first by Honda and Simidu [1917]. Two cylinders C_1 and C_2, each about 3 in. long and $\frac{1}{2}$ in. thick, each with a cone on one end and a concretion on the other end, were arranged as shown in Fig. 9-4. The concretions were

† The reason for this requirement has been discussed in detail by Jakob [1919(*a*)].

machined so that, put together, they formed a hollow space, ⅝ in. long and ¼ in. wide. Inside, a small electric heater was located. Two holes for thermocouples at a distance of less than ½ in. were provided in each cylinder. The body C_1C_2 was located in a copper shell S, and the space between was filled with an insulating material. The whole system then was placed in a porcelain tube which contained hydrogen to prevent oxidation, and the tube was put into an electrical furnace. With this arrangement tests on many samples could be performed in a relatively short time, but the results were not very accurate, owing to the small dimensions and to the radial and other heat losses which were only roughly considered. Schofield [1925] improved the accuracy by doubling the thickness of the samples, quadrupling their length, and placing them in a guard heater, namely a metal tube 3 in. wide, with a heating coil on the middle.

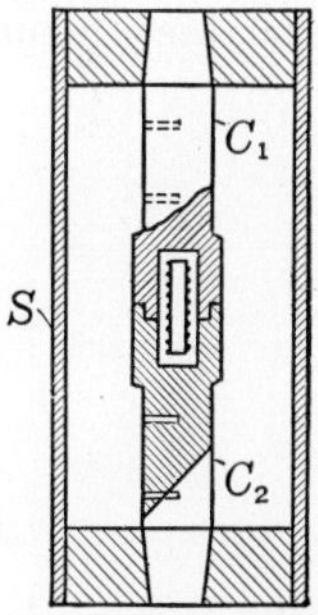

C_1, C_2=Test cylinder
S=Copper shell

FIG. 9-4. Twin-rod device for measurement of the conductivity of metals.

9-4. Single-plate Systems for Thermal-conductivity Tests

For substances available only in smaller sizes, for example, as discs of 4 or 5 in. diameter, the author has designed and used different single-plate arrangements.

The first of these arrangements [Jakob, 1920] was originally designed for the determination of the thermal conductivity of liquids and consisted of a system of plates covered by a Dewar vessel (Fig. 9-5). Because it was also found suitable for the investigation of insulating materials and metals, and because some characteristic items can be learned from this investigation, it will be dealt with in more detail.

Two cylindric copper plates C_1 and C_2, about 4½ in. in diameter, separated by three tiny glass cubes, were placed horizontally on a metal cylinder M. This rested on a grid in a cooling vessel W, through which water was flowing. The liquid under investigation was filled in between the two copper plates and held there by the capillary forces on the edge. Heat energy was produced in an electrical heater H_1 and passed through the liquid layer vertically in the direction from above to below. Thus the temperature decreased and the liquid density increased from above to below, and no convection took place. The temperature drop across the layer was measured by two thermocouples, placed in fine horizontal holes in the copper plates. By a metal ring R of exactly the same outer diameter as the plates C_1 and C_2, the radial radiation of H_1 was reduced, and by R, C_1, C_2, and M a cylindric, smooth, metallic surface was formed, interrupted only by the liquid layer. The heat losses of the whole system were reduced by tilting over it a Dewar vessel D with silver-plated glass walls. The losses above a horizontal plane, laid through the liquid film at half height, could be determined by

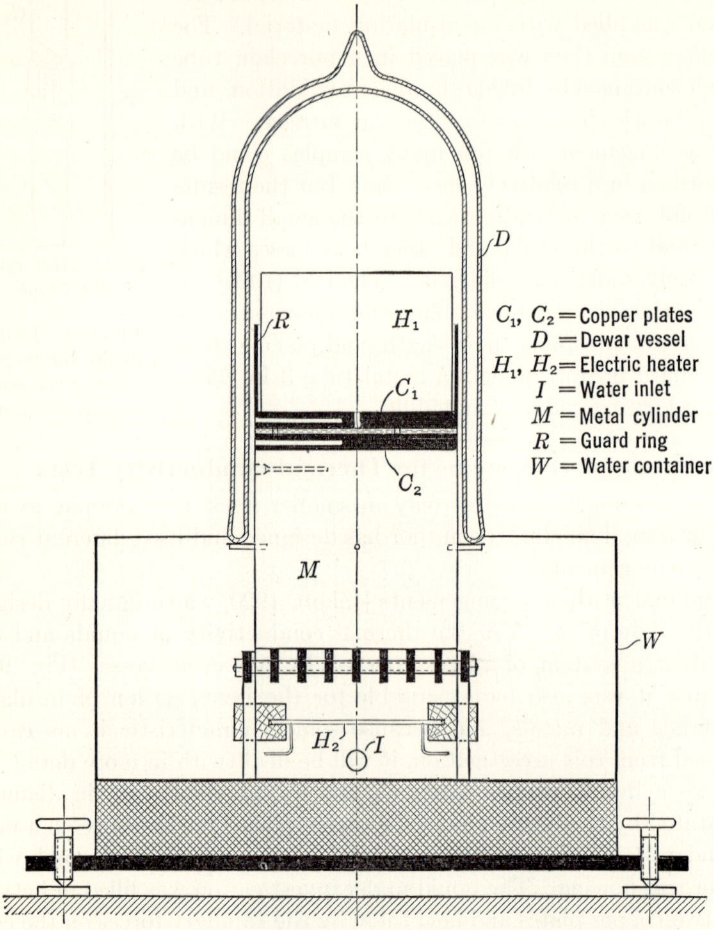

FIG. 9-5. Device for measurement of the conductivity of liquids.

an experiment, in which no liquid was between C_1 and C_2, but just air. This presented a great resistance to any downward heat flow, particularly if M was heated from below almost to the temperature of C_1. An electric heater H_2 was used to adjust the temperature of the water in its flow from the inlet I to the outlet, which is not seen in the figure.

In the experiments for determination of the thermal conductivity of water, the heat loss across D was 0.5% of the total heat flow at 10°C, and 5% at 70°C. Another heat loss was due to evaporation from the edge of the water layer. It was determined mechanically by means of a micrometer needle, which was screwed in from the edges of C_1 and C_2 until it touched the meniscus, or by replacing the evaporated liquid by a measured amount of fresh liquid after the test. Although only a very small mass of water was evaporated, the great latent heat caused a considerable heat loss, for instance, about 2% of the total heat flow at 75°C.

A third correction, smaller than 1%, however, was necessary because some energy passed downward through the inner glass wall of the Dewar vessel. The middle part of Fig. 9–6 represents this wall and the neigh-

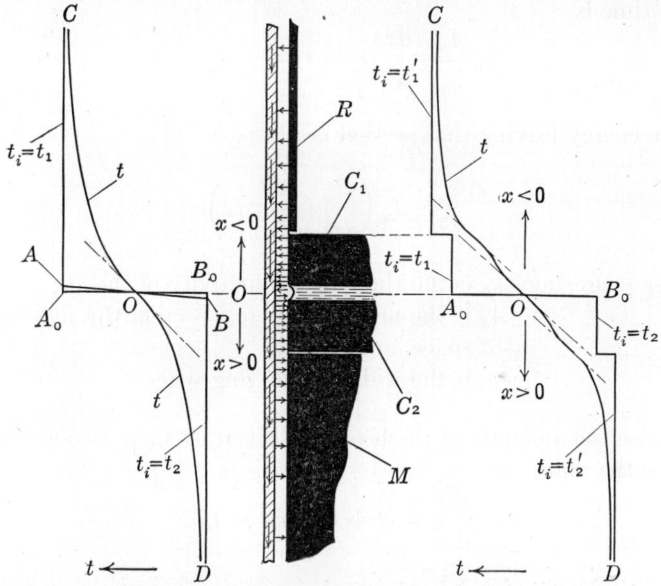

FIG. 9–6. Temperature distribution in inner wall of Dewar vessel D of Fig. 9–5. True temperature distribution in liquid, AB, replaced by A_0B_0. In the graph at the right the temperature drops between R and C_1 and between C_2 and M are considered.

boring parts above and below the water layer. The small arrows indicate how heat is assumed to flow across the air ring space and then in the glass wall downward. Actually, however, the heat will not cross the annular space in exactly horizontal paths. The two graphs at the left and right hand are easily understood, t being the temperature of the glass wall, and t_i the temperature of the metal surfaces inside the Dewar vessel. With increasing negative and positive values of x the temperature t approaches constant values t_1 and t_2, respectively.

The heat loss along the glass wall is given by the equation

$$q = -k_g A_g \left(\frac{dt}{dx}\right)_0 \tag{9-1}$$

where k_g is the thermal conductivity of the glass,
A_g is the area of a cross section through the inner glass wall, and
$(dt/dx)_0$ is the slope of the temperature curve at $x = 0$.

In order to determine that slope, an equation for the curve *COD* is needed.

The heat energy entering a section of the glass wall of the length dx in unit time is

$$k_a \frac{A_a}{s_a} \cdot \frac{dx}{s_a} (t_i - t) - k_g A_g \frac{dt}{dx}$$

and the energy leaving that section is

$$-k_g A_g \left(\frac{dt}{dx} + \frac{d^2t}{dx^2} dx\right)$$

In these expressions k_a is the thermal conductivity of the air,
A_a is the area of the cross section through the ring space, and
s_a is the width of the ring space.

Because the amounts of the heat flow given by these two expressions must be the same

$$\frac{d^2t}{dx^2} = ct + b = c(t - t_i) \tag{9-2}$$

where

$$c = \frac{k_a A_a}{k_g s_a^2 A_g} \tag{9-3}$$

and

$$b = -ct_i \tag{9-4}$$

The general solution of Eq. 9-2 is

$$x = \int \frac{dt}{\sqrt{a + 2\int (ct + b)\, dt}} + d \tag{9-5}$$

a and d being the constants of integration.‡ In our case $c > 0$, because c is defined as a product of positive values. Then the integration yields

$$x = \frac{1}{\pm\sqrt{c}} \ln \pm [b + ct \pm \sqrt{c}\sqrt{a + 2bt + ct^2}\,] + d \tag{9–6}$$

and by differentiation

$$\frac{dt}{dx} = \pm\sqrt{a + 2bt + ct^2} = \overset{(+)}{-}\sqrt{a + ct(t - 2t_i)} \tag{9–7}$$

Only the minus sign can be used, because dt/dx must be negative. In order to determine the constants a and d of Eq. 9–6 the two branches of the curve COD have to be considered. Branch CO may be characterized by the subscript I; branch OD by the subscript II.

For $x = \pm\infty$

$$t = t_i \quad \text{and} \quad \frac{dt}{dx} = 0$$

Hence from Eq. 9–7

$$a_{\mathrm{I}} = ct_1^{\,2} \quad \text{and} \quad a_{\mathrm{II}} = ct_2^{\,2}$$

or generally

$$a = ct_i^{\,2} \tag{9–8}$$

where t_1 and t_2 are values of t_i for $x < 0$ and $x > 0$, respectively.

Further, for $x=0$,

$$t_0 = t_{\mathrm{I}} = t_{\mathrm{II}} \quad \text{and} \quad \left(\frac{dt}{dx}\right)_{\mathrm{I}} = \left(\frac{dt}{dx}\right)_{\mathrm{II}} = \left(\frac{dt}{dx}\right)_0$$

Hence from Eq. 9–7:

$$a_{\mathrm{I}} + ct_0(t_0 - 2t_1) = a_{\mathrm{II}} + ct_0(t_0 - 2t_2)$$

and with Eq. 9–8:

$$ct_1^{\,2} + ct_0(t_0 - 2t_1) = ct_2^{\,2} + ct_0(t_0 - 2t_2)$$

or

$$t_0 = \frac{t_1 + t_2}{2} \tag{9–9}$$

This had to be expected for reasons of symmetry. Substituting now b and a from Eqs. 9–4 and 8 in Eq. 9–6,

$$x = \frac{1}{\pm\sqrt{c}} \ln \pm [2c(t - t_i)] + d \tag{9–10}$$

‡ Another form of the solution is $\theta = Me^{-\sqrt{cx}} + Ne^{\sqrt{cx}}$ with M and N as constants of integration. (See Problem 9–1.)

Substituting here $x = 0$ and $t = t_0 = (t_1 + t_2)/2$, and considering that $t_i = t_1$ for branch I, and $t_i = t_2$ for branch II,

$$\left.\begin{aligned} d_{\mathrm{I}} &= \frac{1}{\mp\sqrt{c}} \ln \pm c(t_2 - t_1) \\ &\text{and} \\ d_{\mathrm{II}} &= \frac{1}{\mp\sqrt{c}} \ln \pm c(t_1 - t_2) \end{aligned}\right\} \tag{9-11}$$

Finally, substituting this in Eq. 9-10,

$$\left.\begin{aligned} x_{\mathrm{I}} &= \frac{1}{+\sqrt{c}} \ln 2 \frac{t - t_1}{t_2 - t_1} \\ &\text{and} \\ x_{\mathrm{II}} &= \frac{1}{-\sqrt{c}} \ln 2 \frac{t - t_2}{t_1 - t_2} \end{aligned}\right\} \tag{9-12}$$

Concerning the signs of this equation, it must be considered that $(t_{\mathrm{I}} - t_1)/(t_2 - t_1) \leqq \frac{1}{2}$ and $(t_{\mathrm{II}} - t_2)/(t_1 - t_2) \leqq \frac{1}{2}$. Therefore the logarithms in Eq. 9-12 become negative. Hence x_{I} and x_{II} are in agreement with Fig. 9-6 only if the signs + and − are chosen as has been done in Eq. 9-12.

Differentiating this equation,

$$\left.\begin{aligned} \left(\frac{dt}{dx}\right)_{\mathrm{I}} &= \sqrt{c}(t - t_1) \\ &\text{and} \\ \left(\frac{dt}{dx}\right)_{\mathrm{II}} &= -\sqrt{c}(t - t_2) \end{aligned}\right\} \tag{9-13}$$

and for $x = 0$ with Eq. 9-9

$$\left(\frac{dt}{dx}\right)_0 = \sqrt{c}\,\frac{t_2 - t_1}{2} \tag{9-14}$$

In the above calculation it has been assumed that the air ring space extends from $x = -\infty$ to $+\infty$. According to Fig. 9-6, which is drawn to scale, corresponding to the mentioned dimensions of the device, 2 in. in both directions are practically sufficient. This amount was chosen as the height of the brass ring R (Fig. 9-5) and as the depth to which the metal cylinder M was immersed into the Dewar vessel.

Substituting now $(dt/dx)_0$ from Eq. 9-14 and c from Eq. 9-3 in Eq. 9-1,

$$q = \sqrt{k_a k_g A_a A_g}\,\frac{t_1 - t_2}{2s_a} \tag{9-15}$$

where $\sqrt{k_a k_g}$ and $\sqrt{A_a A_g}$ are peculiar geometrical mean values.

When the device is to be used for the measurement of the thermal conductivity of solid, non-metallic substances such as glass or ceramics, air spaces between the copper plates C_1, C_2 and the test plate can be avoided by wetting the surfaces with water or oil, or by sealing the plates together by means of paraffin wax or bitumen. When the thickness of the sealing layers is determined by weighing, the temperature drop across them can be calculated from the thickness and the thermal conductivity of the sealing substance. This temperature drop must be deducted from the measured temperature difference between the copper plates C_1 and C_2; it often means a considerable correction.

For measurements at low temperature Jakob [1922] used a modification of the arrangement, which is illustrated by Fig. 9–7. Here the metal cylinder on which the copper plates are placed is entirely inside the Dewar vessel and rests on another copper plate, which is placed on a good insulator. The cylinder is fitted with a small hollow space, into which a double-walled glass tube is introduced. Through this a continuous stream of liquid air is injected into the hollow space. With this arrangement the thermal conductivity of ice has been measured down to −193°F [Jakob and Erk 1929].

The thermal conductivity of metals, which were available as cylinders, 6 in. long, 5 in. in diam., also was determined in the same arrangement. The cylindric samples were located inside the Dewar vessel. In this case the temperature difference was measured by means of thermocouples in radial holes which were drilled into the samples at about 3½ in. distance. The copper plates C_1 and C_2 of Fig. 9–5 could be spared in this case.

The single-plate method with a Dewar vessel, however, can be used only up to moderate temperatures, and with not too large temperature differences between inside and outside. Another single-plate arrangement designed by Jakob [1924] is less simple, but can be used more generally. (See Fig. 9–8.) It consists of a copper plate H, which contains an electric heater; a bronze cooling plate C; a ring heater R; and a top plate T, with a third electrical heating coil. The sample, a disc of 4 in. diameter, is designated by S. This system of plates is surrounded by insulating material. Ten of a total of 16 thermocouples are indicated in the figure; corresponding thermocouples of the heaters H and R or T, are fitted immediately opposite each other; but not all thermocouples of Fig. 9–8 are located in the same vertical plane. The plate S is sealed between H and C, and the temperature drop in the sealing layers is considered as described previously. The parts H, R, and T are adjusted to the same temperature, so that no heat will be lost. If nevertheless some small temperature differences occur, the respective heat losses can

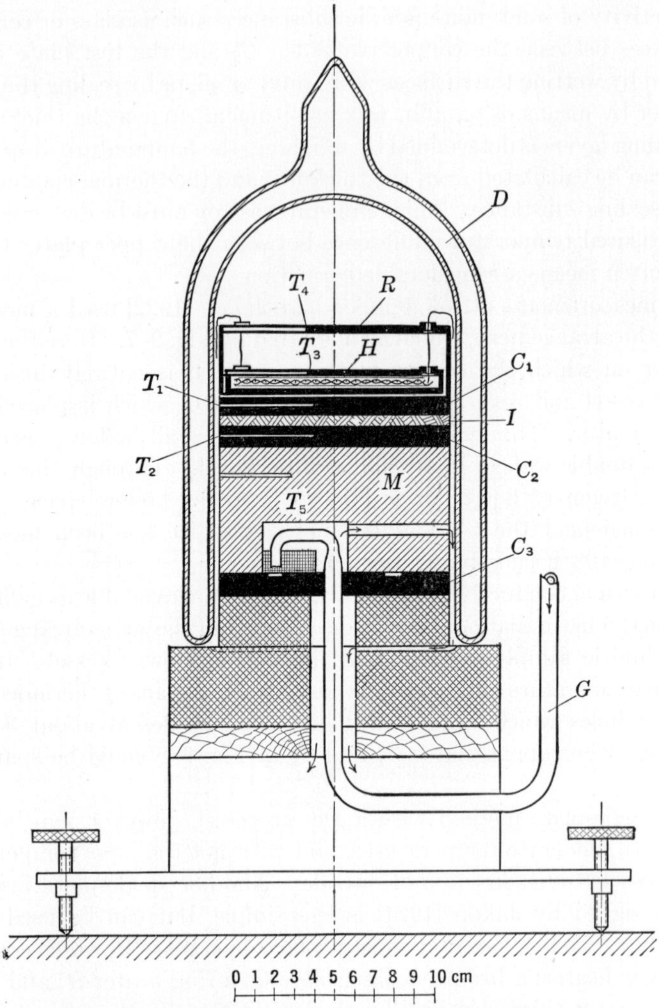

C_1, C_2, C_3 = Copper plates
D = Dewar vessel
G = Double-walled glass tube
H = Electric heater
I = Ice plate (as an example of a test example)
M = Metal cylinder
R = Guard ring
T_1, T_2, T_3, T_4, T_5 = Thermocouples

FIG. 9-7. Device for measurement of the conductivity of liquids and solids at low temperature.

be calculated and taken into account. The accuracy of this method, which was mainly used for testing ceramic materials, is about 1%, even for thin plates and temperatures above 100°C.

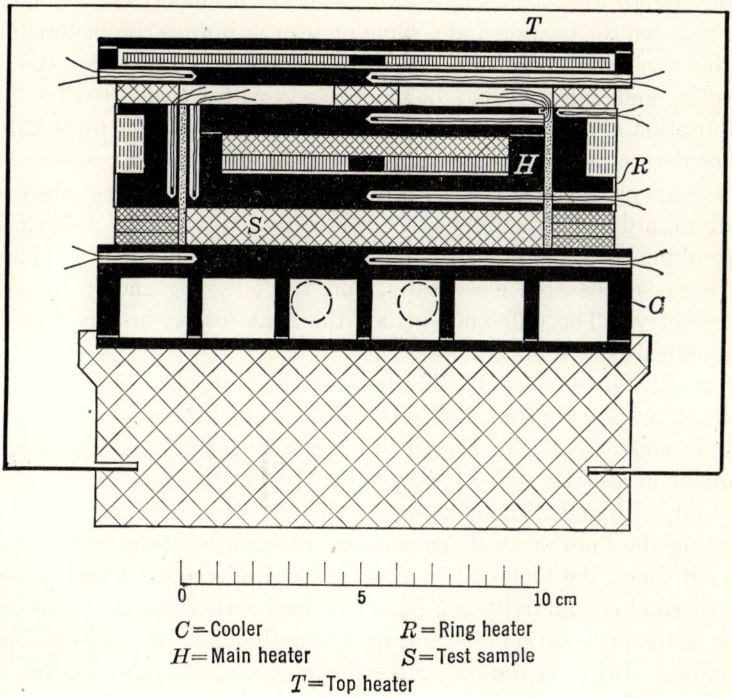

FIG. 9–8. Single-plate device for measurement of the conductivity of non-metallic solids.

Instead of sealing the samples between heater and cooler, Eucken [1911(*a*)] has used a method in which the temperature drops between heater, sample, and cooler can be considered by comparative measurements. In this method, which is particularly useful for small samples, like crystals, the device is subsequently placed in atmospheres of air, carbon dioxide, and hydrogen so that the heat flow must pass thin layers of these gases between the respective surfaces. Owing to the differences in the thermal conductivity of the gases, the total temperature drop will be different and the temperature drop in the gas layers can be eliminated by calculation. The gases are used at vacuum pressures so that convection losses are negligible.

Another modification of the single-plate method as used for liquids is due to Kaye and W. F. Higgins [1928, 1929]. As in the device represented in Fig. 9–5, thin liquid layers were filled into the gap between a heating and cooling plate. The liquid, however, was not held in place by capillary forces, but was free to extend into a small trough, made of asbestos, which surrounded the system like a gallery. The liquid level in the trough was slightly above the liquid layer under test, so that the gap between the heating and cooling plates was always completely filled. In this way, exact measurements on different liquids, such as water, oil, paraffin, and sulfur, could be performed, without any trouble from evaporation of a part of the liquid, in a temperature range up to 210°C.

For determining the thermal conductivity of refractory materials at high temperature, calorimeters, with flowing water as the heat-absorbing medium, are usually employed. Griffiths and Kaye [1923] showed how to combine both calorimetric methods. In their arrangement the incoming heat is measured electrically, and the outgoing energy by water calorimetry. The difference equals the heat losses; in this way ring heaters can be eliminated.

For substances available in the form of thin sheets, an old plate method, originating from Lodge [1878] and called the method of the cut rod, is convenient. It consists in measuring the axial temperature gradient in a metal rod, cut perpendicularly to the axis, and heated at one end, with and without a layer of the test sheet between. By extrapolating the lines of axial temperature distribution from both sides to the cut faces, the temperature drop across the test sheet and herewith the thermal conductivity can be determined rather exactly. The influence of the pressure is important in this method and must be considered carefully. Bringing the rod into different gases, or improving the contact by water, by oil, or by sealing with paraffin wax, may improve the results.

9-5. Comparison of Thermal Conductivities from Isothermal Lines Made Visible

In Sect. 11-2 we shall deal with the method of Ingenhouss [1789], who heated rods covered with thin layers of wax at one end and afterwards measured the length to which melting had proceeded, using the visible line marked at the limit of melting. At this line, obviously, the known melting temperature of wax had existed.

A procedure likewise based on making isothermal lines visible and measuring their location was indicated by Sénarmont [1847, 1847(*a*)].§ It is mainly used for anisotropic substances, particularly crystals. Thin sample plates are covered with wax and heated in radial direction by means of a hot, thin silver tube or cylinder, which is tightly fixed in a hole at the center of the sample. The isothermal limit of melting is

§ For further publications by the same author see Jakob [1926(*a*), pp. 51 and 92].

circular for materials homogeneous in conductivity and elliptical if the thermal conductivity has different values in two directions perpendicular to each other. The method was successfully used to find the influence of magnetization on the thermal conductivity of iron, nickel, or bismuth and the influence of illumination upon the conductivity of selenium. With regard to modifications of the method by Roentgen and other physicists see Jakob [1926(*a*), p. 52]. Only Voigt's [1897, 1898] procedure to compare the conductivities of different isotropic materials will be described in more detail.

Figure 9-9 represents the surface of two plates 1 and 2 of unit thickness, having the conductivities k_1 and k_2, respectively. They are

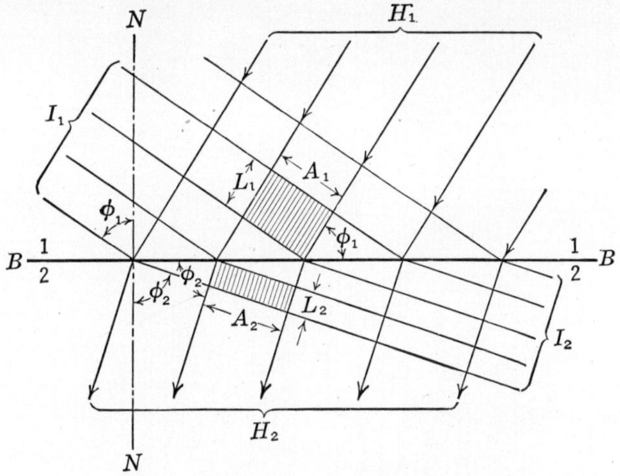

FIG. 9-9. Heat flow oblique to interface of plates in contact.

cemented together along *BB*, and a heat flow may pass through the system along the heat flow lines H_1 and H_2 shown in the figure, which have different directions in the sections 1 and 2 of the systems. The isothermals I_1 and I_2, which, of course, are perpendicular to the flow lines, are inclined by the angles ϕ_1 and ϕ_2 against the normal line *NN*. The temperature difference between any two subsequent isothermals may be the same (Δt). Since the heat flow through the parallelogramms with the sides A_1, L_1 and A_2, L_2 is the same ($q_1' = q_2'$),

$$k_1 \frac{A_1}{L_1} \Delta t = k_2 \frac{A_2}{L_2} \Delta t$$

or

$$\frac{k_1}{k_2} = \frac{\tan \phi_2}{\tan \phi_1} \tag{9-16}$$

Voigt's method is based on this equation. Two congruent plates having the shape of rectangular triangles are cut from the two materials under consideration and cemented together along their largest sides, so that a rectangular plate is formed. The surface of this plate is painted with elaidine acid, some wax and terpentine being added, so that the melting temperature is about 112°F. Then a hot copper block is brought to contact with one side of the plate. The isothermal line, corresponding to the melting temperature, appears very clearly, and the angles ϕ_1 and ϕ_2 can be measured. For consideration of the adaptation of the method to the comparison of the thermal conductivities in different directions in crystals see Jakob [1926(a), p. 92].

CHAPTER 10

Heat Conduction in Steady State, Including Internal Heat Generation

10–1. Unidirectional Heat Flow in a Medium with Uniformly Distributed Heat Sources or Sinks

The probably simplest case of heat conduction with internal heat sources or sinks is represented in Fig. 10–1. Heat energy at the rate q_0

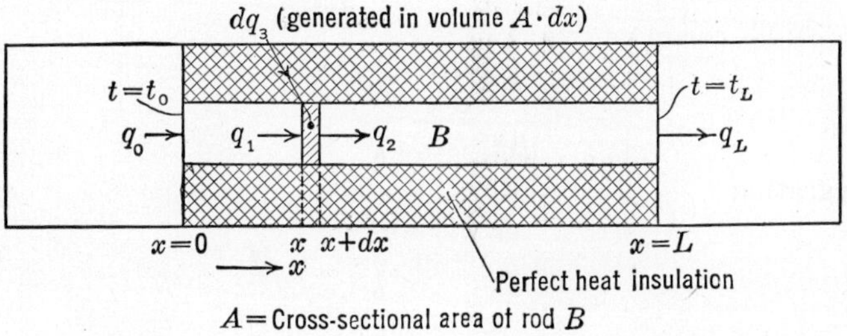

FIG. 10–1. Heat flow in a rod in which heat is generated.

enters a perfectly insulated body B of length L and constant cross-sectional area A. In each unit volume of this rod the heat energy q''' may be produced in unit time. This, for instance, is the case of a wire through which an electric current flows, provided the electric resistivity can be assumed to be independent of temperature; or the case of a chemical reaction, uniformly distributed over the volume of the body, in a range of temperature where the heat of reaction can be considered to be constant; or the case of a metal rod or tube heated by a heating coil wound on the cylinder surface, under the condition that the heat originating in the coil is uniformly distributed over the rod cross section which is practically satisfied, owing to the great thermal conductivity of metals. Finally, there are cases where heat losses occur uniformly over the surface of a metal rod or tube. Then q''' would assume a negative value.

Considering a differential of thickness dx at the distance x from the left end, the incoming heat flow is

$$q_1 = -kA \frac{dt}{dx}$$

The heat flow leaving at the distance $x + dx$ is

$$q_2 = -kA \left(\frac{dt}{dx} + \frac{d^2t}{dx^2} dx \right)$$

and the heat energy originating in the element in unit time may be

$$dq_3 = q''' A \cdot dx$$

Using the heat balance for steady state

$$q_1 + dq_3 = q_2$$

one arrives at

$$q''' A \cdot dx = -kA \frac{d^2t}{dx^2} dx$$

or

$$\frac{d^2t}{dx^2} = -\frac{q'''}{k} \tag{10-1}$$

By integration

$$\frac{dt}{dx} = -\frac{q'''}{k} x + M \tag{10-2}$$

and

$$t = -\frac{q'''}{k} \frac{x^2}{2} + Mx + N = -K \frac{x^2}{2} + Mx + N \tag{10-3}$$

where $K = q'''/k$, and
M and N = the constants of integration.

Using the boundary conditions

$$t = t_0 \quad \text{for} \quad x = 0$$

and

$$t = t_L \quad \text{for} \quad x = L$$

Eq. 10-3 yields

$$N = t_0 \tag{10-4}$$

and

$$M = \frac{t_L - t_0}{L} + \frac{q'''}{k} \cdot \frac{L}{2} \tag{10-5}$$

Finally, by substitution in Eq. 10–3,

$$t = t_0 - \frac{t_0 - t_L}{L} x + \frac{q'''}{k} \cdot \frac{(L - x)x}{2} \tag{10–6}$$

The rate of heat supplied at $x = 0$ is

$$q_0 = -kA \left(\frac{dt}{dx}\right)_{x=0} \tag{10–7}$$

From Eqs. 10–2 and 5

$$\left(\frac{dt}{dx}\right)_{x=0} = M = \frac{t_L - t_0}{L} + \frac{q'''}{k} \cdot \frac{L}{2}$$

Substituting in Eq. 10–7 and rearranging,

$$q_0 + q''' \frac{AL}{2} = k \frac{A}{L} (t_0 - t_L) \tag{10–8}$$

Introducing an equivalent heat rate

$$q_e = q_0 + q''' \frac{AL}{2} \tag{10–9}$$

equal to the heat incoming at $x = 0$, plus half of the heat produced inside the body B, Eq. 10–8 simplifies to

$$q_e = k \frac{A}{L} (t_0 - t_L) \tag{10–10}$$

Thus, the ordinary equation of heat conduction without heat sources or sinks can be used, by adding half of the produced heat to q_0 or subtracting this amount from q_L, the actual rate of heat flow at $x = L$.

As mentioned already, the method can also be used in the case of uniform heat loss from a metal rod. This can be assumed if the temperature difference $(t_0 - t_L)$ is small compared with the temperature difference between surface and environment, or if the total heat loss q_s through the surface is small in comparison to q_0. In this case $q''' = -q_s/AL$ may be substituted in Eq. 10–9.

If the thermal conductivity k of the material of a rod is to be measured and the heat losses can be considered uniform and equal to Aq''' per unit length, Eqs. 10–9 and 10 can be used for the determination of k.

10–2. Relation between Thermal and Electrical Conduction in a Rod

Considering once more Fig. 10–1, it is assumed that by heat reservoirs at both ends of the rod the temperature at these ends is held at t_0 and

t_L. The rod is inserted in an electrical circuit so that a drop in voltage e_L occurs between $x = 0$ and $x = L$.

In the differential section of the rod between x and $x + dx$ heat energy is produced at the rate of $i \cdot de$ in electric units, where i is the electric current and de the potential difference.

Relating the heat production to unit volume,

$$q''' = \frac{i \cdot de}{A \cdot dx} = \frac{de}{dR_e} \cdot \frac{de}{A \cdot dx} \tag{10-11}$$

where dR_e is the electric resistance of the elementary piece of the rod.

Introducing the electric conductivity k_e,

$$q''' = \frac{de}{dx/k_e A} \cdot \frac{de}{A \cdot dx} = k_e \left(\frac{de}{dx}\right)^2$$

and by substitution in Eq. 10–1,

$$k \frac{d^2t}{dx^2} + k_e \left(\frac{de}{dx}\right)^2 = 0 \tag{10-12}$$

a special case of the general differential equation

$$k \left[\frac{\partial^2 t}{\partial x^2} + \frac{\partial^2 t}{\partial y^2} + \frac{\partial^2 t}{\partial z^2}\right] + k_e \left[\left(\frac{\partial e}{\partial x}\right)^2 + \left(\frac{\partial e}{\partial y}\right)^2 + \left(\frac{\partial e}{\partial z}\right)^2\right] = 0 \tag{10-13}$$

This equation originates from Kohlrausch [1872, 1898]. Previously Verdet [see W. Jaeger and Diesselhorst 1900] had already set up an equation for the axial temperature drop in a metal rod inserted in an electrical circuit, for given k, k_e, and h at the surface of the rod.

The further derivation becomes simpler if i instead of e is used as dependent variable. Substituting

$$de = i \cdot dR_e = i \frac{dx}{k_e A}$$

in Eq. 10–11 yields

$$q''' = \frac{i^2}{k_e A^2}$$

Herewith Eq. 10–1 becomes

$$k \frac{d^2t}{dx^2} + \frac{i^2}{k_e A^2} = 0 \quad \text{or} \quad \frac{d^2t}{dx^2} + K = 0 \tag{10-14}$$

where

$$K = \frac{i^2}{k k_e A^2} \tag{10-15}$$

In ranges where $k k_e$ can be considered constant the solution of this differential equation is identical with Eq. 10–3.

If the heat reservoirs at the two ends of the rod B in Fig. 10–1 are held at the same temperature t_0, and the temperature is counted from this temperature as zero level ($\theta = t - t_0$), the solution becomes particularly simple. Let $\theta_m = t_m - t_0$ be the temperature excess in the middle between $x = 0$ and $x = L$.

Then the boundary conditions are

$$\begin{aligned} \theta &= \theta_0 = 0 \quad \text{at} \quad x = 0 \\ \theta &= \theta_m \qquad\ \text{at} \quad x = L/2 \\ \theta &= \theta_0 = 0 \quad \text{at} \quad x = L \end{aligned}$$

Because Eq. 10–3 holds for any zero level of temperature, it can be written as

$$\theta = -K\frac{x^2}{2} + Mx + N \tag{10–16}$$

With the above boundary conditions the equation yields

$$\begin{aligned} 0 &= N \\ \theta_m &= -K\frac{L^2}{8} + M\frac{L}{2} + N \\ 0 &= -K\frac{L^2}{2} + ML + N \end{aligned}$$

Hence

$$\left.\begin{aligned} K &= \frac{8\theta_m}{L^2} \\ M &= \frac{4\theta_m}{L} \\ N &= 0 \end{aligned}\right\} \tag{10–17}$$

By substitution in Eq. 10–16

$$\theta = -\frac{4\theta_m}{L^2}x^2 + \frac{4\theta_m}{L}x \tag{10–18}$$

Further, from Eqs. 10–15 and 17

$$k = \frac{i^2}{k_e A^2 K} = \frac{1}{8k_e}\cdot\frac{L^2}{A^2}\cdot\frac{i^2}{\theta_m}$$

But

$$i = \frac{e_L}{\dfrac{1}{k_e}\cdot\dfrac{L}{A}}$$

Thus

$$k = \frac{k_e}{8}\cdot\frac{{e_L}^2}{\theta_m} \tag{10–19}$$

According to this equation, k is found by measuring the potential difference e_L between the ends of the rod and the temperature difference $\theta_m = t_m - t_0$ between the middle and either end of the rod. The main advantage of the method is that the difficult determination of thermal conductivity is replaced by the simple measurement of electric conductivity.

Therefore this method has been used frequently. Modifications and corrections consider the heat transfer from the surface of the rod to the environment and the dependence of the electric conductivity on temperature. W. Jaeger and Diesselhorst [1900], for instance, showed that the heat loss from the surface of a cylindric rod can be considered by adding to θ_m in Eq. 10–18 a correction term

$$\Delta\theta_m = -\frac{1}{4}\cdot\frac{h}{k}\cdot\frac{L^2}{r}\left(t_a - t_m + \frac{6}{t_m - t_0}\right)$$

where h is the coefficient of heat transfer at the rod surface,
r is the radius of the cylindric rod, and
t_a is the temperature of the surrounding air or any other environment.

In this empirical term, lengths are used in centimeters, and temperatures in centigrades. It is seen that the correction vanishes if $t_m - t_a = 6/(t_m - t_0)$, which can be obtained by a suitable arrangement.*

The influence of heat losses will be dealt with more thoroughly in Sect. 12–2.

Simidu [1917] placed the sample in a hollow cylinder which was heated in such a way that in three cross sections its temperature was the same as that of the sample. By this procedure the heat losses became so much smaller, compared with the axial heat flow, that thinner rods (0.5 cm instead of 1 or 2 cm diameter) and smaller currents (for instance, 10 or 20 amp instead of about 300 amp) could be used.

At very low temperatures (down to −253°C) W. Meissner [1915] operated with rods of only 0.1 to 0.3 cm diameter and 7 cm length. This was possible by using the rods as heaters and resistance thermometers at the same time, according to a suggestion of Diesselhorst [1902].

For use at very high temperature Langmuir [1916] developed a differential equation in which the radiation of glowing wires was considered. By numerical integration of the equation for each individual case he obtained the value of k/k_e.

* The factor $\frac{h}{k}\cdot\frac{L^2}{r}$ is dimensionless; any consistent system of units can be used for its numerical computation.

Another modification of the method was suggested by Mendenhall and applied by Angell [1911]. Whereas, according to Kohlrausch, the heat energy, as well as the electric current which produces it, is flowing in axial direction, and radial heat exchange is prevented by insulation, in the Mendenhall-Angell method only the electrical current is in axial direction, and the heat flows in radial direction outward. Instead of rods, hollow cylinders are used in this method.

Let e = the axial electric potential drop in the cylinder,
i = the electric current,
$2r_1$ = the inner diameter,
$2r_2$ = the outer diameter, and
L = the length.

The following equation is to be used:

$$e \cdot \frac{\pi(r^2 - r_1^2)}{\pi(r_2^2 - r_1^2)} i = -k \cdot 2\pi r \cdot L \frac{dt}{dr} \tag{10–20}$$

The term on the left hand is the rate of heat energy which is produced by the current in a hollow cylinder, within the limits r_1 and r; it must be equal to the rate of heat energy crossing the cylinder area of radius r in radial direction outward (right side of equation). If k = constant, the integration of Eq. 10–20 yields

$$k = \frac{ei}{2\pi L(t_1 - t_2)} \left(\frac{1}{2} - \frac{r_1^2}{r_2^2 - r_1^2} \ln \frac{r_2}{r_1} \right) \tag{10–21}$$

e or i may be eliminated and k_e introduced in this equation, by means of Ohm's law

$$\frac{e}{i} = R_e = \frac{L}{k_e \pi (r_2^2 - r_1^2)}$$

10–3. Heat Conduction in Electrical Coils at Uniform Development of Heat

An important practical case in which heat conduction must be considered in connection with uniform heat production is that of an electrical coil in which heat is generated by the current in the wire.

In a coil of electric resistance R_e a current i causes a voltage drop $e = iR_e$. Thus heat energy is produced at the rate of $ei = e^2/R_e = i^2R_e$. If V is the volume of the coil, including the volume of the insulation of the wires, then the average rate of heat energy developed per unit volume of the coil is

$$q''' = \frac{i^2 R_e}{V} \tag{10–22}$$

The variation of the electric resistivity with temperature presents a very simple means for determining the mean temperature t_m of a coil. Let the cold coil be at a reference temperature t_c. If a very small current, i_c, is passing through, the coil (at the temperature t_c) does not warm up sensibly, and the average resistance of the cold coil, $R_{e,c} = e_c/i_c$, can be obtained by measurement of the current i_c and the potential drop e_c. In the same way the mean resistance of the warm coil, $R_{e,m} = e_m/i_m$, is found from the voltage e_m and the current i_m of operation. But

$$R_{e,m} = R_{e,c}[1 + \epsilon(t_m - t_c)]$$

or

$$t_m = t_c + \frac{1}{\epsilon} \cdot \frac{R_{e,m} - R_{e,c}}{R_{e,c}} \tag{10-23}$$

where ϵ is a temperature coefficient of the electric resistivity of the wire metal defined by either of these equations.

This definition leads to different values of ϵ for different reference temperatures t_c. If t_c, as is usual in physics, is taken as 0°C, then, according to the best experiments for pure copper, $\epsilon = 0.00433\ \mathrm{C}^{-1}$ (in the range of $t_m = 0$°C to $t_m = 200$°C). If, on the other hand, $t_c = 100$°C is used as reference temperature, then the same experiments lead to $\epsilon = 0.00303\ \mathrm{C}^{-1}$. By interpolation it is seen that $\epsilon = 0.004$ is a fair average for $t_c = 25$°C, which in the following calculations will be used as the temperature of "cold" coils unless another value is indicated.

By taking the temperature of a coil in cold state, i.e., without carrying current, as reference point (zero point) of temperature denotation, the calculations become much simpler. This is the reason why Eq. 10–23 is used for the definition of ϵ, and not its special form

$$R_{e,m} = R_{e,0}(1 + \epsilon t_m)$$

with $R_{e,0}$ as resistance at 0°C.

For some pure metals ϵ varies not only with the reference temperature t_c, but also with the temperature t_m. For pure nickel, for instance, according to measurements of Holborn [1921], Eq. 10–23 leads to the values of ϵ given in Table 10–1.

TABLE 10–1

MEAN TEMPERATURE COEFFICIENT ϵ OF THE ELECTRIC RESISTIVITY OF PURE NICKEL

$10^3\epsilon$ [C^{-1}]

t_m [°C]	0	100	200	300
$t_c = 0$°C	5.5	6.7	7.6	8.9
$t_c = 100$°C	4.0	4.6	5.1	6.0

Iron is another metal which behaves in this way. Fortunately most electric coils consist of practically pure copper, where ϵ can be assumed as constant in practical limits. Therefore the mean temperature t_m of a coil under load can be found simply by the measurement of the two resistances $R_{e,c}$ and $R_{e,m}$. The mean temperature, however, is not as important as the highest temperature t_0 occurring inside a coil, and this cannot be measured in a simple way. Vidmar [1918] was the first to show how the difficult measurement of the temperature t_0 at the hottest place can be superseded by the simple measurement of the surface temperature t_s. He directed the attention to the fact that the ratio

$$\phi = \frac{t_m - t_s}{t_0 - t_s} = \frac{\theta_m}{\theta_0} \tag{10–24}$$

is not much different in most kinds of coils. A general treatment of this problem for coils having the shapes of an infinitely wide plate, an infinitely long cylinder, or a sphere has been given by Rogowski [1918].

Starting from Eq. 10–1 for the plate, and from the corresponding differential equations in polar coordinates for cylinder and sphere, Rogowski came to equations of the temperature distribution which can be represented by the unique form

$$\theta = \theta_0 \left(1 - \frac{x^2}{s^2}\right) \tag{10–25}$$

where s = the distance of the surface,

x = the distance of an arbitrary point, from the median plane of the plate, from the center line of the cylinder, and from the center of the sphere, respectively,

$\theta = t - t_s$, and

t = the temperature at the place x.

A common proof of Eq. 10–25 has been given by Jakob [1919]. Let V be the volume between the niveau surfaces $x = 0$ and $x = x$, and A the area of the latter surface, and introduce $V/A = \psi x$, where $\psi = 1$ for the plate, $\psi = \frac{1}{2}$ for the cylinder, and $\psi = \frac{1}{3}$ for the sphere. Then

$$Vq''' = -kA \frac{d\theta}{dx}$$

or

$$\frac{d\theta}{dx} = -\frac{q'''}{k} \psi x$$

Under the assumption that q'''/k = constant, integration yields

$$\theta = M - \frac{q'''}{2k}\psi x^2$$

But $\theta = 0$ for $x = s$, and therefore

$$M = \frac{q'''}{2k}\psi s^2$$

and

$$\theta = \frac{q'''}{2k}\psi\cdot(s^2 - x^2) \tag{10-26}$$

Further, $\theta = \theta_0$ at $x = 0$. Therefore

$$\theta_0 = \frac{q'''}{2k}\psi s^2 \tag{10-27}$$

Combining Eqs. 10-26 and 27 yields Eq. 10-25.

Rogowski further proved that

for an infinitely extended plate, $\phi = 2/3$

for an infinitely long cylinder, $\phi = 2/4$

for a sphere, $\phi = 2/5$

The proof is elementary; for the plate it is given in Chapter V of the elementary book of Jakob and Hawkins [1942].

The integration of Eq. 10-1 was based on the assumption of q'''/k = constant. In fact, the apparent conductivity † k of an electrical coil does not change much with temperature, but q''' does, owing to the considerable temperature differences inside the coil and the rather large value of ϵ for pure metals. The results of this section are restricted to a range of temperature in which the local variation of q'''/k is negligible. Cases where this variation has to be considered will be dealt with in Sect. 10-4.

The case of a coil of rectangular cross section and infinite length is of practical importance. An ordinary coil of rectangular cross section with negligible influence of the curvature behaves as though it were infinitely long in the direction perpendicular to the cross section, since for reasons of symmetry no heat will flow in this direction. This case has been dealt with by Humburg [1909] for equal and different thermal conductivity in the two main directions of the cross section and for equal and different cooling of the four planes which form the surface

† Defined by Eq. 6-17.

of the coil. The simpler case of uniform thermal conductivity and surface temperature has been treated by Jakob [1919], independently of Humburg.

Following Jakob's representation and referring to the symbols in Fig. 10–2, the basic differential equation is

$$\frac{\partial^2\theta}{\partial x^2} + \frac{\partial^2\theta}{\partial y^2} = -\frac{q'''}{k} \tag{10–28}$$

and the boundary conditions are

$$\theta = 0 \quad \text{at} \quad \begin{matrix} x = \pm a \\ y = y \end{matrix} \qquad \theta = 0 \quad \text{at} \quad \left.\begin{matrix} x = x \\ y = \pm b \end{matrix}\right\} \tag{10–29}$$

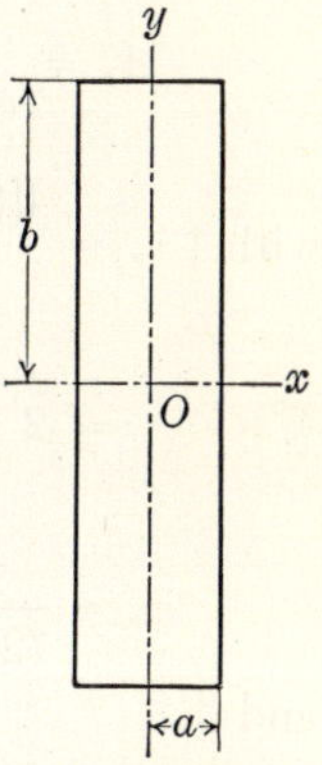

FIG. 10–2. Denotations in the rectangular cross section of an electrical coil.

Like Nusselt's treatment of the cube (see Sect. 8–9), Jakob's procedure is based on Stokes's method of dividing θ into two terms so that

$$\theta = \theta_1 + \theta_2 \tag{10–30}$$

One of these terms can be chosen arbitrarily. Let

$$\theta_1 = -\frac{q'''(x^2 + y^2)}{4k} + C \tag{10–31}$$

This is the equation of the temperature distribution in a circular cross section, and it holds if $x^2 + y^2$ is negligibly small, compared to a^2 and b^2 as occurs in close vicinity to the center O in Fig. 10–2.

By introducing the additional term C in Eq. 10–31 Jakob avoided the difficulty which occurred in Nusselt's derivation where the surface conditions were not fulfilled in the edges of the cube. Here the condition that Eq. 10–29 must hold also in the corners of the cross-sectional area leads to

$$C = \frac{q'''(a^2 + b^2)}{4k}$$

Except for this detail the procedure is entirely similar to that described in Sect. 8–9. It yields the final equation

$$\theta = \frac{q'''}{4k}\Bigg[(a^2 - x^2) + (b^2 - y^2) + \sum_{\nu=0}^{\nu=\infty} A_\nu \cosh(\lambda\pi x)\cdot\cos\left(\frac{2\nu+1}{2b}\pi y\right) + \sum_{\nu=0}^{\nu=\infty} B_\nu \cosh(\mu\pi y)\cdot\cos\left(\frac{2\nu+1}{2a}\pi x\right)\Bigg] = \frac{q'''}{4k}\Phi \tag{10–32}$$

where Φ stands for the expression in brackets,

$$\lambda = \pm \frac{2\nu + 1}{2b} \qquad \mu = \pm \frac{2\nu + 1}{2a}$$

and

$$A_\nu = \frac{b^2}{\cosh (\lambda \pi a)} (v_\nu - w_\nu) \qquad B_\nu = \frac{a^2}{\cosh (\mu \pi b)} (v_\nu - w_\nu)$$

$$\text{with ‡ } v_\nu = \frac{4[(2\nu + 1)^2\pi^2 - 8]}{(2\nu + 1)^3\pi^3} \sin \frac{2\nu + 1}{2} \pi$$

$$= 2 \left\{ \frac{1}{(2\nu + 1)\pi} - \left[\frac{2}{(2\nu + 1)\pi} \right]^3 \right\} 2 \sin \frac{(2\nu + 1)\pi}{2}$$

$$w_\nu = \frac{2}{(2\nu + 1)\pi} 2 \sin \frac{2\nu + 1}{2} \pi$$

and

$$v_\nu - w_\nu = -2 \left[\frac{2}{(2\nu + 1)\pi} \right]^3 2 \sin \left(\frac{2\nu + 1}{2} \pi \right) = \frac{-32}{\pi^3} \cdot \frac{1}{(2\nu + 1)^3} (-1)^\nu$$

By substituting $x = 0$ and $y = 0$ in Eq. 10–32 the following relation is obtained:

$$\theta_0 = \frac{q'''}{4k} \left[(a^2 + b^2) + \left(\sum_{\nu=0}^{\nu=\infty} A_\nu + \sum_{\nu=0}^{\nu=\infty} B_\nu \right) \right] \tag{10–33}$$

The sum of the two first terms is a measure for the maximum temperature of a coil of circular cross section with a diameter $2\sqrt{a^2 + b^2}$ equal to the diagonal of the rectangular cross section. The sums of A_ν and B_ν give the deviation from this temperature.

The mean temperature of the coil is found from the relation

$$\theta_m = \frac{1}{ab} \int_0^a \int_0^b \theta \cdot dx \cdot dy \tag{10–34}$$

By substituting θ from Eq. 10–32 in this equation and integrating,

$$\theta_m = \frac{q'''}{4k} \left[\frac{2}{3} (a^2 + b^2) + \frac{1}{ab} \sum_{\nu=0}^{\nu=\infty} \frac{A_\nu}{\pi^2\lambda^2} \sinh (\lambda \pi a) \cdot \sin \left(\frac{2\nu + 1}{2} \pi \right) \right.$$

$$\left. + \frac{1}{ab} \sum_{\nu=0}^{\nu=\infty} \frac{B_\nu}{\pi^2\mu^2} \sinh (\mu \pi b) \cdot \sin \left(\frac{2\nu + 1}{2} \pi \right) \right] \tag{10–35}$$

‡ In the original paper v_ν was represented in a less simple manner.

It is seen that in θ, θ_0, and θ_m the quantities q''' and k occur only as a factor q'''/k. Thus, as in the former cases, the ratio $\phi = \theta_m/\theta_0$ is independent of the Joulean heat and of the thermal conductivity. It is a mere constant of configuration.

For $b = \infty$, Eq. 10–33 shrinks into

$$(\theta_0)_{b=\infty} = \frac{q'''a^2}{2k} \tag{10–36}$$

But

$$\theta = \frac{q'''}{4k}\Phi$$

Therefore

$$\frac{\theta}{(\theta_0)_{b=\infty}} = \frac{2k}{q'''a^2}\theta = \frac{\Phi}{2a^2} \tag{10–37}$$

Numerical values of this quantity have been calculated and tabulated (see Table 10–2).

TABLE 10–2

Temperature Distribution in Electrical Coils of Rectangular Cross Section

$$\frac{\theta}{(\theta_0)_{b=\infty}} = \frac{2k}{q'''a^2}\theta = \frac{\Phi}{2a^2}$$

b/a	∞	5					1			
y/b	0 to 1	0	$\frac{1}{3}$	$\frac{2}{3}$	$\frac{5}{6}$	1	0	$\frac{1}{3}$	$\frac{2}{3}$	1
$x/a = 0$	1.00	1.00	0.99	0.92	0.72	0	0.59	0.54	0.35	0
$\frac{1}{3}$	0.89	0.89	0.885	0.825	0.65	0	0.54	0.485	0.32	0
$\frac{2}{3}$	0.555	0.555	0.545	0.51	0.415	0	0.35	0.32	0.22	0
1	0	0	0	0	0	0	0	0	0	0

Graphs of the temperature distribution for $b/a = 5$ and $b/a = 1$ are given in Figs. 10–3 and 4.

In Table 10–3 the mean and maximum temperature excesses and their ratio ϕ are represented.

For a graphical representation of the ratio ϕ as a function of the sides ratio of the rectangular cross section see Fig. 10–5. The ordinate axis is met at $\phi = 2/3$, and ϕ approaches this value again for $b/a \to \infty$. In general, ϕ is the same for any couple of ratios $b/a = n$ and $b/a = 1/n$, as easily can be understood. The smallest value occurring for any rec-

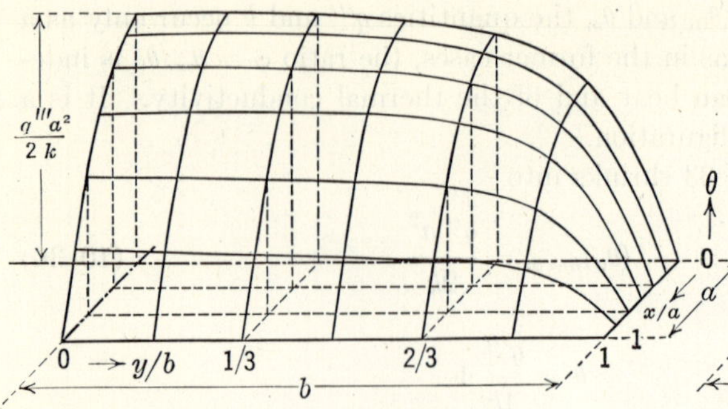

FIG. 10-3. Temperature distribution in an electrical coil of rectangular cross section.

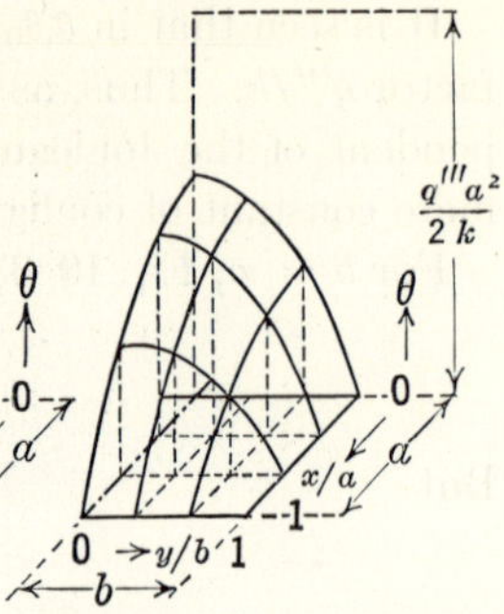

FIG. 10-4. Temperature distribution in an electrical coil of square cross section.

TABLE 10-3

MEAN AND MAXIMUM TEMPERATURE EXCESSES AND THEIR RATIO FOR ELECTRICAL COILS OF RECTANGULAR AND CIRCULAR CROSS SECTIONS

Cross section	Rectangular						Circular
b/a	∞	10	5	2.5	1.5	1	$r = 2a/\sqrt{\pi}$
$(2k/q'''a^2)\theta_m$	0.66···	0.625	0.58			0.28	0.32
$(2k/q'''a^2)\theta_0$	1.00	1.00	1.00			0.59	0.64
$\phi = \theta_m/\theta_0$	0.66···	0.625	0.58	0.52	0.485	0.475	0.50

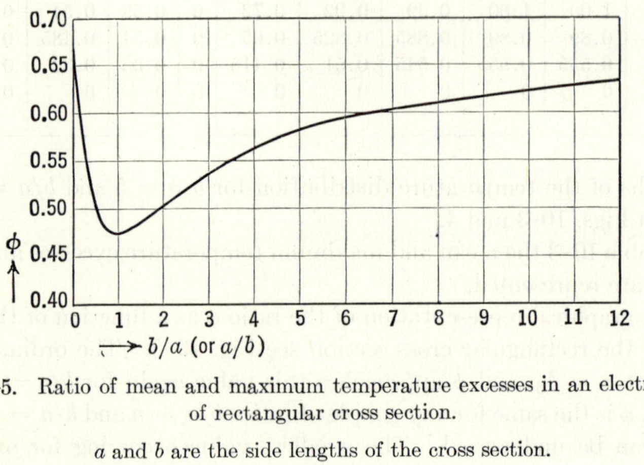

FIG. 10-5. Ratio of mean and maximum temperature excesses in an electrical coil of rectangular cross section.

a and b are the side lengths of the cross section.

tangular cross section is $\phi = 0.475$. It is just 5% smaller than $\phi = 0.5$, which holds for a circular cross section, as may be seen from Table 10–3.

Again it must be kept in mind that q''' actually is not a constant, as has been assumed, but depends on the temperature distribution in the coil. Corrections for this fact, derived for plate and cylinder shapes in Sect. 10–4, may also be used for rectangular cross sections with fair approximation.

10–4. Heat Conduction in Electrical Coils at Non-uniform Development of Heat

The values of ϕ, given in Sect. 10–3, as well as Eqs. 10–25 and 32 for the temperature distribution, are for uniform development of heat in the coil. But, as previously mentioned, heat is not developed uniformly in an ordinary coil. Consider a coil of copper wire with $t_s = 25°C$ and $t_0 = 125°C$. The electric resistance of unit length of the copper wire at the temperature t is

$$R_e' = R_{e,s}'(1 + \epsilon\theta)$$

where $R_{e,s}'$ = that resistance at the temperature t_s, and
$\epsilon = 0.004\ C^{-1}$ (see Sect. 10–3).

Hence

$$R_{e,0}' = 1.4\ R_{e,s}' \text{ at the temperature } t_0 = 125°C$$

But, according to the laws of Ohm and Joule, the heat developed by an electric current is proportional to the resistance. Thus, in the present case, 40% more heat per unit volume is produced at the warmest place inside the coil than just below the surface, where the temperature is equal to t_s.

The influence of this inhomogeneity of heat development has been investigated by Jakob [1943] for the infinitely wide plate, the infinitely long circular cylinder, and the sphere, § and by Th. J. Higgins [1944] for the infinitely long rod of rectangular cross section.

a. The Plane Plate

With $\theta = t - t_s$, the differential equation 10–1 takes the form

$$\frac{d^2\theta}{dx^2} = -\frac{q'''}{k} \tag{10–38}$$

k does not vary much with the temperature and may, therefore, be

§ For detailed derivations of the results given in the following Articles *a*, *b*, and *c* see the original paper.

considered constant. The change of q''' can be considered linear with temperature and may be represented by

$$\frac{q'''}{k} = m + n\theta = m(1 + \epsilon\theta) \tag{10-39}$$

where $m = q_s'''/k \gtreqless 0$,
$n = \epsilon q_s'''/k \gtreqless 0$,

the subscript s referring to surface temperature ($\theta = 0$).

From Eqs. 10-38 and 39

$$\frac{d^2\theta}{dx^2} + n\theta = -m \tag{10-40}$$

For $n > 0$, the reduced equation

$$\frac{d^2\theta}{dx^2} + n\theta = 0$$

has the solution

$$\theta = M \cos (x\sqrt{n}\,) + N \sin (x\sqrt{n}\,)$$

when M and N are the constants of integration.

A particular solution of Eq. 10-40 is

$$\theta = -\frac{m}{n}$$

Hence its general solution is

$$\theta = -\frac{m}{n} + M \cos (x\sqrt{n}\,) + N \sin (x\sqrt{n}\,) \tag{10-40a}$$

Considering the boundary conditions

$$\frac{d\theta}{dx} = 0 \quad \text{for} \quad x = 0$$

and

$$\theta = 0 \quad \text{for} \quad x = s$$

where $2s$ = the thickness of the plate, N becomes zero, and the solution of Eq. 10-40 reduces to

$$\theta = \frac{m}{n}\left[\frac{\cos (x\sqrt{n}\,)}{\cos (s\sqrt{n}\,)} - 1\right] = \frac{1}{\epsilon}\left[\frac{\cos \xi\sigma}{\cos \sigma} - 1\right] \tag{10-41}$$

where, for convenience, the dimensionless quantities $\xi = x/s$ and $\sigma = s\sqrt{n}$ have been introduced.

Substituting $x = 0$,

$$\theta_0 = \frac{m}{n}\left[\frac{1}{\cos(s\sqrt{n})} - 1\right] = \frac{1}{\epsilon}\left[\frac{1}{\cos\sigma} - 1\right] \tag{10–42}$$

and from Eqs. 10–41 and 42

$$\frac{\theta}{\theta_0} = \frac{\cos(x\sqrt{n}) - \cos(s\sqrt{n})}{1 - \cos(s\sqrt{n})} = \frac{\cos(\xi\sigma) - \cos\sigma}{1 - \cos\sigma} \tag{10–43}$$

By substituting θ from Eq. 10–41 in

$$\theta_m = \frac{1}{s}\int_0^s \theta \cdot dx$$

the mean temperature is found to be

$$\theta_m = \frac{m}{n}\left(\frac{\tan(s\sqrt{n})}{s\sqrt{n}} - 1\right) = \frac{1}{\epsilon}\left(\frac{\tan\sigma}{\sigma} - 1\right) \tag{10–44}$$

and

$$\phi = \frac{\theta_m}{\theta_0} = \frac{\sin(s\sqrt{n}) - s\sqrt{n}\cdot\cos(s\sqrt{n})}{1 - \cos(s\sqrt{n})} = \frac{\sin\sigma - \sigma\cdot\cos\sigma}{1 - \cos\sigma} \tag{10–45}$$

According to Eq. 10–41, θ is proportional to m, but, according to Eq. 10–43, the temperature distribution is independent of the value of m.

In a plate with heat sources, according to definition, $\theta > 0$. But for arbitrary values $x < s$, Eq. 10–41 yields positive values of θ only if

$$\cos(x\sqrt{n}) > \cos(s\sqrt{n})$$

and

$$0 < s\sqrt{n} < \frac{\pi}{2} \tag{10–46}$$

At the limit

$$s\sqrt{n} = \frac{\pi}{2} \tag{10–47}$$

θ becomes infinitely great for every value of x. Hence, if the linear Eq. 10–39 were true up to this limit, every point of the plate would be at infinitely high temperature.

Therefore Eq. 10–46 does not mean a physical restriction of the problem. It shows only that already, at a finite value of $s\sqrt{n}$, the temperature at every point, except on the surface, would exceed any finite value.

Substituting n from Eq. 10–47 in Eq. 10–39,

$$q_s''' = \left(\frac{\pi}{2}\right)^2 \frac{k}{\epsilon s^2} = 2.467 \frac{k}{\epsilon s^2} \tag{10–48}$$

Consider a coil of volume V and electric resistance $R_{e,s}$ at temperature t_s. Then, for an electric current i, by definition and Ohm's law,

$$q_s''' = \frac{i^2 R_{e,s}}{V}$$

and by substitution from Eq. 10–48

$$i = \sqrt{2.467 \frac{k}{\epsilon s^2} \cdot \frac{V}{R_{e,s}}} \tag{10–49}$$

This is the current at which the temperature would become infinitely high. To understand this from the energy viewpoint, it must be kept in mind that the increase of the electric resistance inside the coil which is due to the temperature rise will cause an enormous increase of q''' inside the coil, although q_s''' (on the very surface) is supposed to be held at a moderate value.

In the original paper [Jakob 1943] it is shown that

for $\sigma = 0$,

$$\frac{\theta}{\theta_0} = 1 - \xi^2 \tag{10–50}$$

for $\sigma = \pi/2$, Eq. 10–43 yields

$$\frac{\theta}{\theta_0} = \cos\left(\xi \frac{\pi}{2}\right) \tag{10–51}$$

In Fig. 10–6 these two limiting temperature ratios are plotted versus ξ (parabola and cosine curve), and a curve for $\sigma = 1$ is drawn between. The figure shows the surprising result that from zero to infinitely high temperature ($\sigma = 0$ to $\sigma = \pi/2$) the temperature distribution changes only very little.

This means that also the ratio $\phi = \theta_m/\theta_0$ will vary only slightly in the whole range of σ, namely, from $\phi = 2/3$ to $\phi = 2/\pi$. The latter value is only $4\frac{1}{2}\%$ smaller than $2/3$, no matter how thick the plate or how large the temperature coefficient ϵ may be.

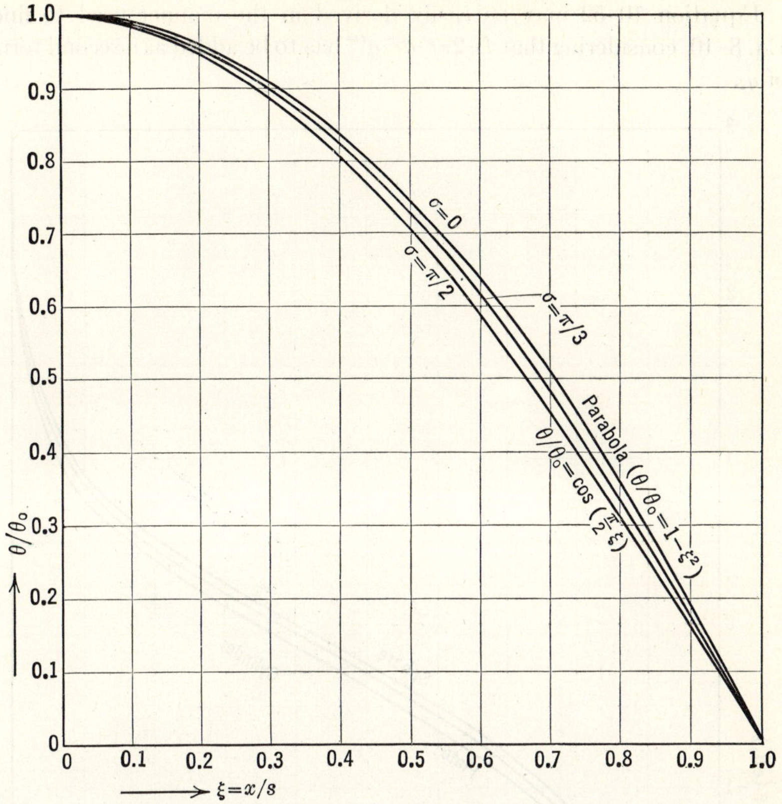

Fig. 10–6. Temperature distribution in a plane plate in which heat is generated at a rate increasing linearly with temperature.

The maximum temperature θ_0, however, according to Eq. 10–42, varies from 0 to ∞, as may be seen from Fig. 10–7, in which $\epsilon\theta_0$ is plotted in logarithmic scale as function of σ/σ_{max}, where $\sigma_{max} = \pi/2$. For $\epsilon = 0$ the temperature excess θ_0 assumes a finite value, given by Eq. 10–27.

b. The Cylinder

For an infinitely long cylinder of radius s, in which heat is developed according to Eq. 10–39, the differential equation that corresponds to Eq. 10–40 for the plate becomes

$$\frac{d^2\theta}{dr^2} + \frac{1}{r} \cdot \frac{d\theta}{dr} + n\theta = -m \tag{10–52}$$

where $\theta = t - t_s$ is the temperature excess at the radial distance r.

Equation 10–52 may be easily derived in the manner used to find Eq. 8–10, considering that $L \cdot 2\pi r \cdot dr \cdot q'''$ has to be added as a second term of q_i.

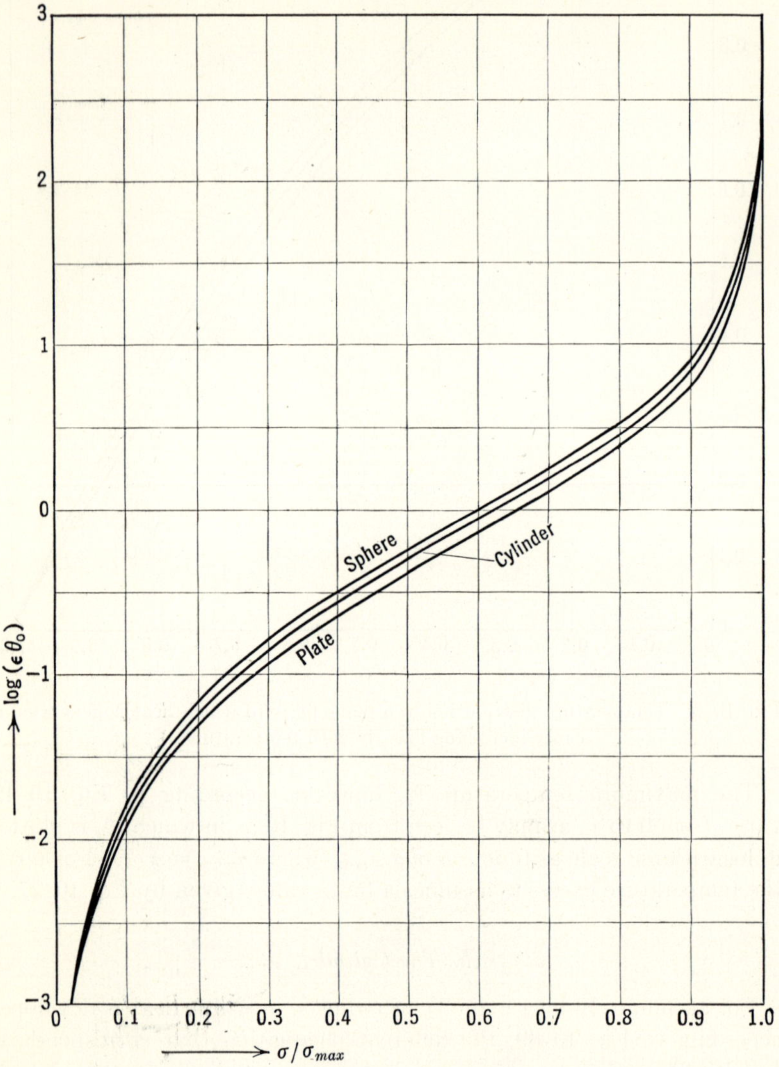

Fig. 10–7. Maximum temperature in bodies whose temperature distribution is given in Figs. 10–6, 8, and 9.

In analogy to Eq. 10–40a the general solution of Eq. 10–52 becomes ||

$$\theta = -\frac{m}{n} + MJ_0(r\sqrt{n}\,) + NY_0(r\sqrt{n}\,) \qquad (10\text{–}52a)$$

where J_0 and Y_0 are the Bessel functions of zero order and first and second kind, respectively, and
M and N are the constants of integration.

Considering the boundary conditions,

$$\frac{d\theta}{dr} = 0 \quad \text{when} \quad r = 0$$

and

$$\theta = 0 \quad \text{when} \quad r = s$$

N becomes zero, and the solution of Eq. 10–52 becomes

$$\theta = \frac{m}{n}\left[\frac{J_0(r\sqrt{n}\,)}{J_0(s\sqrt{n}\,)} - 1\right] = \frac{1}{\epsilon}\left[\frac{J_0(\xi\sigma)}{J_0(\sigma)} - 1\right] \qquad (10\text{–}53)$$

where $\xi = r/s$, and
$\sigma = s\sqrt{n}$.

Substituting $r = 0$,

$$\theta_0 = \frac{m}{n}\left[\frac{1}{J_0(s\sqrt{n}\,)} - 1\right] = \frac{1}{\epsilon}\left[\frac{1}{J_0(\sigma)} - 1\right] \qquad (10\text{–}54)$$

and from Eqs. 10–53 and 54

$$\frac{\theta}{\theta_0} = \frac{J_0(r\sqrt{n}\,) - J_0(s\sqrt{n}\,)}{1 - J_0(s\sqrt{n}\,)} = \frac{J_0(\xi\sigma) - J_0(\sigma)}{1 - J_0(\sigma)} \qquad (10\text{–}55)$$

Substituting θ from Eq. 10–53 in

$$\theta_m = \frac{1}{\pi s^2}\int_0^s \theta \cdot 2\pi r \cdot dr$$

yields the mean temperature

$$\theta_m = \frac{m}{n}\left[\frac{2 \cdot J_1(s\sqrt{n}\,)}{s\sqrt{n} \cdot J_0(s\sqrt{n}\,)} - 1\right] = \frac{1}{\epsilon}\left[\frac{2 \cdot J_1(\sigma)}{\sigma \cdot J_0(\sigma)} - 1\right] \qquad (10\text{–}56)$$

where J_1 is the Bessel function of first order and first kind.

|| A new variable $u = r\sqrt{n}$ may be introduced to bring Eq. 10–52 to the normal form of Bessel's differential equation.

From Eqs. 10–54 and 56

$$\phi = \frac{\theta_m}{\theta_0} = \frac{1}{s\sqrt{n}} \cdot \frac{2\cdot J_1(s\sqrt{n}) - s\sqrt{n}\cdot J_0(s\sqrt{n})}{1 - J_0(s\sqrt{n})}$$

$$= \frac{1}{\sigma} \cdot \frac{2\cdot J_1(\sigma) - \sigma\cdot J_0(\sigma)}{1 - J_0(\sigma)} \tag{10–57}$$

Again, θ is proportional to m, and θ/θ_0 is independent of m. Further again, by definition, $\theta > 0$. But for arbitrary values of $r < s$, Eq. 10–53 yields positive values of θ only if

$$J_0(r\sqrt{n}) > J_0(s\sqrt{n})$$

and

$$0 < s\sqrt{n} < 2.4048 \tag{10–58}$$

the latter value being the argument of the first zero point of the function J_0. This corresponds to the argument $\pi/2$ of the first zero point of the cosine function (see Eq. 10–46).

If

$$s\sqrt{n} = 2.4048 \tag{10–59}$$

then, according to Eq. 10–53, θ becomes infinitely great for every value of r. The rate of heat development at which this occurs can be found in the same way as for the plate, the result being

$$q_s''' = (2.4048)^2 \frac{k}{\epsilon s^2} = 5.783 \frac{k}{\epsilon s^2} \tag{10–60}$$

In analogy to Eqs. 10–50 and 51

for $\sigma = 0$,

$$\frac{\theta}{\theta_0} = 1 - \xi^2 \tag{10–61}$$

and for $\sigma = 2.4048$,

$$\frac{\theta}{\theta_0} = J_0(\xi\cdot 2.4048) \tag{10–62}$$

Figure 10–8 shows curves for these limits, the upper one being the same parabola as in Fig. 10–6. Intermediate curves for the parameters $\sigma = 1$ and $\sigma = 2$ are also drawn in this graph.

By substituting $\sigma = 0$ and $\sigma = 2.4048$ in Eq. 10–57, it is found that ϕ varies from 0.5 to 0.432 only when θ_0 increases from 0 to ∞. For $\epsilon = 0$, Eq. 10–27 again leads to a finite value of θ_0.

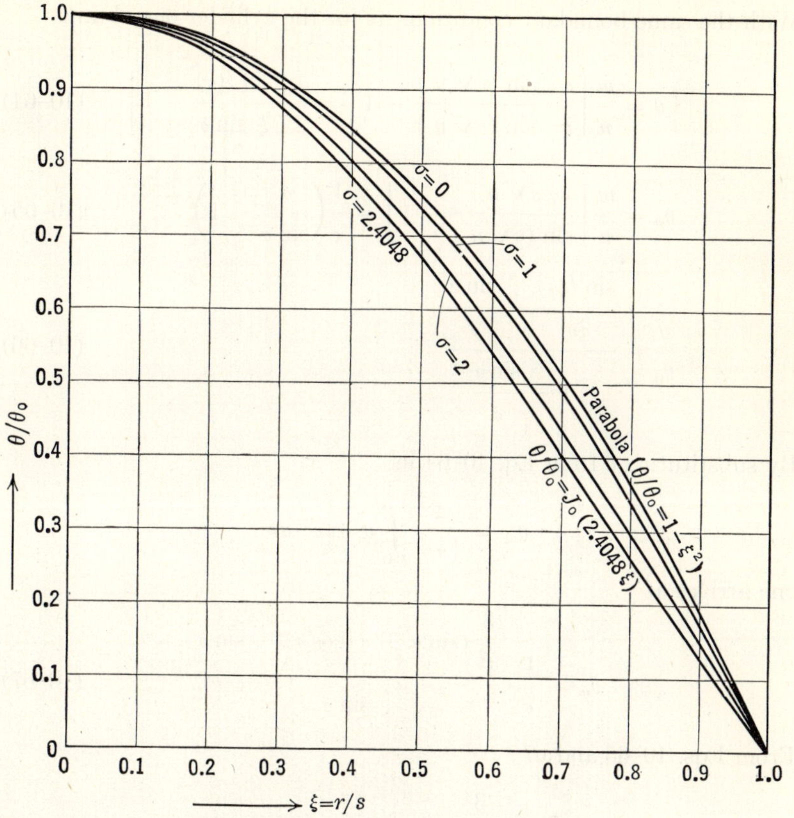

FIG. 10–8. Temperature distribution in a cylinder in which heat is generated at a rate increasing linearly with temperature.

The variation of $\epsilon\theta_0$ with σ/σ_{max} is shown in Fig. 10–7, σ_{max} being equal to 2.4048 in the case of the cylinder.

c. The Sphere

For a spherical coil, again assuming q'''/k according to Eq. 10–39, the differential equation is

$$\frac{d^2\theta}{dr^2} + \frac{2}{r}\frac{d\theta}{dr} + n\theta = -m \qquad (10\text{–}63)$$

Its general solution is

$$\theta = -\frac{m}{n} + \sqrt[4]{n}\left[M\frac{\sin(r\sqrt{n})}{r\sqrt{n}} + N\frac{\cos(r\sqrt{n})}{r\sqrt{n}}\right] \qquad (10\text{–}63a)$$

With the same boundary conditions as for the cylinder one obtains

$$\theta = \frac{m}{n}\left[\frac{s}{r}\cdot\frac{\sin(r\sqrt{n})}{\sin(s\sqrt{n})} - 1\right] = \frac{1}{\epsilon}\left[\frac{\sin(\xi\sigma)}{\xi\cdot\sin\sigma} - 1\right] \tag{10-64}$$

$$\theta_0 = \frac{m}{n}\left[\frac{s\sqrt{n}}{\sin(s\sqrt{n})} - 1\right] = \frac{1}{\epsilon}\left(\frac{\sigma}{\sin\sigma} - 1\right) \tag{10-65}$$

$$\frac{\theta}{\theta_0} = \frac{\dfrac{\sin(\xi\sigma)}{\xi\sigma} - \dfrac{\sin\sigma}{\sigma}}{1 - \dfrac{\sin\sigma}{\sigma}} \tag{10-66}$$

By substituting θ from Eq. 10–64 in

$$\theta_m = \frac{1}{\frac{4}{3}\pi s^3}\int_0^s \theta\cdot 4\pi r^2\cdot dr$$

one arrives at

$$\theta_m = \frac{1}{\epsilon}\cdot\frac{\dfrac{3}{\sigma^2}(\sin\sigma - \sigma\cdot\cos\sigma) - \sin\sigma}{\sin\sigma} \tag{10-67}$$

From Eqs. 10–65 and 67

$$\phi = \frac{\theta_m}{\theta_0} = \frac{\dfrac{3}{\sigma^2}(\sin\sigma - \sigma\cdot\cos\sigma) - \sin\sigma}{\sigma - \sin\sigma} \tag{10-68}$$

As for plate and cylinder, θ is proportional to m, and θ/θ_0 is independent of m.

According to Eq. 10–64, θ is positive only if

$$\frac{\sin(r\sqrt{n})}{r\sqrt{n}} > \frac{\sin(s\sqrt{n})}{s\sqrt{n}}$$

This is the case in the range

$$0 < s\sqrt{n} < \pi \tag{10-69}$$

For

$$s\sqrt{n} = \pi \tag{10-70}$$

which is the first zero point of the function $(\sin\sigma)/\sigma$, the temperature would become infinitely great for every point inside the sphere. This would occur at the following rate of heat development:

$$q_s''' = \pi^2 \frac{k}{\epsilon s^2} = 9.870 \frac{k}{\epsilon s^2} \tag{10–71}$$

For $\sigma = 0$, again,

$$\frac{\theta}{\theta_0} = 1 - \xi^2 \tag{10–61}$$

For $\sigma = \pi$, Eq. 10–66 leads to

$$\frac{\theta}{\theta_0} = \frac{\sin(\xi\pi)}{\xi\pi} \tag{10–72}$$

In Fig. 10–9 curves for the limiting values $\sigma = 0$, $\sigma = \pi$, and two intermediate curves are presented. In these limits ϕ changes from 0.4 to

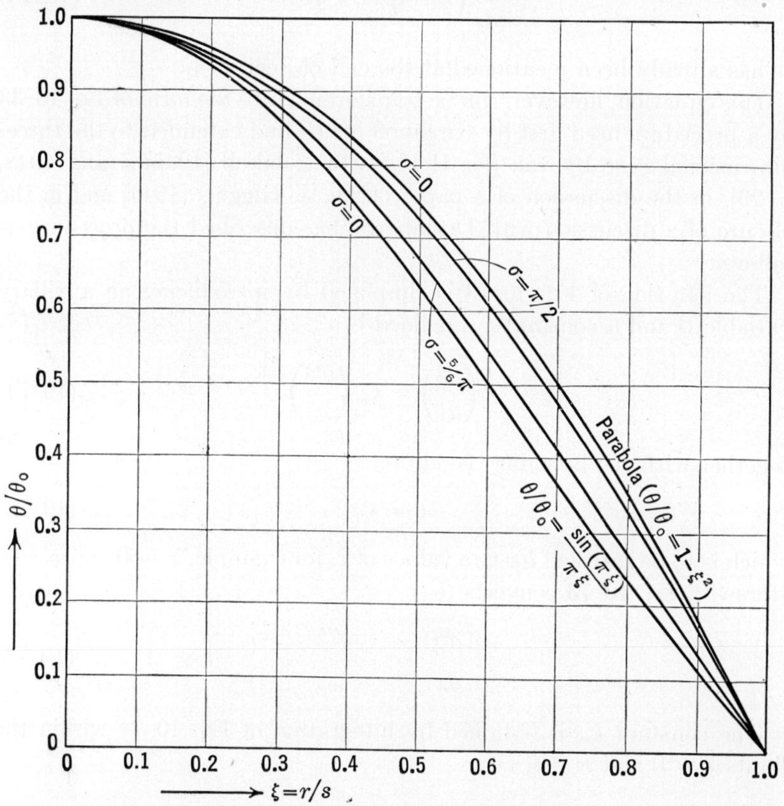

Fig. 10–9. Temperature distribution in a sphere in which heat is generated at a rate increasing linearly with temperature.

0.304, whereas under the assumption of uniform heating ϕ would be invariable and equal to 0.4. The maximum temperature again varies from 0 to ∞, except for $\epsilon = 0$, where again a finite value of θ_0 is found from Eq. 10–27. Figure 10–7 shows how little $\epsilon\theta_0$ for the sphere, drawn versus σ/σ_{max}, where $\sigma_{max} = \pi$, deviates from the corresponding curves for the plate and the cylinder.

d. Influence of the Variation of Thermal Conductivity with Temperature upon the Temperature Distribution

So far it has been assumed that the apparent thermal conductivity of an electrical coil can be considered constant. If the temperature varies appreciably over the cross section, Eq. 10–38 must be replaced by

$$\frac{d}{dx}\left(k\frac{d\theta}{dx}\right) = -q''' \tag{10–73}$$

as has already been mentioned at the end of Sect. 2–4.

This equation, however, can be transformed into the form of Eq. 10–38 by a procedure used first by Kirchhoff [1894] and extended to the three-dimensional case by van der Held [1936]. Jakob [1945(*a*) and 1948, p. 29], in the discussion of a paper of Th. J. Higgins [1944] and in the closure of a discussion with Dusinberre, has described the procedure as follows:

The left side of Eq. 10–73 is simplified by introducing an auxiliary variable Θ and a constant k_m, defined by

$$k\left(\frac{d\theta}{dx}\right) = k_m\left(\frac{d\Theta}{dx}\right) \tag{10–74}$$

together with the boundary condition

$$\theta = \Theta \tag{10–75}$$

which is to be satisfied for two values of x, for example, $x = 0$ and $x = s$. Herewith Eq. 10–73 converts to

$$\frac{d^2\Theta}{dx^2} = -\frac{q'''}{k_m} \tag{10–76}$$

The constant k_m is obtained by integration of Eq. 10–74 within the limits $x = 0$ and $x = s$, i.e.,

$$\int_{\theta=\theta_0}^{\theta=\theta_s} k\cdot d\theta = \int_{\Theta=\theta_0}^{\Theta=\theta_s} k_m\cdot d\Theta = k_m(\theta_s - \theta_0) \tag{10–77}$$

This is valid for any given relationship of k to θ, independent of whether the first integral can be solved analytically or graphically only. Since, in general, θ_0 is not known, but is asked for, it is recommended to use a simple trial-and-error method, namely, to find a first approximation of θ_0 from Eq. 10–42, assuming k to be constant, then to substitute this value of θ_0 in Eq. 10–77, obtain k_m, and repeat the procedure, if necessary.

Jakob has also shown that the same method can be applied for the cylinder and sphere, yielding, respectively:

$$\frac{d^2\Theta}{dr^2} + \frac{1}{r} \cdot \frac{d\Theta}{dr} = -\frac{q'''}{k_m} \tag{10–78}$$

and

$$\frac{d^2\Theta}{dr^2} + \frac{2}{r} \cdot \frac{d\Theta}{dr} = -\frac{q'''}{k_m} \tag{10–79}$$

Obviously, Kirchhoff's method can be employed quite generally. Equations 8–24, 27, and 29, for instance, are cases in which a very simple function for k is used in the left side of Eq. 10–77.

e. Adaption of the Results to Real Coils

Real electrical coils do not have the shape of infinitely wide plates, infinitely long cylinders, and spheres. The relations derived for these, however, come close to those of some real coils. A long, hollow-cylindric coil, for instance, whose inner radius is great compared to the radial thickness of the coil, behaves almost as an infinitely wide plane plate. If the length of such a coil is ten times the thickness, then, according to Table 10–3, ϕ is only about 6% smaller than for infinite length. Therefore the case of the infinitely wide plane plate has considerable practical importance. A hollow-cylindric coil of square cross section behaves almost like an infinitely long cylindric coil of equal cross section, again provided that the inner diameter of the hollow cylindric coil is so great that the curvature can be neglected. According to Table 10–3, ϕ for the square cross section is 5% smaller than for the circular cross section if heat is uniformly produced. Assuming the same difference for non-uniform heat development, the temperature in a hollow cylindric coil of square cross section can be found from the formulas for circular cross section, given in Article *b*. Similarly, the temperatures in hollow-cylindric coils of rectangular cross section can be fairly estimated for the case of non-uniform development of heat. The calculations for spheres show the lowest possible limit of ϕ and give approximations for cylinders of height equal to diameter and for cubes of the same volume.

For rectangular cross section (side lengths $2a$ and $2b$) Th. J. Higgins [1944] has explicitly solved the differential equation. The result is rather involved, compared to the approximation mentioned above. Only Higgins' final equation for the temperature t at any point x, y may be given here:

$$t - t_s = \frac{B}{C^2}(\cos Cx \cdot \sec Ca - 1) - \frac{2B}{a}\sum_{n=1}^{n=\infty}(-1)^{n-1}\frac{\cos M_n x \cdot \cosh K_n y}{K_n{}^2 M_n \cdot \cosh K_n b} \qquad (10\text{-}80)$$

where $B = q_c'''(1 + \epsilon t_s)/k$,

$C^2 = q_c'''\epsilon/k$,

$M_n = \pi(2n - 1)/2a$,

$K_n = +\sqrt{M_n{}^2 - C^2}$.

10-5. Determination of the Apparent Thermal Conductivity of Electrical Coils

For applying the theories dealt with in Sects. 10-3 and 4, the apparent or equivalent thermal conductivity k of the coil material is needed. Very little is known about this. The simplest way to find it will be by direct measurement, using a hollow-cylindric coil or a plane plate, built up from the particular insulated wire, as the sample in one of the standard methods for testing plates or hollow cylinders of heat-insulating material. (See Sects. 9-2 to 4.)

Another experimental method would be to measure θ_m of a cylindric coil, for instance, and determine k from Eq. 10-56 by trial and error.

A third way is to calculate k from the geometrical configuration and the thermal conductivities of the constituents of the coil. An exact theory might be based on the method of conformal mapping; however, it has not been developed as yet. For this reason, an approximative method is recommended. This will be explained with a particular hollow-cylindric coil as example. Temperature measurements made on this type of coil by Rogowski and Vieweg have been compared with the results of the theory [Jakob 1943]. The coil was of square cross section with 44 mm side length and was made from copper wire, 1 mm in diameter, insulated by silk soaked in lacquer. The wire was wound in 40 layers, each including 35.4 windings as an average, the total number of windings being 1416.

For the approximative calculation, the wire of the coil is imagined to be replaced by a conductor of rectangular cross section which has the same cross-sectional area as the wire and an insulation of uniform thickness and of the same cross-sectional area as the actual insulation, including the space between the insulated wires. The ratio of the side lengths of the rectangular cross section of the conductor is further chosen so that the equivalent coil has the same number of windings and the same number of layers as the original coil. Neither the surface of the equivalent coil nor the total length of the conductor and its electrical resistance are different from those of the actual coil. The cross section of some elements of the equivalent coil is represented in Fig. 10–10 in scale; the pertinent lengths are indicated in millimeters.

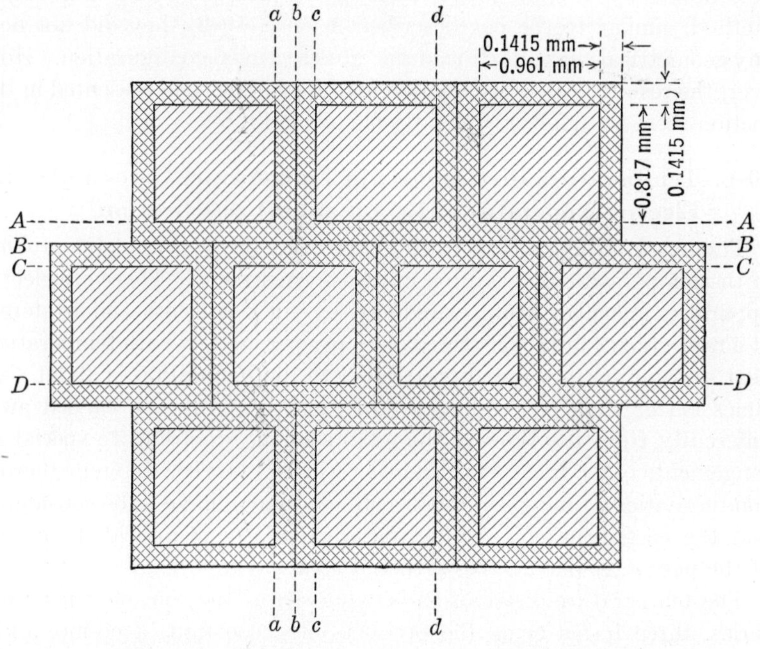

FIG. 10–10. Equivalent cross section of an electric coil.

Now it is assumed that for heat flow in vertical direction the horizontal lines *AA*, *BB*, *CC*, etc., and for heat flow in horizontal direction the vertical lines *aa*, *bb*, *cc*, etc., are isothermals. This allows the calculation of k for vertical and horizontal heat flow (k_v and k_h) if the thermal conductivities k_m and k_i of the metal and the insulation are given. It is necessary only to calculate the thermal conductances of the layers between *AA* and *BB*, between *BB* and *CC*, and between *CC* and *DD* for vertical heat flow; the sum of their reciprocal values (thermal resistances) gives the total thermal resistance of the system in vertical direction.

The procedure in horizontal direction is entirely analogous, using the planes *aa*, *bb*, *cc*, and *dd* (cf. Sects. 2–3 and 6–5 regarding Fig. 6–2).

In this way, assuming $k_m = 3.49$ watt cm^{-1} C^{-1} for copper and $k_i = 0.00143$ watt cm^{-1} C^{-1} for varnished silk, the values $k_v = 0.00556$ and $k_h = 0.00506$ watt cm^{-1} C^{-1} were found. Their difference of about 10% is due to the difference in winding in the two directions. For practical purposes it would be sufficient to consider the coil homogeneous in all directions and to use a mean thermal conductivity of 0.0053 watt cm^{-1} C^{-1}.

In a more exact way the conductivity in different directions has been determined by Richter [1924] and Moore [1927]. Using a graphical method, similar to the one described in Sect. 19–2, they did not need any geometrical simplification of the cross-sectional configuration. However, the results of their calculation and the simpler one presented in this section seem to differ only slightly [Jakob 1943].¶

10–6. Temperature Distribution in Transformer Cores and Other Electric Apparatus of Rectangular Cross Section

Thomas J. Higgins [1945] has also solved the very general problem of the temperature distribution in transformer cores. These are electric apparatus in which heat is produced by eddy currents and hysteresis at a rate which can be considered as increasing linearly with temperature; heat conduction in a direction parallel to the lamination is about sixty times better than perpendicular to it; and heat may be carried away differently from the sides of the cross section, according to special arrangements of oil ducts, etc. In other words, two very different thermal conductivities k_x and k_y, in the directions x and y, have to be considered, and the coefficient of heat transfer h may be a complicated function of the perimeter of the cross-sectional area.

The temperature distribution is expressed as the sum of four infinite series, three terms being the products of a constant, a cosine, and a hyperbolic cosine, the fourth being without the hyperbolic cosine term. The paper further contains a complete numerical example, the result of which is in good agreement with a transformer test taken from the literature. References to fourteen previous papers on the subject are also given in Higgins' paper.

10–7. Heat Conduction in Electric Semi- and Second-class Conductors at Non-uniform Development of Heat

In Sect. 10–4 electric coils have been dealt with, the electric resistance of which increases linearly with temperature. In the present section

¶ See Jakob's closure of the discussion of this paper.

bodies of the same simple form (plate, cylinder, sphere) will be treated which generate Joulean heat at a rate decreasing with rising temperature. That occurs with electric semi-conductors, as arc-lamp carbons, graphite electrodes, silicon carbide resistors, or with conductors of the second class (electrolytes, electrolytical metal salts). In such bodies the maximum temperature at $\xi = 0$ will obviously be smaller than in the previous cases.

In a way similar to that described in Sect. 10–4 for conductors with positive temperature coefficient ϵ, Jakob [1947, 1948] derived the corresponding equations for negative ϵ.

For infinitely wide plates, infinitely long cylinders, and spheres these equations have the same form as Eqs. 10–43, 55, and 66, respectively; however, the functions $\cos(\xi\sigma)$, $J_0(\xi\sigma)$, and $\dfrac{\sin(\xi\sigma)}{\xi\sigma}$ in these equations have to be replaced by $\cosh(\xi\sigma)$, $I_0(\xi\sigma)$, and $\dfrac{\sinh(\xi\sigma)}{\xi\sigma}$, respectively, for any value of ξ between 0 and 1, where I_0 is the symbol for the modified Bessel function of zero order and first kind.

In Figs. 10–11, 12, and 13, θ/θ_0 is plotted for these cases, including the infinite-temperature limits for $\epsilon > 0$, corresponding to Eqs. 10–51, 62, and 72, respectively, which have also been shown in Figs. 10–6, 8, and

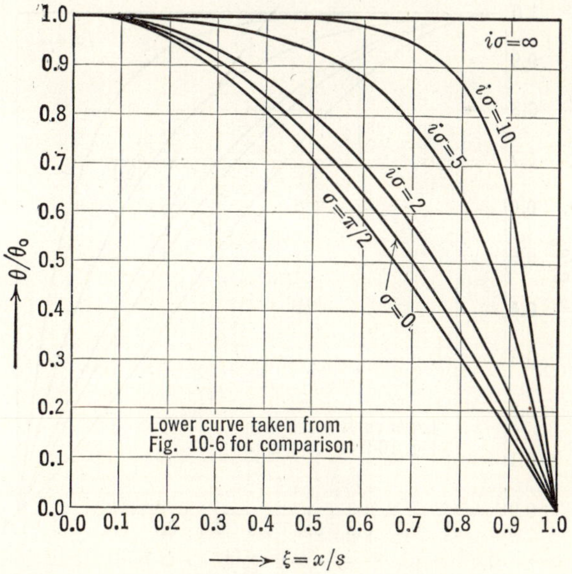

FIG. 10–11. Temperature distribution in a plane plate in which heat is generated at a rate decreasing linearly with increasing temperature.

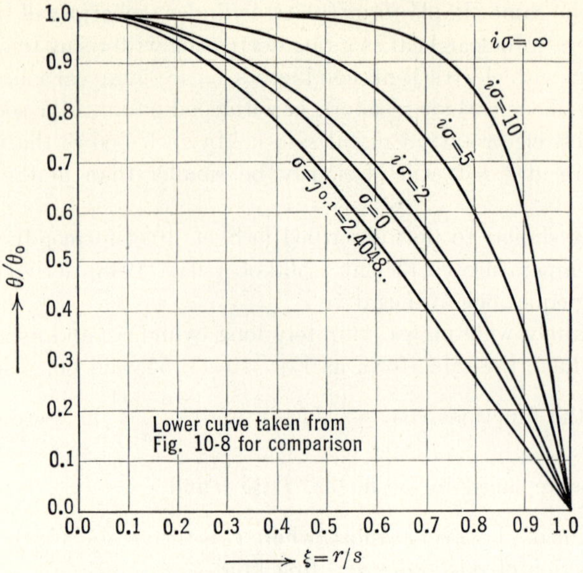

FIG. 10–12. Temperature distribution in a cylinder in which heat is generated at a rate decreasing linearly with increasing temperature.

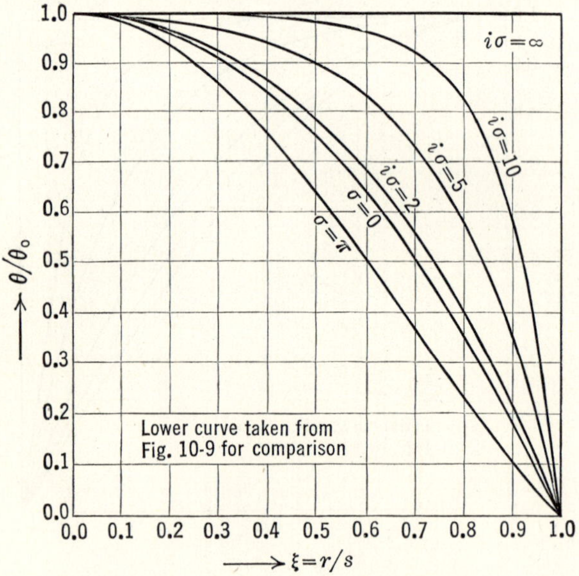

FIG. 10–13. Temperature distribution in a sphere in which heat is generated at a rate decreasing linearly with increasing temperature.

9. The parabola (Eq. 10–50), which is common to plate, cylinder, and sphere and occurs at $\epsilon = 0$, also is plotted in each of Figs. 10–11, 12, and 13.

Only at an advanced stage in the printing of this volume did the author became acquainted with a paper of Wilhelm, W. C. Johnson, and Acton [1943] on catalytic converters which had escaped his attention, probably because its subject seemed to be applications of heat transfer in the chemical industry, to be dealt with in Vol. II. In this paper Jakob's [1947 and 1948] derivations of the temperature distribution in plates and cylinders for $\epsilon < 0$ have been anticipated. Jakob's procedure and graphical representation of the results, however, seem to be simpler and more conspicuous, and only in his papers was the case of spherical arrangement treated. Referring to the cases with $\epsilon > 0$, the priority seems to be Jakob's [1943] because his paper was preprinted before its presentation to the American Society of Mechanical Engineers (December 3, 1942), whereas Wilhelm, Johnson, and Acton delivered their paper on December 29, 1942, in a symposium of the American Chemical Society. It is characteristic for the science of heat transfer that investigators studying such different devices as electric coils and catalytic converters have to use the same methods and arrive at identical results.

10-8. Exothermal and Endothermal Chemical Reactions

The heat development or absorption by chemical reactions is similar to that in electrical coils. In narrow temperature ranges the heat of reaction may be considered as changing linearly with the temperature. Then the equations of Sect. 10–4 or their modifications, described in Sect. 10–7, can be used, and θ_0 will be found.* However, steady-state conditions exist exceptionally only in batch reactions, and in steady-flow reactions the influence of convection in general prevails over that of conduction. Only in such cases as the steady movement of a reacting solid of crushed substance, for instance, in the coal bed on the travelling grid of a furnace, or through a pipe or channel with negligible axial heat conduction or convection, is heat transmitted mainly by conduction in directions perpendicular to the movement of the reacting body, and the temperatures in any given length section are not changing in time.

* For the heat absorption in an endothermic reaction at a rate which increases linearly with the temperature, both θ and θ_0 have negative values, as they will in the case of an endothermic reaction with negative temperature coefficient. See also the graphs in Sect. 10–7. A remarkable feature of Wilhelm, W. C. Johnson, and Acton's [1943] paper is the inclusion of heat transfer from catalysts to a gas in the heat-generation or absorption term without changing the form of the differential equation.

In wider temperature ranges, where q'''/k cannot be considered a linear function of θ, the empirical relation

$$\frac{q'''}{k} = me^{n\theta} \tag{10-81}$$

may be used for exothermal reactions.

For the infinitely extended plate Eqs. 10-38 and 81 yield

$$\frac{d^2\theta}{dx^2} = -me^{n\theta} \tag{10-82}$$

The solution of this equation has been published by Jakob [1933, p. 183].

For the endothermal reactions the minus sign in Eq. 10-38 would have to be replaced by a plus sign, and the integration performed for this case.

10-9. Heating with Electric High-frequency Power

Recently a method of heating has been used in various types of wood processing, such as drying and gluing, in which the heat energy is produced by electric high-frequency power [J. P. Taylor 1943]. Figure

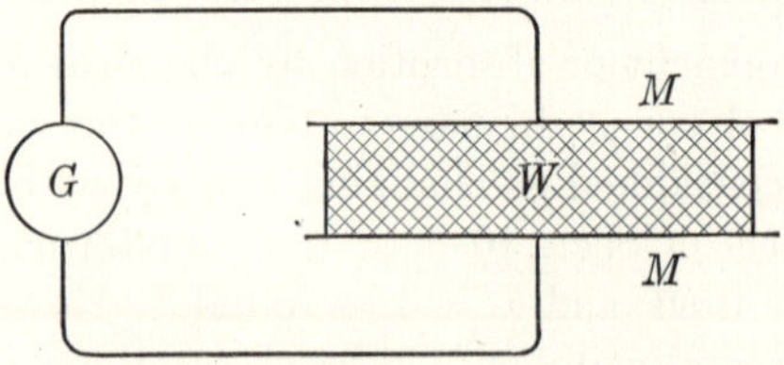

G = High frequency generator
M = Metal electrodes
W = Piece of wood

FIG. 10-14. Electric high-frequency heating of wood.

10-14 is a sketch of the arrangement in its simplest form, Fig. 10-15 gives an electric-circuit diagram, and Fig. 10-16 shows an electric vector diagram to explain the procedure.

The electric high-frequency generator, G (Fig. 10-14), is connected with the metal electrodes, M, which are attached to the piece of wood, W, to be heated. This can be represented schematically (Fig. 10-15) by a capacitance, C_e, paralleled by a resistance, R_e.

In the vector diagram (Fig. 10-16) it is seen that the total current i leads the voltage across the load by something less than 90 degrees.

The reactive current i_C does not contribute to the heating power; only the active current i_R does.

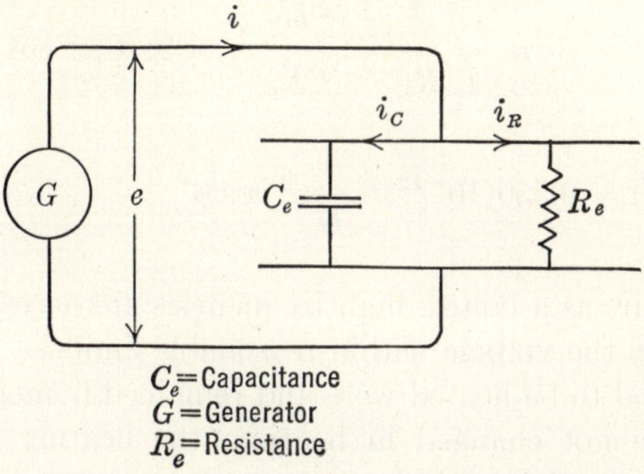

FIG. 10–15. Electric-current diagram of high-frequency heating.

According to Ohm's law $i_R = e/R_e$. Furthermore, $i_C = e/X_e$, where X_e is the capacitive reactance. According to the vector diagram, $\tan(90° - \phi) = i_R/i_C = X_e/R_e$. Since $(90° - \phi)$ is a very small angle, $\tan(90° - \phi) \approx \cos\phi$. Hence $R_e \approx X_e/\cos\phi$.

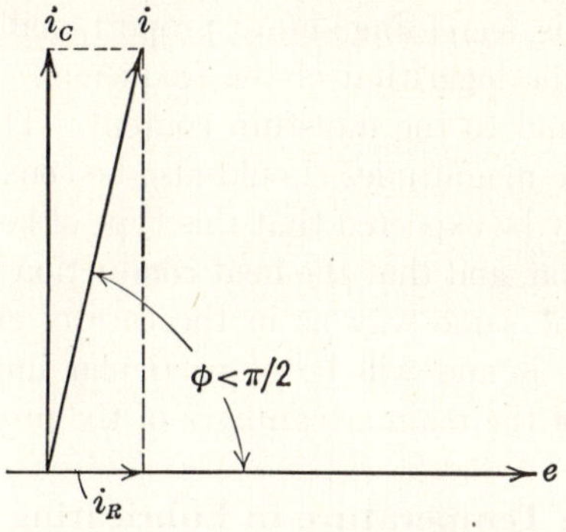

FIG. 10–16. Electric vector diagram of high-frequency heating.

The value of the power factor $\cos\phi$ can easily be measured by means of a radio-frequency bridge. The value of X_e can be calculated from

$$\frac{1}{X_e} = 2\pi f C_e$$

where f = the frequency in sec^{-1},
$C_e = 88.5(10^{-15})\kappa A/D$ in farads,
κ = the dielectric constant of the material,
A = the area of each electrode in square centimeters, and
D = the distance of the electrodes in centimeters.

Combining these relations, the total power in watts required is found to be

$$P = i_R^2 R_e = \frac{e^2}{R_e} = \frac{e^2 \cos \phi}{X_e} = 2\pi f C_e e^2 \cos \phi$$

$$= 0.550(10^{-12}) f\kappa \frac{A}{D} e^2 \cos \phi \tag{10–83}$$

Since (fe^2) occurs as a factor, high frequencies are necessary in practice in order to keep the voltage within reasonable limits.

If the material to be heated were and remained homogeneous, and its properties were not changed in heating, the heating power per unit volume of the material would be

$$q''' = \frac{P}{AD} \tag{10–84}$$

which apparently would be independent of the area A.

On the other hand, as is shown in the quoted paper, the power factor of samples of wood is decreasing almost proportionally with the moisture content and with the logarithm of the frequency. Likewise $(\kappa - 1)$ is virtually proportional to the moisture content. The influence of temperature upon these magnitudes should also be considered.

In general, it may be expected that this type of heating will find more and more application, and that the heat conduction in the material may be dealt with in the same way as in the case of electrical coils. The transient heat flow is and will be of particular importance, since fast heating up is one of the main advantages of the process.

10–10. Maximum Temperature in Lubricating Oil Films

In lubricating oil films heat energy is developed by the friction and conducted to the bearing or journal or to both of them. Since the film is thin compared with the radii of its surfaces, the curvature can be neglected. Nevertheless, the problem is rather complex because the rate of heat development depends on the shearing stress in the oil, and this depends on the dynamic viscosity, the velocity, and its gradient in radial direction; on the other hand, the viscosity is influenced by the temperature, which is varying across the film. Kinsbury [1933] used a graphical method to find the temperature distribution, whereas Muskat and Morgan [1943] made such simplifying assumptions that the differential equation could explicitly be solved.

The theory is based upon the following equations:

$$\mu = f(t) \tag{10-85}$$

$$S_s = \mu \frac{\partial u}{\partial y} \tag{10-86}$$

$$q'' = \frac{1}{J} S_s(u - u_0) = -k \frac{\partial t}{\partial y} \tag{10-87}$$

where μ = the dynamic viscosity of the oil at the temperature t,
k = the thermal conductivity of the oil at the temperature t,
u = the velocity of the oil at the distance y from the bearing surface,
u_0 = the velocity of the oil at the distance y_0, where the temperature has its maximum, t_0,
S_s = the shearing stress due to friction at y,
q'' = the rate of heat flow across the film per unit area at y, and
J = the mechanical equivalent of the thermal unit.

Equation 10–86 is identical with Eq. 3–21. Assuming k = constant, by differentiation and substitution,

$$k \frac{\partial^2 t}{\partial y^2} = -\frac{1}{J} \cdot \frac{\partial}{\partial y}\left(\mu u \frac{\partial u}{\partial y}\right) \tag{10-88}$$

The simplifying assumptions mentioned above are:

1. The velocity distribution across the film is linear, as it would be in a constant-viscosity film.
2. The actual variable viscosity is replaced by an effective mean value μ_m.

Herewith, the solution of Eq. 10–88 becomes

$$\theta = \eta(\theta_j + \Delta t_j - \eta \cdot \Delta t_j) \tag{10-89}$$

where $\eta = y/s$,
s = film thickness,
$\theta = t - t_b$,
$\theta_j = t_j - t_b$,
t_b = temperature of bearing surface,
t_j = temperature of journal surface,
$\Delta t_j = \mu_m u_j^2/2Jk$, and
u_j = surface velocity of journal.

Δt_j is a characteristic temperature difference; its meaning will be explained later on.

According to Eq. 10–89, the maximum temperature excess θ_0 and its position η_0 are

$$\theta_0 = \frac{(\Delta t_j + \theta_j)^2}{4 \cdot \Delta t_j} \tag{10-90}$$

and

$$\eta_0 = \frac{\Delta t_j + \theta_j}{2 \cdot \Delta t_j} \tag{10-91}$$

Hence a maximum of temperature inside the oil film occurs only if $\theta_j < \Delta t_j$.

From Eq. 10–89 by differentiation

$$\frac{d\theta}{d\eta} = \theta_j + \Delta t_j - 2\eta \cdot \Delta t_j \tag{10-92}$$

For $\eta = 0$

$$\left(\frac{d\theta}{d\eta}\right)_b = \theta_j + \Delta t_j \tag{10-93}$$

For $\eta = 1$

$$\left(\frac{d\theta}{d\eta}\right)_j = \theta_j - \Delta t_j \tag{10-94}$$

From these equations and Fig. 10–17 it is seen that for $\Delta t_j = 0$ the straight line a represents the temperature distribution across the film.

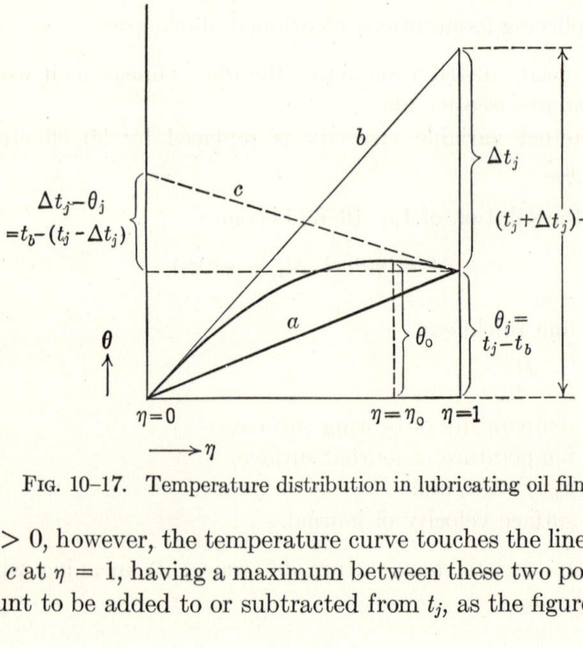

FIG. 10–17. Temperature distribution in lubricating oil film.

For $\Delta t_j > 0$, however, the temperature curve touches the line b at $\eta = 0$ and line c at $\eta = 1$, having a maximum between these two points. Δt_j is an amount to be added to or subtracted from t_j, as the figure shows.

Kinsbury performed the integration of Eq. 10–88 without the simplifications mentioned, for a case where $t_b = 100°F$, $t_j = 140°F$, $u_j = 2000$ in./sec, and found $t_0 = 166.6°F$.

With his value of $k = 323(10^{-6})$ cal sec^{-1} cm^{-1} C^{-1} and an assumed value $\mu_m = 10.3$ cp, as for 140°F, Muskat and Morgan's calculations led to $t_j = 177.2$ and $t_0 = 166.5°F$. The complete agreement of the two values of t_0 is partly accidental because it depends on the assumption of μ_m. Anyway, it is demonstrated that at high speeds the lubricant film may develop very appreciable temperature elevations, even above those of the metal bounding surfaces.

A simpler and more general theory has been developed by Hagg [1944, 1945], starting from Eqs. 10–86, 87, and

$$\mu = me^{-nt} \tag{10–95}$$

where m and n are constants.

Combining these equations,

$$(u_0 - u)\,du = \frac{k}{m}\,e^{nt}\,dt \tag{10–96}$$

and by integration

$$u_0 u - \frac{u^2}{2} = \frac{k}{mn}\,e^{nt} + C \tag{10–97}$$

Eliminating the constant of integration, C, by means of the boundary condition $u = 0$, $t = t_b$,

$$t = \frac{1}{n}\ln\left(e^{nt_b} + \frac{mnu_0}{k}\,u - \frac{mn}{2k}\,u^2\right) \tag{10–98}$$

The value of u_0 in this equation can be obtained by integrating Eq. 10–96 over the thickness of the oil film:

$$\int_0^{u_j}(u_0 - u)\,du = \int_{t_b}^{t_j}\frac{k}{m}\,e^{nt}\,dt$$

or

$$u_0 = \frac{u_j}{2} + \frac{k}{mnu_j}\,(e^{nt_j} - e^{nt_b}) = \frac{u_j}{2} + \frac{k}{nu_j}\left(\frac{1}{\mu_j} - \frac{1}{\mu_b}\right) \tag{10–99}$$

where the subscripts of μ refer to the temperatures with the same subscripts.

By substituting $u = u_0$ in Eq. 10–98,

$$t_0 = \frac{1}{n}\ln\left(e^{nt_b} + \frac{mn}{2k}\,u_0^{\,2}\right) \tag{10–100}$$

If the heat which is axially conducted away by the journal can be neglected, as often occurs under steady-state conditions, the journal can be considered as neither receiving nor giving up heat, so that $u_0 = u_j$ and $t_0 = t_j$. Herewith, from Eq. 10–99,

$$t_j - t_b = \frac{1}{n} \ln \left(1 + \frac{n\mu_b}{2k} u_j^2 \right) \tag{10–101}$$

If $\dfrac{n\mu_b}{2k} u_j^2 \ll 1$, then to a close approximation

$$t_j - t_b = \frac{\mu_b}{2k} u_j^2 \tag{10–102}$$

or

$$t_0 - t_b = \frac{\mu_b}{2k} u_0^2 \tag{10–103}$$

Hagg solved the system of Eqs. 10–86, 87, and 95 also for S_s. Here only one interesting result of the theory about the shearing stress may be mentioned. If a minimum viscosity, μ_0, were fixed, i.e., a maximum temperature t_0 were retained, for instance, by artificial cooling of bearing and journal, then S_s would increase at an increased rate with the velocity u_j and would eventually become infinitely great at a limiting finite value of u_j. Doubtless the assumptions on which the theory is based would break down before this point; however, this case reminds us of the infinite temperature approached in electric coils at finite current (see Sect. 10–4).

Hagg also checked the main features of his theory by measurements of shearing stress and limiting temperatures, t_b and t_j, and by numerical comparison with an example used by Kinsbury.

CHAPTER 11

Heat Conduction in Steady State, Including Heat Dissipation to the Environment

11–1. Restriction to Cases with Constant Surface Coefficient of Heat Transfer

The equations of heat conduction dealt with so far do not include heat losses or gains at the surface of the conducting body. When such a secondary heat flow is to be considered, the surface coefficient of heat transfer h enters the basic equations. An analysis of this coefficient will be presented in the chapters on heat convection. However, there are important cases where the interest is focused on heat conduction, and the surface heat transfer can be simply expressed by Eq. 1–2, with h as a constant factor of given magnitude.

This, for instance, is the procedure in calculating the heat flow in a metal rod or fin protruding from a surface into air at rest or into another fluid at a temperature different from the surface temperature. If, however, forced convection occurs, it must be examined whether h can be assumed as constant.

The cases mentioned of prevailing interest in conduction will be dealt with in the present chapter and in Chapter 12.

11–2. General Equation of the Temperature Distribution in Axial Direction in a Rod Placed in a Medium at Different Temperature

Ingenhouss [1789] demonstrated the differences in the thermal conductivity of solid substances in the following way. He fixed horizontal rods of different metals, all of the same thickness, in the wall of a metal case, so that they protruded by the same length into room air. Each rod was covered with a thin layer of wax. When the vessel was filled with hot water or oil, the wax melted over a greater length on a rod of better conducting material than on a not so well conducting rod. Thus, an order of the conductivities of different materials was found.

Complying with a suggestion made by J. B. Biot in 1816, Despretz [1822, 1828, 1828(*a*)] gave the following theory of the experiment. Referring to Fig. 11-1, let H be a heat source and R a rod of constant cross-sectional area. The temperature differences that occur in the cross

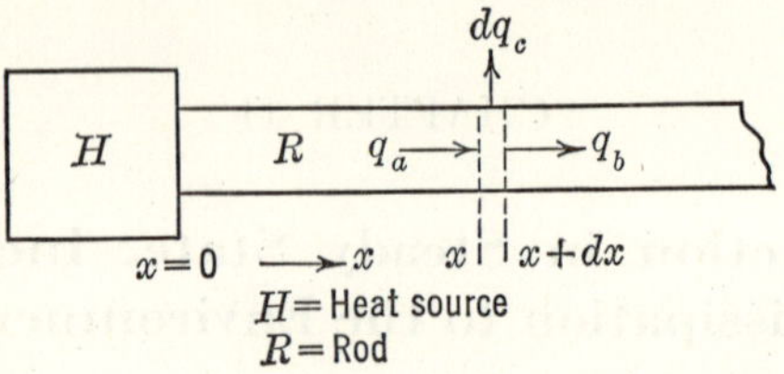

FIG. 11-1. Rod protruding from a heat source.

section are assumed to be negligible compared with those in the axial direction x. Let further:

t_0 = the temperature of the rod at $x = 0$,
t = the temperature of the rod at $x = x$,
t_e = the temperature of the environment of the rod,
$\theta = t - t_e$ = the temperature at $x = x$, counted from t_e as zero level (temperature excess),
A = the area of the cross section of the rod,
C = the circumference (perimeter) of that cross section,
k = the thermal conductivity of the material of the rod, and
h = the surface coefficient of heat transfer between the rod and its environment.

Considering an element of the rod between x and $x + dx$. the rate of incoming heat will be

$$q_a = -kA \frac{dt}{dx} \tag{11-1}$$

The rates of outgoing heat are

$$q_b = -kA \left(\frac{dt}{dx} + \frac{d^2t}{dx^2} dx\right) \tag{11-2}$$

and

$$dq_c = hC(t - t_e)\,dx \tag{11-3}$$

Since in the steady state of heat flow

$$q_a = q_b + dq_c$$

it follows that

$$kA \frac{d^2t}{dx^2} = hC(t - t_e)$$

or

$$kA \frac{d^2\theta}{dx^2} = hC\theta \tag{11-4}$$

Introducing the symbol

$$m = \sqrt{\frac{hC}{kA}} \tag{11-5}$$

the equation assumes the form

$$\frac{d^2\theta}{dx^2} = m^2\theta \tag{11-6}$$

If m is constant, the general solution of Eq. 11-6 becomes

$$\theta = Me^{-mx} + Ne^{mx} \tag{11-7}$$

where M and N are the constants of integration which, as usual, must be determined from the boundary conditions for each special case.

Equation 11-7 may serve different purposes. First, the temperature distribution can be determined; that is, θ dependent on x. Second, substituting this in Eq. 11-3 and integrating, the heat loss q_c along the rod will be found if h is given. Third, the thermal conductivity k can be found if h either is known or can be eliminated from different tests. A simple method for the determination of the ratio of the thermal conductivities of different materials is based upon such elimination. Finally, for rods of any sort of cross section, h can be determined if k is known. These applications will be shown in the following sections.

11-3. Comparison of Thermal Conductivities Based upon the Axial Temperature Distribution in Very Long Rods

The boundary conditions are:

$$t = t_0 \quad \text{or} \quad \theta = \theta_0 \quad \text{at} \quad x = 0$$

and

$$t = t_e \quad \text{or} \quad \theta = 0 \quad \text{at} \quad x = \infty$$

Hence, from Eq. 11-7,

$$M + N = \theta_0 \qquad N = 0 \qquad M = \theta_0$$

and

$$\theta = \theta_0 e^{-mx} \tag{11-8}$$

Let k_{I} and k_{II} be the thermal conductivities of two very long rods, having the same cross section, and h_{I} and h_{II} the respective surface coefficients of heat transfer. The rods protrude from the same heat bath H so that for both

$$\theta = \theta_0 \quad \text{at} \quad x = 0$$

Measuring the distances x_I and x_{II} at which

$$\theta_I = \theta_{II}$$

for instance, by Ingenhouss' wax-melting method, Eq. 11–8 yields the following relationship:

$$\theta_0 e^{-m_I x_I} = \theta_0 e^{-m_{II} x_{II}}$$

or

$$m_I x_I = m_{II} x_{II}$$

where the subscripts I and II refer to the two rods.

Herefrom, using Eq. 11–5,

$$\frac{k_I}{k_{II}} = \frac{h_I}{h_{II}} \left(\frac{x_I}{x_{II}}\right)^2 \tag{11–9}$$

h_I and h_{II} become almost equal if the surfaces of the two rods are made equal, for instance, by covering them with wax or, better, by silver- or chromium-plating and polishing them; the latter treatment reduces the radiant part of heat transfer (see Sect. 7-3), which depends on the temperature, so much that h can be assumed to be practically independent of the temperature. Then

$$h_I = h_{II}$$

and Eq. 11–9 becomes

$$\frac{k_I}{k_{II}} = \left(\frac{x_I}{x_{II}}\right)^2 \tag{11–10}$$

Thus the experiment of Ingenhouss is turned into a quantitative method of comparing thermal conductivites.

A more exact comparison than by measuring the distances x_I and x_{II} at which the temperatures are equal is obtained by measuring the temperatures t_I and t_{II} at the same distance x. Then, from Eq. 11–8,

$$\theta_I = \theta_0 e^{-m_I x}$$

and

$$\theta_{II} = \theta_0 e^{-m_{II} x}$$

Hence, using again Eq. 11–5,

$$\frac{k_I}{k_{II}} = \frac{h_I}{h_{II}} \left(\frac{\ln \theta_{II}/\theta_0}{\ln \theta_I/\theta_0}\right)^2 = \frac{h_I}{h_{II}} \left(\frac{\log \theta_{II}/\theta_0}{\log \theta_I/\theta_0}\right)^2 \tag{11–11}$$

and if

$$h_I = h_{II}$$

$$\frac{k_I}{k_{II}} = \left(\frac{\log \theta_{II}/\theta_0}{\log \theta_I/\theta_0}\right)^2 \tag{11–12}$$

These equations are very simple indeed, but their application is restricted because the results become inaccurate if the rods are not very long.

11–4. Comparison of Thermal Conductivities Based upon the Axial Temperature Distribution in Not Very Long Rods

Since Eq. 11–8 yields no accurate values of k with rods of reasonable length, one has to go back to Eq. 11–7. The best procedure is to take the temperatures at three places, designated by the subscripts 0, 1, and 2 with the same distance between 0 and 1 as between 1 and 2.

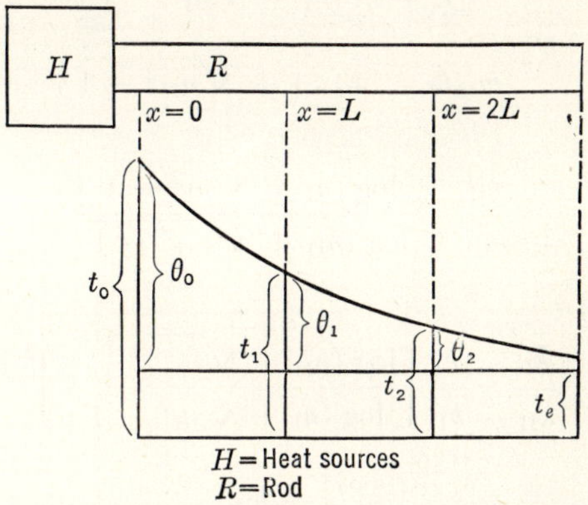

FIG. 11–2. Temperature distribution in a rod protruding from a heat source.

According to Fig. 11–2:

$$t = t_0 \quad \text{or} \quad \theta = \theta_0 \quad \text{at} \quad x = 0$$

$$t = t_1 \quad \text{or} \quad \theta = \theta_1 \quad \text{at} \quad x = L$$

$$t = t_2 \quad \text{or} \quad \theta = \theta_2 \quad \text{at} \quad x = 2L$$

Then, from Eq. 11–7,

$$\theta_0 + \theta_2 = M + N + Me^{-m2L} + Ne^{m2L}$$

$$\theta_1 = Me^{-mL} + Ne^{mL}$$

By division

$$\frac{\theta_0 + \theta_2}{\theta_1} = e^{-mL} + e^{mL} \tag{11–13}$$

Introducing

$$n = \frac{\theta_0 + \theta_2}{2\theta_1} \tag{11–14}$$

then, from Eq. 11–13,

$$e^{mL} = n \overset{+}{(-)} \sqrt{n^2 - 1}$$

or

$$mL = \ln (n + \sqrt{n^2 - 1}) \tag{11–15}$$

Only the plus sign is used because always $e^{mL} > 1$, and $e^{-mL} < 1$, and therefore, from Eqs. 11–13 and 14, $e^{mL} > n$.

Considering again two rods, as described in Sect. 11–3, Eq. 11–15 becomes

$$m_{\text{I}}L = \ln (n_{\text{I}} + \sqrt{n_{\text{I}}^2 - 1})$$

$$m_{\text{II}}L = \ln (n_{\text{II}} + \sqrt{n_{\text{II}}^2 - 1})$$

Hence

$$\frac{m_{\text{I}}}{m_{\text{II}}} = \frac{\log (n_{\text{I}} + \sqrt{n_{\text{I}}^2 - 1})}{\log (n_{\text{II}} + \sqrt{n_{\text{II}}^2 - 1})}$$

Using Eq. 11–5,

$$\frac{k_{\text{I}}}{k_{\text{II}}} = \frac{h_{\text{I}}}{h_{\text{II}}} \left[\frac{\log (n_{\text{II}} + \sqrt{n_{\text{II}}^2 - 1})}{\log (n_{\text{I}} + \sqrt{n_{\text{I}}^2 - 1})}\right]^2 \tag{11–16}$$

If again

$$h_{\text{I}} = h_{\text{II}}$$

then

$$\frac{k_{\text{I}}}{k_{\text{II}}} = \left[\frac{\log (n_{\text{II}} + \sqrt{n_{\text{II}}^2 - 1})}{\log (n_{\text{I}} + \sqrt{n_{\text{I}}^2 - 1})}\right]^2 \tag{11–17}$$

Since n_{I} and n_{II}, by definition, depend only on the measured temperatures t_e, $t_{0,\text{I}}$, $t_{1,\text{I}}$, $t_{2,\text{I}}$, $t_{0,\text{II}}$, $t_{1,\text{II}}$, and $t_{2,\text{II}}$, the ratio $k_{\text{I}}/k_{\text{II}}$ is found from these temperatures by means of Eq. 11–17. This, of course, comes from eliminating the constants of integration M_{I}, N_{I}, M_{II}, and N_{II} by the measurement of the four temperature differences $t_{0,\text{I}} - t_{1,\text{I}}$, $t_{1,\text{I}} - t_{2,\text{I}}$, $t_{0,\text{II}} - t_{1,\text{II}}$, and $t_{1,\text{II}} - t_{2,\text{II}}$.

The temperatures $t_{0,\text{I}}$ and $t_{0,\text{II}}$ may be different; if thermocouples are placed at three points of either rod, it does not matter if the two close to the heating source do not show exactly the same temperature.

In Fig. 11–2 a rod is represented whose zero point $x = 0$ is distant from the surface of the vessel H and whose right end is at $x > 2L$. The method is simplified if the right end is at $x = 2L$ and is held at the same temperature t_0 as the left end. This can be done by two heat sources, H_0 and H_2 (see Fig. 11–3). In this case Eq. 11–7 has the following boundary conditions:

$$t = t_0 \qquad \text{or} \quad \theta = \theta_0 \quad \text{at} \quad x = 0$$

$$t = t_2 = t_0 \quad \text{or} \quad \theta = \theta_0 \quad \text{at} \quad x = 2L$$

Instead of the latter condition the following could be used as well:

$$\frac{dt}{dx} = 0 \quad \text{or} \quad \frac{d\theta}{dx} = 0 \quad \text{at} \quad x = L$$

This is true for reasons of symmetry, as will easily be understood.

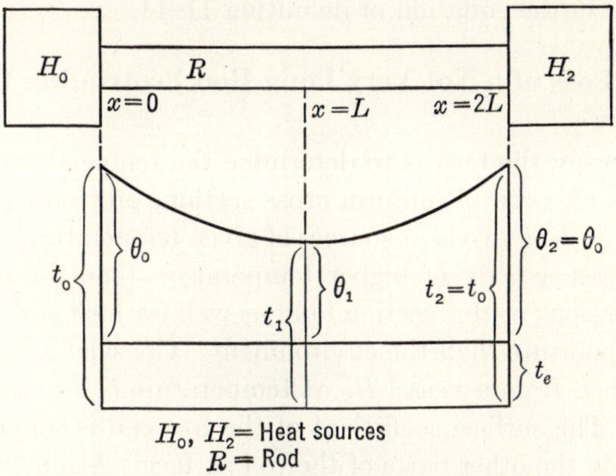

FIG. 11–3. Temperature distribution in a rod connecting two heat sources.

By substituting the boundary conditions and simplifying the resulting expressions,

$$M = \theta_0 \frac{e^{m2L}}{1 + e^{m2L}}, \qquad N = \theta_0 \frac{1}{1 + e^{m2L}}$$

and

$$\theta = \frac{\theta_0}{1 + e^{m2L}} [e^{m(2L-x)} + e^{mx}] \tag{11–18}$$

For $x = L$, by substitution,

$$\theta_1 = \theta_0 \frac{2}{e^{-mL} + e^{mL}} = \frac{\theta_0}{\cosh (mL)} \tag{11–19}$$

Equations 11–14 and 17 hold also for this arrangement, but become simpler, since $n = \theta_0/\theta_1$.

Thus, if the two heaters H_0 and H_2 are kept at the same temperature t_0, it is necessary only to measure this temperature and $t_{1,\mathrm{I}}$ and $t_{1,\mathrm{II}}$, in order to compare the thermal conductivities of two materials I and II.

This method has the further advantage that for the same length of the rods the temperature differences occurring on either rod are smaller than in the system of Fig. 11–2. For this reason the assumptions k = constant and h = constant are more nearly correct.

The latter, in particular, is somewhat questionable because the two rods will not be at exactly the same conditions, even if their surfaces are made exactly equal. The reason is that h depends upon t and θ to some extent. Jakob and Erk [1926] have shown how to correct for this temperature influence by means of a method of finite differences. For a known ratio $k_{\text{I}}/k_{\text{II}}$ the ratio $h_{\text{I}}/h_{\text{II}}$ can be calculated from Eq. 11–16 with fair approximation if n_{I} and n_{II} are found by measuring the temperatures which appear in the equation of definition 11–14.

11–5. Heat Loss of a Not Very Long Rod Protruding from a Heat Source

More generally the task is to determine the temperature distribution and heat loss of a rod of uniform cross section, protruding from a heat source or connecting two heat sources of given temperature. Heat source here means just a body of higher temperature than the environment, and all derivations of this section hold as well for heat sinks, i.e., bodies of lower temperature than the environment. Consider a rod protruding by the length L from a vessel H_0 at temperature t_0 into a fluid of temperature t_e. The surface coefficient of the free end is supposed to be h, the same as on the other parts of the rod surface. Again $\theta = t - t_e$ will be used.

The temperature distribution will be found, starting from Eq. 11–7 with the boundary conditions

$$\theta = \theta_0 \quad \text{at} \quad x = 0 \tag{11–20}$$

and

$$q_1 = hA\theta_1 = -kA\left(\frac{d\theta}{dx}\right)_1 \tag{11–21}$$

where the subscript 1 refers to $x = L$, as subscript 0 refers to $x = 0$.

From Eqs. 11–7 and 20

$$N = \theta_0 - M \tag{11–22}$$

From Eqs. 11–7 and 21

$$h(Ne^{mL} + Me^{-mL}) = -km(Ne^{mL} - Me^{-mL})$$

Substituting N from Eq. 11–22,

$$M = \theta_0 \frac{1}{1 + \dfrac{km - h}{km + h}\, e^{-m2L}} \tag{11–23}$$

Substituting this in Eq. 11–22:

$$N = \theta_0 \frac{1}{1 + \dfrac{km + h}{km - h} e^{m2L}} \tag{11–24}$$

Introducing a new constant, p, by the definition

$$p = \frac{km - h}{km + h} \tag{11–25}$$

and substituting M and N from Eqs. 11–23 and 24 in Eq. 11–7,

$$\theta = \frac{\theta_0}{1 + pe^{-m2L}} (e^{-mx} + pe^{-m2L} e^{mx}) \tag{11–26}$$

This is the equation of temperature distribution.

There are two ways to determine the total heat loss. Either it may be considered that the total heat loss must be covered from the vessel H_0, or the heat losses along the rod and at the free end may be calculated separately and added.

In the first method Eq. 11–1 is used which, applied to $x = 0$, becomes

$$q_0 = -kA \left(\frac{d\theta}{dx}\right)_0$$

From Eq. 11–7

$$\frac{d\theta}{dx} = m(-Me^{-mx} + Ne^{mx}) \tag{11–27}$$

and for $x = 0$

$$\left(\frac{d\theta}{dx}\right)_0 = m(N - M) \tag{11–28}$$

so that

$$q_0 = kmA(M - N) = \sqrt{khAC}(M - N) \tag{11–29}$$

The peculiar term $\sqrt{khAC}$ previously appeared in Eq. 9–15, which was the solution of a problem closely related to the present one.

By substitution of M and N from Eqs. 11–23 and 24 in Eq. 11–29 and some transformation

$$q_0 = \sqrt{khAC} \frac{1 - pe^{-m2L}}{1 + pe^{-m2L}} \theta_0 \tag{11–30}$$

Since $-1 \leqq p \leqq 1$ and $0 \leqq e^{-m2L} \leqq 1$, the fraction in Eq. 11–30 can take any value between 0 and ∞. It will be smaller than 1 when $h < km$, which generally is the case with gases, but may not be with liquids.

The second method starts from Eq. 11–3 (integrated in the limits $x = 0$ and $x = L$) and Eq. 11–1 (applied for $x = L$).

By addition

$$q_0 = \int_0^L hC\theta \cdot dx - kA\left(\frac{d\theta}{dx}\right)_1 \tag{11-31}$$

By substituting θ from Eq. 11–7 and M and N from Eqs. 11–23 and 24, after some transformations, Eq. 11–30 is again obtained, as the reader may check.

A more complex case, concerning an application to thermometry, has been dealt with by Jakob [1939]. If the wires of a thermocouple, which is placed in a well,* are led to the outside, heat is given up to the surrounding air; therefore the inner junction of the thermocouple is cooled and too low a temperature is indicated. If the wires, before being led into the free air, pass first through a metal tube which is in good connection with the well, their heat loss is considerably reduced, and the temperature indication becomes more nearly accurate. Similarly, as in the case dealt with in Sect. 9–4, equations had to be set up for different sections of the measuring device. The result contains two terms p, one for the wires and the other for the guard tube. The general solution is rather clumsy, but the numerical calculation becomes simple.

In the same paper it is shown that a practical suggestion, made by Harper and Brown [1922], in general leads to a result only slightly different from the more nearly correct solution given in Eq. 11–30. These investigators recommend using a corrected length $L_c = L + y_0$ for a fin of rectangular cross section and thickness $2y_0$ and assuming that the heat which actually leaves at the tip of the fin would leave at the added length section and none at the tip ($x = L_c$). Generalizing this suggestion for a rod of cross-sectional area A and circumference C, a corrected length $L_c = L + A/C$ may be used. In this way the problem is reduced to that of a rod with perfectly insulated end surface, the solution of which will be given in Sect. 11–6.

In Jakob's paper just mentioned it is proved that the approximate calculation differs from the exact calculation in the term p only, which is replaced by $e^{-2mA/C}$ or $e^{-2h/km}$ or $e^{2(p-1)/(p+1)}$. For $p = 1$ (that is, $h/km = 0$) both calculations are identical. For $p = 0$ (that is, $h/km = 1$) Eq. 11–30 leads to

$$q_0 = \sqrt{khAC}\theta_0$$

* Langberg [1845] seems to have been the first who inserted thermocouples in holes of rods to measure thermal conductivities.

If, further, $e^{-m2L} = 1$, the approximate calculation yields

$$q_0 = \sqrt{khAC}\,\frac{1 - e^{-2}}{1 + e^{-2}}\,\theta_0$$

that is, a value 24% too small. However, for gases, generally, $h < km$ or $p > 0$, as mentioned in the discussion of Eq. 11–30.

If $p = \frac{1}{2}$ (that is, $h = km/3$), the approximate value is only 3% short of the exact one, and since $e^{-m2L} < 1$, the difference is still smaller. Hence, if $h < km/3$, Harper and Brown's approximation may safely be used.

11–6. Heat Loss of a Rod Connecting Two Heat Sources

Only the case of two heat sources of the same temperature t_0 will be dealt with. Again two methods of derivation are possible, as in Sect. 11–5.

Referring to Fig. 11–3, the temperature distribution is given by Eq. 11–18. For reasons of symmetry

$$q_0 = -q_2$$

the subscript 2 referring to $x = 2L$. Hence, using Eq. 11–3, the total heat loss of the rod becomes

$$2q_0 = hC\int_{x=0}^{x=2L} \theta \cdot dx \tag{11–32}$$

By substituting θ from Eq. 11–18 and performing the integration,

$$2q_0 = 2\sqrt{khAC}\,\frac{e^{mL} - e^{-mL}}{e^{mL} + e^{-mL}}\,\theta_0 = 2\sqrt{khAC}[\tanh{(mL)}]\theta_0 \tag{11–33}$$

As in Sect. 11–5, Eq. 11–1, written for $x = 0$, instead of Eq. 11–3, could have been used for obtaining the result.

Since $d\theta/dx = 0$ at $x = L$, no heat is flowing in the axial direction at this point. Hence Eqs. 11–18 and 19 hold also for a rod of length L, protruding from a source of heat, and being perfectly insulated against heat losses at the free end ($x = L$). The heat loss of such a rod is q_0 and may be calculated from Eq. 11–33.

11–7. General Equations for the Heat Dissipation from Plane Surfaces by Fins

A fin of the kind represented in Fig. 11–4 can be considered a rod of constant cross-sectional area $A = 2y_0Z$ and length L. If, as in the former sections, the thermal conductivity k of the material and the coefficient of heat transfer h can be assumed to be constant, all equations derived for a rod hold also for the fin.

If the thickness of the fin varies with x or the temperature varies along z, equations of a more general character have to be considered. This has been done in a pioneer paper of Harper and Brown [1922] and a very illuminating paper of E. Schmidt [1926]. In both investigations fins of rectangular and wedge-like profile are dealt with; Schmidt has

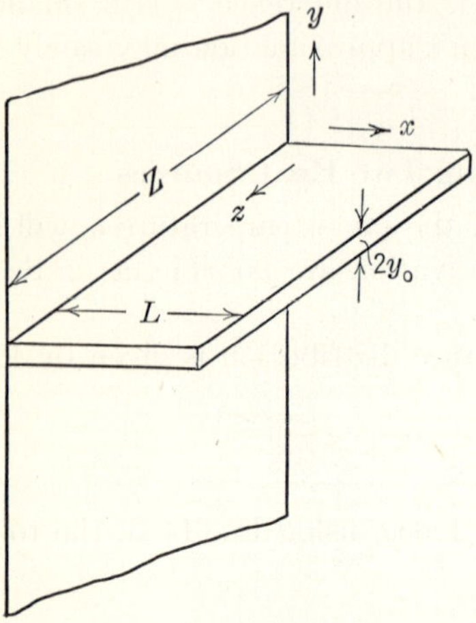

FIG. 11–4. Straight fin.

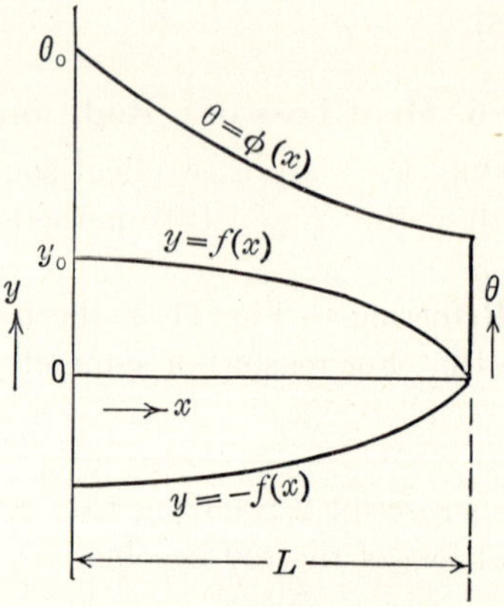

FIG. 11–5. Arbitrary profile of straight fin with temperature distribution.

further analyzed a special case of wedge shape, the triangular cross section, and the case of a fin with constant density of heat flow.

Following his procedure, the case mentioned last will be dealt with first; then the case of constant thickness, $2y_0$, and finally the case of varying thickness, $2y = 2\dfrac{L - x}{L}\, y_0$, will be treated, for both longitudinal and circumferential fins. Schmidt's findings will be completed by some special results from Harper and Brown's paper. In particular, the notion of effectiveness of fins, introduced by these investigators, will be explained and used. In general it will be assumed that the heat flow in the z-direction is negligible.

In Fig. 11–5 a fin profile is represented. It is confined by the symmetrical curves $y = f(x)$ and $y = -f(x)$. The thickness $2y$ is assumed to be small compared to the fin width L. The temperature excess θ is defined as in Sect. 11–2, and the temperature distribution in the direction x may be expressed by the function $\theta = \phi(x)$. The heat flow in direction x, crossing the area $y \cdot z = y$, which is half the cross-sectional area of the fin for unit length $z = 1$, may be denoted by q' at $x = x$, and by q_0' at $x = 0$. Then

$$q' = -ky \frac{d\theta}{dx}$$

and, approximately,†

$$-dq' = h\theta \cdot dx$$

According to these two equations, the differential equation of the temperature distribution in fins of arbitrary profile shape becomes

$$y \frac{d^2\theta}{dx^2} + \frac{dy}{dx} \cdot \frac{d\theta}{dx} - \frac{h}{k}\theta = 0 \tag{11-34}$$

By this equation the two functions $y = f(x)$ and $\theta = \phi(x)$ are linked together.

Provided that all heat is dissipated between $x = 0$ and $x = L$, the boundary conditions are:

$$\theta = \theta_0 \quad \text{for} \quad x = 0$$

$$-ky \frac{d\theta}{dx} = q_0' \quad \text{for} \quad x = 0$$

$$-ky \frac{d\theta}{dx} = 0 \quad \text{for} \quad x = L$$

and

$$h \int_0^L \theta \cdot dx = q_0' \tag{11-35}$$

The last quantity is the total heat dissipation from one side of the fin.

The area

$$a_L = \int_0^L y \cdot dx \tag{11-36}$$

is proportional to the material needed for unit length of the fin. It will be called "semi-profile area."

11-8. The Straight Fin which Requires the Least Material

The question arises for what profile function $f(x)$ the area a_L becomes a minimum, so that the fin needs the least material. Schmidt evaded the use of the calculus of variations and solved this rather difficult problem by the following simple reasoning. For any potential flow a

† It is neglected that for non-rectangular profiles the surface for a length dx is slightly larger than dx.

channel of uniform cross section has the smallest resistance of all channels having the same length and volume. This requires that all heat-flow lines run in the direction x. With temperatures independent of y and homogeneous material the heat flow will also be homogeneous. It follows that a fin of uniform density of heat flow needs the least material. For uniform thermal conductivity this means that $\theta = \phi(x)$ will be a linear equation:

$$\theta = \theta_0 - mx \tag{11-37}$$

Substituting this in Eq. 11-34 and using the boundary condition,‡

$$y = 0 \quad \text{at} \quad x = L$$

the integration yields

$$y = \frac{h}{2k}(x^2 - L^2) - \frac{h\theta_0}{mk}(x - L) \tag{11-38}$$

By substituting θ from Eq. 11-37 in Eq. 11-35 and integrating

$$q_0' = hL\left(\theta_0 - \frac{mL}{2}\right) \tag{11-39}$$

Further, from Eqs. 11-36 and 38,

$$a_L = \frac{h\theta_0 L^2}{2mk} - \frac{h}{3k}L^3 \tag{11-40}$$

In these equations L and m must be determined in such a way that for a given value of heat dissipation, q_0', the area a_L becomes a minimum. Since $\theta \geqq 0$ when $q_0' \geqq 0$, it follows from Eq. 11-37 that $\theta_0 \geqq mL$; otherwise the temperature excess θ would become negative for $x = L$. Substituting $\theta_0/m \geqq L$ in Eq. 11-40, inspection shows that $\theta_0/m = L$; i.e., the smallest possible value of θ_0/m yields the smallest possible value of a_L.

Substituting now $m = \theta_0/L$ in Eq. 11-39,

$$L = \frac{2}{h} \cdot \frac{q_0'}{\theta_0} \tag{11-41}$$

Herewith, the equation of the most favorable profile becomes

$$y = \frac{h}{2k}\left(x - \frac{2}{h} \cdot \frac{q_0'}{\theta_0}\right)^2 \tag{11-42}$$

‡ It is supposed that no heat remains to leave at the tip; then, because equal flow density is supposed, y must be zero at the tip.

Further, from Eqs. 11-37 and 41,

$$\theta = \theta_0 - \frac{h\theta_0^2}{2q_0'}x \tag{11-43}$$

According to these equations, the most favorable profile is confined by parabolas, and the temperature curve is a straight line with $\theta = 0$ at $x = L$, as shown in Fig. 11-6. The main dimensions of the fin profile are given by

$$L = \frac{2}{h} \cdot \frac{q_0'}{\theta_0} \tag{11-44}$$

$$y_0 = \frac{2}{hk}\left(\frac{q_0'}{\theta_0}\right)^2 \tag{11-45}$$

$$a_L = \frac{4}{3h^2k}\left(\frac{q_0'}{\theta_0}\right)^3 \tag{11-46}$$

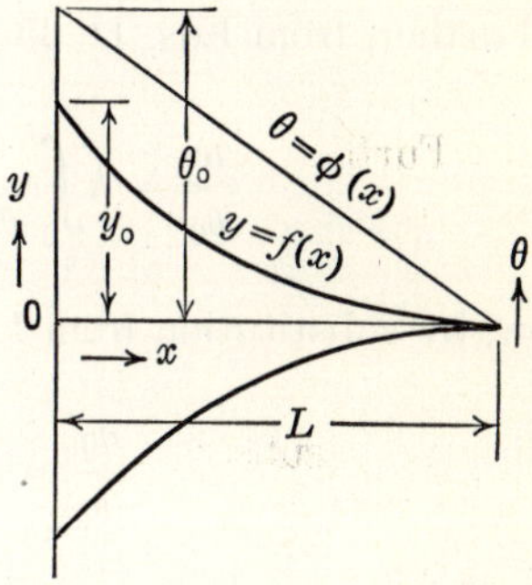

Fig. 11-6. Profile of straight fin requiring least material, with temperature distribution.

It is remarkable that for fixed values of k and h the term q_0'/θ_0 is the only variable in these equations.

11-9. The Straight Fin of Rectangular Profile

The profile represented in Fig. 11-6 is much too acute at the tip for manufacturing and practical use. Considering next the common rectangular profile, the most favorable ratio of width to thickness will be calculated. Taking $y = y_0$ (see Fig. 11-4), Eq. 11-34 becomes

$$\frac{d^2\theta}{dx^2} = \frac{h}{ky_0}\theta \tag{11-47}$$

With the boundary conditions

$$\theta = \theta_0 \quad \text{for} \quad x = 0$$

and

$$\frac{d\theta}{dx} = 0 \quad \text{for} \quad x = L$$

the solution is

$$\theta = \theta_0 \frac{\cosh\left[(x - L)\sqrt{h/ky_0}\,\right]}{\cosh\left(L\sqrt{h/ky_0}\,\right)} \tag{11-48}$$

The second boundary condition, however, holds only if no heat at all is lost at the tip of the fin. The small heat loss that actually will occur at the tip can be corrected approximately by using $L_c = L + y_0$ instead of L, as explained in Sect. 11–5. The symbol L, as used from now on, will include the correction y_0, that is, it will have the meaning of L_c, the subscript being left off for simplicity.

For the rectangular profile, the semi-profile area, defined by Eq. 11–36, becomes

$$a_L = Ly_0 \tag{11–49}$$

Further, from Eqs. 11–35 and 48,

$$\frac{q_0'}{\theta_0} = h\int_0^L \frac{\theta}{\theta_0}\,dx = \sqrt{hky_0}\tanh\left(\sqrt{\frac{h}{ky_0}}\,L\right) \tag{11–50}$$

or, by substitution from Eq. 11–49,

$$\frac{q_0'}{\theta_0} = \sqrt{hky_0}\tanh\left(\sqrt{\frac{h}{k}}\cdot\frac{a_L}{y_0^{3/2}}\right) \tag{11–51}$$

If a_L is given, that thickness y_0 is the most favorable for which q_0'/θ_0 becomes a maximum. In order to find this, the right side of Eq. 11–51 is to be differentiated with respect to y_0, and the result must be put equal to zero. Hence:

$$\tanh u = \frac{3u}{(\cosh u)^2} \quad \text{or} \quad u = \tfrac{1}{6}\sinh(2u) \tag{11–52}$$

where

$$u = \sqrt{\frac{h}{k}}\,a_L y_0^{-3/2} \tag{11–53}$$

By trial and error it is found that the value

$$u = u_f = 1.4192 \tag{11–54}$$

satisfies Eq. 11–52, subscript f standing for "most favorable." The values $\tanh u_f = 0.88944$ and $\cosh u_f = 2.1879$ will be used later on.

According to Eqs. 11–53 and 54, the most favorable thickness of the fin is

$$2y_0 = 2\sqrt[3]{\frac{(h/k)a_L^2}{u_f^2}} \tag{11–55}$$

By substitution in Eq. 11–51

$$\frac{q_0'}{\theta_0} = \sqrt[3]{\frac{h^2ka_L}{u_f}}\tanh u_f \tag{11–56}$$

Using Eqs. 11–49, 54, and 55, the following three relations have been found:

$$L = \frac{u_f}{\tanh u_f} \cdot \frac{1}{h} \cdot \frac{q_0'}{\theta_0} = \frac{1.5957}{h} \cdot \frac{q_0'}{\theta_0} \tag{11–57}$$

$$y_0 = \frac{1}{(\tanh u_f)^2} \cdot \frac{1}{hk}\left(\frac{q_0'}{\theta_0}\right)^2 = \frac{1.2642}{hk}\left(\frac{q_0'}{\theta_0}\right)^2 \tag{11–58}$$

$$a_L = \frac{u_f}{(\tanh u_f)^3} \cdot \frac{1}{h^2k}\left(\frac{q_0'}{\theta_0}\right)^3 = \frac{2.0192}{h^2k}\left(\frac{q_0'}{\theta_0}\right)^3 \tag{11–59}$$

As before, the main dimensions of the fin depend upon k, h, and q_0'/θ_0 only. For the tip of the fin ($x = L$) Eqs. 11–48, 49, 53, and 54 lead to the ratio

$$n = \frac{\theta_L}{\theta_0} = \frac{1}{\cosh u_f} = 0.45706 \tag{11–60}$$

Hence, for the most favorable ratio of width to thickness, the temperature excess at the tip is slightly below half of the excess at the fin root. According to this fact, just two temperature measurements are sufficient to check how far ideal conditions have been approached by a particular choice of fin dimensions.

In the general case, with $u \gtrless u_f$, Eq. 11–48 gives

$$n = \frac{\theta_L}{\theta_0} = \frac{1}{\cosh(\sqrt{h/ky_0}L)} \tag{11–61}$$

From Eqs. 11–50 and 61

$$\frac{q_0'}{\theta_0} = \sqrt{hky_0(1 - n^2)} \tag{11–62}$$

The equations of the main dimensions become

$$L = \frac{\cosh^{-1}(1/n)}{\sqrt{1 - n^2}} \cdot \frac{1}{h} \cdot \frac{q_0'}{\theta_0} \tag{11–63}$$

$$y_0 = \frac{1}{1 - n^2} \cdot \frac{1}{hk}\left(\frac{q_0'}{\theta_0}\right)^2 \tag{11–64}$$

$$a_L = \frac{\cosh^{-1}(1/n)}{\sqrt{(1 - n^2)^3}} \cdot \frac{1}{h^2k}\left(\frac{q_0'}{\theta_0}\right)^3 = f\,\frac{1}{h^2k}\left(\frac{q_0'}{\theta_0}\right)^3 \tag{11–65}$$

where $f = (\cosh^{-1} 1/n)/(\sqrt{1 - n^2}\,)^3$.

For a given value of q_0'/θ_0 only one of the two dimensions L and y_0 can be freely selected, the other one is fixed, according to Eqs. 11–63 and 64. If after such selection n deviates from its optimum value of 0.457, the factor f, and by this the semiprofile area a_L, become greater. In Table 11–1 some values of f are given. It is seen that the minimum

TABLE 11–1

THE FACTOR f OF EQ. 11–65 FOR CALCULATION OF STRAIGHT FINS

n	0.1	0.2	0.3	0.4	0.457	0.5	0.6	0.7	0.8	0.9
f	3.04	2.44	2.16	2.03	2.02	2.025	2.15	2.48	3.20	5.7

is rather flat, so that only greater deviations of n from the most favorable value influence the heat transfer considerably.

Avrami and Little [1942] gave a very general derivation of the temperature distribution in a fin of rectangular cross section. In this derivation all physical quantities are superseded by dimensionless groups, and the heat conduction in the direction y is included. The authors further defined that ratio $(L/y_0)_{opt}$ as an optimum at which the heat output is a maximum for a given value of the Nusselt number hy_0/k. The optimum becomes zero for $hy_0/k \approx 0.8$, which means that only for smaller values of hy_0/k does the fin have a cooling effect; it would have an insulating effect for greater values.

The following values are taken from a table of optimum values given by these authors:

hy_0/k	=100	10	1	0.1	0.01	0.001	0.0001
$(L/y_0)_{opt}$ =	0.955	0.951	0.238	3.75	14.05	46.4	149

The most favorable thickness, however, as defined previously, seems to be a more practical criterion because the minimum of material needed is a more essential item in the design of a fin than the condition where an increase of L no longer improves the cooling effect.

For example, an aluminum fin, $\frac{1}{8}$ in. thick, may be considered, so that $y_0 = 0.00521$ ft and $k = 120$ B hr^{-1} ft^{-1} F^{-1}, and it may be assumed that $h = 8.9$ B hr^{-1} ft^{-2} F^{-1}. Then $hy_0/k = 0.000386$, and according to Eqs. 11–49, 53, and 54 the most favorable length would be $L = 0.376$ ft. According to Avrami and Little, on the other hand, $(L/y_0)_{opt} = 105.2$, yielding $L = 0.548$ ft. This means that a fin of this length gives the maximum heat output at $hy_0/k = 0.000386$. However, it would not be economical to use such a long fin; its cross-sectional area is 46% larger than that of the most favorable fin, and Eq. 11-53 shows that, with such a large cross section, a fin of $y_0 = 0.00672$ ft (instead of 0.00521) and $L = 0.425$ ft (instead of 0.548) would give the greatest heat output q_0' for any given temperature excess θ_0.

The reason why the ratio $(L/y_0)_{opt}$ does not give the most economical fin is that, increasing the length of a fin more and more, the gain in heat transfer eventually becomes so small that it would be a waste of material to design a fin with that ratio.

11-10. The Straight Fins of Triangular and Trapezoidal Profiles

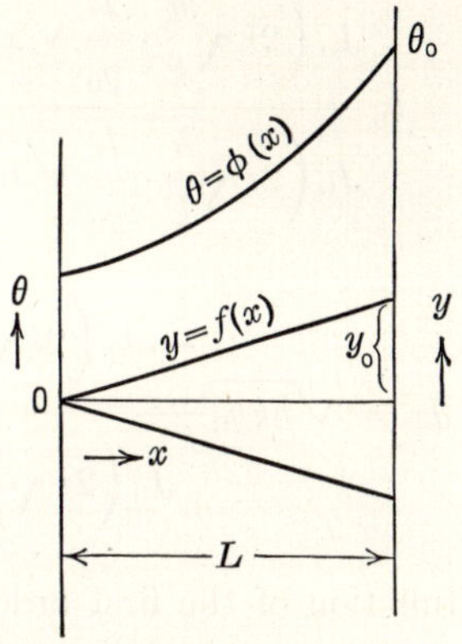

FIG. 11-7. Profile of straight triangular fin with temperature distribution (not to scale).

For a triangular profile the zero point of the system of coordinates may be at the tip of the fin (Fig. 11-7), so that

$$y = \frac{y_0}{L} x$$

By substitution in Eq. 11-34

$$\frac{d^2\theta}{dx^2} + \frac{1}{x} \cdot \frac{d\theta}{dx} - \frac{1}{x} \cdot \frac{h}{k} \cdot \frac{L}{y_0} \theta = 0 \tag{11-66}$$

Introducing $w = \frac{h}{k} \cdot \frac{L}{y_0} x$ as the independent variable, Eq. 11-66 simplifies to

$$\frac{d^2\theta}{dw^2} + \frac{1}{w} \cdot \frac{d\theta}{dw} - \frac{1}{w} \theta = 0 \tag{11-67}$$

Its general solution is

$$\theta = MJ_0(2i\sqrt{w}) + NY_0(2i\sqrt{w}) \tag{11-68}$$

where J_0 and Y_0 are the first and second kind of Bessel's functions of zero order, and M and N are the constants of integration.

According to the theory of Bessel functions, $Y_0 = -\infty$ for $w = 0$. Since in our case θ is positive and real for $x = 0$, the solution simplifies to

$$\theta = MJ_0(2i\sqrt{w}) = M\left[1 + \frac{w}{(1!)^2} + \frac{w^2}{(2!)^2} + \cdots\right] \tag{11-69}$$

Considering the boundary condition $\theta = \theta_0$ for $x = L$, the temperature distribution becomes

$$\theta = \theta_0 \frac{J_0\left(2i\sqrt{\frac{h}{k}\cdot\frac{L}{y_0}}\sqrt{x}\right)}{J_0\left(2i\sqrt{\frac{h}{k}\cdot\frac{L}{y_0}}\sqrt{L}\right)} \tag{11-70}$$

Herewith

$$\frac{q_0'}{\theta_0} = h\int_0^L \frac{\theta}{\theta_0}\,dx = \sqrt{hky_0}\,\frac{-iJ_1\left(2i\sqrt{\frac{h}{k}\cdot\frac{L}{y_0}}\sqrt{L}\right)}{J_0\left(2i\sqrt{\frac{h}{k}\cdot\frac{L}{y_0}}\sqrt{L}\right)} \tag{11-71}$$

where J_1 is the Bessel function of the first order and first kind, which can be represented by

$$-iJ_1(2i\sqrt{w_L}) = \sqrt{w_L}\left[1 + \frac{w_L}{(1!)^2\cdot 2} + \frac{{w_L}^2}{(2!)^2\cdot 3} + \frac{{w_L}^3}{(3!)^2\cdot 4} + \cdots\right] \tag{11-72}$$

where w_L means w for $x = L$. Now those values of L and y_0 have to be found for which q_0'/θ_0 becomes a maximum for a given value

$$a_L = \frac{Ly_0}{2} \tag{11-73}$$

Substituting L from this equation in Eq. 11-71 and introducing

$$u = 4\sqrt{\frac{h}{k}}\,a_L {y_0}^{-3/2}$$

as a new variable, Eq. 11-71 becomes

$$\frac{q_0'}{\theta_0} = \sqrt[3]{\frac{4h^2ka_L}{u}}\cdot\frac{-iJ_1(iu)}{J_0(iu)} \tag{11-74}$$

From the maximum condition $[d(q_0/\theta_0)]/du = 0$ the following equation is obtained:

$$\frac{4}{3} \cdot \frac{1}{u} \cdot \frac{-iJ_1(iu)}{J_0(iu)} = 1 - \left[\frac{-iJ_1(iu)}{J_0(iu)}\right]^2 \tag{11-75}$$

Here from, using Eqs. 11-69 and 72, Schmidt found by trial and error $u_f = 2.6188$ as the most favorable value.

Herewith, finally,

$$L = \frac{1.684}{h} \cdot \frac{q_0'}{\theta_0} \tag{11-76}$$

$$y_0 = \frac{1.6545}{hk}\left(\frac{q_0'}{\theta_0}\right)^2 \tag{11-77}$$

$$a_L = \frac{1.393}{h^2k}\left(\frac{q_0'}{\theta_0}\right)^3 \tag{11-78}$$

The temperature excess at the tip of the fin becomes

$$\theta_L = \theta_0 \frac{J_0(0)}{J_0(iu_f)} = 0.2771\theta_0 \tag{11-79}$$

For a wedge-shaped profile (Fig. 11-7a) the profile equation becomes

$$y = \frac{x(y_0 - y_L)}{L}$$

Again Eq. 11-68 is the solution of the differential equation for the temperature distribution, as has been shown in Harper and Brown's paper. However, here the constant N is different from zero, and instead of

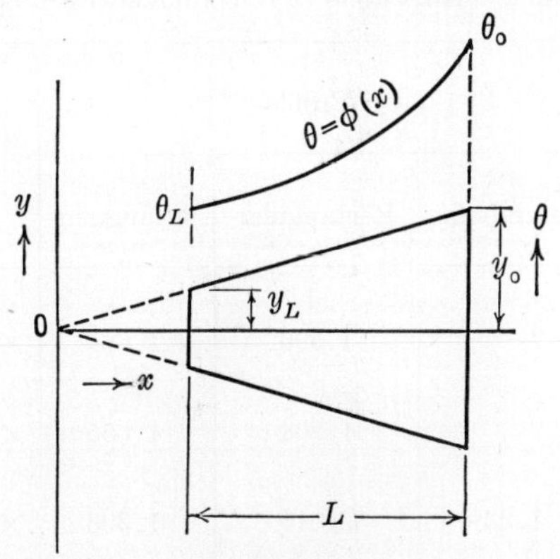

Fig. 11-7a. Profile of straight wedge-shaped fin with temperature distribution (not to scale).

Eq. 11-69 a more complicated equation has been found. It happens that the solution is identical with that for a circular fin of rectangular profile, Eq. 11-94, which will be dealt with in Sect. 11-14. The details of the solution are given in Harper and Brown's paper.

11-11. Comparison of Different Straight Fins

In Fig. 11-8 the three profiles of fins, dealt with in Sects. 11-8 to 10, are represented for equal values of q_0'/θ_0, h, and k. Only one half of

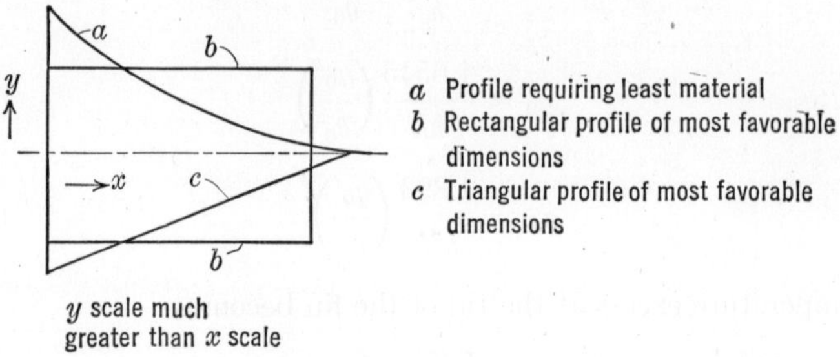

FIG. 11-8. Profiles of three straight fins having equal heat transfer.

either the parabolic or triangular profile is shown, and the thickness of the fins is exaggerated in order to make the sketch as clear as possible.

The most favorable dimensions of the same types of fins are compared in Table 11-2. Also the characteristic temperature ratio θ_L/θ_0 is given

TABLE 11-2

THE MOST FAVORABLE DIMENSIONS OF DIFFERENT TYPES OF STRAIGHT FINS

Dimension	Profile			
	Parabolic	Rectangular	Triangular	
$L =$	2	1.596	1.684	$\times \frac{1}{h} \cdot \frac{q_0'}{\theta_0}$
$y_0 =$	2	1.264	1.655	$\times \frac{1}{hk} \left(\frac{q_0'}{\theta_0}\right)^2$
$a_L =$	1.33···	2.019	1.393	$\times \frac{1}{h^2k} \left(\frac{q_0'}{\theta_0}\right)^3$
$\theta_L/\theta_0 =$	0	0.457	0.277	

in this table. It is seen that the amount of material required increases with the third power of q_0'. Therefore it is recommended to arrange many fins with as small distances as feasible with respect to convection. If the material of the fins is changed, the most favorable width L remains the same, whereas the thickness $2y_0$ changes inversely with the thermal conductivity k. Fins with rectangular profile require about 50% more material than those of parabolic or triangular profile. The triangular is nearly as economic as the profile of minimum material requirement.

An increase of the coefficient of heat transfer h reduces the thickness and, even more, the amount of material that will be needed. But h is increasing with the velocity of the fluid used. Therefore, a different fin thickness should be chosen for small and high velocity; where different velocities occur at the same construction, the fin works under most favorable conditions at only one velocity.

In Table 11–3 the volume V and weight W for different materials are compared for a given value of q_0'/θ_0. Subscript Cu refers to copper.

TABLE 11–3

INFLUENCE OF THE MATERIAL UPON THE SPACE AND WEIGHT REQUIRED FOR FINS AT THE SAME VALUE OF q_0'/θ_0

Metal	k [B hr^{-1} ft^{-1} F^{-1}]	γ [lb ft^{-3}]	V/V_{Cu}	W/W_{Cu}
Copper	222	556	1	1
Aluminum	121	169	1.83	0.556
Iron	30	487	7.33	6.43

The table shows that copper fins require the smallest volume and aluminum fins the least weight. Iron is much less favorable in regard to volume and weight.

11–12. General Equations for the Heat Dissipation from Cylindric Surfaces by Circular Fins

Cylindrical coordinates will be used for circular fins. Let r be the radial distance of a point of the fin from the cylinder axis, r_0 the radius of the surface from which the fin originates (root of the fin), r_L the radius of the tip of the fin. In Fig. 11–9 a fin profile and the temperature distribution along the profile curve are shown schematically.

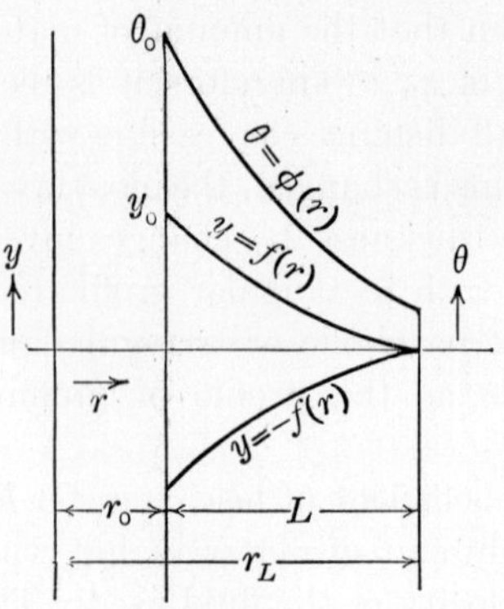

FIG. 11-9. Arbitrary profile of circular fin with temperature distribution (not to scale).

Let q be the heat flow in radial direction through half the cross-sectional area of the fin at $r = r$, and q_0 the heat flow at $r = r_0$. Then, in analogy to the equations given in Sect. 11-7,

$$q = -ky \cdot 2\pi r \frac{d\theta}{dr}$$

and

$$-dq = h\theta \cdot 2\pi r \cdot dr$$

By elimination of q from these equations

$$y \frac{d}{dr}\left(r \frac{d\theta}{dr}\right) + \frac{dy}{dr} r \frac{d\theta}{dr} - \frac{h}{k} r\theta = 0 \tag{11-80}$$

The boundary conditions are

$$\theta = \theta_0 \quad \text{for} \quad r = r_0$$

$$-ky_0 \cdot 2\pi r \frac{d\theta}{dr} = q_0 \quad \text{for} \quad r = r_0$$

$$-ky_L \cdot 2\pi r \frac{d\theta}{dr} = 0 \quad \text{for} \quad r = r_L \S$$

and

$$h \int_{r_0}^{r_L} \theta \cdot 2\pi r \cdot dr = q_0 \tag{11-81}$$

§ In analogy to the procedure with straight fins (see Sect. 11-5) it will be assumed that $r_{L,c} = r_L + A/C$ is used instead of r_L. For convenience the subscript c is omitted above.

The volume

$$V = \int_{r_0}^{r_L} y \cdot 2\pi r \cdot dr \tag{11-82}$$

is proportional to the material needed for the fin. It will be called the "semi-profile volume."

11-13. The Circular Fin Which Requires the Least Material

The reasoning which led to Eq. 11-37 holds also for circular fins. Hence, a minimum of material is needed if

$$\theta = \theta_0 \left(1 - \frac{r - r_0}{r_L - r_0}\right) \tag{11-83}$$

Substituting θ from Eq. 11-83 in Eq. 11-80 and using the boundary condition

$$y = 0 \quad \text{at} \quad r = r_L$$

the integration yields

$$y = \frac{h}{k}\left(\frac{r^2}{3} - \frac{r_L r}{2} + \frac{r_L^3}{6r}\right)$$

or

$$\frac{k}{h} \cdot \frac{y}{r_L^2} = \frac{1}{3}\left(\frac{r}{r_L}\right)^2 - \frac{1}{2} \cdot \frac{r}{r_L} + \frac{1}{6} \cdot \frac{r_L}{r} \tag{11-84}$$

The profile curve according to this equation is represented in Fig. 11-10 with r/r_L and ky/hr_L^2 as coordinates. The curve is similar in

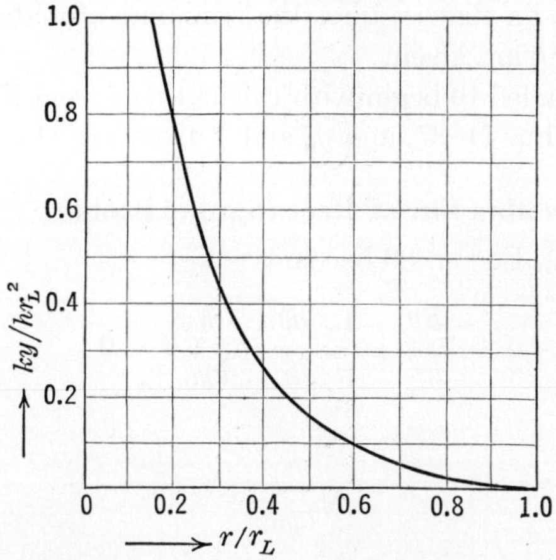

FIG. 11-10. Dimensionless representation of profile of circular fin requiring least material (according to Eq. 11-84).

character to that for the straight fin (see Fig. 11–8), and, in particular, it has also a sharp edge at the tip (horizontal tangent at $r/r_L = 1$).

By substitution of θ from Eq. 11–83 in Eq. 11–81

$$q_0 = h\pi\theta_0 \left(\frac{r_L^2}{3} + \frac{r_L r_0}{3} - \frac{2r_0^2}{3}\right) = h\pi r_0^2\theta_0 \frac{(\rho + 2)(\rho - 1)}{3}$$

$$= h\pi r_0^2\theta_0\sigma \tag{11–85}$$

where

$$\rho = \frac{r_L}{r_0} \tag{11–86}$$

and

$$\sigma = \frac{(\rho + 2)(\rho - 1)}{3} \tag{11–87}$$

From Eqs. 11–86, 84, and 82 the following three relations are obtained:

$$L = r_0(\rho - 1) \tag{11–88}$$

$$y_0 = \frac{h}{k} r_0^2 \frac{(\rho + 2)(\rho - 1)^2}{6} = \frac{h}{k} r_0^2 \chi \tag{11–89}$$

$$V = \frac{h}{k} \pi r_0^4 \frac{(\rho + 1)(\rho - 1)^3}{6} = \frac{h}{k} \pi r_0^4 \psi \tag{11–90}$$

where χ and ψ are functions of ρ which are defined by the two last equations.

From these equations the dimensions L and y_0 and the semi-profile volume V of the fin that requires the minimum of material are found if r_0, q_0, θ_0, h, and k are given.

It is recommended to begin with calculating σ from Eq. 11–85. Then ρ is found from Eq. 11–87, and y_0 and V from Eqs. 11–89 and 90.

11–14. The Circular Fin of Rectangular Profile

Taking $y = y_0$, Eq. 11–80 becomes

$$\frac{d^2\theta}{dr^2} + \frac{1}{r} \cdot \frac{d\theta}{dr} - \frac{h}{ky_0}\theta = 0 \tag{11–91}$$

or, introducing

$$u = \sqrt{\frac{h}{ky_0}} \cdot r = mr \tag{11–92}$$

$$\frac{d^2\theta}{du^2} + \frac{1}{u} \cdot \frac{d\theta}{du} - \theta = 0 \tag{11–93}$$

The general solution of this equation is

$$\theta = MJ_0(iu) + NiH_0(iu) \tag{11-94}$$

where J_0 is the Bessel function of order zero, H_0 is the Hankel function of order zero, and M and N are the constants of integration. For positive and real u the functions $J_0(iu)$ and $iH_0(iu)$ are also positive and real, as is shown in the theory of Bessel functions.

Using the boundary conditions

$$\theta = \theta_0 \quad \text{for} \quad r = r_0$$

and

$$\frac{d\theta}{dr} = 0 \quad \text{for} \quad r = r_L$$

(the latter with the same restriction as made just after Eq. 11-48), the solution becomes

$$\theta = \theta_0 \frac{J_0(imr)H_1(imr_L) + iH_0(imr)iJ_1(imr_L)}{J_0(imr_0)H_1(imr_L) + iH_0(imr_0)iJ_1(imr_L)} \tag{11-95}$$

where J_1 and H_1 are the Bessel and Hankel functions of the first order, respectively. For real and positive u the terms $-iJ_1(iu)$ and $-H_1(iu)$ are also real and positive.

Substituting θ from Eq. 11-95 in Eq. 11-81 and defining σ as in Eq. 11-85,

$$\sigma = \frac{q_0}{h\pi {r_0}^2\theta_0} =$$

$$\frac{2}{mr_0} \cdot \frac{[-H_1(imr_0)][-iJ_1(imr_L)] - [-iJ_1(imr_0)][-H_1(imr_L)]}{[iH_0(imr_0)][-iJ_1(imr_L)] - [J_0(imr_0)][-H_1(imr_L)]} \tag{11-96}$$

where the terms in brackets are all real and positive for real and positive r_0 and r_L.

For given values of r_0, V, h, and k that pair of dimensions $L = r_L - r_0$ and y_0 is to be determined for which σ is a maximum. Schmidt performed this determination, partly by calculation, partly by a graphical trial-and-error method, and represented the result by Fig. 11-11.

This diagram is to be used as follows. From given q_0, h, and r_0 and admitted value θ_0, Eq. 11-96 yields σ. For this value the σ scale on the curve in Fig. 11-11 yields mr_0 and mr_L. By substitution of mr_0 in Eq. 11-92, y_0 is obtained. By subtracting y_0 from $L = r_L - r_0$ the heat loss at the tip of the fin is taken care of (see the paragraph following Eq. 11-48). In other words, the most favorable larger radius will be $r_L - y_0$.

A comparison between rectangular fins and those having a profile according to Eq. 11-84 shows that for the same value of σ, the rectangular fin requires more material, particularly for the higher values of σ.

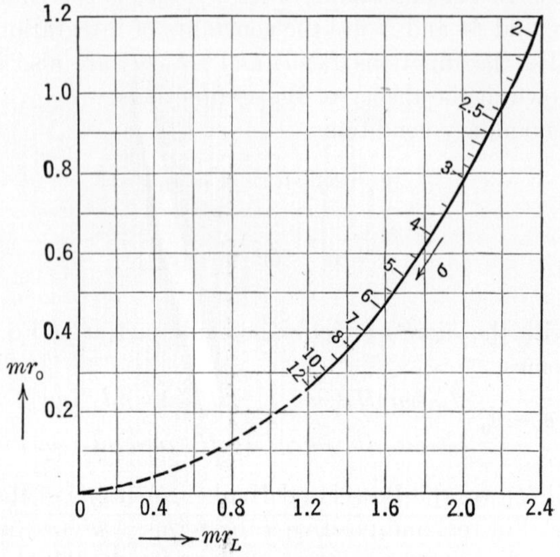

FIG. 11-11. Curve for calculations of dimensions of circular fin of rectangular profile requiring least material.

11-15. Circular Fins of Triangular and Trapezoidal Profiles

It can be supposed that the fin with triangular profile behaves similarly to the fin having a profile which requires the least amount of material. The exact calculation for the triangular profile, however, meets with considerable mathematical difficulties. For this reason Schmidt recommends using the formulas for the straight fin if r_L/r_0 is small, as, for instance, is the case with cylinders of airplane motors.

For greater values of r_L/r_0 the calculations for the profile requiring least material (Eq. 11-84) and for the rectangular profile (see Sect. 11-14) may be performed. Let L_f, $y_{0,f}$, and L_r, $y_{0,r}$ be the characteristic dimensions for these two cases, where the subscripts f and r stand for most favorable fin and rectangular fin, respectively. Those for the triangular profile, L_t and $y_{0,t}$, can be interpolated in the same proportion as has been found for straight fins and is tabulated in Table 11-2. Hence take for a triangular fin

$$L_t = 0.783L_r + 0.217L_f \tag{11-97}$$

$$y_{0,t} = 0.469y_{0,r} + 0.531y_{0,f} \tag{11-98}$$

This yields conservative values for the dimensions of the fin.

Analytical solutions for the circular fin of wedge-shaped (trapezoidal) profile do not seem to be available except for the case where $r \cdot y$ = constant. It is easily understood that this case can be reduced to that of the straight fin of rectangular profile (Sect. 11–9). Comparing a straight rectangular fin of length $z = 2\pi r_0$ with the mentioned circular trapezoidal fin on a cylinder of radius r_0, the cross-sectional areas A of both are constant and equal to $2\pi r_0 y_0$, and the surfaces can be taken as virtually equal if the fins are not too thick.

11–16. The Effectiveness of Fins

Harper and Brown not only derived the essential equations for straight fins of rectangular and triangular cross section some years before Schmidt, but they also extended the analysis to the very complicated case of wedge-shaped cross section, of which the triangular cross section is a special case.

In addition to the tremendous mathematical accomplishments in their paper, there has been introduced a very practical new term which is called the effectiveness of fin surface and can be defined as the ratio of the amount of heat actually dissipated by the fin surface to that which would be dissipated if, without change of h, the fin surface were held at the temperature of the fin base.

Referring, for instance, to Eq. 11–50, the effectiveness for a straight fin of rectangular cross section would be ||

$$\eta_r = \frac{q_0'}{hL\theta_0} = \frac{1}{L}\sqrt{\frac{ky_0}{h}}\tanh\left(L\sqrt{\frac{h}{ky_0}}\right) \tag{11–99}$$

Substituting the dimensionless quantity

$$n = L\sqrt{\frac{h}{ky_0}} \tag{11–100}$$

the following simple form is obtained:

$$\eta_r = \frac{\tanh(n)}{n} \tag{11–101}$$

Equation 11–101 is represented by a curve in Fig. 11–12.

In deriving Eq. 11–50 it was assumed that the influence of heat flow in the direction y (see Fig. 11–5) can be neglected. Harper and Brown analyzed this influence mathematically. In a numerical example dealing

|| Again L may include the correction for the heat loss at the fin tip, so that L is a corrected length which in Sect. 11–9 originally was called L_c.

with extremely unfavorable conditions, namely, steel fins, 0.2 in. thick and 1.2 in. wide, exposed to a wind of about 200 miles per hour, they found that η_r is increased only about ½% compared with the result of Eq. 11–101.

Another remarkable result of their paper is that Eq. 11–101 holds also if θ_0 is not a constant, but varies in the direction z (see Fig. 11–4) according to an arbitrary function $\theta_0 = f(z)$.

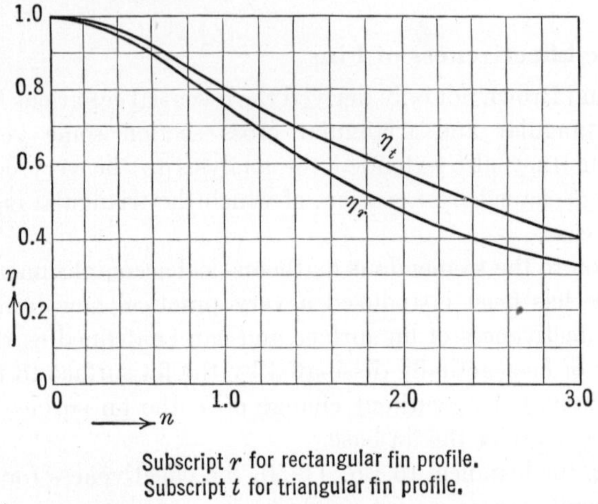

FIG. 11–12. Effectiveness η of straight fins versus $n = L\sqrt{h/ky_0}$.

For the triangular cross section it is shown that Eq. 11–101 is to be replaced by

$$\eta_t = \frac{-\sqrt{2}}{n} \cdot \frac{iJ_1(in\sqrt{2})}{J_0(in\sqrt{2})} \tag{11–102}$$

This is represented by a second curve in Fig. 11–12. It is seen that with increasing n the percentage difference between η_r and η_t becomes rather appreciable.

For the wedge-shaped cross section the ratio of the fin thickness at the tip to the mean thickness y_L/y_m may be called λ, and the effectiveness η_w. Defining a reduction term $\Delta_w\eta$ by the equation

$$\eta_w = \eta_r + \Delta_w\eta \tag{11–103}$$

Harper and Brown obtained the curves which are represented in Figs. 11–13 and 14.

The rectangular circumferential fin is also dealt with in a similar simple way by these authors. They first show that Eq. 11–95 is practically identical to their equation for wedge-shaped straight fins with $\lambda > 1$, as easily can be understood.

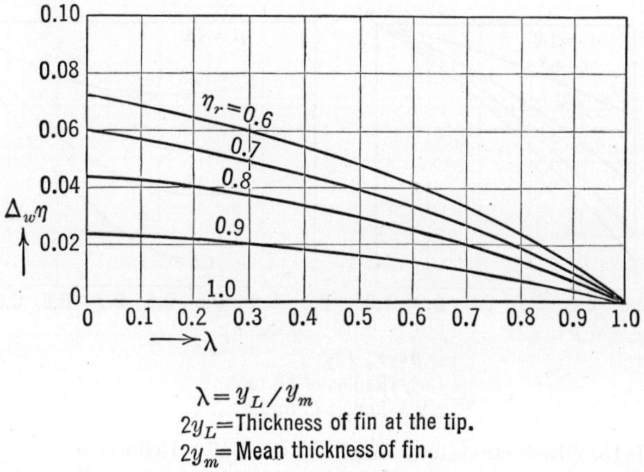

FIG. 11-13. Reduction term $\Delta_w\eta$ for straight fin of wedge-shaped profile (defined by Eq. 11-103) versus λ.

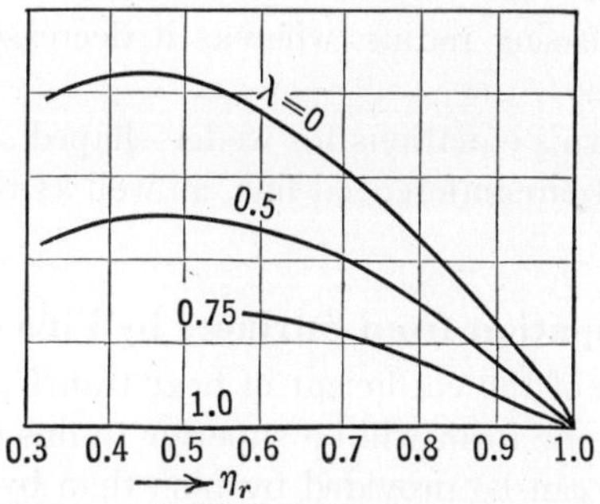

FIG. 11-14. Reduction term $\Delta_w\eta$ for straight fin of wedge-shaped profile (defined by Eq. 11-103) versus η_r.

Same ordinates and designations as in Fig. 11-13.

Introducing ¶ $\rho = r_0/r_L$ and η_c as effectiveness of circumferential fins, a reduction term $\Delta_c\eta$ may be defined by the equation

$$\eta_c = \eta_r + \Delta_c\eta \qquad (11\text{-}104)$$

¶ r_L is the corrected radius, found by adding y_L to the true radius (see p. 235, footnote ‖).

Curves of $\Delta_c\eta$, taken from Harper and Brown's paper, are given in Figs. 11-15 and 16.

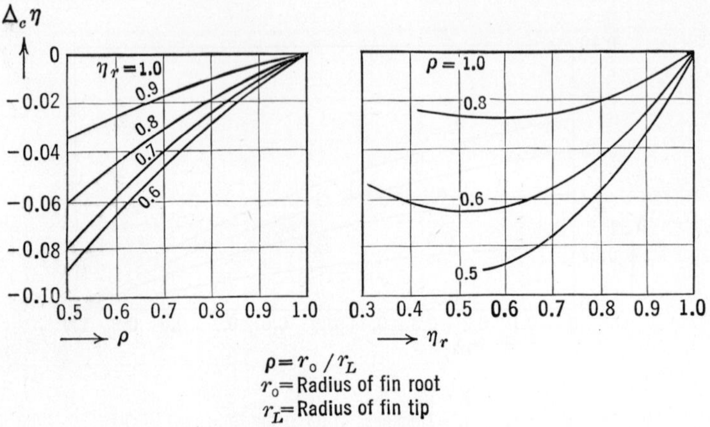

FIG. 11-15. Reduction term $\Delta_c\eta$ for circular fin of rectangular profile (defined by Eq. 11-104) versus ρ.

FIG. 11-16. Reduction term $\Delta_c\eta$ for circular fin of rectangular profile (defined by Eq. 11-104) versus η_r.

It is seen that here the reduction term has a negative value. This is due to the fact that the cross-sectional area of a circumferential fin increases with increasing radius, whereas it decreases in the case of a straight tapered fin.

Harper and Brown's equations for wedge-shaped straight fins include the case of tapered circumferential fins, as well as that of fins of equal thickness.

11-17. Heat Dissipation from Surfaces by Pins or Spines

For equal values of the coefficient of heat transfer, h, extensions of a surface in pin or spine form will be superior to fins because more heat-dissipating surface can be provided by pins than by strips of the same total cross-sectional area.

The case of integral pins of different profiles has been dealt with in a paper of Bueche and Schau [1936]. For conical pins of length L and radius r_0 at the root, for instance, these writers found that

$$\frac{q_0}{k\pi r_0\theta_0} = \frac{r_0}{L}\left[\frac{1}{2}\cdot\frac{2\sqrt{C}J_0(i2\sqrt{C})}{iJ_1(i2\sqrt{C})} + 1\right] = \psi\left(\frac{hr_0}{k}, \frac{r_0}{L}\right) \tag{11-105}$$

where q_0 = the heat energy entering the root of the pin in unit time,

$$C = \frac{2hr_0}{k}\cdot\frac{L}{r_0}\sqrt{1+\left(\frac{r_0}{L}\right)^2}, \text{ and}$$

ψ = a function of the dimensionless groups hr_0/k and r_0/L.

For $r_0/L \to \infty$, the function ψ converges to hr_0/L, and

$$q_0 = h \cdot \pi r_0^2 \theta_0 \tag{11-106}$$

For $r_0/L \to 0$, the function ψ converges to $\sqrt{2hr_0/k}$, and

$$q_0 = \sqrt{2hkr_0}\pi r_0 \theta_0 \tag{11-107}$$

q_0, according to Eq. 11–106, is identical with the heat transfer from a circular disc of radius r_0, and

q_0, according to Eq. 11–107, is the heat transfer from a rod of circular cross section.

For the ranges which practically occur with gases and liquids ($L/r_0 = 0$ to 50, and $hr_0/k = 0.0002$ to 0.075) curves for ψ are represented in Figs. 11–17 and 18.

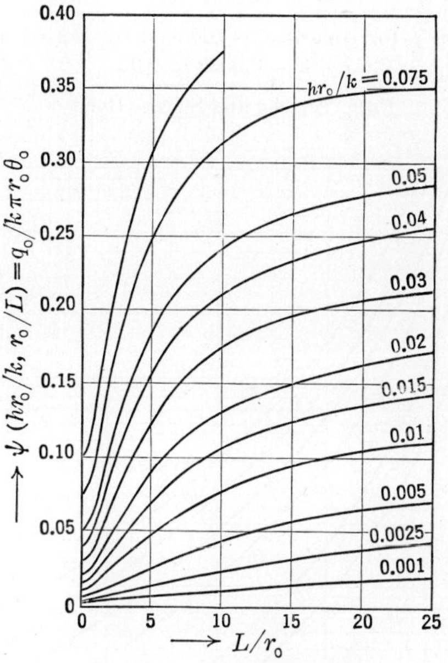

FIG. 11–17. Function ψ for conical pins (defined by Eq. 11–105) versus L/r_0 for $hr_0/k = 0.001$ to 0.1.

From Bueche and Schau [1936].

Using Fig. 11–19, it is possible to find the optimum of L/r_0 as dependent upon hr_0/k and $3G/(\pi r_0^3 \gamma)$ where G is that part of the weight of the wall which belongs to *one* pin, and γ is the specific weight of the pin material.

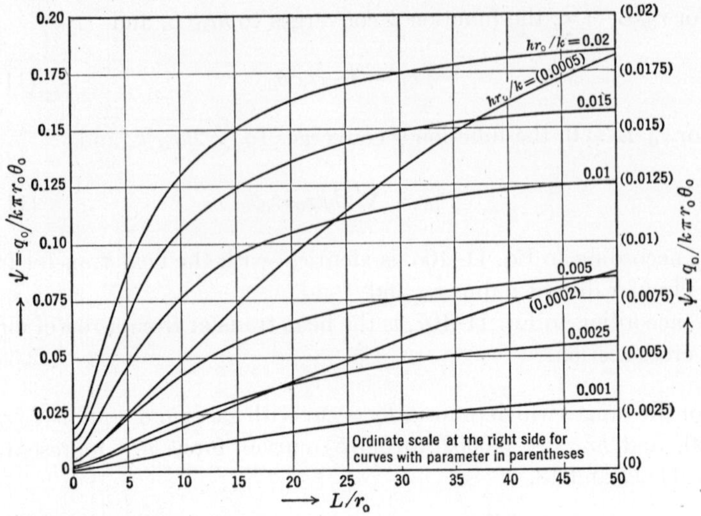

FIG. 11-18. Function ψ for conical pins (defined by Eq. 11-105) versus L/r_0 for $hr_0/k = 0.0002$ to 0.02.

From Bueche and Schau [1936].

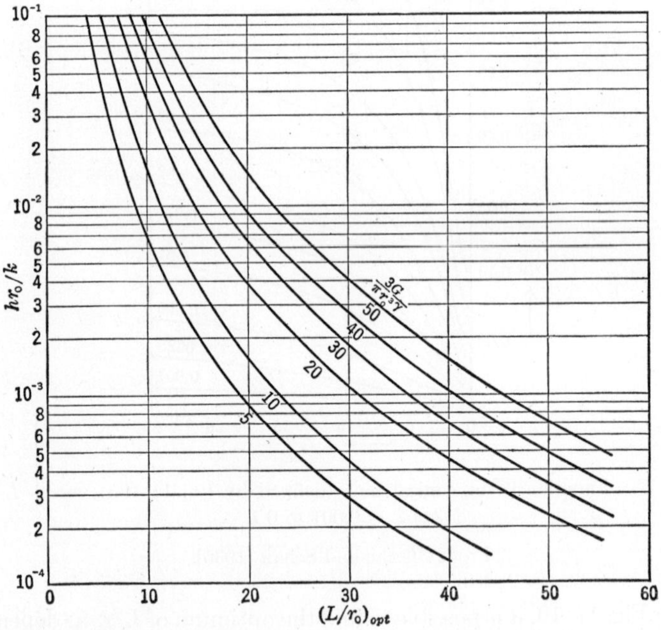

FIG. 11-19. Nusselt number hr_0/k as function of optimum ratio L/r_0 for conical pins.

From Bueche and Schau [1936].

If the extension of the surface consists of integral spines instead of conical pins, the same calculation may be used with good approximation. The main difficulty for pins, as for spines, is to find the pertinent mean value of h. The paper of Bueche and Schau deals also with the change of h along the pin. This, however, is a problem of convection, whereas here, for the time being, only the conduction in rods, fins, and pins has been treated, and h had to be considered a given constant quantity.

11–18. A Generalized Equation for Fins and Spines

In an excellent comprehensive paper, Gardner [1945], following a suggestion of R. D. Douglass [see Sherwood and Reed 1939, p. 210] has shown that the equations derived for any form of fins or spines can be reduced to a generalized Bessel equation, provided the cross-sectional area varies as some power of the distance measured from a given zero point in the direction normal to the base surface.

The heat balance for a differential element of the fin or spine yields the differential equation

$$\frac{d^2\theta}{dx^2} + \left(\frac{1}{A} \cdot \frac{dA}{dx}\right)\frac{d\theta}{dx} - \left(\frac{h}{kA} \cdot \frac{dS}{dx}\right)\theta = 0 \tag{11–108}$$

where S is the surface area of one side of the extended surface from $x = 0$ to $x = x$ (semi-surface area), and
A is the semi-cross-sectional area.

This equation is more general than Eq. 11–34, in which A was replaced by y and S by x, considering a piece of unit length in the z direction.

The generalized Bessel equation, on the other hand, reads

$$x^2\frac{d^2\theta}{dx^2} + [(1 - 2\mu)x - 2\beta x^2]\frac{d\theta}{dx} + [\lambda^2\kappa^2x^{2\lambda} + \beta^2x^2 + \beta(2\mu - 1)x + (\mu^2 - \lambda^2\nu^2)]\theta = 0 \tag{11–109}$$

where β, κ, λ, μ, and ν are constants.

Termwise comparison of Eqs. 11–108 and 109 shows that both have the same form if

$$A = Bx^{1-2\lambda\nu} \tag{11–110}$$

and

$$\frac{dS}{dx} = Ex^{2\lambda(1-\nu)-1} \tag{11–111}$$

where B and E are positive constants.

The boundary conditions are:

$$\theta = \theta_0 \quad \text{when} \quad x = x_0$$

$$\frac{d\theta}{dx} = 0 \quad \text{when} \quad x = L$$

Now a new variable u will be introduced as follows:

$$u = -i\kappa x^{\lambda} = \sqrt{\frac{h}{kA} \cdot \frac{dS}{dx}}\, x \tag{11-112}$$

Since $dS/dx = C$ (half circumference, considering only half the profile),

$$u = mx \tag{11-113}$$

using our former connotation of m.

Herewith, by integration of Eq. 11-109, when ν is an integer or zero,

$$\theta = \theta_0 \left(\frac{u}{u_0}\right)^{\nu} \frac{i^{-\nu}J_{\nu}(iu) + bi^{-\nu}H_{\nu}{}^{(1)}(iu)}{i^{-\nu}J_{\nu}(iu_0) + bi^{-\nu}H_{\nu}{}^{(1)}(iu_0)} \tag{11-114}$$

where $b = \dfrac{i^{1-\nu}J_{\nu-1}(iu_0)}{i^{1-\nu}H_{\nu-1}{}^{(1)}(iu_0)}$

When ν is a fraction,

$$\theta = \theta_0 \left(\frac{u}{u_0}\right)^{\nu} \frac{i^{-\nu}J_{\nu}(iu) + bi^{\nu}J_{-\nu}(iu)}{i^{-\nu}J_{\nu}(iu_0) + bi^{\nu}J_{-\nu}(iu_0)} \tag{11-115}$$

where $b = -\dfrac{i^{1-\nu}J_{\nu-1}(iu_L)}{i^{\nu-1}J_{1-\nu}(iu_L)}$

the subscripts 0 and L referring to $x = 0$ and L, respectively.

The efficiency of the extended surface is given by

$$\eta = \frac{\displaystyle\int_0^{S_L} \theta \cdot dS}{\theta_0 S_L} \tag{11-116}$$

Hence, for $\nu = 0$ or an integer,

$$\eta = \frac{2(1-\nu)}{u_0[1-(u_0/u_L)^{2(1-\nu)}]} \cdot \frac{i^{1-\nu}J_{\nu-1}(iu_0) - bi^{1-\nu}H_{\nu-1}{}^{(1)}(iu_0)}{i^{-\nu}J_{\nu}(iu_0) + bi^{-\nu}H_{\nu}{}^{(1)}(iu_0)} \tag{11-117}$$

and, if ν is a fraction,

$$\eta = \frac{2(1-\nu)}{u_0[1-(u_0/u_L)^{2(1-\nu)}]} \cdot \frac{i^{1-\nu}J_{\nu-1}(iu_0) + bi^{\nu-1}H_{1-\nu}{}^{(1)}(iu_0)}{i^{-\nu}J_{\nu}(iu_0) + bi^{\nu}H_{-\nu}{}^{(1)}(iu_0)} \tag{11-118}$$

For a given fin the exponents of x in Eqs. 11–110 and 111 are known. λ may be eliminated and ν determined for any special case.

Identifying different cases, it is seen that $\nu = \frac{1}{2}$ covers all straight fins or spines of constant cross-sectional area. The wedge-shaped straight fins and the circular fins of constant cross-sectional area are covered by $\nu = 0$. The relation $\nu = -1$ is for conical spines, and $\nu = \frac{1}{3}$ is for circular fins with $y \cdot r =$ constant. Gardner showed further that the straight fin, requiring least material, can be reduced to the case of $\nu = \infty$.

In addition to these cases, different other ones are quoted in Gardner's paper, and many more could be dealt with. Circular fins, requiring least material, or fins of wedge-shaped profile belong to those cases for which the condition formulated in the first sentence of this section is not satisfied [Jakob 1945].

CHAPTER 12

Heat Conduction in Steady State, Including Internal Heat Generation and Heat Losses to the Environment

12–1. Cases to be Dealt with

In Chapter 10 the case of internal heat sources and negligible heat losses was dealt with;* in Chapter 11 heat losses were considered, but sources were excluded. In this chapter some cases will be treated where both heat sources and leakages are to be included.

In determining the thermal conductivity of metals by Kohlrausch's method, the heat losses on the surface of the test sample generally are neglected or roughly estimated (see Sect. 10–2). However, it is not difficult to consider them in a more exact way, as will be shown in Sect. 12–2. It will further be seen that the same procedure yields the correction for axial heat losses in the hollow cylinder method of measuring the thermal conductivity of insulating materials.

12–2. Determination of the Thermal Conductivity of a Metal from the Electric Conduction and the Temperature Drop in a Rod, Considering the Heat Losses

In Sect. 10–2 a method was presented by which the thermal conductivity of a metal is found by carrying an electrical current through a rod and measuring the voltage and the temperature difference between the middle of the rod and its ends. Heat losses at the surface of the rod were assumed to be negligible. Now the same case will be dealt with when heat losses are appreciable.

Using the same symbols as in Sect. 10–2 and considering an element of the rod of thickness dx (Fig. 12–1), the balance between the incoming,

* The transfer of virtually all the generated heat from the surface to the environment, as occurring in electrical apparatus, is not considered a heat "loss."

electrically generated, and outgoing heat energies per unit time becomes

$$q_1 + dq_3 = q_2 + dq_4$$

or

$$-kA\frac{dt}{dx} + q'''A\cdot dx = -kA\left(\frac{dt}{dx} + \frac{d^2t}{dx^2}\,dx\right) + hC(t - t_e)\,dx \qquad (12\text{–}1)$$

where h = the surface coefficient of heat transfer,
C = the circumference (perimeter) of the rod, and
t_e = the temperature of its environment.

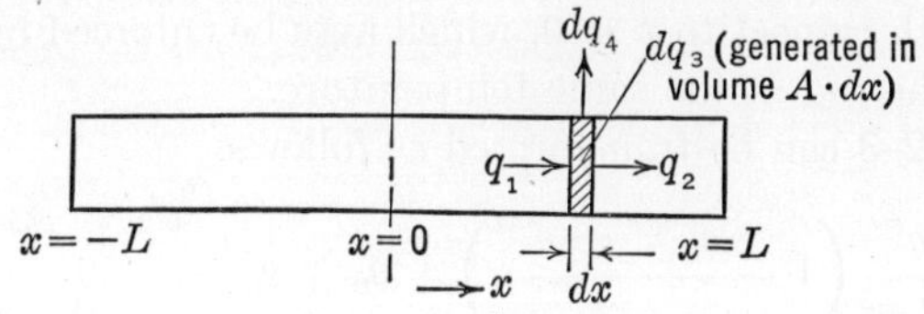

FIG. 12–1. Diagram of heat conduction and generation in a rod and heat loss to environment.

It will be assumed that h is constant. Introducing $\theta = t - t_e$ and $m^2 = (hC)/(kA)$, the heat balance simplifies to

$$\frac{d^2\theta}{dx^2} = m^2\theta - \frac{q'''}{k} \qquad (12\text{–}2)$$

This differential equation may be solved by starting from the reduced equation

$$\frac{d^2\theta}{dx^2} = m^2\theta$$

The solution of this equation is

$$\theta = Me^{-mx} + Ne^{mx}$$

A particular solution of Eq. 12–2 is

$$\theta = \frac{q'''}{km^2}$$

Hence, its general solution is

$$\theta = Me^{-mx} + Ne^{mx} + \frac{q'''}{km^2}$$

Counting x from the middle of the rod (see Fig. 12-1) and denoting the maximum temperature excess by θ_0, the boundary conditions will be

$$\theta = \theta_0 \quad \text{at} \quad x = 0$$

$$\frac{d\theta}{dx} = 0 \quad \text{at} \quad x = 0$$

Herewith:

$$\theta = \left(\theta_0 - \frac{q'''}{km^2}\right)\frac{e^{-mx} + e^{mx}}{2} + \frac{q'''}{km^2} \tag{12-3}$$

Equation 12-3 satisfies Eq. 12-2 and corresponds to the condition of symmetry with respect to $x = 0$, which may be enforced by holding the two ends of the rod at the same temperature.

Equation 12-3 can be transformed as follows:

$$\theta = \frac{q'''}{km^2}\left(1 - \frac{e^{-mx} + e^{mx}}{2}\right) + \theta_0 + \theta_0\frac{e^{-mx} + e^{mx} - 2}{2}$$

$$\theta - \theta_0 = 2\left(\theta_0 - \frac{q'''}{km^2}\right)\left(\frac{e^{mx/2} - e^{-mx/2}}{2}\right)^2$$

$$= 2\left(\theta_0 - \frac{q'''}{km^2}\right)\left(\sinh\frac{mx}{2}\right)^2 \tag{12-4}$$

Substituting $q''' = i^2/(k_eA^2)$, as in Sect. 10-2, and $m^2 = (hC)/(kA)$,

$$\theta - \theta_0 = 2\left(\theta_0 - \frac{i^2}{k_ehAC}\right)\left(\sinh\frac{mx}{2}\right)^2$$

or

$$m = \frac{2}{x}\sinh^{-1}\sqrt{\frac{\theta - \theta_0}{2}\bigg/\left(\theta_0 - \frac{i^2}{k_ehAC}\right)} \tag{12-5}$$

If, by measurement, $\theta = \theta_0$ for $x = 0$ and θ for any distance x, for instance, $\theta = \theta_L$ for $x = L$, have been found, then Eq. 12-5 yields m and k, since A, C, k_e, and h are supposed to be known, and i or the electrical potential drop e in the x direction will also be measured. This equation corresponds to Eq. 10-19 for the case without heat losses.

12-3. Determination of the Thermal Conductivity of an Insulating Material in Hollow-cylinder Form, Considering the Axial Heat Loss in the Cylindric Heating Element

Equations 12-2 and 3 can serve also to determine the axial heat losses in cylindric arrangements for measuring the thermal conductivity as described in Sect. 9–2. In such a device an insulating material M covers a cylindric electrical heating element H (Fig. 12–2). This usually consists of an inner tube I that bears a winding of resistance wire or ribbon W, and an outer tube E that covers this winding. These tubes are electrically insulated from the winding by a thin asbestos sheet or a

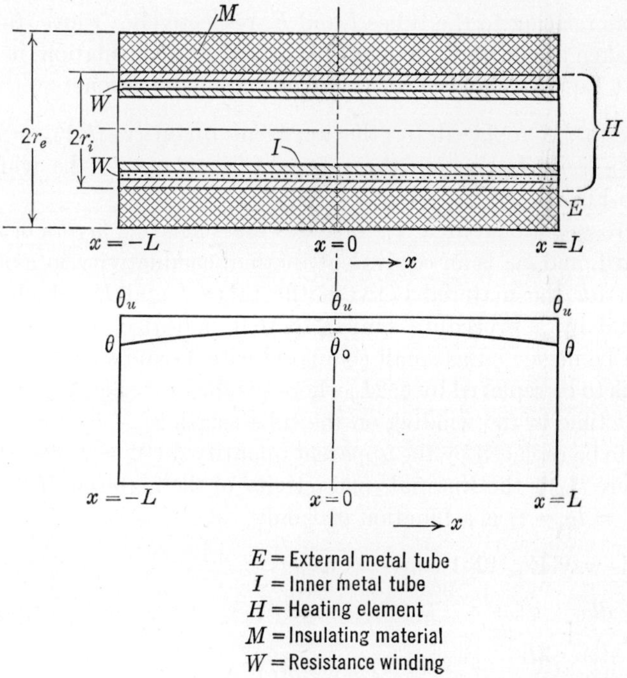

FIG. 12–2. Cylindric shell of insulating material on heating element.

FIG. 12–3. Temperature distribution in the heating element shown in Fig. 12–2.

similar material. If the whole system were infinitely long or perfectly heat insulated at the ends, all the heat produced in the heating element would leave in radial direction, according to Eq. 8–7 or 27, which may be written in the following form:

$$q = -k_M L \cdot 2\pi \frac{t_i - t_e}{\ln r_i - \ln r_e} \tag{12-6}$$

where k_M is the thermal conductivity of the material M, and the subscripts i and e refer to the inner and exterior cylindrical surface of the material M under test.

Since the system is of finite length, some heat will flow in axial direction, so that the main heat flow will not be exactly radial, and a part of

q will be lost. Van Rinsum [1920], who seems to have been the first to deal with this case in an exact manner, found that the deviation of the temperature distribution in the material M from that assumed in Eq. 12–6 is negligible. The axial heat flow in the heating element, however, must be considered. It can be assumed that for any distance x virtually no difference exists between the temperatures t_I and t_E, the subscripts relating to the tubes I and E, respectively. Since, further, t_e can be taken as constant all over the surface of the insulation, it is easily seen that Eq. 12–1 holds with the following modifications:

kA is to be replaced by the apparent quantity $(kA)_a = k_I A_I + k_E A_E + k_W A_W$, where the subscript W relates to the winding W (see Fig. 12–2).

The cross-sectional area A_W includes the insulating layers of asbestos board, and k_W is an equivalent thermal conductivity in axial direction for the material between the tubes I and E, which may be found by a procedure similar to that indicated in Sect. 10–5, or can be neglected as small compared with k_I and k_E.

$q'''A$ is to be replaced by $q/2L$, where q is the heat energy produced per unit time in the winding on the tube length $2L$.

hC is to be replaced by the apparent quantity $(hC)_a = k_M 2\pi/(\ln r_e/r_i)$, where k_M is the thermal conductivity of the material M.

$t = t_i = t_E = t_I$ is a function of x only.

In this way Eq. 12–1 converts to

$$-(kA)_a \frac{dt}{dx} + \frac{q}{2L}\,dx = -(kA)_a\left(\frac{dt}{dx} + \frac{d^2t}{dx^2}\,dx\right) + (hC)_a(t - t_e)\,dx \qquad (12\text{–}7)$$

Introducing $\theta = t - t_e$ and $m_a{}^2 = (hC)_a/(kA)_a$,

$$\frac{d^2\theta}{dx^2} = m_a{}^2\theta - \frac{q}{2L(kA)_a} \qquad (12\text{–}8)$$

The solution may be written in a form analogous to Eq. 12–4:

$$\theta - \theta_0 = 2\left(\theta_0 - \frac{q}{2Lk_aA_am_a{}^2}\right)\left(\sinh\frac{m_ax}{2}\right)^2 \qquad (12\text{–}9)$$

If again θ is measured for a given distance x, then

$$m_a = \sqrt{\frac{(hC)_a}{(kC)_a}} = \sqrt{\frac{k_M 2\pi}{(k_IA_I + k_EA_E + k_WA_W)\ln(r_e/r_i)}} \qquad (12\text{–}10)$$

is the only unknown in Eq. 12–9, and k_M can be found more exactly than from Eq. 12–6. However, in order to obtain m_a from Eq. 12–9, a trial-and-error method has to be used. Introducing

$$\theta_u = \frac{q}{2L(kA)_a m_a^2} = \frac{q}{2L(hC)_a} \tag{12–11}$$

and transforming the last factor in Eq. 12–9 algebraically, the following form is obtained:

$$\theta = (\theta_0 - \theta_u) \cosh (m_a x) + \theta_u \tag{12–12}$$

θ_u, obviously, is that uniform temperature difference $t_u - t_e$ which would exist from $x = -L$ to $x = +L$ (Fig. 12–2), if the heat flow were strictly radial and no axial loss occurred.

Solving for θ_0,

$$\theta_0 = \frac{\theta - \theta_u}{\cosh (m_a x)} + \theta_u \tag{12–13}$$

If the common temperature of the two ends of the rod is changed, the temperature excess θ at a fixed distance x varies. The influence of this variation upon θ_0 can be expressed by

$$\left(\frac{\partial \theta_0}{\partial \theta}\right)_{x=const} = \frac{1}{\cosh (m_a x)} \tag{12–14}$$

or for finite changes

$$\frac{\Delta\theta_0}{\Delta\theta} = \frac{1}{\cosh (m_a x)} \tag{12–15}$$

For not too great distances x the change $\Delta\theta_0$ is a small percentage of $\Delta\theta$.

The procedure for finding the exact value of k_M will be explained with reference to Fig. 12–3. The temperatures θ_0 and θ may be measured. If no axial heat loss occurred, the horizontal line θ_u = constant would be obtained, and k_M could be calculated from Eqs. 12–10 and 11.

θ_u can be found by means of Eq. 12–15, which takes the form

$$\frac{\Delta\theta_0}{\Delta\theta} = \frac{\theta_u - \theta_0}{\theta_u - \theta} = \frac{1}{\cosh (m_a L)} \tag{12–16}$$

and yields

$$\theta_u = \frac{\theta_0 \cosh (m_a L) - \theta}{\cosh (m_a L) - 1} \tag{12–17}$$

Since m_a contains k_M, the value of which is sought, it is recommended to begin with an approximation of Eq. 12–11, namely,

$$\theta_0 = \frac{q}{2L(hC)_a} \tag{12–18}$$

From Eqs. 12–10 and 18 a value of k_M is obtained which may be called $k_{M,1}$ (first approximation of k_M). With this value Eq. 12–17 yields $\theta_{u,1}$ (first approximation of θ_u). Substituting this in Eq. 12–11 yields $k_{M,2}$ (second approximation of k_M). In the same way a third approximation can be found, but the second approximation will be sufficient in all practical cases.

CHAPTER 13

Heat Conduction in Non-periodic Temperature Changes without Heat Losses or Gains

13–1. Limitations in Treating Problems of Heat Conduction in the Unsteady State

The change of temperature in time which occurs in heat-conduction processes is to be determined from Eq. 2–12. The solution of this differential equation for various geometrical configurations and boundary conditions * has always been a favorite field of activity for mathematicians since the days of Fourier. Fourier's series and methods are still the main tools for the treatment of unsteady-state problems of heat conduction.

Here, only the simplest forms of these problems will be dealt with, and even these in a simplified manner that may shock mathematically minded readers. Whereas generally the solutions of the differential equations will be presented, the way of finding these solutions will only occasionally be indicated. However, it may always be seen that differentiation of the final equation leads back to the original differential equation, and that the boundary conditions are satisfied. This verification will generally be the only proof presented. For better mathematical treatment, particularly convergence discussions, mathematical books should be consulted.

According to this limitation of the mathematical treatment, neither the deviation nor the use of the Fourier integral will be shown, although many of the equations presented are based on it. A simple and clear introduction to Fourier's series and integral is given in the book *Heat Conduction* by L. R. Ingersoll, Zobel, and A. C. Ingersoll [1948], to mention only one source of easy information about these important subjects.

* The term boundary conditions, as used in this book, includes the so-called initial conditions. This means that "bounds" are considered to be limitations in time as well as in space.

13-2. General Solution of the Differential Equation of Heat Conduction without Heat Sources

For simplification it will be assumed that the thermal conductivity is the same in all directions, and that no heat is generated or absorbed by the body under consideration. Then Eq. 2-12 simplifies to

$$\frac{\partial t}{\partial \tau} = \alpha \left(\frac{\partial^2 t}{\partial x^2} + \frac{\partial^2 t}{\partial y^2} + \frac{\partial^2 t}{\partial z^2} \right) \tag{13-1}$$

Combining different particular solutions of this equation,

$$t = M + K_1 x + K_2 y + K_3 z + N e^{\kappa\tau + \lambda x + \mu y + \nu z} \tag{13-2}$$

where M, K_1, K_2, K_3, κ, λ, μ, and ν are constants, and

$$\frac{\kappa}{\lambda^2 + \mu^2 + \nu^2} = \alpha$$

This may be checked by substitution of t from Eq. 13-2 in Eq. 13-1 and performance of the differentiations.

There are different types of solutions, all based upon Eq. 13-2; those with imaginary exponents lead to trigonometric functions. The boundary conditions which are needed in a special case will be decisive for the choice of the particular type to be used. In this chapter non-periodic changes of the temperature will be supposed, whereas in Chap. 14 periodic changes will be dealt with.

The famous book of Fourier [1822], which is the basis of the whole phenomenological theory of heat conduction, and the excellent treatise of Carslaw [1921] † have been amply used in these chapters. However, in many instances the author will follow the more simple presentation of the subject in the books of Groeber [1921], Groeber and Erk [1933], and Warburg [1924], who, themselves, based their procedure on that of Fourier. The brief but thorough mathematical survey of the subject by Hobson [1903-1921] has also been consulted.

13-3. Sudden Temperature Change of the Surface of an Infinitely Thick Plate

One of the simplest cases is the unidirectional heat conduction in an infinitely wide plane plate of infinite thickness in the direction x. This coordinate and the heat flow are considered to be positive in the direction

† A new edition of this book has recently been published by Carslaw and J. C. Jaeger [1947].

from the free surface into the interior of the plate.‡ If a homogeneous plane plate of finite width is perfectly heat insulated around its edge, the conduction of heat per unit time and area in the direction perpendicular to the surface is equal to that of an infinitely wide plate. Thus, this and the following section include the case of an infinitely long, perfectly insulated rod. On the other hand, a plate of moderate finite thickness often can be considered as though it were infinitely thick (see also Sect. 14–4). Hence, although in practice there are no infinitely wide and infinitely thick plates, the heat conduction in moderately wide and thick plates may be determined according to the equations of this and the following section. This has been discussed thoroughly by R. L. Perry and Berggren [1944].

The plate is supposed to be initially at the temperature t_i throughout. Now, at the time $\tau = 0$, the temperature on the surface ($x = 0$) is suddenly changed to zero and remains equal to zero at any time τ.

Here the integration leads to the form

$$t = t_i \cdot f(u) \tag{13–3}$$

where

$$u = \frac{x}{2\sqrt{\alpha\tau}} \tag{13–4}$$

and

$$f(u) = \frac{2}{\sqrt{\pi}} \int_0^{\mu} e^{-u^2}\, du \tag{13–5}$$

Expanding the e function in Eq. 13–5 and integrating yields

$$f(u) = \frac{2}{\sqrt{\pi}} \left(\frac{u}{1} - \frac{1}{1!} \cdot \frac{u^3}{3} + \frac{1}{2!} \cdot \frac{u^5}{5} - \cdots \right) \tag{13–6}$$

The function $f(u)$ is known as Gauss's error integral. Numerical values are given in Table 13–1 and plotted versus $x/(2\sqrt{\alpha\tau})$ in Fig. 13–1. It is seen that $f(u) \approx u$ if $u < 0.75$ and $f(u) \approx 1$ if $u > 1.5$.

By substitution of t from Eq. 13–3 in Eq. 13–1 and differentiation it is easily seen that Eq. 13–3 is a solution of Eq. 13–1.

The boundary conditions are:

$$t = t_i \quad \text{at} \quad \tau = 0$$

$$t = 0 \quad \text{at} \quad x = 0 \cdot (\text{for } \tau > 0)$$

‡ The mathematicians use to call this sort of body a plane plate of semi-infinite thickness because the free plane surface ($x = 0$) is considered to be at hand, whereas a plate of infinite thickness would also include the extension from $x = 0$ to $x = -\infty$.

It will be shown that they are satisfied. From Eqs. 13-4 and 6 it follows that $f(u) = 0$ at $x = 0$ for any time $\tau > 0$; this means that the surface of the plate permanently remains at $t = 0$.

TABLE 13-1

VALUES OF GAUSS' ERROR INTEGRAL $f\left(\frac{x}{2\sqrt{\alpha\tau}}\right)$

$\frac{x}{2\sqrt{\alpha\tau}}$	$f\left(\frac{x}{2\sqrt{\alpha\tau}}\right)$	$\frac{x}{2\sqrt{\alpha\tau}}$	$f\left(\frac{x}{2\sqrt{\alpha\tau}}\right)$	$\frac{x}{2\sqrt{\alpha\tau}}$	$f\left(\frac{x}{2\sqrt{\alpha\tau}}\right)$
0.00	0.0000	0.4	0.4284	1.3	0.9340
0.01	0.0113	0.5	0.5205	1.4	0.9523
0.02	0.0226	0.6	0.6039	1.5	0.9661
0.04	0.0451	0.7	0.6778	1.6	0.9763
0.06	0.0676	0.8	0.7421	1.8	0.9891
0.08	0.0901	0.9	0.7969	2.0	0.9953
0.10	0.1125	1.0	0.8427	2.2	0.9981
0.20	0.2227	1.1	0.8802	2.5	0.9996
0.30	0.3286	1.2	0.9103	3.0	1.0000

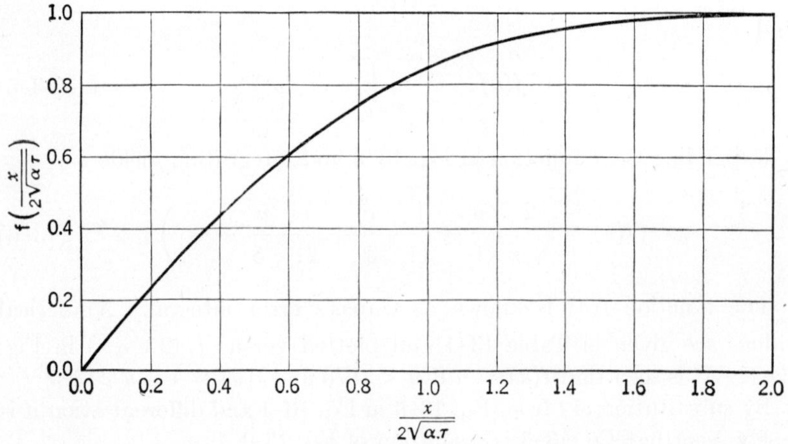

FIG. 13-1. Values of Gauss' error integral.

Furthermore, it will be proved that $f(u) = 1$ at $\tau = 0$ for any place $x > 0$, showing that initially the plate was at the temperature t_i.

Consider r as the radial distance of a point from the zero point in a two-dimensional system of rectangular coordinates y and z, so that:

$$r^2 = y^2 + z^2 \tag{13-7}$$

According to the rules of calculus,

$$\int_{y=-\infty}^{y=+\infty}\int_{z=-\infty}^{z=+\infty} e^{-(y^2+z^2)}\,dy\,dz = \int_{-\infty}^{+\infty} e^{-y^2}\,dy \int_{-\infty}^{+\infty} e^{-z^2}\,dz$$

$$= \left(\int_{-\infty}^{+\infty} e^{-y^2}\,dy\right)^2 \tag{13-8}$$

On the other hand, using polar coordinates with the surface element $2\pi r \cdot dr$, and covering again the entire plane by integration,

$$\int_{r=0}^{r=\infty} e^{-r^2}\cdot 2\pi r\cdot dr = \pi \tag{13-9}$$

Since the results of Eqs. 13-8 and 9 must be identical,

$$\int_{-\infty}^{+\infty} e^{-y^2}\,dy = \sqrt{\pi}$$

and

$$\int_0^{\infty} e^{-y^2}\,dy = \frac{\sqrt{\pi}}{2} \tag{13-10}$$

Hence, Eq. 13-5 yields $f(u) = 1$ for $u = \infty$, which, according to Eq. 13-4, occurs at $\tau = 0$.

There is an unsteadiness at $x = 0$ and $\tau = 0$, due to the sudden change of the surface temperature; but for any couple of variables $x > 0$ and $\tau > 0$ the function $f(u)$ has a steady (analytical) character.

If the temperature of the surface is changed to $t = t_0$ instead of $t = 0$ at $\tau = 0$, then in Eq. 13-3 the values t and t_i have to be replaced by $\theta = t - t_0$, and $\theta_i = t_i - t_0$, respectively, as is easily understood, so that

$$\frac{\theta}{\theta_i} = f(u) \tag{13-11}$$

The heat flow in the direction x becomes

$$q = -kA\frac{\partial\theta}{\partial x} = -kA\frac{\partial\theta}{\partial u}\cdot\frac{\partial u}{\partial x} = -\frac{kA\theta_i e^{-x^2/4\alpha\tau}}{\sqrt{\pi\alpha\tau}} \tag{13-12}$$

The heat flow entering the plate is found by taking $x = 0$ and is

$$q_0 = -\frac{kA\theta_i}{\sqrt{\pi\alpha\tau}} \tag{13-13}$$

and the total heat energy entering in the time interval from $\tau = 0$ to $\tau = \tau$ becomes

$$Q_0 = \int_{\tau=0}^{\tau=\tau} q_0 \cdot d\tau = -2kA\theta_i \sqrt{\frac{\tau}{\pi\alpha}} \qquad (13\text{–}14)$$

From Eq. 13–11, furthermore,

$$\frac{\theta_i - \theta}{\theta_i} = \frac{t - t_i}{t_0 - t_i} = 1 - f(u) \qquad (13\text{–}15)$$

This means that a sudden change of the surface temperature from t_i to t_0 causes a change of the temperature t at any point, equal to the fraction $[1 - f(u)]$ of $(t_0 - t_i)$. This fraction $(t - t_i)/(t_0 - t_i)$ is something like an efficiency ratio, with numerical values between 0 and 1 only.

13–4. Sudden Start of Heating the Plane Surface of an Infinitely Thick Plate

Another boundary condition has to be considered if a source of heat energy suddenly begins and then permanently continues to deliver a constant amount of heat per unit time to the free surface of an infinitely thick plate.§ This, for instance, occurs if a plate at moderate temperature is suddenly exposed to a source of radiation of much higher temperature.

Referring to the differential equation 13–1, the boundary conditions become

1. $t = 0$ at $\tau = 0$
2. $\dfrac{\partial t}{\partial x} = -c,$ or $v = 0$ at $x = 0$
3. $t = 0$ at $x = \infty$

where c is a constant, and a new variable v is introduced, defined by

$$v = c + \frac{\partial t}{\partial x} \qquad (13\text{–}16)$$

From Eqs. 13–1 and 16

$$\frac{\partial t}{\partial \tau} = \alpha \frac{\partial v}{\partial x}$$

and by differentiation,

$$\frac{\partial}{\partial x}\left(\frac{\partial t}{\partial \tau}\right) = \frac{\partial v}{\partial \tau} = \alpha \frac{\partial^2 v}{\partial x^2} \qquad (13\text{–}17)$$

§ Referring to the notion "infinitely thick plate," see the introductory remarks in Sect. 13–3.

According to Eqs. 13–3 to 5, this differential equation and the second boundary condition are satisfied by the form

$$v = N\frac{\sqrt{\pi}}{2}f(u) = c + \frac{\partial t}{\partial x} \tag{13–18}$$

where N is a constant.

From Eq. 13–18 by integration

$$cx + t + \phi(\tau) = \int N\frac{\sqrt{\pi}}{2}f(u)\cdot dx \tag{13–19}$$

where ϕ is an arbitrary function of time.

But

$$\int f(u)\cdot dx = x\cdot f(u) - \int x\frac{\partial f(u)}{\partial x}\,dx$$

By substituting in Eq. 13–19

$$cx + t + \phi(\tau) = N\left(x\int_0^u e^{-u^2}\,du + \sqrt{\alpha\tau}e^{-u^2}\right) \tag{13–20}$$

From the first boundary condition ($t = 0$ at $\tau = 0$), considering that $u = \infty$ for $\tau = 0$ and using Eq. 13–10,

$$cx + \phi(\tau)_{\tau=0} = Nx\frac{\sqrt{\pi}}{2}$$

Hence $\phi(\tau)_{\tau=0} = 0$, and $N = 2c/\sqrt{\pi}$.

Substituting this in Eq. 13–20,

$$t + \phi(\tau) = \frac{2c}{\sqrt{\pi}}\left(x\int_0^u e^{-u^2}\,du + \sqrt{\alpha\tau}e^{-u^2}\right) - cx$$

Substituting $t = 0$ for $x = \infty$ (third boundary condition) and considering again Eq. 13–10,

$$\phi(\tau) = cx - cx = 0$$

Hence, $\phi(\tau)$ is identical zero for any value of τ, and

$$t = \frac{2c}{\sqrt{\pi}}\left(x\int_0^u e^{-u^2}\,du + \sqrt{\alpha\tau}e^{-u^2}\right) - cx \tag{13–21}$$

Considering that $u = x/2\sqrt{\alpha\tau}$,

$$t = 2c\sqrt{\alpha\tau}u\left(\frac{2}{\sqrt{\pi}}\int_0^u e^{-u^2}\,du + \frac{e^{-u^2}}{\sqrt{\pi}u} - 1\right) \tag{13–22}$$

If $x = 0$,

$$t = 2c\sqrt{\frac{\alpha\tau}{\pi}}$$

showing that the surface temperature is increasing proportionally to $\sqrt{\tau}$. Hence, obviously, the present problem is identical with that of heat conduction in a thick plate for a temperature increase proportional to $\sqrt{\tau}$. The case of temperature increase proportional to the time will be dealt with in Sect. 13–7.

For $\tau = \infty$, Eq. 13–22 yields

$$t = -cx$$

Obviously, the linear temperature drop which permanently exists at $x = 0$, according to the second boundary condition, extends into the plate with increasing time and, finally, dominates, whereas the e functions of Eqs. 13–21 and 22 are vanishing.

13–5. Sudden Temperature Changes of One or Both Surfaces of a Plane Plate of Finite Thickness

For a plane plate of finite thickness, $2s$, a more general case of unidirectional heat conduction will be dealt with than for the infinitely thick plate. It will be assumed that the initial temperature distribution across the plate is given by an arbitrary function $t = f(x)$, and the two free faces are suddenly brought to, and then permanently held at, $t = 0$.

In this case, the boundary conditions of Eq. 13–1 are

$$\left.\begin{array}{lll} 1. & t = f(x) \quad \text{at} & \tau = 0 \\ 2. & t = 0 \quad \text{at} & x = 0 \\ 3. & t = 0 \quad \text{at} & x = 2s \end{array}\right\} \tag{13–23}$$

An appropriate solution of Eq. 13–1 is:

$$t = \sum_{\nu=1}^{\nu=\infty} N_\nu e^{-\nu^2\pi^2\alpha\tau/4s^2} \sin\frac{\nu\pi x}{2s} \tag{13–24}$$

By differentiation it may be seen that this equation satisfies Eq. 13–1, and by substitution of $x = 0$ or $x = 2s$ it follows that it is in agreement with the second and third boundary condition.

Substituting the first boundary condition, the equation becomes

$$\sum_{\nu=1}^{\nu=\infty} N_\nu \sin \frac{\nu\pi x}{2s} = f(x)$$

and introducing $\xi = \pi x/2s$,

$$\sum_{\nu=1}^{\nu=\infty} N_\nu \sin(\nu\xi) = \phi(\xi) \tag{13-25}$$

where $\phi(\xi) = f(x)$.

This, however, is a Fourier series without cosine terms. From the theory of these series it is known that

$$N_\nu = \frac{1}{\pi}\int_{\xi=0}^{\xi=2\pi} \phi(\xi)\sin(\nu\xi)\cdot d\xi$$

Because $\phi(-\xi) = -\phi(\xi)$ and $\sin(\nu\xi) = -\sin(-\nu\xi)$,

$$N_\nu = \frac{2}{\pi}\int_{\xi=0}^{\xi=\pi} \phi(\xi)\sin(\nu\xi)\cdot d\xi$$

Reconverting from ξ to x, as independent variable, and considering that $x = 2s$ for $\xi = \pi$,

$$N_\nu = \frac{1}{s}\int_{x=0}^{x=2s} f(x)\sin\frac{\nu\pi x}{2s}\cdot dx \tag{13-26}$$

Hence, the complete solution is given by Eqs. 13–24 and 26.

If, initially, the entire plate is at $t = 0$, and at the time $\tau = 0$ one surface ($x = 2s$) is suddenly heated to, and then permanently held at, $t = t_2$, whereas the other surface ($x = 0$) is held at $t = 0$, then the boundary conditions are

$$\left.\begin{array}{lll} 1. & t = 0 \quad \text{at} & \tau = 0 \\ 2. & t = 0 \quad \text{at} & x = 0 \\ 3. & t = t_2 \quad \text{at} & x = 2s \end{array}\right\} \tag{13-27}$$

Here, as in Eqs. 8–35 and 10–30, the temperature may be considered as composed of two parts:

$$t = t_{\text{I}} + t_{\text{II}} \tag{13-28}$$

The first part may be the following particular integral of Eq. 13–1:

$$t_{\text{I}} = t_2 \frac{x}{2s} \tag{13-29}$$

which is the temperature distribution in the steady state, corresponding to the second and third boundary conditions.

Since Eq. 13–1 has to satisfy also the first boundary condition, t_{II} obviously must satisfy the boundary conditions

$$\left.\begin{aligned} &1. \quad t_{II} = -t_2(x/2s) \quad \text{at} \quad \tau = 0 \\ &2. \quad t_{II} = 0 \quad \text{at} \quad x = 0 \\ &3. \quad t_{II} = 0 \quad \text{at} \quad x = 2s \end{aligned}\right\} \tag{13–30}$$

Herewith $t = t_I + t_{II}$ satisfies all three conditions of Eq. 13–27. But the boundary conditions of Eq. 13–30 are a special case of those of Eq. 13–23 with $f(x) = -t_2(x/2s)$.

Therefore t_{II} is given by Eqs. 13–24 and 26. Further according to Eq. 13–30,

$$f(x) = -t_2 \frac{x}{2s}$$

Substituting this in Eq. 13–26,

$$N_\nu = -\frac{t_2}{2s^2} \int_0^{2s} x \sin \frac{\nu \pi x}{2s} \, dx$$

and by integration ||

$$N_\nu = \frac{2t_2}{\nu \pi} \cos (\nu \pi)$$

Hence

$$N_1 = -\frac{2}{\pi} t_2$$

$$N_2 = \frac{1}{2} \cdot \frac{2}{\pi} t_2$$

$$N_3 = -\frac{1}{3} \cdot \frac{2}{\pi} t_2$$

$$\cdots\cdots\cdots$$

Herewith from Eqs. 13–24, 28, and 29

$$t = t_2 \left[\frac{x}{2s} - \frac{2}{\pi} \left(e^{-\pi^2 \alpha \tau / 4s^2} \sin \frac{\pi x}{2s} - \frac{1}{2} e^{-4\pi^2 \alpha \tau / 4s^2} \sin \frac{2\pi x}{2s} + \frac{1}{3} e^{-9\pi^2 \alpha \tau / 4s^2} \sin \frac{3\pi x}{2s} - \cdots \right) \right] \tag{13–31}$$

This solution includes the case of initial temperature t_i of the whole plate and sudden change of one surface to the temperature t_2, whereas the

|| Considering that $\int y \cdot \sin y \cdot dy = \sin y - y \cdot \cos y$.

second one is held at t_i. It is necessary only to replace every value of t in the derivation of Eq. 13–31 by $\theta = t - t_i$.

13–6. Sudden Temperature Change of the Surface of a Body at Originally Uniform Temperature

In this section will be treated the case of a plane plate of finite thickness, of an infinitely long cylinder, of a sphere and some other bodies which initially are at uniform temperature $t = t_i$ and whose surface temperature suddenly (at the time $\tau = 0$) is changed to and then permanently held at t_s.

Introducing $\theta = t - t_s$, obviously $\theta_i = t_i - t_s$ and $\theta_s = 0$. Therefore, the case dealt with is identical with that of sudden change of the surface temperature to zero, if only a temperature scale in terms of θ, instead of t, is considered. θ is the temperature excess above surface temperature, but for simplicity it will be called temperature in this section.

a. The Infinitely Wide Plate

For the plane plate this case is covered already by Eqs. 13–24 and 26. The latter, using the θ scale and $f(x) = \theta_i$, simplifies to

$$N_\nu = \frac{2\theta_i}{\nu\pi}(1 - \cos \nu\pi) \qquad (13\text{–}32)$$

Hence

$$N_1 = \frac{4}{\pi}\theta_i \qquad N_2 = 0$$

$$N_3 = \frac{1}{3}\cdot\frac{4}{\pi}\theta_i \qquad N_4 = 0$$

$$N_5 = \frac{1}{5}\cdot\frac{4}{\pi}\theta_i \qquad N_6 = 0$$

$$\cdots\cdots\cdots\cdots$$

and

$$\theta = \frac{4}{\pi}\theta_i\left(e^{-\pi^4\alpha\tau/4s^2}\sin\frac{\pi x}{2s} + \frac{1}{3}e^{-9\pi^2\alpha\tau/4s^2}\sin\frac{3\pi x}{2s} + \frac{1}{5}e^{-25\pi^2\alpha\tau/4s^2}\sin\frac{5\pi x}{2s} + \cdots\right) \qquad (13\text{–}33)$$

At the beginning ($\tau = 0$)

$$\theta = \frac{4}{\pi}\theta_i\left(\sin\frac{\pi x}{2s} + \frac{1}{3}\sin\frac{3\pi x}{2s} + \frac{1}{5}\sin\frac{5\pi x}{2s} + \cdots\right) \qquad (13\text{–}34)$$

Since at $\tau = 0$ the temperature of the plate is supposed to be θ_i the sum of the series in parentheses must be equal to $\pi/4$ for any value of x within the limits $2s > x > 0$.

For $\tau = \infty$, every point in the plate will be at $\theta = 0$, since the two surfaces are held at this temperature.

When dealing with symmetrical bodies, it is generally more suitable to count distances from the symmetry plane, axis, or center as zero point. For the plate a median plane would be the locus of $x = 0$. Then, instead of Eq. 13–33, the following equation would have been obtained:

$$\theta = \frac{4}{\pi}\theta_i\left(e^{-\pi^2\alpha\tau/4s^2}\cos\frac{\pi x}{2s} - \frac{1}{3}e^{-9\pi^2\alpha\tau/4s^2}\cos\frac{3\pi x}{2s} + \cdots\right) \tag{13–35}$$

Here, for symmetry reasons, cosine terms appear, instead of the sine terms of Eq. 13–33.

Equation 13–35 may be written in a more general form as

$$\theta = \theta_i \sum_{\nu=1}^{\nu=\infty} \frac{2}{\dfrac{2\nu-1}{2}\pi\sin\dfrac{2\nu-1}{2}\pi} e^{-\left(\frac{2\nu-1}{2}\pi\right)^2\alpha\tau/s^2}\cos\left(\frac{2\nu-1}{2}\pi\frac{x}{s}\right) \tag{13–36}$$

The reason for some details of this representation is to bring the equation in analogy to that for a cylindric body, as will be explained in Article *b*.

b. The Infinitely Long Cylinder

The second symmetrical body to be dealt with is the infinitely long cylinder of radius s.

The result is in close analogy to Eq. 13–36. It will be given without proof by the equation

$$\theta = \theta_i \sum_{\nu=1}^{\nu=\infty} \frac{2}{\beta_\nu J_1(\beta_\nu)} e^{-\beta_\nu^2\alpha\tau/s^2} J_0\left(\beta_\nu\frac{r}{s}\right) \tag{13–37}$$

where J_0 = the Bessel function of first kind and zero order,
J_1 = the Bessel function of first kind and first order, and
β_ν = the νth zero place of J_0.

The Bessel functions J_0 and J_1 are closely related to the cosine and sine function, respectively, and the zero places of J_0, called here β_ν, are in analogy with the zero places of the cosine function $(2\nu - 1)\pi/2$ (see Eq. 13–36). This is the reason why $(2\nu - 1)\pi/2$ and $\sin\,(2\nu - 1)\pi/2$

have been introduced in Eq. 13–36. The latter function alternates from $+1$ to -1, producing the alternating signs of Eq. 13–35. The function $J_1(\beta_\nu)$ in Eq. 13–37, however, not only changes the sign, but decreases in its numerical value at the same time, as does $J_0\left(\beta_\nu \frac{r}{s}\right)$ for any fixed value of r/s.

c. The Sphere

The temperature distribution in a sphere will also be given without the proof:

$$\theta = \theta_i \sum_{\nu=1}^{\nu=\infty} \frac{2}{-\cos(\nu\pi)} e^{-(\nu\pi)^2 \alpha\tau/s^2} \frac{\sin\left(\nu\pi \frac{r}{s}\right)}{\nu\pi \frac{r}{s}} \tag{13–38}$$

Since $\cos(\nu\pi)$ is alternating between -1 and $+1$, it can be shown that Eq. 13–38 is in perfect analogy to Eq. 13–36, considering that in sphere functions $(\sin \phi)/\phi$ is the equivalent of $\cos \phi$ of linear functions and of $J_0(\phi)$ of cylinder functions. Generally, equations for cylinder and sphere can directly be written down in analogy to the equations for plates (as Eqs. 13–37 and 38 have the same pattern as Eq. 36). However, it is not always simple to find the correct analogy. For instance, the denominator of any term on the right side of Eq. 13–37 can be expressed by

$$-\left[\frac{dJ_0\left(\beta_\nu \frac{r}{s}\right)}{d\frac{r}{s}}\right]_{\frac{r}{s}=1} = \beta_\nu J_1(\beta_\nu) \tag{13–39}$$

Thus, from the third factor under the sum sign of Eq. 13–38, the denominator of the first factor is found as follows:

$$-\left[\frac{d\dfrac{\sin\left(\nu\pi \frac{r}{s}\right)}{\nu\pi \frac{r}{s}}}{d\frac{r}{s}}\right]_{\frac{r}{s}=1} = -\cos(\nu\pi) \tag{13–40}$$

By analogy, the respective denominators in Eq. 13–36 must be

$$-\left[\frac{d\cos\left(\dfrac{2\nu-1}{2}\pi\dfrac{x}{s}\right)}{d\dfrac{x}{s}}\right]_{\frac{x}{s}=1} = (2\nu-1)\frac{\pi}{2}\sin\frac{2\nu-1}{2}\pi \qquad (13\text{–}41)$$

Using these analogies and proceeding in the right way may save much work in handling these functions and equations.

Williamson and L. H. Adams [1919], using Eq. 13–38, calculated θ/θ_i for different values of r/s and $\alpha\tau/s^2$. Table 13–2 is taken from their paper.

TABLE 13–2

Temperature Ratio θ/θ_i for a Sphere

(According to Eq. 13–38)

r/s	1000 $\alpha\tau/s^2$									
	0	4	16	36	64	100	196	256	400	∞
0	1	1.000	1.000	0.994	0.910	0.707	0.288	0.160	0.0386	0
0.05	1	1.000	1.000	0.994	0.908	0.705	0.287	0.159	0.0385	0
0.25	1	1.000	1.000	0.979	0.858	0.647	0.260	0.144	0.0347	0
0.33···	1	1.000	0.999	0.961	0.813	0.600	0.239	0.132	0.0319	0
0.5	1	1.000	0.990	0.875	0.676	0.474	0.184	0.102	0.0246	0
0.66···	1	1.000	0.906	0.679	0.473	0.316	0.119	0.0661	0.0160	0
0.75	1	0.993	0.792	0.531	0.354	0.232	0.0869	0.0480	0.0116	0
0.95	1	0.394	0.179	0.103	0.0644	0.0411	0.0152	0.0084	0.0020	0
1	1	0	0	0	0	0	0	0	0	0

d. Different Bodies

The same authors calculated numerical values of θ_0/θ_i for the infinitely wide plate, the infinitely long cylinder, the sphere, and some other bodies. Their results are represented in Table 13–3 and Fig. 13–2.

This table and this figure include values for the infinitely long square bar, the cube, and a cylinder of equal length and diameter. In the same paper the complete equations of the temperature distribution in all these bodies are presented; further those for the cylinder of arbitrary length L, the rectangular parallelepiped of any side lengths ("brick-shaped"), and a special case of the parallelepiped, namely, an infinitely

TABLE 13-3

TEMPERATURE RATIO θ_0/θ_i FOR DIFFERENT SYMMETRIC BODIES

$\alpha\tau/s^2$	Plate of Infinite Width	Square Bar of Infinite Length	Cube	Cylinder of Infinite Length	Cylinder of Equal Length and Diameter	Sphere
0	1	1	1	1	1	1
0.032	1.000	1.000	1.000	0.999	0.999	0.998
0.080	0.975	0.951	0.927	0.918	0.895	0.828
0.100	0.949	0.901	0.856	0.848	0.805	0.707
0.160	0.846	0.715	0.605	0.627	0.530	0.409
0.240	0.702	0.493	0.346	0.399	0.280	0.187
0.320	0.578	0.334	0.193	0.252	0.145	0.085
0.800	0.177	0.031	0.006	0.016	0.003	0.001
1.600	0.025	0.001	0.000	0.000	0.000	0.000
3.200	0.000	0.000	0.000	0.000	0.000	0.000

NOTE. s is the radius or the half-side length of the cross section.

long rod of rectangular cross section for any side lengths of the cross section.

The case of a hollow cylinder of infinitely thick wall is of the same importance for the approximate calculation of the transient ¶ conduction through the walls of a tube as is the case of the semi-infinite plate with respect to the plate of infinite thickness. An infinitely long circular cylindric hole is supposed to exist in an infinitely extended body of initial temperature t_i. Let s and r, respectively, be the radius of the hole and any greater distance from, and normal to, the cylinder axis. The surface of the hole is suddenly (at time $\tau = 0$) brought to, and then permanently held at, temperature t_s. Let, further, $\theta = t - t_i$, and $\theta_s = t_s - t_i$.

Carslaw and J. C. Jaeger [1940] found the analytical solutions for θ/θ_s and $q_s/(kL\theta_s)$ as two integrals, containing Bessel functions, where q_s is the rate of heat flow entering the body from the cylindric surface over an axial length L. J. C. Jaeger and Clarke [1942] tabulated the dimensionless quantity $q_s/(kL\theta_s)$. Another dimensionless quantity is $\alpha Q_s/(kLs^2\theta_s)$, where Q_s is the total heat energy which has entered the body in a section of length L at the time τ. Both dimensionless groups have been tabulated also by R. L. Perry and Berggren [1944]. Table 13-4 is taken from their paper.

¶ "Transient" is a notation for non-periodic changing.

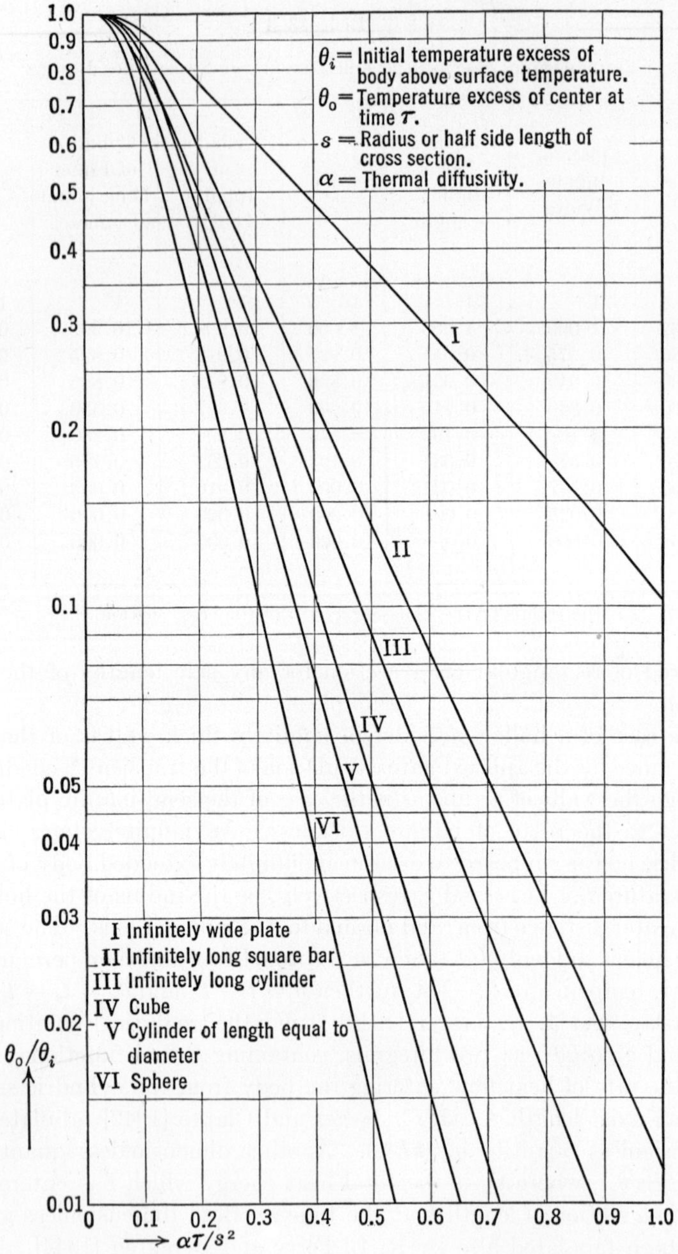

FIG. 13–2. Change of center temperature of different bodies caused by sudden change of surface temperature.

TABLE 13–4

DIMENSIONLESS GROUPS, CHARACTERIZING THE TRANSIENT HEAT FLOW FROM A PLANE OR CYLINDRIC SURFACE INTO AN INFINITELY EXTENDED MEDIUM

$\alpha\tau/s^2$	Group $q_s/(kL\theta_s)$, Characteristic for the Rate of Heat Flow Entering:		Group $\alpha Q_s/(kLs^2\theta_s)$, Characteristic for Total Heat Flow Entering:	
	Semi-infinite Plane Plate	Hollow Cylinder of Infinitely Large Outer Diameter [1]	Semi-infinite Plane Plate	Hollow Cylinder of Infinitely Large Outer Diameter [1]
0.001	112.08	115.2	0.2242	0.226
0.005	50.13	52.9	0.5013	0.516
0.01	35.45	38.51	0.7090	0.737
0.02	25.06	28.09	1.0026	1.060
0.05	15.85	18.83	1.5854	1.727
0.1	11.21	14.13	2.2417	2.529
0.2	7.926	10.776	3.1706	3.740
0.4	5.605	8.373	4.4843	5.613
0.6	4.576	7.288	5.4918	7.164
0.8	3.963	6.633	6.3411	8.547
1.0	3.545	6.180	7.0898	9.822
1.5	2.894	5.467	8.6829	12.72
2.0	2.506	5.029	10.026	15.35
2.5	2.242	4.726	11.209	17.79
3.0	2.047	4.499	12.280	20.09
3.5	1.895	4.321	13.265	22.30

[1] The first two values of Column 3 and the values of Column 5 have been determined graphically.

The values for the semi-infinite plate which are shown for comparison are calculated for the same area $A = 2\pi sL$, for which q_s and Q_s are calculated in the case of the cylindric surface. Substituting this in Eqs. 13–13 and 14, and adjusting the symbols and the sign, the following expressions are obtained:

$$\frac{q_s}{kL\theta_s} = \frac{2\sqrt{\pi}}{\sqrt{\dfrac{\alpha\tau}{s^2}}}$$

and

$$\frac{\alpha Q_s}{kLs^2\theta_s} = 4\sqrt{\pi} \cdot \sqrt{\frac{\alpha\tau}{s^2}}$$

These equations have been used for the computation of Columns 2 and 4 of Table 13–4.

Carslaw and J. C. Jaeger [1940] have also derived an equation for θ/θ_s when the cylinder wall is of finite thickness, and this equation is reproduced in R. L. Perry and Berggren's [1944] paper. The equation is very complex and includes slowly converging series. Therefore, generally, graphical, numerical, or experimental solutions must be used to treat practical cases (see Chapters 18, 19, and 20). The exact results, as given in Tables 13–1 and 4, may be used to shorten and simplify the numerical or graphical procedure (see Sect. 19–5).

13–7. Temperature Increase Proportional to Time on the Surface of a Body at Originally Uniform Temperature

Williamson and Adams further developed the equations for heating a body from an initial overall temperature t_i under the assumption that the surface temperature is increasing proportionally to the time. Again all sorts of bodies, mentioned in Sect. 13–6, are taken into consideration. Numerical solutions, however, are given only for the infinitely extended plate, the infinitely long cylinder, and the sphere; moreover the numerical data refer only to pieces of one sort of glass and to one particular time rate of temperature increase at the surface. The reason for this restriction was that the purpose of the investigation was

TABLE 13–5

θ/θ_s for an Infinitely Wide Plate whose Surface Temperature Increases Proportionally to Time ($\theta_s = c\tau$)

x/s	$\alpha\tau/s^2$								
	0.08	0.16	0.20	0.32	0.40	0.80	1.60	2.00	4.00
0	0.004	0.045	0.074	0.170	0.231	0.464	0.694	0.752	0.875
0.33···	0.030	0.104	0.142	0.245	0.305	0.522	0.728	0.779	0.889
0.5	0.082	0.194	0.238	0.346	0.402	0.594	0.770	0.814	0.906
0.66···	0.214	0.353	0.398	0.498	0.545	0.698	0.830	0.862	0.930
0.8	0.418	0.548	0.586	0.664	0.699	0.802	0.889	0.911	0.955
1	1	1	1	1	1	1	1	1	1

to precalculate the temperature changes in the manufacture of optical glass.

TABLE 13-6

θ/θ_s FOR AN INFINITELY LONG CYLINDER WHOSE SURFACE TEMPERATURE INCREASES PROPORTIONALLY TO TIME ($\theta_s = c\tau$)

r/s	$\alpha\tau/s^2$			
	0.08	0.16	0.32	0.80
0	0.016	0.123	0.354	0.691
0.33···	0.054	0.191	0.420	0.725
0.5	0.122	0.287	0.505	0.768
0.66···	0.268	0.443	0.628	0.828
0.8	0.470	0.621	0.755	0.888
1	1	1	1	1

However, it is easy to generalize the results given in the tables of Williamson and Adams. From Tables 13-5, 6, and 7 [see Jakob, 1933, pp. 204 and 205] the temperature excess, $\theta = t - t_i$, of any point of the

TABLE 13-7

θ/θ_s FOR A SPHERE WHOSE SURFACE TEMPERATURE INCREASES PROPORTIONALLY TO TIME ($\theta_s = c\tau$)

r/s	$\alpha\tau/s^2$				
	0.016	0.08	0.16	0.32	0.80
0	0.00	0.054	0.219	0.506	0.792
0.33···	0.00	0.090	0.290	0.560	0.815
0.5	0.00	0.162	0.385	0.626	0.844
0.66···	0.02	0.312	0.529	0.722	0.884
0.8	0.14	0.516	0.686	0.819	0.925
1	1	1	1	1	1

body above the initial temperature $t_i (\theta_i = 0)$ of the body can be found for an arbitrary instant of time τ, if the rate of temperature increase at the surface, $(\partial\theta/\partial\tau)_s$, is a given constant c, so that $\theta_s = (\partial\theta/\partial\tau)_s\tau = c\tau$.

13-8. Sudden Temperature Change of the Environment of Some Symmetrical Bodies at Originally Uniform Temperature (Groeber's Charts)

The most general case that will be treated in this chapter involves the sudden temperature change of the environment of some symmetrical bodies, particularly the infinitely wide plane plate, the infinitely long cylinder, and the sphere. More specifically, it will be assumed that the body which initially is at the temperature t_i suddenly is brought into an environment at temperature t_e. Here, of course, the coefficient of heat transfer h enters the equations. But, since the interest is focused on the heat conduction and h is assumed to be constant, these cases, like those in Chapter 11, are dealt with in the present part of the book, which is devoted to heat-conduction problems (see the remarks in Sect. 11-1).

The solutions of the respective differential equations will be presented in the form chosen by Groeber [1925], and his procedure in the derivation of these solutions [1921] will be followed and abridged only as far as possible.

Charts for the practical use of the solutions have been published by Gurney and Lurie [1923], Groeber [1925], Beuken and Paschkis [1937], Heisler [1947], and others. Those of Groeber are reproduced in this section, those of Beuken and Paschkis in Sect. 20-9, those of Heisler in Sects. 13-9 and 20-9.

a. The Infinitely Wide Plate

Introducing $\theta = t - t_e$ as the temperature excess of a point over the environment temperature t_e, Eq. 13-1 becomes

$$\frac{\partial \theta}{\partial \tau} = \alpha \frac{\partial^2 \theta}{\partial x^2} \tag{13-42}$$

with the boundary conditions

1. $\theta = \theta_i \quad \text{at} \quad \tau = 0$

2. $\dfrac{\partial \theta}{\partial x} = -b\theta \quad \text{at} \quad x = s$

3. $\dfrac{\partial \theta}{\partial x} = b\theta \quad \text{at} \quad x = -s$

where

$$b = \frac{h}{k} \tag{13-43}$$

Particular integrals of Eq. 13–42 are

$$\theta = Me^{-m^2\alpha\tau} \sin (mx)$$

and

$$\theta = Ne^{-n^2\alpha\tau} \cos (nx) \tag{13–44}$$

as may be checked by differentiation.

For symmetry reasons, $M = 0$. Substituting θ from Eq. 13–44 in the second or third boundary condition,

$$\frac{ns}{bs} = \cot (ns) \tag{13–45}$$

This is a transcendental equation, the meaning of which Fourier made clear by the following geometrical construction (see Fig. 13–3):

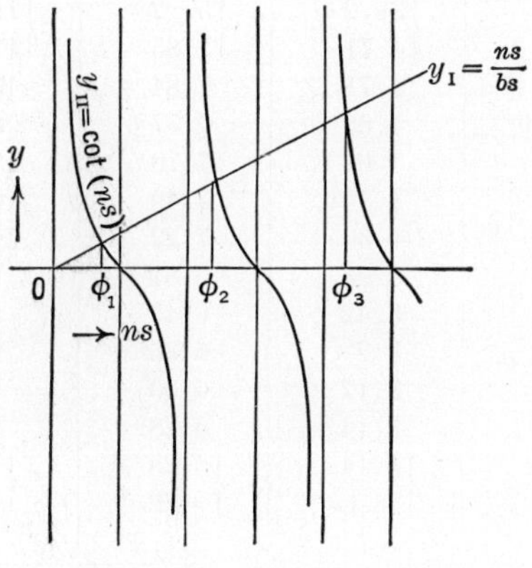

FIG. 13–3. Graphical interpretation and solution of Eq. 13–45.

Using ns as abscissa and y as ordinate, the figure shows a straight line $y_{\mathrm{I}} = ns/bs$ and a system of curves $y_{\mathrm{II}} = \cot (ns)$ for a given value of bs. The infinite number of intersecting points of the curves y_{I} and y_{II} determines an infinite number of "roots" of Eq. 13–45, i.e., of values of $ns > 0$, for which the equation is satisfied. The νth root may be called ϕ_ν.

Table 13–8 shows the first five roots of this equation as calculated by Groeber. With these values the general solution becomes

$$\theta = \sum_{\nu=1}^{\nu=\infty} N_\nu e^{-\phi_\nu{}^2\alpha\tau/s^2} \cos \left(\phi_\nu \frac{x}{s}\right) \tag{13–46}$$

According to the first boundary condition of Eq. 13–42,

$$\theta_i = \sum_{\nu=1}^{\nu=\infty} N_\nu \cos\left(\phi_\nu \frac{x}{s}\right) \tag{13–47}$$

which again is a Fourier series. This time, however, the series is not so simple as in Eq. 13–25, where ν were the integers from 1 to ∞, because here $\phi_\nu = \frac{2\nu - 1}{2}\pi$ holds only if $bs = \infty$, and $\phi_\nu = \nu\pi$ only if $bs = 0$.

TABLE 13–8

ROOTS ϕ_ν OF EQ. 13–45 (SUDDEN TEMPERATURE CHANGE OF ENVIRONMENT OF PLANE PLATE)

bs	ϕ_1	ϕ_2	ϕ_3	ϕ_4	ϕ_5
∞	$\pi/2 =$ 1.57	$3\pi/2 =$ 4.71	$5\pi/2 =$ 7.85	$7\pi/2 =$ 11.00	$9\pi/2 =$ 14.15
1000	1.57	4.71	7.84	10.98	14.13
100	1.56	4.66	7.77	10.88	14.00
50	1.54	4.62	7.70	10.78	13.87
20	1.50	4.49	7.49	10.51	13.55
10	1.43	4.30	7.22	10.20	13.22
4	1.26	3.93	6.81	9.78	12.87
1	0.86	3.42	6.43	9.52	12.65
0.5	0.65	3.29	6.36	9.47	12.61
0.1	0.31	3.17	6.30	9.43	12.57
0.01	0.10	3.14	6.28	9.42	12.57
0	0.00 $= 0 \cdot \pi$	3.14 $= 1\pi$	6.28 $= 2\pi$	9.42 $= 3\pi$	12.57 $= 4\pi$

For any other value of bs the arguments of the cosine terms are other prescribed values, as given in Table 13–8.

By a procedure similar to that for simple Fourier series, here the coefficients are found to be

$$N_\nu = \frac{\int_{-s}^{+s} \theta_i \cos\left(\phi_\nu \frac{x}{s}\right) \cdot dx}{\int_{-s}^{+s} \cos^2\left(\phi_\nu \frac{x}{s}\right) \cdot dx} = \theta_i \frac{2 \sin \phi_\nu}{\phi_\nu + \sin \phi_\nu \cos \phi_\nu} \tag{13–48}$$

By substitution in Eq. 13–46

$$\frac{\theta}{\theta_i} = \sum_{\nu=1}^{\nu=\infty} \frac{2 \sin \phi_\nu}{\phi_\nu + \sin \phi_\nu \cos \phi_\nu} e^{-\phi_\nu^2 \alpha\tau/s^2} \cos\left(\phi_\nu \frac{x}{s}\right) \tag{13–49}$$

Since the quantities ϕ_ν depend on bs only, Eq. 13–49 can be considered a special case of an equation

$$\frac{\theta}{\theta_i} = \Phi\left(bs, \frac{\alpha\tau}{s^2}, \frac{x}{s}\right) \qquad (13\text{–}50)$$

which contains only dimensionless groups as variables, and in which Φ represents the function given as an infinite sum in Eq. 13–49.

Equation 13–35 is a special case of Eq. 13–49 which occurs if $h = \infty$.

Equation 13–49 can be used to calculate, for any arbitrary instant of time, $\tau > 0$, numerical values of the maximum temperature excess θ_0 in the plate, its surface temperature excess θ_s, and the excess of heat content H of a piece of the plate, having the area A, between $x = 0$ and $x = s$, over its heat content at the temperature of environment. If the initial heat content of this half of the plate was H_i, then the heat given up by either surface of the plate in the time interval from $\tau = 0$ to $\tau = \tau$ may be expressed by

$$Q = H_i - H \qquad (13\text{–}51)$$

The temperature excesses θ_0 and θ_s are found by substituting $x = 0$ and $x = s$, respectively, in Eq. 13–49.

The heat loss Q, finally, is identical with the decrease of the heat capacity of the respective volume of the plate. Hence, assuming $\rho c_p =$ constant, θ from Eq. 13–49 is to be substituted in

$$Q = A\rho c_p \int_{x=0}^{x=s} (\theta_i - \theta)\, dx \qquad (13\text{–}52)$$

Considering Eq. 13–51 the initial heat content is given by

$$H_i = As\rho c_p \theta_i \qquad (13\text{–}53)$$

In this way the following three equations for an infinitely wide plate have been obtained:

$$\frac{\theta_0}{\theta_i} = \sum_{\nu=1}^{\nu=\infty} \frac{2 \sin \phi_\nu}{\phi_\nu + \sin \phi_\nu \cos \phi_\nu} e^{-\phi_\nu{}^2 \alpha\tau/s^2} = \Phi_0\left(bs, \frac{\alpha\tau}{s^2}\right) \qquad (13\text{–}54)$$

$$\frac{\theta_s}{\theta_i} = \sum_{\nu=1}^{\nu=\infty} \frac{2 \sin \phi_\nu}{\phi_\nu + \sin \phi_\nu \cos \phi_\nu} e^{-\phi_\nu{}^2 \alpha\tau/s^2} \cos \phi_\nu = \Phi_s\left(bs, \frac{\alpha\tau}{s^2}\right) \qquad (13\text{–}55)$$

$$\frac{H}{H_i} = 1 - \sum_{\nu=1}^{\nu=\infty} \frac{2 \sin^2 \phi_\nu}{\phi_\nu{}^2 + \phi_\nu \sin \phi_\nu \cos \phi_\nu} (1 - e^{-\phi_\nu{}^2 \alpha\tau/s^2})$$

$$= \Phi_H\left(bs, \frac{\alpha\tau}{s^2}\right) \qquad (13\text{–}56)$$

where Φ_0, Φ_s, and Φ_H are the functions represented by the respective infinite sums. Each can take values between 0 and 1 only and represents a sort of efficiency of cooling, in regard to either the diminution of the

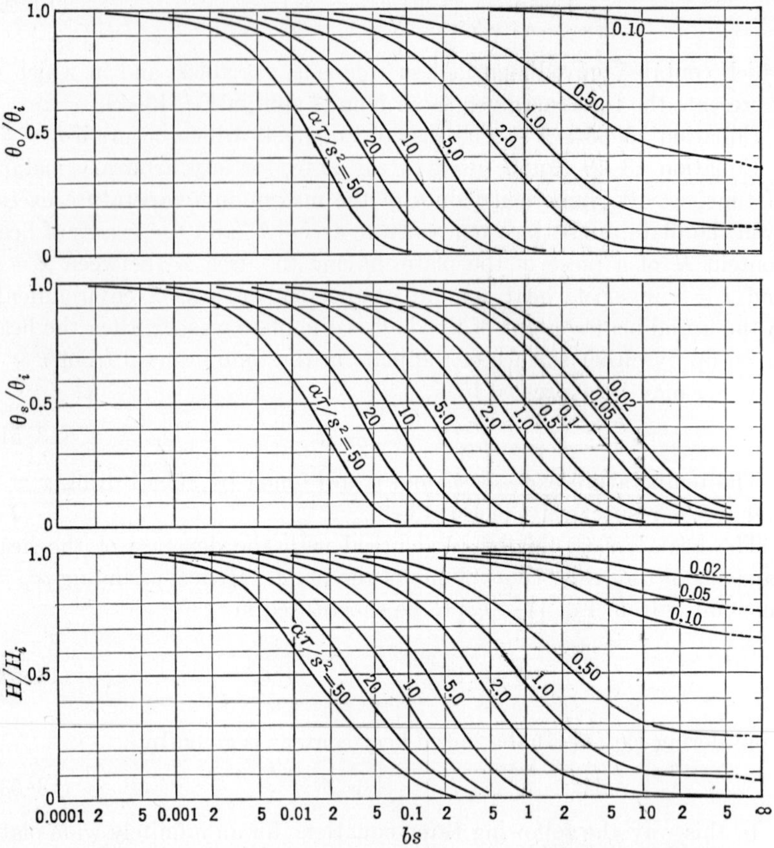

Fig. 13-4. θ_0/θ_i versus bs for plane plate, according to Eq. 13-54.

Fig. 13-5. θ_s/θ_i versus bs for plane plate, according to Eq. 13-55.

Fig. 13-6. H/H_i for plane plate, versus bs according to Eq. 13-56.

Figures from Groeber [1925].

temperature excess of the plate when placed into a colder environment, or the diminution of the heat contents.

Instead of Eq. 13-56, the ratio of the heat loss from one side of the plate to the original heat content may be expressed by

$$\frac{Q}{H_i} = \frac{H_i - H}{H_i} = 1 - \Phi_H\left(bs, \frac{\alpha\tau}{s^2}\right) \tag{13-57}$$

Groeber calculated numerical values of the functions Φ_0, Φ_s and $(1 - \Phi_H)$ for different values of bs in the range of 0.0001 to ∞, and for $\alpha\tau/s^2$ between 0.02 and 50. These are tabulated and represented by graphs in his paper [1925]. Figures 13–4 to 6 are reproductions of these graphs; in Fig. 13–6, however, Φ_H values are plotted instead of his $(1 - \Phi_H)$ values.

These charts allow one to find immediately the temperature in the median plane and at the surface of a plane plate and its heat loss for an arbitrary time without any calculation, if only the numerical values of the two dimensionless groups, bs and $\alpha\tau/s^2$, have been calculated.

Equations 13–49, 54, 55, and 56 obviously hold also, and Figs. 13–4 to 6 may be used for a plane plate which is perfectly insulated on one side and suddenly cooled on the other one, if x is counted from the insulated surface as zero and s is considered as the thickness of this plate.

b. The Infinitely Long Cylinder

In perfect analogy to the procedure used for the plate, equations for an infinitely long cylinder and for a sphere of radius s have been derived, where θ is the temperature excess of a point at distance r from the center line of the cylinder or the center of the sphere over the environment temperature t_e.

The differential equation for the cylinder is:

$$\frac{\partial\theta}{\partial\tau} = \alpha\left(\frac{\partial^2\theta}{\partial r^2} + \frac{1}{r}\frac{\partial\theta}{\partial r}\right) \tag{13–58}$$

and the boundary conditions are

1. $$\theta = \theta_i \quad \text{at} \quad \tau = 0$$

2. $$\frac{\partial\theta}{\partial r} = -b\theta \quad \text{at} \quad r = s$$

Assume

$$\theta = Ne^{-n^2\alpha\tau}f(nr) \tag{13–59}$$

to be a particular integral of Eq. 13–58. The form of the function $f(nr)$ is found by substitution of θ from Eq. 13–59 in Eq. 13–58. This yields

$$\frac{d^2f(nr)}{dr^2} + \frac{1}{r}\cdot\frac{df(nr)}{dr} + n^2\cdot f(nr) = 0 \tag{13–60}$$

and introducing $z = nr$

$$\frac{d^2f(z)}{dz^2} + \frac{1}{z}\cdot\frac{df(z)}{dz} + f(z) = 0 \tag{13–61}$$

which is Bessel's differential equation of zero order. Hence, Eq. 13–59 becomes

$$\theta = Ne^{-n^2\alpha\tau}J_0(nr) \tag{13–62}$$

For symmetry reasons no particular solution with $Y_0(nr)$ as factor can exist.

Again, n cannot take arbitrary values, but only those compatible with the second boundary condition of Eq. 13–58. From this condition and Eq. 13–62:

$$\frac{ns}{bs} = \frac{J_0(ns)}{J_1(ns)} \tag{13–63}$$

in close analogy to Eq. 13–45. The roots of this transcendental equation may be called χ_ν. Some values of χ_ν, as calculated by Groeber, are represented in Table 13–9. From Eq. 13–63, as from Table 13–9, it is

TABLE 13–9

Roots χ_ν of Eq. 13–63 (Sudden Temperature Change of Environment of Cylinder)

bs	χ_1	χ_2	χ_3	χ_4	χ_5
∞	2.405	5.520	8.654	11.792	14.931
50	2.35	5.41	8.48	11.56	
20	2.29	5.26	8.25	11.27	
10	2.17	5.03	7.96	10.94	
4	1.906	4.60	7.52	10.54	
1.0	1.253	4.08	7.16	10.27	
0.5	0.940	3.96	7.09	10.22	
0.1	0.443	3.86	7.03	10.19	
0.05	0.315	3.85	7.02	10.18	
0	0.000	3.832	7.016	10.174	13.324

seen that the zero points ("roots") of J_0 belong to $bs = \infty$, and those of J_1 belong to $bs = 0$. Substituting χ_ν in Eq. 13–62, the general solution becomes

$$\theta = \sum_{\nu=1}^{\nu=\infty} N_\nu e^{-\chi_\nu^2\alpha\tau/s^2} J_0\left(\chi_\nu \frac{r}{s}\right) \tag{13–64}$$

The coefficients N_ν are found, as those of Fourier series, and substituted in Eq. 13-64. The result is:

$$\frac{\theta}{\theta_i} = \sum_{\nu=1}^{\nu=\infty} \frac{2J_1(\chi_\nu)}{\chi_\nu[J_0^2(\chi_\nu) + J_1^2(\chi_\nu)]} e^{-\chi_\nu^2 \alpha\tau/s^2} J_0\left(\chi_\nu \frac{r}{s}\right) \tag{13-65}$$

or

$$\frac{\theta}{\theta_i} = \mathrm{X}\left(bs, \frac{\alpha\tau}{s^2}, \frac{r}{s}\right) \tag{13-66}$$

where X stands for the sum in Eq. 13-65. For $h = \infty$, Eq. 13-65 simplifies to the form of Eq. 13-37.

Also the further procedure is analogous to that for the plane plate. In particular

$$Q = H_i - H = L\rho c_p \int_{r=0}^{r=s} (\theta_i - \theta) 2\pi r \cdot dr \tag{13-67}$$

and

$$H_i = L\pi s^2 \rho c_p \theta_i \tag{13-68}$$

The final equations corresponding to Eqs. 13-54 to 57 are

$$\frac{\theta_0}{\theta_i} = \sum_{\nu=1}^{\nu=\infty} \frac{2}{\chi_\nu} \cdot \frac{J_1(\chi_\nu)}{J_0^2(\chi_\nu) + J_1^2(\chi_\nu)} e^{-\chi_\nu^2 \alpha\tau/s^2} = \mathrm{X}_0\left(bs, \frac{\alpha\tau}{s^2}\right) \tag{13-69}$$

$$\frac{\theta_s}{\theta_i} = \sum_{\nu=1}^{\nu=\infty} \frac{2}{\chi_\nu} \cdot \frac{J_1(\chi_\nu)}{J_0^2(\chi_\nu) + J_1^2(\chi_\nu)} e^{-\chi_\nu^2 \alpha\tau/s^2} J_0(\chi_\nu) = \mathrm{X}_s\left(bs, \frac{\alpha\tau}{s^2}\right) \tag{13-70}$$

$$\frac{H}{H_i} = 1 - \sum_{\nu=1}^{\nu=\infty} \frac{4}{\chi_\nu^2} \cdot \frac{J_1^2(\chi_\nu)}{J_0^2(\chi_\nu) + J_1^2(\chi_\nu)} (1 - e^{-\chi_\nu^2 \alpha\tau/s^2})$$

$$= \mathrm{X}_H\left(bs, \frac{\alpha\tau}{s^2}\right) \tag{13-71}$$

$$\frac{Q}{H_i} = 1 - \mathrm{X}_H\left(bs, \frac{\alpha\tau}{s^2}\right) \tag{13-72}$$

Figures 13-7 to 9 give a graphical representation of Eqs. 13-69 to 71.

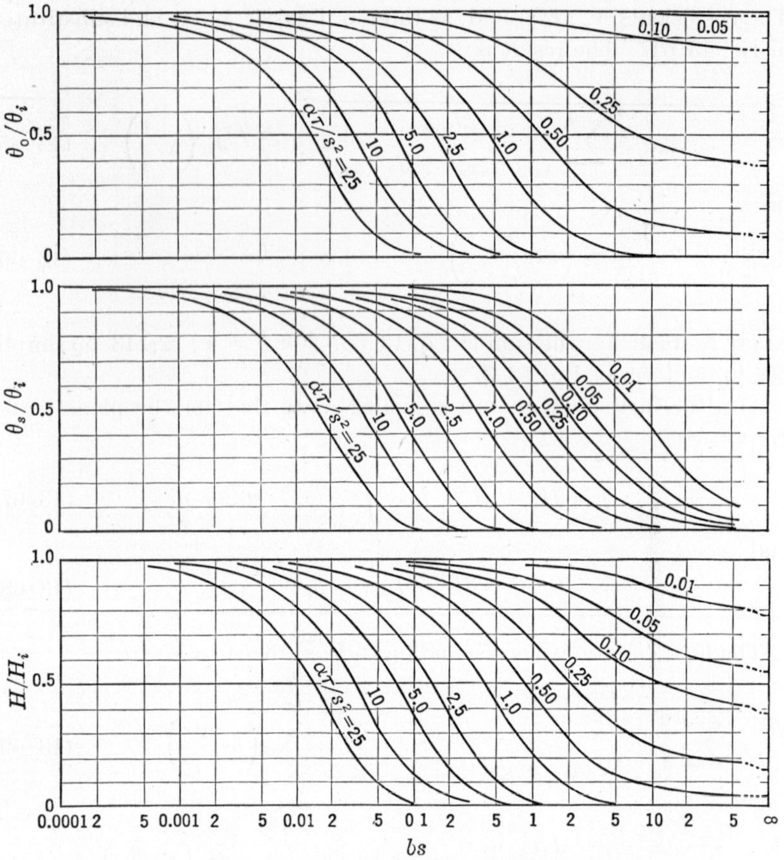

FIG. 13–7. θ_0/θ_i versus bs for cylinder, according to Eq. 13–69.
FIG. 13–8. θ_s/θ_i versus bs for cylinder, according to Eq. 13–70.
FIG. 13–9. H/H_i versus bs for cylinder, according to Eq. 13–71.
Figures from Groeber [1925].

c. The Sphere

Here the differential equation is

$$\frac{\partial\theta}{\partial\tau} = \alpha\left(\frac{\partial^2\theta}{\partial r^2} + \frac{2}{r}\cdot\frac{\partial\theta}{\partial r}\right) \tag{13–73}$$

The boundary conditions are

1. $$\theta = \theta_i \quad \text{at} \quad \tau = 0$$

2. $$\frac{\partial\theta}{\partial r} = -b\theta \quad \text{at} \quad r = s$$

Assuming that

$$\theta = Ne^{-n^2\alpha\tau}g(nr) \tag{13-74}$$

where g is a function of nr, and substituting this in Eq. 13–73 yields

$$\frac{d^2g(nr)}{dr^2} + \frac{2}{r} \cdot \frac{dg(nr)}{dr} + n^2g(nr) = 0 \tag{13-75}$$

with the particular solution

$$g(nr) = \frac{\sin(nr)}{nr}$$

Hence, Eq. 13–74 becomes

$$\theta = Ne^{-n^2\alpha\tau}\frac{\sin(nr)}{nr} \tag{13-76}$$

A second solution contains $\cos(nr)/nr$, which is an uneven function. For symmetry reasons the integration constant M, with which it would have to be multiplied, must be equal to zero.

The second boundary condition to Eq. 13–73 leads to

$$bs = 1 - ns\cot(ns) \tag{13-77}$$

with roots ψ_ν given in Table 13–10. Equation 13–77 can be written as

TABLE 13–10

ROOTS ψ_ν OF EQ. 13–77 (SUDDEN TEMPERATURE CHANGE OF ENVIRONMENT OF SPHERE)

bs	ψ_1	ψ_2	ψ_3	ψ_4
∞	$\pi =$ 3.14	$2\pi =$ 6.28	$3\pi =$ 9.42	$4\pi =$ 12.57
50	3.08	6.12	9.24	12.31
20	2.98	5.98	8.98	12.00
10	2.84	5.72	8.66	11.65
4	2.45	5.23	8.20	11.25
1.0	1.57	4.71	7.85	10.99
0.5	1.17	4.60	7.79	10.95
0.1	0.54	4.52	7.74	10.91
0.05	0.39	4.51	7.72	10.91
0	0	4.49	7.72	10.90

well in the form

$$\frac{ns}{bs} = \frac{\dfrac{\sin (ns)}{ns}}{-\dfrac{d}{d(ns)}\left[\dfrac{\sin (ns)}{ns}\right]}$$

in perfect analogy to Eq. 13–45 if this is written in the form

$$\frac{ns}{bs} = \frac{\cos (ns)}{-\dfrac{d}{d(ns)} \cos (ns)}$$

Substituting ψ_ν in Eq. 13–76, the general solution becomes

$$\frac{\theta}{\theta_i} = \sum_{\nu=1}^{\nu=\infty} 2 \frac{\sin \psi_\nu - \psi_\nu \cos \psi_\nu}{\psi_\nu - \sin \psi_\nu \cos \psi_\nu} e^{-\psi_\nu{}^2 \alpha\tau/s^2} \frac{\sin \left(\psi_\nu \dfrac{r}{s}\right)}{\psi_\nu \dfrac{r}{s}} \tag{13–78}$$

For $h = \infty$ this equation reduces to Eq. 13–38. Finally, in analogy to Eqs. 13–54 to 57 and 69 to 72,

$$\frac{\theta_0}{\theta_i} = \sum_{\nu=1}^{\nu=\infty} 2 \frac{\sin \psi_\nu - \psi_\nu \cos \psi_\nu}{\psi_\nu - \sin \psi_\nu \cos \psi_\nu} e^{-\psi_\nu{}^2 \alpha\tau/s^2} = \Psi_0 \left(bs, \frac{\alpha\tau}{s^2}\right) \tag{13–79}$$

$$\frac{\theta_s}{\theta_i} = \sum_{\nu=1}^{\nu=\infty} 2 \frac{\sin \psi_\nu - \psi_\nu \cos \psi_\nu}{\psi_\nu - \sin \psi_\nu \cos \psi_\nu} e^{-\psi_\nu{}^2 \alpha\tau/s^2} \frac{\sin \psi_\nu}{\psi_\nu} = \Psi_s \left(bs, \frac{\alpha\tau}{s^2}\right) \tag{13–80}$$

$$\frac{H}{H_i} = 1 - \sum_{\nu=1}^{\nu=\infty} \frac{6}{\psi_\nu{}^3} \cdot \frac{(\sin \psi_\nu - \psi_\nu \cos \psi_\nu)^2}{\psi_\nu - \sin \psi_\nu \cos \psi_\nu} (1 - e^{-\psi_\nu{}^2 \alpha\tau/s^2})$$

$$= \Psi_H \left(bs, \frac{\alpha\tau}{s^2}\right) \tag{13–81}$$

$$\frac{Q}{H_i} = 1 - \Psi_H \left(bs, \frac{\alpha\tau}{s^2}\right) \tag{13–82}$$

where H_i is the excess of the initial heat content of the sphere above that at environment temperature t_e.

Equations 13–79 to 81 are represented by Figs. 13–10 to 12.

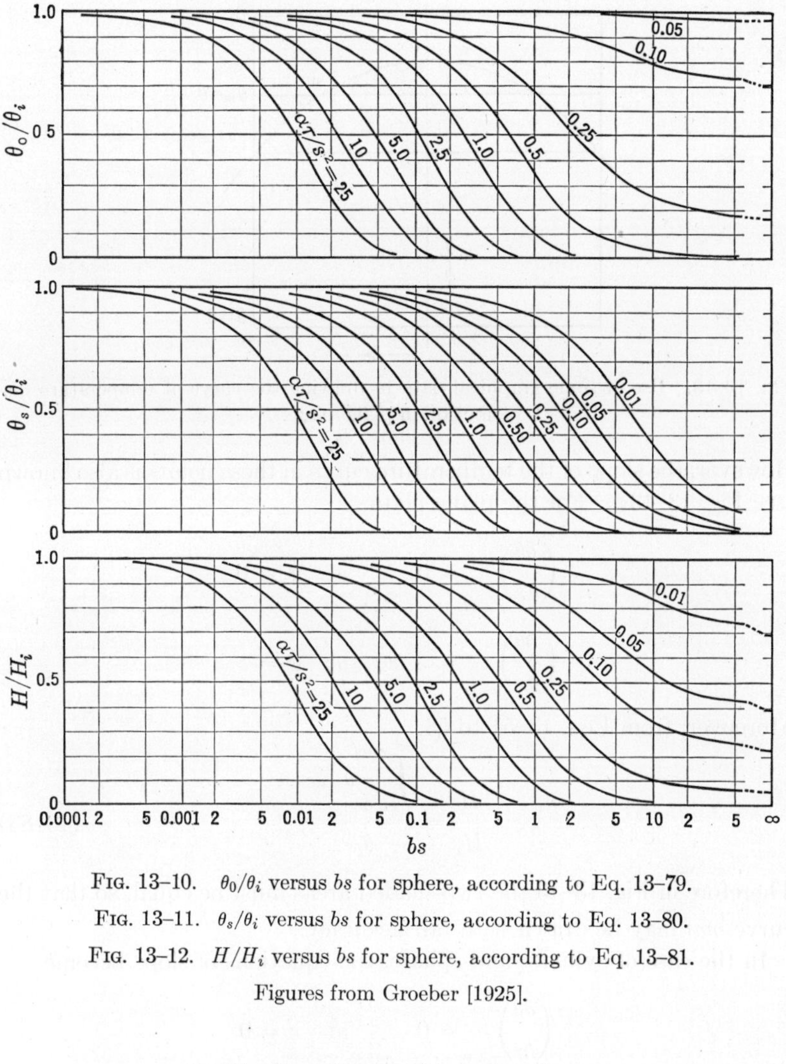

FIG. 13-10. θ_0/θ_i versus bs for sphere, according to Eq. 13-79.
FIG. 13-11. θ_s/θ_i versus bs for sphere, according to Eq. 13-80.
FIG. 13-12. H/H_i versus bs for sphere, according to Eq. 13-81.
Figures from Groeber [1925].

d. The Use of the Φ, X, and Ψ Charts

One advantage of the charts, given in Figs. 13-4 to 12, is their independence of dimensions. Moreover, ratios of the temperatures and heat capacities are presented as functions of only two simple parameters, bs and $\alpha\tau/s^2$, so that numerical calculations are reduced to a minimum.

Concerning the temperature distribution in the bodies under consideration, it might seem that two points only, given by the maximum temperature, θ_0, and the surface temperature, θ_s, would not suffice.

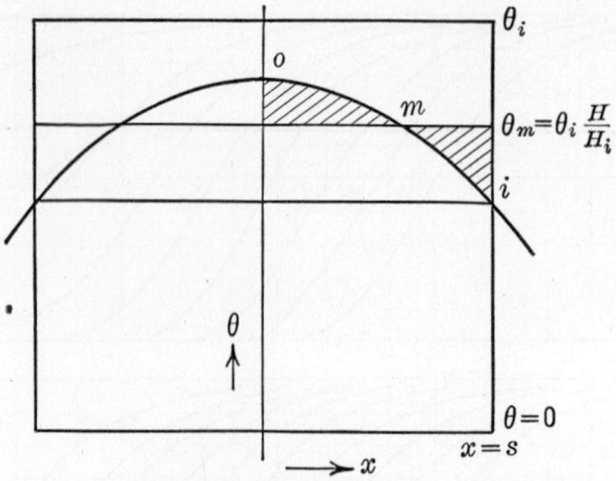

FIG. 13-13. Use of some graphical data in drawing the curve of temperature distribution in plane plate.

However, the slope of the temperature curve in these points is also known (see Fig. 13-13). For the plane plate

$$\left(\frac{\partial \theta}{\partial x}\right)_{x=0} = 0 \quad \text{at} \quad x = 0$$

$$\left(\frac{\partial \theta}{\partial x}\right)_{x=s} = -b\theta_s \quad \text{at} \quad x = s$$

Moreover, from Eqs. 13-51 to 53,

$$\frac{H}{H_i} = \frac{\int_{x=0}^{x=s} \theta \cdot dx}{\theta_i s} \tag{13-83}$$

Therefore in Fig. 13-13 the two shaded areas must be equal, so that the curve $\overline{omi}$ may be drawn with fair accuracy.

In the case of cylinder and sphere, the equations of slope become

$$\left(\frac{\partial \theta}{\partial r}\right)_{r=0} = 0 \quad \text{at} \quad r = 0$$

and

$$\left(\frac{\partial \theta}{\partial r}\right)_{r=s} = -b\theta_s \quad \text{at} \quad r = s$$

For the cylinder Eq. 13-83 is to be replaced by

$$\frac{H}{H_i} = \frac{\int_{r=0}^{r=s} \theta \cdot 2\pi r \cdot dr}{\theta_i \cdot \pi s^2} = \frac{\int_{r=0}^{r=s} \theta \cdot d(r^2)}{\theta_i s^2} \tag{13-84}$$

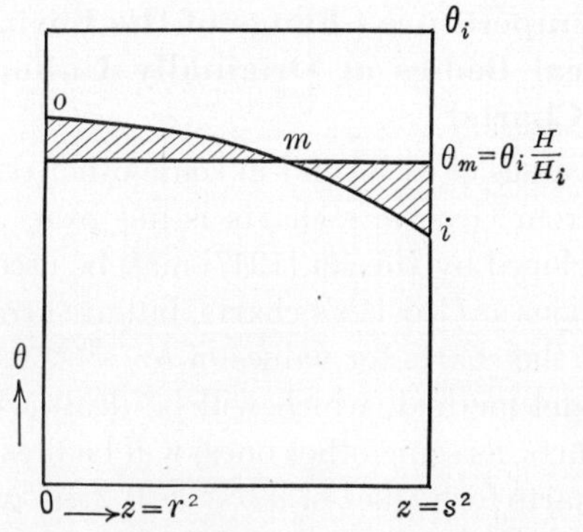

FIG. 13-14. Use of some graphical data in drawing the curve of temperature distribution in cylinder.

In Fig. 13-14 r^2 is taken as abscissa, and θ as ordinate. Equation 13-84 will be satisfied if $\overline{omi}$ is drawn so that the two shaded areas become equal. At $r^2 = 0$ the curve should be horizontal, as in Fig. 13-13, but at $r^2 = s^2$, the tangent must be taken equal to $(d\theta/dz)_{r=s}$, where $z = r^2$, that is,

$$\left(\frac{\partial \theta}{\partial z}\right)_{r=s} = \left(\frac{\partial \theta}{\partial r} \cdot \frac{dr}{dz}\right)_{r=s} = -\frac{b\theta_s}{2s}$$

For spheres Eq. 13-83 is to be replaced by

$$\frac{H}{H_i} = \frac{\int_{r=0}^{r=s} \theta \cdot 4\pi r^2 \cdot dr}{\theta_i \cdot 4\pi s^3/3} = \frac{\int_{r=0}^{r=s} \theta \cdot d(r^3)}{\theta_i s^3} \tag{13-85}$$

Here a diagram like Fig. 13-12 may be used, with $z = r^3$ as abscissa. In this case the slope at $r = s$ becomes

$$\left(\frac{\partial \theta}{\partial z}\right)_{r=s} = -\frac{b\theta_s}{3s^2}$$

When herewith a curve like that in Fig. 13-14 is found, it may be converted to one with r as abscissa. The slope $(\partial\theta/\partial r)_{r=s}$ of this curve should be equal to $-b\theta_s$. This is a check of the accuracy of the construction.

Instead of taking θ as ordinate, it is more convenient to use θ/θ_i. The mean coordinate then becomes H/H_i instead of $\theta_i \cdot H/H_i$. Also for abscissae dimensionless quantities (x/s, r^2/s^2, r^3/s^3) can be used.

13–9. Sudden Temperature Change of the Environment of Some Symmetrical Bodies at Originally Uniform Temperature (Heisler's Charts)

For very large values of $\alpha\tau/s^2$ and in some other regions interpolation or extrapolation from Groeber's charts is not exact enough. In these ranges charts developed by Heisler [1947] may be used. They are based on the same equations as Groeber's charts, but are arranged in a different manner. Most of the charts for values of $\alpha\tau/s^2 < 0.2$ were determined by an electric-model method, which will be dealt with in Chapter 20; therefore these charts, as some other ones, will be presented in Sect. 20–9 only. Heisler's charts for values of $\alpha\tau/s^2 > 0.2$ are given in Figs. 13–15 to 21, however, with the same symbols as are used in Sect. 13–8.

In Figs. 13–15, 16, and 17, which are restricted to $bs > 0.01$, the ratio θ_0/θ_i is plotted versus $\alpha\tau/s^2$, with $1/bs$ as parameter for plane, cylinder, and sphere, respectively. In Figs. 13–18, 19, and 20 values of θ/θ_0 for these three cases are plotted versus $1/bs$ with x/s or r/s as parameter. Multiplying any value θ_0/θ_i from the first set of figures with a value θ/θ_0 from a corresponding figure of the second set yields θ/θ_i for the particular values of $\alpha\tau/s^2$, bs, and x/s (or r/s).

Figure 13–21, finally, is a graph of θ_0/θ_i versus $m(\alpha\tau/s^2)bs = m\alpha b\tau/s$, where $m = 1$, 2, and 3 for plate, cylinder, and sphere, respectively. This figure is valid for $\alpha\tau/s^2 > 0.2$ and $bs < 0.01$. For $m\alpha\tau/s^2 > 100$ the ratio θ_0/θ_i reduces to $1 - m\alpha b\tau/s$.

Heisler has further proved that the following remarkable relationship exists. If q_s'' is a heat-flux density permanently produced at $x = s$ (or $r = s$), starting at $\tau = 0$, the ordinates in Figs. 13–15 to 17 present values of $1 - h(\theta_0 - \theta_i)/q_s''$ instead of θ_0/θ_i. This can be used in high-frequency induction heating of metal bodies, where the heating is actually concentrated in layers below the surface, having depths of about 0.020 in. at a frequency of $0.2(10^6)$ cycles per second and 0.001 in. at $5(10^6)$ cycles per second. In this case the ordinates of Figs. 13–18 to 20 are values of $[1 - h(\theta - \theta_i)/q_s'']/[1 - h(\theta_0 - \theta_i)/q_s'']$. From this ratio the temperature excess θ at any point, for instance, $\theta = \theta_s$, can be determined if q_s'' is given, and vice versa.

Finally it may only be mentioned that Heisler has also developed a relatively simple graphical procedure to determine the temperature change at any place in a short cylinder which is suddenly exposed to an ambient medium at a temperature different from the initial temperature of the cylinder. This is an important case of two-dimensional flow which often occurs in engineering practice.

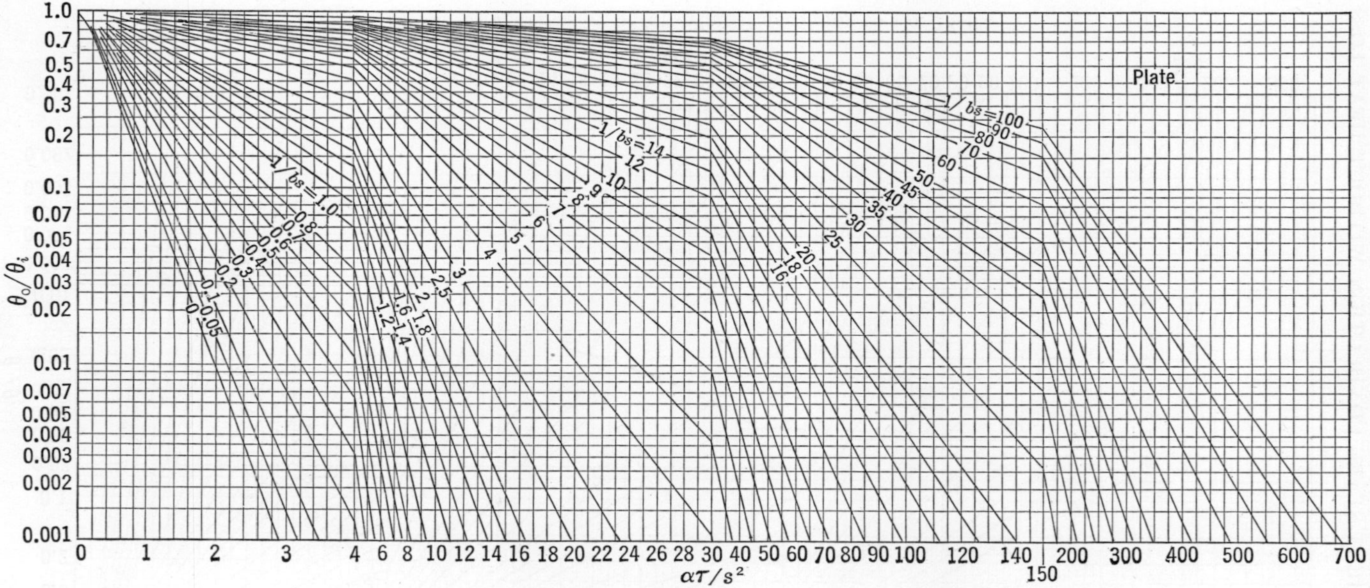

FIG. 13–15. θ_0/θ_i versus $\alpha\tau/s^2$ for plane plate according to Eq. 13–54.

Courtesy Dr. M. P. Heisler [1947].

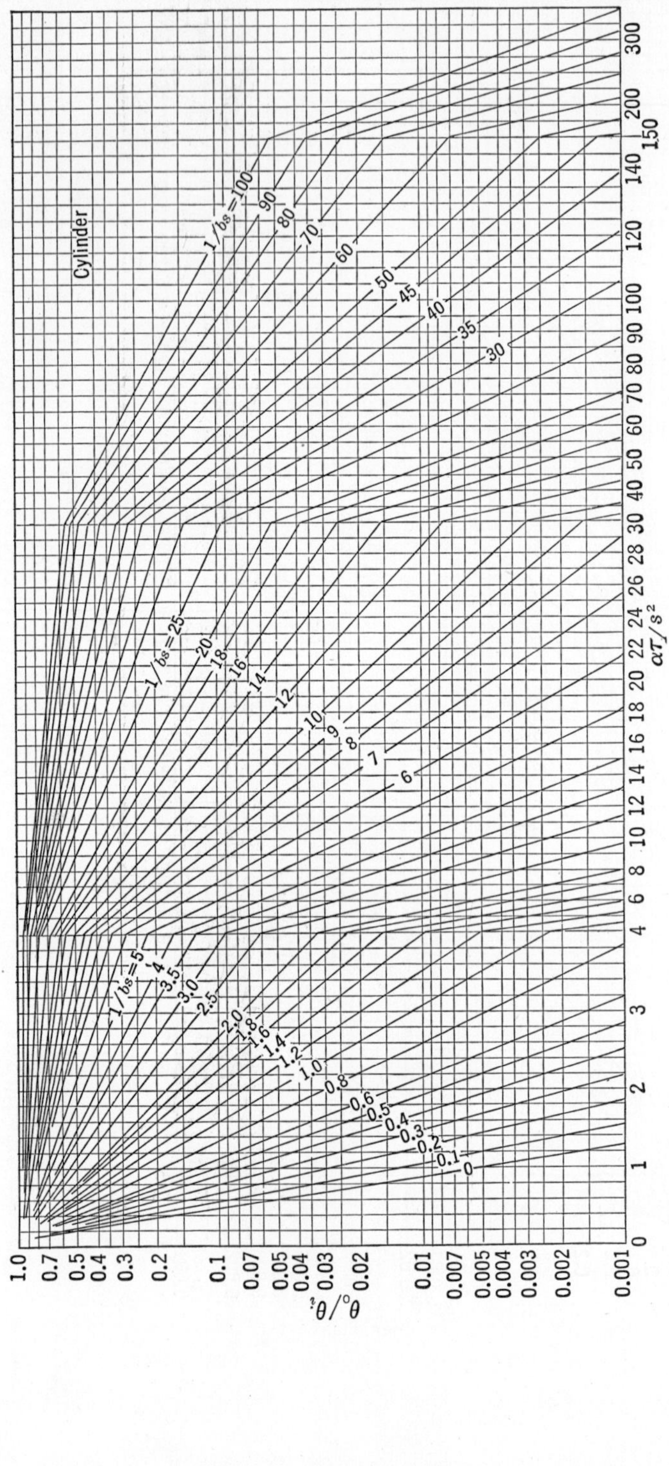

FIG. 13–16. θ_0/θ_i versus $\alpha\tau/s^2$ for cylinder according to Eq. 13–69.

Courtesy Dr. M. P. Heisler [1947].

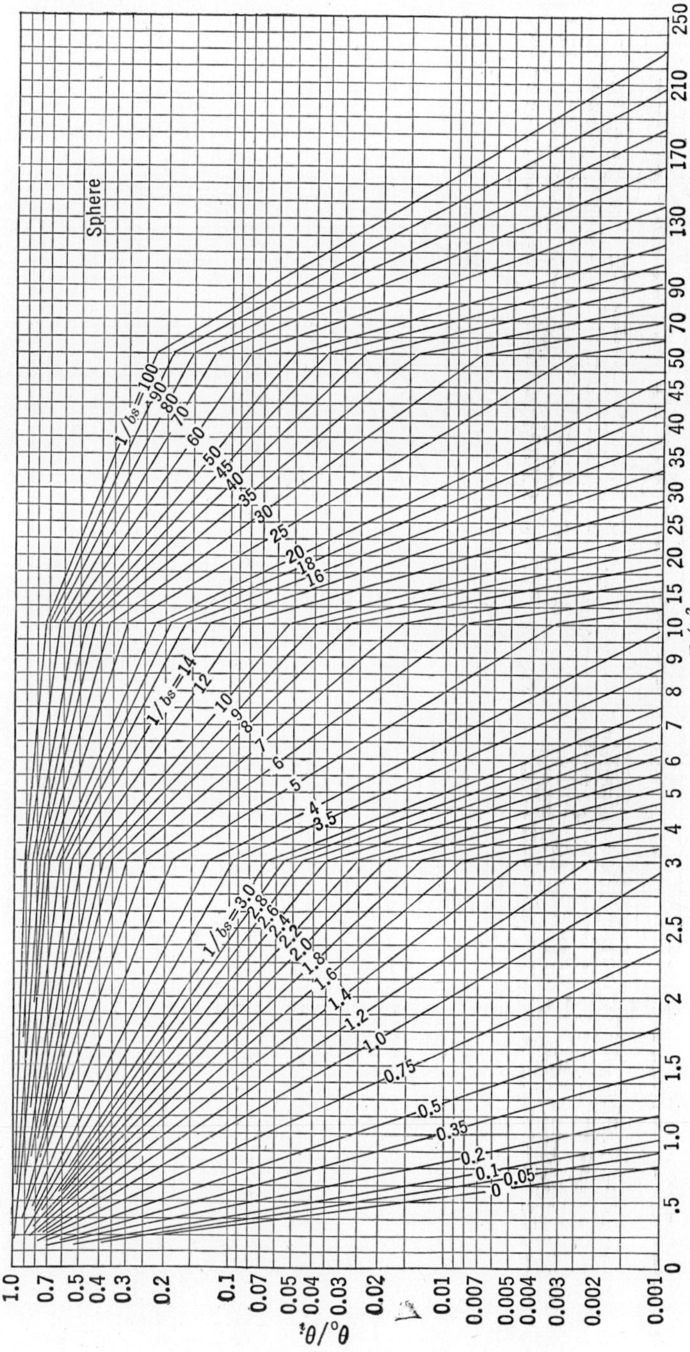

FIG. 13-17. θ_0/θ_i versus $\alpha\tau/s^2$ for sphere according to Eq. 13-79. Courtesy Dr. M. P. Heisler [1947].

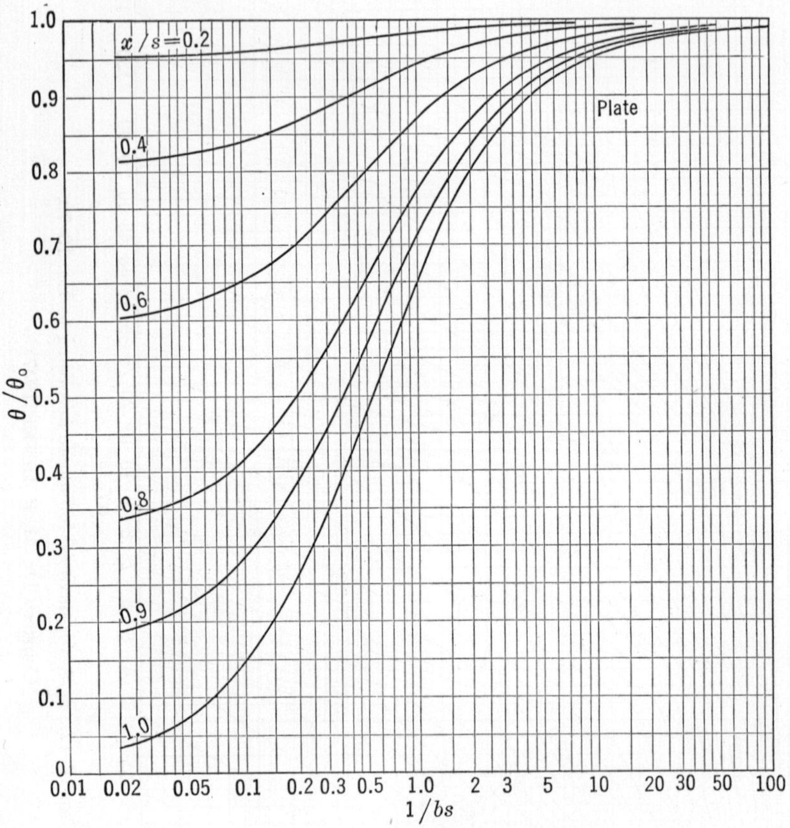

FIG. 13-18. θ/θ_0 versus $1/bs$ for plane plate from Eq. 13-49. Courtesy Dr. M. P. Heisler [1947].

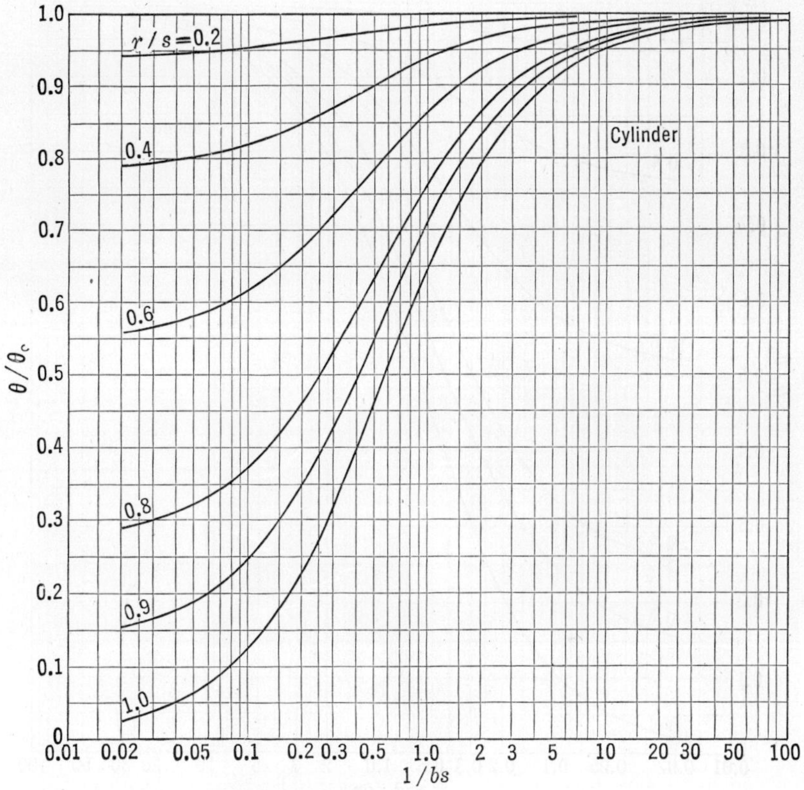

FIG. 13–19. θ/θ_0 versus $1/bs$ for cylinder from Eq. 13–65.
Courtesy Dr. M. P. Heisler [1947].

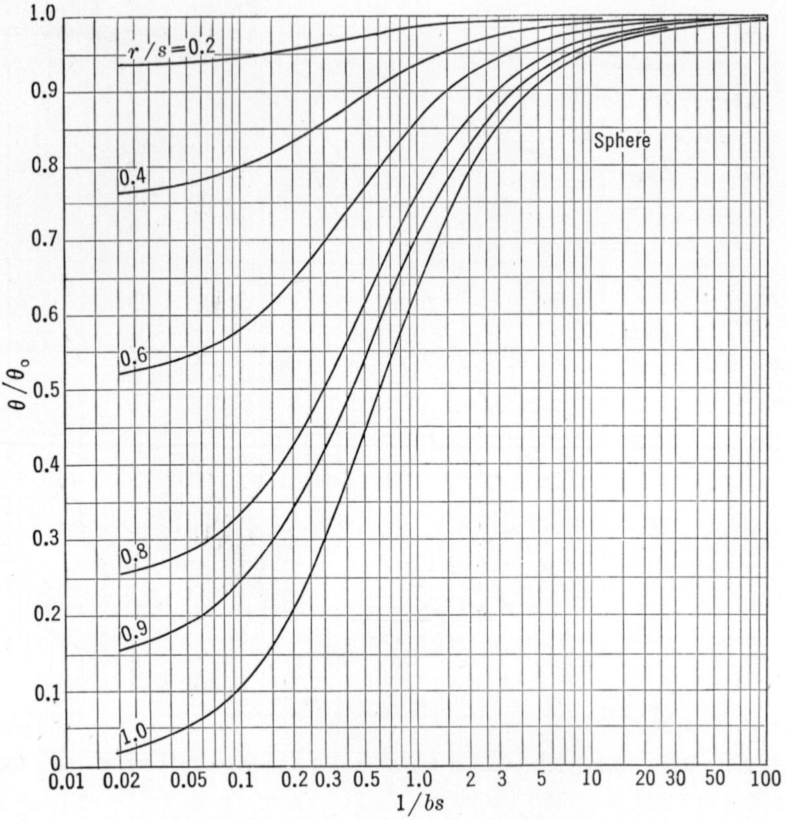

FIG. 13-20. θ/θ_0 versus $1/bs$ for sphere from Eq. 13-78.
Courtesy Dr. M. P. Heisler [1947].

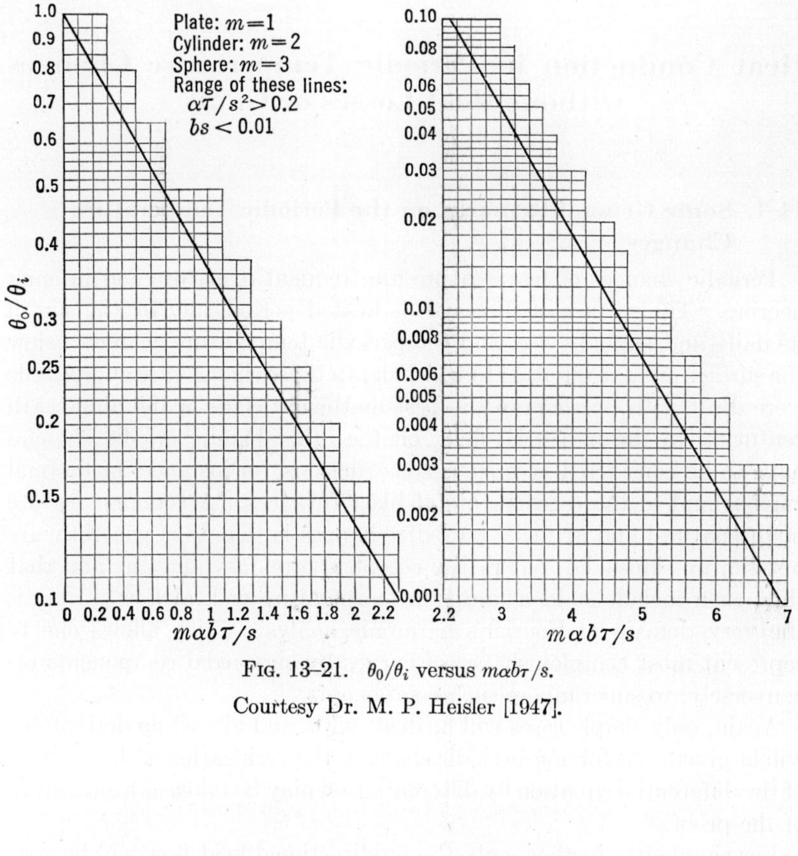

FIG. 13–21. θ_0/θ_i versus $mab\tau/s$.

Courtesy Dr. M. P. Heisler [1947].

CHAPTER 14

Heat Conduction in Periodic Temperature Changes without Heat Losses or Gains

14–1. Some General Remarks on the Periodic Temperature Changes

Periodic changes of temperature are frequent in nature and in engineering. The surface of the earth is heated periodically by the sun in its daily and annual cycle, and the periodic temperature changes below the surface and the question to which depth these changes are appreciable were dealt with by some of the best mathematicians of the nineteenth century. In the industrial field, on the other hand, periodic changes have to be considered in various constructions, for instance, in thermal engines and in the regenerators of blast furnaces. A common feature of the natural and artificial periodic changes is that they generally are not simply sinusoidal, but rather complex, often asymmetric, and that they even include sudden jumps in temperature or heat flow. Here is the very domain of Fourier's harmonic analysis which allows one to represent most complex shapes of curves by sinusoidal components or, conversely, to superimpose such components.

Again, only simple cases will be dealt with, and almost no derivations will be given. As for non-periodic changes, the verification of the solution of the differential equation by differentiation may be taken as a substitute of the proof.

For simplicity, further, only the unidirectional heat flow will be considered, so that the pertinent differential equation is

$$\frac{\partial \theta}{\partial \tau} = \alpha \frac{\partial^2 \theta}{\partial x^2} \tag{14–1}$$

where θ is the temperature excess of a point x at the time τ over the mean temperature t_m of any cycle in the periodic process.

The most complicated cases dealt with here will be those of periodic changes of the environment temperature of an infinitely thick plate

(Sect. 14–3) and of the surface temperature of a plate of finite thickness (Sects. 14–4 and 14–5). Regarding the much more difficult case of an infinitely long cylinder exposed to temperature fluctuations of its environment, reference is made to a paper of Groeber [1928].

14–2. Periodic Changes of the Surface Temperature of an Infinitely Thick Plate

It will be assumed that the temperature of the free surface ($x = 0$) of an infinitely thick plate changes with time according to the equation

$$\theta_{0,\tau} = N \cdot \cos \frac{2\pi\tau}{\tau_c} = N \cdot \cos(\omega\tau) \tag{14–2}$$

where τ_c is the time needed for one full cycle, and
$\omega/2\pi = 1/\tau_c$ is the frequency of the oscillation.

If τ is counted from an instant of time where all but cyclic temperature changes have been damped down to virtually zero, it is obvious that also in the interior of the plate the temperature will change with the same frequency, $\omega/2\pi$.

In the particular integral of Eq. 14–1,

$$\theta_p = e^{nx} e^{\alpha n^2 \tau} \tag{14–3}$$

(cf. Eq. 13–2 *) no real exponent of the second e function satisfies the postulate of damping the non-cyclic changes. Only an imaginary exponent $i\omega\tau$ can do it, since, according to the algebra of complex values,

$$e^{i\omega\tau} = \cos(\omega\tau) + i \sin(\omega\tau) \tag{14–4}$$

Comparison of Eqs. 14–3 and 4 leads to

$$n^2 = \frac{i\omega}{\alpha} \quad \text{or} \quad n = \pm\sqrt{\frac{i\omega}{\alpha}} \tag{14–5}$$

But

$$\sqrt{i} = \pm \frac{1 + i}{\sqrt{2}}$$

Hence

$$n = \pm(1 + i)\sqrt{\frac{\omega}{2\alpha}} = \pm(1 + i)m$$

where, for simplicity,

$$m = \sqrt{\frac{\omega}{2\alpha}} \tag{14–6}$$

has been introduced.

* The symbol λ of that equation is here replaced by n because λ will be used for wavelength.

By substitution in Eq. 14-3, and then using Eq. 14-4,

$$\theta_p = e^{\pm(1+i)mx}e^{i\omega\tau} = e^{\pm mx}e^{i(\omega\tau \pm mx)}$$

$$= e^{\pm mx}[\cos(\omega\tau \pm mx) + i \sin(\omega\tau \pm mx)] \qquad (14\text{-}7)$$

θ_p may be split into two parts:

$$\theta_p = \theta_{p,1} + i\theta_{p,2}$$

where

$$\theta_{p,1} = e^{\pm mx} \cos(\omega\tau \pm mx) \qquad (14\text{-}8)$$

and

$$\theta_{p,2} = e^{\pm mx} \sin(\omega\tau \pm mx)$$

These are particular integrals of Eq. 14-1, and so are any values $N\theta_{p,1}$ and $M\theta_{p,2}$, where N and M are arbitrary constants, as may be checked by differentiation. However, only $N\theta_{p,1}$ is in agreement with the boundary condition for $x = 0$ (Eq. 14-2), which means that $M = 0$. Moreover, the plus signs are in disagreement with the following physical fact. If a very thick plate which originally was at temperature t_m is periodically heated and cooled by changing the surface temperature according to a sine function, oscillating around t_m, the temperature excess, θ, on a point of the plate at distance x from the surface will never exceed the amplitude $\theta_{0,a}$ at $x = 0$. But, with a positive exponent of the e function in Eq. 14-8, it would approach an infinite value for $x \to \infty$.

Hence, the solution of Eq. 14-1 has the unique form

$$\theta = \theta_{0,a}e^{-mx} \cos(\omega\tau - mx) = \theta_{x,a} \cos(\omega\tau - mx) \qquad (14\text{-}9)$$

where $\theta_{0,a}$ = the amplitude of the temperature difference $t_0 - t_m$,
t_0 = the temperature at $x = 0$,
t_m = the mean temperature,
$\theta_{x,a}$ = the amplitude of the temperature difference $t - t_m$, and
t = the temperature at the distance x from the free surface.

The following features of the process can be recognized from Eq. 14-9:

1. Neither the frequency, $\omega/2\pi$, nor the sinusoidal form of the temperature oscillation changes with x; only its amplitude decreases with increasing x according to the following e function:

$$\theta_{x,a} = \theta_{0,a}e^{-mx} \qquad (14\text{-}10)$$

2. With increasing x, the amplitude of oscillation is delayed; it occurs at the time

$$\tau = \frac{mx}{\omega} = \sqrt{\frac{1}{2\omega\alpha}}\,x \qquad (14\text{-}11)$$

3. The temperature has a maximum at $\tau = 0$ and $x = 0$. The next maximum at $x = 0$ occurs if $\tau = \tau_c$. At this instant a maximum occurs also at a place $x = \lambda$, where $\omega\tau_c = m\lambda$, i.e., at the distance

$$\lambda = \frac{\omega\tau_c}{m} = \sqrt{2\omega\alpha\tau_c} = 2\pi\sqrt{2\frac{\alpha}{\omega}} \tag{14–12}$$

At $\tau = \nu\tau_c$ the maximum has reached the point $x = \nu\lambda$, if ν is a positive integer. Hence the distance λ represents the "wavelength" of a temperature wave which moves into the plate. Because the temperature maximum always needs a time interval, τ_c, to proceed by a distance λ, its velocity is:

$$v = \frac{\lambda}{\tau_c} = \frac{\omega}{m} = \sqrt{2\omega\alpha} \tag{14–13}$$

It is remarkable that this velocity depends upon the frequency $\omega/2\pi$ of the temperature fluctuation and upon α, which therefore was often called temperature conductivity before the name thermal diffusivity was more generally adopted.

Figure 14–1 shows the temperature distribution at two instants τ_1 and τ_2, where $\tau_2 > \tau_1$. The full line represents the temperature distribution at the instant τ_1, the dotted line the distribution at τ_2. By these two curves the moving of the temperature wave into the plate is pictured. The dash-dot lines represent $\theta = \theta_{x,a} = \theta_{0,a}e^{-mx}$ and $\theta = -\theta_{x,a}$. Obviously, each of the curves, $\tau =$ constant, will touch the line $\theta = \theta_{x,a}$, whenever the cosine in Eq. 14–9 equals 1, i.e., at $x = \omega\tau/m + (2\nu - 2)\pi/m$, and the line $\theta = -\theta_{x,a}$, when it equals -1, i.e., at $x = \omega\tau/m + (2\nu - 1)\pi/m$. All other points of the curve, $\tau =$ constant, will be between these two limiting lines, which, therefore, form envelopes of the system of curves, $\tau =$ constant.

Of course, the maxima and minima of the curves, $\tau =$ constant, lie inside the enveloping curves. Their locus is given by the equation $\partial\theta/\partial x = 0$, or substituting θ from Eq. 14–9,

$$\theta_{0,a}me^{-mx}[\sin(\omega\tau - mx) - \cos(\omega\tau - mx)] = 0$$

Hence, maxima or minima occur either when $x = \infty$ (which is self-evident) or if $\tan(\omega\tau - mx) = 1$. The latter equation yields

$$\omega\tau - mx = \frac{5 - 4\nu}{4}\pi$$

or

$$x = \frac{\omega\tau}{m} - \frac{5 - 4\nu}{4} \cdot \frac{\pi}{m}$$

where ν is any positive integer.

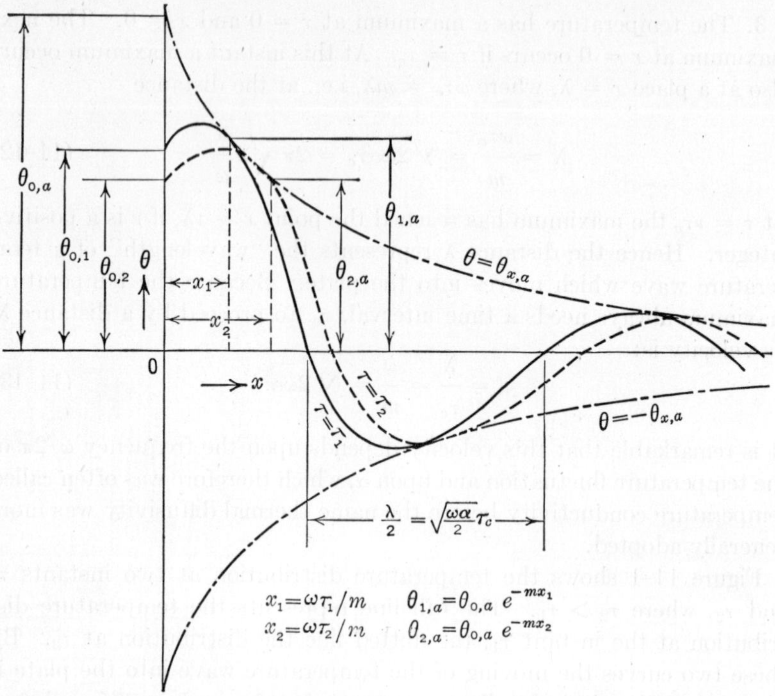

FIG. 14–1. Temperature distribution at two different instants in infinitely thick plate whose surface temperature changes periodically.

If ν is increased by 2, x increases by λ, as can easily be verified. The distance between the places where a temperature wave, at τ = constant, touches the envelope and has its maximum is $\frac{1}{4} \cdot \frac{\pi}{m}$.

According to Eq. 14–10, the amplitude of the temperature wave decreases with an e function of the depth of penetration, x. The "logarithmic decrement" is $m = \sqrt{\omega/2\alpha}$. This means a stronger reduction of the amplitude for higher frequency and smaller thermal diffusivity.

Also from Eq. 14–10, the depth at which the amplitude $\theta_{x,a}$ has been reduced to the μth part of $\theta_{0,a}$ is:

$$x_\mu = \frac{1}{m} \ln \mu = \frac{2.3026}{m} \log \mu = 1.30\sqrt{\alpha\tau_c} \log \mu \qquad (14\text{–}14)$$

The depth at which the amplitude is damped to 1% of that of the surface is

$$x_{\mu=100} = 2.60\sqrt{\alpha\tau_c}$$

A wall of this thickness and even a somewhat thinner wall can obviously be considered as virtually infinitely thick. It is remarkable that the value x_μ depends on the product of α and τ_c. Therefore the equations for an infinitely thick plate may be applied to the rather thin cylinder wall of a combustion engine, although for metals α is large, because τ_c usually is extraordinarily small.

In addition to the temperature change and distribution, it is of interest to know the heat flow entering or leaving the plate at any instant, and particularly the heat stored in the plate during one half of a cycle and restored to the environment during the second half of the cycle.

The heat flow entering by an area A of the surface is

$$q = -kA\left(\frac{\partial\theta}{\partial x}\right)_{x=0}$$

Substituting θ from Eq. 14–9, performing the differentiation, and finally substituting $x = 0$ yield

$$q = kA\theta_{0,a}\sqrt{\frac{\omega}{\alpha}}\cos\left(\omega\tau + \frac{\pi}{4}\right) \tag{14–15}$$

Although the amplitude of heat flow, $kA\theta_{0,a}\sqrt{\omega/\alpha}$, is greater for high than for small frequency, $\omega/2\pi$, more heat is stored at small frequency, as will now be shown.

q is positive in the limits $\omega\tau + (\pi/4) = -\pi/2$ and $+\pi/2$, i.e., from $\tau = -3\pi/4\omega$ to $+\pi/4\omega$. Integrating Eq. 14–15 in these limits yields the following heat energy stored in half a cycle:

$$H = kA\theta_{0,a}\sqrt{\frac{\omega}{\alpha}}\int_{-3\pi/4\omega}^{\pi/4\omega}\cos\left(\omega\tau + \frac{\pi}{4}\right)\cdot d\tau = \frac{2}{\sqrt{\omega}}\sqrt{\rho c_p k}A\theta_{0,a}$$

$$= \sqrt{\frac{2}{\pi}}\sqrt{\rho c_p k}\sqrt{\tau_c}A\theta_{0,a} \tag{14–16}$$

14–3. Periodic Change of the Temperature of a Medium in Contact with the Plane Surface of an Infinitely Thick Plate

This is a case of particular interest in mechanical engineering. In an internal-combustion engine, for instance, the gas temperature undergoes cyclic changes, and it is important to learn how far the cylinder wall follows these changes. The high temperatures used in combustion engines do not damage the cylinder wall, only because of the fact that the temperature changes are greatly damped in the transfer of heat, as will be shown in this section.

Let the temperature of the environment fluctuate according to the equation

$$\theta_e = \theta_{e,a} \cos \frac{2\pi\tau}{\tau_c} = \theta_{e,a} \cos (\omega\tau) \tag{14-17}$$

where $\theta_{e,a}$ is the amplitude of the excess of the temperature of environment above its mean temperature, $t_{e,m}$.

Introducing again $b = h/k$ (see Eq. 13-43), the integration of Eq. 13-1 has led to the following equation of the temperature distribution in the plate [Hobson 1903-1921, p. 186]:

$$\theta = \theta_{e,a}\eta e^{-mx} \cos (\omega\tau - mx - \epsilon) \tag{14-18}$$

where $\eta = \sqrt{\dfrac{1}{1 + 2m/b + 2m^2/b^2}}$,

$$\epsilon = \tan^{-1} \frac{1}{1 + b/m},$$

and

$$\frac{b}{m} = b\sqrt{\frac{\alpha\tau_c}{\pi}} = \frac{h}{\sqrt{\rho c_p k}}\sqrt{\tau_c}.$$

Numerical values of the functions η and ϵ, which depend only on b/m, are given in Table 14-1.

TABLE 14-1

VALUES OF η AND ϵ, AS DEFINED AND USED IN EQS. 14-18 TO 21 FOR PERIODIC CHANGE OF TEMPERATURE

$\frac{m}{b}$	$\frac{b}{m}$	η	$\frac{360}{2\pi}\epsilon$	$\frac{m}{b}$	$\frac{b}{m}$	η	$\frac{360}{2\pi}\epsilon$
∞	0	0	45°	1	1	0.447	26° 34′
100	0.01	0.00704	44° 43′	0.5	2	0.632	18° 26′
50	0.02	0.0140	44° 26′	0.2	5	0.830	9° 28′
20	0.05	0.0345	43° 36′	0.1	10	0.905	5° 12′
10	0.1	0.0673	42° 16′	0.05	20	0.951	2° 43′
5	0.2	0.128	39° 48′	0.02	50	0.980	1° 07′
2	0.5	0.277	33° 41′	0.01	100	0.990	0° 34′
1	1	0.447	26° 34′	0	∞	1	0°

Comparison of Eqs. 14-17 and 18 shows the following facts:

1. At the surface of the plate ($x = 0$) the amplitude of the temperature fluctuation becomes

$$\theta_{0,a} = \theta_{e,a}\eta \tag{14-19}$$

Since $\eta \leqq 1$, the amplitude at the surface is smaller than in the environing medium, and η is the fraction to which the amplitude is reduced by the heat transfer to the plate.

2. In the depth x the amplitude becomes

$$\theta_{x,a} = \theta_{e,a}\eta e^{-mx} = \theta_{0,a}e^{-mx} \tag{14-20}$$

By this equation the present case is reduced to the former one (see Eq. 14-9).

3. At $x = 0$ and $\tau = 0$, the temperature excess may be called $\theta_{0,0}$. From Eqs. 14-18 and 19

$$\theta_{0,0} = \theta_{0,a} \cos(-\epsilon)$$

Since at $x = 0$ and $\tau = \tau$

$$\theta_{0,\tau} = \theta_{0,a} \cos(\omega\tau - \epsilon) \tag{14-21}$$

it is evident that only at the instant $\tau = \epsilon/\omega$ does the temperature at the surface have its first maximum. The ratio ϵ/ω is the time lag of the amplitude of oscillation at $x = 0$ and $(mx + \epsilon)/\omega$ is the time lag at $x = x$, counted from the previous amplitude of the temperature of environment.

The heat energy stored in half a period is found from Eq. 14-16, where $\theta_{0,a}$ is to be substituted from Eq. 14-19.

14-4. Periodic Change of the Temperature of Both Surfaces of a Plane Plate of Finite Thickness

The thickness of the plate may be $2s$, and the zero point of the coordinate x be located in the median plane of the plate because the temperatures are symmetrical to this plane. The temperature of both surfaces is supposed to change according to the equation

$$\theta_s = \theta_{s,a} \cos \frac{2\pi\tau}{\tau_c} = \theta_{s,a} \cos(\omega\tau) \tag{14-22}$$

where $\theta_{s,a}$ is the amplitude of the temperature excess at the surface above the mean temperature, t_m.

The temperature distribution has been derived by Groeber [1921]. The result of this derivation can be expressed by the equation:

$$\theta = \theta_{s,a} \sqrt{\frac{f_c^2(\xi\sigma) + f_s^2(\xi\sigma)}{f_c^2(\sigma) + f_s^2(\sigma)}} \cos\left[\omega\tau + \tan^{-1}\frac{f_c(\sigma)f_s(\xi\sigma) - f_s(\sigma)f_c(\xi\sigma)}{f_c(\sigma)f_c(\xi\sigma) + f_s(\sigma)f_s(\xi\sigma)}\right]$$

$$= \theta_{s,a}\zeta \cos(\omega\tau + \phi) \tag{14-23}$$

where $\omega\tau = 2\pi(\tau/\tau_c)$,
$\xi = x/s$,

$$\sigma = ms = \sqrt{\frac{\pi}{\alpha\tau_c}}\, s,$$

$f_c(z) = \cos(z)\cosh(z)$,
$f_s(z) = \sin(z)\sinh(z)$, and
ζ and ϕ are two functions defined by Eq. 14–23.

It is seen that $\theta/\theta_{s,a}$ is a function Ω of three dimensionless groups:

$$\frac{\theta}{\theta_{s,a}} = \Omega\left(\frac{x}{s}, \frac{\tau}{\tau_c}, \frac{\alpha\tau_c}{s^2}\right) \tag{14–24}$$

Tables of the values of $f_c(z)$ and $f_s(z)$ from $z = 0$ to $z = 10$, in intervals of 0.02, are given in a book of Hohenemser and Prager [1933]. Groeber has calculated numerical values of the function ζ; they are represented in Table 14–2.

TABLE 14–2

VALUES OF THE FUNCTION ζ, DEFINED BY EQ. 14–23 FOR PERIODIC CHANGE OF TEMPERATURE

σ	ξ								
	0	$\frac{1}{8}$	$\frac{2}{8}$	$\frac{3}{8}$	$\frac{4}{8}$	$\frac{5}{8}$	$\frac{6}{8}$	$\frac{7}{8}$	1
0	1	1	1	1	1	1	1	1	1
0.5	0.98	0.98	0.98	0.98	0.98	0.98	0.98	0.99	1
1.0	0.77	0.77	0.77	0.78	0.79	0.81	0.85	0.91	1
1.5	0.47	0.47	0.47	0.48	0.52	0.58	0.68	0.83	1
2.0	0.27	0.27	0.28	0.30	0.36	0.45	0.58	0.77	1
4.0	0.04	0.04	0.05	0.08	0.13	0.22	0.37	0.64	1
8.0	0.00	0.00	0.01	0.01	0.02	0.05	0.14	0.36	1
∞	0	0	0	0	0	0	0	0	..

Again, the temperature distribution in the wall at subsequent instants of time has the character of a wave moving into the plate. In Fig. 14–2 the lines of constant σ are envelopes of such waves. The wave existing at the time $\tau = 0.146\tau_c$ is represented in the figure. The equation of its envelope is $\sigma = 4.0$; the ratio,† $\theta_{s,\tau}/\theta_{s,a} = 0.61$. It is remarkable that

† From Eq. 14–23: $\theta_{s,\tau}/\theta_{s,a} = \sqrt{1}[\cos(2\pi\tau/\tau_c) + 0] = 0.61$.

in the median plane,‡ $\theta_{0,\tau}/\theta_{s,a} = 0.04$. For symmetry reasons the temperature wave must have the slope zero at this point.

In the case of the infinitely thick plate the wavelength was independent of x; for the plate of finite thickness it increases with x, as may be seen from the figure. It is further seen that for $\sigma \gtreqless 4$ either half of the plate

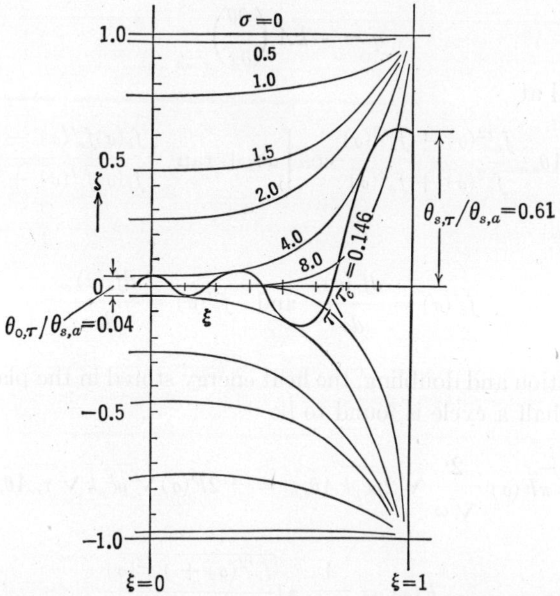

FIG. 14-2. Temperature distribution at any instant in a plate of finite thickness, both surfaces of which suffer the same periodic temperature change. From Groeber [1921].

can be dealt with as though it were infinitely thick, because no appreciable temperature change occurs any more at $x \approx 0$.

It should be mentioned also that the case treated in this section is identical with that of a plate of thickness s with one perfectly insulated surface, the other one undergoing a sinusoidal temperature oscillation.

14-5. Periodic Storage and Release of Heat Energy by a Plane Plate of Finite Thickness

The periodic temperature change of the two surfaces of a plate causes a cyclic storage of heat energy in the plate. This is of particular interest in the design of the regenerators which serve for the periodic storage and release of heat; the so-called stove, used as the air preheater of a

‡ From Table 14-2: $\sigma = 4$ for $\xi = 0$.

blast furnace, is one of the most important types of these devices. The theory will be given in Vol. II of this book; the present section deals with only one feature of the complex picture of the regenerator.

Groeber started from a modification of Eq. 14–23. He substituted θ from this modified equation in the heat-flow equation

$$q = -kA\left(\frac{\partial \theta}{\partial x}\right)_{x=s}$$

and arrived at

$$q = -kmA\theta_{s,a}\frac{f_c'^2(\sigma) + f_s'^2(\sigma)}{f_c^2(\sigma) + f_s^2(\sigma)}\cos\left[\omega\tau + \tan^{-1}\frac{f_c(\sigma)f_s'(\sigma) - f_s(\sigma)f_c'(\sigma)}{f_c(\sigma)f_c'(\sigma) + f_s(\sigma)f_s'(\sigma)}\right] \tag{14-25}$$

where

$$f_c'(\sigma) = \frac{df_c(\sigma)}{d\sigma} \quad \text{and} \quad f_s'(\sigma) = \frac{df_s(\sigma)}{d\sigma}$$

By integration and doubling, the heat energy stored in the plate of thickness $2s$ in half a cycle is found to be

$$H = \mp\sqrt{\pi}F(\sigma)\frac{2}{\sqrt{\omega}}\sqrt{2\rho c_p k}A\theta_{s,a} = \mp 2F(\sigma)\sqrt{\rho c_p k}\sqrt{\tau_c}A\theta_{s,a} \tag{14-26}$$

where

$$F(\sigma) = \frac{1}{\sqrt{\pi}}\sqrt{\frac{f_c'^2(\sigma) + f_s'^2(\sigma)}{f_c^2(\sigma) + f_s^2(\sigma)}}$$

This function is represented in Fig. 14–3. It has a maximum at $\sigma = 1.2$. Hence, for given values of ρ, c_p, k, $\theta_{s,a}$, and τ_c, a maximum of heat energy will be stored in half a period if the thickness of the wall, $2s$, which, according to Eq. 14–23, equals $2\sigma\sqrt{\alpha\tau_c/\pi}$, is made equal to $2(1.2)\sqrt{\alpha\tau_c/\pi} = 1.35\sqrt{\alpha\tau_c}$.

The figure shows further that a deviation of $\pm 20\%$ from the most favorable value of σ or s does not appreciably reduce the storage capacity of the wall.

For $\sigma \to \infty$, Eq. 14–26, obviously, will reduce to Eq. 14–16, i.e., $F(\sigma)$ becomes equal to $\sqrt{2/\pi} = 0.798$. However, according to Fig. 14–3, the storage capacity of a wall is practically equal to that of an infinitely thick wall if only $\sigma > 2$; i.e., $s > 2\sqrt{\alpha\tau_c/\pi} = 1.13\sqrt{\alpha\tau_c}$, provided that but one surface of the wall undergoes a temperature fluctuation according to Eq. 14–22. If this equation holds for both surfaces, the double thickness is necessary.

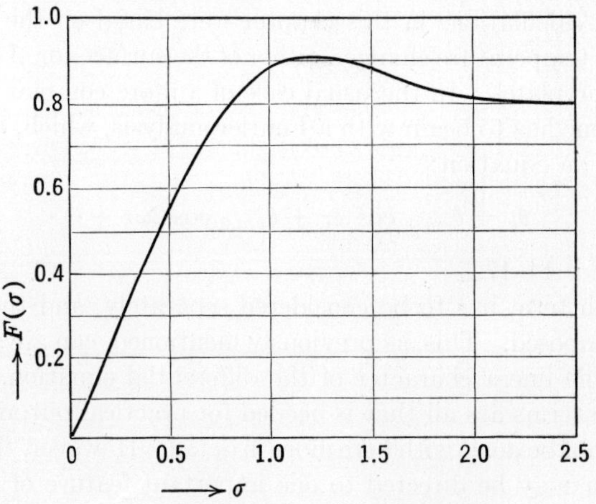

FIG. 14–3. Function $F(\sigma)$, characteristic for periodic storage and release of heat in plate of finite thickness (see Eq. 14–26).

From Groeber and Erk [1933].

A modification of Eq. 14–26 is:

$$H = (2s)A\rho c_p(2\theta_{s,a})\frac{\sqrt{\pi}}{2\sigma}F(\sigma) = H_{s,a}\Omega_H\left(\frac{\alpha\tau_c}{s^2}\right) \qquad (14\text{–}27)$$

where $H_{s,a}$ = the heat energy which could be stored if the whole plate were heated from $-\theta_{s,a}$ to $+\theta_{s,a}$, and

$\Omega_H = (\sqrt{\pi}/2\sigma)F(\sigma)$ = the fraction of $H_{s,a}$ that actually can be stored while the surface temperature changes in these limits.

Groeber has performed the numerical calculation. Table 14–3 contains his results, slightly modified by the present author.

TABLE 14–3

THE FUNCTION Ω_H, DEFINED BY EQ. 14–27 FOR PERIODIC STORAGE OF HEAT ENERGY

$\alpha\tau_c/s^2$	Ω_H	$\alpha\tau_c/s^2$	Ω_H	$\alpha\tau_c/s^2$	Ω_H	$\alpha\tau_c/s^2$	Ω_H	$\alpha\tau_c/s^2$	Ω_H
0	0	0.4	0.26	1.1	0.44	1.8	0.60	5.0	0.90
0.05	0.09	0.5	0.28	1.2	0.47	1.9	0.62	6.0	0.92
0.10	0.12	0.6	0.31	1.3	0.49	2.0	0.64	7.0	0.94
0.15	0.15	0.7	0.34	1.4	0.52	2.5	0.71	8.0	0.95
0.20	0.18	0.8	0.36	1.5	0.54	3.0	0.77	9.0	0.96
0.25	0.20	0.9	0.39	1.6	0.56	3.5	0.82	10.0	0.97
0.30	0.23	1.0	0.42	1.7	0.58	4.0	0.86	∞	1

So far, all calculations in this chapter were based on the assumption of sinusoidal temperature changes, either of the surfaces or of the environment of plane plates. In the usual case of a more complex form of the oscillation, one has to begin with a Fourier analysis, which, for instance, may lead to an equation:

$$\theta_e = \theta_{e,a,1} \cos \omega\tau + \theta_{e,a,2} \cos 2\omega\tau + \cdots \tag{14–28}$$

instead of Eq. 14–17.

Then, each term has to be considered separately, and the waves are to be superimposed. This, as previously mentioned, can always be done, because of the linear character of the differential equation. Generally two or three terms are all that is needed for practical purposes.

This will not be dealt with here more in detail. However, the attention of the reader may be directed to one important feature of the picture: Equations 14–14 and 16 show that the depth of penetration of a wave and the heat energy stored in a plate are both proportional to $\sqrt{\tau_c}$ in the simplest case, and it is not much different in more complicated cases. Hence, the higher harmonics penetrate less deeply and contribute less energy to the heat storage than the basic wave. This, frequently, means a further simplification of the situation in so far as the higher harmonics, in general, are less effectual and need not be considered so carefully as the basic wave.

CHAPTER 15

Heat Conduction at Variable Temperature, Including Internal Heat Generation and Heat Dissipation to the Environment

15–1. Uniform and Constant Heat Generation

This chapter deals with some complex cases of heat conduction. In addition to the items mentioned in the title, absorption of heat, such as may occur in endothermic reactions, and heat gains from the environment also can be included just by changing the respective signs.

First, the relatively simple case of homogeneous and unvariable heat generation will be treated in the way chosen by Awbery in probably the first paper on this subject [1927]. In particular, his derivation of the equations for the symmetrical heat conduction in a sphere will be shown.

The differential equation 2–10, from which Awbery started, may be written in the form

$$\frac{\partial t}{\partial \tau} = \alpha \cdot \nabla^2 t + G \tag{15–1}$$

where t = the temperature of a point x, y, z,
∇^2 = the Laplacian operator, and
G = $q'''/\rho c_p$ and is supposed to be constant.

If $G = 0$, using Eq. 13–2,

$$t = \Sigma N e^{\kappa\tau + \lambda x + \mu y + \nu z} \quad \text{(Fourier's equation)} \tag{15–2}$$

where κ, λ, μ, and ν are constants, and $\alpha(\lambda^2 + \mu^2 + \nu^2) = \kappa$.

If $\partial t/\partial \tau = 0$,

$$t = M - \frac{G}{6\alpha}(x^2 + y^2 + z^2) \quad \text{(Poisson's equation)} \tag{15–3}$$

where M is a constant of integration.

By combination of Eqs. 15–2 and 3 the general solution of Eq. 15–1 becomes

$$t = M - \frac{G}{6\alpha}(x^2 + y^2 + z^2) + \Sigma N e^{\kappa\tau + \lambda x + \mu y + \nu z} \tag{15-4}$$

For heat conduction in a sphere, symmetrical to the center, Eq. 15–1 becomes

$$\frac{\partial t}{\partial \tau} = \frac{\alpha}{r^2} \cdot \frac{\partial}{\partial r}\left(r^2 \frac{\partial t}{\partial r}\right) + C \tag{15-5}$$

or, introducing

$$u = rt \tag{15-6}$$

as a new variable,

$$\frac{\partial u}{\partial \tau} = Gr + \alpha \frac{\partial^2 u}{\partial r^2} \tag{15-7}$$

The solution of this equation is

$$u = H + Kr - \frac{Gr^3}{6\alpha} + \Sigma e^{-n^2\alpha\tau}[M \cos(nr) + N \sin(nr)]$$

where H and K are constants.

By substitution from Eq. 15–6

$$t = \frac{H}{r} + K - \frac{Gr^2}{6\alpha} + \frac{1}{r}\Sigma e^{-n^2\alpha\tau}[M \cos(nr) + N \sin(nr)] \tag{15-8}$$

Since the temperature at the center of the sphere will be finite,

$$H = M = 0$$

Hence

$$t = K - \frac{Gr^2}{6\alpha} + \frac{1}{r}\Sigma N e^{-n^2\alpha\tau} \sin(nr) \tag{15-9}$$

Assuming now that the whole sphere initially is at the temperature t_i and that, at $\tau = 0$, its surface ($r = s$) is brought to, and then held at temperature t_s,

$$t_i = K - \frac{Gr^2}{6\alpha} + \frac{1}{r}\Sigma N \sin(nr) \tag{15-10}$$

$$t_s = K - \frac{Gs^2}{6\alpha} + \frac{1}{s}\Sigma N e^{-n^2\alpha\tau} \sin(ns) \tag{15-11}$$

The surface temperature t_s is supposed to be constant for $\tau > 0$. This requires

$$ns = \nu\pi$$

where ν is any integer.* Hence

$$K = t_s + \frac{Gs^2}{6\alpha}$$

Introducing $\theta = t - t_s$, Eq. 15–10 becomes

$$r\theta_i + \frac{Gr}{6\alpha}(s^2 - r^2) = \Sigma N \sin\frac{\nu\pi r}{s} \tag{15–12}$$

The constants N, obviously, are the coefficients of a Fourier series and may be represented by

$$N_\nu = \frac{2}{s}\int_0^s \left[rt_i - \frac{Gr}{6\alpha}(s^2 - r^2)\right]\sin\frac{\nu\pi r}{s}\,dr$$

which Awbery showed to reduce to

$$N_\nu = (-1)^\nu\left[\frac{2Gs^3}{\nu^3\pi^3\alpha} - \frac{2st_i}{\nu\pi}\right] \tag{15–13}$$

Hence the final equation for the temperature excess becomes

$$\theta = \frac{G(s^2 - r^2)}{6\alpha} + \frac{1}{r}\Sigma N_\nu e^{-\nu^2\pi^2\alpha\tau/s^2}\sin\frac{\nu\pi r}{s} \tag{15–14}$$

with N_ν to be taken from Eq. 15–13.

When $\tau = \infty$,

$$\theta = t - t_s = \frac{G(s^2 - r^2)}{6\alpha} = \frac{q'''}{6k}(s^2 - r^2) \tag{15–15}$$

This is identical with Eq. 10–26 for a sphere under steady-state conditions. By differentiation

$$\left(\frac{\partial\theta}{\partial r}\right)_{\tau=\infty} = -\frac{2Gr}{6\alpha}$$

and the heat flow through the surface becomes

$$(q_s)_{\tau=\infty} = -kA\left(\frac{\partial\theta}{\partial r}\right)_s = \frac{k\cdot 4\pi s^2\cdot 2Gs}{6\alpha} = \frac{4}{3}s^3q'''$$

in agreement with our original assumption.

* It will be noticed that the symbol ν, as used from now on, has nothing to do with the same symbol as used in Eqs. 15–2 and 4.

When $r = 0$,

$$\frac{1}{r}\sin\frac{\nu\pi r}{s} = \frac{\nu\pi}{s}$$

and

$$\theta = \frac{Gs^2}{6\alpha} + \Sigma N_\nu \frac{\nu\pi}{s} e^{-\nu^2\pi^2\alpha\tau/s^2} \tag{15-16}$$

This is the temperature excess in the center of the sphere. When $\tau = \infty$, it becomes

$$\theta = \frac{q'''s^2}{6k} \tag{15-17}$$

which is identical with Eq. 10–27 for a sphere.

When $G = 0$, Eq. 15–14 reduces to

$$\theta = -\frac{2s}{\pi r}\theta_i \Sigma \frac{1}{\nu} e^{-\nu^2\pi^2\alpha\tau/s^2} \sin\frac{\nu\pi r}{s} \tag{15-18}$$

This equation is identical with Eq. 13–38.

For a sphere at an initial temperature distribution as given by Eq. 15–15, and suddenly brought into an environment at unvariable temperature t_e, Awbery obtains the following equation:

$$t - t_e = \frac{(s^2 - r^2)G}{6\alpha} + \frac{ksG}{3h\alpha} + \frac{1}{r}\Sigma N_\nu e^{-\psi_\nu^2\alpha\tau/s^2} \sin\psi_\nu \frac{r}{s} \tag{15-19}$$

where ψ_ν = the νth root of Eq. 13–77, tabulated in Table 13–10,

$$N_\nu = \frac{2s^2hF}{\psi_\nu} \cdot \frac{[\psi_\nu^2k^2 + (hs - k)^2]^{1/2}}{\psi_\nu^2k^2 + hs(hs - k)}, \text{ and}$$

$$F = t_s - t_e - (ksG/3h\alpha).$$

Considering Eqs. 13–43 and 77, it can be proved that the given expression for N_ν is identical with

$$N_\nu = \frac{2sF}{\psi_\nu} \cdot \frac{\sin\psi_\nu - \psi_\nu\cos\psi_\nu}{\psi_\nu - \sin\psi_\nu\cos\psi_\nu} \tag{15-20}$$

Using the values of ψ_ν given in Table 13–10, the difficulty in the numerical calculation, regretted by Awbery, disappears.

If $G = 0$, the initial temperature becomes equal to t_s and $F = t_s - t$. In this case, Eq. 15–19, with N_ν according to Eq. 15–20, becomes identical with Eq. 13–78, as can easily be checked.

15-2. Non-uniform and Variable Heat Generation in an Infinite Solid

The most general treatment of the problem of transient heat conduction, including non-uniform and variable heat generation in the conducting body, is that of Paterson [1941]. His procedure and results will be presented in an abridged manner in this and the next section.

Instead of Eq. 15-1 the differential equation takes the form

$$\frac{\partial t}{\partial \tau} = \alpha \cdot \nabla^2 t + G(x, y, z, \tau) \tag{15-21}$$

(for any values of x, y, z, and $\tau > 0$) where

$$\nabla^2 t = \frac{\partial^2 t}{\partial x^2} + \frac{\partial^2 t}{\partial y^2} + \frac{\partial^2 t}{\partial z^2}$$

and G is a function representing that part of the temperature change with time which is due to the generation of heat inside the body.

The initial temperature distribution may be given by the function

$$t = \phi(x, y, z) \quad \text{at} \quad \tau = 0 \tag{15-22}$$

As in the case of steady heat flow in coils, the heat generation will be considered a linear function of the temperature, viz.,

$$G(x, y, z, \tau, t) \equiv f(x, y, z, \tau) + t \cdot g(\tau) \tag{15-23}$$

where f and g are symbols of functions.

If the heat generation is independent of the temperature t,

$$G(x, y, z, \tau, t) = f(x, y, z, \tau) \tag{15-24}$$

For a given point x, y, z, τ, using x', y', z', and τ' as the respective variables, the solution of Eq. 15-21 becomes

$$t = \frac{1}{(4\pi\alpha\tau)^{3/2}} \int_{-\infty}^{\infty}\int_{-\infty}^{\infty}\int_{-\infty}^{\infty} \phi(x', y', z') e^{-\frac{(x-x')^2+(y-y')^2+(z-z')^2}{4\alpha\tau}} dx' \cdot dy' \cdot dz' +$$

$$\int_0^{\tau} \frac{1}{[4\pi\alpha(\tau-\tau')]^{3/2}} \int_{-\infty}^{\infty}\int_{-\infty}^{\infty}\int_{-\infty}^{\infty} f(x', y', z', \tau') e^{-\frac{(x-x')^2+(y-y')^2+(z-z')^2}{4\alpha(\tau-\tau')}} dx' \cdot dy' \cdot dz' \cdot d\tau'$$

$$= t_1 + t_2 \tag{15-25}$$

The first term (t_1) represents the solution when no heat is generated; the second one (t_2) is the contribution due to the generation of heat. For convenience, the second term may be expressed by

$$t_2 = \int_0^{\tau} \frac{\partial t_2}{\partial \tau'} d\tau' \tag{15-26}$$

If the heat generation is also dependent on the temperature t, Eq. 15-23, instead of 15-24, has to be considered. However, by substituting $t = \theta(x, y, z, \tau)e^{\int_0^\tau g(\tau')d\tau'}$, this case can be reduced to the former one, as is shown in Paterson's paper. Herewith, the solution of Eq. 15-21 becomes:

$$t = e^{\int_0^\tau g(\tau')d\tau'} \left[t_1 + \int_0^\tau e^{-\int_0^{\tau'} g(\tau')d\tau'} \frac{\partial t_2}{\partial \tau'} d\tau' \right] \tag{15-27}$$

The corresponding solutions for two-dimensional or linear conduction can easily be deduced from Eq. 15-27, by omitting z' or y' and z', their differentials, and the corresponding integrals in Eq. 15-25.

If f and g are constant in space and time, Eq. 15-27 simplifies to

$$t = \frac{f}{g}(e^{g\tau} - 1) + \frac{e^{g\tau}}{(4\pi\alpha\tau)^{3/2}} \int_{-\infty}^{\infty}\int_{-\infty}^{\infty}\int_{-\infty}^{\infty} \phi(x', y', z') e^{-\frac{(x-x')^2+(y-y')^2+(z-z')^2}{4\alpha\tau}} dx' \cdot dy' \cdot dz' \tag{15-28}$$

For the case of spherical symmetry, instead of Eqs. 15-22, 23, 27, and 28, the following equations are to be used for any fixed point r at time τ, r' and τ' being the corresponding variables.

$$t = \phi(r) \quad \text{at} \quad \tau = 0 \tag{15-29}$$

$$G(r, \tau, t) = f(r, \tau) + t \cdot g(\tau) \tag{15-30}$$

$$t = \frac{e^{\int_0^\tau g(\tau')\,d\tau'}}{r} \left\{ \frac{1}{(4\pi\alpha\tau)^{1/2}} \int_0^\infty r'\phi(r')\left[e^{-\frac{(r-r')^2}{4\alpha\tau}} - e^{-\frac{(r+r')^2}{4\alpha\tau}}\right] dr' + \int_0^\tau \frac{e^{-\int_0^{\tau'} g(\tau')\,d\tau'}}{[4\pi\alpha(\tau-\tau')]^{1/2}} \int_0^\infty r' f(r', \tau')\left[e^{-\frac{(r-r')^2}{4\alpha(\tau-\tau')}} - e^{-\frac{(r+r')^2}{4\alpha(\tau-\tau')}}\right] dr' \cdot d\tau' \right\} \tag{15-31}$$

$$t = \frac{f}{g}(e^{g\tau} - 1) + \frac{e^{g\tau}}{r(4\pi\alpha\tau)^{1/2}} \int_0^\infty r' \cdot \phi(r')\left[e^{-\frac{(r-r')^2}{4\alpha\tau}} - e^{-\frac{(r+r')^2}{4\alpha\tau}}\right] dr' \tag{15-32}$$

15-3. Non-uniform and Variable Heat Generation in a Bounded Solid Body

When the solid is in any way bounded, a surface condition has to be considered. If, as in most practical cases, this condition can be expressed by the vanishing of a linear function of the temperature t and of the normal temperature gradient $\partial t/\partial n$, at the surface, the problem can be reduced to the one without generation of heat.

In a certain range (R) of x, y, z, comprised by the solid, the differential equation takes the form given by Eq. 15–21 with Eqs. 15–22 and 23 and completed by the boundary condition

$$\left(\frac{\partial t}{\partial n}\right)_s = t \cdot \chi_s(x, y, z, \tau) - \psi_s(x, y, z, \tau) \qquad (15\text{–}33)$$

where the subscript s relates to the surface, n stands for normal to the surface, and is positive in inward direction, and χ_s and ψ_s are functions of time and space.

The meaning of the last equation may become clear by considering the following three cases:

a. The surface temperature is described.

b. There is no heat loss through the surface.

c. There is a heat loss into the environment at temperature $t_e(x, y, z, \tau)$, where t_e is a function of time and space.

In the first case

$$t = \frac{\psi_s}{\chi_s}$$

which is a function of space and time.

The second case means that

$$\chi_s = \psi_s = 0$$

so that, indeed, $(\partial t/\partial n)_s = 0$.

The third case may be expressed by

$$\chi_s = \frac{\psi_s}{t_e} = \frac{h}{k} \quad \text{or} \quad k\left(\frac{\partial t}{\partial n}\right)_s = t - t_e$$

Paterson, dealing with this most general case of heat conduction, again removes the term $t \cdot g(t)$ by substituting $\theta = v \cdot e^{\int_0^\tau g(\tau') \cdot d\tau'}$. By two more, ingenious substitutions, he reduces the general solution to that of the corresponding general problem without generation of heat. Then he applies this to the special cases of a semi-infinite plate and sphere, the free surface of the plate being at $x = 0$, that of the sphere at $r = s$.

The solutions given for each of the three surface conditions mentioned above are long expressions, but presented in a perspicuous form, ready for special applications.

15-4. Transient Heat Transfer in Cables

The generation of heat in cables is not confined to the copper core, but includes eddy-current losses in the metal sheath of the cable and dielectric losses in the insulation. Hence, even the steady-state temperature distribution in radial direction is not simple. Transient-state heat development is very complex. Here only a couple of papers will be referred to which contain valuable information on this subject.

The theory of oil-filled cables has been dealt with by Shanklin and Buller [1931]. They start from the ordinary differential equation for radial heat flow, express it in the so-called Heaviside form (see Sect. 17-6), arrive at operational equations, and solve them by use of Heaviside's expansion theorem. The following special cases are dealt with, cable in still air, in an air blast, under water, under ground (in ducts): and buried directly in the earth. A very elegant graphical way of solving the equations for abrupt and other load variations of a cable has been shown by Miller and Wollaston [1933]. The equation of the temperature distribution is obtained in the usual way of integration of the differential equation; it appears as an infinite series of Bessel functions of the first and second kinds, both of order zero. The ratio of the corresponding coefficients of these two kinds of functions, as occurring in the solution, changes in the range $\pm\infty$ which is highly inconvenient in graphical work. By a simple transformation this ratio is converted to a coefficient having the range ± 1. The roots of the Bessel functions are determined as intersections of two curves, as is usual in the graphical solution. However, since the intersections are invariant against any transformations of the intersecting lines, the lines are distorted by appropriate transformation so that very convenient ranges and angles of intersection occur, and the procedure is reduced to the routine handling of two charts which are contained in the paper.

CHAPTER 16

Some Applications of the Unsteady-state Equations

16–1. General Remarks on the Use of the Unsteady-state Equations

The equations dealt with in Chaps. 13 to 15 can be used either to find the thermal diffusivity of materials from temperature changes in space and time, or to calculate the temperature variation if the diffusivity is known.

The greater part of this chapter will be devoted to the determination of the diffusivity. Since ρ and c_p are known or can be measured by simple methods, the experimental determination of α also yields k.

The first to use such a method was Ångstroem [1861, 1863, 1864], who produced periodic changes by alternately heating and cooling metal rods in air from one end. About the same time Neumann [1862] showed how to find α by air cooling of rods, rings, spheres, and cubes which had been heated before on one spot. Since that time numerous modifications of these methods have been applied.

Their advantage is that generally a short time is sufficient for an experiment. Further, heat losses and gains have less influence on the result the faster the temperature changes. This follows from the differential equation of the temperature distribution. Considering a rod in an environment and using the same symbols as in Sects. 2–4 and 11–2, the differential equation is

$$\rho c_p A \frac{\partial \theta}{\partial \tau} = kA \frac{\partial^2 \theta}{\partial x^2} - hC\theta \qquad (16\text{–}1)$$

or

$$\frac{\partial \theta}{\partial \tau} = \alpha \frac{\partial^2 \theta}{\partial x^2} - \frac{h}{\rho c_p} \cdot \frac{C}{A} \theta = \alpha \frac{\partial^2 \theta}{\partial x^2} - \beta\theta \qquad (16\text{–}2)$$

where $\beta = \dfrac{h}{\rho c_p} \cdot \dfrac{C}{A}$. Hence, indeed, the influence of h becomes smaller and smaller with increasing $\partial\theta/\partial\tau$.

It is essential to perform the measurements under conditions where $\alpha\, \partial^2\theta/\partial x^2$ is great, i.e., where the isochrones have a small radius of curvature, so that the last term of Eq. 16–2 has a small influence only.

Another advantage of these methods is that, α being a merely kinematic quantity, no heat-energy measurements are needed. The determination of the specific heat, of course, generally requires an energy measurement; this, however, does not concern us here and is simpler, in most cases, since temperature differences can be held smaller than in heat-transfer experiments, where temperature differences are essential for the procedure. The main handicap of the unsteady-state methods is to define and satisfy the boundary conditions and to measure the varying temperature with sufficient accuracy.

As an example of the reverse procedure, viz., determination of θ from α, the cooling of apples in ships as calculated by Awbery (see Sect. 15–1) will be dealt with.

16–2. Combination of Measurements in Steady and Transient State

A method going back to Forbes [1864, 1867] shows clearly the characteristic features of the transient-state methods.

One end ($x = 0$) of a rod of length L (Fig. 16–1) is held at temperature t_0 by a temperature bath, the other end is perfectly heat insulated. The rod, provided with thermocouples, is placed horizontally in an environment of temperature t_e. The temperature curve a is found by measurement under steady-state conditions. Now the bath is taken away, and the whole rod is heated up to t_0 and left to cool in the air (Curve b). The amount by which the heat energy in an element of length dx decreases in unit time is

$$dq = -\rho c_p A \frac{\partial t}{\partial \tau}\, dx$$

By integration,

$$\Delta q = \int_{x=x}^{x=L} dq = -\rho c_p A \int_{x=x}^{x=L} \frac{\partial t}{\partial \tau}\, dx = \rho c_p A F \tag{16–3}$$

where

$$F = -\int_{x=x}^{x=L} \frac{\partial t}{\partial \tau}\, dx$$

This is equivalent to the shaded area between the x axis and the Curve c in Fig. 16–1, and Δq, obviously, is the rate of the heat flow from the rod to the environment between $x = x$ and $x = L$. Now, Forbes supposed that the rate of heat flow from an element of length dx to the environment is the same under steady- and transient-state conditions if only t is the same. Hence

$$\Delta q = -kA\left(\frac{dt}{dx}\right)_{x=x} = kAE \tag{16–4}$$

since the term in the middle is the amount of heat which passes the cross section at $x = x$ in the direction from left to right and is given up

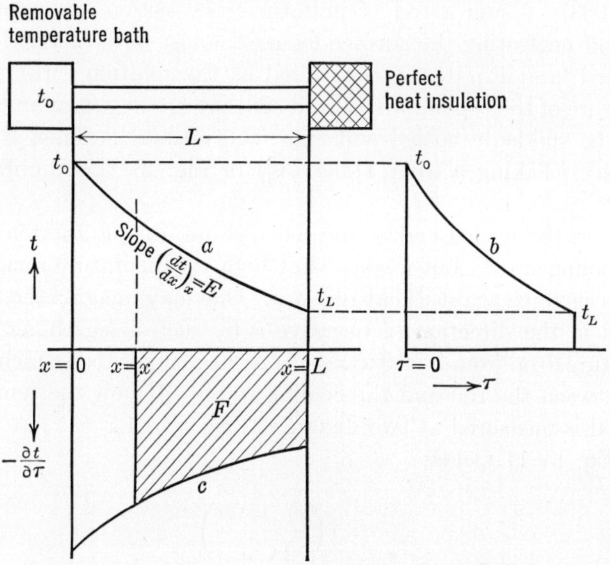

FIG. 16–1. Determination of thermal diffusivity by combination of temperature measurements in steady and transient state.

to the environment in unit time between $x = x$ and $x = L$. For simplification $E = -(dt/dx)_{x=x}$ has been introduced.

By comparison of Eqs. 16–3 and 4,

$$\alpha = \frac{k}{\rho c_p} = \frac{F}{E} \tag{16–5}$$

From the derivation it is seen that, from the arithmetic values E of the gradient and F of the area indicated in Fig. 16–1, the thermal conductivity k can be found by means of Eq. 16–5 for any temperature $t = t_0$ to $t = t_L$.

However, Forbes' assumption that the convection on any element dx is the same, whether $(\partial t/\partial x)_{x=x} = -E$ (Curve a) or zero (case of Curve b) will hold only approximately, since in the former case the environing medium will have a horizontal velocity component in the direction x in addition to the vertical movement which controls the second case.

16–3. Sudden Change of the Temperature of One End of a Rod

Neumann's method has been simplified by Kirchhoff and Hansemann [1881], who cooled a cube from an initial uniform temperature by sprinkling cold water on one of its surfaces. Their procedure has been still simplified and improved by Schulze [1898], Grueneisen [1900], and Giebe [1903]. Using a rod of uniform cross section and considerable length and neglecting the surface losses, the last term of Eq. 16–2 may be omitted, and Eq. 13–11 considered as the solution. Let t_e be the temperature of the cooling water, and measure $\theta = t - t_e$ at any distance x from the suddenly cooled end and at any time τ. Then Eq. 13–11 yields $f(u)$. Taking u from Table 13–1 or Fig. 13–1, one obtains $\alpha = x^2/4\tau u^2$.

However, the cooled end of the rod will not assume the temperature of the cooling water, but a somewhat higher temperature owing to the surface resistance against heat transfer. One may imagine the rod to be extended in the direction of negative x by such a length Δx that the temperature drop from $x = 0$ to $x = -\Delta x$ just equals that which actually exists between the rod and the cooling water. If now the temperature of the rod is measured at two distances $x = x_1$ and $x = x_2$ at the same time τ, Eq. 13–11 yields:

$$\frac{\theta_1}{\theta_i} = f\left(\frac{x_1 + \Delta x}{2\sqrt{\alpha\tau}}\right)$$

$$\frac{\theta_2}{\theta_i} = f\left(\frac{x_2 + \Delta x}{2\sqrt{\alpha\tau}}\right)$$

The left-hand sides being known by measurement, $(x_1 + \Delta x)/(2\sqrt{\alpha\tau})$ and $(x_2 + \Delta x)/(2\sqrt{\alpha\tau})$ can be taken from Fig. 13–1, and Δx and α may be calculated from these two values.

So far, the heat loss of the rod, perpendicular to its axis, has been neglected. It can be considered, using the general solution of Eq. 16–2,

$$\frac{\theta}{\theta_i} = \frac{1}{2}\left[e^{x\sqrt{\beta/\alpha}} f\left(\frac{x}{2\sqrt{\alpha\tau}} + \sqrt{\beta\tau}\right) + e^{-x\sqrt{\beta/\alpha}} f\left(\frac{x}{2\sqrt{\alpha\tau}} - \sqrt{\beta\tau}\right)\right] \quad (16\text{–}6)$$

For $h = 0$, this reduces to Eq. 13–11.

Instead of sprinkling the end of the rod with water, it may be heated at a constant rate by radiation from a glowing platinum sheet. This frees the method from the resistance against heat transfer, mentioned above. Here the boundary conditions are

$$\theta = \theta_i \quad \text{at} \quad \tau = 0$$

$$\frac{\partial \theta}{\partial x} = -c \quad \text{at} \quad x = 0$$

and the solution of the differential equation becomes

$$\theta = \frac{c}{2}\sqrt{\frac{\alpha}{\beta}}\left[e^{-x\sqrt{\beta/\alpha}}f\left(\frac{x}{2\sqrt{\alpha\tau}} - \sqrt{\beta\tau}\right) - e^{x\sqrt{\beta/\alpha}}f\left(\frac{x}{2\sqrt{\alpha\tau}} + \sqrt{\beta\tau}\right)\right] \tag{16–7}$$

To handle Eqs. 16–6 and 7, they may be expanded to powers of β, restriction to the first power being sufficient. α/β can be found by means of the Despretz method. Diesselhorst [1903–1921, pp. 224–227] has given a general representation of the original papers.

Another interesting method has been developed by L. Lorenz [1881]. It is based on the following reasoning. If at constant temperature of environment a rod is first heated on one end and then is left to cool, each mean temperature t_m which occurs in heating at a time τ' will occur again in cooling at a time τ''. Under the assumption that the heat loss of the rod to the environment of temperature t_e depends only on $\theta_m = t_m - t_e$, and therefore can be expressed by a function $\phi(\theta_m)$, it can be eliminated from two heat balances for the incoming, leaving, and stored amounts of heat energy per unit time, set up for the instants τ' and τ''. The result is:

$$\alpha = LA\,\frac{\left(\dfrac{\partial\theta_m}{\partial\tau}\right)_{\tau'} - \left(\dfrac{\partial\theta_m}{\partial\tau}\right)_{\tau''}}{\left(\dfrac{\partial\theta}{\partial x}\right)_0' - \left(\dfrac{\partial\theta}{\partial x}\right)_L' - \left(\dfrac{\partial\theta}{\partial x}\right)_0'' + \left(\dfrac{\partial\theta}{\partial x}\right)_L''} \tag{16–8}$$

where $\theta = t - t_e$, the prime and double prime signs refer to the instants τ' and τ'', and the subscripts 0 and L to $x = 0$ and $x = L$, respectively.

By means of thermocouples, inserted in the rod at distances of 2 cm, Lorenz measured and represented θ as function of x and τ.

16–4. Periodic Change of the Temperature of One End of a Rod

Ångstroem, in his pioneer paper, started from Eq. 16–2. Its solution can be written in the form

$$\theta = \sum_{\nu=0}^{\nu=\infty} N_\nu e^{-n_\nu x} \cos\left(\frac{2\nu\pi\tau'}{\tau_c} - m_\nu x - \epsilon_\nu\right) \tag{16–9}$$

where

$$0 < \tau' = \frac{\tau}{\nu} < 1$$

and m_ν and n_ν are defined by the equations:

$$m_\nu n_\nu = \frac{\nu\pi}{\alpha\tau_c} \tag{16-10}$$

$$n_\nu{}^2 - m_\nu{}^2 = \frac{\beta}{\alpha} \tag{16-11}$$

The constants N_ν and ϵ_ν are given by the temperature as function of time. Assume θ to be measured at two distances, x and $x + L$, from one end of the rod, and represented by a Fourier analysis, as:

$$\theta_x = \sum_{\nu=0}^{\nu=\infty} a_\nu \cos\left(\frac{2\nu\pi\tau'}{\tau_c} + b_\nu\right) \tag{16-12}$$

$$\theta_{x+L} = \sum_{\nu=0}^{\nu=\infty} f_\nu \cos\left(\frac{2\nu\pi\tau'}{\tau_c} + g_\nu\right) \tag{16-13}$$

From Eqs. 16–9, 12, and 13

$$\frac{a_\nu}{f_\nu} = e^{n_\nu L} \tag{16-14}$$

$$b_\nu - g_\nu = m_\nu L \tag{16-15}$$

Equations 16–14 and 15 yield n_ν and m_ν for any value of ν. Substituting these in Eqs. 16–10 and 11 yields α and β.

It is seen that this method can be used with any sort of periodic heating and cooling of one end of the rod.

It is preferable to use a less general but more convenient procedure, as R. W. King [1915] did by changing the temperature of one end according to a cosine function:

$$\theta_{x=0} = \theta_m + N \cos \omega\tau \tag{16-16}$$

He further found an arrangement, in which with sufficient accuracy

$$\theta_{x=\infty} = 0 \tag{16-17}$$

In this case, the solution of the differential equation 16–2 can be written as

$$\theta = \theta_m e^{-x\sqrt{\beta/\alpha}} + Ne^{-nx} \cos(\omega\tau - mx - \epsilon) \tag{16-18}$$

where

$$m = \frac{1}{\alpha}\sqrt{\frac{-\beta + \sqrt{\beta^2 + \omega^2}}{2}} \tag{16–19}$$

and

$$n = \frac{1}{\alpha}\sqrt{\frac{\beta + \sqrt{\beta^2 + \omega^2}}{2}} \tag{16–20}$$

Again, as in Ångstroem's procedure, α and β are the two unknown quantities. King eliminates the latter one as follows:

According to Eq. 14–13, a heat wave of length λ is travelling in the rod at the velocity

$$v = \frac{\lambda}{\tau_c} = \frac{2\pi}{m\tau_c} = \frac{L}{\Delta\tau} \tag{16–21}$$

where $\Delta\tau$ is the time in which the wave proceeds by λ in the direction x.

Substituting m from Eq. 16–19,

$$\beta = \frac{v^2}{4\alpha} - \frac{4\pi^2\alpha}{\tau_c^2 v^2} \tag{16–22}$$

Assuming now, that α and β are independent of frequency, and operating at two periods $\tau_{c,1}$ and $\tau_{c,2}$, Eq. 16–22 leads to:

$$\frac{v_1^2}{4\alpha} - \frac{4\pi^2\alpha}{\tau_{c,1}^2 v_1^2} = \frac{v_2^2}{4\alpha} - \frac{4\pi^2\alpha}{\tau_{c,2}^2 v_2^2}$$

or

$$\alpha = \frac{\tau_{c,1}\tau_{c,2}v_1v_2}{4\pi}\sqrt{\frac{v_1^2 - v_2^2}{\tau_{c,2}^2 v_2^2 - \tau_{c,1}^2 v_1^2}} \tag{16–23}$$

Figure 16–2 is a diagram of King's arrangement. An electric heating coil H is fed from E over the resistor R, and connected parallel to a variable section of the resistor r. The variation is made sinusoidal by means of a rotating disc D of suitable shape. The sample S is a wire heated by H and cooled by the surrounding air. It is extended so much to the right side that Eq. 16–7 can be used. Two thermocouples t' and t'' are placed on the sample at a distance L from each other. Their electromotive forces occurring if r is in its medium position are compensated against the batteries B' and B'' by means of the resistors R' and R'' and the galvanometers G' and G''.

The observation, then, consists only in measuring how much later G'' goes through zero than G' for the periods $\tau_{c,1}$ and $\tau_{c,2}$. From these time lags $\Delta\tau_1$ and $\Delta\tau_2$ the values of v_1 and v_2 are found by means of Eq. 16–21.

King measured α on copper and tin wires 10 to 20 in. long and 0.1 in. thick. The distance L was about 1 in. and the periods $\tau_{c,1}$ and $\tau_{c,2}$ were about 130 and 290 sec, respectively.

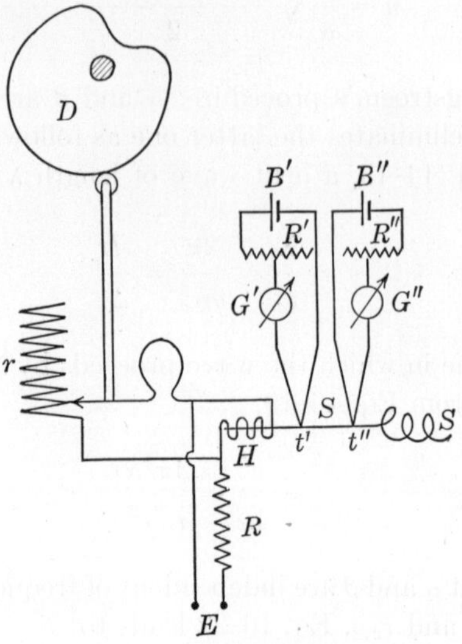

FIG. 16–2. Arrangement for determination of thermal diffusivity of metals by periodic heating of a wire.

B', B'' = Electric batteries	H = Electric heating coil
D = Rotating disc	R, R', R'', r = Resistors
E = Source of electricity	S = Test sample
G', G'' = Galvanometers	t', t'' = Thermocouples

16–5. Periodic Change of the Temperature of the Surface of a Cylinder

In measuring the thermal conductivity of humid materials, a difficulty occurs by the migration of humidity due to the changes of temperature. Tanasawa [1935] evaded this difficulty by employing a periodic heat flow. A sinusoidal heat wave was produced on the surface of a cylindric specimen, and the temperature was measured at different radial distances. Using temperature differences of less than ½F and frequencies of $\frac{1}{40}$ to $\frac{1}{80}$ sec only, the shifting of the humidity was negligible. An interesting result was that the thermal diffusivity of foundry sand first increased and then decreased with increasing humidity, with a maximum at a humidity of 10% by volume. The reason was that, starting with dry sand, k first rose fast and then slowly to about five times its original value, whereas c_p increased at a linear rate.

16-6. Cooling of Spherical Bodies Containing Uniformly Distributed Heat Sources

Some decades ago, ship loads of apples, going from Australia to England, deteriorated from a disease, called "brown heart," which occurred under insufficient cooling conditions. Apples, when placed on shipboard, are usually warm and have to be cooled down to be carried in cold storage. They also generate heat by their respiration. The question arose to what extent their temperature is higher after a given time in the refrigerated chamber than the temperature would be if they generated no heat, the latter temperature being well known. This was the problem which induced Awbery's [1927] theoretical investigation, dealt with in Sect. 15-1. Awbery first assumed that the apples are initially at uniform temperature t_i and that their skins assume the temperature t_s immediately when they are introduced into the hold. Since the center of each apple is at the highest temperature, the value there is the only one of interest in the practical problem.

For a finite value of G, Eq. 15-16, with N_ν according to Eq. 15-13, would give the temperature at $r = 0$. If $G = 0$, the first term of Eq. 15-16 and the term with G in Eq. 15-13 vanish. Hence, in the former case the temperature is higher by

$$\Delta t = \frac{Gs^2}{6\alpha} + \Sigma(-1)^\nu \frac{2Gs^2}{\nu^2\pi^2\alpha} e^{-\nu^2\pi^2\alpha\tau/s^2} \tag{16-24}$$

Assuming that $s = 4$ cm as radius of an apple,

$$G = 0.0000133 \text{ C/sec},$$
$$\alpha = 0.00155 \text{ cm}^2\text{/sec},$$

it was found that the temperature effect Δt of the heat generation was very small, namely, only 0.0232 C when, after about 2 hours, equilibrium was reached. Obviously G is much too small to influence the result appreciably.

For more powerful reactions than those occurring in an apple or for electric heating, this will be different.

CHAPTER 17

Special Mathematical Methods Used in Heat Conduction

17–1. Mathematical Methods under Consideration

So far, only straight solutions of the differential equations of heat conduction have been dealt with. By differentiation it could be seen that the differential equations were actually satisfied, and the number of constants of integration showed that the solution was general. In some cases the attention was directed to the analogy between plate, cylinder, and sphere which allowed one to write down the solution for all three shapes if the solution for one of them was given (see, for instance, Sects. 8–8 and 13–6, Article *c*). This is already a special method as understood in the title of this chapter; however, it will not be dealt with in more detail.

Another method is based on the principle of similarity; it will be briefly considered in Sect. 17–2. The procedure in conformal mapping will be dealt with only by means of one simple illustrative example (Sect. 17–3). The use of fictitious sources and images (Sect. 17–4) and of a moving system of coordinates (Sect. 17–5) will be treated somewhat more in detail. Sections 17–6 and 17–7, finally, are devoted to the very elements of operational calculus. It will just be indicated how Heaviside's method and the Laplace transform may be used for the solution of heat-conduction problems. As for conformal mapping, one has to study mathematical books to become more familiar with the theory and application of these powerful methods.

17–2. Similarity in Heat Conduction

Much use will be made of the principle of similarity in the chapters on convection. What does similarity mean in heat conduction?

For simplification heat conduction in one direction only will be assumed. Then the differential equations are

$$\frac{\partial \theta_{\mathrm{I}}}{\partial \tau_{\mathrm{I}}} = \alpha_{\mathrm{I}} \frac{\partial^2 \theta_{\mathrm{I}}}{\partial x_{\mathrm{I}}^2} \tag{17-1}$$

and

$$\frac{\partial \theta_{\mathrm{II}}}{\partial \tau_{\mathrm{II}}} = \alpha_{\mathrm{II}} \frac{\partial^2 \theta_{\mathrm{II}}}{\partial x_{\mathrm{II}}^2} \tag{17-2}$$

where θ is a temperature excess, for instance, above temperature of environment, and the subscripts I and II refer to two systems which are to be "similar." This comprises:

1. Similarity of configuration, expressed by

$$x_{\mathrm{II}} = C_\lambda x_{\mathrm{I}} \tag{17-3}$$

and, for instance, $L_{\mathrm{II}} = C_\lambda L_{\mathrm{I}}$, where C_λ is a constant and L a characteristic length of the arrangement.

2. Similarity of time intervals, expressed by

$$\tau_{\mathrm{II}} = C_\tau \tau_{\mathrm{I}} \tag{17-4}$$

and, for instance, $\tau_{0,\mathrm{II}} = C_\tau \tau_{0,\mathrm{I}}$, where τ_0 is a characteristic time interval, for instance, the time of a cycle.

3. Similarity of temperature excesses, expressed by

$$\theta_{\mathrm{II}} = C_\theta \theta_{\mathrm{I}} \tag{17-5}$$

4. Similarity of diffusivities

$$\alpha_{\mathrm{II}} = C_\alpha \alpha_{\mathrm{I}} \tag{17-6}$$

A criterion for similarity may be that the solutions of the differential equations 17-1 and 2 can be expressed by the same function Φ:

$$\Phi(x_{\mathrm{I}}, L_{\mathrm{I}}, \tau_{\mathrm{I}}, \theta_{\mathrm{I}}, \alpha_{\mathrm{I}}) = 0 \tag{17-7}$$

and

$$\Phi(x_{\mathrm{II}}, L_{\mathrm{II}}, \tau_{\mathrm{II}}, \theta_{\mathrm{II}}, \alpha_{\mathrm{II}}) = 0 \tag{17-8}$$

The question arises whether C_λ, C_τ, C_θ, and C_α are all arbitrary constants. By substitution from Eqs. 17-3 to 6 in Eq. 17-2,

$$\frac{C_\theta}{C_\tau} \cdot \frac{\partial \theta_{\mathrm{I}}}{\partial \tau_{\mathrm{I}}} = \frac{C_\alpha C_\theta \alpha_{\mathrm{I}}}{C_\lambda^2} \cdot \frac{\partial^2 \theta_{\mathrm{I}}}{\partial x_{\mathrm{I}}^2}$$

Comparison with Eq. 17-1 shows that

$$\frac{C_\theta}{C_\tau} = \frac{C_\alpha C_\theta}{C_\lambda^2} \quad \text{or} \quad \frac{C_\alpha C_\tau}{C_\lambda^2} = 1 \tag{17-9}$$

Again, substituting from Eqs. 17-3 to 6,

$$\frac{\alpha_{II}\tau_{0,II}}{L_{II}^2} = \frac{\alpha_I\tau_{0,I}}{L_I^2} = N_{Fo} \tag{17-10}$$

This means that the differential equations are similar if $\alpha\tau_0/L^2$ has a fixed value. This is a dimensionless group for which the name "Fourier number" is usual.*

More generally, $\alpha_{II}\tau_{II}/x_{II}^2 = \alpha_I\tau_I/x_I^2$ for any corresponding lengths and times in the two systems.

Another relationship can be found from the boundary condition

$$h_I\theta_{s,I} = -k_I\left(\frac{\partial\theta_I}{\partial x_I}\right)_s \tag{17-11}$$

which means that in the steady state the heat given up by unit area of surface (subscript s) to the environment must be equal to that conducted to the surface from the inside of the body under consideration.

For heat flow in any direction, the derivative $(\partial\theta_I/\partial x_I)_s$ would have to be superseded by $(\partial\theta_I/\partial n_I)_s$, where n is the direction normal to the surface.

From Eq. 17-11

$$\left(\frac{\partial\theta_I}{\partial x_I}\right)_s = -\frac{h_I}{k_I}\theta_{s,I} = -b_I\theta_{s,I} \tag{17-12}$$

where $b_I = h_I/k_I$ is a characteristic ratio. For the second system, by analogy,

$$\left(\frac{\partial\theta_{II}}{\partial x_{II}}\right)_s = -b_{II}\theta_{s,II} \tag{17-13}$$

Introducing C_β by

$$b_{II} = C_\beta b_I \tag{17-14}$$

and substituting from Eqs. 17-3 and 5,

$$\frac{C_\theta}{C_\lambda}\left(\frac{\partial\theta_I}{\partial x_I}\right)_s = C_\beta b_I C_\theta\theta_I$$

By comparison with Eq. 17-12

$$\frac{C_\theta}{C_\lambda} = C_\beta C_\theta \quad \text{or} \quad C_\lambda C_\beta = 1 \tag{17-15}$$

* Instead of the noun number the terms modulus and group are also used for this and other dimensionless expressions.

Substituting again from Eqs. 17–3 and 14,

$$b_{II}L_{II} = b_I L_I \tag{17–16}$$

This is a second dimensionless group which must be constant if similarity is to exist. Further, geometrical similarity requires

$$\frac{L_{II}}{L_I} = C_\lambda \tag{17–17}$$

if one characteristic length is sufficient to define the geometrical configuration. This is the case for a sphere or a cube or an infinitely long cylinder. For a cylinder of diameter D and length L, on the other hand, in addition to Eq. 17–17 the following would be needed:

$$\frac{D_{II}}{D_I} = C_\lambda$$

or

$$\frac{L_{II}}{D_{II}} = \frac{L_I}{D_I}$$

Finally, of course, the similarity of temperature excesses is prescribed by

$$\frac{\theta_{II}}{\theta_{s,II}} = \frac{\theta_I}{\theta_{s,I}}$$

From the above it is seen that some of the variables of Eqs. 17–7 and 8 appear only in combination with other ones, so that the general form of these equations would be

$$\Phi\left(\frac{x}{L}, \frac{\alpha\tau}{L^2}, \frac{\theta}{\theta_s}, bL\right) = 0 \tag{17–18}$$

provided that only one characteristic length L has to be considered.

Therefore it is not accidental that in the equations of heat conduction, dealt with in the former chapters, just the above dimensionless groups occur.

On the other hand, if tests should be made to find the temperature distribution or the heat conduction in geometrically similar bodies of any complex figuration, Eq. 17–18 presents the advantage of reducing considerably the number of experiments needed, as will be explained when dealing with the principle of similarity in convection (see Sects. 21–4, 23–1 and 2).

17-3. Conformal Mapping

The differential equation

$$\frac{\partial^2\theta}{\partial x^2} + \frac{\partial^2\theta}{\partial y^2} + \frac{\partial^2\theta}{\partial z^2} = 0 \tag{17-19}$$

which is known as Laplace's equation, has a very general meaning from a physical and mathematical viewpoint. It covers the so-called potential flow, diffusion, electrical and magnetic flow, heat conduction, and other forms of physical processes. Ingenious experimental methods are based upon this relation, as will be shown in Chapter 20. In the present section the fact that the so-called analytical functions of complex variables satisfy Laplace's equation will be applied to simplify problems of heat conduction. The method used for this purpose is called conformal mapping. For the theory of this method any books on advanced calculus, potential theory, and theory of functions may be consulted. At this point only the mathematical meaning of the procedure will be briefly explained and illustrated by a simple practical problem.

The method is based on certain relationships of complex quantities to each other, a few of which will be employed in the following presentation of the method. A complex variable

$$z = x + iy \tag{17-20}$$

can be considered as a vector (0z) drawn from the zero point of a cartesian system of coordinates (Fig. 17-1). This system may be called the z plane.

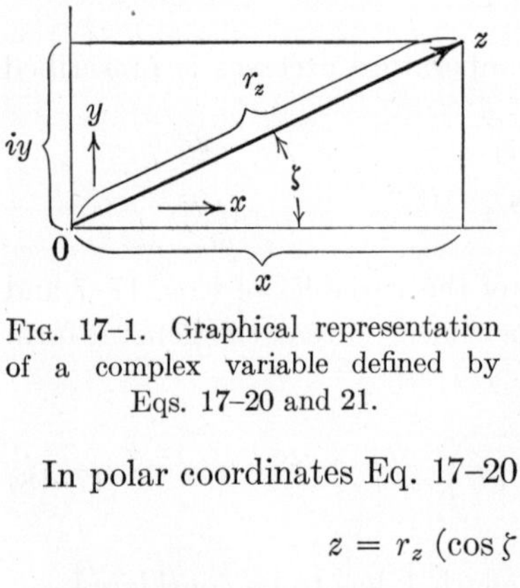

FIG. 17-1. Graphical representation of a complex variable defined by Eqs. 17-20 and 21.

In polar coordinates Eq. 17-20 can be expressed by

$$z = r_z(\cos\zeta + i\sin\zeta) = r_z e^{i\zeta} \tag{17-21}$$

where r_z is the scalar value of z, and $e^{i\zeta}$ is the vector of unit magnitude in the direction oz.

If values of a complex variable w are coordinated to those of another complex variable z by a prescription (like $w = \sqrt[n]{z}$), w is called a complex function of z, which may be expressed by

$$w = w(z)$$

By means of this function the points of the w plane are referred to the points of the z plane, or, in other words, the points of the one are "mapped" upon the other plane. If the function w has a unique derivative dw/dz, it is called an analytical function. It can be expressed as the sum of a real and an imaginary part:

$$w = u + iv \tag{17-22}$$

where both u and v are real functions of x and y.

Considering Eq. 17-20, it can be shown that each of the variables u, v, and w satisfies Laplace's equation, and that

$$\frac{\partial u}{\partial x} = \frac{\partial v}{\partial y} \tag{17-23}$$

$$\frac{\partial u}{\partial y} = -\frac{\partial v}{\partial x} \tag{17-24}$$

These two equations are known as Cauchy-Rieman's equations.

The relation of a complex analytic function w to the complex variable z on which it depends allows a simple and important interpretation. As x and y can be considered to be rectangular coordinates in a plane, the two constituents u and v of w (see Eq. 17-22) may be considered rectangular coordinates in another plane, called the w plane. In this way each point of the z plane is uniquely coordinated to a point of the w plane. Each line or surface in one plane corresponds to a line or surface in the other one.

It is further shown in the theory of functions that, for corresponding lines at corresponding points of the two planes, ratios of any differential lengths, as well as angles which may occur, are not changed in mapping if only $dw/dz \neq 0$; or, as it is usually expressed, mapping is similar in the smallest parts, i.e., conformal and isogonal. The method making use of this fact is called conformal mapping.

By choosing

$$u(x, y) = M \tag{17-25}$$

and

$$v(x, y) = N \tag{17-26}$$

and varying the constants M and N, two systems of curves may be defined which intersect each other perpendicularly. If subsequent curves of these systems always differ by $\Delta u = 1$ and $\Delta v = 1$, respectively, then Eqs. 17-25 and 26 represent a cartesian system in the w plane, the horizontals and verticals of which form squares of side length 1. The curves in the z plane which correspond to the horizontal and vertical lines in the w plane also intersect each other perpendicularly according to the above statement on conformal mapping.

In this way each square of unit side length in the w plane is mapped in the z plane as a mesh with right angles at its four corners. Now, each side of the unit squares in the w plane may be divided in n parts, and n^2 squares may be formed by drawing parallel lines to the axes of coordinates. Mapping these smaller squares, meshes are formed which, according to the definition of conformal mapping, will approach squares if n approaches infinity. Therefore a mesh in the z plane which corresponds to a square of side lengths $\Delta u = 1$ and $\Delta v = 1$ in the w plane can be considered as built up of n^2 squares if the unit square of the w plane is divided in n^2 squares, provided that $n^2 \to \infty$. However, it is of paramount importance for the application of the method of conformal mapping to the solution of heat-conduction problems and similar problems in other branches of physics, that the curvilinear squares, as the meshes of finite size may be called, can be considered as built up of exactly the same number n^2 of true squares, though n^2 will be an infinitely great number.

The procedure will now be shown on the basis of a simple mathematical example, which then will be applied for solving a special problem of heat conduction. Let the equation of transformation be

$$w = z^2 \tag{17-27}$$

By substitution from Eqs. 27–20 and 22

$$u + iv = (x + iy)^2 = x^2 + 2ixy - y^2$$

Herefrom, using also Eqs. 17–25 and 26,

$$u = x^2 - y^2 = M \tag{17-28}$$

$$v = 2xy = N \tag{17-29}$$

By differentiation it can be seen that Cauchy-Rieman's Eqs. 17–23 and 24 are satisfied.

Using Eq. 17–21 and a corresponding one for w,

$$z = x + iy = r_z e^{i\zeta}$$

$$w = u + iv = r_w e^{i\omega}$$

Equation 17–27 converts to

$$r_w e^{i\omega} = r_z^{\,2} e^{2i\zeta}$$

Herefrom, according to the algebra of complex quantities,

$$r_w = r_z^{\,2} \tag{17-30}$$

and

$$\omega = 2\zeta \tag{17-31}$$

By means of these two equations, the curvilinear squares of Fig. 17–3 have been derived from the squares of Fig. 17–2.

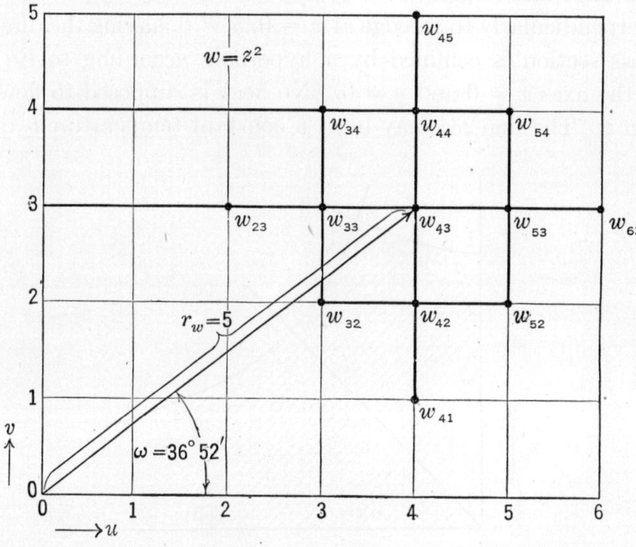

FIG. 17–2. Squares in complex w plane defined by Eqs. 17–27, 28, and 29.

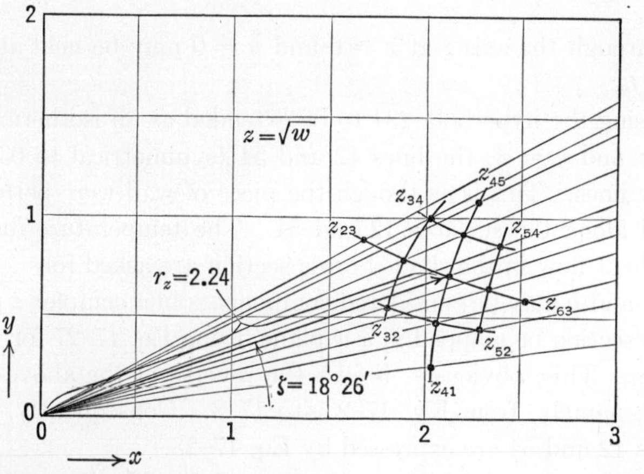

FIG. 17–3. Curvilinear squares in complex z plane, corresponding to the squares in the w plane shown in Fig. 17–2.

For instance, point w_{43} has the coordinates $u = 4$ and $v = 3$ and the distance $r_w = 5$ from the zero point of the system. Hence $\omega = 36°52'$. It follows that $r_z = \sqrt{5} = 2.24$, and $\zeta = \omega/2 = 18°26'$, etc.

The corresponding heat-conduction problem can be stated as follows:

Figure 17–4 may represent a cross section through a piece of a wall, taken perpendicularly to an edge at $x = 0$, $y = 0$, having the direction z. The cross section is confined by a hyperbola according to Eq. 17–29 and by the axes $x = 0$ and $y = 0$. No heat is supposed to flow in the direction z. The line $\overline{234}$ may be at a constant temperature t_3, and the

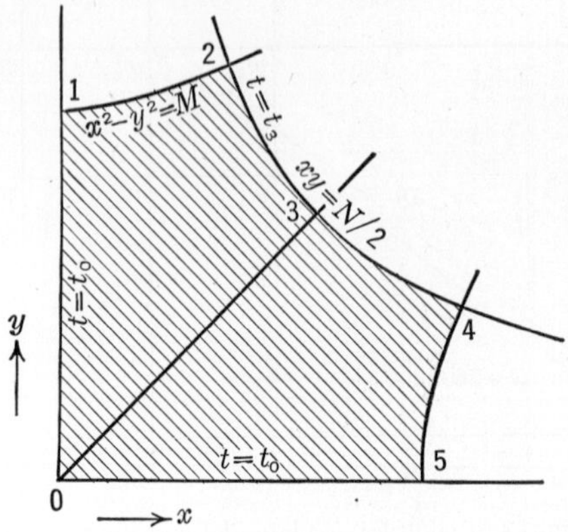

FIG. 17–4. Hyperbolic isothermals in the corner of a wall.

planes through the axis z at $x = 0$ and $y = 0$ may be held at the temperature t_0.

Supposing the hyperbola $\overline{234}$ to be extended as an isothermal line up to $x = \infty$ and $y = \infty$, the lines $\overline{12}$ and $\overline{54}$ (symmetrical to $\overline{03}$) may be heat-flow lines. This is as though the piece of wall were perfectly heat insulated along the surfaces $\overline{12}$ and $\overline{54}$. The temperature distribution and the heat flow in the shaded cross section are asked for.

Now x and y may be considered components in a complex z plane and the cross section be mapped in a w plane, using Eq. 17–27 for the transformation. This, obviously, is just the reverse of the above-described mapping, namely, from Fig. 17–2 into 17–3. Consequently, the heat-flow lines $\overline{12}$ and $\overline{54}$ are expressed by Eq. 17–28.

The resulting system is shown in Fig. 17–5. The lines $\overline{234}$ and $\overline{105}$ in this figure may be supposed to be held at the same temperatures $t = t_3$ and $t = t_0$ as the corresponding lines in Fig. 17–4. Since, as emphasized above, the shaded area between these lines in each of these figures can be divided up in the same number n^2 of very small squares, confined by isothermals and flow lines, the heat conduction through these areas between the two lines mentioned is the same in both figures. The time rate of thermal conduction in Fig. 17–5 is

$$q' = k \frac{u_5 - u_1}{v_3 - v_0} (t_3 - t_0) \tag{17–32}$$

where the subscripts correspond to the points with the same numbers in the figure, and the body 12345 is supposed to have the thickness 1 perpendicular to the shaded area. The rate of heat conduction for the

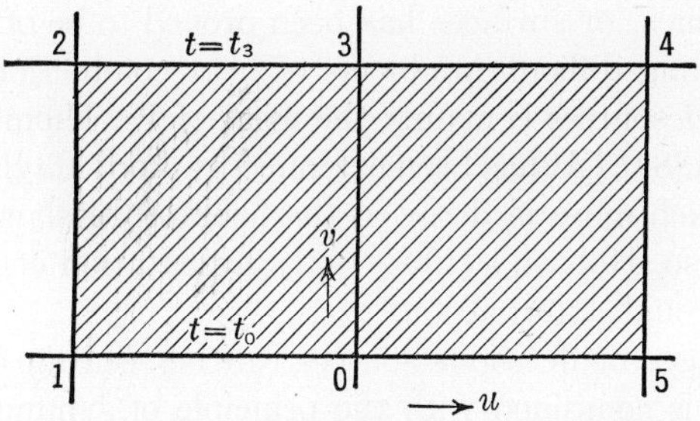

FIG. 17–5. Isothermals of Fig. 17–4 converted to straight lines by conformal mapping.

corresponding body of thickness 1 in Fig. 17–4 has the same value q', as calculated from Eq. 17–32.

It is seen, from this example, how the relatively complicated case of heat flow in Fig. 17–4 is reduced to the extremely simple case of Fig. 17–5 and Eq. 17–32.

Various applications to more complicated cases are found in the literature; for instance, the heat flow in a wall of constant thickness around a corner is calculated in a paper of E. Schmidt and Sellschopp [1932]. More often than for heat-conduction problems, the method of conformal mapping has been used in other fields of science where potential flow occurs (fluid flow, electrical and magnetic flow). These solutions may be used in corresponding heat-flow problems. Rothe, Ollendorff, and Pohlhausen [1933, p. 129 ff.], for instance, have determined the magnetic field on poles of a dynamo. Their solution directly applies to the case of heat flow around the corner of a wall of very different thickness on the two sides of the corner. The main and often serious difficulty of the method is the appropriate choice of an equation of transformation, such as Eq. 17–27 in the simple example presented above.

Often the procedure is combined with the use of the concept of point and line sources or images, which will be dealt with in the following section.

17-4. The Method of Fictitious Sources and Images

If heat is discharged by a source inside or on the surface of a body, either in a steady or unsteady state, the temperature distribution and variation can be calculated from the differential equation of heat conduction and the boundary conditions. If, on the other hand, temperature differences and changes are observed, one may search for sources of heat that have caused or can be imagined to have caused these differences or variations. A mathematical method which describes a given temperature field as though it were caused by heat sources and sinks concentrated in single points, lines, or surfaces has been proved to be of great heuristic value for handling difficult problems of heat conduction. The theory of fictitious heat sources is mainly the work of W. Thomson (later Lord Kelvin) [1880, 1884].† It has been extended by Lord Rayleigh [1911] and is presented much more in detail in the book of Carslaw [1921] already mentioned. Also, Hobson's [1903–1921] presentation of the main equations gives thorough information.

The advantages of fictitious sources can be seen most clearly if the source method is combined with the principle of symmetry, as used in electrostatics and hydrodynamics. It is known also as the method of images. The simplest application of this principle is to find the distribution of the temperature in a medium limited by planes. The medium is imagined to be extended beyond these boundaries, and the temperature distribution is determined by fictitious sources in the original space and symmetrically located sinks in the extention of the space. Also this method has been described thoroughly by Carslaw (*l.c.*) and Hobson (*l.c.*).

a. Use of Fictitious Sources and Images under Steady-state Conditions

The principle may be explained by an example for steady-state conditions dealt with more briefly in Carslaw's book and probably in earlier publications.

Referring to Fig. 17–6, let a cable or insulated wire be imbedded horizontally in the soil which is considered to be infinitely extended below its plane surface $y = 0$. The Joulean heat, q', released per unit length of the cable and the thermal conductivity, k, of the soil are given. The temperature distribution in the soil and on the surface of the cable is asked for.

If the radius, r_c, of the cable is small compared to the depth, s, at which it is buried, it can be assumed that close around it the heat flow is

† Some of the ideas and solutions of this theory, however, were already known to Fourier and Poisson, 60 years before.

homogeneous and in radial direction and that each point of the cable surface is at the same temperature t_c. Moreover, this temperature and heat flow can be imagined as produced by a homogeneous source of heat springing from the center line of the cable. The rate of heat produced in and flowing radially from unit length of this so-called line source may be equal to q' and be called the strength of the source. It is further supposed that the volume actually taken by that cable is filled by soil of the conductivity k, as the space around the cable.

The excess of the temperature t of any point, $y < 0$, in the soil over the surface temperature t_s of the soil at $y = 0$ may be called θ.

If the medium around the cable were infinitely extended and not only semi-infinitely, as the soil is, which is limited by the plane $y = 0$, the temperature distribution would be

$$\theta_f = \theta_c - \frac{q'}{2\pi k} \ln \frac{r}{r_c} \qquad (17\text{–}33)$$

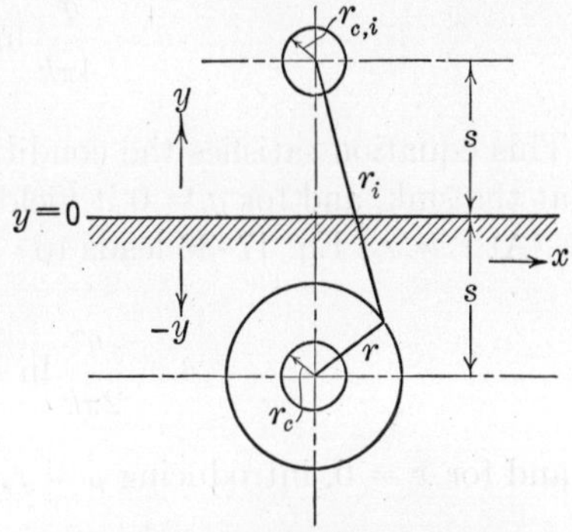

r_c= Radius of cable.
$r_{c,i}$=r_c=Radius of cable image.

FIG. 17–6. Configuration of fictitious heat source and image for determination of temperature distribution around a cable imbedded in soil.

It is easily seen that this formula cannot be applied to points of the semi-infinite medium at greater values of the radius r. At $y = 0$, for instance, r varies with x, so that θ_f would likewise vary instead of remaining equal to zero for arbitrary values of x. The subscript f has been added to indicate that this is not the true, but a fictitious, temperature.

The method of images, as applied to this example, consists in supposing the medium to be infinite and an image of the cylinder of radius r_c to exist, symmetrical to the plane $y = 0$. The temperature excess is supposed to be $\theta_i = -\theta_f$ at points, $y > 0$, symmetrical to those where the temperature θ_f exists. Then the image of the cylindrical heat source would be a sink of heat with $\theta_{c,i} = -\theta_{c,f}$ at the cylinder surface, and for any point of the soil surface $y = 0$ a negative temperature excess θ_i would neutralize the positive excess θ_f. Again, a line source of strength $-q'$ (that is, a sink) is the image of the line source q'.

Corresponding to Eq. 17–33:

$$\theta_i = -\theta_c + \frac{q'}{2\pi k} \ln \frac{r_i}{r_c} \qquad (17\text{–}34)$$

By superposition of the two temperature excesses

$$\theta = \theta_f + \theta_i = \frac{q'}{2\pi k}\left(\ln\frac{r_i}{r_c} - \ln\frac{r}{r_c}\right)$$

But

$$r = \sqrt{x^2 + (s+y)^2} \quad \text{and} \quad r_i = \sqrt{x^2 + (s-y)^2} \tag{17-35}$$

and by substitution

$$\theta = \frac{q'}{4\pi k}\ln\frac{x^2 + (s-y)^2}{x^2 + (s+y)^2} \tag{17-36}$$

This equation satisfies the conditions $\theta = \infty$ at the source and $\theta = -\infty$ at the sink, and for $y = 0$ it yields $\theta = 0$, as required.

At $r = r_c$, Eq. 17–36 leads to

$$\theta = \frac{q'}{2\pi k}\ln\frac{\sqrt{x^2 + (s-y)^2}}{r_c} \tag{17-37}$$

and for $x = 0$, introducing $\rho = r_c/2s$,

$$\theta_{min} = \frac{q'}{2\pi k}\ln\frac{1-\rho}{\rho} \tag{17-38}$$

$$\theta_{max} = \frac{q'}{2\pi k}\ln\frac{1+\rho}{\rho} \tag{17-39}$$

where θ_{min} is for $y = r_c - s$, and θ_{max} for $y = -r_c - s$. From Eqs. 17–38 and 39

$$\theta_{max} - \theta_{min} = \frac{q'}{2\pi k}\ln\frac{1+\rho}{1-\rho} = \frac{q'}{2\pi k}2\rho\left(1 + \frac{1}{3}\rho^2 + \frac{1}{5}\rho^4 + \cdots\right) \tag{17-40}$$

If $r_c \ll 2s$

$$\theta_{max} - \theta_{min} = \frac{q'}{2\pi k}\cdot\frac{r_c}{s} \tag{17-41}$$

Take, for instance, a cable of 1 in. diameter, buried at a depth $s = 6$ ft in dry soil ($k = 0.15$ B hr^{-1} ft^{-1} F^{-1}) and releasing $q' = 10$ B ft^{-1} hr^{-1}. Since $\rho = r_c/2s = 0.00348$, Eqs. 17–38 and 41 yield:

$$\theta_{min} = \frac{10}{2\pi 0.15}\ln\frac{1-0.00348}{0.00348} = 60.0\text{ F}$$

and

$$\theta_{max} - \theta_{min} \approx \frac{10}{\pi 0.15}0.00348 = 0.074\text{ F}$$

Obviously the excess of the surface temperature of the cable above that of the soil surface varies only from about 60.1 to 60.0 all around the cable.

Without going further into the details, the heat-flow lines between the line source and its image, the line sink, are shown in Fig. 17–7. The flow

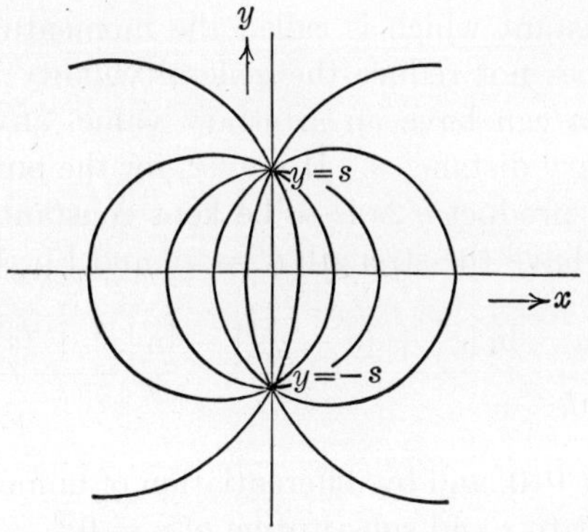

FIG. 17–7. Heat-flow lines between line source and its image.

lines are circles through the source and sink points $x = 0$, $y = -s$ and $x = 0$, $y = s$. For the proof see the book of Rothe, Ollendorff, and Pohlhausen [1933, p. 106].

If the source and sink shift closer and closer to one another, they eventually unite to a so-called doublet (double source, dipole) as shown in Fig. 17–8. In this case Eq. 17–36 would lead to $\theta = 0$ all over the

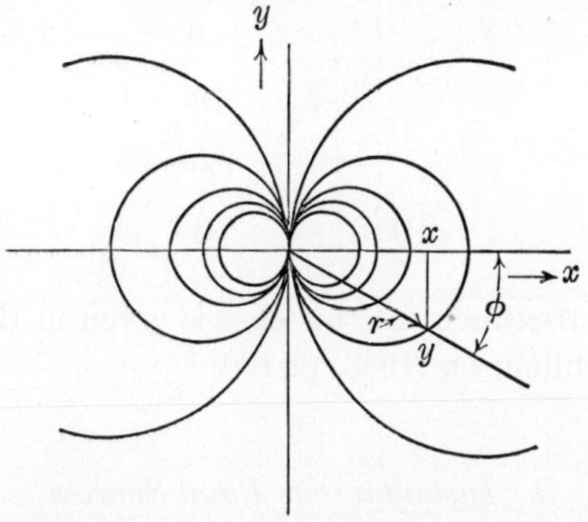

FIG. 17–8. Double line source, also called doublet or dipole, with heat-flow lines.

x, y plane, which would mean that the source and sink just cancel one another's action. The sink, however, is fictitious, and heat actually will flow into the semi-infinite space, $y < 0$. Hence, this cannot be the right solution. In the theory of functions it is shown that, in order to obtain the true solution, the approach $s \to 0$ may be performed under the condition:

$$q'2s = m \tag{17-42}$$

where m is a constant which is called the momentum of the doublet. This condition does not reduce the general validity of the source-sink concept. Since m can have an arbitrary value, any strength q' may exist at an arbitrary distance s. However, for the purpose of approaching the limit, the product $q'2s$ is to be kept constant, so that at $s = 0$ the source would have the strength $q' = \infty$, and Eq. 17-36 yields:

$$\theta = \frac{m}{8\pi k} \cdot \frac{\ln [x^2 + (s - y)^2] - \ln [x^2 + (s + y)^2]}{s}$$

If $s \to 0$, then $\theta = 0/0$, and by differentiation of numerator and denominator with respect to s and substitution of $s = 0$:

$$\theta = -\frac{m}{2\pi k} \cdot \frac{y}{x^2 + y^2} = \frac{m}{2\pi k} \cdot \frac{1}{r} \sin \phi \tag{17-43}$$

Referring to Fig. 17-8, it is seen that for a direction given by an angle ϕ, the temperature θ takes the following values:

x	y	r	θ
0	0	0	∞
		∞	0
0			$-\frac{m}{2\pi k y}$
$\gtrless 0$	0		0

A more general treatment of this case is given in the book of Rothe, Ollendorff, and Pohlhausen [1933, p. 109].

b. *Instantaneous Point Sources*

Consider an infinitely extended body, homogeneous in its thermal properties and at temperature t_e. A spherical part of this body, of radius s, is supposed to be suddenly brought to the temperature t_i at the time $\tau = 0$. It is asked how this sphere will be cooled by its environment.

The differential equation of radial heat conduction in a sphere is

$$\frac{\partial \theta}{\partial \tau} = \alpha \left(\frac{\partial^2 \theta}{\partial r^2} + \frac{2}{r} \cdot \frac{\partial \theta}{\partial r} \right) \tag{17-44}$$

where $\theta = t - t_e$.

Substituting

$$\zeta = \theta r \tag{17-45}$$

Eq. 17–44 becomes

$$\frac{\partial \zeta}{\partial \tau} = \alpha \frac{\partial^2 \zeta}{\partial r^2} \tag{17-46}$$

The boundary conditions are:

$$\zeta = \theta_i r \quad \text{when} \quad \tau = 0 \quad \text{and} \quad 0 < r \leqq s$$

$$\zeta = 0 \quad \text{when} \quad \tau = 0 \quad \text{and} \quad r > s$$

$$\zeta = 0 \quad \text{when} \quad r = 0$$

The substitution of ζ is a standard method to convert the equation for a sphere into that for a plane plate. For the latter the general solution is known to be: ‡

$$\zeta = \frac{\theta_i}{2\sqrt{\pi \alpha \tau}} \int_0^s r' [e^{-\frac{(r-r')^2}{4\alpha\tau}} - e^{-\frac{(r+r')^2}{4\alpha\tau}}]\, dr' \tag{17-47}$$

where ζ belongs to a particular radius r which is considered as constant, whereas the variable radius over which the integration is performed is denoted by r'. Expanding the integrand in powers of r' and assuming s to be small, the following approximate solution is obtained:

$$\theta = \frac{\theta_i s^3}{6\sqrt{\pi \alpha^3 \tau^3}} e^{-\frac{r^2}{4\alpha\tau}} \left[1 + \left(\frac{r^2}{\alpha\tau} - 6 \right) \frac{s^2}{40\alpha\tau} \right] \tag{17-48}$$

The initial heat content of the sphere (more exactly, the excess of heat content above that at temperature t_e) is introduced as

$$H_i = \tfrac{4}{3}\pi s^3 \rho c_p \theta_i \tag{17-49}$$

Herewith Eq. 17–48 becomes

$$\theta = \frac{H_i}{\rho c_p (4\pi\alpha\tau)^{3/2}} e^{-\frac{r^2}{4\alpha\tau}} \left[1 + \left(\frac{r^2}{\alpha\tau} - 6 \right) \frac{s^2}{40\alpha\tau} \right] \tag{17-50}$$

‡ For derivation see, for instance, Carslaw [1921, Chapter III, Sects. 16 and 18].

If $r \to 0$ at H_i = constant,

$$\theta = \frac{H_i}{\rho c_p (4\pi\alpha\tau)^{3/2}} e^{-\frac{r^2}{4\alpha\tau}} \tag{17–51}$$

where $r^2 = x^2 + y^2 + z^2$.

If $\tau \to 0$,

$$\theta = 0 \quad \text{at} \quad r > 0$$

$$\theta = \infty \quad \text{at} \quad r = 0$$

Therefore the temperature distribution, given by Eq. 17–51, can be considered as caused by an instantaneous point source of strength H_i, sprung at the point $x = 0$, $y = 0$, $z = 0$ at the time $\tau = 0$. The dimension of H_i is the British thermal unit in the British system of units.§

If the heat source originates at the point $x = x'$, $y = y'$, $z = z'$, instead of $x = 0$, $y = 0$, $z = 0$, Eq. 17–51 converts to:

$$\theta = \frac{H_i}{\rho c_p (4\pi\alpha\tau)^{3/2}} e^{-\frac{(x-x')^2+(y-y')^2+(y-y')^2}{4\alpha\tau}} \tag{17–52}$$

This is the fundamental solution of the equation of conduction in three dimensions.

Consider, now, the case that in a finite volume V of the infinitely extended body the temperature suddenly is increased by

$$\theta_i = f(x, y, z) \tag{17–53}$$

Then an infinite number of heat sources can be assumed to exist in the volume V, each, however, contributing only a differential, dH_i, to heating up the body. Here

$$dH_i = (dV)\rho c_p \theta_i \tag{17–54}$$

corresponds to Eq. 17–49, and H_i in Eq. 17–52 is to be replaced by dH_i. Furthermore,

$$\theta = \theta_i \quad \text{at} \quad r = \sqrt{x'^2 + y'^2 + z'^2} \quad \text{for} \quad \tau \to 0, \quad \text{instead of} \quad \theta = \infty$$

Each of the infinitely many heat sources in the volume V contributes to the temperature at any point x, y, z at any time τ. Hence Eq. 17–52 generalizes to

$$\theta = \int_V \frac{dH_i}{\rho c_p (4\pi\alpha\tau)^{3/2}} e^{-\frac{(x-x')^2+(y-y')^2+(z-z')^2}{4\alpha\tau}}$$

$$= \frac{1}{(4\pi\alpha\tau)^{3/2}} \int_V \theta_i e^{-\frac{(x-x')^2+(y-y')^2+(z-z')^2}{4\alpha\tau}} dV \tag{17–55}$$

§ It seems to be much better to call H_i the strength of the point source than the value $H_i/(\rho c_p)$, as Carslaw does. The physical dimension of Carslaw's "strength" would be [F ft^3].

It is seen that, from the instantaneous heat sources, given by Eqs. 17–53 and 54, the temperature distribution over the infinitely extended body at any time can be calculated by Eq. 17–55. It should be kept in mind that the content of the volume V is suddenly heated up to the temperature θ_i, but not held at this temperature.

c. Continuous Point Sources

If a point source, existing at $x = 0$, $y = 0$, $z = 0$, does not act instantaneously only, but continues from $\tau = 0$ to $\tau = \tau$, variable, however, with a total capacity of

$$H_i = \int_0^\tau \phi(\tau')\, d\tau' \tag{17–56}$$

then, by integration,

$$\theta = \frac{1}{\rho c_p (4\pi\alpha)^{3/2}} \int_0^\tau \phi(\tau') \frac{e^{-\frac{r^2}{4\alpha(\tau-\tau')}}}{(\tau - \tau')^{3/2}}\, d\tau' \tag{17–57}$$

This distribution of temperature at time τ is due to a rate of heat originating from the source and having a mean value:

$$q_{i,m} = \frac{1}{\tau} \int_0^\tau \phi(\tau')\, d\tau' \tag{17–58}$$

If

$$\phi(\tau') = \text{constant} = q_i \tag{17–59}$$

then Eq. 17–57 becomes

$$\theta = \frac{q_i}{\rho c_p (4\pi\alpha)^{3/2}} \int_0^\tau \frac{e^{-\frac{r^2}{4\alpha(\tau-\tau')}}}{(\tau - \tau')^{3/2}}\, d\tau'$$

$$= \frac{q_i}{4\rho c_p (\pi\alpha)^{3/2}} \int_{\eta=1/\sqrt{\tau}}^{\eta=\infty} e^{-\frac{r^2}{4\alpha}\eta^2}\, d\eta \tag{17–60}$$

where

$$\eta = \frac{1}{\sqrt{\tau - \tau'}}$$

For $\tau \to \infty$ this simplifies to

$$\theta = \frac{q_i}{4\pi k r} \tag{17–61}$$

which is the temperature distribution in steady state for a sphere (Eq. 8–36).

d. *Instantaneous Spherical Surface Sources*

A spherical shell of inner radius s and outer radius $s + \Delta s$ may be surrounded by an infinitely extended body of the same material at temperature t_e. At the time $\tau = 0$ the shell is suddenly brought to temperature t_i. Again, by definition, $\theta = t - t_e$. The initial heat content of the shell is

$$H_i = \tfrac{4}{3}\pi[(s + \Delta s)^3 - s^3]\rho c_p \theta_i \tag{17-62}$$

If

$$\Delta s \to 0 \quad \text{while} \quad H_i = \text{constant}$$

then, in analogy to Eq. 17–51, the following relation can be derived:

$$\theta = \frac{H_i}{8\pi^{3/2} s r \rho c_p (\alpha\tau)^{1/2}} [e^{-\frac{(r-s)^2}{4\alpha\tau}} - e^{-\frac{(r+s)^2}{4\alpha\tau}}] \tag{17-63}$$

This is the temperature excess due to an instantaneous spherical surface source of strength H_i. (See footnote §, p. 338.)

e. *Continuous Spherical Surface Sources*

Again using the definitions of H_i and $q_{i,m}$ given by Eqs. 17–56 and 58 and assuming q_i = constant, according to Eq. 17–59, the temperature distribution due to a continuous spherical surface source has been found to be

$$\theta = \frac{q_i}{4\pi^{3/2} s r \rho c_p \sqrt{\alpha}} \int_{1/\sqrt{\tau}}^{\infty} \frac{1}{\eta^2} [e^{-\frac{(r-s)^2\eta^2}{4\alpha}} - e^{-\frac{(r+s)^2\eta^2}{4\alpha}}]\, d\eta \tag{17-64}$$

where

$$\eta = \frac{1}{\sqrt{\tau - \tau'}}$$

For $\tau \to \infty$, again Eq. 17–61 holds when $r > s$, whereas for $r < s$:

$$\theta = \frac{q_i}{4\pi k s} \tag{17-65}$$

This equation has occurred already as Eq. 8–36.

f. *Instantaneous Line Sources*

If the temperature is dependent on two coordinates, x and y, and independent of the third one, z, one has a two-dimensional heat flow in an infinitely wide plate, the thickness of which in the direction z may equal L. The plate may be perfectly insulated on its two surfaces.

In any plane, z = constant, of this plate a square with the sides $x = \pm s$ and $y = \pm s$ may be at the initial temperature t_i, the surroundings originally being at the temperature t_e all over the plate up to $x = \infty$ and $y = \infty$. The temperature excess $\theta = t - t_e$ at any point of the plate at a time $\tau > 0$ is asked for.

It can be expressed by

$$\theta = \frac{\theta_i}{4\pi\alpha\tau} \int_{-s}^{+s} \int_{-s}^{+s} e^{-\frac{(x-x')^2+(y-y')^2}{4\alpha\tau}} dx'\, dy' \tag{17-66}$$

The initial excess of heat contents of the piece at initial temperature t_i is

$$H_i = 4s^2 L\rho c_p \theta_i \tag{17-67}$$

It is supposed to remain the same while $s \to 0$. At this limit

$$\theta = \frac{H_i}{4\pi Lk\tau} e^{-\frac{x^2+y^2}{4\alpha\tau}} \tag{17-68}$$

This is the temperature due to an instantaneous line source of strength H_i which is uniformly distributed over the length L of the coordinate $x = 0$, $y = 0$.

If this line is located at $x = x'$, $y = y'$ (instead at $x = 0$, $y = 0$),

$$\theta = \frac{H_i}{4\pi Lk\tau} e^{-\frac{(x-x')^2+(y-y')^2}{4\alpha\tau}} \tag{17-69}$$

In complete analogy to the procedure, given by Eqs. 17–53 to 55 for point sources, an infinite number of line sources of differential strength dH_i may spring in a volume V, bordered by two planes $z = z_1$ and $z = z_2$, where $z_2 - z_1 = L$. The result is similar to Eq. 17–55.

g. Instantaneous Cylindrical Surface Sources

Introducing

$$H_i = \pi[(s + \Delta s)^2 - s^2]L\rho c_p \theta_i \tag{17-70}$$

in analogy to Eq. 17–62, and supposing

$$\Delta s \to 0 \quad \text{while} \quad H_i = \text{constant}$$

the following equation is obtained:

$$\theta = \frac{H_i}{8\pi^2 Lk\tau} e^{-\frac{s^2+r^2}{4\alpha\tau}} \int_0^{2\pi} e^{\frac{sr}{2\alpha\tau}\cos\phi} d\phi \tag{17-71}$$

According to Carslaw, this may be transformed to

$$\theta = \frac{H_i}{4\pi Lk\tau} e^{-\frac{s^2+r^2}{4\alpha\tau}} J_0\left(\frac{isr}{2\alpha\tau}\right) \tag{17-72}$$

where J_0 is the Bessel function of zero order, and $i = \sqrt{-1}$.

Continuous line and cylindrical surface sources may be handled in analogy to Articles *c* and *e*.

h. Instantaneous Plane Sources

In an infinitely extended medium a plane plate between $x = s$ and $x = -s$ may initially be at temperature t_i and everything else be at temperature t_e. Then:

$$\theta = \frac{\theta_i}{2\sqrt{\pi\alpha\tau}} \int_{-s}^{+s} e^{-\frac{(x-x')^2}{4\alpha\tau}} dx' \tag{17-73}$$

Let again $s \to 0$, whereas

$$H_i'' = 2s\rho c_p \theta_i = \text{constant} \tag{17-74}$$

Then:

$$\theta = \frac{H_i''}{2\rho c_p \sqrt{\pi\alpha\tau}} e^{-\frac{(x-x')^2}{4\alpha\tau}} \tag{17-75}$$

This is the equation for the temperature distribution in linear heat flow, which Fourier already had interpreted as due to a plane source.

In analogy to Eqs. 17-53 and 55,

$$\theta = \frac{1}{\sqrt{4\pi\alpha\tau}} \int_{x=x_1}^{x=x_2} \theta_i e^{-\frac{(x-x')^2}{4\alpha\tau}} dx \tag{17-76}$$

where

$$\theta_i = f(x) \tag{17-77}$$

This describes the initial temperature distribution between $x = x_1$ and $x = x_2$.

Groeber [1921, pp. 62 and 66] has directed attention to the fact that for $x_1 = -\infty$ and $x_2 = +\infty$, Eq. 17-76 is more general than a solution which can be found by means of Fourier integrals, because the latter are restricted by the condition

$$\int_{-\infty}^{+\infty} f(x) \cdot dx < \infty \tag{17-78}$$

whereas Eq. 17-77 is free of this restriction.

For the application of doublets to transient-state problems see Rothe, Ollendorff, and Pohlhausen [1933, p. 156].

17-5. Transformation from a Fixed to a Moving System of Coordinates for Moving Heat Sources

The theory of heat flow due to a moving source is of considerable importance and has broad application in the general treatment of heat flow in metals, for instance, in arc welding, surface hardening, and continuous casting or quenching.

Rosenthal [1946] dealt thoroughly with the theory of moving heat sources and their applications. His procedure will be followed in this section. It is based upon the following experimental consideration by Bornefeld [1933]. If a heat source is moved at constant velocity on a sufficiently large body, a quasi-stationary state soon occurs. This means that an observer stationed at the source and moving along with it would fail to notice any change in the temperature around him as the source moves along.

For a fixed system without internal heat sources, Eq. 2-10 becomes

$$\alpha\left(\frac{\partial^2 t}{\partial x^2}+\frac{\partial^2 t}{\partial y^2}+\frac{\partial^2 t}{\partial z^2}\right)=\frac{\partial t}{\partial \tau} \tag{17-79}$$

If the properties of the material are considered as independent of temperature, assume that heat at the rate q is supplied by a point source moving along the x axis with a constant speed v. To determine the temperature distribution around the heat source, a moving system of coordinates will be used with the origin at the point source.

Figure 17-9 illustrates the moving point source. After a time τ has elapsed, the origin of the system will have been displaced from 0′ to 0,

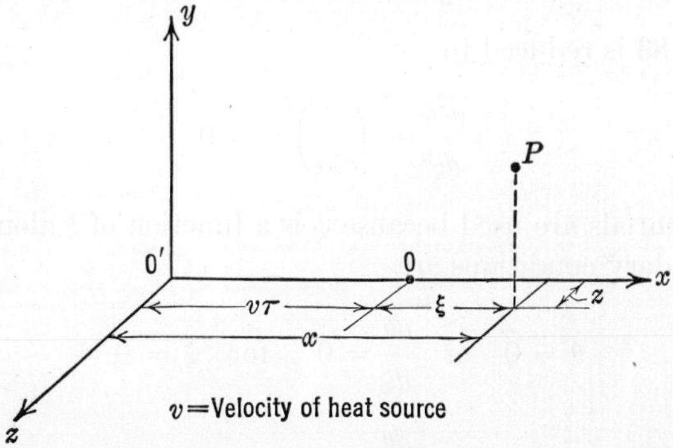

FIG. 17-9. Denotations referring to a moving point source of heat.

a distance of $v\tau$. The coordinates of some point P may be x, y, z with respect to the fixed system and ξ, y, z in the moving coordinate system.

Introducing $\xi = x - v\tau$, Eq. 17–79 becomes

$$\alpha\left(\frac{\partial^2 t}{\partial \xi^2} + \frac{\partial^2 t}{\partial y^2} + \frac{\partial^2 t}{\partial z^2}\right) = -v\frac{\partial t}{\partial \xi} + \frac{\partial t}{\partial \tau} \tag{17–80}$$

Since in the moving system $\partial t/\partial \tau = 0$, Eq. 17–80 turns into the equation of the quasi-stationary state:

$$\frac{\partial^2 t}{\partial \xi^2} + \frac{\partial^2 t}{\partial y^2} + \frac{\partial^2 t}{\partial z^2} = -\frac{v}{\alpha}\cdot\frac{\partial t}{\partial \xi} \tag{17–81}$$

This equation yields the temperature excess

$$\theta = t - t_i = e^{-\frac{v}{2\alpha}\xi}\phi(\xi, y, z) \tag{17–82}$$

where t_i = the initial temperature of the solid, and
ϕ = a function to be determined for various geometrical configurations.

Substituting t from Eq. 17–82 in Eq. 17–81,

$$\frac{\partial^2 \phi}{\partial \xi^2} + \frac{\partial^2 \phi}{\partial y^2} + \frac{\partial^2 \phi}{\partial z^2} - \left(\frac{v}{2\alpha}\right)^2\phi = 0 \tag{17–83}$$

a. Linear Flow of Heat without Losses (Plane Source)

If heat is considered to flow in the ξ direction only,

$$\frac{\partial \theta}{\partial y} = 0; \qquad \frac{\partial \theta}{\partial z} = 0$$

and Eq. 17–83 is reduced to

$$\frac{d^2 \phi}{d\xi^2} - \left(\frac{v}{2\alpha}\right)^2\phi = 0 \tag{17–84}$$

Total differentials are used because ϕ is a function of ξ alone.

The boundary conditions are

$$\theta = 0 \quad \text{or} \quad \frac{d\theta}{d\xi} = 0 \qquad \text{for} \quad \xi = \pm\infty \tag{17–85}$$

$$-k\frac{d\theta}{d\xi} = q'' \quad \text{for} \quad \xi = 0 \tag{17–86}$$

where q'' is the rate of heat produced per unit area by a plane source at $\xi = 0$.

The general solution of Eq. 17–84 is

$$\phi = Me^{-\frac{v}{2\alpha}\xi} + Ne^{\frac{v}{2\alpha}\xi} \tag{17–87}$$

Substituting ϕ into Eq. 17–82,

$$\theta = Me^{-\frac{v}{\alpha}\xi} + N \tag{17–88}$$

and by differentiation

$$\frac{d\theta}{d\xi} = \left(-M\frac{v}{\alpha}\right)e^{-\frac{v}{\alpha}\xi} \tag{17–89}$$

Using Eq. 17–86,

$$M = \frac{\alpha}{k}\cdot\frac{q''}{v} = \frac{q''}{\rho c_p v} \tag{17–90}$$

Now the two ranges, $\xi < 0$ and $\xi > 0$, must be dealt with separately.

For $\xi < 0$:

Since $d\theta/d\xi = 0$ for $\xi = -\infty$, it follows from Eq. 17–89 that $M = 0$, so that in this region the temperature distribution is given by

$$\theta = N \tag{17–91}$$

For $\xi > 0$:

Since $\theta = 0$ for $\xi = +\infty$, Eq. 17–88 yields

$$N = 0$$

Therefore, in this range,

$$\theta = Me^{-\frac{v}{\alpha}\xi} \tag{17–92}$$

Equations 17–91 and 92 must lead to the same temperature at $\xi = 0$. Therefore

$$M_{\xi \geqq 0} = N_{\xi \leqq 0}$$

Hence, from Eq. 17–90,

$$N = \frac{q''}{\rho c_p v} \tag{17–93}$$

According to Eqs. 17–91 to 93:

For $\xi < 0$

$$\theta = \frac{q''}{\rho c_p v} = \theta_{max} \tag{17–94}$$

For $\xi > 0$

$$\theta = \frac{q''}{\rho c_p v}e^{-\frac{v}{\alpha}\xi} = \theta_{max}e^{-\frac{v}{\alpha}\xi} \tag{17–95}$$

Obviously, $q''/\rho c_p v$ is the temperature excess (θ_{max}) at the heat source. Equation 17–94 shows that this maximum value of the temperature, imparted to the solid, remains after the source moves on to a new position, provided that there are no losses as has been assumed.

The temperature distribution of this case is indicated by the line $\overline{abc}$ in Fig. 17–10.

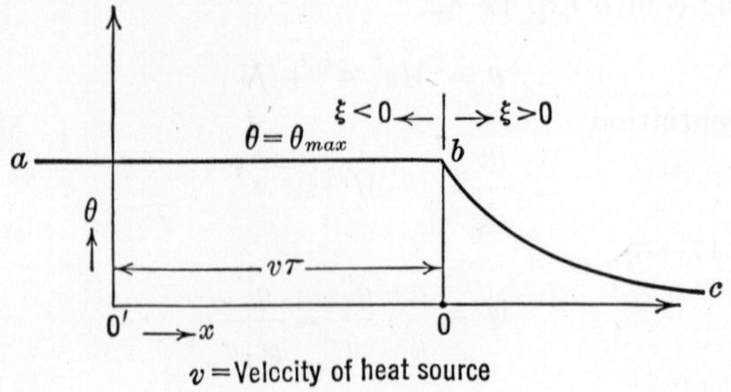

FIG. 17–10. Temperature distribution for linear flow of heat, originating from moving heat source, without losses.

It is seen that at $\xi = 0$ a discontinuity in the heat flow exists, all heat flowing in the positive ξ direction and nothing in the opposite direction.

b. *Linear Flow of Heat with Losses*

Consider now a rod of constant cross-sectional area A and assume temperature differences in the cross section to be negligible and its initial temperature to be equal to environment temperature. Then the above solution can be made to include surface losses. In this case the differential equation of linear heat flow with surface losses becomes

$$\frac{\partial \theta}{\partial \tau} = \alpha \left(\frac{\partial^2 \theta}{\partial x^2} - m^2 \theta \right) \tag{17–96}$$

where, as formerly, $m^2 = hC/kA$ (see Sect. 11–2).

For the quasi-stationary state:

$$\frac{d^2 \phi}{d\xi^2} - \left[\left(\frac{v}{2\alpha} \right)^2 + m^2 \right] \phi = 0 \tag{17–97}$$

which corresponds to Eq. 17–84 without losses.

The general solution of this differential equation is

$$\phi = Me^{-\sqrt{\left(\frac{v}{2\alpha}\right)^2 + m^2}\,\xi} + Ne^{\sqrt{\left(\frac{v}{2\alpha}\right)^2 + m^2}\,\xi} \tag{17–98}$$

By substitution in Eq. 17–82, which is also a solution of Eq. 17–97,

$$\theta = Me^{-\left(\sqrt{\left(\frac{v}{2\alpha}\right)^2 + m^2} + \frac{v}{2\alpha}\right)\xi} + Ne^{\left(\sqrt{\left(\frac{v}{2\alpha}\right)^2 + m^2} - \frac{v}{2\alpha}\right)\xi} \tag{17–99}$$

Using again the condition $\theta = 0$ at $\xi = \pm\infty$:

For $\xi < 0$

$$M = 0$$

and

$$\theta = Ne^{\left(\sqrt{\left(\frac{v}{2\alpha}\right)^2 + m^2} - \frac{v}{2\alpha}\right)\xi} \tag{17-100}$$

For $\xi > 0$

$$N = 0$$

and

$$\theta = Me^{-\left(\sqrt{\left(\frac{v}{2\alpha}\right)^2 + m^2} + \frac{v}{2\alpha}\right)\xi} \tag{17-101}$$

At the source, $\xi = 0$, the value of θ is a maximum and equal for both temperature distributions. Hence

$$M_{\xi \geqq 0} = N_{\xi \leqq 0} = \theta_{max}$$

Using Eqs. 17–100 and 101, it follows that at $\xi = 0$ the heat flow in the positive ξ direction is

$$q_1'' = k\theta_{max}\left(\sqrt{\left(\frac{v}{2\alpha}\right)^2 + m^2} + \frac{v}{2\alpha}\right) \tag{17-102}$$

and in the negative direction

$$q_2'' = k\theta_{max}\left(\sqrt{\left(\frac{v}{2\alpha}\right)^2 + m^2} - \frac{v}{2\alpha}\right) \tag{17-103}$$

As in Case *a*, there is a jump in the heat flow equal to

$$q_1'' - q_2'' = \frac{kv}{\alpha}\theta_{max} \tag{17-104}$$

The value of θ_{max} can be found by considering that the total heat output of the source per unit time and unit area is

$$q'' = q_1'' + q_2'' \tag{17-105}$$

From Eqs. 17–102, 103, and 105

$$\theta_{max} = \frac{q''}{2k\sqrt{\left(\frac{v}{2\alpha}\right)^2 + m^2}} \tag{17-106}$$

The temperature distribution is sketched in Fig. 17–11, considering that the source supplies more heat to the region ahead of it than to the region behind it.

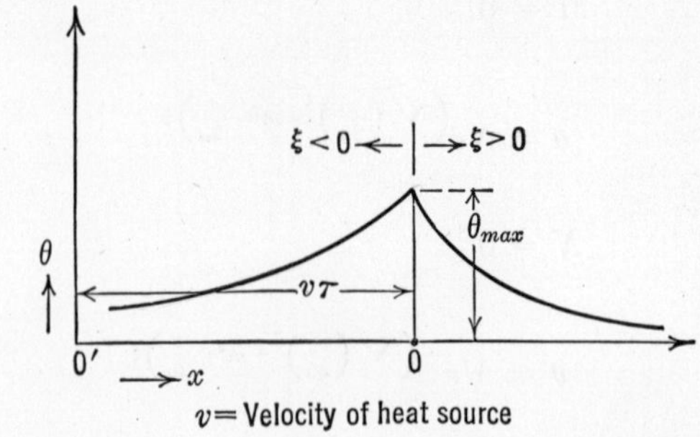

FIG. 17–11. Temperature distribution for linear flow of heat, originating from moving heat source, with losses.

c. Two-dimensional Flow of Heat without Losses (Line Source)

If heat flows in the y as well as the ξ direction, a uniform line source of heat output q' per unit time and unit length is considered lying in the z axis of the ξ, y, z system of coordinates and moving along the x axis. Since no heat flow is assumed in the z direction, $\partial\theta/\partial z = 0$, and Eq. 17–83 becomes

$$\frac{\partial^2 \phi}{\partial \xi^2} + \frac{\partial^2 \phi}{\partial y^2} - \left(\frac{v}{2\alpha}\right)^2 \phi = 0 \tag{17–107}$$

with the boundary conditions

$$\left.\begin{aligned} \frac{\partial\theta}{\partial\xi} &= 0 \quad \text{for} \quad \xi = \pm\infty \\ \frac{\partial\theta}{\partial y} &= 0 \quad \text{for} \quad y = \pm\infty \end{aligned}\right\} \tag{17–108}$$

If a cylinder of radius $r = \sqrt{\xi^2 + y^2}$ is drawn around the line source, cylindrical coordinates may be used with the boundary conditions

$$\frac{\partial\theta}{\partial r} = 0 \quad \text{at} \quad r = \infty \tag{17–109}$$

and

$$-2\pi k r \frac{\partial\theta}{\partial r} = q' \quad \text{at} \quad r = 0 \tag{17–110}$$

From the boundary conditions and the symmetrical form of Eq. 17–107 with respect to ξ and y, it follows that ϕ is a function of r alone, and the equation is converted to

$$\frac{d^2\phi}{dr^2} + \frac{1}{r} \cdot \frac{d\phi}{dr} - \left(\frac{v}{2\alpha}\right)^2 \phi = 0 \tag{17–111}$$

The general solution of this equation is

$$\phi = MI_0\left(\frac{vr}{2\alpha}\right) + NK_0\left(\frac{vr}{2\alpha}\right) \tag{17–112}$$

where I_0 is the modified Bessel function of the first kind, zero order, and K_0 is the modified Bessel function of the second kind, zero order.

Substituting in Eq. 17–82,

$$\theta = e^{-\frac{v}{2\alpha}\xi}\left[MI_0\left(\frac{vr}{2\alpha}\right) + NK_0\left(\frac{vr}{2\alpha}\right)\right] \tag{17–113}$$

Since $\theta \to 0$ for $r \to \infty$ (taking $\xi = 0$ for convenience),

$$\theta = 0 = MI_0(\infty) + NK_0(\infty)$$

But

$$I_0\left(\frac{vr}{2\alpha}\right) \to \infty, \quad \text{as} \quad r \to \infty$$

$$K_0\left(\frac{vr}{2\alpha}\right) \to 0, \quad \text{as} \quad r \to \infty$$

Hence, $M = 0$ and

$$\theta = Ne^{-\frac{v}{2\alpha}\xi}K_0\left(\frac{vr}{2\alpha}\right) \tag{17–114}$$

The function $K_0(vr/2\alpha)$ tends to $\ln r$ as $r \to 0$. Hence, $r(dK_0/dr)$ tends toward 1, and, because $\xi = 0$, the boundary condition of Eq. 17–110 is fulfilled, so that

$$N = \frac{q'}{2\pi k}$$

and

$$\theta = \frac{q'}{2\pi k} e^{-\frac{v}{2\alpha}\xi}K_0\left(\frac{vr}{2\alpha}\right) \tag{17–115}$$

To satisfy Eq. 17–109, it must be noted that $K_0(vr/2\alpha) \to \sqrt{\alpha\pi/vr}\, e^{-\frac{vr}{2\alpha}}$ as $r \to \infty$.

d. Two-dimensional Flow of Heat with Surface Losses (Thin Plate)

If the thickness s of a plate is so small that the gradient $\partial\theta/\partial z$ can be neglected, the equations of two-dimensional flow may be made to include surface losses.

Let h_1 and h_2 be the coefficients of heat transfer for the two surfaces, and

$$b_1 = \frac{h_1}{k} \quad \text{and} \quad b_2 = \frac{h_2}{k}$$

Further, for a very thin plate,

$$\frac{C}{A} = \frac{2}{s}$$

Therefore

$$\frac{h_1}{k} \cdot \frac{C}{A} = \frac{2b_1}{s} \quad \text{and} \quad \frac{h_2}{k} \cdot \frac{C}{A} = \frac{2b_2}{s}$$

Herewith, by definition,

$$m^2 = \frac{h_1 + h_2}{2k} \cdot \frac{C}{A} = \frac{b_1 + b_2}{s}$$

Therefore the differential equation, corresponding to Eq. 17-111, becomes

$$\frac{d^2\phi}{dr^2} + \frac{1}{r} \cdot \frac{d\phi}{dr} - \left[\left(\frac{v}{2\alpha}\right)^2 + m^2\right]\phi = 0 \qquad (17\text{-}116)$$

The solution of this equation is

$$\theta = \frac{q'}{2\pi k} e^{-\frac{v}{2\alpha}\xi} K_0 \left[\sqrt{\left(\frac{v}{2\alpha}\right)^2 + m^2}\, r\right] \qquad (17\text{-}117)$$

where $q' = q/s$ and q is the rate of heat dissipated by a source of length s.

e. Three-dimensional Heat Flow (Point Source)

If heat flows in the ξ, y, and z directions from a point source at O moving in the direction x, the function ϕ depends only upon the radius r, where $r = \sqrt{\xi^2 + y^2 + z^2}$.

In analogy to Eq. 17–111,

$$\frac{d^2\phi}{dr^2} + \frac{z}{r} \cdot \frac{d\phi}{dr} - \left(\frac{v}{2\alpha}\right)^2 \phi = 0 \tag{17–118}$$

Using

$$\frac{d^2\phi}{dr^2} + \frac{z}{r} \cdot \frac{d\phi}{dr} = \frac{1}{r} \cdot \frac{d^2(r\phi)}{dr^2}$$

Eq. 17–118 becomes

$$\frac{d^2(r\phi)}{dr^2} - \left(\frac{v}{2\alpha}\right)^2 r\phi = 0 \tag{17–119}$$

with the general solution

$$r\phi = Me^{-\frac{vr}{2\alpha}} + Ne^{\frac{vr}{2\alpha}} \tag{17–120}$$

By substitution in Eq. 17–82,

$$\theta = e^{-\frac{v}{2\alpha}\xi} \frac{Me^{-\frac{vr}{2\alpha}} + Ne^{\frac{vr}{2\alpha}}}{r} \tag{17–121}$$

The boundary conditions are

$$\left.\begin{aligned} \theta = 0 \quad \text{or} \quad \frac{\partial\theta}{\partial r} = 0 \quad \text{for} \quad r = \infty \\ \text{and} \qquad\qquad \\ -4\pi k r^2 \frac{\partial\theta}{\partial r} = q \quad \text{for} \quad r = 0 \end{aligned}\right\} \tag{17–122}$$

where q is the rate of heat flow from the source.

In using the first boundary condition (taking $\xi = 0$ for convenience), it must be considered that $e^{\frac{vr}{2\alpha}}$ tends to infinity at a much greater rate than r alone, so that

$$\lim_{r \to \infty} \frac{e^{\frac{vr}{2\alpha}}}{r} = \infty$$

To satisy the condition $\theta = 0$,

$$N = 0$$

and Eq. 17–121 simplifies to

$$\theta = Me^{-\frac{v}{2\alpha}\xi} \frac{e^{-\frac{vr}{2\alpha}}}{r} \tag{17–123}$$

Since

$$\lim_{r \to 0} \left(-r^2 \frac{\partial \theta}{\partial r} \right) = M$$

the second boundary condition yields

$$M = \frac{q}{4\pi k}$$

and Eq. 17-123 becomes

$$\theta = \frac{q}{4\pi k} e^{-\frac{v}{2\alpha}\xi} \frac{e^{-\frac{vr}{2\alpha}}}{r} \tag{17-124}$$

This is the equation of the temperature distribution around the point source in an infinite medium. It may be used when a welding electrode (point source) is moved in a straight line on the plane surface of a thick metal body, considered semi-infinite and homogeneous. Neglecting the heat losses from the free surface, the heat flow into the semi-infinite medium is just half of that called q in the equation. Hence, the double of the energy of the electrode would have to substitute as q in Eq. 17-124 in order to obtain θ in the body under consideration.

In the paper of Rosenthal various other metallurgical applications of the foregoing theory are described, and the results of experiments are presented which check the theory. Furthermore, the method of images is applied in connection with moving sources. Concerning this material the original paper should be consulted.

A more recent paper of Rosenthal and Cameron [1947] deals with the temperature distribution in a cylinder heated by a point source moving along its axis. This problem arose in connection with the temperature distribution produced in a gun barrel by firing a bullet.

17-6. Sudden Temperature Changes Dealt with by Means of Heaviside's Operational Calculus

Operational calculus as first developed by Heaviside [1893] has proved to be a very powerful tool for solving engineering problems. This method, originally created for the treatment of problems of transient electric flow, enables us to solve certain problems of unsteady-state heat flow in a particularly quick and easy way.

For readers not familiar with Heaviside's calculus, some of its basic principles and their application to heat-flow problems will be presented here. No attempt will be made to give any rigorous mathematical proofs. For a more thorough study, reference is made to the books of Berg [1929], Carslaw and J. C. Jaeger [1941], and Churchill [1944].

a. Heaviside's Operator p

The use of operators is very common in mathematics. In the expression

$$y = \cos (x)$$

for instance, the symbol cos stands for a trigonometric operator which is applied to the operand x. The inverse operator $\cos^{-1}$ applied to y, yields x according to the equation

$$\cos^{-1} (y) = \cos^{-1} (\cos x) = x$$

Heaviside defined an operator p and the corresponding inverse operator $1/p$ by the equations:

$$p[f(\tau)] = \frac{d}{d\tau} f(\tau) \tag{17–125}$$

and

$$\frac{1}{p} [f(\tau)] = \int_0^\tau f(\tau)\, d\tau \tag{17–126}$$

The first of these operations yields the same result as differentiation, which means applying the well-known operator $d/d\tau$ to a continuous function $f(\tau)$. In the case of p, however, the operand $f(\tau)$ may be a continuous or discontinuous function of τ. Similarly, the operator $1/p$ is the same as the well-known integration operator $\int$, except that $1/p$ is applicable to both continuous and discontinuous functions, whereas the operator $\int$ is applicable only to continuous functions of τ.

It can be easily proved that the operators p and $1/p$ are inverse of each other. Either operation followed by the other leaves the original function unchanged.

$$p \left\{ \frac{1}{p} [f(\tau)] \right\} = \frac{d}{d\tau} \left[\int_0^\tau f(\tau)\, d\tau \right] = f(\tau)$$

$$\frac{1}{p} \left\{ p[f(\tau)] \right\} = \int_0^\tau \frac{d}{d\tau} f(\tau)\, d\tau = f(\tau)$$

b. Heaviside's Unit Function

The definition of a unit function was a great step towards simplification of problems in which an electromotive force or a temperature is suddenly applied. The unit function will be denoted here by the symbol

H, instead of the symbol **1** used by Heaviside. It is defined as a discontinuous function of time τ (Fig. 17–12) according to the equations

$$\left.\begin{aligned} H(\tau) &= 0 \quad \text{for} \quad \tau < 0 \\ H(\tau) &= 1 \quad \text{for} \quad \tau > 0 \end{aligned}\right\} \tag{17–127}$$

At $\tau = 0$, the unit function suddenly jumps from 0 to 1 and so has all the values between 0 and 1.

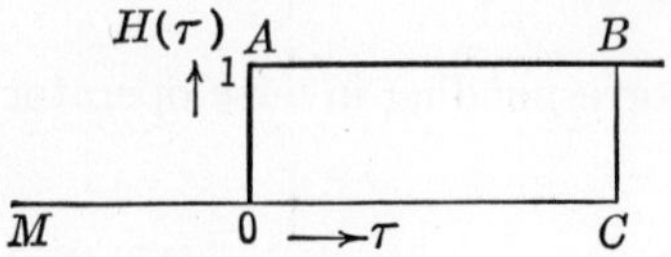

FIG. 17–12. Heaviside's unit function.

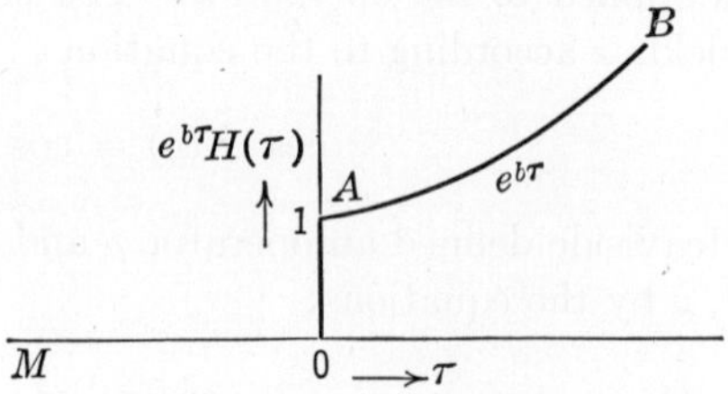

FIG. 17–13. Discontinuous function in terms of Heaviside's unit function.

Other functions of time can be easily expressed in terms of the unit function. Figure 17–13, for example, shows a discontinuous function $e^{b\tau}H(\tau)$ which is obtained by multiplying $e^{b\tau}$ by $H(\tau)$. Thus the unit function offers a unique method of defining discontinuous functions by single mathematical expressions. Ordinarily $e^{b\tau}H(\tau)$ would have to be expressed by the two equations

$$\begin{aligned} f(\tau) &= 0 \quad \text{for} \quad \tau < 0 \\ f(\tau) &= e^{b\tau} \quad \text{for} \quad \tau > 0 \end{aligned}$$

In Fig. 17–13, the line $\overline{MO}$ stands for the first of these equations, and the curve $\overline{AB}$ for the second one, while $e^{b\tau}H(\tau)$ is represented by $\overline{MOAB}$.

c. Operational Equations

In the subsequent analysis the symbol H will be used instead of $H(\tau)$ for the sake of simplicity.

Consider the operator $1/p$ acting on a function of time. say on the unit function H. Then by definition (see Eq. 17–126)

$$\frac{1}{p}H = \int_0^\tau H\,d\tau = \tau H$$

The result of the integration is given by the area $OABC = \tau H$ in Fig. 17–12. Though $H = 1$ for $\tau > 0$, the symbol H will be retained to draw attention to the fact that H is discontinuous at $\tau = 0$.

The operation $1/p$ may be repeated:

$$\frac{1}{p}\left(\frac{1}{p}H\right) = \frac{1}{p^2}H = \frac{1}{p}\tau H = \int_0^\tau \tau H \cdot d\tau = \frac{\tau^2}{2}H$$

Repeating this operation n times,

$$\frac{\tau^n}{n!}H = \frac{1}{p^n}H \tag{17–128}$$

This equation indicates a method of converting a function of time τ into a function of p. In general:

$$f(\tau)H = \phi(p)H \tag{17–129}$$

As $(1/p^n)H$ means n times integrating H, so $\phi(p)$ means a certain operation on H. The expression $\phi(p)$ is an algebraic function of the operator p, which, when operating on the time function H, transforms it into the time function $f(\tau)H$.

Equations of this nature are called operational equations. Equation 17–129 is entirely symbolic. According to the Heaviside calculus, it can also be expressed in the form

$$\phi(p) = p\int_0^\infty e^{-p\tau}f(\tau)\,d\tau \tag{17–130}$$

The integral in this equation is known as the Laplace transform, which will be dealt with briefly in Sect. 17–7.

For illustration, one operational equation will be derived:

$$\frac{p}{p+b}H = \frac{1}{1+\dfrac{b}{p}}H = \left(1 - \frac{b}{p} + \frac{b^2}{p^2} - \frac{b^3}{p^3}\cdots\right)H$$

$$= \left(1 - b\tau + \frac{b^2\tau^2}{2!} - \frac{b^3\tau^3}{3!}\cdots\right)H$$

Hence

$$\frac{p}{p+b}H = e^{-b\tau}H \tag{17–131}$$

Similarly, it can be proved that

$$\frac{p}{p-b}H = e^{b\tau}H \tag{17–132}$$

It is seen that ordinary algebra is used in dealing with the operator p.

In general, the method transforms differential equations with constant coefficients into algebraic equations which are much easier to handle.

In the solution of unsteady-state problems by operational calculus, the first step is to set up the differential equations as usual. The differential equations are then written in the form of operational expressions. The last step is to evaluate these operational expressions by using one or more of the theorems given below.

d. Heaviside's Expansion Theorem

When the operator $\phi(p)$ in Eq. 17–129 can be expressed as a rational fraction with no multiple roots in the denominator, it may be written as the sum of partial fractions, each one of which may be separately evaluated. For example:

$$\frac{p}{p^2 - 5p + 6} H = \left(\frac{p}{p-3} - \frac{p}{p-2} \right) H = \frac{p}{p-3} H - \frac{p}{p-2} H$$

and using Eq. 17–132,

$$\frac{p}{p^2 - 5p + 6} H = (e^{3\tau} - e^{2\tau}) H$$

e. Heaviside's Superposition Theorem

If $\phi(p)H$ in Eq. 17–129 has the form $\phi_1(p) f_2(\tau) H$, i.e., if the operand is $f_2(\tau)H$ instead of the simple unit function H, then, according to the same equation, $f_2(\tau)$ can be replaced by its equivalent $\phi_2(p)$. This is a simple case of the superposition theorem, yielding

$$\phi_1(p) \cdot f_2(\tau) H = \phi_1(p) \cdot \phi_2(p) H \tag{17–133}$$

For example, consider $\dfrac{1}{p+b} e^{-b\tau} H$.

From Eq. 17–131

$$e^{-b\tau} H = \frac{p}{p+b} H$$

and using the superposition theorem,

$$\frac{1}{p+b} e^{-b\tau} H = \frac{1}{p+b} \cdot \frac{p}{p+b} H = \frac{p}{(p+b)^2} H$$

But

$$\frac{p}{(p+b)^2}H = \left(\frac{1}{p} - \frac{2b}{p^2} + \frac{3b^2}{p^3} \cdots\right)H$$

$$= \left(\tau - 2b\frac{\tau^2}{2!} + 3b^2\frac{\tau^3}{3!} \cdots\right)H$$

Hence

$$\frac{1}{p+b}\,e^{-b\tau}H = \tau e^{-b\tau}H$$

It can be proved that the order of $\phi_1(p)$ and $\phi_2(p)$ can be reversed, but the order of the operators with respect to the time function has to be maintained. For example,

$$\frac{1}{p+b} \cdot \frac{p}{p+b}H = \frac{p}{p+b} \cdot \frac{1}{p+b}H$$

But

$$\frac{1}{p+b}\,e^{-b\tau}H \neq e^{-b\tau}\frac{1}{p+b}H$$

The operators themselves follow all the rules of common algebra, but their order with respect to the time function must not be changed.

f. Irrational Operators

In most of the cases operational expressions involving irrational operators can be handled by expanding them in a series of $1/p^n$, as was done for rational operators in the derivation of Eq. 17–131. Sometimes, however, it becomes necessary to apply the Laplace transform.

For example, it may be required to determine $f(\tau)$ in Eq. 17–129 when $\phi(p) = p^{1/2}$. Using Eq. 17–130,

$$p^{-1/2} = \int_0^\infty e^{-p\tau} f(\tau)\,d\tau \qquad (17\text{–}134)$$

It will be proved that Eq. 17–134 is satisfied by

$$f(\tau) = \frac{1}{\sqrt{\pi\tau}}$$

Substituting this value in the equation and further replacing τ by x^2 and $x\sqrt{p}$ by y:

$$\int_0^\infty e^{-p\tau}\frac{1}{\sqrt{\pi\tau}}\,d\tau = \frac{2}{\sqrt{\pi}}\int_0^\infty e^{-px^2}\,dx = \frac{2}{\sqrt{\pi}}\,(p)^{-1/2}\int_0^\infty e^{-y^2}\,dy \qquad (17\text{–}135)$$

According to Eq. 13–10, the integral has the value $\sqrt{\pi}/2$. Hence Eq. 17–135 is identical with Eq. 17–134, and Eq. 17–129 becomes

$$p^{1/2}H = \frac{1}{\sqrt{\pi\tau}}H \tag{17–136}$$

A similar procedure leads to

$$p^{-1/2}H = 2\sqrt{\frac{\tau}{\pi}}\,H \tag{17–137}$$

g. Some Operational Formulas

For convenience a few operational formulas are given below, which may be needed for applying Heaviside's method to heat-transfer problems.

$$p^{1/n}H = \frac{\tau^n}{n!}H \tag{17–128}$$

$$\frac{1}{p+b}H = \frac{1}{b}(1 - e^{-b\tau})H \tag{17–138}$$

$$\frac{p}{p+b}H = e^{-b\tau}H \tag{17–131}$$

$$p^{1/2}H = \frac{1}{\sqrt{\pi\tau}}H \tag{17–136}$$

$$p^{-1/2}H = 2\sqrt{\frac{\tau}{\pi}}\,H \tag{17–137}$$

$$\sqrt{p+b}H = \frac{e^{-b\tau}}{\sqrt{\pi\tau}} + b\int_0^\tau \frac{e^{-b\tau}}{\sqrt{\pi\tau}}\,d\tau \tag{17–139}$$

$$e^{\sqrt{bp}}H = \left[1 + \sqrt{\frac{b}{\pi}}\left(\tau^{-1/2} - \frac{1}{2}\cdot\frac{b}{3!}\tau^{-3/2} + \frac{1\cdot 3}{2\cdot 2}\cdot\frac{b^2}{5!}\tau^{-5/2}\cdots\right)\right]H \tag{17–140}$$

$$e^{-\sqrt{bp}}H = \left[1 - \sqrt{\frac{b}{\pi}}\left(\tau^{-1/2} - \frac{1}{2}\cdot\frac{b}{3!}\tau^{-3/2} + \frac{1\cdot 3}{2\cdot 2}\cdot\frac{b^2}{5!}\tau^{-5/2}\cdots\right)\right]H \tag{17–141}$$

$$\frac{bp}{p^2+b^2}H = (\sin b\tau)H \tag{17–142}$$

$$\frac{b^2}{p^2+b^2}H = (\cos b\tau)H \tag{17–143}$$

h. Application of Heaviside's Method to a Heat-conduction Problem

A perfectly insulated rod, extending from $x = 0$ to infinity, and being originally at zero temperature, is suddenly touched at $x = 0$ by a hot body of constant temperature t_0. Develop expressions for the temperature distribution in the rod in terms of x and τ, the rate of heat q_0 entering the rod, and the total quantity of heat Q_0 entering the rod in time τ.

SOLUTION: Since x is independent of time τ, the operator p can be used like a constant. Consequently the Laplace equation can be written in the following form of an ordinary differential equation:

$$\frac{d^2t}{dx^2} - \frac{1}{\alpha}pt = 0$$

and its solution becomes

$$t = M_1 e^{-\sqrt{(p/\alpha)}x} + N_1 e^{+\sqrt{(p/\alpha)}x} \tag{17–144}$$

where M_1 and N_1 are arbitrary constants.

At $x = \infty$, $t = 0$. Hence $N_1 = 0$. At $x = 0$, $t = t_0H$. Hence $M_1 = t_0H$.

Using $t = t_0H$ and not $t = t_0$ takes care of the discontinuity at $\tau = 0$. With these boundary values Eq. 17–144 becomes

$$t = e^{-\sqrt{(p/\alpha)}x}t_0H$$

Using Eq. 17–141, for $x > 0$,

$$t = t_0\left[1 - \frac{x}{\sqrt{\pi\alpha}}\left(\tau^{-1/2} - \frac{1}{2}\cdot\frac{x^2/\alpha}{3!}\tau^{-3/2} + \frac{1\cdot 3}{2\cdot 2}\cdot\frac{(x^2/\alpha)^2}{5!}\tau^{-5/2}\cdots\right)\right] \tag{17–145}$$

The rate of heat entering the rod may be determined by considering that

$$\frac{dq}{dx} = -\rho c_p A p t$$

Differentiating a second time,

$$\frac{d^2q}{dx^2} = -\rho c_p A p\frac{dt}{dx} = \frac{1}{\alpha}pq$$

The general solution of this equation is

$$q = M_2 e^{-\sqrt{(p/\alpha)}x} + N_2 e^{+\sqrt{(p/\alpha)}x}$$

At $x = \infty$, $q = 0$. Hence $N_2 = 0$. At $x = 0$, $q = q_0H$. Hence $M_2 = q_0H$. Herewith

$$q = e^{-\sqrt{(p/\alpha)}x} q_0 H \tag{17-146}$$

But

$$q = -kA \frac{dt}{dx}$$

Substituting t from Eq. 17–145,

$$q = kA \sqrt{\frac{p}{\alpha}}\, e^{-\sqrt{p/\alpha}x} t_0 H \tag{17-147}$$

and comparing Eqs. 17–146 and 147,

$$q_0 = kAt_0 \sqrt{\frac{p}{\alpha}}\, H \tag{17-148}$$

Using Eq. 17–136,

$$q_0 = \frac{kAt_0}{\sqrt{\pi\alpha\tau}} \tag{17-149}$$

Finally:

$$Q_0 = \int_0^\tau q_0 \cdot d\tau = \frac{1}{p}\, q_0 \tag{17-150}$$

Substituting from Eq. 17–148,

$$Q_0 = \frac{kAt_0 \sqrt{\dfrac{p}{\alpha}}}{p}\, H = \frac{kAt_0}{\sqrt{\alpha}} \cdot \frac{1}{\sqrt{p}}\, H$$

and using Eq. 17–137,

$$Q_0 = 2kAt_0 \sqrt{\frac{\tau}{\pi\alpha}} \tag{17-151}$$

Notice that at the time of contact ($\tau = 0$), q_0 is infinite, although Q_0 is a finite quantity.

Equations 17–149 and 151 are virtually the same as Eqs. 13–13 and 14; moreover, Eq. 17–151 could have been obtained directly from Eqs. 17–149 and 150 by integration. However, the author wanted to demonstrate the procedure of the Heaviside calculus in all details in the present simple example.

17-7. Application of the Laplace Transform to Heat-conduction Problems

An operational procedure, similar to Heaviside's calculus and possibly more general, is now preferably used for the solution of complicated heat-conduction problems. It is based on the Laplace transform, which was defined by Eq. 17–130. Only the simplest steps of the mathematical technique can be presented in this section.

a. Some Simple Laplace Transforms

If $F(\tau)$, a function of τ, is multiplied by $e^{-s\tau}$ and integrated from 0 to ∞, a function of s is obtained, which may be expressed by $f(s)$ and is called the Laplace transform of $F(\tau)$. Using the symbol L for this operation,

$$LF(\tau) = \int_0^\infty e^{-s\tau}F(\tau)\,d\tau = f(s) \qquad (17\text{–}152)$$

where s is a positive number large enough to make the integral converge.||

Consider, for example,

$$F(\tau) = e^{-b\tau}$$

Its Laplace transform is

$$Le^{-b\tau} = \int_0^\infty e^{-s\tau}e^{-b\tau}\,d\tau = \frac{1}{s+b} \qquad (17\text{–}153)$$

Some other important transforms are:

$$L\frac{\tau^n}{n!} = \frac{1}{s^{n+1}} \qquad (17\text{–}154)$$

where n is a positive integer

$$L \sin b\tau = \frac{b}{s^2 + b^2} \qquad (17\text{–}155)$$

$$L \cos b\tau = \frac{s}{s^2 + b^2} \qquad (17\text{–}156)$$

$$L\left[1 - \sqrt{\frac{b}{\pi}}\left(\tau^{-1/2} - \frac{1}{2}\cdot\frac{b}{3!}\tau^{-3/2} + \frac{1\cdot 3}{2\cdot 2}\cdot\frac{b^2}{5!}\tau^{-5/2}\cdots\right)\right] = \frac{1}{s}e^{-\sqrt{b/s}} \qquad (17\text{–}157)$$

$$L\frac{1}{\sqrt{\pi\tau}} = s^{-1/2} \qquad (17\text{–}158)$$

$$L2\sqrt{\frac{\tau}{\pi}} = s^{-3/2} \qquad (17\text{–}159)$$

|| Equation 17–152 may be obtained from Eq. 17–130 by replacing p by s and $\phi(p)/p$ by $f(s)$.

It should be noticed that Laplace transforms are similar to Heaviside's operational expressions given in Sect. 17–6, Article *g*. Dividing any of Heaviside's operational expressions by p and then replacing p by s yield the corresponding Laplace transform.

b. Some Theorems about the Laplace Transform

Some important theorems may be presented without proof.

1. If $LF(\tau) = f(s)$ and $F'(\tau)$ and $F''(\tau)$ are the first and second derivatives of $F(\tau)$, respectively, then

$$LF'(\tau) = s \cdot f(s) - F(+0) \tag{17–160}$$

where $F(+0)$ stands for the value of $F(\tau)$ which is approached when τ approaches 0 from the positive direction. Also

$$LF''(\tau) = s^2 f(s) - sF(+0) - F'(+0) \tag{17–161}$$

This theorem converts the derivatives of $F(\tau)$ to algebraic functions of s so that differential equations with τ as independent variable can be transformed into algebraic equations with s as variable.

2. If

$$f_1(s) = LF_1(\tau) \quad \text{and} \quad f_2(s) = LF_2(\tau)$$

then

$$f_1(s) + f_2(s) = L[F_1(\tau) + F_2(\tau)] \tag{17–162}$$

and

$$f_1(s) \cdot f_2(s) = \int_0^\tau F_1(\lambda)F_2(\tau - \lambda)\, d\lambda = \int_0^\tau F_1(\tau - \lambda)F_2(\lambda)\, d\lambda \tag{17–163}$$

The latter part of this theorem is called the convolution.

3. If

$$f(s) = LF(\tau)$$

then

$$f(s + b) = Le^{-b\tau}F(\tau) \tag{17–164}$$

and

$$e^{-bs}f(s) = LF(\tau + b) \tag{17–165}$$

4. If

$$LF(\tau) = f(s)$$

then

$$L\tau \cdot F(\tau) = -\frac{d}{ds} f(s) \tag{17–166}$$

In general

$$L\tau^n F(\tau) = (-1)^n \frac{d^n}{ds^n} f(s) \tag{17–167}$$

Heaviside's expansion theorem, given in Sect. 17–6, Article *d*, is also frequently used in the evaluation of Laplace transforms.

c. A Simple Example of Application

The transformation of a differential equation into an algebraic equation will be shown for the case of the Fourier-Poisson equation for unsteady-state heat conduction.

$$\frac{\partial t}{\partial \tau} = \alpha \frac{\partial^2 t}{\partial x^2} \tag{17–168}$$

where $t = t(x, \tau)$ is a function of x and τ, and the boundary conditions are

$$t(x, 0) = 0, \quad \text{i.e.,} \quad t = 0 \quad \text{at} \quad \tau = 0$$

$$t(0, \tau) = t_0, \quad \text{i.e.,} \quad t = t_0 \quad \text{at} \quad x = 0$$

$$t(\infty, \tau) = 0, \quad \text{i.e.,} \quad t = 0 \quad \text{at} \quad x = \infty$$

Replacing $F(\tau)$ by the form $t(x, \tau)$, Eq. 17–160 becomes

$$L\frac{\partial t}{\partial \tau} = Lt'(x, \tau) = s \cdot t(x, s) - t(x, 0) = s \cdot t(x, s) \tag{17–169}$$

where $t(x, s)$ is the Laplace transform of $t(x, \tau)$ with respect to τ.¶

From Eq. 17–152,

$$L\alpha \frac{\partial^2 t(x, \tau)}{\partial x^2} = \int_0^\infty \alpha \frac{\partial^2}{\partial x^2} e^{-s\tau} t(x, \tau)\, d\tau$$

Assuming that the function $e^{-s\tau}t(x, \tau)$ satisfies the conditions under which the order of differentiation with respect to x and integration with respect to τ can be interchanged, and again considering Eq. 17–152,

$$L\alpha \frac{\partial^2 t(x, \tau)}{\partial x^2} = \alpha \frac{\partial^2}{\partial x^2} \int_0^\infty e^{-s\tau} t(x, \tau)\, d\tau = \alpha \frac{\partial^2}{\partial x^2} t(x, s) \tag{17–170}$$

The only variable in the last term is x, since s is only a suitable positive number. Hence

$$L\alpha \frac{\partial^2 t(x, \tau)}{\partial x^2} = \alpha \frac{d^2}{dx^2} t(x, s) \tag{17–171}$$

¶ It should be noticed that using the same function symbol t does not mean that $t(x, \tau)$ and $t(x, s)$ have the same form.

Writing Eq. 17–168 in the form

$$L \frac{\partial t}{\partial \tau} = L\alpha \frac{\partial^2 t}{\partial x^2} \tag{17–172}$$

and substituting from Eqs. 17–169 and 171,

$$s \cdot t(x, s) = \alpha \frac{d^2}{dx^2} t(x, s) \tag{17–173}$$

Such a "subsidiary equation" may be solved by applying Laplace transform with respect to x, or, if it is simple, the usual methods for solving ordinary differential equations may be employed. In the present case the solution is given in Eq. 11–7, which may be written in the form

$$t(x, s) = Me^{-\sqrt{s/\alpha}x} + Ne^{+\sqrt{s/\alpha}x} \tag{17–174}$$

Since $t(\infty, \tau) = 0$, Eq. 17–152 yields $Lt(\infty, \tau) = 0$ and $t(\infty, s) = 0$. Hence $N = 0$. Further $t(0, \tau) = t_0$. Again applying Eq. 17–152,

$$t(0, s) = \frac{t_0}{s}$$

and therefore

$$M = \frac{t_0}{s}$$

Herewith Eq. 17–174 becomes

$$t(x, s) = \frac{t_0}{s} e^{-\sqrt{s/\alpha}x} \tag{17–175}$$

Using Eq. 17–157,

$$t(x, \tau) = t_0 \left[1 - \frac{x}{\sqrt{\pi\alpha}} \left(\tau^{-1/2} - \frac{1}{2} \cdot \frac{x^2/\alpha}{3!} \tau^{-3/2} + \frac{1 \cdot 3}{2 \cdot 2} \cdot \frac{(x^2/\alpha)^2}{5!} \tau^{-5/2} \cdots\right)\right] \tag{17–176}$$

which is identical with Eq. 17–145.

Concerning the so-called inversion theorem and other features of the theory of the Laplace transform, the mathematical books of Carslaw and J. C. Jaeger [1941], Churchill [1944], or others should be consulted.

CHAPTER 18

Numerical Methods for the Solution of Heat-conduction Problems

18–1. Numerical Methods under Consideration

Except for the simplest geometric shapes and boundary conditions the analytical solution of the differential equation of heat conduction is a most cumbersome if not impossible job. Approximative solutions may be obtained by numerical, graphical, or experimental methods. Chapters 18 to 20 will be devoted to these methods. In the present chapter first a numerical method for steady-state problems will be dealt with, the roots of which may be found in Cross's [1932] analysis of continuous frames by distributing fixed-end moments. Independently and with essential modifications it has been developed by Southwell [see Christopherson and R. V. Southwell 1938], who called it the relaxation method, but did not use it for heat-flow problems. The application of the method in this field is due to Emmons [1943, 1944].

For steady-state and composite rectangular areas Frocht [1946] developed a simple and fast method of numerical solutions, directly based on Laplace's equation. It would transgress the scope of this book to reproduce from the original paper the tables computed by Frocht for use in the calculations.

For unsteady state Emmons employed a straightforward numerical method which shortly afterwards was aptly generalized by Dusinberre [1945]. In its use for transient heat conduction it is closely related to the Binder-Schmidt graphical method (see Chapter 19).

18–2. Application of the Relaxation Method to Steady-state Conduction Problems

For introduction into the method of relaxation, consider two-dimensional conduction as indicated by the pattern of Fig. 18–1. The heat is assumed to flow in the directions of x and y only, through equal rods of length $\Delta x = \Delta y$, which form a network in the y, x plane. Each point represents a sink for the heat conducted to it by the rods meeting there. The conduction through any rod is assumed to be equal to the conduction in a square of side length Δx. For instance, the heat flow in rod $\overline{10}$ supersedes the conduction indicated by the parallel arrows in the figure. The fields of conduction in the directions $\overline{10}$, $\overline{20}$, $\overline{30}$, and $\overline{40}$ partly superimpose one another. There is no objection against this procedure, which is substantially the same as that used in the derivation of the differential equation of heat conduction in Sect. 2-4.

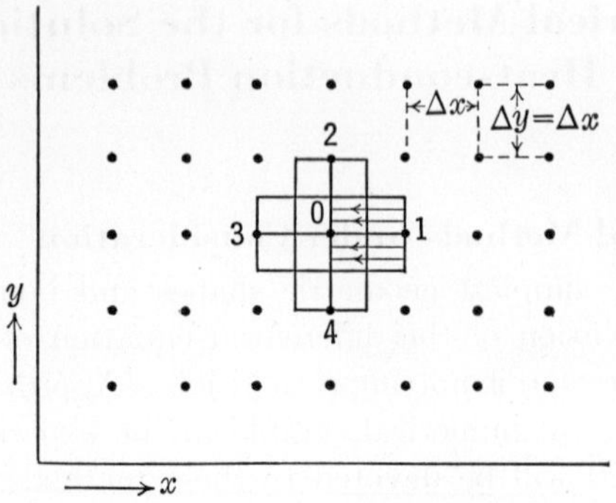

FIG. 18-1. Replacing homogeneous medium by fictitious rods as elements in the relaxation treatment.

Hence, for a body of unit length in the direction z, the conductive heat flow to the point 0 becomes

$$q_0' = k(t_1 + t_2 + t_3 + t_4 - 4t_0) \tag{18-1}$$

where k is the thermal conductivity of the material, and the subscripts belong to the five points involved.

In the steady state of heat flow without internal sources or sinks of heat, $q_0' = 0$. If now the five temperatures have been assumed, as a guess, a positive or negative value of q_0' may be found from Eq. 18-1. The relaxation method consists in correcting this erroneous value by a reduction (or increase) of the heat flow from each of the four points around point 0 by the amount $q_0'/4$. By this correction, however, the heat balances made up for each of the points 1, 2, 3, and 4 will be influenced, and now these balances will have to be corrected. The trick of the procedure consists in beginning the correction with that point of the whole system where the numerical value of q_0' obtained by Eq. 18-1 is greatest. According to this equation, the subsequent disturbance at each of the four neighboring points becomes only one quarter of that first correction. The method will be understood more readily from an example. The same one used by Emmons has been selected. However, the calculation will be explained more in detail, since Emmons' presentation is too brief for easy understanding. Figure 18–2 shows a cross section (in the y, x plane) through a furnace, the heat loss of which is to be calculated for 1 ft length in the z direction. Steady-state conditions with 500°F and 100°F inner and outer surface temperature, respectively, may be assumed; the thermal conductivity of the wall material is 1.5 B hr^{-1} ft^{-1} F^{-1}.

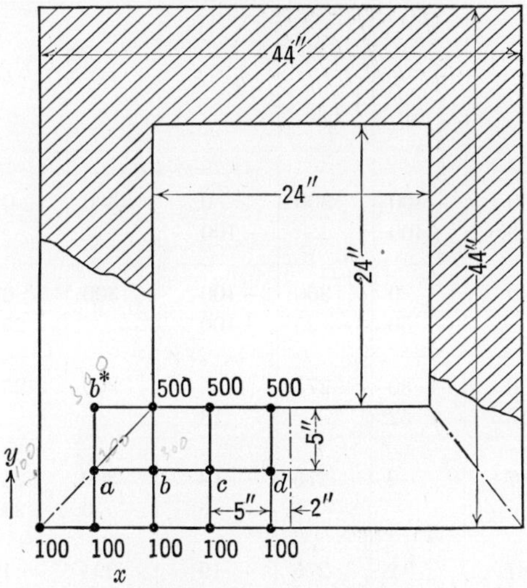

FIG. 18–2. Cross section through furnace wall with fictitious grid of rods as elements in the relaxation treatment.

In one corner of the wall, a grid with rods of 5 in. length is shown. The temperatures in the points a, b, c, and d are to be calculated. The result will be a first approximation only, since the meshes of the system are very wide.

Table 18–1 contains the step-wise computation, starting (step 0) from the assumption of $t = 300$°F at each of the four indicated points. Each step consists of the correction of the temperature of one point and corresponding change of (q_0'/k) for this and the neighboring points. For instance, since Eq. 18–1 yields $q_0'/k = -400$ for point a, a correction of $-400/4 = -100$ is made in step 1. The correction for point a is doubled in steps 2, 5, and 9, in consideration of the influence from point b^* to a (see Fig. 18–2), which is the same as that from b to a. The points at the inner and outer surface (see Fig. 18–2) are not corrected because their temperatures are fixed by boundary conditions in the present case. It will be noticed that the calculation consists in nothing but dividing the extreme value of q_0'/k by 4 and adding the

TABLE 18-1

COMPUTATION OF TEMPERATURE DISTRIBUTION IN THE WALL SHOWN BY FIG. 18-2 (FIRST APPROXIMATION IN RELAXATION TREATMENT)

Point		*a*		*b*		*c*		*d*	
Step No.	Result of Step No.	t	q_0'/k	t	q_0'/k	t	q_0'/k	t	q_0'/k
0	0	300	−400	300	0	300	0	300	0
1		−100	+400		−100				
	1	200	0	300	−100	300	0	300	0
2			− 50	− 25	+100		−25		
	2	200	− 50	275	0	300	−25	300	0
3		− 13	+ 52		− 13				
	3	187	2	275	− 13	300	−25	300	0
4					− 6	− 6	+24		−6
	4	187	2	275	− 19	294	− 1	300	−6
5			− 10	− 5	+ 20		− 5		
	5	187	− 8	270	+ 1	294	− 6	300	−6
6		− 2	+ 8		− 2				
	6	185	0	270	− 1	294	− 6	300	−6
7					− 2	− 2	+ 8		−2
	7	185	0	270	− 3	292	+ 2	300	−8
8							− 2	− 2	+8
	8	185	0	270	− 3	292	0	298	0
9			− 2	− 1	+ 4		− 1		
	9	185	− 2	269	+ 1	292	− 1	298	0
10		− 1	+ 4		− 1				
	10	184	+ 2	269	0	292	− 1	298	0
11		+ 0.5	− 2		− 0.5				
	11	184.5	0	269	− 0.5	292	− 1	298	0

resulting value to the temperature of the particular point and to the values of q_0'/k of the neighboring points. For this reason, a table like the present one can be calculated in less than half an hour if only the pattern has been fixed.

Table 18–1 shows that the temperatures obtained by the fifth step differ only by 2 F from the final result, obtained in 11 steps, and that the total correction of point *a* amounts to 115 F, whereas the true temperature at *d* is only 2 F below the original guess.

A second approximation of the temperatures at *a*, *b*, *c*, and *d* may be obtained by inserting points *e* to *m* in the centers of the meshes (see Fig. 18–3) and continuing the method, starting from the first-approximation temperatures as given in the last line of Table 18–1.

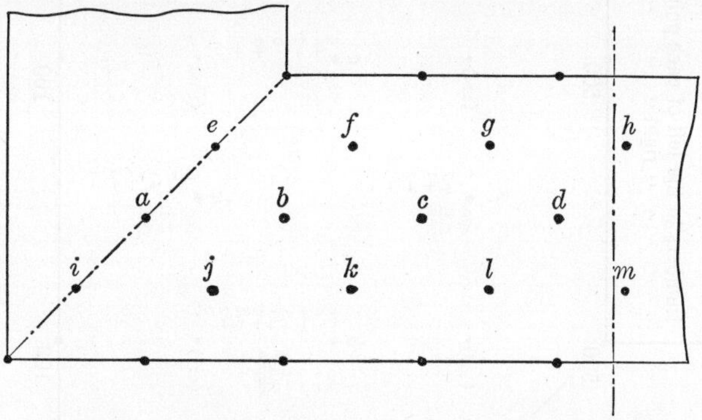

FIG. 18–3. Cross section through wall corner with points for second approximation of relaxation treatment of the case indicated by Fig. 18–2.

This time the procedure is shown in the simplified form as used by Emmons (Fig. 18–4). The assumed temperatures *t* and their subsequent corrections form a column at the left of each point; the values of q_0'/k form a column at the right side. The corrections are continued until the lowest values of q_0'/k deviate by only 0.5 or 1 from zero, showing a perfect balance all over the cross section. Then the temperature columns are added, with the second-approximation "corrected temperatures" as the result. Finally, interpolated temperatures in parentheses are added for the final calculation.

The temperatures according to the first and second approximation are shown in Table 18–2. It is seen that only the temperatures at *b* and *e* required an appreciable correction in this second approximation.

In filling up Table 18–1 and Fig. 18–4 no "overshooting" of temperature correction, as recommended by Emmons, has been used. However, it is clear that if the tendency of change is recognized, "overshooting" may abbreviate the procedure considerably.

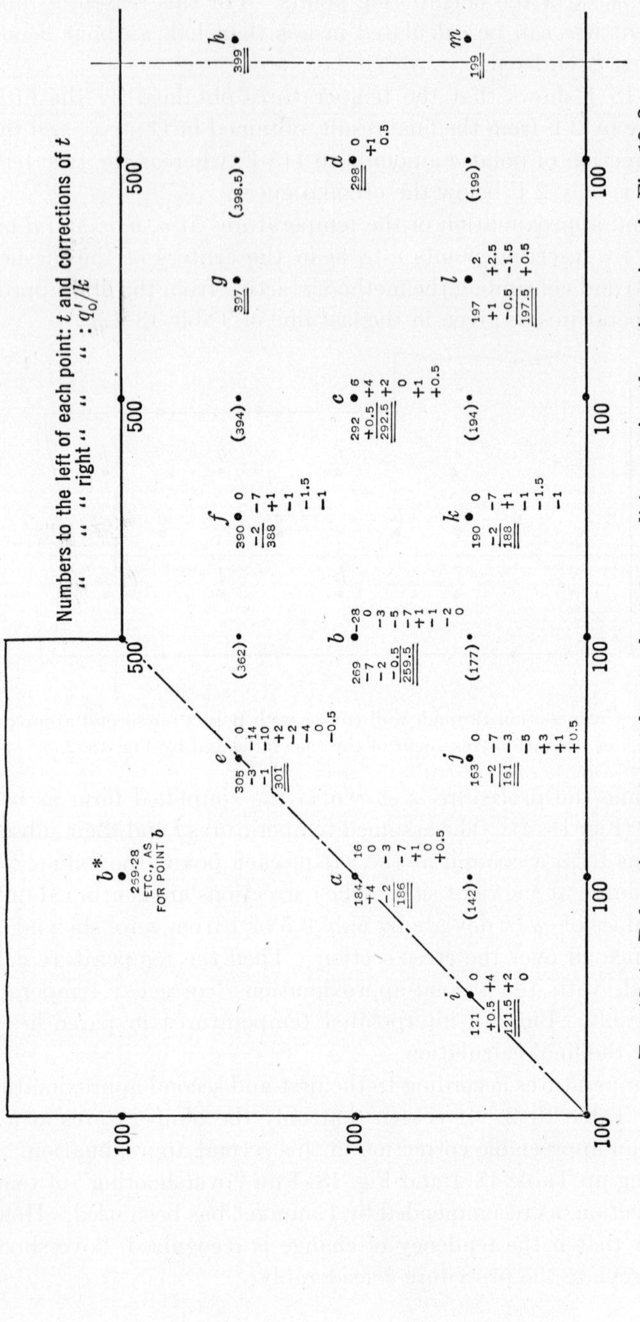

FIG. 18–4. Relaxation treatment for steady-state conditions, using the points given in Fig. 18–3.

TABLE 18–2

TEMPERATURES AT THE POINTS OF FIGS. 18–3 AND 4 (APPROXIMATIONS IN RELAXATION TREATMENT)

Point	*a*	*b*	*c*	*d*	*e*	*f*	*g*	*h*	*i*	*j*	*k*	*l*	*m*
Original assumption	300	300	300	300									
First approximation	184	269	292	298	305	390	397	399	121	163	190	197	199
Second approximation	186	259.5	292.5	298	301	388	397	399	121.5	161	188	197.5	199

The total heat flow is found by summing up the rates of heat flowing through the rods which end at the outer surface (see Fig. 18–2).

Denoting these sums by q_{I}' and q_{II}', respectively, the first approximation yields:

$$q_{\mathrm{I}}' = 8k[(500 - t_b) + (500 - t_c) + 0.9(500 - t_d)]$$

and

$$q_{\mathrm{II}}' = 8k[(t_a - 100) + (t_b - 100) + (t_c - 100) + 0.9(t_d - 100)]$$

The factor 8 occurs because only one eighth of the wall has been considered in the previous calculations. The factor 0.9 in the last term is necessary because the distance from point d to the vertical symmetry line equals only 2 in. Substituting $k = 1.5$ and the final temperatures from Table 18–1, $q_{\mathrm{I}}' = 7450$ and $q_{\mathrm{II}}' = 7484$ are obtained. The values should be equal. Their average is $q' = 7467$ B hr^{-1} ft^{-1}.

For the second approximation, deviating slightly from Emmons' procedure, 5 rods starting inside and 8 ending outside, including rods through the interpolated points in Fig. 18–4, are considered as carriers of the heat flow. In this way

$$q_{\mathrm{I}}' = 8(k)590.8 \quad \text{and} \quad q_{\mathrm{II}}' = 8(k)609.7$$

were obtained with an average of $q' = 7203$ B hr^{-1} ft^{-1}, i.e., 3½% less than according to the first approximation.

The method of relaxation is not restricted to plane surfaces. For curved surfaces, it is necessary only to fill the cross-sectional area by a grid, as indicated in Fig. 18–5, with boundaries that follow the curved boundaries as closely as possible. By taking the mesh width smaller, any approximation may be reached; however, a rather rough approximation of the boundaries will be sufficient in most practical cases.

In less complicated cases, for instance, cylindric boundaries, Emmons recommends first bringing the boundaries into a more convenient shape by conformal mapping.

By modifying Eq. 18–1, the method may also be adapted to bodies which are inhomogeneous in k. For instance, it would be simple to solve

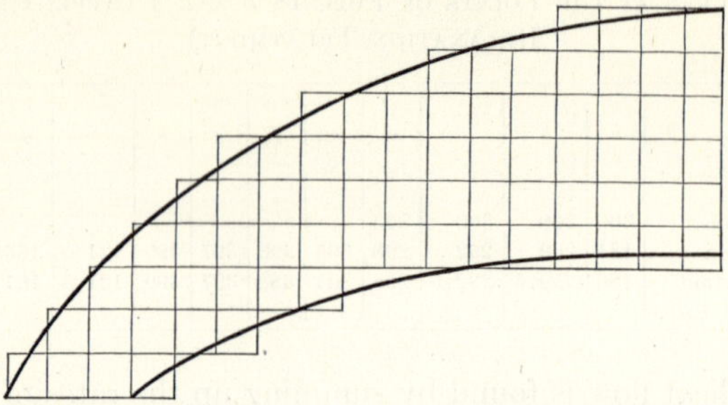

FIG. 18–5. Relaxation grid for a wall with curved surfaces.

the former problem for a pattern with different thermal conductivities in the directions x and y.

Also three-dimensional problems can be treated by relaxation. Every point may be considered as surrounded by six points, if a square lattice is used, or by eight points, if a body-centered square lattice is used. Some results for a cubical furnace of the same cross section as shown in Fig. 18–2 are given in Emmons' [1943] first-mentioned paper.

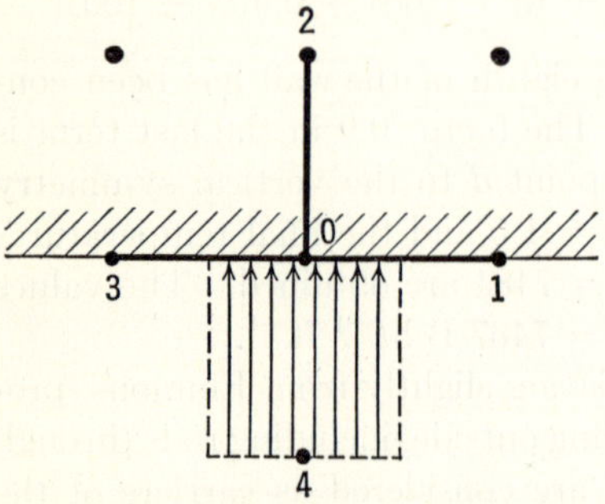

FIG. 18–6. Consideration of environment temperature in the relaxation treatment.

If in the boundary conditions environment temperatures are given instead of surface temperatures, the procedure may be modified as follows:

In Fig. 18–6 the points 0, 1, and 3 are supposed to be on the surface, the point 2 inside, and point 4 outside in an imagined extension of the medium below the surface. Then, for the surface, Eq. 18–1, obviously, changes to

$$q_0' = k(t_1 + t_2 + t_3 - 3t_0) + h\cdot\Delta x(t_e - t_0) \qquad (18\text{–}2)$$

where t_e = the temperature of the environment, and
h = the surface coefficient of heat transfer.

One may use, as well, Eq. 18–1, with a fictitious temperature

$$t_4 = t_0 + \frac{h}{k}\,\Delta x(t_e - t_0) \qquad (18\text{–}3)$$

If h is the same everywhere on the surface, $h \cdot \Delta x/k$ is a constant factor. However, the method is not restricted to this assumption, and this may be a great advantage in dealing with cases where h varies. It would be enormously difficult, if possible at all, to treat such a case analytically.

Also, the influence of inner heat sources, as in electrical coils or chemically reacting bodies, is covered by Eq. 18–1. Here q_0' is the rate of heat flow absorbed by a piece of unit of thickness and area $(\Delta x)^2$ around point 0 of Fig. 18–1. If, according to our method of denotations, q_0''' is the rate of heat developed by unit volume around point 0, then Eq. 18–1 turns into

$$q_0' = k(t_1 + t_2 + t_3 + t_4 - 4t_0) + (\Delta x)^2 q_0''' \qquad (18\text{–}4)$$

The only difference in the procedure shown in Table 18–1 or Fig. 18–4 is to correct t for every point of the lattice in such a way that q_0'/k, according to Eq. 18–4, becomes zero. The merit of the relaxation method will be recognized best if a case is considered where this value varies with t and possibly with k, as in electrical coils or under chemical reactions. This makes the analytical treatment very difficult, as has been demonstrated in Chapter 10. Using the relaxation method, q_0'''/k may be any temperature function, given, for instance, by a complicated curve. Then one has just to take q_0'''/k from the curve for each point of the grid and at each step, and compensate for this value by correcting the temperatures as shown above.

The case where heat is produced in a very thin layer, as in frictional processes, can be dealt with similarly to the case of surface heat transfer (see Fig. 18–6).

18–3. Numerical Solutions of Unsteady-state Conduction Problems for One-dimensional Heat Flow

In unsteady-state problems the temperature distribution in the body under consideration is generally given at a certain instant τ, and the change of the distribution is asked for.

Following Dusinberre's [1945] procedure, the conducting body may be divided in cubical blocks of side length Δx, with centers at 1, 2, 3, etc., as represented in Fig. 18–7. Regarding k and c_p as constant and the heat flow q as one-dimensional, the temperatures in

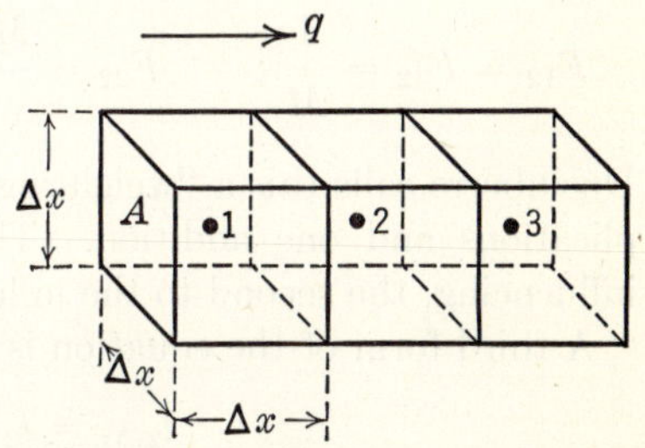

FIG. 18–7. Configuration of three adjacent blocks, used in numerical method for linear unsteady-state conduction problems.

vertical planes through 1, 2, 3, parallel to plane A, may be t_1, t_2, t_3 at the time τ and $t_{1,\Delta\tau}$, $t_{2,\Delta\tau}$, $t_{3,\Delta\tau}$ after a small time interval $\Delta\tau$.

It will be assumed that this time interval is so small that

1. the initial temperature gradients $(t_1 - t_2)/\Delta x$ and $(t_2 - t_3)/\Delta x$ can be used during this interval,
2. the temperature at point 2 is not affected by any change at the left of cube 1 and at the right of cube 3, and
3. the heat capacity of cube 2 can be computed, using the temperature at point 2.

Under these assumptions the following heat balance for block 2 is obtained:

$$k\frac{(\Delta x)^2}{\Delta x}(t_1 - t_2)\,\Delta\tau + k\frac{(\Delta x)^2}{\Delta x}(t_3 - t_2)\,\Delta\tau = \rho c_p(\Delta x)^3(t_{2,\Delta\tau} - t_2) \qquad (18\text{–}5)$$

It yields

$$t_{2,\Delta\tau} - t_2 = \frac{t_1 - 2t_2 + t_3}{M}$$

or

$$t_{2,\Delta\tau} = \frac{t_1 + (M-2)t_2 + t_3}{M} \qquad (18\text{–}6)$$

where

$$M = \frac{\rho c_p(\Delta x)^2}{k\cdot\Delta\tau} = \frac{(\Delta x)^2}{\alpha\cdot\Delta\tau} \qquad (18\text{–}7)$$

Equation 18–6 defines an "averaging" procedure. It involves three steps, namely, one multiplication, one addition, and one division.

Equation 18–6 can also be expressed in the following way:

$$t_{2,\Delta\tau} = F_{12}t_1 + F_{22}t_2 + F_{32}t_3 \qquad (18\text{–}8)$$

where

$$F_{12} = F_{32} = \frac{1}{M}; \qquad F_{22} = \frac{M-2}{M}; \qquad \text{and} \qquad \Sigma F = 1 \qquad (18\text{–}9)$$

Dusinberre calls this a "weighting" procedure. It involves three multiplications and one addition. The first subscript of F refers to the influencing, the second to the influenced, temperature.

A third form of the equation is

$$t_{2,\Delta\tau} = t_2 + C_{12} + C_{32} \qquad (18\text{–}10)$$

where

$$C_{12} = \frac{t_1 - t_2}{M} \quad \text{and} \quad C_{32} = \frac{t_3 - t_2}{M} \qquad (18\text{–}11)$$

This is called a "correction" procedure. It involves two subtractions, two divisions, and one algebraic addition.

Considering a case of perfect insulation at the left side of block 1 and at the right side of block 2, the second term in Eq. 18–5 vanishes, and

$$t_1 - t_2 = M(t_{2,\Delta\tau} - t_2) \tag{18–12}$$

Since both blocks are equal, the largest change which could occur would make

$$t_{2,\Delta\tau} = t_{1,\Delta\tau} = \frac{t_1 + t_2}{2}$$

Hence, in general,

$$t_{2,\Delta\tau} \lesseqgtr \frac{t_1 + t_2}{2} \quad \text{and} \quad M \gtreqless 2 \tag{18–13}$$

For $M = 2$, from Eq. 18–6,

$$t_{2,\Delta\tau} = \frac{t_1 + t_3}{2} \tag{18–14}$$

The graphical method of Binder-Schmidt (see Chapter 19) is based on this assumption.

By inspection of Eq. 18–9 it follows that $F_{22} \geqq 0$ for $M \geqq 2$, so that all "weights" F in Eq. 18–8 become positive. Otherwise, $t_{2,\Delta\tau}$ would oscillate under some conditions, and the solution may not converge.

For $M = 3$, Eq. 18–6 yields

$$t_{2,\Delta\tau} = \frac{t_1 + t_2 + t_3}{3} \tag{18–15}$$

which is particularly convenient for calculations.

From Eqs. 18–6 and 7 it follows that for larger values of M the time interval $\Delta\tau$ becomes smaller, and the convergence correspondingly slower.

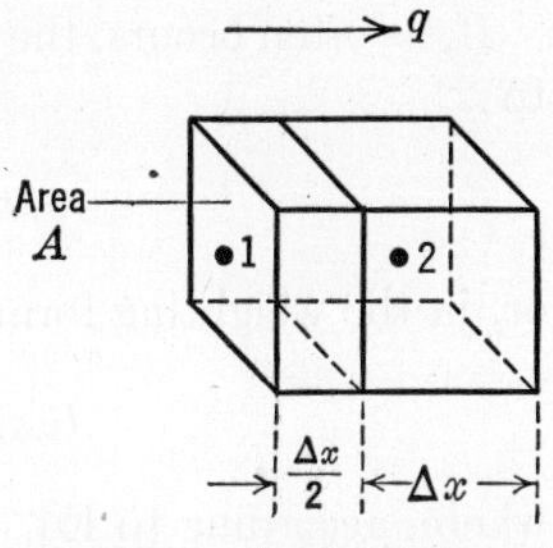

FIG. 18–8. Configuration of one and one-half adjacent blocks, used in numerical method for linear unsteady-state conduction problems, including surface heat transfer.

At a surface where convection occurs, an arrangement according to Fig. 18–8 may be considered, with an environment temperature t_0 at the left of area A. Here the heat balance for the half-block 1 becomes

$$h(\Delta x)^2(t_0 - t_1)\,\Delta\tau + k\frac{(\Delta x)^2}{\Delta x}(t_2 - t_1)\,\Delta\tau = \rho c_p \frac{(\Delta x)^3}{2}(t_{1,\Delta\tau} - t_1) \tag{18–16}$$

In addition to M a second modulus, defined by

$$N = h\frac{\Delta x}{k} \tag{18-17}$$

may be introduced in Eq. 18–16. Hence

$$N(t_0 - t_1) + (t_2 - t_1) = \frac{M}{2}(t_{1,\Delta\tau} - t_1)$$

or

$$t_{1,\Delta\tau} = \frac{2Nt_0 + (M - 2N - 2)t_1 + 2t_2}{M} \tag{18-18}$$

or in the "weighting" form,

$$t_{1,\Delta\tau} = F_{01}t_0 + F_{11}t_1 + F_{21}t_2 \tag{18-19}$$

where

$$F_{01} = \frac{2N}{M}; \qquad F_{11} = \frac{M - 2N - 2}{M}; \qquad F_{21} = \frac{2}{M} \tag{18-20}$$

The condition $F_{11} \geqq 0$ requires that

$$M \geqq 2N + 2 \tag{18-21}$$

At a well-insulated surface h, N, and F_{01} may approximate zero.

If, as often occurs, the initial condition is $t_0 = t_1$, then Eq. 18–18 leads to

$$t_{1,\Delta\tau} - t_1 = \frac{2}{M}(t_2 - t_1) \tag{18-22}$$

or, in the weighting form,

$$t_{1,\Delta\tau} = (F_{01} + F_{11})t_1 + F_{21}t_2 \tag{18-23}$$

where, according to Eq. 18–20,

$$F_{01} + F_{11} = \frac{M - 2}{M}; \qquad F_{21} = \frac{2}{M} \tag{18-24}$$

Further, in analogy to Eq. 18–13,

$$t_{1,\Delta\tau} \leqq \frac{t_1 + 2t_2}{3} \quad \text{and} \quad M \geqq 3 \tag{18-25}$$

considering that the difference in mass of blocks 1 and 2 influences the approach to an equilibrium temperature $t_{1,\Delta\tau} = t_{2,\Delta\tau}$.

If N is large, Eq. 18–21 causes M to become large, and many steps will be required. In this case Eq. 18–18 may be brought into the form

$$t_{1,\Delta\tau} - t_1 = \frac{2N}{M}(t_0 - t_1) - \frac{2}{M}(t_1 - t_2) \qquad (18\text{–}26)$$

or for the lower limit $M = 2N + 2$,

$$t_{1,\Delta\tau} - t_1 = (t_0 - t_1) - \frac{2}{M}(t_0 - t_2) \qquad (18\text{–}27)$$

In both equations the second term on the right side will be small and sometimes negligible. The calculation for the blocks 2, 3, etc., may be performed with a different, and much smaller, value of M.

For a composite wall with different thermal diffusivities α_1 and α_3 to the left and right sides of point 2 in Fig. 18–7 one possible choice would be

$$\frac{(\Delta x)_1}{(\Delta x)_3} = \sqrt{\frac{\alpha_1}{\alpha_3}}$$

Then, since $\Delta\tau$ should be the same for both sides,

$$M_1 = M_3$$

However, it may be more convenient to have $(\Delta x)_1 = (\Delta x)_3$ and therefore to choose

$$\frac{M_1}{M_3} = \frac{\alpha_3}{\alpha_1} \qquad (18\text{–}28)$$

Then, in order to find $t_{2,\Delta\tau}$, the present author recommends to use k_1 in the first term of Eq. 18–5, k_3 in the second term, and $(\rho_1 c_{p1} + \rho_3 c_{p3})/2$ instead of ρc_p in the third term.

In Dusinberre's paper it is also shown how to proceed if k and c_p vary independently with temperature. (See also footnote, p. 379.)

18–4. Numerical Solutions of Unsteady-state Conduction Problems for Two-dimensional Heat Flow

Figure 18–9 represents adjacent cubical blocks of depth L. Considering heat flow only in the x and y directions, as has been done in Sect. 18–2 with respect to Fig. 18–1, the following heat balance, similar to Eq. 18–5, is found by equating the sum of the amounts of heat conducted from the outer blocks to the block 0 and the heat stored in this block:

$$k \frac{(\Delta x)^2}{\Delta x} (t_1 + t_2 + t_3 + t_4 - 4t_0)\,\Delta\tau = \rho c_p (\Delta x)^3 (t_{0,\Delta\tau} - t_0)$$

or

$$t_{0,\Delta\tau} = \frac{t_1 + t_2 + t_3 + t_4 + (M - 4)t_0}{M} \tag{18-29}$$

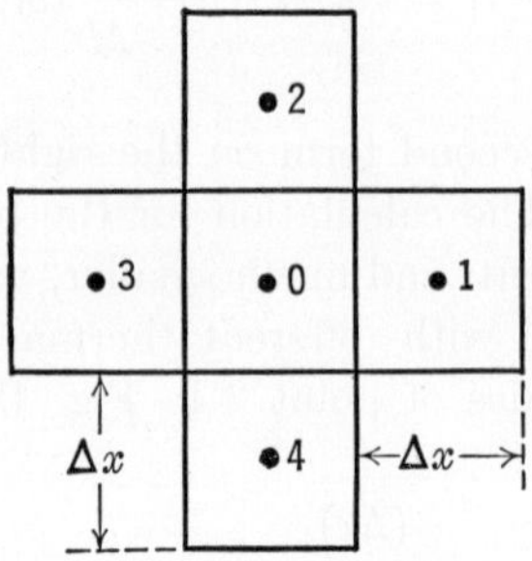

FIG. 18-9. Configuration of five adjacent blocks used in numerical method for two-dimensional unsteady-state conduction problems.

The greatest allowable change in t_0 would occur if $t_{0,\Delta\tau}$ assumed the average of the surrounding temperatures. Hence

$$M \geqq 4 \quad \text{and preferably} \quad M = 5$$

For $M = 4$

$$t_{0,\Delta\tau} = \frac{t_1 + t_2 + t_3 + t_4}{4} \tag{18-30}$$

For $M = 5$

$$t_{0,\Delta\tau} = \frac{t_0 + t_1 + t_2 + t_3 + t_4}{5} \tag{18-31}$$

As an example of the application of Eq. 18-30, again Emmons' illustration (Fig. 18-2) is used, this time with the assumption that at the time $\tau = 0$ the temperature of the wall is 100°F throughout, and that the inner surface is suddenly brought to, and then held at, 500°F. Furthermore, $\alpha = 0.01$ sq ft/hr and $\Delta x = 0.208$ ft may be given. Equation 18-7 yields $\Delta\tau = 1.083$ hr, and Fig. 18-10, taken from Emmons' first paper, shows the temperatures in the time intervals $\Delta\tau$, as calculated from those in the surrounding points by means of Eq. 18-30.

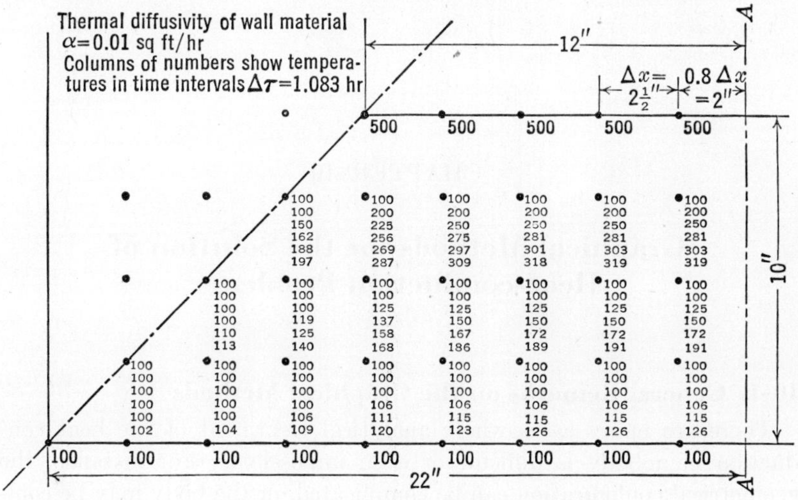

FIG. 18–10. Numerical solution of an unsteady-state problem (furnace wall as in Fig. 18–2).

Surface conditions and inner heat sources or sinks, including those due to changes of phase or chemical constitution, can be taken care of, as in the steady-state and one-dimensional heat flow, without considerable complications.

Three-dimensional problems can be handled in a way similar to those for two dimensions.

NOTE (REFERRING TO THE LAST PARAGRAPH OF SECT. 18–3)

Just before this book was to appear, a paper of Boelter and Tribus [1949] was published in which practical work sheets for the numerical solution of unidirectional heat-conduction problems are proposed. These are based upon the use of $M = 2$ and essentially consist of matrices with temperatures at time $\tau = 0$ written for intervals Δx in the first row and temperatures at $\tau = \Delta\tau,\ 2\cdot\Delta\tau,\ \cdots\ n\cdot\Delta\tau$ in the following rows, the columns belonging to values $x = 0,\ x = \Delta x,\ x = 2\cdot\Delta x,\ \cdots\ m\cdot\Delta x$, etc. According to Eq. 18–14, the temperature at $t_{m,n}$ in any row and column is the average of $t_{m-1,n-1}$ and $t_{m+1,n-1}$ in the preceding row. Each of these values is the average of two values of the next preceding row. In this way the matrix is built up and simple calculating rules are given. It is also shown how walls composed of materials having different diffusivities α_{m-1} and α_{m+1} to the left and right from the mth column can be handled. In averaging temperatures from its two neighbor columns, each of these temperatures is multiplied by a weighting factor depending upon the diffusivity in the particular column. Surface heat transfer is handled in a similar manner.

CHAPTER 19

Graphical Methods for the Solution of Heat-conduction Problems

19-1. General Remarks on the Graphical Methods

There are many cases where analytical treatment of the heat conduction in a body is difficult or even impossible. For instance, the geometrical configuration can be complicated, or the body may be composed of materials of different thermal conduction, or the thermal conduction of its parts may have another than linear dependence on temperature. Also, the temperature distribution in a body at an instant under consideration can be complicated, or the temperature of environment and the surface coefficient of heat transfer may change in time. Finally, analytical treatment may be possible by a Fourier analysis or other means, but the accuracy needed may be not great enough to make it worth while to perform the analysis, and a less exact, but easier, ready result will be preferred. For such cases some very expedient graphical methods have been developed many years before the numerical methods were used for the solution of heat-conduction problems. They yield only approximative results; however, an accuracy can be obtained which is sufficient for most practical purposes.

In this chapter only one of the procedures for steady-state heat flow will be presented. The main emphasis will be laid on the method of finite differences for unsteady-state problems.

The steady-state method dealt with in the following section has been developed by Lehmann [1909] for the determination of a two-dimensional electrical flow. However, both electrical- and thermal-flow patterns being governed by the same laws of potential theory, this method can be and has been used for thermal conduction as well.

The method based on the calculus of finite differences was indicated first by Binder [1911; see also Jakob 1943(*a*)], then developed anew and more in detail by E. Schmidt [1924], and further improved and extended by Nessi and Nisolle [1929]. The following presentation of the subject is based mainly on the work of these men.

19-2. Construction of Heat-flow Lines and Isothermals

Figure 19-1 shows isothermal and adiabatic lines, which form a net with meshes of approximately square form in the x, y plane. The lines are curved; however, they intersect each other under right angles. Each mesh represents the cross section of a prismatic bar with isothermal and

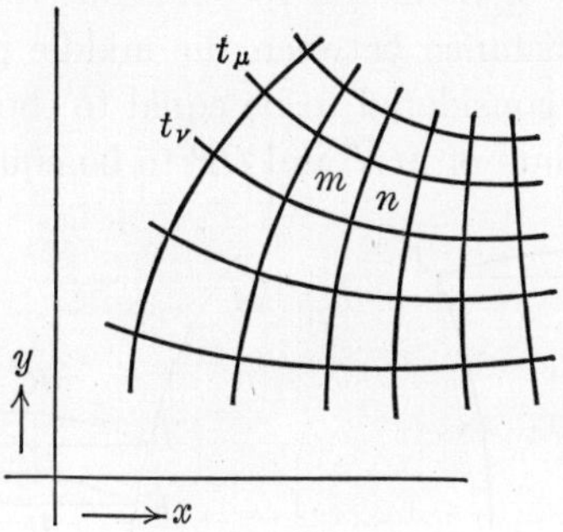

FIG. 19-1. Net formed by heat-flow lines and isothermals.

adiabatic surfaces; the length of each prism in the direction z of the system of coordinates may be equal to unity. The thermal conductivity, k, may be the same in all prisms.

The rates of heat flow in adjacent prisms, m and n, of unit length will be

$$q_m' = k \frac{D_m}{L_m} (t_\mu - t_\nu) \tag{19-1}$$

$$q_n' = k \frac{D_n}{L_n} (t_\mu - t_\nu) \tag{19-2}$$

where D_m and D_n are the mean distances of the adiabatics, confining the meshes m and n, respectively; L_m and L_n are the mean distances of the isothermals which form two borders of the meshes m and n; and t_μ and t_ν are the temperatures of these isothermals.

If

$$\frac{D_m}{L_m} = \frac{D_n}{L_n} = 1 \tag{19-3}$$

the meshes may be called curvilinear squares, as in Sect. 17-3. For these, according to Eqs. 19-1 and 2,

$$q_m' = q_n'$$

Hence, two curvilinear squares with the same isothermal at heat entrance have also identical isothermals where the heat leaves, provided q' is the same for both. This yields a useful check in the construction of curve systems like that of Fig. 19-1.

To prove whether a mesh can be assumed to be a curvilinear square, as defined by Eq. 19-3, Lehmann recommends the construction shown in Fig. 19-2. The distance between the middle points of the straight lines $\overline{MO}$ and $\overline{NP}$ is considered to be equal to the distance D, that between the middle points of $\overline{MN}$ and $\overline{OP}$ to be equal to the length L.

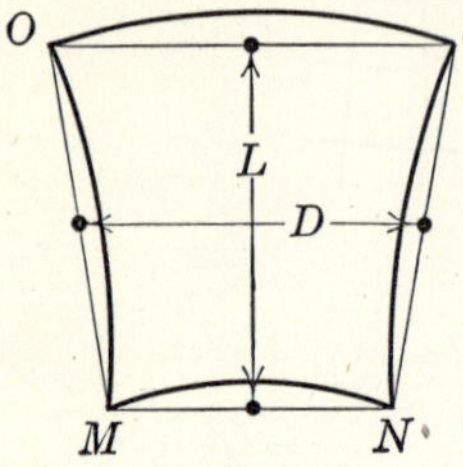

FIG. 19-2. Curvilinear square.

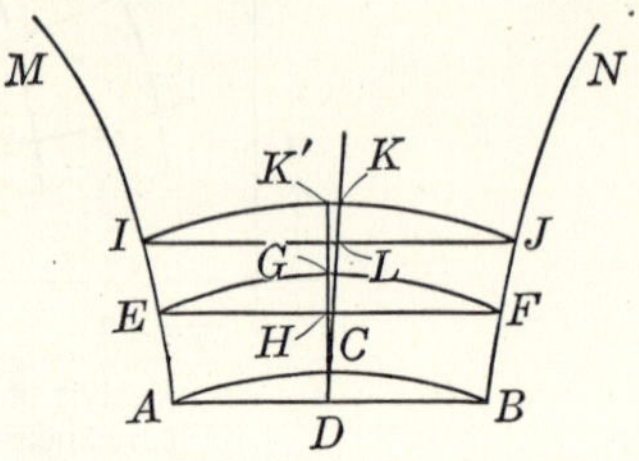

FIG. 19-3. Construction to divide the space between adiabatics and isothermals into curvilinear squares.

In order to insert curvilinear squares in a system, with boundaries $\overline{MA}$, $\overline{AB}$, and $\overline{BN}$ (Fig. 19-3), where $\overline{AB}$ is the curved line between A and B, the following procedure is recommended:

1. Draw a straight line $\overline{AB}$ and make $\overline{AD} = \overline{DB}$.
2. Draw $\overline{DC}$ perpendicular to $\overline{ADB}$.
3. Find a straight line $\overline{EF}$ by trial and error so that $\overline{AE} = \overline{BF} = (\overline{EF})/4$.
4. Extend $\overline{DC}$ to G so that also $\overline{CG} = (\overline{EF})/4$.
5. Draw $\overline{EGF}$ so that the angles at E and F are 90°.
6. Find H as intersection of $\overline{CG}$ and $\overline{EF}$.
7. Extend $\overline{DG}$ to K' so that $\overline{CK'} = \overline{EH}$.
8. Make $\overline{EI} = \overline{FJ} = \overline{GK'}$.
9. Draw $\overline{IK'J}$ so that the angles at I and J are 90°.
10. Make $\overline{IL} = \overline{LJ}$.
11. Draw $\overline{LK}$ perpendicular to $\overline{ILJ}$.
12. Draw a curve $\overline{CK}$ so that the angles at C and K are 90°. Then $IACK$ and $KCBJ$ are two curvilinear squares between $\overline{MA}$ and $\overline{NB}$.

More than two could be inserted in a similar way. Proceeding from $\overline{IJ}$ as basis, as under (1) from $\overline{AB}$, will yield another couple of curvilinear squares in the space inside $MIKJN$.

Various modifications may be used in order to improve the result. Consider, for instance, curves $\overline{MA}$, $\overline{AB}$, $\overline{BN}$ (Fig. 19-4), and curvilinear squares in between, constructed from $\overline{AB}$ as basis, as shown above. Then, because, according to Eq. 19-3, only the geometrical configuration is essential, and not the direction of heat flow, the construction can be checked, using $\overline{AG}$ as basis and proceeding from left to right, or from $\overline{BH}$, proceeding from right to left, instead of using $\overline{AB}$ and proceeding from the bottom to the top. In this way, lines deviating somewhat from $\overline{CFI}$ and $\overline{DFE}$ may be found, and a better approximation will be obtained from the different results.

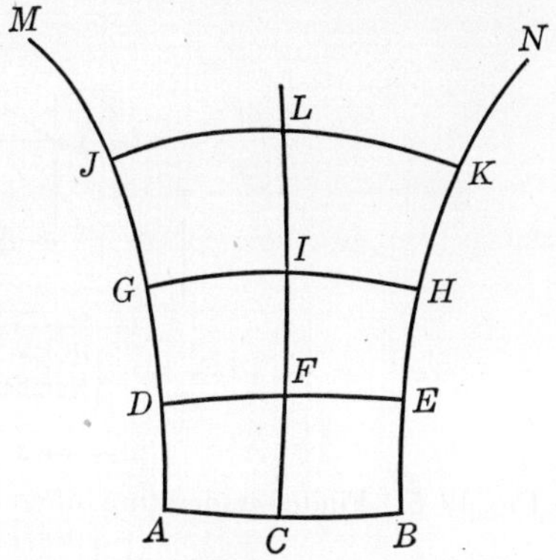

FIG. 19-4. Curvilinear squares inserted between two heat-flow lines.

Generally, if a system of isothermals and flow lines is to be inserted inside $JABK$ (see Fig. 19-4), it is recommended to begin by dividing it in two parts by a flow line as shown above. If $\overline{CFIL}$ has been found to be this line, by the described trial-and-error procedure, then the space inside $JACL$ may also be divided in two parts by another flow line, or the space inside $ABED$ may be bisected by an isothermal line, and so forth, always re-checking one by another.

Finally, it may be mentioned that, in a temperature field as represented by Fig. 19-1, under the assumption of Eqs. 19-1 to 3, either the heat flow or the temperature drop, whichever is unknown, can be found just by counting the number of curvilinear squares in both directions.

19-3. The Method of Finite Differences for Heat Conduction through Homogeneous Plane Plates

Again we start from the differential equation of unidirectional heat conduction

$$\frac{\partial t}{\partial \tau} = \alpha \frac{\partial^2 t}{\partial x^2} \tag{19-4}$$

A corresponding equation with finite differences instead of the differentials is:

$$\frac{\Delta_\tau t}{\Delta \tau} = \alpha \frac{\Delta_x{}^2 t}{(\Delta x)^2} \tag{19-5}$$

where the subscripts τ and x indicate that the temperature differences, Δt and $\Delta^2 t$, are to be taken at varying τ and x, respectively.

$\Delta\tau$ and Δx are used as unit changes in time and distance. The method is based on step-wise proceeding by these unit differences.

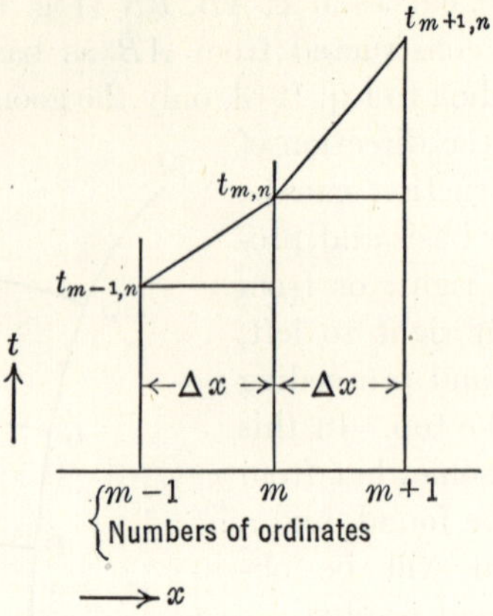

FIG. 19-5. Finite temperature differences in a direction x at a fixed instant of time.

If $t_{m,n}$ denotes the temperature at a distance, $m \cdot \Delta x$, from any starting point, and at a time, $n \cdot \Delta\tau$, counted from any zero point of time, then, as may be seen from Fig. 19-5,

$$\begin{aligned}\Delta_\tau t &= t_{m,n+1} - t_{m,n}\\ \Delta_x t &= t_{m,n} - t_{m-1,n}\\ \Delta_x{}^2 t &= (t_{m+1,n} - t_{m,n}) - (t_{m,n} - t_{m-1,n})\\ &= t_{m+1,n} - 2t_{m,n} + t_{m-1,n}\end{aligned}$$

Hence, Eq. 19-5 may be written as a formula of recursion:

$$t_{m,n+1} - t_{m,n} = \alpha \frac{\Delta\tau}{(\Delta x)^2}(t_{m+1,n} - 2t_{m,n} + t_{m-1,n}) \tag{19-6}$$

By this equation the temperature at a place $x = m \cdot \Delta x$ can be calculated for the instant $\tau = (n + 1)\,\Delta\tau$ if the distribution close to that place x is known at a previous instant $\tau = n \cdot \Delta\tau$. The three temperatures given in Fig. 19-5 are sufficient for this purpose.

Consider a temperature distribution at time $\tau = n \cdot \Delta\tau$, as given in Fig. 19-6 by the points a, b, c, d, e, f, and g. Connecting the points c and e by a straight line,

$$\begin{aligned}\overline{dd'} &= \tfrac{1}{2}(t_{m+1,n} + t_{m-1,n}) - t_{m,n}\\ &= \tfrac{1}{2}(t_{m+1,n} - 2t_{m,n} + t_{m-1,n})\end{aligned} \tag{19-7}$$

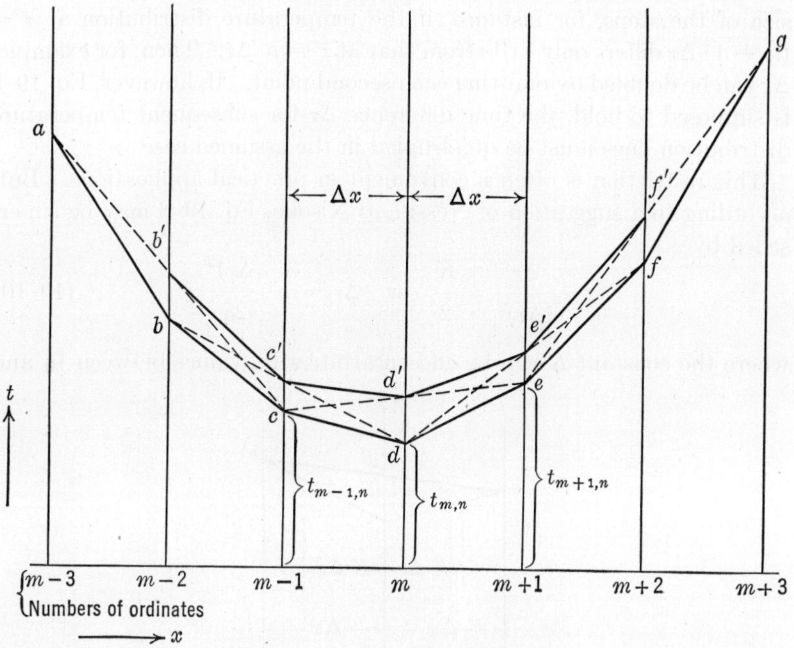

FIG. 19-6. Graphical determination of temperatures in a direction x at instants of time differing by finite increments.

Multiplying this equation by the constant factor $2\alpha \cdot \Delta\tau/(\Delta x)^2$, the increase of temperature at the place $m \cdot \Delta x$ in a time interval $\Delta\tau$ is obtained, as given by Eq. 19-6. Moreover, if such values of Δx and $\Delta\tau$ are chosen that

$$\alpha \frac{\Delta\tau}{(\Delta x)^2} = \frac{1}{2} \quad \text{or} \quad \Delta\tau = \frac{(\Delta x)^2}{2\alpha} \tag{19-8}$$

which is always possible, then

$$\overline{dd'} = t_{m,n+1} - t_{m,n}$$

$$\overline{md'} = t_{m,n+1}$$

and Eq. 19-6 simplifies to

$$t_{m,n+1} = \tfrac{1}{2}(t_{m+1,n} + t_{m-1,n}) \tag{19-9}$$

In the same manner as d', the points b', c', e', and f' have been found. The broken line $\overline{b'c'd'e'f'}$ represents the temperature distribution at the time $\tau = (n + 1)\,\Delta\tau$. Proceeding in the same way, the lines of temperature in equal steps of $\Delta\tau$ may be constructed.

It may be desirable, from the viewpoint of drawing, to change the size of the steps, for instance, if the temperature distribution at $\tau = (n+1)\,\Delta\tau$ differs only little from that at $\tau = n \cdot \Delta\tau$. Then, for example, Δx can be doubled by omitting each second point. If, however, Eq. 19–8 is supposed to hold, the time difference $\Delta\tau$ for subsequent temperature distribution lines must be quadrupled in the assumed case.

This restriction is often inconvenient in practical applications. But, according to a suggestion of Nessi and Nisolle, Eq. 19–8 may be superseded by

$$\alpha \frac{\Delta\tau}{(\Delta x)^2} = \frac{K}{2} \quad \text{or} \quad \Delta\tau = K\frac{(\Delta x)^2}{2\alpha} \tag{19–10}$$

where the constant K can be chosen arbitrarily, values between $\frac{1}{4}$ and

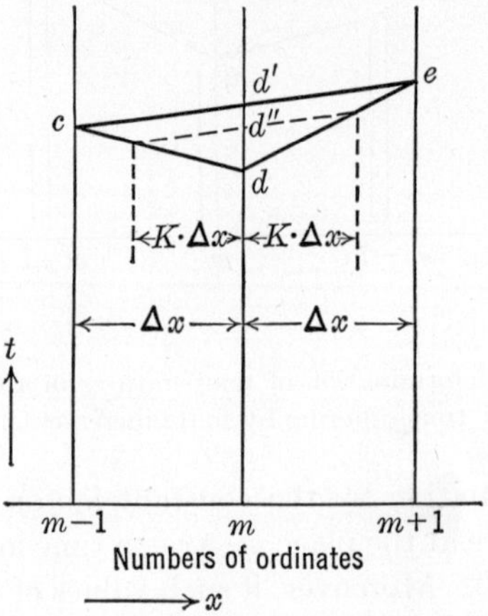

FIG. 19–7. Reduction of the time interval $\Delta\tau$ between subsequent temperature lines in the graphical method of finite differences.

1 being preferable in most cases. Referring to Fig. 19–7, where c, d, e, and d' are the same points as in Fig. 19–6,

$$\overline{dd''} = K(\overline{dd'}) = \frac{K}{2}\,(t_{m+1,n} - 2t_{m,n} + t_{m-1,n})$$

$$= \alpha \frac{\Delta\tau}{(\Delta x)^2}\,(t_{m+1,n} - 2t_{m,n} + t_{m-1,n}) \tag{19–11}$$

and $\overline{md''} = t_{m,n+1}$. Here, according to Eq. 19–10, $\Delta\tau$ is K times greater than before. Therefore, the time interval is controlled by the arbitrary factor K.

If a sudden temperature change is assumed to occur in the system, for instance, a change from t_i to t_s at the free surface of the semi-infinite plate, then first the question arises from what surface temperature in the range from t_i to t_s the graphical construction should be started.

According to Eq. 13-15,

$$\frac{t - t_i}{t_s - t_i} = 1 - f\left(\frac{x}{2\sqrt{\alpha\tau}}\right)$$

Now, for $x = \Delta x$ and $\tau = \Delta\tau$, using $\Delta\tau$ according to Eq. 19-8,

$$\frac{x}{2\sqrt{\alpha\tau}} = \frac{1}{\sqrt{2}} = 0.7071$$

and from Table 13-1:

$$\frac{t_{1,1} - t_i}{t_s - t_i} = 0.3173$$

where $t_{1,1}$ is the temperature at $m = 1$ and $n = 1$. Also, further on, the first of double subscripts will indicate the value of m, and the second one the value of n. Using finite differences and beginning with $t = t_s$ at $x = 0$ and $\tau = 0$,

$$t_{1,1} = \tfrac{1}{2}(t_{0,0} + t_{2,0}) = \tfrac{1}{2}(t_s + t_i)$$

or

$$\frac{t_{1,1} - t_i}{t_s - t_i} = \frac{\frac{1}{2}(t_s + t_i) - t_i}{t_s - t_i} = 0.5$$

to be compared with the value 0.3173 above.

This is not satisfactory. Therefore R. L. Perry and Berggren [1944] recommend to use the temperature $(t_i + t_s)/2$ as $t_{0,0}$. This yields

$$t_{1,1} = \frac{1}{2}\left(\frac{t_i + t_s}{2} + t_i\right)$$

or

$$\frac{t_{1,1} - t_i}{t_s - t_i} = \frac{\frac{1}{2}\left(\frac{t_i + t_s}{2} + t_i\right) - t_i}{t_s - t_i} = 0.25$$

which is still a rather poor approximation.

A better approximation, close to the surface, is obtained by using smaller steps, for instance, $\Delta x' = (\Delta x)/2$ and correspondingly $\Delta\tau' = (\Delta\tau)/4$, instead of Δx and $\Delta\tau$ in Eq. 19-8. Assuming $t_{0,0} = (t_i + t_s)/2$ as before, but $t_{0,n} = t_s$ for $n > 1$, and following the geometry of the construction algebraically through four steps up to $\Delta\tau = 4\cdot\Delta\tau'$ (see Fig. 19-8),

$$\frac{t_{1,4} - t_i}{t_s - t_i} = \frac{5}{8} = 0.625$$

This is to be compared with the result of Eq. 13–15, taking $x = \Delta x' = (\Delta x)/2$, $\tau = 4 \cdot \Delta\tau' = \Delta\tau$, and, therefore,

$$\frac{x}{2\sqrt{\alpha\tau}} = \frac{0.7071}{2} = 0.35355$$

This, according to Table 13–1, corresponds to $(t - t_i)/(t_s - t_i) = 0.617$, showing that the graphical method approaches the analytical solution

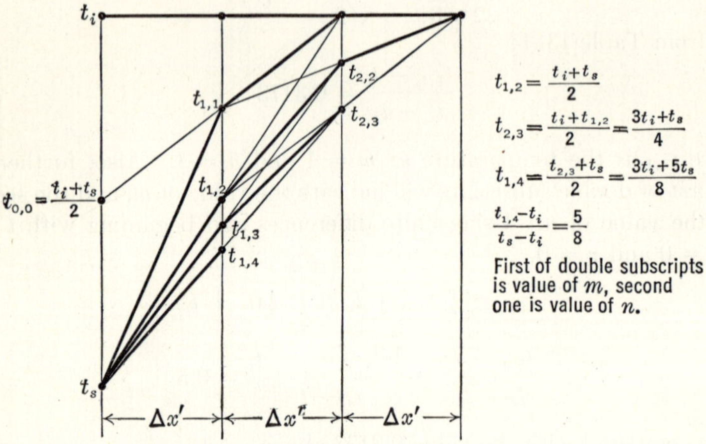

FIG. 19–8. Recommended start of the finite-differences treatment at a surface where the temperature suddenly changes.

satisfactorily. The advantage of this modification is even more conclusively shown by comparing $(t_{2,4} - t_i)/(t_s - t_i) = 5/16 = 0.3125$ with the corresponding analytically determined value 0.3173.

19–4. The Method of Finite Differences for Heat Conduction through Composite Plane Plates

Nessi and Nisolle's procedure is particularly appropriate for composite walls. The lower part of Fig. 19–9 shows the cross section of a wall consisting of two layers I and II with the thermal conductivities and diffusivities k_{I}, α_{I} and k_{II}, α_{II}, respectively. In the upper part is shown a construction of the temperature t_2' at a time $\tau' = (n + 1)\,\Delta\tau$ from t_1, t_2, t_3 at the time $\tau = n \cdot \Delta\tau$. Assume that

$$\alpha_{\text{I}} \frac{\Delta\tau_{\text{I}}}{(\Delta x_{\text{I}})^2} = \alpha_{\text{II}} \frac{\Delta\tau_{\text{II}}}{(\Delta x_{\text{II}})^2} = \frac{1}{2} \tag{19–12}$$

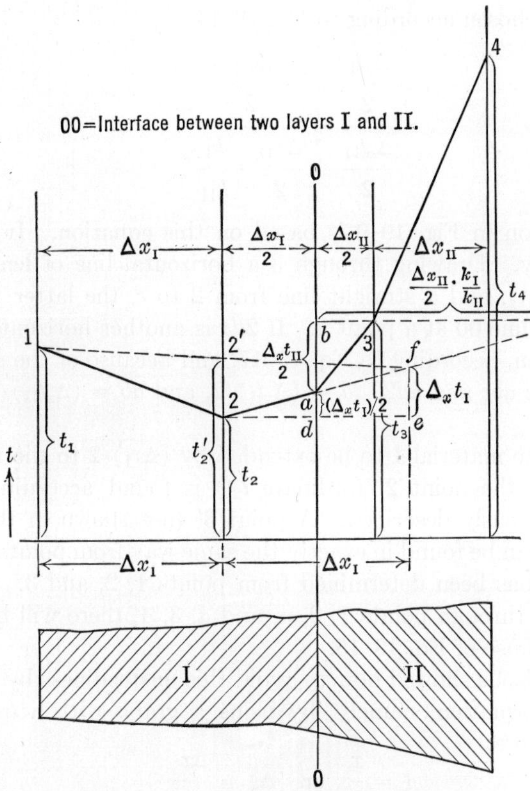

FIG. 19–9. Graphical method of finite differences applied to a composite wall.

Since the temperature distribution is wanted for the same time instants in both wall sections,

$$\Delta\tau_I = \Delta\tau_{II} = \Delta\tau \tag{19–13}$$

Then

$$\frac{\Delta x_I}{\Delta x_{II}} = \sqrt{\frac{\alpha_I}{\alpha_{II}}} \tag{19–14}$$

At the interface the heat flow is the same for both sections, so that

$$-k_I \frac{\partial t_I}{\partial x_I} = -k_{II} \frac{\partial t_{II}}{\partial x_{II}} \tag{19–15}$$

or

$$k_I \frac{\Delta_x t_I}{\Delta x_I} = k_{II} \frac{\Delta_x t_{II}}{\Delta x_{II}} \tag{19–16}$$

If $\Delta x_{\text{I}}/\Delta x_{\text{II}}$ is chosen according to Eq. 19–14,

$$\frac{\dfrac{\Delta_x t_{\text{I}}}{2}}{\dfrac{\Delta_x t_{\text{II}}}{2}} = \frac{\dfrac{\Delta x_{\text{I}}}{2}}{\dfrac{\Delta x_{\text{II}}}{2} \cdot \dfrac{k_{\text{I}}}{k_{\text{II}}}} \tag{19–17}$$

The construction in Fig. 19–9 is based on this equation. It is partly self-explanatory. Drawing through 3 a horizontal line of length $\overline{bc} = (\Delta x_{\text{II}}/2) \cdot (k_{\text{I}}/k_{\text{II}})$, and a straight line from 2 to c, the latter intersects the separating line $\overline{00}$ at a point a. If $\overline{2de}$ is another horizontal line of length Δx_{I}, then, according to Eq. 19–17, and because of the similarity of the triangles abc and $ad2$, $\overline{da} = (\Delta_x t_{\text{I}})/2$, and $\overline{ab} = (\Delta_x t_{\text{II}})/2$. Also, $\overline{ef} = \Delta_x t_{\text{I}}$.

Imagining the material I to be extended by $(\Delta x_{\text{I}})/2$ to the right side of the line $\overline{00}$, the point $2'$ (ordinate t_2') is found according to the procedure previously described. A point $3'$ (not shown in the figure) above point 3 can be found in exactly the same way from points 2, 3, and 4, as point $2'$ has been determined from points 1, 2, and 3. Drawing smooth curves through points 1, 2, a, and a, 3, 4, there will be a bend at point a, because of Eq. 19–15.

Nessi and Nisolle showed how to avoid this unsteadiness by a mathematical transformation, namely, by using a graph with a quantity ξ as abscissa, where:

$$\xi = \frac{x}{k} \quad \text{or} \quad \Delta\xi = \frac{\Delta x}{k} \tag{19–18}$$

Substituting $k \cdot \Delta\xi$ for Δx in Eq. 19–8,

$$\Delta\tau = \frac{\rho c_p k}{2} (\Delta\xi)^2 \tag{19–19}$$

This means superseding α in the differential equation by $1/\rho c_p k$. In fact, by substituting $x = k\xi$ in Eq. 19–4,

$$\frac{\partial t}{\partial \tau} = \alpha \frac{\partial}{\partial x}\left(\frac{\partial t}{\partial \xi} \cdot \frac{d\xi}{dx}\right) = \frac{\alpha}{k} \cdot \frac{\partial^2 t}{\partial \xi^2} \cdot \frac{d\xi}{dx} = \frac{1}{\rho c_p k} \cdot \frac{\partial^2 t}{\partial \xi^2}$$

Furthermore, from Eq. 19–15, substituting $x_{\text{I}} = k_{\text{I}}\xi_{\text{I}}$ and $x_{\text{II}} = k_{\text{II}}\xi_{\text{II}}$,

$$k_{\text{I}} \frac{\partial t_{\text{I}}}{\partial \xi_{\text{I}}} \cdot \frac{d\xi_{\text{I}}}{dx_{\text{I}}} = k_{\text{II}} \frac{\partial t_{\text{II}}}{\partial \xi_{\text{II}}} \cdot \frac{d\xi_{\text{II}}}{dx_{\text{II}}}$$

or

$$\frac{\partial t_{\text{I}}}{\partial \xi_{\text{I}}} = \frac{\partial t_{\text{II}}}{\partial \xi_{\text{II}}} \tag{19–20}$$

Hence, the temperature distribution at an initial time $\tau = 0$ would be represented by a curve without bend at the separating plane between the sections I and II. However, according to Eq. 19–19, different time intervals would have to be used in the sections I and II of the wall, namely,

$$\Delta\tau_{\text{I}} = \frac{\rho_{\text{I}} c_{p\text{I}} k_{\text{I}}}{2} (\Delta\xi_{\text{I}})^2$$

and

$$\Delta\tau_{\text{II}} = \frac{\rho_{\text{II}} c_{p\text{II}} k_{\text{II}}}{2} (\Delta\xi_{\text{II}})^2$$

This would be very inconvenient, if not prohibitive, since, beginning with a smooth temperature curve at $\tau = 0$, later on the left-hand branch of the temperature distribution (section I in Fig. 19–9) would belong to another instant of time than the right-hand branch (section II) and would have to be reduced, after each time step, to the same value of τ before continuing with the graphical construction. For this reason, Nessi and Nisolle recommend to use Eq. 19–10 and take

$$\Delta\tau_{\text{I}} = K_{\text{I}} \frac{\rho_{\text{I}} c_{p\text{I}} k_{\text{I}} (\Delta\xi_{\text{I}})^2}{2}$$

and

$$\Delta\tau_{\text{II}} = K_{\text{II}} \frac{\rho_{\text{II}} c_{p\text{II}} k_{\text{II}} (\Delta\xi_{\text{II}})^2}{2}$$

Choosing further,

$$\frac{K_{\text{I}}}{K_{\text{II}}} = \frac{\rho_{\text{II}} c_{p\text{II}} k_{\text{II}} (\Delta\xi_{\text{II}})^2}{\rho_{\text{I}} c_{p\text{I}} k_{\text{I}} (\Delta\xi_{\text{I}})^2} = \frac{\alpha_{\text{I}}}{\alpha_{\text{II}}} \left(\frac{\Delta x_{\text{II}}}{\Delta x_{\text{I}}}\right)^2 \tag{19–21}$$

the time intervals become equal, namely,

$$\Delta\tau_{\text{I}} = \Delta\tau_{\text{II}} = \Delta\tau \tag{19–22}$$

Moreover, choosing

$$\frac{K_{\text{I}}}{K_{\text{II}}} = \frac{\alpha_{\text{I}}}{\alpha_{\text{II}}}$$

the initial partitions Δx_{I} and Δx_{II} would also become equal, namely,

$$\Delta x_{\text{I}} = \Delta x_{\text{II}} = \Delta x \tag{19–23}$$

so that now the composite wall could be considered a simple wall with only one sort of partitions, Δx, for space, and one sort of intervals, $\Delta\tau$, for time. However, it is more practicable to choose Δx_{I} and Δx_{II} different and so that each section of a composite wall can be divided into an integral number of partitions. This can also be done by the choice of appropriate values of $K_{\rm I}$ and $K_{\rm II}$, according to Eq. 19–21.

19–5. The Method of Finite Differences for Heat Conduction through Cylindric Shells

Nessi and Nisolle also showed first how the method of finite differences can be applied to radial heat conduction in cylindric and spherical bodies.

The differential equation for cylinders is

$$\frac{\partial t}{\partial \tau} = \alpha \frac{\partial^2 t}{\partial r^2} + \frac{\alpha}{r} \cdot \frac{\partial t}{\partial r} \tag{19–24}$$

Introducing

$$\xi = \ln r, \quad \text{that is,} \quad r = e^{\xi} \tag{19–25}$$

the following equation is obtained:

$$\frac{\partial t}{\partial \tau} = \frac{\alpha}{r^2} \cdot \frac{\partial^2 t}{\partial \xi^2} \tag{19–26}$$

or, written as an equation of finite differences,

$$\frac{\Delta_\tau t}{\Delta \tau} = \frac{\alpha}{r^2} \cdot \frac{{\Delta_x}^2 t}{(\Delta \xi)^2} \tag{19–27}$$

Now, as for the plane plate,

$$\frac{\alpha}{r^2} \cdot \frac{\Delta \tau}{(\Delta \xi)^2} = \frac{K}{2} \tag{19–28}$$

Since

$$d\xi = \frac{dr}{r}$$

or

$$\Delta \xi = \frac{\Delta r}{r}$$

by substitution in Eq. 19–28,

$$\Delta \tau = K \frac{(\Delta r)^2}{2\alpha} = K \frac{(e^{\xi}\, \Delta \xi)^2}{2\alpha} \tag{19–29}$$

$\Delta \tau$ is constant if Δr is constant. Hence $\Delta \xi$ varies inversely to r. If a graph with t as function of r is given, then it has to be transformed into one with $\xi = \ln r$ as abscissa and partitions of distance $\Delta \xi = (\Delta r)/r$.

Again, Δr and $\Delta \tau$ may be changed in the run of the construction, and an appropriate ratio of $\Delta \tau/(\Delta r)^2$ may be selected by assuming a proper value of K.

If, on the other hand, Δr were made proportional to r, and K selected proportional to $1/r^2$, then $\Delta \xi$ and $\Delta \tau$ would remain constant. The unit interval, $\Delta \tau$, however, would be fixed in this case, which may be less convenient for the graphical representation of subsequent temperature distributions.

Composite cylindrical shells may be dealt with by introducing

$$\left.\begin{aligned} \xi_{\mathrm{I}} &= \frac{\ln r_{\mathrm{I}}}{k_{\mathrm{I}}} \\ \xi_{\mathrm{II}} &= \frac{\ln r_{\mathrm{II}}}{k_{\mathrm{II}}} \end{aligned}\right\} \qquad (19\text{–}30)$$

$$\cdots\cdots$$

R. L. Perry and Berggren [1944] thoroughly treated the case of a hollow cylinder, originally at uniform temperature, whose inner surface is suddenly changed to, and then kept at, a different temperature. Thermal conditions near the inner surface are practically independent of the presence of the outer surface until the temperature rise at that surface becomes appreciable. Hence, the authors recommend to begin with calculating the change of temperature at the radial distance of the outer surface, assuming an infinitely thick cylinder. A table and a graph suitable for this purpose are included in their paper.

Moreover, since the temperature gradient close to the inner surface often is very steep, they recommend to make use of the analytical solution known for the semi-infinite plate, considering that for the innermost part of a hollow cylinder the curvature can be neglected with good approximation. This is discussed in detail in the paper mentioned.

Another recommendation may be taken from the same paper. In carrying out the graphical solution of a hollow-cylinder problem, the successive points become confusingly close together when they are still an appreciable distance from the steady-state line. In this region, the rate of approach to steady state is proportional to the difference between temperatures for finite and infinite times. Thus the differences between the temperatures at steady state and those at several of the last graphically determined points can be plotted on semi-logarithmic paper and extrapolated to give the differences at later times. These temperature difference-versus-time curves have the same slope for different radial positions.

19-6. The Method of Finite Differences for Heat Conduction through Spherical Shells

The treatment of spherical shells is analogous to that of the cylindrical arrangement. Here the differential equation is

$$\frac{\partial t}{\partial \tau} = \alpha \frac{\partial^2 t}{\partial r^2} + \frac{2\alpha}{r} \cdot \frac{\partial t}{\partial r} \tag{19-31}$$

Introducing

$$\xi = -\frac{1}{r} \tag{19-32}$$

the following equations are obtained:

$$\frac{\partial t}{\partial \tau} = \frac{\alpha}{r^4} \cdot \frac{\partial^2 t}{\partial \xi^2} = \alpha \xi^4 \frac{\partial^2 t}{\partial \xi^2} \tag{19-33}$$

and

$$\frac{\Delta_\tau t}{\Delta \tau} = \frac{\alpha}{r^4} \cdot \frac{\Delta_\xi{}^2 t}{(\Delta \xi)^2} \tag{19-34}$$

Now, as for plate and cylinder,

$$\frac{\alpha}{r^4} \cdot \frac{\Delta \tau}{(\Delta \xi)^2} = \frac{K}{2} \tag{19-35}$$

Since

$$d\xi = \frac{dr}{r^2} \quad \text{and} \quad \Delta \xi = \frac{\Delta r}{r^2}$$

by substitution in Eq. 19-35

$$\Delta \tau = K \frac{(\Delta r)^2}{2\alpha} = K \frac{(\Delta \xi)^2}{2\alpha \xi^4} \tag{19-36}$$

Again, the radius is to be divided into equal parts Δr, and the original graph is to be transformed, this time with $\xi = -1/r$ as abscissa. From here on, the procedure is entirely analogous to that described for cylindrical shells.

For composite spherical shells the following transformation may be used:

$$\left.\begin{aligned} \xi_{\mathrm{I}} &= -\frac{1}{r_{\mathrm{I}} k_{\mathrm{I}}} \\ \xi_{\mathrm{II}} &= -\frac{1}{r_{\mathrm{II}} k_{\mathrm{II}}} \\ &\cdots\cdots\cdots \end{aligned}\right\} \tag{19-37}$$

19-7. The Method of Finite Differences for Heat Conduction in Different Bodies, Including the Heat Transfer on the Surfaces

One particular advantage of the method of finite differences is that variations of the temperature of environment, t_e, and of the surface coefficient of heat transfer, h, can be taken care of without much complication.

Consider a plane wall (Fig. 19-10) extending from the surface plane, $x = s$, to the right and exposed to a medium of temperature $t = t_e$

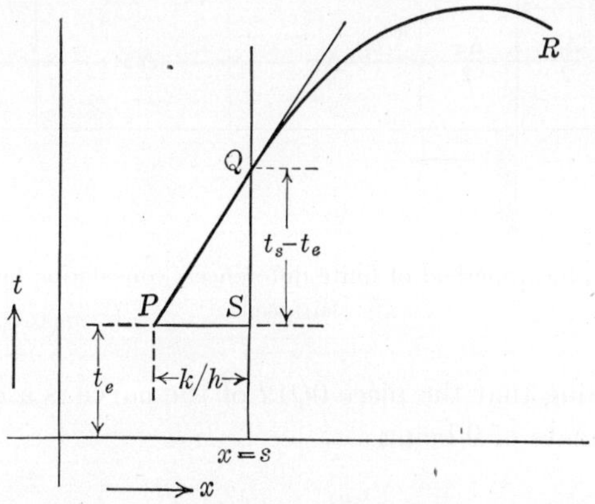

FIG. 19-10. Replacement of a fluid in contact with a wall surface by an extension of the wall into the fluid space.

at $x < s$. Since, per unit surface area and unit time, the same amount of heat, $-q''$, is flowing from the wall to the surface as from the surface to the environment:

$$-q'' = k\left(\frac{\partial t}{\partial x}\right)_s = h(t_s - t_e) \tag{19-38}$$

The tangent $\overline{PQ}$ to the curve $\overline{QR}$, the horizontal line $\overline{PS}$ at ordinate $t = t_e$, and the vertical distance $\overline{SQ}$ form a triangle in which, according to Eq. 19-38, $\overline{PS} = k/h$. It is as though the plate were extended from $x = s$ to the left-hand side, and $\overline{PQR}$ can be considered as the temperature distribution at the instant $\tau = \tau$. Thus, the construction according to the method of finite differences may be extended to the left-hand side.

In Fig. 19-11 the temperature distribution in a wall at the time instant τ is given by the curve through the points Q, 1, 2, 3, 4. It may be extended to a point 0, outside the wall, at a distance $\Delta x/2$ from the sur-

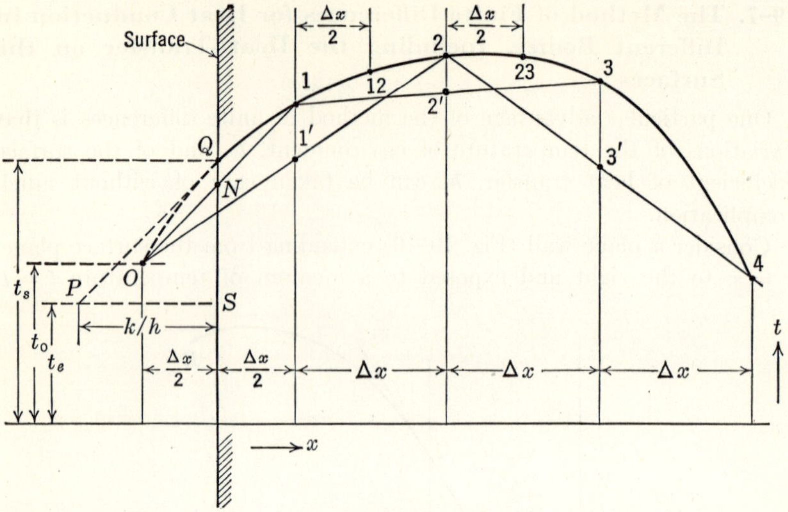

FIG. 19–11. Graphical method of finite differences, considering heat transfer at the surface.

face, by assuming that the piece $\overline{0Q12}$ of the curve is a curve of second order, the ordinate of 0 being

$$t_0 = \frac{8t_s - 6t_1 + t_2}{3} \tag{19–39}$$

as will be shown later on. It will also be shown that the vertical distances $\overline{RQ}$ (see Fig. 19–12) and $\overline{NQ}$ can be represented by the equations

$$\overline{RQ} = \frac{5(t_1 - t_s) - (t_2 - t_1)}{3} \tag{19–40}$$

and

$$\overline{NQ} = \frac{2(t_1 - t_0) - (t_2 - t_0)}{3} \tag{19–41}$$

A part of Fig. 19–11 is reproduced as Fig. 19–12 on a larger scale and with some new notations. The equation of the curve $\overline{0Q12}$ may be

$$z = ay + by^2 \tag{19–42}$$

and the scale of y may be chosen so that $y_{0.5} = 0.5$; $y_1 = 1$; $y_2 = 2$.

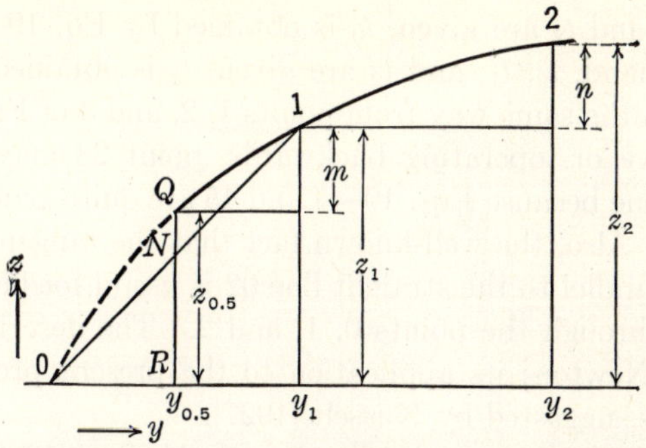

FIG. 19-12. A detail of the construction in Fig. 19-11.

From the last two data

$$a = \frac{4z_1 - z_2}{2}$$

and

$$b = \frac{z_2 - 2z_1}{2}$$

and by substitution

$$\overline{RQ} = \frac{6z_1 - z_2}{8} \tag{19-43}$$

Substituting

$$z_1 = \overline{RQ} + m$$

and

$$z_2 = \overline{RQ} + m + n$$

one obtains

$$\overline{RQ} = \frac{5m - n}{3} \tag{19-44}$$

which is identical with Eq. 19-40.

Since

$$\overline{NQ} = \overline{RQ} - \frac{z_1}{2}$$

by substitution from Eq. 19-43,

$$\overline{NQ} = \frac{2z_1 - z_2}{8} \tag{19-45}$$

which is identical with Eq. 19-41.

Finally

$$t_0 = t_s - \overline{RQ}$$

Substituting RQ from Eq. 19-40, one obtains Eq. 19-39.

So, if t_s, t_1, and t_2 are given, t_0 is obtained by Eq. 19–39 or 40. If, on the other hand, t_0, t_1, and t_2 are given, t_s is obtained by means of Eq. 19–41. In the same way from points 1, 2, and 3 of Fig. 19–11 point 12 of the curve or, operating backwards, point 23 may be obtained. This can be done because Eqs. 19–44 and 45 are quite general corollaries of Eq. 19–42. Also, the well-known fact that the tangent of curve $\overline{012}$ in point 1 is parallel to the straight line $\overline{02}$ is useful for finding the best-fitting curve through the points 0, 1, and 2. The described procedure goes back to Newton; its application to the present problem of finite differences was suggested by Nusselt [1927].

The points 1′, 2′, 3′ are obtained as described in Sect. 19–3. Connecting these points by a curve, a point (12)′ below point (12) will be found, and graphical extension of curve 2′(12)′1′ to the left will yield a point Q' below Q. Finally, a point 0′ is obtained by drawing from 1′ a parallel to $Q'P$. Then the construction is continued with 0′ as starting point.

A simpler, good procedure is to draw a straight line $\overline{P1}$ intersecting the ordinate $x = -\Delta x/2$ at a point little distant from 0 of Fig. 19–11, and to continue with the polygonal distribution without smoothing it by curves.

It can easily be recognized from Fig. 19–11 what effect it has when the pole P is moving because of changes of k/h or of the distance $\overline{QS}$.

If the procedure of Nessi and Nisolle with $\xi = x/k$ is used, then substitution in Eq. 19–38 shows that the distance k/h in the t, x chart (Fig. 19–10) is to be replaced by $1/h$ in a corresponding t, ξ chart.

For cylinders and spheres, the symbol x in Fig. 19–10 and in Eq. 19–38 has to be replaced by r.

Substituting, then, $\xi = \ln r$, the equation converts to

$$\frac{k}{s}\left(\frac{\partial t}{\partial \xi}\right)_s = h(t_s - t_e) \tag{19–46}$$

Hence the distance, $\overline{PS} = k/h$, in the t, x chart is to be superseded by a distance, $k/(hs)$, in a corresponding t, ξ chart for cylinders.

Introducing $\xi = -1/r$, Eq. 19–38 becomes

$$\frac{k}{s^2}\left(\frac{\partial t}{\partial \xi}\right)_s = h(t_s - t_e) \tag{19–47}$$

Hence, for the sphere, $\overline{PS}$ of Fig. 19–10 is to be replaced by $k/(hs^2)$ in the t, ξ chart.

CHAPTER 20

Experimental Analogy Methods for the Solution of Heat-conduction Problems

20–1. General Remarks on the Experimental Analogy Methods

Fourier's differential equation is not restricted to heat conduction; and, particularly, that special case of the equation which is known as Laplace's equation holds for all cases which can be treated by the potential theory. In Chapter 28 the application of Fourier's equation to problems of mechanical diffusion will be dealt with. In this chapter it will be shown how experimental arrangements for the determination of potential differences in electrical and magnetical fields, or of electrical currents in suitable combinations of electrical resistors and capacitors, can be used to solve problems of heat conduction.

The analogy between thermal and electrical conduction was used many years ago, for instance, by, and even before, Kohlrausch (see Sect. 10–2). But Langmuir, Adams, and Meikle [1913] seem to have been the first who made practical use of the similarity between a heat flow-temperature field and an electrical flow-voltage field of the same geometrical configuration. In the next year (1914) Jakob used the similarity of thermal and magnetic fields for testing insulating materials in the Physikalisch-Technische Reichsanstalt. Because of World War I this method was not published until years later [Jakob 1922(*a*)]. Again, a year afterward, Nusselt [1915] showed how from the analogy between thermal and electrostatic fields the thermal conductivity of materials in arbitrary form can be determined.

Whereas these investigators used the analogy methods for the steady state of heat conduction only, the Dutch engineer Beuken [1936, 1936(*a*)] showed how to apply the thermal- and electrical-flow analogy to problems of unsteady heat flow. Owing to the mathematical difficulties of this sort of problems, Beuken's experimental procedure has found particular interest and application.

20-2. Basic Equations of Analogy Methods for Steady-state Problems

The methods mentioned for steady state are based on the following general reasoning.

In a homogeneous body B (Fig. 20-1) of thermal conductivity k, two sections of the surface, having the areas A_1 and A_2, may be at constant temperatures (t_1 and t_2, respectively), and everywhere else the surface may be perfectly heat insulated. Then, if $t_1 > t_2$, heat will flow from A_1 to A_2 according to Laplace's equation:

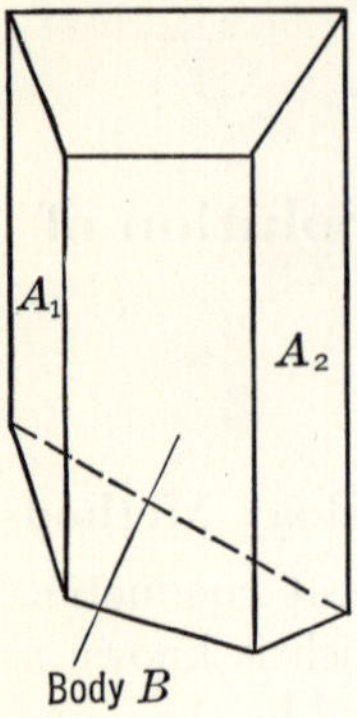

FIG. 20-1. Homogeneous body B with isothermal surfaces A_1 and A_2.

$$\frac{\partial^2 t}{\partial x^2} + \frac{\partial^2 t}{\partial y^2} + \frac{\partial^2 t}{\partial z^2} = 0 \tag{20-1}$$

The rate of heat flow is

$$q = -k \int_{A_1} \frac{\partial t}{\partial n} \, dA_1 \tag{20-2}$$

where n is the direction normal to a differential element dA_1, counted as positive for decreasing temperature. The integration is to be performed over the area A_1.

Introducing a new variable

$$\phi = \frac{t - t_2}{t_1 - t_2} \tag{20-3}$$

Eq. 20-1 is transformed to

$$\frac{\partial^2 \phi}{\partial x^2} + \frac{\partial^2 \phi}{\partial y^2} + \frac{\partial^2 \phi}{\partial z^2} = 0 \tag{20-4}$$

with the boundary conditions

$$\phi = 1 \quad \text{for} \quad t = t_1$$

and

$$\phi = 0 \quad \text{for} \quad t = t_2$$

whatever values these temperatures may have. Likewise, Eq. 20-2 is transformed to

$$q = k(t_1 - t_2) \int_{A_1} - \frac{\partial \phi}{\partial n} \, dA_1 \tag{20-5}$$

Hence, ϕ does not depend on t_1 and t_2, but only on the shape of the body B, and the same holds for the integral in Eq. 20-5. This has the physical dimension of a length; it can be considered as the ratio of a mean effective area A_m to a mean effective length L_m of the body B and may be expressed by

$$\lambda_c = \int_{A_1} -\frac{\partial \phi}{\partial n}\, dA_1 \qquad (20\text{–}6)$$

Since the characteristic length * λ_c has nothing to do with temperature and heat flow, but solely with the geometrical configuration of the body B, it can be determined by any suitable experimental method which is based on Eqs. 20–1 and 2, whatever the meaning of t, q, and k may be, if only the medium under consideration is in a shape similar to body B. The reason is that Eqs. 20–1 to 5 hold also if in a second arrangement all lengths in the body are m times greater. Assuming now that t_1 and t_2 are left unchanged, $\partial\phi/\partial n$ at any place of the surface A_1 in the second arrangement will be numerically equal to $1/m$ of that gradient at a corresponding place in the original set-up, and the integration in Eq. 20–5 or 6 will have to be extended over a surface m^2 times as large. Hence λ_c and q will be m times greater than their values in the original set-up. This is easily understood by imagining two similar spherical shells. For the same temperature difference $t_1 - t_2$ will q obviously be doubled in a similar shell of double size.

For a complex shape it would be difficult, if possible at all, to find λ_c analytically. The essence of the experimental analogy methods which are dealt with in this section is to determine λ_c by other than thermal measurements. Then, from Eq. 20–5, either the thermal conductivity k can be obtained, if this is asked for, or for a planned thermal construction, similar to and in general much larger than the model, the heat flow q can be precalculated if k is known. Different ways of procedure will be shown.

20–3. Liquid or Solid Electrical Conductor as Model for Steady-state Problems

Langmuir, Adams, and Meikle constructed containers of such form that an electrolyte, filled in, took the shape of parts of furnace walls (edges, corners) under consideration. Each model was made of glass and copper plates, fastened together at their edges by cement, the copper-plate surfaces imitating the areas A_1 and A_2 of the original object (see Fig. 20–1). A standard decimeter cube with two opposite sides of copper, filled with the same electrolyte, was used as a standard resistance and placed in series with the model.

* Langmuir and coworkers called it shape factor. This denotation is reserved for dimensionless factors in this book (see, for instance, Eq. 8–13).

Obviously the current i in the electrolyte corresponds to the heat flow q, and the potential drop Δe between the electrodes to the temperature drop $\Delta t = t_1 - t_2$ between the areas A_1 and A_2 of the original body. Hence the equations for the heat flow in the original and for the electrical current in a model, with all lengths in the scale $m{:}1$ to the original, are

$$q = k \frac{A_c}{L_c} \Delta t = k\lambda_c \,\Delta t \tag{20–7}$$

and

$$i = k_e \frac{m^2 A_c}{m L_c} \Delta e = m k_e \lambda_c \,\Delta e \tag{20–8}$$

By comparison,

$$q = \frac{1}{m} \cdot \frac{k}{k_e} \cdot \frac{\Delta t}{\Delta e} i = \frac{1}{m} \cdot \frac{k}{k_e} \cdot \frac{\Delta t}{R_e} \tag{20–9}$$

where R_e is the electrical resistance of the electrolyte in the model. In the experiments described the measurement of the electrical conductivity of the electrolyte, k_e, was not necessary because the characteristic length for the standard resistance was known. With subscript *st* for standard resistance, obviously:

$$i = k_e(\lambda_c)_{st}\,(\Delta e)_{st} \tag{20–10}$$

and by substitution in Eq. 20–9

$$q = \frac{k}{m} (\lambda_c)_{st} \frac{(\Delta e)_{st}}{\Delta e} \Delta t \tag{20–11}$$

In this way, the actual measurement was reduced to that of two voltages $(\Delta e)_{st}$ and Δe. The authors found a saturated solution of copper sulfate containing 0.5% by volume of sulfuric acid the most suitable electrolyte. The current density used was about 5 milliamperes per square centimeter. The accuracy of their experiments was limited to about 1 to 3% because of the so-called "transfer resistance," a sort of polarization, amounting to about 0.05 volt, which, unlike ordinary polarization, reverses instantly on reversing the current.

In order to eliminate this source of error Nukiyama and Yosikata [1930] used high-frequency current in water as the electrolyte.

Boelter, Cherry, and H. A. Johnson [1942, p. IV–34] describe how to find potential and flow lines in an electrolyte by means of a microphone hummer connected to the copper electrodes as generator and an earphone attached to two exploring probes. These are insulated wires with free metallic ends ("needles") and may be brought to different places in the electrolyte. Wherever the sound strength has its minimum, the two needles are at places of equal potential.

It might be possible, as well, to measure the potential differences between one electrode and a single probe at different places of the electrolyte; however, the "transfer resistance" already mentioned would have to be considered.

Langmuir and his coworkers represented a part of their results by empirical formulas. One of these gives the characteristic length for the heat flow across coherent plane walls of thickness L with square edges and corners. Referring to Fig. 20–2,

ΣA_1 is the total area of the wall, except the areas for the square edges (and corners, if any).

ΣE is the total length of all square edges.

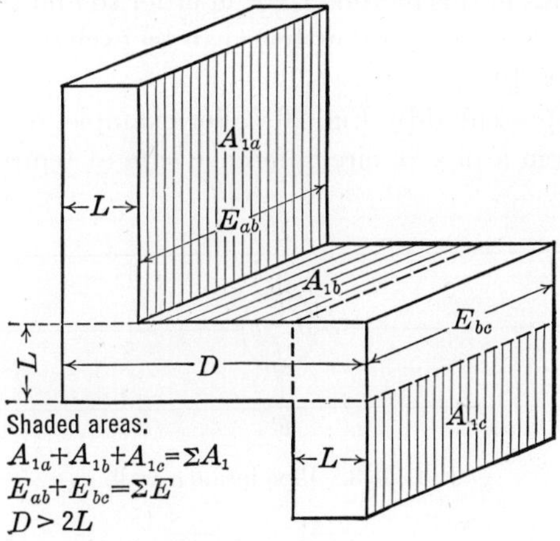

FIG. 20–2. Wall with square edges.

Calling, further, n the number of square corners (none occurring in the figure), the formula becomes

$$\lambda_c = \frac{\Sigma A_1}{L} + 0.54 \cdot \Sigma E + 0.15nL \qquad (20\text{–}12)$$

The first term is the characteristic length for the wall, edges being cut out as indicated in Fig. 20–2, and corners, if any, cut out in a similar way. The second and third terms of the equation have been found by the experiments with the electrolyte. These terms may also be used separately for each square edge or corner in a wall. Denoting the distances of subsequent edges or corners by D (see Fig. 20–2), only constructions for which $D > 2L$ are included in Eq. 20–12; those with $D < 2L$ are dealt with separately in Langmuir's paper.

If the heat flow is only two-dimensional, Awbery and Schofield [1929] recommend to cut thin sheets of known electric conductivity into a shape similar to that of the heat-conducting body and to measure either the electrical potential at different places or the total electrical resistance R_e of these sheets. With this method the heat flow from heating plates, pipes, etc., which are embedded in a wall at given surface temperature can be determined. Schofield, later on [1930, 1931], developed the method in detail and compared it with the mathematical or graphical treatment of the same problems. In order to satisfy the requirement of uniform potential of A_1 and A_2, which is more difficult than with the electrodes in an electrolyte, Schofield soldered copper bars to those lines of the sheet models where the original is at the temperatures t_1 and t_2. It is advantageous in this method that, in order to find the characteristic length λ_c, equipotentials and flow lines can be exchanged for the reason indicated in Sect. 19–2.

In the case represented by Fig. 20–3, for example, one wants to know the heat flow from a pipe of circumference $\overline{bdfh}$ at temperature t_1 inside

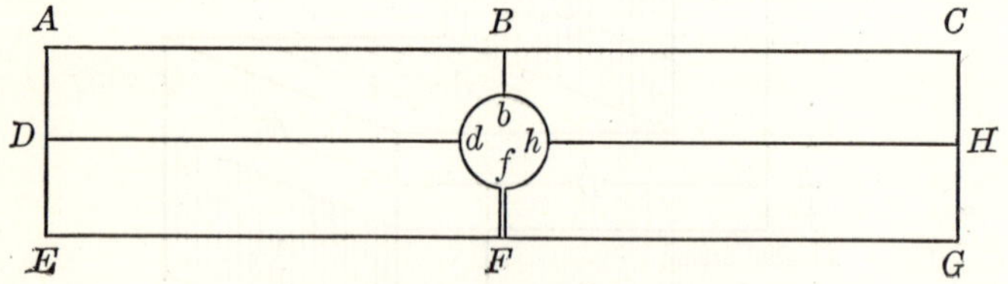

FIG. 20–3. Pipe inside a wall.

a wall to the wall surfaces $\overline{ABC}$ and $\overline{EFG}$, both at temperature t_2, whereas the sides $\overline{ADE}$ and $\overline{CHG}$ are supposed to be perfectly heat insulated. In the sheet model it would not be easy to bring each point of the lines $\overline{AC}$ and $\overline{EG}$ to the same electrical potential, and the involved electric resistances would be inconveniently small. Schofield, however, cut the sheet along $\overline{fF}$ and used the two lines $\overline{fF}$ as poles for the electric current; i.e., he used the flow line $\overline{fF}$ of the original as potential line in the model. In addition to the elimination of the two difficulties mentioned a check of the accuracy of the method was possible by measuring the potential along $\overline{Dd}$ and $\overline{Hh}$ by a potential probe, as used in the electrolyte method. If Δe is the potential drop from the left to the right line at $\overline{fF}$, then, for symmetry reasons, the potential drop up to $\overline{dD}$ should be $(\Delta e)/4$ and that up to $\overline{hH}$ equal to $3(\Delta e)/4$ for any points of these lines.

Considering unit length of the original system in the direction perpendicular to the drawing plane of Fig. 20-3 and model sheets of the thickness s, the heat flow from $\overline{bdfh}$ to $\overline{AC}$ and $\overline{EG}$ may be called q', and the electric current from the one line at $\overline{fF}$ over bB to the second line at $\overline{fF}$ may be i. It will be shown that, instead of Eqs. 20-7 to 9, the three following are to be used:

$$q' = k \frac{A_c}{L_c} \Delta t = k\lambda_c \, \Delta t \tag{20-13}$$

$$i = k_e \frac{mL_c}{mA_c} s \cdot \Delta e = k_e \frac{s}{\lambda_c} \Delta e \tag{20-14}$$

$$q' = \frac{kk_e s \cdot \Delta e \cdot \Delta t}{i} \tag{20-15}$$

In Eq. 20-13, A_c may be considered as a length, the effective cross-sectional area for the heat flow being equal to $(A_c \cdot 1)$. Because of the exchange of potential and flow lines the mean effective cross section in Eq. 20-14 is $s \cdot L_c$, and the mean effective path length is A_c. In the drawing plane of Fig. 20-3 the model may have an arbitrary linear scale $m{:}1$ with respect to the original, since both lengths, L_c and A_c, in Eq. 20-14 are to be multiplied by m. The papers of Schofield show an excellent agreement between experiment and calculation in the cases mentioned and some others.

An electrical network which no longer has any geometrical similitude with the object under consideration may be used as a model. Wilhelm, W. C. Johnson, Wynkoop, and Collier [1948] describe a network which they developed for analyzing a cylindrical catalytic converter. The procedure was to divide a radial section of the bed of granules into a convenient number of concentric annuli. The annuli were chosen with a thickness in the axial direction so small that, in effect, the analysis was for a radial cross-sectional area. Electrical resistances proportional to the various thermal resistances in the annuli were assembled into a network structure. Electric current was introduced in each element in proportion to the heat generated by chemical reaction. Because the rate of reaction depends upon the temperature, the current to each element had to be adjusted according to the voltage at the element. After balance had been achieved, the radial voltage distribution described the corresponding temperature distribution of the converter. This procedure was repeated at a series of positions or slices in the axial direction. Details of the electrical arrangement and mathematical procedure may be studied in the original paper.

20–4. Electromagnet as Model for Steady-state Problems

The problem for which the author used the magnetic analogy was to determine the heat flow represented by the curved full lines in Fig. 20–4. The figure gives the boundary lines of a cross section through a twin-plate arrangement (see Sect. 9–3) broken off at $\overline{EADJ}$. The heater was placed between $\overline{AB}$ and $\overline{CD}$, the coolers above $\overline{EF}$ and below $\overline{IJ}$.

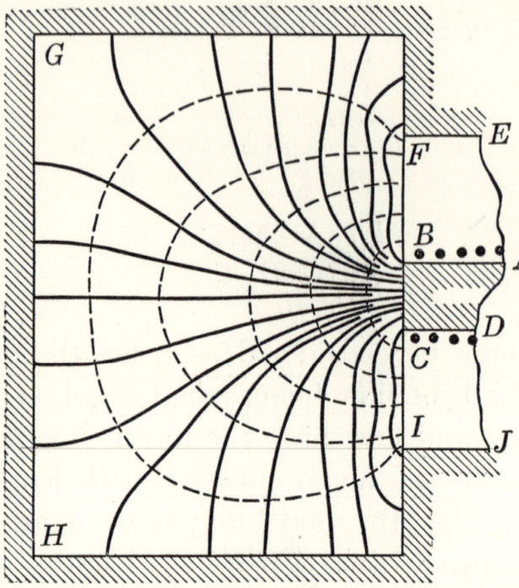

FIG. 20–4. Electromagnetic model for determination of heat flow through a space of rectanglar cross section.

ABCD = Iron plate with electrical winding

EFGHIJ = Inner contour of transformer steel sheets

(Iron plate and transformer sheets are in close contact at the right side; not shown in the figure)

BFGHIC = Rectangular cross section with magnetic flow and niveau lines

Between $\overline{AB}$ and $\overline{EF}$, and between $\overline{CD}$ and $\overline{IJ}$, were the plates of the material the thermal conductivity of which was to be measured. There was no guard ring; however, the space bounded by the line $\overline{BCIHGFB}$ was filled with a heat-insulating material. The line $\overline{ABCD}$ was at temperature t_1; the line $\overline{EFGHIJ}$ was at temperature t_2.

The model consisted of an iron plate placed inside of $\overline{ABCD}$, with an electric winding as indicated in the figure, and a package of transformer steel sheets outside of $\overline{EFGHIJ}$. The right-side ends of the iron plate and the package of sheets were brought to close contact, so that an iron-closed electromagnet was formed. A cardboard covered with iron filing was placed in the space $\overline{BCIHGFB}$, and the magnet was excited by direct current. Photos were taken of the iron-file chains occurring on the cardboard, and the curved full lines in Fig. 20–4 were drawn along these magnetic flow lines. The lines $\overline{BC}$, $\overline{FGHI}$, and the dotted lines are lines of constant magnetic potential. From this system λ_c was calculated by a graphical method similar to that presented in Sect. 19–2. As far as the author knows, this was the first time that isothermal and heat-flow lines were determined experimentally by an analogy method.

20–5. Electrical Capacitor as Model for Steady-state Problems

Nusselt, having used the spherical and cubical shell, with heater inside and cooler outside, to measure the thermal conductivity of an insulating material, was anxious to extend this loss-free method to test samples, filling the space between an inner rectangular heating plate and a rectangular cooling case with a rectangular cooling plate as top. The heat flow q, the outer surface temperature t_1 of the heater, and the inner surface temperature t_2 of the coolers were directly measured. In order to determine the characteristic length λ_c, defined by Eq. 20–7, Nusselt used an electrical capacitor, composed of two wooden cases, the smaller of which was placed inside the larger one. The inner case had the same outer-surface dimensions as the electrical heater of the original set-up; the other one had inner surfaces equal in dimensions to the corresponding surfaces of the original cooling system. Covering the two surfaces with tin foil, a capacitor with air as dielectric was formed.

The analogy is based on the fact that the differential equation of the potential e in a dielectric and the equation of the electrical charge Q_e are similar to Eqs. 20–1 and 2, namely,

$$\frac{\partial^2 e}{\partial x^2} + \frac{\partial^2 e}{\partial y^2} + \frac{\partial^2 e}{\partial z^2} = 0 \tag{20–16}$$

and

$$Q_e = -\frac{\kappa}{4\pi}\int \frac{\partial e}{\partial n}\, dA_1 \tag{20–17}$$

where κ = the dielectric constant which is dimensionless in the electrostatic system of units, and

n = the direction, normal to the inner surface, so that the potential decreases with increasing n.

Equation 20–17 is found as follows:

If Q_e'' is the electrical charge of unit area of the surface 1, then the total charge is

$$Q_e = \int_{A_1} Q_e'' \cdot dA_1 \tag{20–18}$$

But by definition,

$$Q_e = C_e(e_1 - e_2) \tag{20–19}$$

where C_e is the capacitance of the condenser. Hence

$$C_e(e_1 - e_2) = \int_{A_1} Q_e''\cdot dA_1 \tag{20–20}$$

In electrostatics it is shown that

$$Q_e'' = -\frac{\kappa}{4\pi}\cdot\frac{\partial e}{\partial n} \tag{20–21}$$

Substituting Q_e'' from Eq. 20–21 in Eq. 20–20 and equating this to Eq. 20–19, Eq. 20–17 is obtained.

Introducing now

$$\phi = \frac{e - e_2}{e_1 - e_2} \tag{20–22}$$

in complete analogy to Eq. 20–3, comparison with Eq. 20–6 yields

$$\lambda_c = \frac{4\pi C_e}{\kappa} \tag{20–23}$$

Since $\kappa = 1$ for vacuum and virtually also for air,

$$\lambda_c = 4\pi C_e \tag{20–24}$$

Substituting this in Eqs. 20–5 and 6,

$$k = \frac{q}{(t_1 - t_2)4\pi C_e} \tag{20–25}$$

In the electrostatic system of units C_e has the dimension of a length. It has to be measured for the model capacitor by a standard method of electrostatics. It may be mentioned that before Nusselt the analogy method with a capacitor had already been envisaged by Northrup [1913] in an enlightening paper entitled "Some Aspects of Heat Flow."

20–6. Other Possibilities of Models for Steady-state Problems

It does not seem that the equation of the pressure drop in a porous medium has been used so far as analogous to the Laplace equation. However, it could, considering that the pressure drop corresponds to the temperature drop and the fluid flow to the heat flow, provided that the flow is not turbulent. The author encountered problems where it might have been useful to calculate the fluid flow by means of thermal-conduction measurements. The flow of fluids through porous plates, in piles of sand or coal, or in the soil where the pressure at two or more surfaces is given may be precalculated by making a small model of similar configuration from a substance of known thermal conductivity; this then should be heated and cooled at the corresponding surfaces, and the heat flow and temperature drop measured.

Generally, any fluid flow under potential flow condition could be used for imitating the heat conduction, and there are still other possibilities in different fields of physics where the Laplace equation occurs.

20-7. Basic Equations of an Analogy Method for Unsteady-state Problems

A linear non-steady heat flow, as expressed by Fourier's equation

$$\frac{\partial t}{\partial \tau} = \alpha \frac{\partial^2 t}{\partial x^2} \tag{20-26}$$

can be imitated by an electrical model as shown in Fig. 20-5. Let one pole of a source of electricity with variable voltage e be directly connected

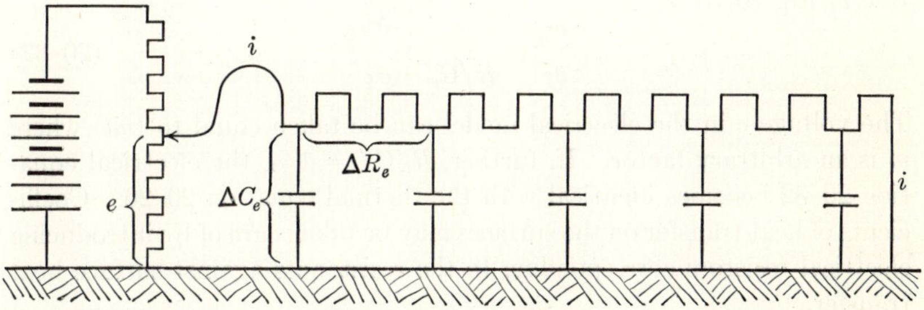

FIG. 20-5. Model using electric resistors and condensers for determination of unsteady heat flow in a wall.

with the ground and the other pole over an induction-free resistance, the instantaneous current being i. The resistance consists of many equal sections, $\Delta R_e = R_e' \cdot \Delta x$, where Δx is a small distance and R_e' the electrical resistance per unit length. The connecting points between subsequent resistors ΔR_e are connected with the ground over small condensers of equal capacitance, $\Delta C_e = C_e' \cdot \Delta x$, where C_e' is the capacitance per unit length. If at a place x the potential is changed by $\Delta_\tau e$ in a time interval $\Delta\tau$, the elementary capacitor will be charged by the quantity $(\Delta i \cdot \Delta\tau)$. Using further the subscript x for local change at time τ, then, according to electrophysics,

$$i = \frac{\Delta_x e}{R_e' \cdot \Delta x} \tag{20-27}$$

$$\Delta i = (C_e' \cdot \Delta x) \frac{\Delta_\tau e}{\Delta\tau} \tag{20-28}$$

Obviously, R_e' corresponds to the thermal resistance R' per unit length, and C_e' corresponds to the product $C_pA = \rho c_p A$ because Eq. 20–28 is equivalent to the equation

$$\Delta q = (\rho c_p A \cdot \Delta x) \frac{\Delta_\tau t}{\Delta \tau} \tag{20-29}$$

where Δq is the heat energy stored in unit time if the temperature is increased by $\Delta_\tau t$ in the time interval $\Delta \tau$.

Reducing the finite differences in Eqs. 20–27 and 28 to differentials,

$$i = \frac{1}{R_e'} \cdot \frac{\partial e}{\partial x} \tag{20-30}$$

$$\frac{\partial e}{\partial \tau} = \frac{1}{C_e'} \cdot \frac{\partial i}{\partial x} \tag{20-31}$$

Differentiating Eq. 20–30 with respect to x and substituting the derivative in Eq. 20–31,

$$\frac{\partial e}{\partial \tau} = \frac{1}{R_e'C_e'} \cdot \frac{\partial^2 e}{\partial x^2} \tag{20-32}$$

The voltage e in the electrical model can be taken equal to (mt) where m is an arbitrary factor. If, further, $R_e'C_e' = 1/\alpha$, the electrical equation 20–32 becomes identical with the thermal equation 20–26. Coefficients of heat transfer on the surfaces may be taken care of by introducing electrical resistors corresponding to the resistances against surface heat transfer.

That this analogy has been applied only 20 years later than the electric analogy for steady state is due to the difference of the numerical values of $R_e' \cdot C_e'$ and $1/\alpha$ in practical arrangements. In the words of Richards [1913], who seems to have first recognized this difficulty, "An electric charge is imparted to a body in a fraction of a second, whereas the amount of heat required to bring the body to thermal equilibrium is very great and it takes minutes and sometimes hours to obtain that state."

Considering insulating materials, as used in furnace design, α may be of the order of magnitude of 0.003 cm^2/sec. Then $R_e'C_e'$ would have to be $1/0.003 \approx 330$ ohm·farad/sq cm. Selecting, now, for instance, $R_e' = 10^6$ ohm/cm, then $C_e' = 0.33(10^{-3})$ farad/cm $= 330$ μfarad/cm is required. These are much higher values than could be realized in a practicable model. However, remembering that in problems of transient conduction α occurs only in the combination $N_{Fo} = \alpha\tau/x^2$, Beuken recognized that a reduction of α has no influence on the result if τ is increased in the same proportion.

Thus, a model with $1/R_e'C_e' = 200\alpha$ for time intervals of $\frac{1}{200}$ of those in the original may be made. Then 1 hour of heating in a furnace wall will be imitated by 18 seconds of charging the capacitors in the model. As convenient resistances and capacitances per unit length, 10^5 ohms/cm and 16.5 μfarad/cm are recommended. Considering elements of the thermal set-up of 3-cm length, resistors of $0.3(10^6)$ ohms and capacitors of about 50 μfarad may be arranged.

20–8. Permanent Models for Unsteady-state Problems

On this basis Beuken has constructed different models in the Nederlandsch Instituut voor Electrowarmte en Electrochemie, the last one [1939] † as a permanent device with five units, each with a resistor, variable from 100 to 1,111,000 ohms, and with a capacitor, variable from 0.1 to 111 μfarad. In the United States a much larger permanent model has been erected at Columbia University by Paschkis, who previously did valuable work in this field in cooperation with Beuken [see Beuken and Paschkis 1937; Paschkis and Beuken 1938]. The Columbia model [Paschkis and Baker 1942] contains 15 sections, each with decade resistors for 100 to 1,111,000 ohms and condensers of 152 μfarad available in steps of 0.1 μfarad. These are said to be accurate within $\pm 1\%$, their insulation resistance being at least 15,000 megohms per microfarad. The voltage in this model can be varied from 0 to 286 volts; currents between 0.1 and 30 milliamperes can be set and held constant by electron tubes. Instead of electrostatic voltmeters, as used by Beuken, amplifiers in combination with recording millivoltmeters are employed, only 10^{-9} ampere being drawn from the model by this arrangement. Also an ampere-hour meter, combined with an amplifier, is provided.

20–9. Actual Solution of Unsteady-state Problems by the Model Method

With his first model Beuken solved a number of practical problems occurring in the operation of furnaces. Of general importance is Beuken and Paschkis' determination of a function which they called uniformity ratio and which may be denoted by ϕ_u or χ_u, for infinitely wide planes and infinitely long cylinders, respectively. They were concerned with the heating up of plates or cylinders, originally at a uniform temperature which they took as the zero level of temperature. Temperatures in this scale may be expressed by a prime sign; so the initial temperature of the body to be heated would be $t_i' = 0$. If it is brought into a furnace at the constant temperature t_e', then, at a time τ, the largest temperature

† In a recent paper on Beuken's model Gelissen [1948] gives a thorough bibliography about its use by Beuken and his coworkers and successors.

difference occurring in the sample is the difference between the temperature at the surface and in the center, $t_s' - t_0'$. In metallurgical operation, small differences, $t_s' - t_0'$, may be required on the end of a heating process; this is the practical importance of the functions which, according to the authors' definition, would be:

$$\phi_u \text{ (or } \chi_u) = \frac{t_s' - t_0'}{t_s'} \tag{20-33}$$

It seems to be preferable to reduce this formula to the usual temperature scale. Calling t_i the initial temperature of the body, then obviously

$$t_s' = t_s - t_i$$

$$t_0' = t_0 - t_i$$

$$t_e' = t_e - t_i$$

Hence

$$\phi_u \text{ (or } \chi_u) = \frac{t_s - t_0}{t_s - t_i} \tag{20-34}$$

The temperature excess over environment temperature t_e (which here is the temperature in the oven) will be introduced as:

$$\theta = t - t_e$$

Then

$$\phi_u \text{ (or } \chi_u) = \frac{\theta_s - \theta_0}{\theta_s - \theta_i} \tag{20-35}$$

or with the denotations Φ and X of Eqs. 13-54, 55, 69, and 70:

$$\phi_u = \frac{\Phi_s - \Phi_0}{\Phi_s - 1} \tag{20-36}$$

$$\chi_u = \frac{X_s - X_0}{X_s - 1} \tag{20-37}$$

It is seen that the functions ϕ_u and χ_u can be calculated from Figs. 13-4, 5, 7, and 8. However, if $(t_s - t_0)$ is small, the accuracy of these figures is not sufficient; for this reason it is convenient to complete Groeber's graphs by the values ϕ_u and χ_u, found experimentally by Beuken and Paschkis (Figs. 20-6 and 7).‡

‡ Dr. Paschkis kindly put original-size charts of these figures at the author's disposal.

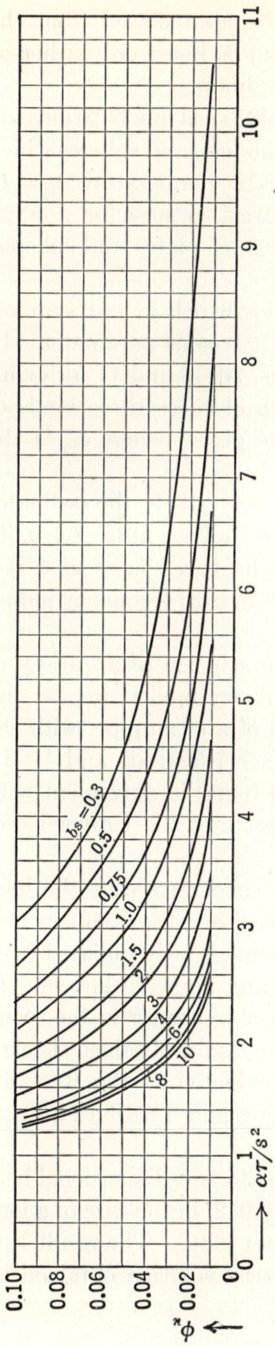

FIG. 20-6. Function ϕ_u defined by Eq. 20-36 and determined by model experiments. (Heating up of plane plates.) Courtesy Dr. V. Paschkis.

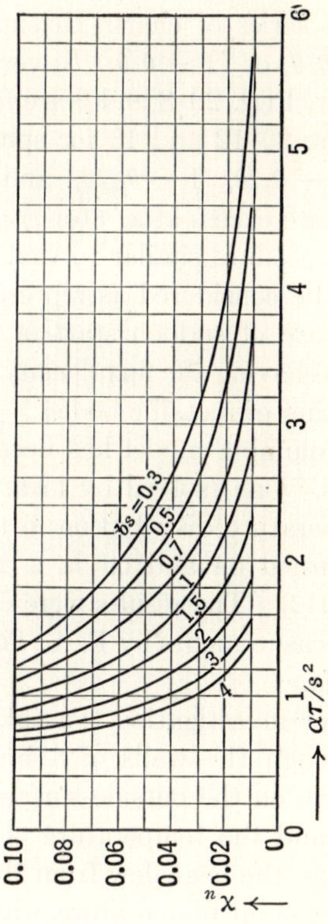

FIG. 20-7. Function χ_u defined by Eq. 20-37, and determined by model experiments. (Heating up of cylinders.) Courtesy Dr. V. Paschkis.

The efficiency of the model method can be appreciated from the authors' claim that their experiments took about 50 hours only, whereas the exact calculation would have required 5000 hours.

Heisler [1947] has used the same permanent electrical model to develop charts for the temperature change in plates, cylinders, and spheres which are suddenly exposed to an environment at a temperature $t_e \neq t_i$. These charts are valid for $\alpha\tau/s^2 < 0.20$, in contrast to those for $\alpha\tau/s^2 > 0.20$, dealt with in Sect. 13-9. The charts for θ_s of plates and spheres, however, were obtained by calculation.

In this series of charts Heisler plotted ratios which in our symbols are $1 - \theta/\theta_i$ or $(1 - \theta/\theta_i)/bs$ versus $\alpha\tau/s^2$ with $1/bs$ as parameter. In particular, Figs. 20-8 and 9 are for plates, Figs. 20-10 and 11 for cylinders, Figs. 20-12 and 13 for spheres. In each of these three kinds of bodies $1 - \theta_s/\theta_i$, $1 - \theta_{1/2}/\theta_i$, and $1 - \theta_o/\theta_e$ are given, where $\theta_{1/2}$ is the temperature excess at x/s (or r/s) $= 1/2$.

In a procedure similar to that described in Sect. 13-9, the ordinates can also be considered as representing $h(\theta - \theta_i)/q_s''$ or $h(\theta - \theta_i)/q_s''bs$ for the case of sudden start of a permanent heating source at $x = s$ (or $r = s$) having the flux density q_s'', as occurs in high-frequency induction heating of metallic bodies.

The Columbia model has been used for the solution of a variety of problems. One (solved by means of a preliminary model) was to find the temperature distribution in the insulation of a steampipe, with the steam turned on and off in a given time cycle [Paschkis and H. D. Baker 1942]. The results were compared and found in agreement with actual measurements by R. L. Perry and Berggren at the University of California.

Another investigation [Paschkis 1942] was concerned with the heat flow through the walls of a building, where either the temperature-time curve on the outside surface or the amount of heat radiated from the sun and the temperature of the surrounding air were known. In both cases, the heat flow from the inside surface to the air in the room, held at constant temperature, was desired. The results obtained checked closely with direct thermal measurements conducted at the Research Laboratory of the American Society of Heating and Ventilating Engineers.

Finally, the application of the model method to two-dimensional heat flow has been shown by Avrami and Paschkis [1942] in the case of a long rectangular bar placed with a long side on a hot plate. The result is in good agreement with the available mathematical solution obtained by Avrami and Little [1942].

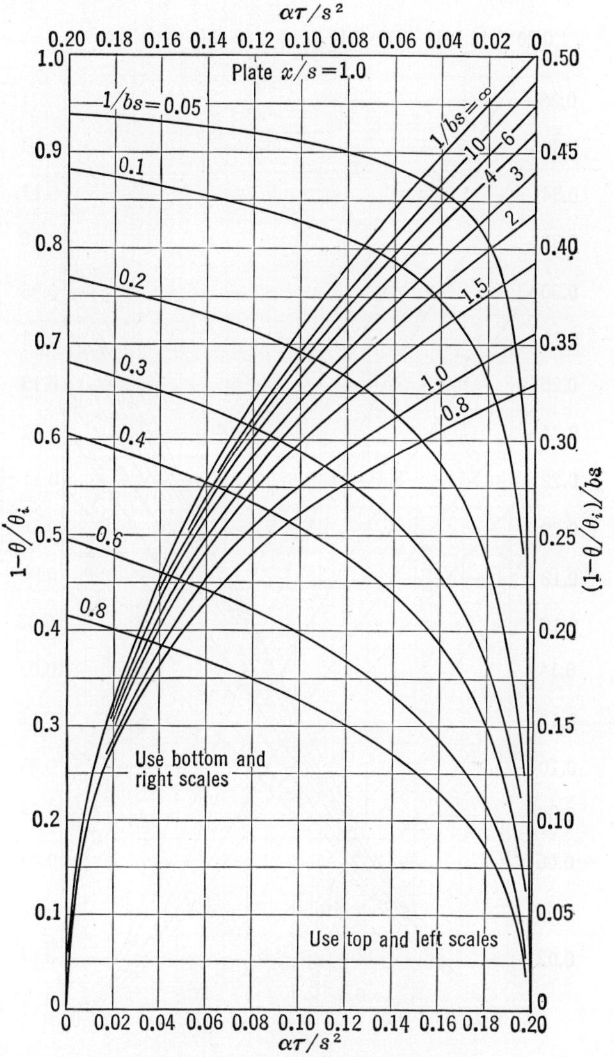

FIG. 20–8. Temperature change on the surface of plane plate after sudden change of environment temperature, for $\alpha\tau/s^2 < 0.20$.

Courtesy Dr. M. P. Heisler [1947].

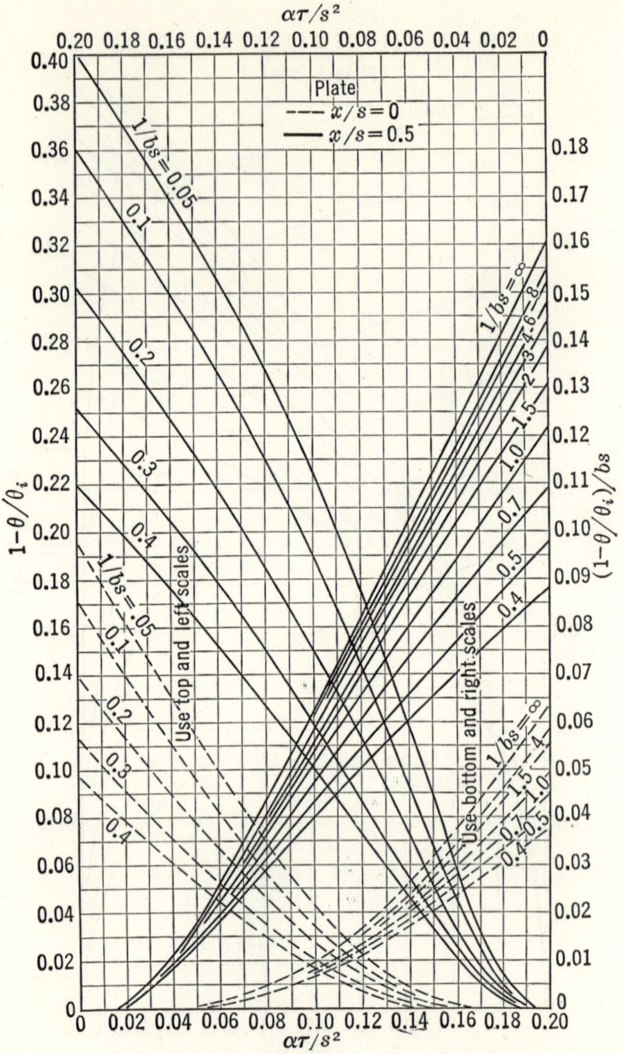

FIG. 20–9. Temperature change in the interior of plane plate after sudden change of environment temperature, for $\alpha\tau/s^2 < 0.20$.

Courtesy Dr. M. P. Heisler [1947].

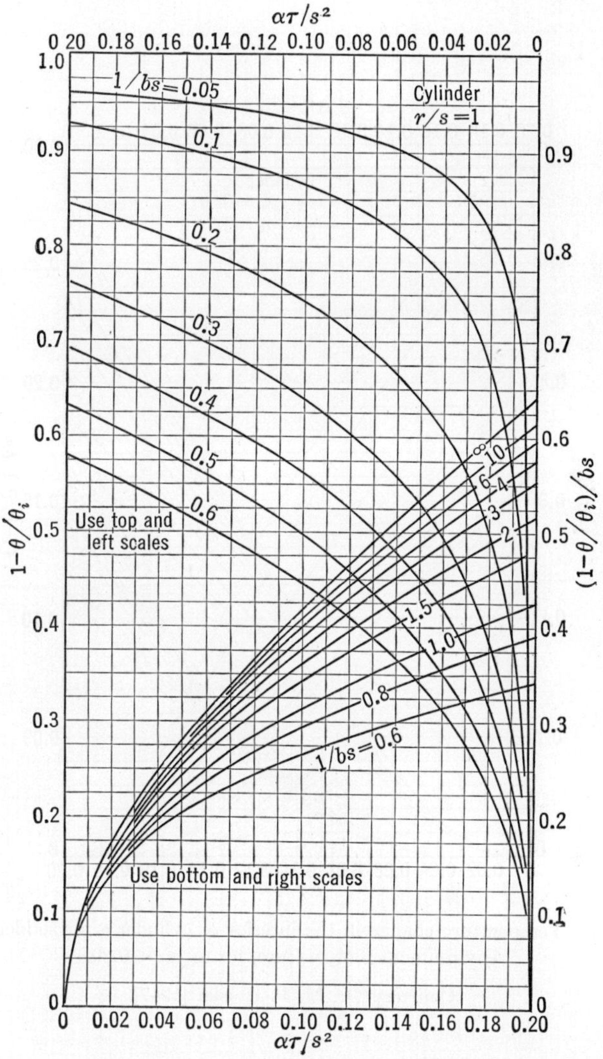

FIG. 20–10. Temperature change on the surface of cylinder after sudden change of environment temperature, for $\alpha\tau/s^2 < 0.20$.

Courtesy Dr. M. P. Heisler [1947].

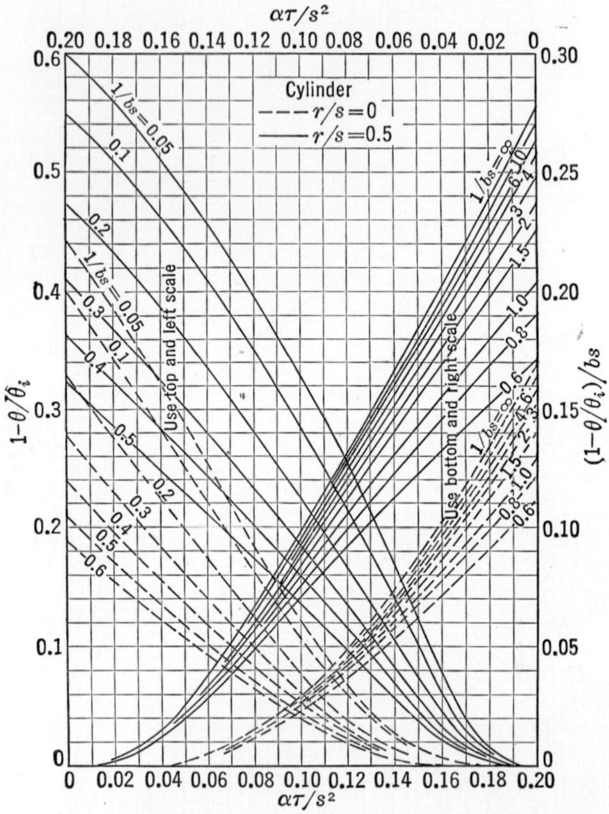

FIG. 20–11. Temperature change in the interior of cylinder after sudden change of environment temperature, for $\alpha\tau/s^2 < 0.20$.

Courtesy Dr. M. P. Heisler [1947].

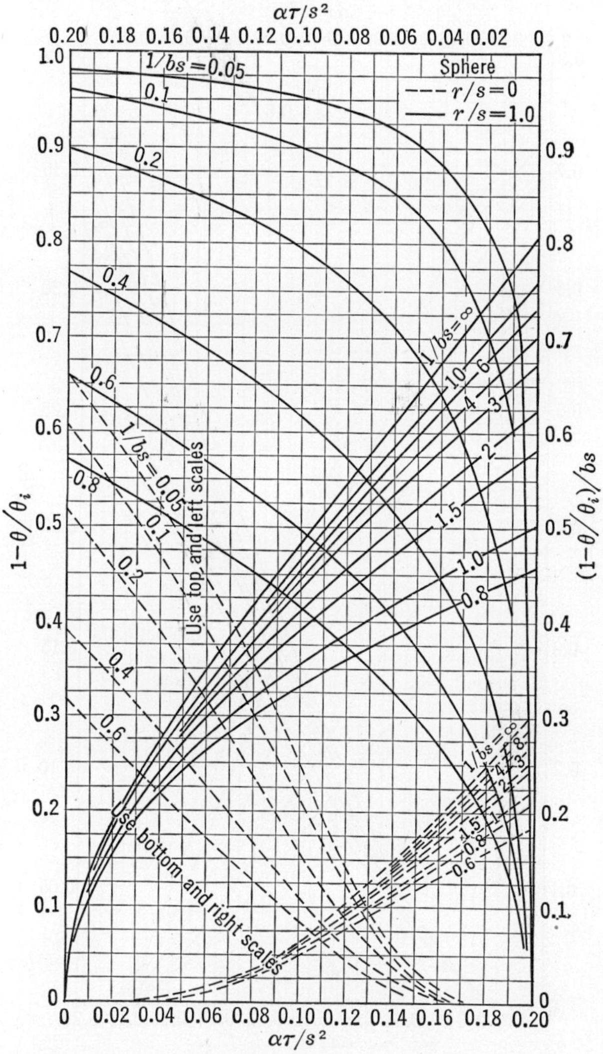

Fig. 20–12. Temperature change on the surface and in the center of sphere after sudden change of environment temperature, for $\alpha\tau/s^2 < 0.20$.

Courtesy Dr. M. P. Heisler [1947].

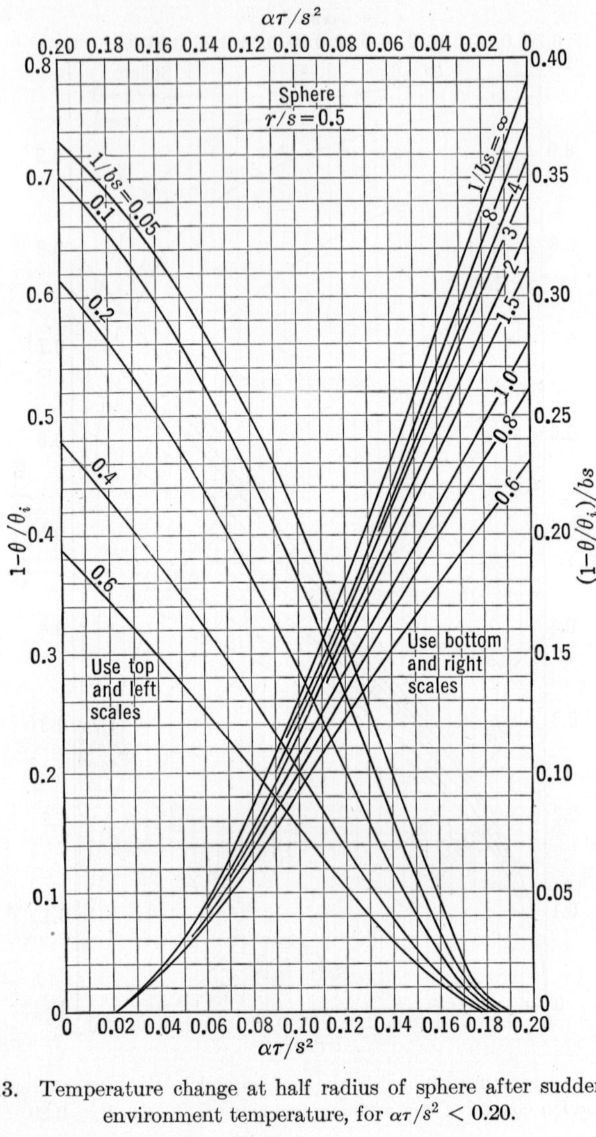

FIG. 20–13. Temperature change at half radius of sphere after sudden change of environment temperature, for $\alpha\tau/s^2 < 0.20$.

Courtesy Dr. M. P. Heisler [1947].

PART D. HEAT CONVECTION WITHOUT CHANGE OF PHASE OR CONSTITUTION

CHAPTER 21

Some General Features of Heat Convection

21–1. The Different Kinds of Flow and Convection

Flow and temperatures are the main characteristics of heat convection. Experience has shown that two different patterns of flow occur. Considering, for instance, the flow of a fluid through a straight tube, it is known that, at sufficiently small velocity, the fluid moves in axial direction only. In this and related cases one speaks of laminar or streamline flow. Under certain conditions, in particular, when the velocity is increased and reaches an amount fixed by the kinematic viscosity of the fluid and the configuration of the walls, mixing of the fluid layers occurs, because of an oscillatory motion of the particles normal to the main flow direction. This kind of flow is called turbulent flow. Except for a transitory stage, the flow is well defined, although the details of its mechanism are not fully recognized as yet.

The heat transfer under laminar flow conditions is mainly due to the thermal conduction, i.e., to the molecular heat exchange between the layers of the fluid. The heat transfer in turbulence, on the other hand, is mainly due to the oscillatory motions of super-molecular portions of the fluid; heat energy is carried to and fro by such particles, and the thermal conductivity plays a secondary role only in this mechanism. It is not surprising that the heat transfer is better, often of a higher order of magnitude than that in streamline flow. One of the dominating methods used is to split the field of heat transfer into a turbulent section, which is supposed to exist in the bulk of the fluid, and a streamline layer close to the walls.

Referring to the driving forces of convection, it is usual to distinguish between two groups, natural or free convection, and artificial or forced convection. Free convection is caused by differences in the specific weight of a substance, due to differences in temperature. The lighter parts which generally are those of higher temperature move upwards;

the heavier ones fall down under the action of gravity. Forced convection occurs if pressure differences are applied, as in the flow through a tube. In both cases laminar or turbulent motion may occur.

21-2. The Mean Velocity and Temperature of a Fluid

Some peculiarities and difficulties occurring in heat transfer by convection have been mentioned already in Sects. 3-1 and 3.

One of the most intriguing is the definition of the temperature of the fluid to or from which heat is transferred. Theory and experience made it evident that at the very surface the temperature of a fluid is not different from the surface temperature, except for gases at rather low vacuum pressure. The temperature at considerable distance from the surface is generally used as fluid temperature; however, the temperature of the fluid at the surface or the mean of these temperatures is also often introduced in empirical equations.

In the case of pipes or channels, the temperature in the center line would satisfy best the requirement of considerable distance from the wall. However, a mean temperature is more characteristic for the storage and convection of heat. Generally the so-called cup-mixing temperature is used, i.e., the temperature which any quantity of the fluid, passing the cross-sectional area of the pipe or channel in a time interval, would assume if it were collected and perfectly mixed in a cup.

Often the heat picked up or given up by the fluid between two cross-sectional areas is needed; in this case no others than the mixing temperatures can be used.

The procedure to find this average is similar to that of determining the average for an arbitrary velocity distribution over the cross section.

In Part a of Fig. 21-1 the curve v shows the velocity distribution over the diameter of a tube, which may be given from a traverse measurement with Pitot tubes. Flow in axial direction only and symmetry around the axis are assumed. The time rate of volume displacement, $\dot{V}$, is found by multiplying the cross-sectional area of an annulus of inner radius r and width dr with v and integrating in the limits $r = 0$ and $r = s$, where $2s$ is the inner diameter of the tube. Hence

$$\dot{V} = \int_{r=0}^{r=s} v \cdot 2\pi r \cdot dr \tag{21-1}$$

If v is given as a function of r, this may be found analytically. Otherwise, two graphical methods can be used. Either $(v \cdot r)$ may be plotted as ordinate over r as abscissa (Part b of Fig. 21-1), or v may be plotted

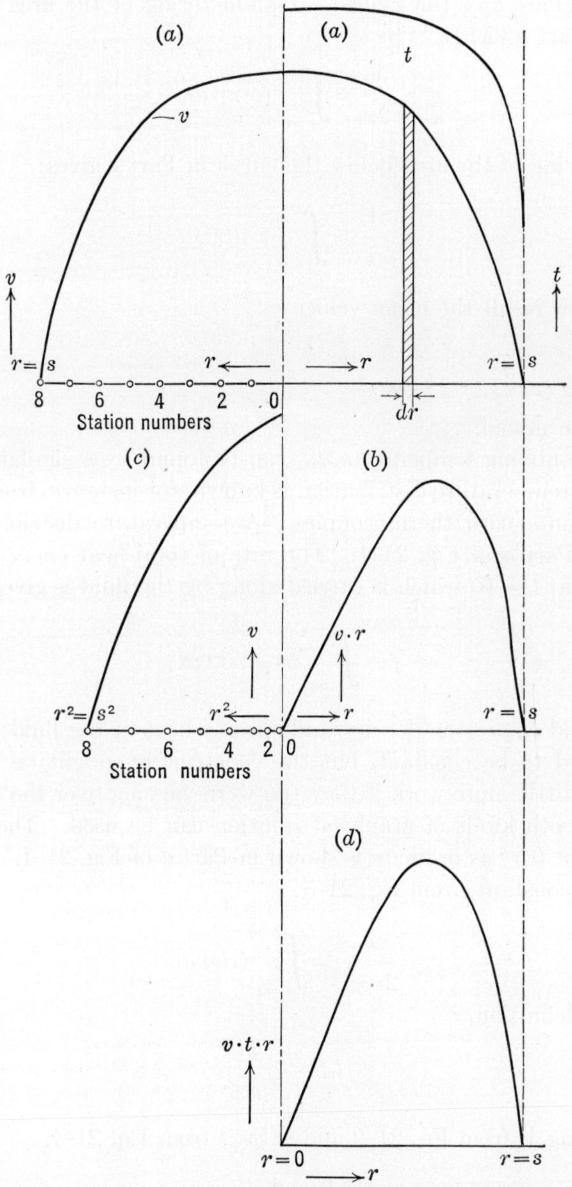

FIG. 21–1. Procedures to find the means of velocities and temperatures from curves of their distribution.

versus r^2 (Part *c* of the figure). Planimetering of the area below the curve in Part *b* yields

$$\frac{\dot{V}}{2\pi} = \int_{r=0}^{r=s} (v \cdot r) \cdot dr$$

Planimetering of the area below the curve in Part *c* gives:

$$\frac{\dot{V}}{\pi} = \int_{r=0}^{r=s} v \cdot d(r^2)$$

From either result the mean velocity

$$v_m = \frac{\dot{V}}{\pi s^2} \tag{21-2}$$

can be determined.

The cup-mixing temperature, t_m, can be found by a similar construction if the temperature distribution is known, for instance, from traverse measurements with thermocouples. A temperature distribution, t, is shown in Part *a* of Fig. 21–1. The rate of total heat energy (counted from zero at $t = 0$) which is carried along by the fluid is given by

$$q = \int_{r=0}^{r=s} v\rho c_p t \cdot 2\pi r \cdot dr \tag{21-3}$$

where ρ and c_p are the density and specific heat of the fluid; these will be assumed to be constant, but the construction might be performed with only little more work if they too were varying over the radius.

Again, both kinds of graphical solution can be used. The one with the product (vtr) as ordinate is shown in Part *d* of Fig. 21–1. Considering ρc_p as constant, from Eq. 21–3:

$$\frac{q}{2\pi\rho c_p} = \int_{r=0}^{r=s} (vtr)\, dr \tag{21-4}$$

Now, by definition,

$$t_m = \frac{q}{\dot{V}\rho c_p} \tag{21-5}$$

Substituting $\dot{V}$ from Eq. 21–2 and $q/(\rho c_p)$ from Eq. 21–4,

$$t_m = \frac{2}{s^2 v_m} \int_{r=0}^{r=s} (vtr)\, dr \tag{21-6}$$

The integral is found by planimetering the area below the curve in Part *d* of Fig. 21–1.

Sometimes, however, the mean has to be taken over the cross-sectional area A. This mean can be represented as follows:

$$t_{m,A} = \frac{1}{\pi s^2} \int_{r=0}^{r=s} t \cdot 2\pi r \cdot dr$$

$$= \frac{2}{s^2} \int_0^s tr \cdot dr \tag{21–6a}$$

21–3. The Mean Temperature Difference between Two Fluids Separated by a Wall

Another question is: What average temperature difference should be taken if a fluid receives heat energy from a surface of uniform temperature when flowing over this surface? A more general case will be dealt with which includes the answer to this question.

Two fluids are assumed to flow on both sides of a separating wall. If the thermal resistance of the wall is negligible and the time rates of mass, entrance, and exit temperatures of the two fluids are given, it is proposed to find what may suitably be considered the mean temperature difference between them.

Considering parallel flow and denoting by y the area covered by the flow and by t_{I} and t_{II} the mean temperatures of the warmer and colder

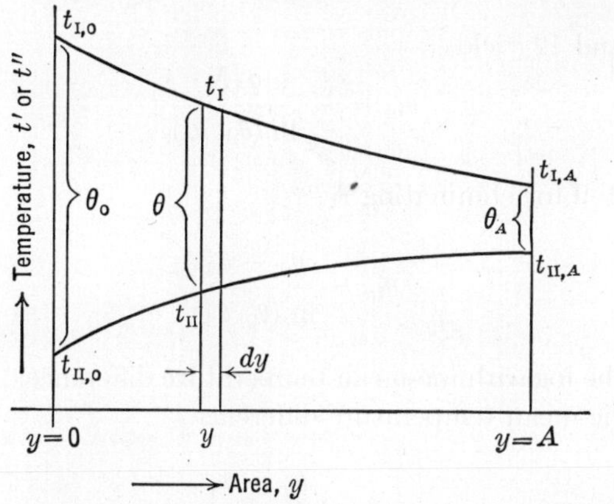

FIG. 21–2. Temperature distributions in two fluids flowing parallel on the two sides of a wall.

fluid, respectively (Fig. 21–2), then for an element dy of the surface, according to Eq. 3–4,

$$dq = U(t_{\text{I}} - t_{\text{II}})\, dy \tag{21–7}$$

Denoting the rates of mass flow of the two fluids by $\dot{m}_{\mathrm{I}}$ and $\dot{m}_{\mathrm{II}}$, respectively, and their specific heats by c_{I} and c_{II}, respectively,

$$dq = -\dot{m}_{\mathrm{I}} c_{\mathrm{I}} \cdot dt_{\mathrm{I}} = \dot{m}_{\mathrm{II}} \cdot c_{\mathrm{II}} \cdot dt_{\mathrm{II}} \tag{21–8}$$

or

$$d(t_{\mathrm{I}} - t_{\mathrm{II}}) = -dq \cdot N \tag{21–9}$$

where

$$N = \frac{1}{\dot{m}_{\mathrm{I}} c_{\mathrm{I}}} + \frac{1}{\dot{m}_{\mathrm{II}} c_{\mathrm{II}}} \tag{21–10}$$

From Eqs. 21–7 and 9, introducing $\theta = t_{\mathrm{I}} - t_{\mathrm{II}}$,

$$\frac{d\theta}{\theta} = -UN \cdot dy$$

and by integration from $y = 0$ to $y = A$

$$\frac{1}{N} \ln \frac{\theta_A}{\theta_0} = -UA \tag{21–11}$$

Defining the mean temperature difference θ_m by

$$\int_{y=0}^{y=A} dq = q_A = UA\theta_{\ln} \tag{21–12}$$

Eqs. 21–11 and 12 yield

$$\theta_{\ln} = -\frac{q_A N}{\ln (\theta_A/\theta_0)}$$

Using Eq. 21–9 for eliminating N,

$$\theta_{\ln} = \frac{\theta_0 - \theta_A}{\ln (\theta_0/\theta_A)} \tag{21–13}$$

$\theta_{\ln}$ is called the logarithmic-mean temperature difference. It is seen that the arithmetic-mean temperature difference

$$\theta_a = \frac{\theta_0 + \theta_A}{2} \tag{21–14}$$

cannot, generally, be used instead of $\theta_{\ln}$ in Eq. 21–12. However, from Eq. 21–13,

$$\theta_{\ln} = \theta_0 \frac{1 - (\theta_A/\theta_0)}{\ln (\theta_0/\theta_A)}$$

Substituting

$$\ln\frac{\theta_0}{\theta_A} = \ln\frac{1+\dfrac{\theta_0-\theta_A}{\theta_0+\theta_A}}{1-\dfrac{\theta_0-\theta_A}{\theta_0+\theta_A}}$$

$$= 2\left[\frac{\theta_0-\theta_A}{\theta_0+\theta_A}+\frac{1}{3}\left(\frac{\theta_0-\theta_A}{\theta_0+\theta_A}\right)^3+\frac{1}{5}\left(\frac{\theta_0-\theta_A}{\theta_0+\theta_A}\right)^5+\cdots\right]$$

the following equation is obtained:

$$\theta_{\ln} = \frac{\dfrac{\theta_0+\theta_A}{2}}{1+\dfrac{1}{3}\left(\dfrac{\theta_0-\theta_A}{\theta_0+\theta_A}\right)^2+\dfrac{1}{5}\left(\dfrac{\theta_0-\theta_A}{\theta_0+\theta_A}\right)^4+\cdots} \tag{21–15}$$

This equation shows the relationship between the logarithmic and arithmetic mean. If $(\theta_0 - \theta_A)/(\theta_0 + \theta_A)$ approaches zero, the two means approach each other.

If $\theta_0/\theta_A = 2$, that is, $(\theta_0 - \theta_A)/(\theta_0 + \theta_A) = 1/3$, the denominator of Eq. 21–15 becomes 1.04. Hence, for $\theta_0/\theta_A \lesseqgtr 2$, the arithmetic mean differs from the logarithmic mean by less than 4%.

In the case of counterflow of the two fluids the last term of Eq. 21–8, as well as that of Eq. 21–10, has to be taken with a negative sign, because both t_{I} and t_{II} decrease in the flow direction of the warmer fluid.

Calling the temperature differences on the two ends of the separating wall θ_{max} and θ_{min}, where $\theta_{max} > \theta_{min}$, the following equation holds for parallel and counterflow:

$$\theta_{\ln} = \frac{\theta_{max} - \theta_{min}}{\ln\,(\theta_{max}/\theta_{min})} \tag{21–16}$$

This equation holds also for the special case defined by the first sentence of this section. It is as though the coefficient of heat transfer h_{I} on the side of the warmer fluid and $m_{\text{I}}c_{\text{I}}$ were infinitely great. Then t_{I} would equal the temperature of the surface in contact with the colder fluid, and Eq. 21–7 would simplify to

$$dq = h(t_{\text{I}} - t_{\text{II}})\,dy$$

where h is the coefficient of transfer between the surface and the colder fluid, in fact, the only fluid actually occurring in this case.

In the derivation of Eq. 21–16 it was assumed that U is constant. For linear variation of U with θ, Colburn [1933] has derived the following formula:

$$q = A \frac{U_2\theta_1 - U_1\theta_2}{\ln (U_2\theta_1/U_1\theta_2)} \tag{21–17}$$

where the subscripts 1 and 2 refer to the two ends of the separating wall. Again, for the simpler case of one surface at t_{I} and one fluid at t_{II}, the same equation may be used, replacing U_1 and U_2 by h_1 and h_2, respectively.

21–4. Introduction to the Methods of Similarity

a. Three Kinds of Approach

Straight solutions of the differential equations, presented in Sects. 3–4, 6, and 7, have been found only in a few cases and only under certain simplifications of the considered problems. However, by employing the principle of similarity to these equations, a multitude of important results and much insight into the complex mechanisms of convection processes have been obtained. The greater part of the chapters dealing with convection will be devoted to this principle and to that procedure of its application which is known as dimensional analysis.

In Sect. 17–2 the principle of similarity was described as applied to heat conduction. In the present section the different methods of procedure will be briefly characterized and illustrated by some cases of flow of fluids which play a role in convection. The application of the principle to heat convection itself will be demonstrated in Chapters 23 and 24.

The concept of general similarity has its origin in the geometric similarity and may, therefore, be explained by the most simple example from geometry. Two triangles, characterized by the subscripts I and II, are called similar if correspondingly situated sides have the same length ratio $a_{\text{II}}/a_{\text{I}}$. An application would be to find, from the given three sides a_{I}, b_{I}, and c_{I} of one triangle, the lengths b_n and c_n of the second and third side of many similar triangles, characterized by the numbers n = II, III, IV $\cdots$ if the lengths a_n of their first side are given. Instead of separately measuring each of these second and third sides, as a person without knowledge of geometry might do, it is sufficient and much simpler to calculate the lengths b_n and c_n by multiplying b_{I} and c_{I} by the ratios a_n/a_{I}.

The conversion of this procedure to dynamics is due to Newton. The first who applied it to heat transfer was Nusselt. In a brief analysis of the different kinds of procedures which can and have been used, Zahm

[1928] distinguishes between the integral, differential, and dimensional methods which are attributed to Newton, Stokes (or Helmholtz), and Rayleigh, respectively.

Newton defined dynamically similar systems as those satisfying the following three equations:

$$\frac{s_{II}}{s_I} = a; \qquad \frac{\tau_{II}}{\tau_I} = b; \qquad \frac{m_{II}}{m_I} = c$$

where s, τ, and m stand for any length, time interval, and mass occurring in the two systems I and II, and a, b, and c are factors of proportionality.

Instead of $m_{II}/m_I = c$ one may use the ratio of corresponding forces

$$\frac{F_{II}}{F_I} = \frac{m_{II}s_{II}/\tau_{II}^2}{m_I s_I/\tau_I^2} = \frac{ac}{b^2}$$

Stokes started from the differential equations of flow and the condition that all kinds of forces involved must have the same factors of proportionality in the two compared systems. The procedure is similar to that described in Sect. 17–2. It will be demonstrated in this section by a basic example.

In Rayleigh's method, finally, it is assumed that the relations between the different physical quantities can be expressed in the form of sums of power functions. Generally only one term of the sum is used. For instance:

$$Q_r = CQ_1^{\alpha}Q_2^{\beta}Q_3^{\gamma}\cdots \tag{21–18}$$

where C is a dimensionless constant, Q_1, Q_2, Q_3 $\cdots$ are the quantities upon which a required quantity Q_r is supposed to depend, and α, β, γ $\cdots$ are constant exponents. The method consists of converting this equation in an equation of dimensionless groups G_A, G_B, G_C $\cdots$ of the form

$$G_R = KG_A^{a}G_B^{b}C_C^{c}\cdots \tag{21–19}$$

where K is a constant and a, b, c $\cdots$ are constant exponents. If these constants are found by experiments for one set of groups G_A, G_B, G_C $\cdots$ they are supposed to be valid for any other sets with arbitrary values G_A', G_B', G_C' $\cdots$ for corresponding cases. The procedure which is known as dimensional analysis will also be demonstrated on a particular case.

b. Conditions for Similarity of Flow

As an example of the use of the differential method the conditions of similarity of flow are to be determined. In general, the answer is that three kinds of similarity are necessary, namely, geometrical, kinematic, and dynamic similarity.

Geometrical similarity means that the surfaces in contact with the fluid are to be geometrically similar. Infinitely long tubes, for instance, having perfectly smooth inner surfaces satisfy this condition for inside flow. In the present example the flow of fluids inside such tubes will be considered.

Kinematic similarity exists if all combinations of products and ratios of lengths and time intervals which occur in the Navier-Stokes equations have the same proportionality factors in the systems which are supposed to be similar.

For flow in the horizontal direction, x, for instance, Eqs. 3–14, 19, and 20 lead to

$$\rho\left(\frac{\partial v_x}{\partial \tau} + v_x \frac{\partial v_x}{\partial x}\right) = -\frac{\partial p}{\partial x} + \frac{4}{3}\mu \frac{\partial^2 v_x}{\partial x^2} \tag{21–20}$$

If D is the diameter of the tube and v_m the mean velocity, then kinematic similarity requires that in all similar systems

$$v_x \frac{\partial v_x}{\partial x} = C_1 \frac{{v_m}^2}{D} \tag{21–21}$$

and

$$\frac{\partial^2 v_x}{\partial x^2} = C_2 \frac{v_m}{D^2} \tag{21–22}$$

where C_1 and C_2 are dimensionless constants found for *one* of these systems.

Dynamic similarity exists if the ratio of the different forces (inertia forces and frictional forces in the present case) is the same in all systems under comparison. Therefore

$$\rho v_x \frac{\partial v_x}{\partial x} = C_3 \mu \frac{\partial^2 v_x}{\partial x^2} \tag{21–23}$$

where C_3 is a dimensionless constant, found for *one* of these systems.

By substitution from Eqs. 21–21 and 22,

$$\frac{v_m D}{\mu/\rho} = \frac{v_m D}{\nu} = \frac{C_2 C_3}{C_1} = C \tag{21–24}$$

Hence the flow in tubes is similar if the dimensionless group $v_m D/\nu$ has a constant value.* In honor to Osborne Reynolds, who was the

* A similar, though more elementary, derivation will be found in an excellent paper on the Reynolds number by Bakhmeteff [1936(*a*)].

first to apply the theory of similarity to the flow of viscous fluids (in 1883), this group is called Reynolds number. It will be denoted by the symbol N_{Re}.

Reynolds was also the first to state that under certain conditions the flow in a tube suddenly changes from the streamline form to turbulence; he further concluded, from the theory of similarity, that this will always occur at the same value of N_{Re}, and he found that critical value, $(N_{Re})_{cr}$, by experiments; the most reliable value seems to be 2300.

c. Pressure Drop in Smooth Tubes

It is self-explanatory that the term $\partial_p/\partial x$ in Eq. 21–20 must stand in a constant ratio to $\rho v_x(\partial v_x/\partial x)$ for similar systems. However, the case of the pressure drop in a tube will now be used to explain and illustrate the procedure in the third similarity method, which has been mentioned above and called dimensional analysis.

Starting from Eq. 21–18, the first step is to state or guess upon which quantities, Q_1, Q_2, Q_3 $\cdots$ the required quantity, Q_r, is depending. The required quantity here is the pressure drop per unit length of the tube, $\Delta p/L$, assuming that the pressure drop is proportional to the length. The quantities on which $\Delta p/L$ is supposed to depend are the inner diameter D of the tube, the velocity v_m, the density ρ, and the dynamic viscosity μ of the fluid. Then Eq. 21–18 takes the form

$$\frac{\Delta p}{L} = CD^\alpha v_m{}^\beta \rho^\gamma \mu^\delta \tag{21–25}$$

Now, it is known that in any physical equation the sum of the exponents of the basic units must be the same on the right and left side. In order to find relationships between the exponents a case may be assumed where all symbols, except the exponents, are equal to unity. Operating in a system with unit mass as one of the three basic units, only the physical dimensions Λ (for length), T (for time), and M (for mass) would be left in Eq. 21–25. Hence

$$\frac{\mathrm{M}}{\Lambda^2\mathrm{T}^2} = \Lambda^\alpha \cdot \frac{\Lambda^\beta}{\mathrm{T}^\beta} \cdot \frac{\mathrm{M}^\gamma}{\Lambda^{3\gamma}} \cdot \frac{\mathrm{M}^\delta}{\Lambda^\delta \mathrm{T}^\delta} \tag{21–26}$$

Now, equating on both sides, the sum of the exponents

$$\begin{aligned} &\text{of } \Lambda: \quad -2 = \alpha + \beta - 3\gamma - \delta \\ &\text{of T:} \quad -2 = \qquad -\beta \qquad\quad - \delta \\ &\text{of M:} \quad\;\;\; 1 = \qquad\qquad\quad \gamma + \delta \end{aligned}$$

From these three equations by algebra,

$$\beta = \alpha + 3$$

$$\gamma = \alpha + 2$$

$$\delta = -\alpha - 1$$

By substitution in Eq. 21–25 and combination of quantities having the same exponent into dimensionless groups, an equation of the form of Eq. 21–19 is obtained:

$$\frac{\Delta p}{\rho v_m{}^2/2} = 2C\frac{L}{D}\left(\frac{v_m D\rho}{\mu}\right)^{\alpha+1} \qquad (21\text{–}27)$$

This is the general equation of the pressure drop in a smooth circular tube whether the flow is laminar or turbulent.

By comparison with Eq. 21–26 it is seen that only one of the four exponents is needed. The reason is that each of the basic dimensions Λ, T, and M yields one equation of determination so that three unknowns can be eliminated. Since in general these exponents have to be determined by experiments, the use of dimensional analysis means a considerable simplification and reduction of work.

It is further seen that the required quantity, which in dimensionless form is expressed as the ratio of the pressure drop to the kinetic energy of unit volume, depends only on the length ratio L/D and on the Reynolds number.

For streamline flow the unknown α can be found by a balance of forces, namely,

$$\left(\frac{\pi}{4}D^2\right)\Delta p = -(\pi D)L\mu\left(\frac{dv}{dr}\right)_{r=D/2} \qquad (21\text{–}28)$$

where r is the radial distance from the center line, and $\mu(dv/dr)_{r=D/2}$ is the frictional shearing stress on the inner surface of the tube (see Eq. 3–21). The minus sign is necessary if Δp is wanted as a positive quantity.

According to Eq. 21–28, Δp is proportional to μ. This is compatible with Eq. 21–27 only if $\alpha = -2$. Hence

$$\Delta p = 2C\frac{L}{D}\cdot\frac{\mu}{v_m D\rho}\cdot\frac{\rho v_m{}^2}{2} \qquad (21\text{–}29)$$

Determining one or the other exponent by using some special physical facts, as has been done by applying Eq. 21–28, is a procedure characteristic for dimensional analysis which will be used also in subsequent chapters.

For laminar flow, it is also possible to find the constant C from physical considerations. It can be assumed that a balance of forces, similar to Eq. 21–28, exists at any radius r:

$$(\pi r^2)\,\Delta p = -(2\pi r)L\mu\frac{dv}{dr} \tag{21–30}$$

By integration within the limits $r = 0$ and $r = r$,

$$v = v_0 - \frac{\Delta p}{4\mu L}r^2 \tag{21–31}$$

where v_0 is the velocity in the center line ($r = 0$).

Integrating further within the limits $r = 0$ and $r = s = D/2$,

$$v_0 = \frac{\Delta p}{4\mu L}s^2 \tag{21–32}$$

From Eqs. 21–31 and 32,

$$v = v_0\left(1 - \frac{r^2}{s^2}\right) \tag{21–33}$$

Defining the mean velocity by

$$v_m = \frac{1}{\pi s^2}\int_{r=0}^{r=s} v\cdot 2\pi r\cdot dr \tag{21–34}$$

and substituting v from Eq. 21–33:

$$v_m = \frac{v_0}{2} \tag{21–35}$$

Finally, substituting v_0 from Eq. 21–35 in Eq. 21–32 and replacing $2s$ by D,

$$\frac{\Delta p}{\rho v_m{}^2/2} = 64\frac{L}{D}\cdot\frac{\mu}{v_m D\rho} = \frac{64}{N_{Re}}\cdot\frac{L}{D} \tag{21–36}$$

This is Poiseuille's law, and by comparison with Eq. 21–29 it is seen that $C = 32$.

In general, the exponents as well as the constant C have to be determined by direct experiments. For turbulent flow in a smooth tube, for instance, $\alpha = -\frac{5}{4}$ and $C = 0.3164$ have been found to hold for Reynolds numbers up to about 200,000. Substituting these in Eq. 21–27,

$$\frac{\Delta p}{\rho v_m{}^2/2} = \frac{0.3164}{\sqrt[4]{N_{Re}}}\cdot\frac{L}{D} \tag{21–37}$$

The dimensionless equations 21–36 and 37 are very useful in dealing with forced heat convection. Here, however, they served only to demonstrate the procedure of dimensional analysis.

It should be understood that the first prerequisite for successful use of dimensional analysis is to know what quantities are involved in a case in order to set up a right starting equation, as Eq. 21–25 in the present case. Otherwise entirely wrong results may be obtained. The quantities involved may be known from intuition and experience; however, starting from a well-founded differential equation, using Stokes's method, is the safest procedure. In the present case, Stokes's differential equations show that the quantities used in Eq. 21–25 are the right ones. Anyway, caution is necessary if one operates with dimensional analysis, instead of starting from the differential equations.

d. Pressure Drop in Rough Tubes

How important it is to include all influential variables in a set-up for dimensional analysis, like Eq. 21–25, becomes evident by comparison of the pressure drop in rough tubes with the corresponding result for smooth tubes. A general form of the pressure-drop equation is

$$\frac{\Delta p}{\rho v_m^2/2} = \phi \frac{L}{D} \tag{21–38}$$

where ϕ is called the coefficient of tube resistance or friction.

A graph of ϕ for smooth and one kind each of rough and corrugated tubes is given in Fig. 21–3. The transition from laminar to turbulent

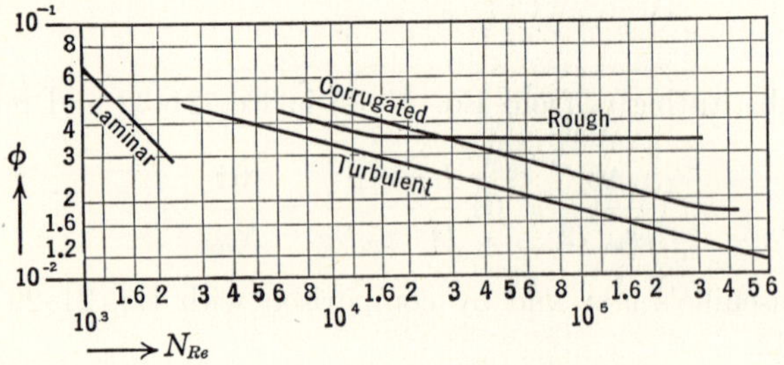

FIG. 21–3. Friction coefficient ϕ for smooth, rough, and corrugated tubes. From Jakob [1933*a*].

flow requires a jump. For smooth tubes, according to Eqs. 21–36 and 37,

$$\phi = \frac{64}{N_{Re}} \text{ for laminar flow}$$

and

$$\phi = \frac{0.3164}{\sqrt[4]{N_{Re}}} \text{ for turbulent flow}$$

when

$$N_{Re} < 200{,}000$$

With rough tubes the following behavior has been revealed by experience:

1. The critical Reynolds number, $(N_{Re})_{cr}$, is about the same as for smooth tubes.

2. At the same Reynolds number the pressure drop is greater with rough tubes.

3. The coefficient ϕ cannot be expressed by a simple power function of N_{Re}. With increasing Reynolds number, ϕ approaches a constant value. Hence Δp eventually becomes proportional to v_m^2, instead of being throughout proportional to $v_m^{1.75}$. On the other hand, another dimensionless term, d/D, enters the equation, where d is a length dimension, characteristic for the roughness elements. If, for instance, the roughness elements are small hemispherical extensions of the surface, d may be their diameters, or if the elements are half-tetrahedrons, d may be the diagonal or side length or ratio of volume and surface area, etc. Nikuradse [1933] obtained well-defined roughnesses in a tube by thoroughly sieved sand grains of average diameter d which, by means of a special technique, were bedded in a very thin layer of lacquer on the inner tube surface. The ratio d/s was used as a measure of the roughness and varied from $\frac{1}{15}$ to $\frac{1}{507}$. The experiments showed that for that sort of rough tubes

$$\phi = \frac{1}{[1.74 - 2 \log (d/s)]^2} \tag{21–39}$$

It may be mentioned that Jakob [1940], using Nikuradse's data, found that in a smooth tube the local velocity at a distance from the wall $y_m = 0.24s$ equals the mean velocity. This is valid in the range of $N_{Re} = 3000$ to 4,500,000.

Introducing the dimensionless independent variable

$$\xi = \sqrt{\frac{\phi}{2}} \cdot \frac{2sv_m}{\nu} \tag{21–40}$$

$\xi = 1$ virtually represents perfect smoothness, and $\xi > 100$, perfect roughness.

For $\xi = 1$, y_m decreased from $0.24s$ to $0.23s$, for $\xi > 100$, y_m decreased from $0.23s$ to $0.22s$, if d/s decreased from $\frac{1}{15}$ to $\frac{1}{507}$.

Hence, for both smooth and rough tubes $y_m = 0.23s$ is a fair value for that distance from the surface at which $v = v_m$. This is in line with the fact, mentioned above, that $(N_{Re})_{cr}$ is virtually independent of the roughness.

21-5. The Boundary-layer Concept

From the fact that the pressure drop in rough tubes eventually becomes proportional to the velocity square it can be concluded that the effect of roughness is to stop completely the moving particles which fluctuate between the core of the fluid and a layer close to the wall. This interaction apparently plays an essential role concerning distribution of velocity and pressure drop. Whereas in the laminar flow in a tube the velocity distribution is parabolic, turbulence has the effect of mixing the fluid so well that the velocity decreases only slightly between the center line and places close to the tube surface, as would be the case in a frictionless fluid. However, in a boundary layer on the surface the velocity is bound to decrease steeply to the value zero, which it assumes at the surface. Such velocity slope is not compatible with considerable cross movements because they would equalize the velocity. Hence the fluid is streaming in axial direction without sensible flow oscillations by turbulence; i.e., the flow close to the wall is laminar and shows the characteristic velocity slope caused by viscosity.

Having clearly recognized this dual behavior of the fluid in the core and on the wall, Prandtl [1905] simplified it by the following idealization. He considered the actually viscous fluid as free from friction in the core, though turbulent under the action of vortices coming from the wall, and as viscous only in a thin boundary layer on the surface.

Like the theory of similarity, the boundary-layer concept is used as a substitute for the difficult or impossible solution of the Navier-Stokes equations. It has been proved to be one of the most efficient tools in fluid mechanics, as well as in heat transfer and related fields. In this section, only a case of isothermal flow will be treated to illustrate the use of this principle. Referring to Fig. 21-4, which schematically represents the distribution of the velocity v over a tube diameter, it is seen that the velocity increases almost proportionally to y up to v_b at the distance from the wall $y = b$. Prandtl [1910], in one of his first papers on this subject, assumed $y \leqq b$ to be the region of the boundary layer and $y \geqq b$ the region of the turbulent core. Actually, as he stated in that paper, there is a transition region in between where viscosity and convection are of similar influence. However, he neglected this field of transition to simplify the mathematical treatment.

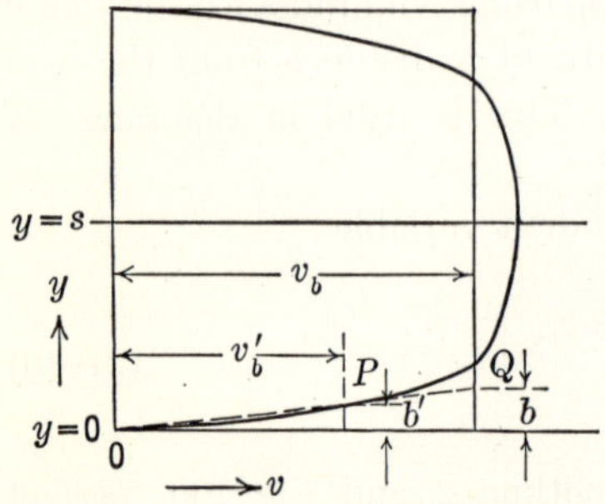

FIG. 21-4. Distribution of velocity with indication of boundary layer.

A balance of the forces acting in the boundary layer is given as

$$-\pi r^2 \frac{dp}{dx} = \mu 2\pi r \frac{dv}{dy}$$

Substituting

$$\frac{dv}{dy} = \frac{v_b'}{b'}$$

(see Fig. 21–4), the following relation is obtained:

$$\frac{dp}{dx} = -\frac{2\mu v_b'}{b'r}$$

But, according to Eq. 21–37,

$$\frac{dp}{dx} = -\phi \frac{1}{2s} \rho \frac{v_b^2}{2} \tag{21–41}$$

where $\phi = 0.3164/\sqrt[4]{N_{Re}}$.

Equating,

$$b' = \frac{8\mu v_b'}{\phi \rho v_b^2} \tag{21–42}$$

According to Blasius, it is easily found of which order the thickness of the boundary layer is. Extending the straight line $\overline{OP}$ to Q,

$$\frac{b}{b'} = \frac{v_b}{v_b'}$$

Substituting b' from Eq. 21–42,

$$b = \frac{8\mu}{\phi \rho v_b} \tag{21–43}$$

or

$$\frac{b}{2s} = \frac{8}{\phi} \cdot \frac{\nu}{2sv_b} = \frac{8}{\phi} \cdot \frac{1}{N_{Re}} \tag{21–44}$$

Because ϕ depends on N_{Re} only, the ratio $b/2s$, which is characteristic for the thickness of the boundary layer, depends also only on N_{Re}. Substituting $\phi = 0.3164/\sqrt[4]{N_{Re}}$, the following is obtained:

For $N_{Re} =$ 10,000 $b/2s = 1/40$
For $N_{Re} =$ 100,000 $b/2s = 1/225$

Hence the boundary layer is very thin indeed in most practical cases.

21-6. Universal Relations for the Velocity Distribution

In Chapter 24 it will be shown that the temperature distribution across a flowing fluid is either entirely or nearly similar to the velocity distribution so that, if the latter is known, the heat transfer by convection can be readily determined. Therefore it is gratifying and important that very general relations for the velocity distribution across the boundary layer and the core of the flow have been set up.

Prandtl [1921] made the assumption that the velocity distribution close to a surface depends upon the physical properties of the fluid in that region, but is independent of other magnitudes, particularly distance of other surfaces and average or maximum velocity. Thus, the distribution of the velocity v in the direction y perpendicular to the surface would depend on ρ, μ, and the shear stress $S_{s,y=0}$ at the wall, which for simplicity will be termed S. The most practical formulation is obtained if the following two dimensionless magnitudes are used: †

$$v^* = v\sqrt{\frac{\rho}{S}} \tag{21-45}$$

and

$$y^* = \frac{y}{\nu}\sqrt{\frac{S}{\rho}} \tag{21-46}$$

Then, according to Prandtl's postulate, a universal relation should exist, which can be expressed by

$$v^* = \Phi(y^*) \tag{21-47}$$

Nikuradse [1932] verified this by experiments in the range from N_{Re} = 4000 to 3,240,000. Figure 21-5 is his representation of Eq. 21-47 in a semi-logarithmic system of coordinates.

The range of experiments begins with $y^* = 10$. At low values the velocity distribution is represented by the curve A. This curve meets the two straight lines B and C at $y^* \approx 150$. Line B is drawn with particular consideration of the points close to the tube axis, line C giving preference to the points close to the surface. Nikuradse represented these lines by the following equations:

$$v^* = 5.5 + 5.75 \log_{10} y^* \quad \text{for } B \tag{21-48}$$

and

$$v^* = 5.84 + 5.52 \log_{10} y^* \quad \text{for } C \tag{21-49}$$

† The quantity $\sqrt{S/\rho}$ is the so-called friction velocity, and y^* can be considered a Reynolds number. See Bakhmeteff [1936, 1936(*a*)], whose nomenclature, however, is different.

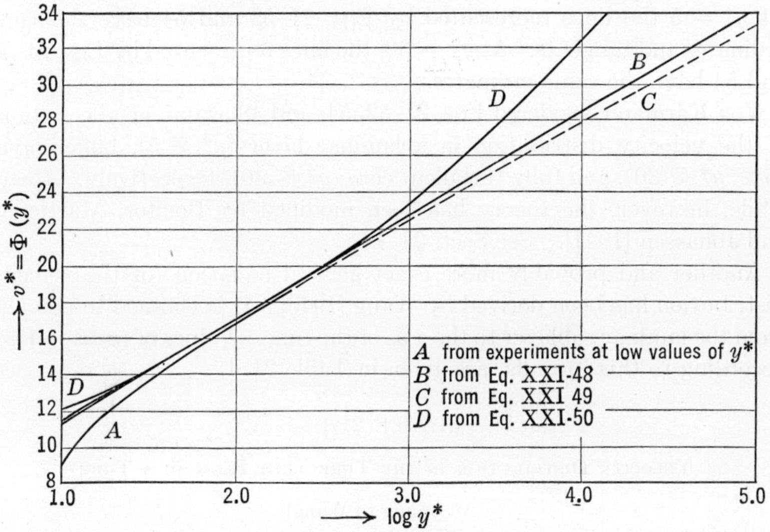

FIG. 21–5. v^* versus y^*, defined by Eqs. 21–45 and 46, respectively. (Dimensionless representation of velocity distribution close to a surface.)

The range of Blasius' equation and von Kármán's seventh root relation is represented by the curve D and the equation

$$v^* = 8.74(y^*)^{1/7} \tag{21–50}$$

According to the experiments, it is restricted to the range of $\log y^* \approx 1.6$ to 2.6, that is, $y^* \approx 40$ to 400.

At $y = 0$ or $y^* = 0$

$$S = \mu \frac{dv}{dy} \quad \text{or} \quad \Phi'(y^*) = \frac{dv^*}{dy^*} = 1 \tag{21–51}$$

Von Kármán [1939] assumed that this holds up to $y^* = 5$, where $v^* = 5$ would occur. In this range, then, according to Eq. 21–51,

$$v^* = y^* \tag{21–52}$$

Von Kármán further replaced curve D in the range from $y^* = 5$ to $y^* = 30$ by a straight line, as will be dealt with in Sect. 24–4.

The equations of this line are:

$$\Phi'(y^*) = \frac{dv^*}{dy^*} = \frac{5}{y^*} \tag{21–53}$$

and

$$v^* = 5\left[1 + \ln\left(\frac{y^*}{5}\right)\right] \tag{21–54}$$

At $y^* = 5$ the lines represented by Eqs. 21–52 and 54 have the same ordinates and tangents. At $y^* = 30$, the lines represented by Eqs. 21–48 and 54 have the same ordinates.

Von Kármán considered Eqs. 21–52, 54, and 48 as universal equations of the velocity distribution in a laminar layer ($y^* \lesseqgtr 5$), buffer layer ($5 \lesseqgtr y^* \lesseqgtr 30$), and fully turbulent core ($y^* \gtreqless 30$), respectively. Meanwhile, however, the theory has been modified by Boelter, Martinelli, and Jonassen [1941]. See Sect. 24–4.

Another and probably more exact general equation for the velocity distribution has been derived by Wang [1946]. It is claimed to be valid from the laminar sublayer to the axis of the tube. Values of $(v_0/v - 1)v^*$, according to this formula, are given in Table 21–1.

TABLE 21–1

VELOCITY DISTRIBUTION IN THE TURBULENT FLOW IN A TUBE

(According to Wang)

r/s	$(v_0/v - 1)v^*$	r/s	$(v_0/v - 1)v^*$
0.00	0.00	0.60	2.49
0.02	0.01	0.70	3.34
0.04	0.04	0.80	4.49
0.10	0.15	0.85	5.28
0.20	0.43	0.90	6.37
0.30	0.80	0.93	7.31
0.40	1.26	0.96	8.75
0.50	1.81	0.98	10.50

21–7. The Equivalent Diameter

The cylindric tube is the most convenient duct for any sort of flow, as well as for theoretical treatment. However, in many instances, other than circular tubes must be used for reasons of design and materials. Then all calculations concerning flow and convection become much more complicated and difficult, if possible at all. To evade such difficulties the concept of an equivalent circular tube is used, i.e., a tube which would present the same resistance against flow or would secure the same heat transfer as the duct actually used under equal or comparable conditions. The diameter of an equivalent tube is termed equivalent diameter and may be denoted by D_e.

Considering that dynamic similarity exists if the ratio of inertia forces and frictional forces is the same (see Article b of Sect. 21–4), and that the former are proportional to the cross-sectional area A of a duct and the latter proportional to its circumference (perimeter) C, the equivalent diameter is usually defined by

$$D_e = \frac{4A}{C} \tag{21-55}$$

because for a circular tube of inner diameter D, according to this definition,

$$D_e = \frac{4 \cdot \pi D^2/4}{\pi D} = D$$

This primarily holds when the pressure drop is wanted; however, because of the existing similarity between momentum transfer and heat transfer, which will be dealt with in Chapter 24, it is also used in problems of heat transfer by convection.

For the flow of a liquid in an open channel where no appreciable friction occurs at the free surface, C has to be replaced by the "wetted" part of the circumference, C_w. In the same way a difference is often made between the "heated" and "unheated" part of the circumference in cases where the heat transfer is restricted to a part of the perimeter. However, it has been found in some such cases that to use the wetted part of the circumference, C_w, gives better results than the use of the heated part, C_h, when $C_h \neq C_w$.

TABLE 21-2

EQUIVALENT DIAMETER D_e AND CONSTANT N OF PRESSURE-DROP EQUATION 21-56 FOR DIFFERENT SHAPES OF CROSS SECTION

Shape of cross-sectional area	$2b/2a$	$D_e/2b$	N
Circle (diameter $D = 2b$)		1	64
Ellipse (major axis $2a$, minor axis $2b$)	0.7	1.17	65
	0.5	1.30	68
	0.3	1.44	73
	0.2	1.50	76
	0.1	1.55	78
Rectangle (long side $2a$, short side $2b$)	0.5	1.3	62
	0.333	1.50	69
	0.25	1.60	73
	0.2	1.67	76
	0.1	1.82	85
	0	2.00	96
Equilateral triangle (side length $2b$)		0.58	53
Square (side length $2b$)		1	57
Annular ring (width $2b$)		2	96

Experimentation has shown that for non-circular ducts Eq. 21–36 is to be replaced by:

$$\frac{\Delta p}{\rho v_m{}^2/2} = N\frac{L}{D_e} \cdot \frac{\mu}{\rho D_e v_m} \tag{21–56}$$

where N is an individual constant for each shape of the cross section. For some configurations D_e and N are given in Table 21–2; those of N have been calculated by Davies and White [1929], using an approximative method of Boussinesq [1868]. The constant N is much smaller for square and triangular than for circular cross-sectional area. However, for the same velocity and cross-sectional area the smallest pressure drop occurs with circular tubes. The values of N given in the table have partly been proved by experiments also.

For turbulence, Eq. 21–37 remains unchanged if D is replaced by D_e. This has been proved for various shapes by Nikuradse [1930]. It may be noted that the critical Reynolds number for non-circular cross section was found to be somewhat smaller than for cylindric tubes.

CHAPTER 22

Analytical Solutions of the Convection Equations

22–1. Free Heat Convection

Possibly the first of the few practicable straightforward solutions of the convection equations under steady-state conditions is due to L. Lorenz [1881]. It deals with the heat transfer from a vertical surface at uniform and constant temperature t_s to a colder gas in contact with it when gravity is the only force acting on that gas. Lorenz correctly recognized that the gas close to the surface streams straight upwards, and he assumed that the horizontal streams were negligible and that consequently the pressure was constant in a horizontal plane, whereas it changed from one horizontal layer to the next. This, obviously, is the case of streamline flow on a vertical wall. He further assumed that the environment temperature t_e at an infinite distance from the wall was constant and that the properties of the gas ρ, μ, c_p, and k were independent of the temperature. Setting up a heat balance for a differential section of the flowing fluid and integrating, Lorenz obtained the following equation:

$$h_{c,m} = C \sqrt[4]{\frac{g\rho^2 c_p k^3 \theta_s}{\mu L T_e}} \tag{22–1}$$

where $h_{c,m}$ = the mean coefficient of heat transfer by convection,
L = the height of the surface,
$\theta_s = t_s - t_e$ = its temperature excess above environment,
T_e = the temperature of environment in the absolute temperature scale, and
C = a constant for which Lorenz found the value 0.548.

This equation means a triumph of the classical theory, having revealed for the first time the complex nature of the coefficient of heat transfer by free convection, h, in a form which has been proved valid with good approximation through more than half a century.

One particular consequence of this equation is that the rate of heat exchange, q, is proportional to $\theta_s^{5/4}$. This has been proved in numerous experiments, beginning with a publication by Dulong and Petit [1817], who arrived at an exponent 1.23, instead of 1.25.

The derivation of Eq. 22–1 will not be given, since more exact solutions are presently known. Nusselt [1909, p. 81] directed attention to the fact that, according to experiments with air, the equation becomes less exact at temperature excesses below 20 F and fails entirely when θ_s approaches zero. Avoiding some oversimplifications made by Lorenz, Nusselt and Juerges [1928] succeeded in essentially improving the theory. The most elaborate improvement, however, is due to E. Schmidt and Beckmann [1930], in cooperation with Pohlhausen, who found a way to integrate the differential equations set up by these authors. The experimental part of their work consisted in the determination of the velocities and temperatures close to the surface of electrically heated vertical plates of about 5 and 20 in. height. The temperatures were measured by a thermocouple with wires of 0.015 mm diameter, and the air velocities by a quartz thread, 0.02 mm diameter, 10 to 20 mm long. The deflections of the thread by the influence of the flow, which can be considered a measure for the air velocity, were observed by a microscope; the calibration showed that the deflection is not proportional to the velocity.

At different heights, x, above the lower edge of a plate and at horizontal distances, y, from the plate, the vertical velocity components, v_x, and the temperatures, t, were found to be distributed as indicated in Figs. 22–1 and 2, which refer to the 5-in. plate. Both v_x and t increase with x, and for constant x the velocity has a maximum at some distance from the wall.

In developing a theory of the process, the authors started from the Navier-Stokes equations for the flow directions x and y, the continuity equation for incompressible fluids,* and the Fourier equation for a moving fluid. For simplification the following temperature function was introduced:

$$\phi = \frac{t - t_e}{t_s - t_e} = \frac{\theta}{\theta_s} = \frac{T - T_e}{T_s - T_e} \tag{22–2}$$

Then, from Eqs. 3–14, 18, and 20,

$$v_x \frac{\partial v_x}{\partial x} + v_y \frac{\partial v_x}{\partial y} = \nu \left(\frac{\partial^2 v_x}{\partial x^2} + \frac{\partial^2 v_x}{\partial y^2} \right) + g \frac{T_s - T_e}{T_e} \phi \tag{22–3}$$

$$v_x \frac{\partial v_y}{\partial x} + v_y \frac{\partial v_y}{\partial y} = \nu \left(\frac{\partial^2 v_y}{\partial x^2} + \frac{\partial^2 v_y}{\partial y^2} \right) \tag{22–4}$$

* A gas in free convection can be considered virtually uncompressed.

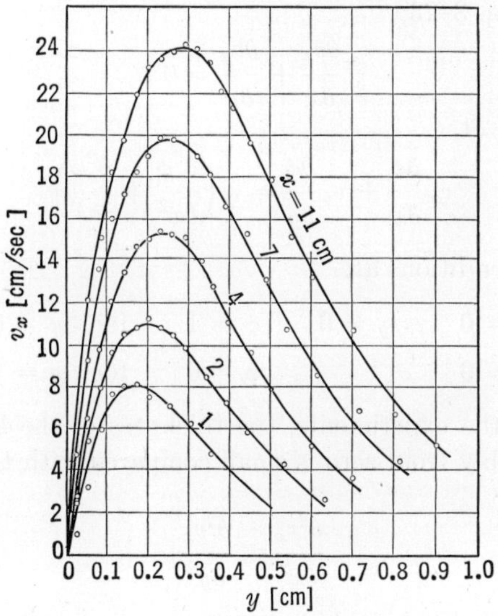

FIG. 22–1. Velocity field for air in free convection on a vertical plate. From E. Schmidt and Beckmann [1930].

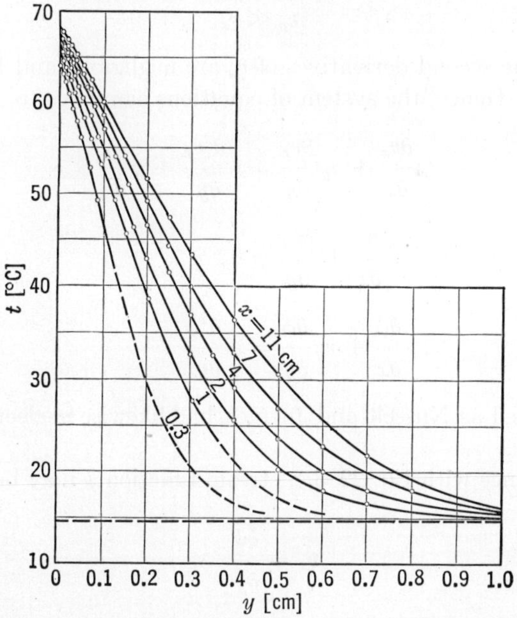

FIG. 22–2. Temperature field for air in free convection on a vertical plate. From E. Schmidt and Beckmann [1930].

Further, from Eq. 3–23,

$$\frac{\partial v_x}{\partial x} + \frac{\partial v_y}{\partial y} = 0 \tag{22-5}$$

and, from Eq. 3–24,

$$v_x \frac{\partial \phi}{\partial x} + v_y \frac{\partial \phi}{\partial y} = \alpha \left(\frac{\partial^2 \phi}{\partial x^2} + \frac{\partial^2 \phi}{\partial y^2} \right) \tag{22-6}$$

The boundary conditions are

$$\left.\begin{array}{llll} v_x = 0 & v_y = 0 & \phi = 1 & \text{for} \quad y = 0 \\ v_x = 0 & & \phi = 0 & \text{for} \quad y = \infty \end{array}\right\} \tag{22-7}$$

According to the experiments, the thickness of the layer in which v and t differ sensibly from zero is small compared with the height of the plate. Hence

$$\frac{\partial^2 v_x}{\partial x^2} \ll \frac{\partial^2 v_x}{\partial y^2}$$

and

$$\frac{\partial^2 \phi}{\partial x^2} \ll \frac{\partial^2 \phi}{\partial y^2}$$

Further, obviously,

$$v_y \ll v_x$$

Therefore, the second derivatives of v_y are negligible, and Eq. 22–4 can be omitted. Hence, the system of equations simplifies to

$$v_x \frac{\partial v_x}{\partial x} + v_y \frac{\partial v_x}{\partial y} = \nu \frac{\partial^2 v_x}{\partial y^2} + g \frac{\theta_s}{T_e} \phi \tag{22-8}$$

$$\frac{\partial v_x}{\partial x} + \frac{\partial v_y}{\partial y} = 0 \tag{22-5}$$

$$v_x \frac{\partial \phi}{\partial x} + v_y \frac{\partial \phi}{\partial y} = \alpha \frac{\partial^2 \phi}{\partial y^2} \tag{22-9}$$

Lorenz, as well as Nusselt and Juerges, had likewise neglected $v_x \cdot \partial v_x / \partial x$ and $v_x \cdot \partial \phi / \partial x$.

In accordance with Eq. 22–5, a stream function ψ may be introduced, defined by

$$v_x = \frac{\partial \psi}{\partial y} \tag{22-10}$$

and

$$v_y = -\frac{\partial \psi}{\partial x} \tag{22-11}$$

By substitution in Eqs. 22–8 and 9,

$$\frac{\partial\psi}{\partial y}\cdot\frac{\partial^2\psi}{\partial x\cdot\partial y}-\frac{\partial\psi}{\partial x}\cdot\frac{\partial^2\psi}{\partial y^2}=\nu\frac{\partial^3\psi}{\partial y^3}+g\frac{\theta_s}{T_e}\phi \tag{22–12}$$

and

$$\frac{\partial\psi}{\partial y}\cdot\frac{\partial\phi}{\partial x}-\frac{\partial\psi}{\partial x}\cdot\frac{\partial\phi}{\partial y}=\alpha\frac{\partial^2\phi}{\partial y^2} \tag{22–13}$$

In order to solve this system of partial differential equations, the mathematician E. Pohlhausen suggested introducing the new variable

$$\xi = c\frac{y}{\sqrt[4]{x}} \tag{22–14}$$

where

$$c=\sqrt[4]{\frac{g\theta_s}{4\nu^2 T_e}}$$

and to replace the functions ϕ and ψ by two other functions χ and ζ, defined by

$$\chi(\xi)=\phi(x,y) \tag{22–15}$$

and

$$\zeta(\xi)=\frac{\psi(x,y)}{4\nu c\sqrt[4]{x^3}} \tag{22–16}$$

Substituting these in Eqs. 22–12 and 13, Pohlhausen succeeded in replacing the partial differential equations by the following more simple system of two ordinary equations, in which single, double, and triple primes stand for the first, second, and third derivatives with respect to ξ:

$$\zeta'''+3\zeta\cdot\zeta''-2(\zeta')^2+\chi=0 \tag{22–17}$$

and

$$\chi''+\frac{3\nu}{\alpha}\zeta\chi'=0 \tag{22–18}$$

with the boundary conditions: †

$$\left.\begin{array}{lllll}\zeta=0 & \zeta'=0 & \chi=1 & \text{for} & \xi=0\\ \text{and}\\ \zeta'=0 & \zeta''=0 & \chi=0 & \text{for} & \xi=\infty\end{array}\right\} \tag{22–19}$$

† When $y = 0$, from Eq. 22–7: $v_x = 0$ and $\phi = 1$.
Hence from Eq. 22–15: $\chi = 1$, and from Eq. 22–10: $\partial\psi/\partial y = 0$.
Then from Eqs. 22–14 and 16: $\zeta' = 0$.
Further from Eq. 22–7: $v_y = 0$.
Hence from Eq. 22–11: $\partial\psi/\partial x = 0$, and herewith from Eq. 22–16: $\zeta = 0$.
When $y = \infty$, from Eq. 22–7: $v_x = 0$ and $\phi = 0$.
Hence from Eq. 22–15: $\chi = 0$, and from Eqs. 22–10 and 16: $\zeta' = 0$.
Finally, since $\partial v_x/\partial y = 0$ at $y = \infty$, from Eq. 22–10: $\partial^2\psi/\partial y^2 = 0$, and from Eq. 22–16: $\zeta'' = 0$.

Now the velocity components and temperatures can be expressed as functions of x, χ, and ζ, the latter ones being functions of the variable ξ, which is defined by Eq. 22–14 as a function of x and y. Thus

$$v_x = \frac{\partial\psi}{\partial y} = 4\nu c\sqrt[4]{x^3}\,\frac{\partial\xi}{\partial y}\,\zeta' = 4\nu c^2\sqrt{x}\,\zeta' \tag{22–20}$$

$$v_y = -\frac{\partial\psi}{\partial x} = -\nu c\left[\frac{1}{\sqrt[4]{x}}\,\zeta - c\,\frac{y}{\sqrt{x}}\,\zeta'\right] \tag{22–21}$$

$$\phi(x, y) = \chi \tag{22–15}$$

From Eqs. 22–20 and 15 it follows that different horizontal cuts through the systems of experimental curves, given in Figs. 22–1 and 2, must lead to unique curves for $\zeta' = v_x/(4\nu c^2\sqrt{x}\,)$ and for $\chi = \theta/\theta_s$ as ordinates when ξ is used as abscissa. Schmidt and Beckmann checked this, and Pohlhausen performed a numerical integration of the differential equations 22–17 and 18, having expanded their terms into badly convergent series and starting from the experimentally determined values of ζ'' and χ' at $y = 0$ (i.e., at $\xi = 0$). (See Fig. 22–3.) ‡

The curves ζ' and χ of this figure are supposed to supersede the systems of velocity curves in Fig. 22–1 and of temperature curves in Fig. 22–2, respectively. These two curves are to be considered as the solution of the partial differential equations 22–5, 8, and 9. Curve χ has been closely verified by the experiments. Curve ζ', however, deviates by about 10% from the corresponding experimental curves obtained with the plate of 5 in. height, and considerably more from those for the 20-in. plate. This is probably due to the effect of a starting section below the 20-in. plate and to beginning turbulence. Fortunately it is sufficient to know the temperature distribution, and particularly the temperature slope at $y = 0$, in order to determine the heat transfer on a surface. Considering that all but the radiated heat is transferred by conduction through a very thin layer adjacent to the surface, the following identity exists:

$$h_c\theta_s = -k\left(\frac{\partial\theta}{\partial y}\right)_{y=0} = -k\theta_s\left(\frac{\partial\phi}{\partial y}\right)_{y=0} \tag{22–22}$$

where h_c is the local coefficient of heat transfer by convection. By substitution from Eqs. 22–15 and 14,

$$h_c = -k\sqrt[4]{\frac{g\theta_s}{4\nu^2 T_e x}}\,\chi_0' \tag{22–23}$$

where the subscript 0 stands for $\xi = 0$.

‡ A table, from which this figure has been developed, is given in the paper.

Since ν/α is the only parameter in Eq. 22–18, the derivative in Eq. 22–23 depends only upon this ratio. For air and other diatomic gases, ν/α is almost independent of the temperature. Assuming $\nu/\alpha = 0.733$,

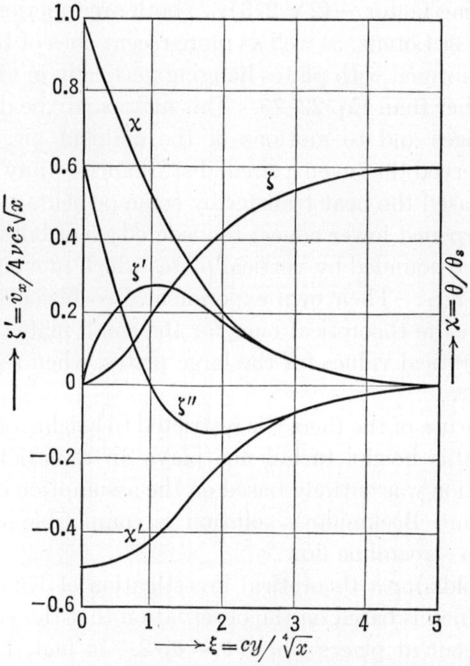

FIG. 22–3. Solution of the differential equations 22–17 and 18 for $\nu/\alpha = 0.733$, including dimensionless representation of velocity field ζ' and temperature field χ for air in free convection on a vertical plate.

From E. Schmidt and Beckmann [1930].

Pohlhausen obtained $\chi_0' = -0.508$ under the conditions of Schmidt and Beckmann's experiments. From Eq. 22–23, by integration over the height L of the plate,

$$h_{c,m} = -\frac{2\sqrt{2}}{3} \cdot \frac{k}{L} \sqrt[4]{\frac{gL^3\theta_s}{\nu^2 T_e}}\, \chi_0' \qquad (22\text{–}24)$$

and substituting $\chi_0' = -0.508$,

$$h_{c,m} = 0.479 \frac{k}{L} \sqrt[4]{\frac{gL^3\theta_s}{\nu^2 T_e}} \qquad (22\text{–}25)$$

This may be compared with Lorenz' result of the year 1881, by converting Eq. 22–1 into the form

$$h_{c,m} = C \frac{k}{L} \sqrt[4]{\frac{gL^3\theta_s}{\nu^2 T_e}} \sqrt[4]{\frac{\nu}{\alpha}} = 0.513 \frac{k}{L} \sqrt[4]{\frac{gL^3\theta_s}{\nu^2 T_e}} \qquad (22\text{–}26)$$

where the last term is obtained by substituting $C = 0.548$, as calculated by Lorenz, and $\nu/\alpha = 0.733$, as used above. Equation 22–26 differs from Eq. 22–24 only in the factor $0.548\sqrt[4]{\nu/\alpha}$, which is 7% larger than the corresponding factor $-(2\sqrt{2}/3)\chi_0'$; both are functions of ν/α alone.

Experiments of Lorenz, as well as more recent ones of R. Weise [1935], which were performed with plates hanging vertically in air, led to values about 25% higher than Eq. 22–25. This may partly be due to the influence of the edges and to motions of the ambient air. Schmidt and Beckmann observed that even a small disturbance by any motion around the edges increased the heat transfer by some percentages. Their small plates had sharpened lower edges; the side edges of both the small and large plates were bounded by vertical plates which prevented convection around these edges. Their own experimental coefficients are in perfect agreement with the theoretical ones for the small plates and about 4% above the theoretical values for the large plates, when ν is taken at surface temperature.

However, the use of the theory is restricted to heights of not more than 2 ft. Above this height turbulence plays an essential role, whereas Lorenz' derivation was entirely based on the assumption of laminar flow, and Schmidt and Beckmann's solution is compatible with but slight deviations from streamline flow.

The same holds for a theoretical investigation of Kimball and W. J. King [1932], which is based on the observation that the velocity maxima of Fig. 22–1 occur at places where $\theta = \theta_s/2$. In fact, Fig. 22–3 shows that the maximum of ζ' occurs close to $\chi = 0.5$, and this maximum corresponds to a maximum of $\partial v_x/\partial y$ for any fixed value of x. The authors might have also remarked that the maximum occurs close to $\xi = 1$ so that, according to Eq. 22–14, $y = \sqrt[4]{x/c}$. This is a very much simpler expression than the one which the authors find for the "equivalent Langmuir film." The authors further assume that the temperature drop between the wall and the locus of the maximum is linear which is a fair approximation of the curve χ in Fig. 22–3. Finally it is supposed that the heat carried upwards by convection is divided into two equal parts by a surface containing the points where maximum velocity occurs. The merit of this last assumption is questionable, although the authors found their results, based on this and the previous statements, in satisfactory agreement with Schmidt and Beckmann's observations.

For horizontal cylinders Hermann [1936] followed and extended the procedure of Schmidt and Beckmann and found velocity and temperature distributions which are rather similar to those for vertical plates. Instead of the heights x above the lower edge of the plate the angle between any radius of the cross-sectional circle and its vertical radius was introduced. In the calculations again the results of Pohlhausen's numerical integrations were used.

22–2. Forced Heat Convection in Laminar Flow through a Tube or Duct at Uniform Surface Temperature

Whereas the velocity distribution of a fluid passing through a circular tube in streamline motion has been found to be utterly simple (Eq. 21–33), it is enormously difficult to calculate the heat flow to such fluid from the wall, when the wall is kept at uniform temperature and the fluid enters the tube at a different temperature. The problem has only been solved neglecting the variations of the thermal properties of the fluid with temperature. Graetz [1885] found a mathematical solution which 25 years later was rediscovered by Nusselt [1910]. Groeber [1921] and Jakob [1933, p. 235], partly in cooperation with H. Eck, recalculated and corrected the constants of Graetz's equation. An excellent review of Graetz's and related papers has been given by Drew [1931].

Transformation of the components of Eq. 3–24 in the directions y and z to polar coordinates, for steady state, axial symmetry, and flow in the direction x only, yields

$$v_x \frac{\partial t}{\partial x} = \alpha \left(\frac{\partial^2 t}{\partial r^2} + \frac{1}{r} \cdot \frac{\partial t}{\partial r} + \frac{\partial^2 t}{\partial x^2} \right) \tag{22–27}$$

This equation could also have been directly written down by superimposing the terms with x (taken from Eq. 3–24) and those with r (taken from Eq. 8–10). Considering that the temperature curve for axial direction will be much less curved than the radial one, $\partial^2 t/\partial x^2$ may be neglected. Since, further, the influence of temperature upon viscosity is neglected, the velocity distribution is given by Eq. 21–33, and Eq. 22–27 simplifies to

$$2v_m \left[1 - \left(\frac{r}{s} \right)^2 \right] \frac{\partial \theta}{\partial x} = \alpha \left(\frac{\partial^2 \theta}{\partial r^2} + \frac{1}{r} \cdot \frac{\partial \theta}{\partial r} \right) \tag{22–28}$$

where v_m is the mean velocity§, s is the radius of the inner surface of the tube, and $\theta = t - t_s$. This equation can also be derived from a heat balance for a cylinder ring, co-axial with the tube, of length dx and thickness dr, considering radial thermal conduction and axial heat convection by the fluid [see Nusselt 1910].

§ This mean is taken over r in a plane through the center line.

In order to find a particular integral of Eq. 22–28 a separation of the influences of x and r may be tried by the form

$$\theta = X \cdot R \tag{22–29}$$

where X is a function of x and R a function of r. Substituting this in Eq. 22–28,

$$\frac{2\rho c_p v_m}{kX} \cdot \frac{dX}{dx} = \frac{\dfrac{d^2R}{dr^2} + \dfrac{1}{r}\dfrac{dR}{dr}}{[1 - (r/s)^2]R}$$

This must be valid for any value of r and x. Therefore each side can be taken equal to an arbitrary figure $-1/b^2$. Then

$$\frac{dX}{dx} = -\frac{k}{2b^2 \rho c_p v_m} X \tag{22–30}$$

and

$$\frac{d^2R}{dr^2} + \frac{1}{r} \cdot \frac{dR}{dr} + \frac{1}{b^2}\left[1 - \left(\frac{r}{s}\right)^2\right] R = 0 \tag{22–31}$$

The integral of Eq. 22–30 is

$$X = Ne^{-\frac{\alpha}{2b^2 v_m} x} \tag{22–32}$$

where N is the constant of integration.

By substitution in Eq. 22–29,

$$\theta = Ne^{-\frac{\alpha}{2b^2 v_m} x} R \tag{22–33}$$

Substituting $r/b = u$ and $s/b = \beta$ in Eq. 22–31,

$$\frac{d^2R}{du^2} + \frac{1}{u} \cdot \frac{dR}{du} + \left(1 - \frac{u^2}{\beta^2}\right) R = 0 \tag{22–34} \|$$

A particular integral of this equation is

$$R(u) = \sum_{n=0}^{n=\infty} B_{2n} u^{2n} \tag{22–35}$$

where $B_0 = 1$.

$$B_2 = -\frac{1}{2^2}.$$

$$B_{2n} = \frac{1}{(2n)^2}\left(\frac{1}{\beta^2} B_{2n-4} - B_{2n-2}\right).$$

‖ It may be noted that Bessel's differential equation has a term β^2/u^2 instead of u^2/β^2 in the above equation.

The series is convergent for any value of u, when $\beta > 1$. For $r = s$, by definition, $u = \beta$ and $\theta = 0$. Hence, according to Eq. 22–33, $R = 0$. In other words, the constants β are the roots of the equation

$$R(\beta) = 0 \tag{22–36}$$

The nth root may be denoted by β_n, and a subscript n is used for terms belonging to β_n. Nusselt's calculation led to the following values:¶

$$\beta_0 = 2.705 \qquad \beta_1 = 6.66 \qquad \beta_2 = 10.3$$

Introducing $w = r/s$ and using Eq. 22–33, the general solution of the differential equation becomes

$$\theta = \sum_{n=0}^{n=\infty} N_n e^{-\frac{\beta_n^2}{2}\cdot\frac{\alpha}{sv_m}\cdot\frac{x}{s}} R_n(w, \beta) \tag{22–37}$$

The term R_n in this equation, belonging to $\beta = \beta_n$, satisfies the differential equation

$$\frac{d^2R}{dw^2} + \frac{1}{w}\cdot\frac{dR}{dw} + \beta^2(1 - w^2)R = 0 \tag{22–38}$$

which is found by substituting $r/s = w$ in Eq. 22–31.

Introducing the boundary condition $\theta = \theta_0$ at $x = 0$,

$$\theta_0 = \sum_{n=0}^{n=\infty} N_n R_n \tag{22–39}$$

Graetz determined the coefficients N_n by a procedure similar to that used in a Fourier analysis. He first proved mathematically the following two theorems:

$$\int_0^1 R_n R_m w(1 - w^2)\, dw = 0 \tag{22–40}$$

and

$$\int_0^1 R_n R_n w(1 - w^2)\, dw = \frac{1}{2\beta_n}\left[\left(\frac{\partial R}{\partial \beta}\right)_n \frac{dR_n}{dw}\right]_{w=1} \tag{22–41}$$

where the integer $m \neq n$.

¶ β_0 is in excellent agreement with the values calculated by Graetz and Drew; β_1 is close to Drew's value; β_2 had not been calculated before. Lee, Nelson, Cherry, and Boelter [1938] further indicate $\beta_3 = 14.671$. Their value β_2 is 3.7% larger than Nusselt's value. It does not seem that β_3 has been checked by anyone else.

Multiplying both sides of Eq. 22-39 by $R_n w(1 - w^2)\,dw$ and integrating,

$$\theta_0 \int_0^1 R_n w(1 - w^2)\,dw = N_n \int_0^1 R_n^2 w(1 - w^2)\,dw \qquad (22\text{-}42)$$

Rewriting Eq. 22-38 for $R = R_n$,

$$R_n w(1 - w^2)\,dw = -\frac{1}{\beta_n^2}\left(w \cdot d\frac{dR_n}{dw} + dR_n\right) = -\frac{1}{\beta_n^2}\,d\left(w\frac{dR_n}{dw}\right)$$

Integrating,

$$\int_0^1 R_n w(1 - w^2)\,dw = -\frac{1}{\beta_n^2}\left[\left(w\frac{dR_n}{dw}\right)_{w=1} - \left(w\frac{dR_n}{dw}\right)_{w=0}\right]$$

$$= -\frac{1}{\beta_n^2}\left(\frac{dR_n}{dw}\right)_{w=1} \qquad (22\text{-}43)$$

From Eqs. 22-41, 42, and 43

$$N_n = -\frac{2\theta_0}{\beta_n\left(\dfrac{\partial R}{\partial \beta}\right)_{n,w=1}} \qquad (22\text{-}44)$$

$R_{n,w=1}$ and $(\partial R/\partial \beta)_{n,w=1}$ can be calculated from Eq. 22-35, considering that $u = \beta w = \beta$.

In this way, for the first three terms ($n = 0$, 1, and 2), respectively,

$$\left(\frac{\partial R}{\partial \beta}\right)_{n,w=1} = -0.50082;\ 0.371;^{*}\ \text{and}\ -0.505$$

Substituting these values in Eq. 22-44, respectively,

$$\frac{N_n}{\theta_0} = 1.477;\ -0.810;\ \text{and}\ 0.385$$

The present author checked these values analytically and by graphical integration of Eq. 22-42, using the values of R_0, R_1, and R_2 calculated by Nusselt and represented in Table 22-1. The results of the graphical procedure are

$$\frac{N_n}{\theta_0} = 1.478;\ -0.808;\ \text{and}\ 0.384$$

in excellent agreement with the corresponding values given above.†

* In Graetz's original paper 1.12290 is given, because of an error in the calculation.

† Lee, Nelson, Cherry, and Boelter's [1938] value of N_3/θ_0 seems to be erroneous, possibly because of a printing error. They further indicate $N_4/\theta_0 = -0.479$, which is unlikely.

The values of Table 22–1 are represented in Fig. 22–4. The right end of curve R_2 has been drawn differently from Nusselt's original representation for reasons to be explained later on.

TABLE 22–1

NUMERICAL VALUES OF FUNCTION $R_n(r/s,\ \beta_n)$, DEFINED BY EQ. 22–35 AND USED FOR THE CALCULATION OF FORCED CONVECTION IN LAMINAR FLOW THROUGH A TUBE

r/s	R_0	R_1	R_2
0	1	1	1
0.1	0.9818	0.8923	0.753
0.2	0.9290	0.6067	0.206
0.3	0.8456	0.2367	−0.290
0.4	0.7382	−0.1062	−0.407
0.5	0.6147	−0.3399	−0.204
0.6	0.4833	−0.4317	0.104
0.7	0.3506	−0.3985	0.278
0.8	0.2244	−0.3051	0.278
0.9	0.1069	−0.1637	0.144
1.0	0	0	0

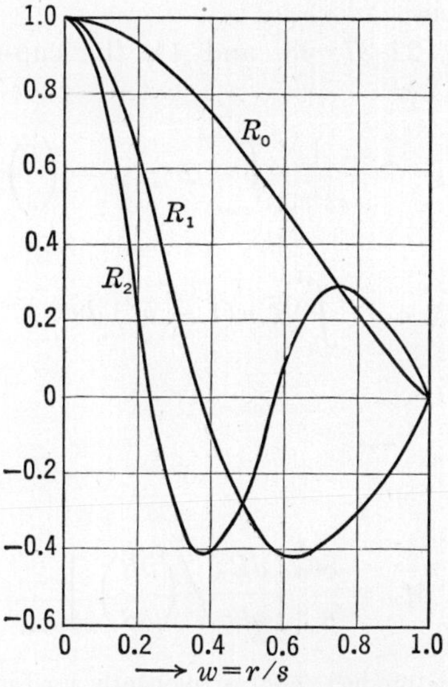

FIG. 22–4. Graphical representation of the functions R_0, R_1, and R_2 defined by Eq. 22–35. (Forced convection in laminar flow through a tube.)

By substitution of the values of N_n and β_n in Eq. 22–37,

$$\frac{\theta}{\theta_0} = 1.477e^{-3.658\frac{\alpha}{sv_m}\cdot\frac{x}{s}}R_0\left(\frac{r}{s}\right) - 0.810e^{-22.178\frac{\alpha}{sv_m}\cdot\frac{x}{s}}R_1\left(\frac{r}{s}\right)$$

$$+ 0.385e^{-53.05\frac{\alpha}{sv_m}\cdot\frac{x}{s}}R_2\left(\frac{r}{s}\right) - \cdots$$

$$= 1.477e^{-m_0x}R_0 - 0.810e^{-m_1x}R_1 + 0.385e^{-m_2x}R_2 - \cdots \tag{22–45}$$

In the last line the following term has been introduced for convenience:

$$m_n = \frac{\beta_n{}^2}{2}\cdot\frac{\alpha}{s^2v_m} = \frac{\beta_n{}^2}{2}\cdot\frac{\pi k}{C_p\dot{V}} = \frac{\beta_n{}^2}{2}\cdot\frac{\pi k}{c_p\dot{m}} \tag{22–46}$$

where‡ $C_p = \rho c_p$,
$\dot{V} = \pi s^2 v_m$ = rate of volume flow, and
$\dot{m} = \rho\dot{V}$ = rate of mass flow.

It is seen that the temperature distribution depends only on r/s and a dimensionless group, $\alpha/sv_m\cdot x/s$. According to Fig. 22–4, the sum in Eq. 22–45 is not always oscillating. However, the exponents are rapidly decreasing so that the three terms given above are sufficient for most numerical calculations.

From Eqs. 21–6, 22–37, 43, and 44, the cup-mixing temperature becomes

$$\theta_m = \sum_{n=0}^{n=\infty} N_n e^{-m_nx}\frac{1}{\pi s^2 v_m}\int_{r=0}^{r=s} 2v_m\left[1 - \left(\frac{r}{s}\right)^2\right]2\pi rR_n\cdot dr$$

$$= 4\sum_{n=0}^{n=\infty} N_n e^{-m_nx}\int_0^1 R_n w(1 - w^2)\,dw$$

$$= \sum_{n=0}^{n=\infty} M_n e^{-m_nx} \tag{22–47}$$

where

$$M_n = \frac{8\theta_0}{\beta_n{}^3}\left[\frac{dR_n}{dw}\Big/\left(\frac{\partial R}{\partial \beta}\right)_n\right]_{w=1}$$

‡ Instead of $(c_p\dot{m})$ the product $(c_p w)$ is frequently used in literature, where w is the time rate of weight. In this case c_p is related to unit force (instead of to unit mass).

M_n can be calculated from β_n and Eq. 22–35. Jakob [1933, p. 235] preferred to use the equation

$$M_n = 4N_n \int_0^1 R_n w(1 - w^2)\, dw \tag{22–48}$$

which is obtained from Eqs. 22–43 and 44, since he had previously determined by a graphical method both the coefficient N_n and the integral.

The following three values were obtained:

$$\frac{M_n}{\theta_0} = 0.820;\, 0.0972\colon 0.0135$$

These three values are in excellent agreement with Groeber's results, the first two also with Drew's values, and the first one with Graetz's value.§ A direct comparison with Nusselt's results is not possible because he used the local-mean temperature instead of the cup-mixing temperature; however, an indirect comparison will be made later on.

By substitution of our values of M_n in Eq. 22–47,

$$\frac{\theta_m}{\theta_0} = 0.820e^{-m_0 x} + 0.0972e^{-m_1 x} + 0.0135e^{-m_2 x} \cdots \tag{22–49}$$

Often a temperature function, similar to the one defined by Eq. 22–2, is used, namely,

$$\phi = \frac{t_0 - t_m}{t_0 - t_s} = 1 - \frac{\theta_m}{\theta_0} \tag{22–50}$$

The total heat exchange between fluid and surface from $x = 0$ to $x = x$ can be obtained in two different ways, either from

$$q = -\int_{x=0}^{x=x} k \cdot 2\pi s \left(\frac{\partial \theta}{\partial r}\right)_{r=s} dx \tag{22–51}$$

or from

$$q = \pi s^2 \cdot v_m \rho c_p (\theta_0 - \theta_m) \tag{22–52}$$

By differentiation of Eq. 22–45 the rate of local heat transfer per unit area is obtained as follows:

$$q'' = -k\left(\frac{\partial \theta}{\partial r}\right)_{r=s} = k\frac{\theta_0}{s}[1.499e^{-m_0 x} + 1.078e^{-m_1 x} + 0.358e^{-m_2 x} + \cdots] \tag{22–53}$$

§ His second value is 0.0325, due to the mentioned error of calculation (see footnote *, p. 454).

It decreases rapidly with increasing x. By substitution in Eq. 22–51 and integration,

$$q = 2\pi k\theta_0 \left[\frac{1.499}{m_0}(1 - e^{-m_0 x}) + \frac{1.078}{m_1}(1 - e^{-m_1 x}) + \frac{0.3581}{m_2}(1 - e^{-m_2 x}) + \cdots \right] \tag{22–54}$$

On the other hand, substituting θ_m from Eq. 22–49 in Eq. 22–52,

$$q = \pi s^2 v_m \rho c_p \theta_0 (1 - 0.820e^{-m_0 x} - 0.0972e^{-m_1 x} - 0.0135e^{-m_2 x} - \cdots) \tag{22–55}$$

Substituting m_n from Eq. 22–46 in the denominators of Eq. 22–54 and identifying the respective terms with e functions in Eqs. 22–54 and 55, it follows that

$$\frac{4}{\beta_0^2} \cdot 1.499 = 0.820; \qquad \frac{4}{\beta_1^2} \cdot 1.078 = 0.0972; \qquad \frac{4}{\beta_2^2} \cdot 0.3581 = 0.0135$$

This is in agreement with Nusselt's original values || of β_n. Hence the values 1.078 and 0.358, given in Eq. 22–53, seem to be correct, whereas 1.114 and 0.503, respectively, given in Nusselt's paper, were erroneous. Nusselt probably obtained these coefficients by taking $(dR_n/dr)_{r=s}$ from his graph, having drawn the right ends of the curves R_1 and R_2 differently from those in our Fig. 22–4.

Since, also, the sum of the terms without e function in Eqs. 22–54 and 55 must be equal, it follows that

$$0.820 + 0.0972 + 0.0135 + \cdots = 1$$

Obviously the terms with $n > 2$ still contribute 0.068 to this sum. This is only another expression of the fact, which can be concluded from Eq. 22–55, that for $x \to \infty$,

$$q_{x=\infty} = \pi s^2 v_m \rho c_p \theta_0 \tag{22–56}$$

Equation 22–56, by the way, is self-explanatory, since the total excess of heat energy of the incoming fluid will have been given up to the tube wall for $x = \infty$.

The local coefficient of heat transfer, h, is found by substituting q'' from Eq. 22–53 and θ_m from Eq. 22–49 in

$$q'' = h\theta_m \tag{22–57}$$

|| Checked by the present author.

It starts ¶ with $h = \infty$ at $x = 0$, and with $x \to \infty$ it approaches the limit

$$h = \frac{1.499}{0.820} \cdot \frac{k}{s} = 1.828 \frac{k}{s} \tag{22-58}$$

Different definitions of a mean coefficient of heat transfer are thinkable. The following one is based on the log-mean temperature difference $\theta_{\ln}$:

$$h_{\ln} = \frac{q}{2\pi s x \theta_{\ln}} = \frac{\pi s^2 v_m \rho c_p}{2\pi s x} (\theta_0 - \theta_m) \frac{\ln(\theta_0/\theta_m)}{\theta_0 - \theta_m} = \frac{\rho c_p s v_m}{2x} \ln(\theta_0/\theta_m) \tag{22-59}$$

where q is substituted from Eq. 22–52, and θ_m is to be calculated from Eq. 22–49. Since Eq. 21–16 is based on the assumption of U = constant, $h_{\ln}$ is that apparent, unvariable coefficient of heat transfer * which would yield the same heat transfer as actually is obtained owing to the true coefficient, which varies over the length according to Eqs. 22–57 and 53.

Before he succeeded in solving the problem of heat transfer of Poiseuillean distribution of velocity, Graetz [1883] had dealt with the simpler case of uniform velocity over the cross section. It led to the following solution:

$$\frac{\theta}{\theta_0} = \sum_{n=1}^{n=\infty} \frac{2}{\beta_n J_1(\beta_n)} e^{-\beta_n^2 \frac{\alpha}{s v_0} \cdot \frac{x}{s}} J_0\left(\beta_n \frac{r}{s}\right) \tag{22-60}$$

where J_0 and J_1 are the Bessel functions of zero and first order,
β_n is the nth root of the equation $J_0(\beta) = 0$, and
v_0 is the uniform flow velocity.

Details about this solution are given in the mentioned paper of Drew [1931]. Of course, uniform velocity distribution does not occur in real fluids flowing through a tube. However, immediately past the inlet the distribution will be nearly uniform when the fluid has entered at uniform velocity. Further, when liquids are heated or gases are cooled in laminar flow, the velocity has a distribution somewhere between the parabolic and uniform pattern. In these cases the viscosity is smallest

¶ Obviously $1.499 + 1.078 + 0.358 \cdots = \infty$, whereas the sum $\frac{4}{\beta_0^2} 1.499 + \frac{4}{\beta_1^2} 1.078 + \frac{4}{\beta_2^2} 0.358 + \cdots = 1$, as proved in the discussion of Eq. 22–55.

* It should not be overlooked that in this definition the Graetzian temperature distribution is assumed, which can be established only when h is varying with x, whereas $h_{\ln}$ is assumed to be constant. This is the reason why $h_{\ln}$ is a fictitious quantity only, but a quite practicable one, in the writer's opinion.

close to the wall; therefore the velocity (Curve a in Fig. 22-5) has a steeper slope than the parabolic curve b, whereas for cooling of liquids or heating of gases the temperature gradient close to the wall is more flat (Curve c). Hence, Eq. 22-60 represents a limit for case a.

This equation of Graetz was rediscovered, many years later, by another scientist, Lévêque [1928], as Eq. 22-45 had been by Nusselt [1910]. Lévêque's paper further included a useful supplement to the theory for Poiseuillean velocity distribution. According to Eq. 22-50, obviously

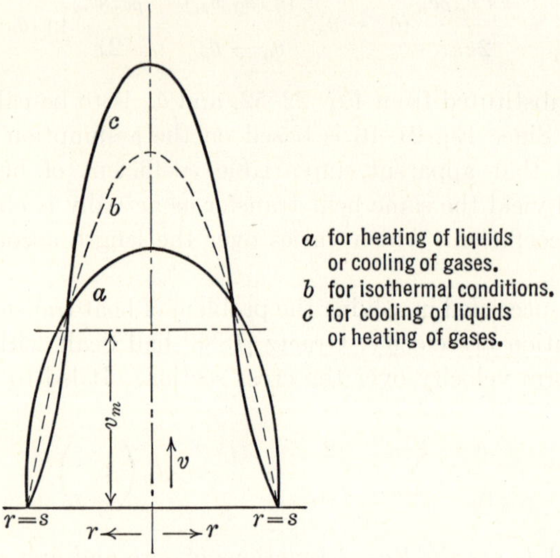

Fig. 22-5. Velocity distributions in a tube.

$t_m \to t_s$ when $x \to \infty$, and $t_m = t_0$ and $\theta_m/\theta_0 = 1$ for $x = 0$. According to Eq. 22-49, however, the first three terms yield only 0.9307 when $x = 0$. Hence more terms would have to be determined, which would require excessive labor. Lévêque, having found a solution for flat plates, used this as an asymptotic approximation, considering that, according to Eq. 22-45, $s^2 \to \infty$ has the same effect on the result as $x \to 0$. As shown in Drew's [1931] analytical paper, Lévêque arrived at

$$\frac{t_0 - t_m}{(t_0 - t_s) + 0.6(t_{s,0} - t_{s,L})} = 6.47\left(\frac{\pi k L}{4c_p\dot{m}}\right)^{2/3} = 5.55\left(\frac{c_p\dot{m}}{kL}\right)^{-2/3} \qquad (22\text{-}61)$$

where $t_{s,0}$, t_s, and $t_{s,L}$ are the surface temperatures at $x = 0$, $x = x$, and $x = L$, respectively, assuming a linear temperature change of the surface temperature with x. If $t_{s,L} = t_s = t_{s,0}$, the left side of Eq. 22-61 represents a limit for ϕ as defined by Eq. 22-50.

In Fig. 22-6, ϕ is plotted versus $c_p\dot{m}/kL$.

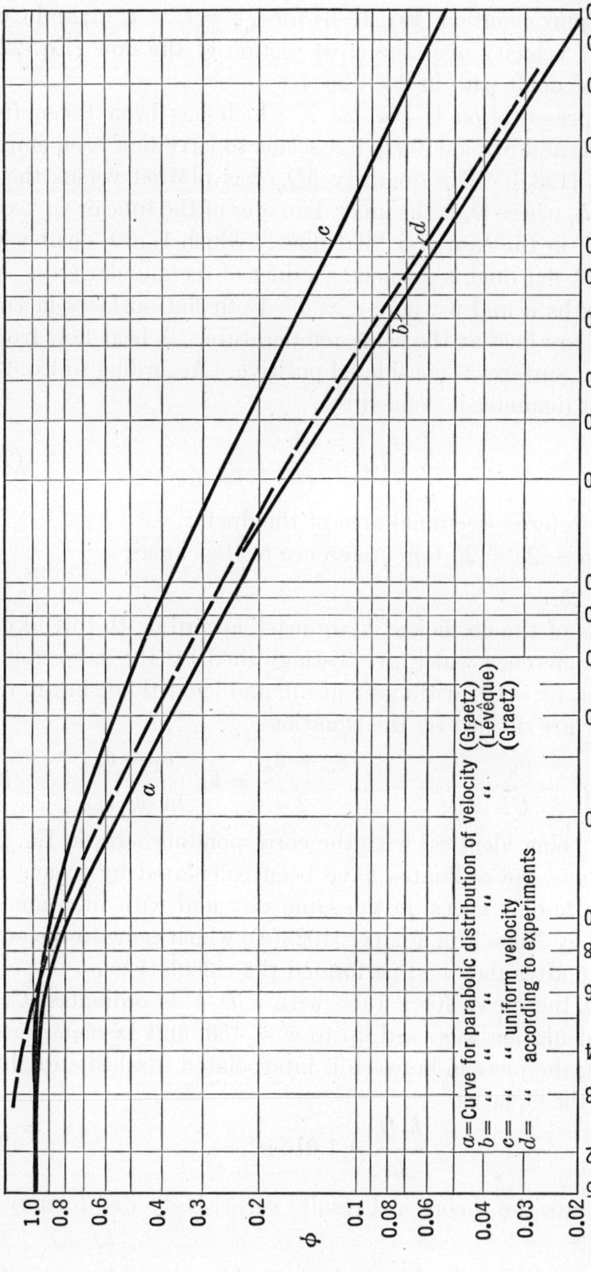

FIG. 22–6. Heat transfer in a tube for linear flow.

Curve a is for parabolic velocity distribution (Eq. 22–49); Curve b is Lévêque's approximation (Eq. 22–61 for $t_{s,0} = t_s = t_{s,L}$); and Curve c is for constant velocity over the cross section of the flow (Eq. 22–60). Curve d will be dealt with in Sect. 26–1.†

Another representation is Fig. 22–7, which has been taken from a paper of Norris and Streid [1940], but seems to have first been employed by McAdams [1931]. The quantity hD_e/k is plotted versus the ratio $\Phi = 4c_p\dot{m}/\pi kL$, where D_e is the inner diameter of the tube or an "equivalent diameter" in the case of a "flat duct," which is also dealt with by the authors. A flat duct is defined as a duct of rectangular cross section with side lengths a and b where $a \gg b$. Both flat surfaces in contact with the fluid are held at the same temperature. A heat flow from the fluid to the flat surfaces is considered positive. According to Eq. 21–55, the equivalent diameter is defined by

$$D_e = \frac{4A}{C} \tag{22-62}$$

where $A = a \cdot b$ (cross-sectional area of the duct),
$C = 2a + 2b \approx 2a$ (circumference of that area),

so that $D_e \approx 2b$.

Three kinds of the coefficient h are used, according to three kinds of temperature difference which are distinguished by the subscripts 0, a, and ln for inlet ($x = 0$), arithmetic mean, and logarithmic mean, respectively. They are defined by the equation

$$q'' = \frac{q}{CL} = h_0\theta_0 = h_a \frac{\theta_0 + \theta_m}{2} = h_{\ln} \frac{\theta_0 - \theta_m}{\ln(\theta_0/\theta_m)} \tag{22-63}$$

the last term being identical with the corresponding term in Eq. 22–59.

For flat ducts the ordinates have been calculated by means of the method of finite differences, in the same way and with the same result as previously by Nusselt in a paper [1923(a)] which came to the authors' attention only after they had performed the calculations.

Concerning the curve for a tube with h_aD/k as ordinates, Graetz's method of calculation was used up to $\Phi \approx 150$, and Lévêque's method for $\Phi > 2000$; the piece in between is interpolated (dashed straight line) according to the equation

$$\frac{h_aD}{k} = 1.615\Phi^{1/3} \tag{22-64}$$

which represents the theoretical results surprisingly well for any value $\Phi > 10$.

† The attention of the reader may also be directed to a thorough analysis of superimposed free and forced viscous convection in vertical pipes, including useful tables and charts by Martinelli and Boelter [1942].

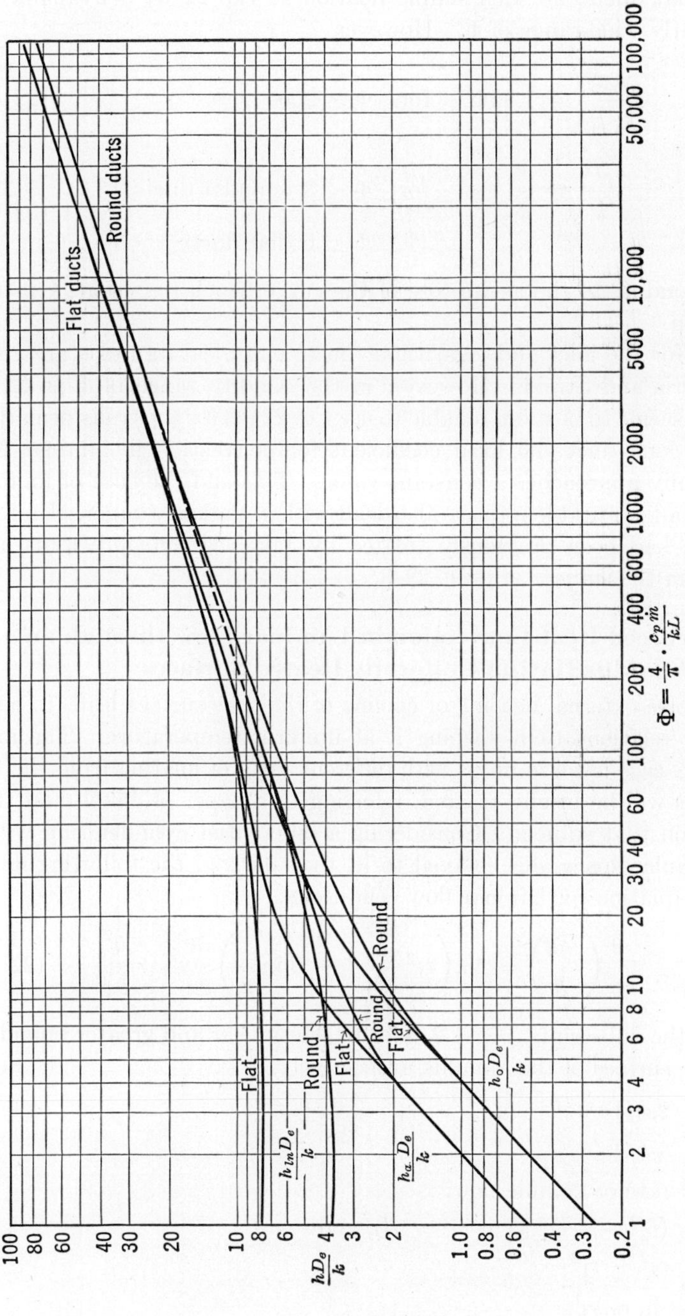

FIG. 22–7. Heat transfer by forced convection in ducts.

For flat ducts, no such simple relation as Eq. 22–64 is available for a similarly wide range of Φ. However,

$$\frac{h_a D_e}{k} = 1.85\Phi^{1/3} \quad \text{for} \quad \Phi > 2000 \tag{22-65}$$

$$\frac{h_a D_e}{k} = \frac{\Phi}{2} \quad \text{for} \quad \left.\begin{array}{l}\Phi < 3 \text{ (circular ducts)} \\ \Phi < 6 \text{ (flat ducts)}\end{array}\right\} \tag{22-66}$$

corresponding to the asymptote of the curve which represents Graetz's equation.

Also for the inlet and logarithmic mean empirical formulas are given by Norris and Streid. However, in the validity range of Eqs. 22–64 to 66 it seems to be more reliable to use Eq. 22–63 for the transformation.

In a long duct the local coefficient for convection in laminar flow eventually approaches a constant value. This and the case of uniform heat input per unit length are also dealt with in the same paper. Uniform heating had been previously treated by Jakob [1939(*a*), 1940(*a*)] and will form the subject of Sect. 22–3.

22–3. Forced Heat Convection in Laminar Flow through an Annulus Having Uniformly Heated Surfaces

Uniform external heating or cooling of the tube surface leads to much simpler relations than keeping it at uniform temperature. The more complex case of an annulus with different rates of heating from the two surfaces will be briefly treated, referring to a paper of Jakob and Rees [1941] on that subject. Considering a differential cylinder ring within the annular space and co-axial to it, they derived the following differential equation for laminar flow conditions:

$$\frac{\partial}{\partial r}\left(r\frac{\partial t}{\partial r}\right) = Nr\left(r_1^2 - r^2 + B\ln\frac{r}{r_1}\right)\frac{\partial t}{\partial x} - r\frac{\partial^2 t}{\partial x^2} \tag{22-67}$$

where the subscripts 1 and 2 refer to the smaller and greater boundary surface of the annulus, respectively,

$$N = \frac{2\rho c_p \dot{V}}{\pi k M},$$

$\dot{V}$ = rate of volume flow,

$M = (r_2^2 - r_1^2)(r_2^2 + r_1^2 - B)$, and

$$B = \frac{r_2^2 - r_1^2}{\ln r_2/r_1}.$$

In the case of uniform heating or cooling of the fluid from either or both boundary surfaces of the annulus it can be supposed that the heat energy is absorbed or given up by the fluid in such a manner that at any sufficiently great distance x from the entrance and at any radial distance r from the center line the temperature change per unit length is constant, so that

$$\frac{\partial t}{\partial x} = C$$

and

$$\frac{\partial^2 t}{\partial x^2} = 0$$

Hence Eq. 22–67 simplifies to:

$$\frac{d}{dr}\left(r\frac{dt}{dr}\right) = NC\left({r_1}^2 r - r^3 + Br \cdot \ln\frac{r}{r_1}\right) \tag{22–68}$$

Since N is constant, Eq. 22–68 is a mere conduction equation. Integration and some transformation yield the following general result:

$$\frac{t_2 - t_1}{CL} = \frac{t_2 - t_1}{\Delta t} = \frac{\rho c_p \dot{V}}{2\pi k L}\mathrm{X} \tag{22–69}$$

with

$$\mathrm{X} = \frac{1}{(1-y^2)(1+y^2-z)}\left\{(1-y^2)\left[\frac{3}{4}(1+y^2) - z\right]\right.$$
$$\left. + y^2 \ln y(y^2 - z) + (\ln y)[1 - y^2(y^2 - z) - z]P_1\right\}$$

where $t_2 - t_1$ = the temperature difference between the surfaces 2 and 1 at any distance $x >$ starting length,
$\Delta t = CL$ = the increase of temperature at any fixed radius r over a length $\Delta x = L$, in the region of $x >$ starting length,
$y = r_1/r_2$,
$z = (y^2 - 1)/\ln y$,
$P_1 = q_1/q$,
q_1 = the rate of heat input at the surface 1, and
q = the sum of the two rates of heat input at the surfaces 1 and 2.

Figure 22–8 is a plot of the shape factor X. The curve $P_1 = 0$ represents the case of heating from surface 2 only, and $P_1 = 1$ is for heating from surface 1 only. $P_1 < 0$ is for cooling of surface 1; $P_1 > 1$ for cooling of surface 2. The singular point $\mathrm{X} = 0.75$ at $y = 0$ and $P_1 = 0$

refers to a tube, as limiting case of an annulus. All other cases at $y = 0$ yield $X = \pm\infty$ because finite amounts of heat transferred by an infinitely thin wire must give rise to an infinitely great temperature drop.

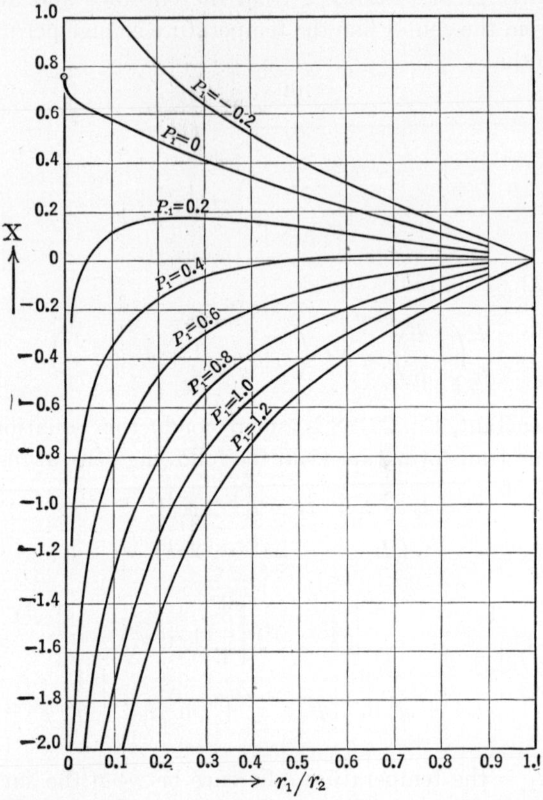

FIG. 22-8. Shape factor X defined by Eq. 22-69. (Forced convection in laminar flow through an annulus.)

Defining coefficients of heat transfer h_{12} and h_{21} by

$$q_1 = h_{12}\Delta x \cdot 2\pi r_1(t_1 - t_2) \tag{22-70}$$

and

$$q_2 = h_{21}\Delta x \cdot 2\pi r_2(t_2 - t_1) \tag{22-71}$$

it follows from Eq. 22-69 that

$$X_{P_1=0} = \frac{k}{h_{21}r_2} \tag{22-72}$$

and

$$X_{P_1=1} = \frac{k}{h_{12}r_1} \tag{22-73}$$

For a cylindric tube, heated from outside ($X_{P_1=0} = 0.75$), Eq. 22-72 leads to

$$t_2 - t_1 = 0.75 \frac{q_2}{2\pi k \cdot \Delta x} \tag{22-74}$$

The case of a plane gap heated from one side, for instance, from surface 1, is included in the system by the relationship $s = r_2 - r_1 > 0$, $r_1/r_2 = 1$, where s is the width of the gap. The following is obtained:

$$\frac{h_{12}s}{k} = 2 \tag{22-75}$$

as is explained in the original paper of Jakob and Rees [1941, p. 629].

This case had previously been dealt with, more in detail, by Jakob [1939(*a*)]. Taking again x in the direction of flow, and y perpendicular across the gap (from $y = 0$ to $y = s$), the velocity distribution becomes

$$v = 4v_{max}\left(\frac{y}{s} - \frac{y^2}{s^2}\right) = 6v_m\left(\frac{y}{s} - \frac{y^2}{s^2}\right) \tag{22-76}$$

where v_{max} and v_m are the maximum and mean velocity, respectively. Furthermore, the following equation for the temperature distribution was obtained:

$$t = t_1 - K\left(\frac{y^4}{2s^3} - \frac{y^3}{s^2} + y\right) \tag{22-77}$$

where subscript 1 refers to the heating surface, $K = q/(k \cdot \Delta x \cdot \Delta z)$, and the rate of heat flow q is taken for the area $\Delta x \cdot \Delta z$. Because one-sided heating has been supposed, $dt/dy = 0$ for $y = s$. (See Fig. 22-9.)

If both sides are heated uniformly and equally, the temperature distribution has been found to be

$$t = t_1 - K\left(\frac{y^4}{2s^3} - \frac{y^3}{s^2} + \frac{y}{2}\right) \tag{22-78}$$

with a minimum $t_{y=s/2} = t_1 - \frac{5}{32}Ks$.

Furthermore,

$$\frac{h_{12}s}{k} = \frac{h_{21}s}{k} = \frac{16}{5} \tag{22-79}$$

where h_{12} is defined by

$$\frac{q}{2} = h_{12} \cdot \Delta x \cdot \Delta z (t_1 - t_{y=s/2})$$

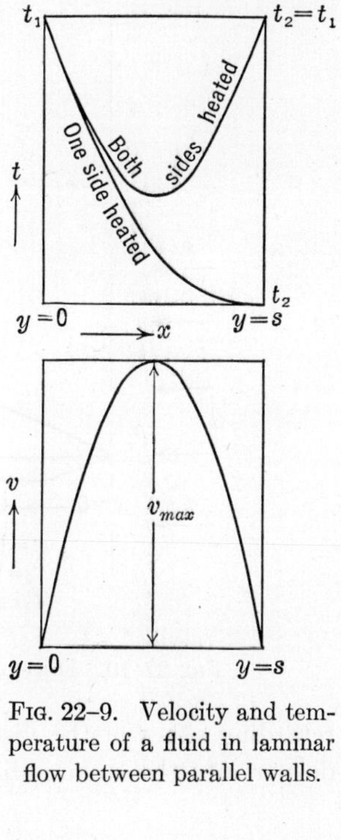

FIG. 22-9. Velocity and temperature of a fluid in laminar flow between parallel walls.

As in Eq. 22–75, h_{12} depends on s and k only, and not upon the absolute value of the flow velocity or range of heat exchange. However, it depends on the velocity distribution in the cross section, which is a function of the geometrical configuration and the boundary conditions.

22–4. Forced Heat Convection in Laminar Flow of a Fluid Parallel to a Surface

Let an unbounded fluid at uniform velocity meet the leading edge of a very thin plane plate and proceed flowing parallel to that plate; only one side of the plate may be considered.

Referring to Sect. 21–5, it is understood that at the leading edge the fluid is suddenly caused to adhere to the solid surface. As a result a thin layer is formed in which the viscosity forces of the fluid bring about a change of velocity. The development of that layer is schematically indicated in Fig. 22–10. As the stream moves over the surface, the

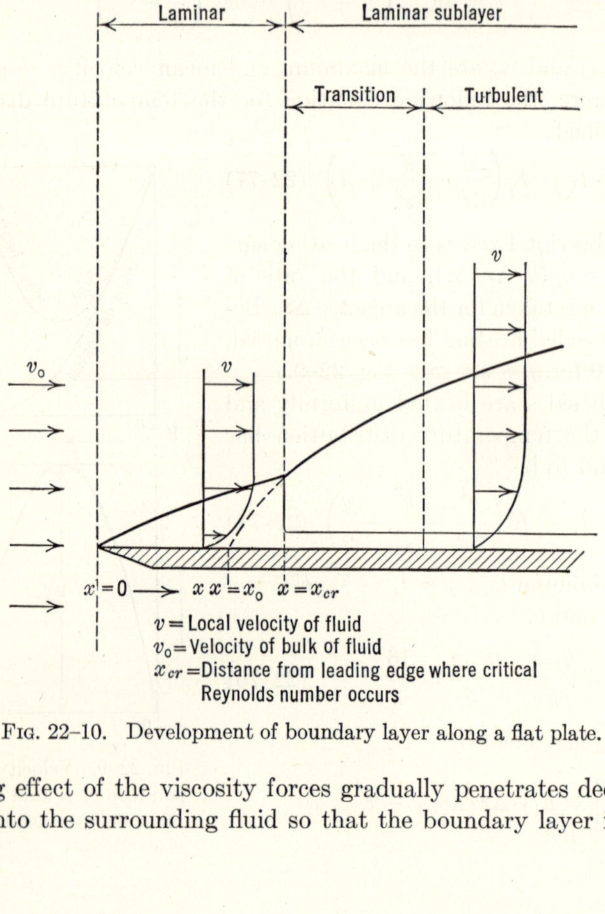

Fig. 22–10. Development of boundary layer along a flat plate.

retarding effect of the viscosity forces gradually penetrates deeper and deeper into the surrounding fluid so that the boundary layer increases in thickness. For some distance the flow in that layer is laminar; in a transition range it gradually becomes turbulent; in the turbulent stage, finally, the thickness of the layer increases more rapidly than it did in the laminar range. Within the boundary layer, however, a very thin laminar film still exists next to the surface, being a permanent feature of any fully established flow.

The boundary layer of a free stream parallel to a surface can be characterized by a dimensionless Reynolds number xv_0/ν, where x is the distance from the leading edge in the stream direction, $x = 0$, in Fig. 22–10, and v_0 is the unaltered velocity. The transition from laminar to turbulent flow is not sharp; it depends upon original disturbances in the turbulent stream ahead of the surface.

Because of the analogy between momentum transfer and heat transfer, the mode of heat transmission depends upon whether the flow in the boundary layer is laminar or turbulent. A difference in the coefficient of heat transfer by convection can also be expected if heating does not start at the leading edge, but is preceded by a hydrodynamical starting section.

Direct solutions of the convection equations for a laminar stream parallel to a plane plate are due to Pohlhausen [1921]. Again only cases of steady-state heat flow are treated. Furthermore, it is assumed that the velocities are small compared to the sound velocity, and ν and k are small and of the same order of magnitude. Finally it is considered that virtually no pressure drop occurs in this sort of unbounded flow.

Then, Eqs. 3–14, 19, 20, 23, and 24 simplify ‡ to

$$v_x \frac{\partial v_x}{\partial x} + v_y \frac{\partial v_x}{\partial y} = \nu \frac{\partial^2 v_x}{\partial y^2} \tag{22–80}$$

$$\frac{\partial v_x}{\partial x} + \frac{\partial v_y}{\partial y} = 0 \tag{22–5}$$

$$v_x \frac{\partial t}{\partial x} + v_y \frac{\partial t}{\partial y} = \alpha \frac{\partial^2 t}{\partial y^2} + \frac{\mu}{J\rho c_p}\left(\frac{\partial v_x}{\partial y}\right)^2 \tag{22–81}$$

where, as previously, x is taken in the direction of the flow, and y is the perpendicular distance from the surface. The last term replaces $q'''/(\rho c_p)$ of Eq. 3–24. The factor $\mu(\partial v_x/\partial y)^2$ in that term represents the power per unit volume due to the frictional force $\mu \cdot \partial v_x/\partial y$, which is converted to a rate of heat flow by means of the mechanical equivalent of heat J.

‡ For details see Pohlhausen's paper.

The boundary conditions of the above equations are

$$v_x = 0 \qquad v_y = 0 \qquad \text{for} \quad y = 0$$

$$v_x = v_0 \qquad \qquad \text{for} \quad y = \infty$$

where v_0 is the velocity of the undisturbed free flow.

Pohlhausen's paper deals with two important cases:

a. Uniform surface temperature.

b. External perfect heat insulation of the plane plate.

a. Uniform Surface Temperature

In this case the following boundary conditions must be added:

$$t = t_s \quad \text{for} \quad y = 0$$

$$t = t_0 \quad \text{for} \quad y = \infty$$

where t_0 is the undisturbed uniform temperature of the arriving stream. Further, the last term of Eq. 22–81 may be neglected, since the rate of frictional heat is generally small compared to the heat flow perpendicular to the surface.

Again the law of continuity (Eq. 22–5) is satisfied by introducing a stream function ψ, defined by

$$v_x = \frac{\partial \psi}{\partial y} \tag{22–10}$$

and

$$v_y = -\frac{\partial \psi}{\partial x} \tag{22–11}$$

Furthermore, the following new variables are introduced:

$$\xi = \frac{1}{2}\sqrt{\frac{v_0}{\nu}} \cdot \frac{y}{\sqrt{x}} \tag{22–82}$$

$$\zeta(\xi) = \frac{\psi}{\sqrt{\nu v_0}} \cdot \frac{1}{\sqrt{x}} \tag{22–83}$$

$$\phi(\xi) = \frac{t - t_s}{t_0 - t_s} = \frac{\theta}{\theta_0} \tag{22–84}$$

The last formula is analogous to Eqs. 22–2 and 50.

By substitution in Eqs. 22–10 and 11,

$$v_x = \tfrac{1}{2} v_0 \zeta' \tag{22-85}$$

and

$$v_y = \frac{1}{2}\sqrt{\nu v_0} \cdot \frac{1}{\sqrt{x}} (\xi\zeta' - \zeta) \tag{22-86}$$

The partial differential equations 22–80 and 81 convert into the ordinary differential equations

$$\zeta''' + \zeta \cdot \zeta'' = 0 \tag{22-87}$$

$$\phi'' + \frac{\nu}{\alpha} \zeta \cdot \phi' = 0 \tag{22-88}$$

with the boundary conditions

$$\zeta = 0 \qquad \zeta' = 0 \qquad \phi = 0 \qquad \text{for} \quad \xi = 0$$

$$\zeta' = 2 \qquad \phi = 1 \qquad \text{for} \quad \xi = \infty$$

Equation 22–87 was solved, and values of ζ, ζ', and ζ'' were calculated by Blasius [1908] (see Table 22–2 and Fig. 22–11). Herewith

TABLE 22–2

Solution of Eq. 22–87 for Laminar Flow on a Vertical Surface

ξ	ζ	ζ'	ζ''
0	0	0	1.33
0.2	0.04	0.29	1.33
0.4	0.13	0.55	1.32
0.6	0.26	0.80	1.27
0.8	0.44	1.04	1.18
1.0	0.66	1.27	1.05
1.2	0.93	1.47	0.90
1.4	1.24	1.64	0.70
1.6	1.58	1.77	0.53
1.8	1.96	1.88	0.39
2.0	2.37	1.94	0.27
2.2	2.8	1.98	0.17
2.4		2.00	0.10
2.6		2.00	0.04
2.8		2.00	0.00
3.0		2.00	0.00
3.2		2.00	0.00

Pohlhausen solved Eq. 22–88. According to this equation, the temperature gradient in the direction y at the surface ($y = 0$, $\xi = 0$) solely depends on ν/α, as was the case with respect to Eq. 22–18.

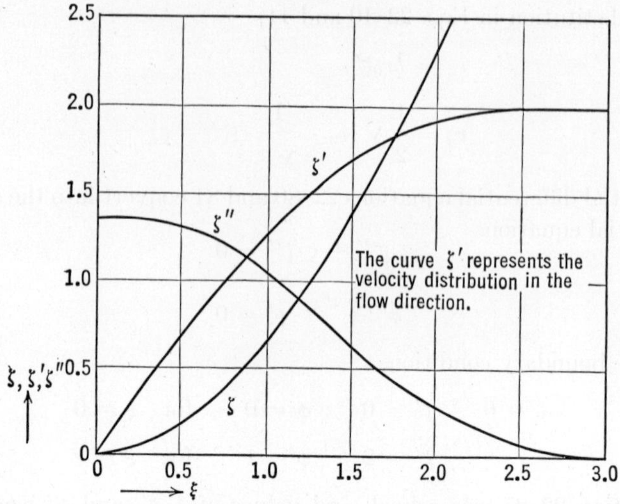

FIG. 22–11. Values of the variable ζ and its first and second derivatives occurring in Eqs. 22–85 to 88. (Forced convection in laminar flow parallel to a surface.)

Table 22–3 contains values of $\phi_{\xi=0}'$ calculated by Pohlhausen.§ He represented them approximately by the equation

$$\phi_{\xi=0}' = 0.664 \sqrt[3]{\frac{\nu}{\alpha}} \tag{22–89}$$

TABLE 22–3

VALUES OF FUNCTIONS, DEFINED BY EQS. 22–84 AND 96, FOR LAMINAR FLOW ON PLANE SURFACES

ν/α	$\phi_{\xi=0}'$	δ
0.6	0.552	0.77
0.7	0.585	0.835
0.8	0.614	0.895
0.9	0.640	0.95
1.0	0.664	1.00
1.1	0.687	1.05
7.0	1.29	2.515
10.0	1.46	2.965
15.0	1.67	3.535
100	.....	6.7
1000	.....	12.9

§ The first six rows cover the range of gases, ν/α being equal to ⅔ for monatomic gases and about 1.1 for four-atomic gases. The next three rows are for water of about 20, 10, and 0°C.

Using the values of the table, the heat flow from a plate of length L and width W to the fluid is readily determined as

$$q = -kW \int_{x=0}^{x=L} \left(\frac{\partial t}{\partial y}\right)_{y=0} dx = \frac{k}{2} \sqrt{\frac{v_0}{\nu}}\, W(t_s - t_0)\phi_{\xi=0}' \int_{x=0}^{x=L} x^{-\frac{1}{2}}\, dx$$

$$= k \sqrt{\frac{v_0 L}{\nu}}\, W(t_s - t_0)\phi_{\xi=0}' \qquad (22\text{–}90)$$

Employing Eq. 22–89,

$$q = 0.664k \sqrt{\frac{v_0 L}{\nu}} \sqrt[3]{\frac{\nu}{\alpha}}\, W(t_s - t_0) \qquad (22\text{–}91)$$

This equation is in good agreement with the result of direct measurements on air by Jakob and Dow [1946].

Another method to determine the heat transfer between a surface and a fluid in parallel flow to it will be described in Sect. 22–6 for turbulent flow. It is based on an empirical formula for the velocity distribution in the boundary layer, an equation for the thickness of the layer, and Reynolds' theorem of similarity between temperature and velocity distribution. The same method might be used for the laminar range, starting from an approximate, but rather exact, solution of the differential equation of the boundary layer by Pohlhausen [1921(*a*)], which led to

$$v_x = \left(\frac{2y}{b} - \frac{2y^3}{b^3} + \frac{y^4}{b^4}\right) v_0 \qquad (22\text{–}92)$$

and

$$b = 5.83x \sqrt{\frac{\nu}{v_0 x}} \qquad (22\text{–}93)$$

This method has not been used so far, since Eq. 22–91 was considered sufficient. Jakob and Dow [1946], however, employed it to estimate how much a curvature of the surface will influence the heat transfer.

Accustomed to the simple parabolic velocity distribution in cylindric tubes, one might think that a simpler form than Eq. 22–92 could be applied for plane surfaces. However, all the attempts made by Boussinesq, King, and Aichli, before Pohlhausen, and by Lévêque afterward, failed because of oversimplified assumptions. Lévêque, for instance, assumed a linear velocity distribution, instead of Eq. 22–92. These mathematical attacks on the problem have been thoroughly described by Drew [1931].

b. Perfectly Heat-insulated Plane Plate

This is the important case of a flat-plate thermometer, as used for measurement of the temperature in air streams, etc.

Here the two additional boundary conditions are

$$\frac{\partial t}{\partial y} = 0 \quad \text{for} \quad y = 0$$

$$t = t_0 \quad \text{for} \quad y = \infty$$

The general procedure is the same as in case *a*. The heat of friction, however, cannot be neglected; in fact, friction is the only source of heat in this case. Hence, Eq. 22–88 is not valid any more.

Instead of $\phi(\xi)$, as defined by Eq. 22–84, the following new variable is introduced: ||

$$\chi(\xi) = \frac{J\rho c_p(t - t_0)}{\rho {v_0}^2/2} = \frac{2Jc_p(\theta - \theta_0)}{{v_0}^2} \tag{22–94}$$

where v_0 is the velocity of flow at some considerable distance from the surface. By this definition the kinetic energy which disappears when the unit volume of the fluid is stopped adiabatically is brought into a relation to the energy which appears as an equivalent in the heating of the fluid from t_0 to t.

The differential equation with the variables ξ, ζ, and χ becomes ¶

$$\zeta''' + \zeta\cdot\zeta'' = 0 \tag{22–87}$$

and

$$\chi'' + \frac{\nu}{\alpha}\zeta\chi' + \frac{1}{2}\cdot\frac{\nu}{\alpha}(\zeta'')^2 = 0 \tag{22–95}$$

with the boundary conditions

$$\zeta = 0 \quad \zeta' = 0 \quad \chi' = 0 \quad \text{for} \quad \xi = 0$$

$$\zeta' = 2 \quad \chi = 0 \quad \text{for} \quad \xi = \infty$$

The boundary conditions of cases *a* and *b* differ in the last term only. Whereas in case *a* the temperature $t = t_s =$ constant, for $y = 0$ and any value x, the surface temperature varies with x in case *b* and

|| This variable is one quarter of the corresponding function in Pohlhausen's [1921] paper. If the British gravitational system is used, it is to be expressed in ft lb/B, c_p in B slug^{-1} F^{-1}, and v_0 in ft/sec.

¶ The third term of Eq. 22–95 differs from the corresponding one of Pohlhausen because of the difference in the definition of χ.

$(\partial t/\partial y)_{y=0} = 0$ over all the length. $\phi = 1$ and $\chi = 0$, however, are only different expressions for the same fact, namely, that $t = t_0$ for $x = 0$ or $y = \infty$.

From ζ, which is known from Blasius' calculations, χ can be calculated numerically. Denoting by t_s the indication of a flat-plate thermometer, the temperature of the fluid may be expressed by

$$t_0 = t_s - \Delta = t_s - \frac{{v_0}^2}{2Jc_p}\,\delta \tag{22-96}$$

where Δ is the correction of the indication of the flat thermometer, and δ is defined by Eq. 22-96, both quantities being functions of ν/α only.

Values of δ are given in Table 22-3, the first nine according to Pohlhausen's own calculation,* the last two for very viscous liquids as calculated by Eckert and W. Weise [1941].

The temperature increase, corresponding to an adiabatic compression, would be

$$t_{ad} - t_0 = \frac{{v_0}^2}{2Jc_p} \tag{22-97}$$

Hence, according to Eq. 22-96,

$$\Delta = (t_{ad} - t_0)\,\delta \tag{22-98}$$

For a gas with $\nu/\alpha = 1$, according to Table 22-3,

$$\delta = 1$$

so that

$$t_0 = t_{ad} - \Delta \tag{22-99}$$

Comparing Eqs. 22-96 and 99,

$$t_s = t_{ad}$$

It may be assumed that the gas under consideration has the same specific heat as air. Then, at $v_0 = 100$ ft/sec, Eq. 22-97 yields $t_{ad} - t_0 = 0.84$ F as temperature excess of the surface above the temperature of the undisturbed fluid.

For air with $\nu/\alpha \approx 0.7$, $\delta = 0.835$. Again, at $v_0 = 100$ ft/sec and using Eq. 22-99, the temperature excess becomes

$$\Delta = 0.84(0.835) = 0.7 \text{ F}$$

According to Eqs. 22-97 and 98, the temperature excess of the surface is increasing as the square of the velocity.

* According to the difference in the definition, our values are four times smaller than those of Pohlhausen.

By increasing ν/α, the function δ takes higher values, and so the temperature excess Δ also becomes greater. This correction plays a considerable role in the thermometry of fast-moving fluids.

The influence of compressibility has been dealt with more exactly by Emmons and Brainerd [1941], in particular, for velocities of the order of sound velocity. They found, for instance, that Eq. 22-96 takes the following, more general, form:

$$t_s = t_0\{1 + 0.5\sigma[\eta(N_{Pr})^{1/2} + (1 - \eta)N_{Pr}]\} \qquad (22\text{-}100)$$

where $\sigma = (\kappa - 1)(N_{Ma})_0^2$,

$\kappa = c_p/c_v$,

$(N_{Ma})_0 = v_0/c$ = Mach number for the undisturbed flow velocity,

c = sound velocity, and

$\eta = 1 - 0.1 \log_{10}(2\sigma + 0.4)$.

22-5. Forced Heat Convection in Turbulent Flow through a Tube at Uniform Surface Temperature

According to the complicated flow conditions, theoretical determination of the heat transfer to or from a fluid in turbulent flow through a tube is still more difficult than for laminar flow. Latzko [1921] has devoted to this subject a thorough theoretical investigation. Only a few basic items of his work can be dealt with here.† He started from Prandtl's and von Kármán's equations on shearing stress, velocity distribution in turbulent flow, and "turbulent conductivity," a sort of apparent conductivity which includes the effect of turbulent mixing movements. This concept will be dealt with more in detail in connection with the application of the principle of similarity to forced convection problems. Also the explanation of the analogy between heat transfer and impulse transportation, as used in Latzko's paper, will have to be postponed.

Using the items mentioned above, Latzko came to the following differential equation for the temperature distribution in a circular tube when turbulence is fully established:

$$\frac{\partial}{\partial r}\left[r\left(\frac{s^2 - r^2}{2s}\right)^{6/7}\frac{\partial \theta}{\partial r}\right] = Kr\left[1 - \left(\frac{r}{s}\right)^2\right]^{1/7}\frac{\partial \theta}{\partial x} \qquad (22\text{-}101)$$

where x = the axial distance of a point of the fluid from the inlet,

r = the radial distance of that point from the surface,

s = the radius of the inner tube surface,

† For readers who would study the original paper it may be mentioned that $4g$ in Latzko's Eq. 12a should be replaced by 49, and that in his Fig. 7A, using any consistent units, the ordinates are values of $3600K$ and not values of K, as is erroneously indicated there. The following equations require the use of consistent units.

θ = the temperature excess of a point over surface temperature,

$$K = \frac{8}{7} \cdot \frac{v_m^{1/4}(2s)^{33/28}}{0.199\nu^{1/4}},$$

v_m = the mean velocity, and
ν = the kinematic viscosity.

Assuming a solution of the form

$$\theta = X \cdot R \doteq e^{-mx} \cdot R \qquad (22\text{–}102)$$

and substituting it in Eq. 22–101, this equation converts to

$$\frac{d}{dr}\left[r\left(\frac{s^2 - r^2}{2s}\right)^{6/7} \frac{dR}{dr}\right] = -mKRr\left[1 - \left(\frac{r}{s}\right)^2\right]^{1/7}$$

Introducing

$$u = \left[1 - \left(\frac{r}{s}\right)^2\right]^{1/7}$$

the following differential equation is obtained:

$$\frac{d}{du}\left[(1 - u^7)\frac{dR}{du}\right] = -49mK\left(\frac{s}{2}\right)^{6/7} u^7 R = -\omega u^7 R \qquad (22\text{–}103)$$

with the boundary conditions

$$R = 0 \qquad \text{at} \quad u = 0$$

$$\frac{dR}{du} = \text{finite} \quad \text{at} \quad u = 1$$

ω is a dimensionless number defined by the equation. Solutions of Eq. 22–103 which at the same time satisfy the boundary conditions exist only for fixed values of ω. By a method of approximation, restricted to three terms, assuming further uniform temperature distribution at $x = 0$, Latzko eventually arrived at

$$\left.\begin{aligned} \frac{\theta}{\theta_0} = {} & 1.129e^{-m_1 x}(0.9544u - 0.0212u^3 + 0.0668u^5) \\ & -0.180e^{-m_2 x}(-0.7472u - 4.275u^3 + 6.022u^5) \\ & +0.048e^{-m_3 x}(20.34u - 54.80u^3 + 35.47u^5) \end{aligned}\right\} \qquad (22\text{–}104)$$

where $m_n = \beta_n \dfrac{1}{D}\sqrt[4]{\dfrac{\nu}{v_m D}}$,

$D = 2s$ = the tube diameter,
$\beta_1 = 0.1510$, $\beta_2 = 2.844$, and $\beta_3 = 29.42$.

Latzko further derived the following equation for a local coefficient of heat transfer, which is based on the local mean temperature instead of the cup-mixing temperature:

$$h = 0.0346 v_m \rho c_p \left(\frac{\nu}{v_m D}\right)^{1/4} \cdot \frac{1.078 e^{-m_1 x} + 0.134 e^{-m_2 x} + 0.980 e^{-m_3 x}}{0.970 e^{-m_1 x} + 0.024 e^{-m_2 x} + 0.006 e^{-m_3 x}} \tag{22-105}$$

Because of the finite temperature difference between surface and entering fluid, h is starting with an infinitely great value. This cannot be seen from the approximative equation 22-105 with three terms only. In the exact solution, the numerator would supposedly approach infinity for $x = 0$. However, as soon as x becomes finite, the fast increase of the exponents with n causes a rapid decrease of h. When x approaches infinity, the following limit is obtained:

$$h_{min} = 0.0384 v_m \rho c_p \left(\frac{\nu}{v_m D}\right)^{1/4} \tag{22-106}$$

Latzko further showed that his formulas are in good agreement with experimental results of Nusselt's.

To determine the heat transfer in a starting section where turbulence is not fully established, he had to derive first an expression for the velocity distribution. He used the boundary-layer concept, referring to Fig. 22-12. Let the fluid enter a tube with a velocity v_0 uniform over the

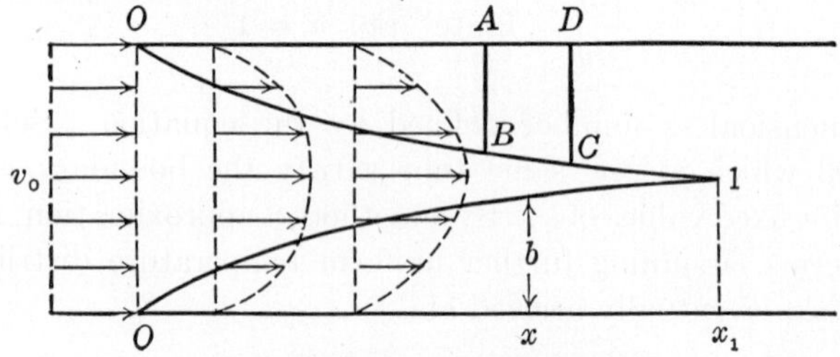

FIG. 22-12. Development of velocity distribution and boundary layer in a tube.

inlet cross section. The layers close to the wall are retarded by the influence of inner friction, as is indicated by velocity arrows in the figure. This retarding effect grows in radial direction toward the tube axis, until the boundary layer reaches the axis. Up to this point there is a core of undisturbed fluid. Considering the equilibrium of forces in an element, $ABCD$, of the boundary layer and some additional conditions as for continuity, pressure drop, relation between shearing stress and velocity distribution, the following equation for the thickness b of the boundary layer was obtained:

$$\xi = \frac{b}{s} = 1.41\chi - 0.048\chi^2 + 0.168\chi^3 \qquad (22\text{–}107)$$

where

$$\chi = \left(\frac{\nu}{v_m D}\right)^{1/5} \left(\frac{x}{D}\right)^{4/5}$$

The starting length x_1 up to the point 1, where the difference between undisturbed core and boundary layer vanishes, is determined by the condition $\xi = 1$ or $\chi = 0.746$, and becomes

$$x_1 = 0.693D \left(\frac{vD}{\nu}\right)^{1/4} \qquad (22\text{–}108)$$

Equations for velocity and temperature distribution were likewise developed, and finally, by means of a heat balance, two equations for the coefficient of heat transfer were derived, one belonging to that starting section in which turbulence is only partly established (h_p), and another one for fully established turbulence (h_f). The result, though analytically clumsy, is presented in a handy graphical form in the paper. It is notable that the coefficient of heat transfer is somewhat slower in approaching the minimum value, given by Eq. 22–106.

The first of the two equations mentioned is

$$h_p = Bv_m\rho c_p \left(\frac{\nu}{v_m D}\right)^{1/4} = Bv_m\rho c_p (N_{Re})^{-1/4} \qquad (22\text{–}109)$$

where B = a function of ξ.

Some values of χ and B, taken from graphs in the paper, are presented in Table 22–4.

For the range of fully established turbulence

$$h_f = 0.03461 v_m \rho c_p \left(\frac{\nu}{v_m D}\right)^{1/4} \frac{0.969e^{-m_1 x} + 0.038e^{-m_2 x}}{0.873e^{-m_1 x} + 0.0068e^{-m_2 x}} \qquad (22\text{–}110)$$

which for $x = \infty$ simplifies to Eq. 22–106.

Summarizing, the following cases must be distinguished:

1. Hydrodynamic and thermal states fully established (Eq. 22–106).

2. Hydrodynamic state fully established, because of starting section at inlet temperature; thermal state unestablished (Eq. 22–105).

3. Hydrodynamic and thermal states unestablished with hydrodynamic and thermal actions starting at the same point (Eqs. 22–109 and 110).

Major differences in the experimental results of different reliable observers are due to different starting conditions which had not been recognized and considered before Latzko's analysis. This, however,

TABLE 22–4

NUMERICAL VALUES OF FUNCTIONS OCCURRING IN EQS. 22–107 AND 109 FOR CONVECTION IN STARTING SECTION OF TUBE

ξ	χ	$3600 \cdot B$
0	0	∞
0.05	0.025	258
0.1	0.045	216
0.15	0.07	198
0.2	0.10	186
0.25	0.135	179
0.3	0.17	173
0.35	0.205	168
0.4	0.24	164
0.45	0.28	161
0.5	0.32	158
0.55	0.36	155
0.6	0.40	153
0.65	0.445	151
0.7	0.485	149
0.75	0.53	148
0.8	0.57	146
0.85	0.615	145
0.9	0.655	144
0.95	0.70	143
1.0	0.746	142

will have to be checked more in detail by experiments. It may be improved, for instance, in so far as it is based on von Kármán's so-called "seventh-root rule" of velocity distribution which more recently has been found to hold only in a restricted range of Reynolds numbers.

22–6. Forced Heat Convection in Turbulent Flow Parallel to a Plane Plate at Uniform Temperature

This case is also treated in the same paper of Latzko [1921]. It is supposed that the plate is so thin that the leading edge does not affect the arriving stream. Instead of this a streamlined leading edge may be used [see Jakob and Dow 1946]. Neglecting the heating effect of friction (as in Sect. 22–4, case *a*) and assuming uniform pressure, the velocity field and temperature field are similar, as again will be anticipated and dealt with more in detail later on.

According to von Kármán [1921], the velocity distribution and thickness of the boundary layer in the turbulent range may be expressed by

$$v_x = v_0 \left(\frac{y}{b}\right)^{1/7} \tag{22–111}$$

and

$$b = 0.366x \left(\frac{\nu}{v_0 x}\right)^{1/5} \tag{22–112}$$

with the denotations used in Sect. 22–4. Then, for similarity reasons,

$$\theta = \theta_0 \left(\frac{y}{b}\right)^{1/7} \tag{22–113}$$

where θ_0 is the temperature excess of the undisturbed stream above surface temperature.

Referring to Fig. 22–13, which represents a boundary layer in development, the heat balance of an element of length dx and unit width in the

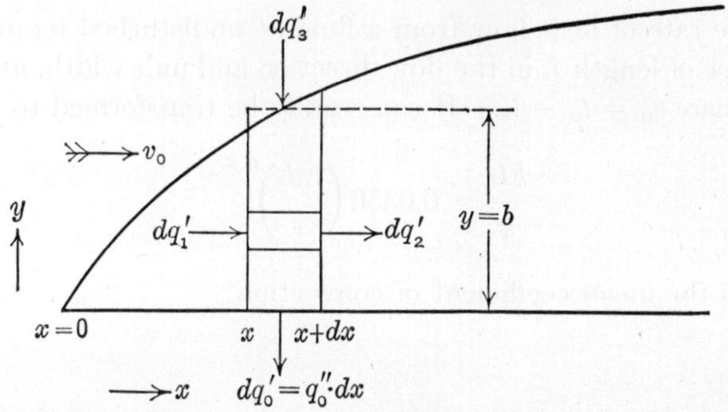

FIG. 22–13. Heat flow through a differential element of a boundary layer.

direction z (perpendicular to the drawing plane) will be set up. The prime and double prime signs may refer to unit width in the direction z, and unit area in the x, z plane, respectively. Then the following rates of heat flow to and from the element occur:

$dq_0' = q_0'' \cdot dx$ will leave the volume element.

$q_1' = \int_{y=0}^{y=b} v\rho c_p \theta \cdot dy$ will enter from the left.

$q_2' = \int_{y=0}^{y=b} v\rho c_p\theta \cdot dy + \left(\frac{\partial}{\partial x} \int_{y=0}^{y=b} v\rho c_p\theta \cdot dy \right) dx$ will leave to the right.

$dq_3' = \left(\frac{\partial}{\partial x} \int_{y=0}^{y=b} v \cdot dy \right) \rho c_p\theta_0 \cdot dx$ enters because of the feeding of the volume of the layer from above (increase of volume by influx at $\theta = \theta_0$).

Since

$$q_1' + dq_3' = dq_0' + q_2'$$

where $q_2' - q_1'$ is a differential of the order of magnitude of dq_0' and dq_3',

$$\frac{\partial}{\partial x} \int_0^b v\rho c_p\theta \cdot dy = -\rho c_p\theta_0 \frac{\partial}{\partial x} \int_0^b v \cdot dy + q_0'' = 0 \qquad (22\text{–}114)$$

By substitution from Eqs. 22–111, 112, and 113 and integration,

$$q_0'' = 0.0285\rho c_p v_0\theta_0 \left(\frac{\nu}{v_0 x} \right)^{1/5} \qquad (22\text{–}115)$$

and

$$q_0' = \int_{x=0}^{x=L} q_0'' \cdot dx = 0.0356\rho c_p v_0\theta_0 L \left(\frac{\nu}{v_0 L} \right)^{1/5} \qquad (22\text{–}116)$$

This is the rate of heat flow from a fluid at undisturbed temperature t_0 to a surface of length L in the flow direction and unit width, at temperature t_s, where $\theta_0 = t_0 - t_s$. It can readily be transformed to

$$\frac{hL}{k} = 0.0356 \left(\frac{v_0 L}{\nu} \right)^{0.8} \frac{\nu}{\alpha} \qquad (22\text{–}117)$$

where h is the mean coefficient of convection.

CHAPTER 23

Similarity Relations for Heat Convection

23–1. Heat Transfer in Free Convection

Nusselt [1909(*a*), 1910(*a*), 1915(*a*)] seems to have first employed the principle of similarity to the field of heat transfer, in two papers, one devoted to forced convection in tubes and the other to free convection in general.

In this section some basic relations on heat transfer in free convection will be derived from the principle of similarity, using both the differential and the dimensional method (see Sect. 21–4).

a. Differential Method

Referring to the differential method, Nusselt's original procedure will be followed. He started from the following equations:

$$dq = -k\frac{\partial t}{\partial n}dA = h\theta_s \cdot dA \tag{23–1}$$

$$\rho\left(\frac{\partial v_x}{\partial \tau} + v_x\frac{\partial v_x}{\partial x} + v_y\frac{\partial v_x}{\partial y} + v_z\frac{\partial v_x}{\partial z}\right) = -\frac{\partial p}{\partial x} + \mu\left(\frac{\partial^2 v_x}{\partial x^2} + \frac{\partial^2 v_x}{\partial y^2} + \frac{\partial^2 v_x}{\partial z^2}\right) \tag{23–2}$$

$$\rho\left(\frac{\partial v_y}{\partial \tau} + v_x\frac{\partial v_y}{\partial x} + v_y\frac{\partial v_y}{\partial y} + v_z\frac{\partial v_y}{\partial z}\right) = -\frac{\partial p}{\partial y} + \mu\left(\frac{\partial^2 v_y}{\partial x^2} + \frac{\partial^2 v_y}{\partial y^2} + \frac{\partial^2 v_y}{\partial z^2}\right) \tag{23–3}$$

$$\rho\left(\frac{\partial v_z}{\partial \tau} + v_x\frac{\partial v_z}{\partial x} + v_y\frac{\partial v_z}{\partial y} + v_z\frac{\partial v_z}{\partial z}\right)$$
$$= -\frac{\partial p}{\partial z} + \mu\left(\frac{\partial^2 v_z}{\partial x^2} + \frac{\partial^2 v_z}{\partial y^2} + \frac{\partial^2 v_z}{\partial z^2}\right) + g\beta\rho\theta \quad (23\text{–}4)$$

$$\frac{\partial v_x}{\partial x} + \frac{\partial v_y}{\partial y} + \frac{\partial v_z}{\partial z} = 0 \quad (23\text{–}5)$$

$$\rho c_p\left(\frac{\partial \theta}{\partial \tau} + v_x\frac{\partial \theta}{\partial x} + v_y\frac{\partial \theta}{\partial y} + v_z\frac{\partial \theta}{\partial z}\right) = k\left(\frac{\partial^2 \theta}{\partial x^2} + \frac{\partial^2 \theta}{\partial y^2} + \frac{\partial^2 \theta}{\partial z^2}\right) \quad (23\text{–}6)$$

where $\theta = t - t_e$, the temperature of environment being termed t_e. All these equations are known to us from Chapter 3 (see Eqs. 3–14, 17, 19, 20, 23, 24, and 25). However, it will be noticed that the fluid is considered incompressible so that for gases only moderate changes of pressure in the heat-transfer process are covered. Thermal expansibility, however, represented by β in Eq. 23–4, is included; in fact, it yields the driving force for free convection.

The variable n and the heat flow dq in Eq. 23–1 are considered positive in the direction from the surface into the fluid.

The boundary conditions of the system of equations are

$$v_x = 0; \quad v_y = 0; \quad v_z = 0; \quad \theta = 0 \quad \text{at} \quad n = \infty$$
$$v_x = 0; \quad v_y = 0; \quad v_z = 0; \quad \theta = \theta_s \quad \text{at} \quad n = 0$$

A case may be considered where the geometrical configuration is defined by one distance, for instance, the diameter D of an infinitely long, horizontal cylinder. According to the general definition of similarity given in Sect. 21–4, the following equations must be valid if the two similar systems are denoted by the subscripts 1 and 2 and the Greek letters without subscript designate constants:

$$\left.\begin{aligned}
x_2 &= \lambda x_1; \quad y_2 = \lambda y_1; \quad z_2 = \lambda z_1; \quad D_2 = \lambda D_1 \\
\tau_2 &= \zeta \tau_1 \\
\theta_2 &= \phi \theta_1; \quad \theta_{s,2} = \phi \theta_{s,1} \\
v_{x,2} &= \omega v_{x,1}; \quad v_{y,2} = \omega v_{y,1}; \quad v_{z,2} = \omega v_{z,1} \\
g_2 &= \xi g_1 \\
p_2 &= \psi p_1 \\
\beta_2 &= \chi \beta_1 \\
\rho_2 &= \sigma \rho_1 \\
\mu_2 &= \eta \mu_1 \\
c_{p,2} &= \epsilon c_{p,1} \\
k_2 &= \kappa k_1
\end{aligned}\right\} \quad (23\text{–}7)$$

Writing, now, any equation of our system, for instance, Eq. 23–4 for cases 1 and 2,

$$\rho_1\left(\frac{\partial v_{z,1}}{\partial \tau_1} + v_{x,1}\frac{\partial v_{z,1}}{\partial x_1} + v_{y,1}\frac{\partial v_{z,1}}{\partial y_1} + v_{z,1}\frac{\partial v_{z,1}}{\partial z_1}\right)$$

$$= -\frac{\partial p_1}{\partial z_1} + \mu_1\left(\frac{\partial^2 v_{z,1}}{\partial {x_1}^2} + \frac{\partial^2 v_{z,1}}{\partial {y_1}^2} + \frac{\partial^2 v_{z,1}}{\partial {z_1}^2}\right) + g_1\beta_1\rho_1\theta_1 \quad (23\text{–}8)$$

$$\rho_2\left(\frac{\partial v_{z,2}}{\partial \tau_2} + v_{x,2}\frac{\partial v_{z,2}}{\partial x_2} + v_{y,2}\frac{\partial v_{z,2}}{\partial y_2} + v_{z,2}\frac{\partial v_{z,2}}{\partial z_2}\right)$$

$$= -\frac{\partial p_2}{\partial z_2} + \mu_2\left(\frac{\partial^2 v_{z,2}}{\partial {x_2}^2} + \frac{\partial^2 v_{z,2}}{\partial {y_2}^2} + \frac{\partial^2 v_{z,2}}{\partial {z_2}^2}\right) = g_2\beta_2\rho_2\theta_2 \quad (23\text{–}9)$$

Substituting from Eq. 23–7 in Eq. 23–9,

$$\sigma\rho_1\left[\frac{\omega}{\zeta}\cdot\frac{\partial v_{z,1}}{\partial \tau_1} + \frac{\omega^2}{\lambda}\left(v_{x,1}\frac{\partial v_{z,1}}{\partial x_1} + v_{y,1}\frac{\partial v_{z,1}}{\partial y_1} + v_{z,1}\frac{\partial v_{z,1}}{\partial z_1}\right)\right]$$

$$= -\frac{\psi}{\lambda}\cdot\frac{\partial p_1}{\partial \zeta_1} + \frac{\eta\omega}{\lambda^2}\mu_1\left(\frac{\partial^2 v_{z,1}}{\partial {x_1}^2} + \frac{\partial^2 v_{z,1}}{\partial {y_1}^2} + \frac{\partial^2 v_{z,1}}{\partial {z_1}^2}\right) + \xi\chi\sigma\phi g_1\beta_1\rho_1\theta_1 \quad (23\text{–}10)$$

Equations 23–8 and 10 must be identical. This occurs only if

$$\frac{\sigma\omega}{\zeta} = \frac{\sigma\omega^2}{\lambda} = \frac{\psi}{\lambda} = \frac{\eta\omega}{\lambda^2} = \xi\chi\sigma\phi \quad (23\text{–}11)$$

Equations 23–2, 3, and 5, when treated as has just been done with Eq. 23–4, lead also to some of the relations given by Eq. 23–11. Equation 23–6, however, yields

$$\frac{\sigma\epsilon\phi}{\zeta} = \frac{\sigma\epsilon\omega\phi}{\lambda} = \frac{\kappa\phi}{\lambda^2} \quad (23\text{–}12)$$

Equations 23–11 and 12 show that the following five relations exist * between the eleven quantities, λ, ζ, ϕ, ω, ξ, ψ, χ, σ, η, ϵ, and κ.

$$\eta = \lambda\omega\sigma \quad (23\text{–}13)$$

$$\zeta = \frac{\lambda}{\omega} \quad (23\text{–}14)$$

$$\psi = \omega^2\sigma \quad (23\text{–}15)$$

$$\eta^2 = \lambda^3\phi\xi\sigma^2\chi \quad (23\text{–}16)$$

$$\kappa = \eta\epsilon \quad (23\text{–}17)$$

* One more relation might be expected. However, there is one identity involved in Eqs. 23–11 and 12. Equation 23–14 can be obtained by equating the first two terms of either Eq. 23–11 or 12.

Because of the existence of these five conditions, only six out of the eleven quantities mentioned can be arbitrarily chosen.

Substitutions from Eq. 23–7 in these equations yield a number of similarity conditions. According to Eq. 23–13,

$$\frac{\rho_1 L_1 v_1}{\mu_1} = \frac{\rho_2 L_2 v_2}{\mu_2} \tag{23–18}$$

where L stands for any considered distance, and v for any considered velocity. This is known to us already from Article *b* of Sect. 21–4 and means that similarity exists only when any Reynolds number, defined in one system, is equal to the corresponding number in the second system.

Equation 23–14 leads to the condition

$$\frac{\tau_1 v_1}{L_1} = \frac{\tau_2 v_2}{L_2} \tag{23–19}$$

It is self-explanatory that this equation is always satisfied for steady-state conditions, which will be assumed for the time being.

Equation 23–15 yields

$$\frac{\rho_1 {v_1}^2}{p_1} = \frac{\rho_2 {v_2}^2}{p_2} \tag{23–20}$$

This means that occurring kinetic energies and pressures must have the same ratio in similar systems.† In free convection, kinetic energies are generally neglected; they can as well be considered to be virtually equal all over the system; this would hold also for the corresponding pressures.

Hence, Eq. 23–18 is not new to us, and Eqs. 23–19 and 20 are not used presently. Equations 23–16 and 17, on the other hand, reveal two important new relations:

$$\frac{g_1 \beta_1 {\rho_1}^2 {D_1}^3 \theta_{s,1}}{{\mu_1}^2} = \frac{g_2 \beta_2 {\rho_2}^2 {D_2}^3 \theta_{s,2}}{{\mu_2}^2} = N_{Gr} \tag{23–21}$$

$$\frac{\mu_1 c_{p,1}}{k_1} = \frac{\mu_2 c_{p,2}}{k_2} = N_{Pr} \tag{23–22}$$

They show that there are two dimensionless groups, called Grashof number (N_{Gr}) and Prandtl number (N_{Pr}), which must have the same values

† The name Euler number has been suggested for this ratio [Thomas 1943].

in both systems, as Reynolds number (N_{Re}) must, if similarity is to exist.‡

The Grashof number can also be written in the following form:

$$N_{Gr} = \frac{g\beta D^3\theta_s}{\nu^2} \tag{23–23}$$

and for ideal gases, according to Eq. 3–18, it becomes

$$N_{Gr} = \frac{gD^3\theta_s}{\nu^2 T_0} \tag{23–24}$$

Another form of the Prandtl number is

$$N_{Pr} = \frac{\nu}{\alpha} \tag{23–25}$$

Returning now to Eq. 23–1, a twelfth similarity ratio ι may be introduced, defined by

$$h_2 = \iota \cdot h_1$$

Then:

$$-k_1 \frac{\partial t_1}{\partial n_1} = h_1\theta_{s,1}$$

and

$$-k_2 \frac{\partial t_2}{\partial n_2} = h_2\theta_{s,2}$$

or

$$-\frac{\kappa\phi}{\lambda} k_1 \frac{\partial t_1}{\partial n_1} = \iota\phi h_1\theta_{s,1}$$

which is possible only if

$$\frac{\kappa}{\lambda} = \iota$$

‡ Groeber [1921], who first used the name Grashof number for the quantity defined by Eq. 23–21, suggested also (l.c., p. 122) to use the two first letters of this and corresponding names as symbols in equations, i.e., *Gr, Re*, etc. This has widely been accepted. However, according to the general principles of letter symbol standardization, formulated by the American Standards Association [1943]: "A letter symbol is a single character, with subscripts or superscripts if required, used to designate a physical magnitude in mathematical equations and expressions. Two or more symbols together always represent a product." Hence this association has suggested N_{Gr}, N_{Re}, etc., as symbols for the dimensionless groups. The present author does not know why Groeber suggested the name of Grashof for this dimensionless group. The other names used to characterize dimensionless groups, also some which will be mentioned later on, are chosen to honor men who have had merit in the development of fluid mechanics and heat transmission.

Using $D_2/D_1 = \lambda$ as above, it follows that

$$\frac{h_1D_1}{k_1} = \frac{h_2D_2}{k_2} = \frac{hD}{k} = N_{Nu} \tag{23-26}$$

By this equation another dimensionless magnitude, N_{Nu}, is defined, which is called Nusselt number.

Now, Eqs. 23-1 to 6 have all been used, and the result is that a system of only three variables (N_{Gr}, N_{Pr}, and N_{Nu}) has been left, two of which may be considered as independent variables in writing

$$N_{Nu} = \Phi(N_{Gr}, N_{Pr}) \tag{23-27}$$

The meaning is that each fixed pair of values, N_{Gr} and N_{Pr}, determines a series of similar cases all of which have the same Nusselt number, N_{Nu}. All imaginable cases, which may be thought of as points in a quadrant of a plane with coordinates $N_{Gr} > 0$ and $N_{Pr} > 0$, determine one unique surface in the space with positive values of N_{Nu} as third coordinate.

In other words, having chosen a value of N_{Pr}, for instance, $N_{Pr} = 0.74$, which comprises all diatomic gases at atmospheric pressure and 212°F, then N_{Nu} is a unique function of N_{Gr}, and the result will be the same, whatever the single quantities in Eq. 23-23 are, if only N_{Gr} is the same. This is a tremendous accomplishment which is due to the principle of similarity. It has been checked by a great number of experiments with different geometrical dimensions, substances, temperatures, etc., as will be shown later on.

It may be noticed that Eq. 23-27 covers also the free convection on a vertical wall if only the height L of the wall instead of D is introduced in the definition of the Grashof number. Hence, it is not surprising, though satisfying, that the result of Lorenz' derivation can be brought into this form by rewriting Eq. 22-26 as

$$N_{Nu} = C\sqrt[4]{N_{Gr} \cdot N_{Pr}} \tag{23-28}$$

whereas the Schmidt-Beckmann-Pohlhausen equation 22-23, replacing x by L, can be expressed by

$$N_{Nu} = \sqrt[4]{N_{Gr}} \cdot \chi_0'(N_{Pr}) \tag{23-29}$$

b. Dimensional Method

Referring to the use of dimensional analysis for the same free convection problem, the description given in the elementary book of Jakob and

Hawkins [1942, p. 67] will be followed here. The procedure is analogous to that used in Article *c* of Sect. 21–4.

First, a guess is made of which quantities h may depend on, and the following equation is set up:

$$h = CL^a\theta_s{}^b\rho^f\mu^i\beta^j c_p{}^m g^n k^r \tag{23–30}$$

where C is a constant.

The following symbols for the basic units are used:

H for heat energy.
T for time.
Λ for length.
M for mass.
Θ for temperature.

Writing an equation of the units only, analogous to Eq. 21–26,

$$\mathrm{HT^{-1}\Lambda^{-2}\Theta^{-1}} =$$
$$\mathrm{\Lambda^a\Theta^b(M\Lambda^{-3})^f(MT^{-1}\Lambda^{-1})^i\Theta^{-j}(HM^{-1}\Theta^{-1})^m(T^{-2}\Lambda)^n(HT^{-1}\Lambda^{-1}\Theta^{-1})^r}$$

Identifying the sums of the exponents to the left and right,

For H: $1 = m + r$
" T: $-1 = -i - 2n - r$
" Λ: $-2 = a - 3f - i + n - r$
" M: $0 = f + i - m$
" Θ: $-1 = b - j - m - r$

By means of these equations five exponents may be expressed in terms of the three others, say j, m, n:

$$a = -1 + 3n$$
$$b = j$$
$$f = 2n$$
$$i = m - 2n$$
$$r = 1 - m$$

Substituting these in Eq. 23–30, rearranging, and collecting the terms containing like exponents,

$$\frac{hL}{k} = C\left(\frac{g\rho^2L^3}{\mu^2}\right)^n(\beta\theta_s)^j\left(\frac{\mu c_p}{k}\right)^m \tag{23–31}$$

This equation can be further simplified if the role of g and β in Eqs. 23–1 to 6 is considered.

Since g and β appear only in the form of a product $(g\beta)$ in these equations, they must have the same exponent. Therefore Eq. 23–31 becomes

$$\frac{hL}{k} = C\left(\frac{g\beta\rho^2L^3\theta_s}{\mu^2}\right)^n\left(\frac{\mu c_p}{k}\right)^m = C\left(\frac{g\beta L^3\theta_s}{\nu^2}\right)^n\left(\frac{\nu}{\alpha}\right)^m \tag{23–32}$$

or

$$N_{Nu} = C(N_{Gr})^n(N_{Pr})^m \tag{23–33}$$

which is a special form of the general equation 23–27.

For ideal gases, according to Eq. 3–18, $\beta = 1/T_e$. Hence

$$N_{Gr} = \frac{gL^3\theta_s}{\nu^2T_e} \tag{23–34}$$

From Eq. 23–28 it is seen that for gases, according to the old theory of Lorenz,

$$m = n = \tfrac{1}{4}$$

The constants C, n, and m have to be determined by experiments. Theoretically, for either laminar or turbulent flow, three experiments under different conditions should be sufficient to find these three constants and so would allow the determination of h for all possible changes of the eight independent variables of Eq. 23–30. However, a great number of experiments has been performed, some at a time when the theory of similarity was not yet known. For a comparison of this theory with the results of actual measurements see Chapters 25 and 26.

23–2. Heat Transfer in Forced Convection

In deriving the similarity relations which govern processes of forced convection, Nusselt [1910(a)] used the differential method, as he did when dealing with free convection. Here dimensional analysis will be used for the derivation.

The equation of forced convection of heat inside a tube of diameter D is assumed to be of the form

$$h = Cv^aD^b\mu^fk^j\rho^mc_p{}^n \tag{23–35}$$

where C is a constant and v is the mean velocity in axial direction.

In the same way as before, when dealing with Eq. 23–30,

$$\mathrm{HT^{-1}\Lambda^{-2}\Theta^{-1}} \\ = (\mathrm{T^{-1}\Lambda})^a\Lambda^b(\mathrm{MT^{-1}\Lambda^{-1}})^f(\mathrm{HT^{-1}\Lambda^{-1}\Theta^{-1}})^j(\mathrm{M\Lambda^{-3}})^m(\mathrm{HM^{-1}\Theta^{-1}})^n$$

Identifying the sums of the exponents,

$$
\begin{aligned}
\text{For H:} &\quad 1 = j + n \\
\text{" T:} &\quad -1 = -a - f - j \\
\text{" } \Lambda\text{:} &\quad -2 = a + b - f - j - 3m \\
\text{" M:} &\quad 0 = f + m - n \\
\text{" } \Theta\text{:} &\quad -1 = -j - n
\end{aligned}
$$

and by algebra, considering that the first and last of these equations are identical,

$$
\begin{aligned}
a &= m \\
b &= m - 1 \\
f &= n - m \\
j &= 1 - n
\end{aligned}
$$

Substituting in Eq. 23–35,

$$\frac{hD}{k} = C\left(\frac{vD\rho}{\mu}\right)^n \left(\frac{\mu c_p}{k}\right)^m \tag{23–36}$$

or

$$N_{Nu} = C(N_{Re})^n(N_{Pr})^m \tag{23–37}$$

The analogy between Eqs. 23–33 and 37 is striking. The Grashof number plays a role in free convection similar to that of the Reynolds number in forced convection. In both cases, the Prandtl number, a mere property of the substance, has an additional influence. The constants C, m, and n, of course, are not the same in both cases. But, again, theoretically three experiments would be sufficient to cover the influence of the six independent variables of Eq. 23–35. Since similarity exists only if the kind of flow is the same, different constants will occur in the laminar and turbulent region, as was the case for free convection.

In the special case where $m = n$, the two dimensional groups in Eqs. 23–36 and 37 are combined in one group, defined by

$$N_{Pe} = N_{Re} \cdot N_{Pr} = \frac{vD}{\alpha} \tag{23–38}$$

which is called Péclet number in honor of the French physicist Péclet.

Another combined group is the Graetz number, which for circular cross section is defined by

$$N_{Gz} = \frac{\pi}{4} N_{Pe} \cdot \frac{D}{L} \tag{23–39}$$

This group occurred in the exponents of Graetz's equation 22–37 in the form $\frac{v_m s}{\alpha} \cdot \frac{s}{x}$, which would be equal to $\frac{1}{\pi} N_{Gz}$ if x is considered the characteristic length.

A more general definition of this group, not restricted to circular cross section of the channel, is

$$N_{Gz} = \frac{\dot{m} c_p}{kL} \tag{23–40}$$

where $\dot{m}$ is the time rate of mass flow.

Using the Graetz number, Eq. 22–59 could be written as the following relation between dimensionless groups:

$$\frac{h_{\ln} 2s}{k} = \frac{v_m s}{\alpha} \cdot \frac{s}{x} \ln \frac{\theta_0}{\theta_m}$$

or

$$N_{Nu} = \frac{1}{\pi} N_{Gz} \cdot \ln \frac{\theta_0}{\theta_m} \tag{23–41}$$

According to Eqs. 23–38 and 39, N_{Gz} can be considered a function of N_{Re} and N_{Pr}. Hence, Eq. 23–41 is in agreement with Eq. 23–37. Referring to the additional ratio D/L see Sect. 23–5.

Brinn, Friedman, Gluckert, and Pigford [1948] directed attention to the interesting fact that the transfer of heat to a fluid in rod-like flow in a tube is mathematically identical to unsteady-state heating of a long cylinder with negligible surface resistance. Considering that the time of passage through the tube is $\tau = V/\dot{V}$; $V = \pi s^2 L$; and $\dot{V} = \dot{m}/\rho$, Eq. 23–40 can also be written in the form

$$N_{Gz} = \frac{\rho c_p \cdot \pi s^2}{k\tau} = \frac{\pi s^2}{\alpha \tau} = \frac{\pi}{N_{Fo}} \tag{23–40a}$$

where N_{Fo} is the Fourier number. Herewith Eq. 22–60 becomes identical with Eq. 13–37, as claimed by the authors.

23–3. Interrelation between the Equations for Free and Forced Convection

Brown and Marco [1942, p. 111] directed attention to a relation between Eqs. 23–33 and 37, which helps to make the similarity of these two equations understandable. The present author has slightly modified their deduction to the following reasoning.

Considering the heat transfer by free convection from a vertical surface of height L at temperature t_s to the air of environment at temperature t_e, it may be calculated as though the boundary layer, driven by the buoyant force, were moving upward with a mean velocity v_m. Assuming the whole temperature drop perpendicular to the surface to be concentrated in the boundary layer and its mean temperature to be $t_f = (t_s + t_e)/2$, the buoyancy force for a piece of the layer of unit horizontal cross-sectional area and unit height would be

$$\frac{\Delta p}{L} = g\beta\rho_f \frac{t_s - t_e}{2}$$

For unit cross-sectional area and height L

$$\Delta p = g\beta\rho_f L \frac{t_s - t_e}{2}$$

Combining this with Eq. 23–37, but considering that in the flow parallel to plates the path length of flow, L, counted from the leading edge, is the characteristic length for the dimensionless groups,

$$N_{Nu} = C(N_{Re})^n(N_{Pr})^m = C\left(\frac{v_f L}{\nu}\right)^n (N_{Pr})^m = C\left(\frac{g\beta L^3\theta}{\nu^2}\right)^{n/2}(N_{Pr})^m$$
$$= C(N_{Gr})^{n/2}(N_{Pr})^m \tag{23–42}$$

Hence, the square root of the Grashof number can be considered a special case of the Reynolds number. This theoretical result is in satisfactory agreement (at least in the order of magnitude) with the results of experiments.

23–4. A Graphical Interpretation of Nusselt's Number

The meaning of Nusselt's number can easily be understood from the following deduction.§

Consider a surface at homogeneous temperature t_s and a fluid in contact with the surface, having the temperature t_e (environment temperature) at a great distance from the surface. Then the heat flow can be expressed either by using a coefficient of heat transfer h or as thermal conduction through a thin film on the surface. Hence

$$h(t_s - t_e) = -k_f\left(\frac{\partial t}{\partial y}\right)_{y=0} \tag{23–43}$$

where k_f = the thermal conductivity of the fluid in the film, and
t = the temperature at distance y from the surface.

§ Slight modification of a personal communication of Dr. R. C. Martinelli to the author.

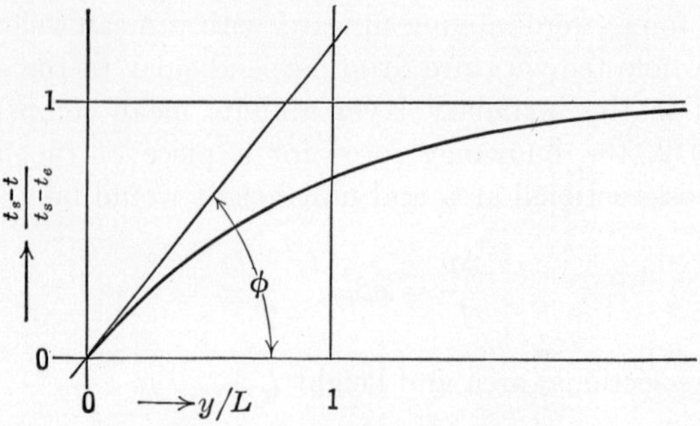

Fig. 23–1. Graphic interpretation of the Nusselt number according to Eq. 23–45.

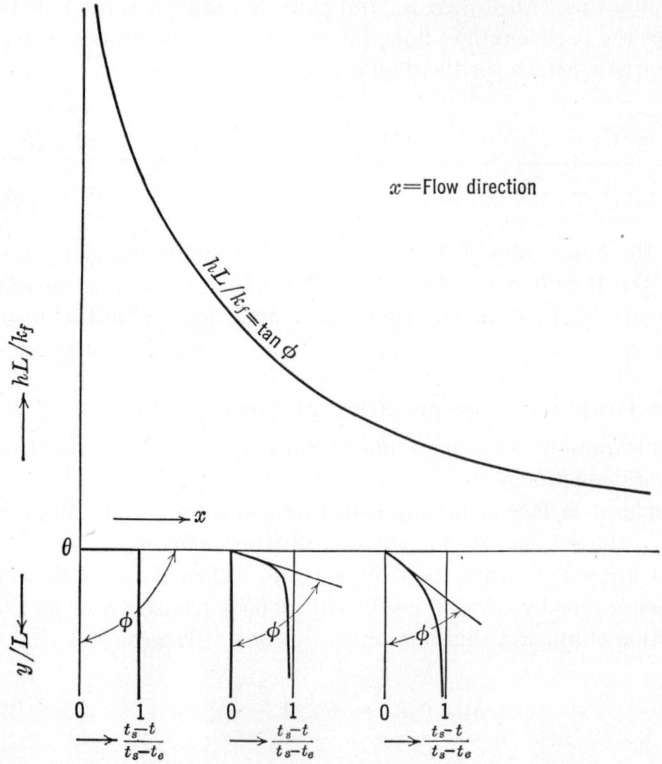

Fig. 23–2. Graphic interpretation of the change of the Nusselt number in the flow of a fluid parallel to a plane surface.

Introducing a characteristic length L, by simple algebra,

$$\frac{hL}{k_f} = -\frac{\left(\frac{\partial t}{\partial y}\right)_{y=0}}{(t_s - t_e)/L} = \left[\frac{\partial\left(\frac{t_s - t}{t_s - t_e}\right)}{\partial(y/L)}\right]_{y=0} \tag{23-44}$$

or

$$N_{Nu} = \tan\phi \tag{23-45}$$

since the first term in Eq. 23-44 is a Nusselt number, and the last term is equal to $\tan\phi$ in Fig. 23-1. Hence || the Nusselt number is nothing but the ratio of the temperature gradient $\partial t/\partial y$ at $y = 0$, to the constant average gradient $(t_s - t_e)L$.

If, for instance, the surface is a plane plate and a fluid is flowing parallel to this plate in the direction x (see Fig. 23-2), then the temperature distribution in the direction y and the temperature gradient,

$$\left(\frac{\partial t}{\partial y}\right)_{y=0} = \left[\frac{(t_s - t_e)}{L}\right] \tan\phi,$$

change as indicated in the figure. Hence the Nusselt number starts with an infinite value at $x = 0$ and decreases with increasing x, as may be seen from the figure.

23-5. More Complex Cases of Similarity

So far only such geometrically simple cases as an infinitely long cylinder of diameter D or tube of inner diameter D or infinitely wide wall of height L have been considered. If more than one geometrical dimension is necessary to describe the configuration, ratios of these dimensions have to be added as dimensionless groups. The heat transfer to a fluid, passing a tube of diameter D and length L, for instance, could be described by an equation

$$N_{Nu} = C(N_{Re})^n(N_{Pr})^m\left(\frac{D}{L}\right)^i \tag{23-46}$$

instead of Eq. 23-37.

Here four experiments would theoretically be sufficient to find the constants C, n, m, and i for a given type of flow (laminar or turbulent).

There are other items which make the situation more complex. So far, it has been assumed that the physical properties of the fluid are independent of pressure and temperature. Fortunately the influence of pressure on μ and k is small in most practical cases. On the other hand,

|| The minus sign comes from the definition of the heat-flow direction and is, therefore, irrelevant in the present interpretation of the Nusselt number.

both the dynamic and kinematic viscosity decrease very much with increasing temperature for liquids, and increase considerably for gases. The influence of temperature on the thermal conductivity and diffusivity of gases is quite similar to that on the two kinds of viscosity. For liquids, k and α do not change much with the temperature, except for water. Many details have been given in Chapters 5 and 6. Complete similarity would exist only if, at any distance from the surface or in the flow direction, the viscosities and the other properties had the same ratios for fluids under comparison; otherwise new dimensionless groups would have to be introduced.

In particular, it will be understood that there is a lack of similarity in the direction of decreasing temperature if gases and liquids are compared, notwithstanding exact geometrical similarity of the configuration, due to the opposite trend of change of viscosity with temperature. Another case where similarity is deficient is that of heating or cooling by a surface under otherwise similar conditions. A liquid flowing through a tube, for instance, is less viscous close to the hot surface than in the colder core when heated from outside, whereas just the opposite is the case when the liquid is cooled.

The addition of a dimensionless group $(\mu_s/\mu)^j$ has been proved as a practical means of taking care of these counteracting effects, where μ_s and μ are the viscosities at surface and core temperature, respectively. This would mean an extension of Eq. 23–46 to the form

$$N_{Nu} = C(N_{Re})^n(N_{Pr})^m\left(\frac{D}{L}\right)^i\left(\frac{\mu_s}{\mu}\right)^j \qquad (23\text{–}47)$$

23–6. The Influence of Temperature upon the Similarity Relations for Gases

Restricting the consideration to gases, Nusselt [1915(a)] succeeded in extending the validity of Eq. 23–27 to temperature differences for which the properties of the gases cannot be assumed as constant any more.

In the dimensionless groups in this equation Nusselt uses mean values of the properties, taken over the boundary layer, denoted by the subscript m, and defined as follows:

$$\frac{\theta_s}{T_m} = \int_{T_0}^{T_s}\frac{dT}{T} = \ln\frac{T_s}{T_0}$$

or

$$T_m = \frac{\theta_s}{\ln(T_s/T_0)} \qquad (23\text{–}48)$$

$$\mu_m = \frac{1}{\theta_s}\int_{T_0}^{T_s} \mu \cdot dT; \quad c_{p,m} = \frac{1}{\theta_s}\int_{T_0}^{T_s} c_p \cdot dT; \quad k_m = \frac{1}{\theta_s}\int_{T_0}^{T_s} k \cdot dT \tag{23–49}$$

$$\rho_m = \frac{1}{\theta_s}\int_{T_0}^{T_s} \rho \cdot dT = \frac{1}{\theta_s}\int_{T_0}^{T_s} \rho_0 T_0 \frac{dT}{T} = \rho_0 \frac{T_0}{T_m} \tag{23–50}$$

It is further assumed that the influence of temperature on μ, c_p, and k can be taken care of by empirical equations of the form

$$\mu = \mu_0 \left(\frac{T}{T_0}\right)^{\eta}; \quad c_p = c_{p_0}\left(\frac{T}{T_0}\right)^{\epsilon}; \quad k = k_0\left(\frac{T}{T_0}\right)^{\kappa} \tag{23–51}$$

where η, ϵ, and κ are constants (of course different from those defined by Eq. 23–7).

Nusselt also derived an exact expression for this case, starting from differential equations which are more general than Eqs. 23–2 to 6 in so far as ρ, μ, and k are considered dependent on temperature.

Comparing again two similar systems 1 and 2, obviously,

$$\phi = \frac{T_2}{T_1} = \frac{T_{0,2}}{T_{0,1}} = \frac{T_{s,2}}{T_{s,1}}$$

or

$$\frac{T_{s,1}}{T_{0,1}} = \frac{T_{s,2}}{T_{0,2}} = N_T \tag{23–52}$$

where N_T is a constant.

The differential method of similarity further reveals that N_{Nu} depends on two more dimensionless groups, namely, on a gravity term ¶

$$\frac{g\rho^2 D^3}{\mu^2} = N_G \tag{23–53}$$

and the Prandtl number

$$\frac{\mu c_p}{k} = N_{Pr} \tag{23–22}$$

Hence

$$N_{Nu} = \mathrm{X}(N_G, N_{Pr}, N_T) \tag{23–54}$$

where X is a function symbol.

This is a relationship which can be used for any temperature difference. It is a more general equation for gases than Eq. 23–27, and first of all it is seen that the Grashof number, as defined by Eq. 23–34, is split in two parts N_G and N_T. It is further seen that in Eq. 23–54 the

¶ This, instead of the Grashof number, occurs because for ideal gases in Eq. 23–7 $\chi = 1/\phi$, which simplifies Eqs. 23–11 and 16.

properties ρ, μ, c_p, and k can be taken either at T_s or T_0, because any of these quantities can be converted from one to the other temperature by multiplication with a power function of T_s/T_0, and since N_{Nu} depends on N_T, the equation remains satisfied.

Equation 23–27, when used with mean values of the gas properties, defined by Eqs. 23–48 to 51, actually satisfies Eq. 23–54 and can therefore be used for any temperature differences. This has been checked experimentally as will be shown later on. However, it does not eliminate the lack of similarity between the heating and cooling of geometrically similar systems. (See Sect. 23–5.)

In the same paper Nusselt [1915(a)] has also dealt with the case of forced convection, including the influence of gravity, i.e., free convection. In this case the velocity at great distance from the surface is not zero any more, but v_0, and for two similar cases:

$$\omega = \frac{v_{0,2}}{v_{0,1}}$$

This leads to the Reynolds number as influential quantity, and the general equation of heat transfer becomes

$$N_{Nu} = \Psi[(N_{Re})_0, N_{Pr}, N_{Gr}] \tag{23–55}$$

where Ψ is a function symbol, and the subscript 0 indicates that the Reynolds number at the undisturbed velocity v_0 has to be taken.

This again holds only for such temperature differences as do not sensibly affect the properties of the substance.

For gases this restriction does not exist for the form

$$N_{Nu} = \psi[(N_{Re})_0, N_G, N_T, N_{Pr}] \tag{23–56}$$

If v_0 is great, the influence of gravity vanishes, and Eq. 23–56 simplifies to

$$N_{Nu} = \Psi[(N_{Re})_0, N_T, N_{Pr}] \tag{23–57}$$

Finally, unsteady-state cases also can be included by making use of Eq. 23–19. For convenience the expressions in this equation may be divided by those of Eq. 23–18, yielding the dimensionless group

$$\frac{\mu}{\rho} \cdot \frac{\tau}{L^2} = N_\tau \tag{23–58}$$

which must be the same in similar cases. This leads to the relation

$$N_{Nu} = \Omega[(N_{Re})_0, N_G, N_T, N_{Pr}, N_\tau] \tag{23–59}$$

where Ω is the function symbol. Let, for instance, the surface be initially at T_i and suddenly be brought to T_s and kept at that temperature, i.e.,

$$T = T_i \text{ (including } T_s = T_i) \text{ at } \tau = 0$$

Then Eq. 23-59 expresses the fact that the coefficient of heat transfer depends also on the time which has elapsed since the start of the temperature change. This is not self-explanatory, indeed could not be expected at all. Nusselt claims to have discovered and explained this phenomenon first [1914].

Equation 23-59 is restricted to ideal gases; however, it would not present difficulties to develop an equation of that sort without this restriction.

CHAPTER 24

Similarity between Thermal and Dynamical Transfer, without and with the Boundary-layer Concept

24–1. Similarity Relations between Heat Transfer and Exchange of Impulse

So far, similarity between different cases of heat transfer has been dealt with. In this section similarity relations between different components of the convection process will be developed which are based on Reynolds' [1874] idea of proportionality between the velocity and temperature field near a heat-exchanging surface. Reynolds made it clear that such proportionality should include the effect of both molecular diffusion and mechanical mixing.

The first to apply this idea to convection was Prandtl [1910]. Considering the heat transfer between a horizontal tube and a fluid flowing through it under negligible compressibility, he compared the differential equations of the velocity field and temperature field in the steady state, the first being a special case of Eqs. 3–14, 19, 20, the second one being based on Eq. 3–24:

$$\frac{\partial v}{\partial \tau} + v\frac{\partial v}{\partial x} = -\frac{1}{\rho}\cdot\frac{dp}{dx} + \frac{\mu}{\rho}\cdot\frac{\partial^2 v}{\partial x^2} \tag{24–1}$$

$$\frac{\partial \theta}{\partial \tau} + v\frac{\partial \theta}{\partial x} = \frac{q'''}{\rho c_p} + \frac{k}{\rho c_p}\cdot\frac{\partial^2 \theta}{\partial x^2} \tag{24–2}$$

where v is the velocity in the x direction,

θ is the temperature excess of the fluid above surface temperature, and

q''' is the rate of heat generation in uniformly distributed heat sources per unit volume.

These two equations can be considered similar if the following conditions are satisfied:

$$\theta = Cv \quad \text{and} \quad \theta_m = Cv_m \tag{24-3}$$

$$\frac{q'''}{c_p} = -C\frac{dp}{dx} \tag{24-4}$$

$$\frac{k}{c_p} = \mu \quad \text{or} \quad N_{Pr} = 1 \tag{24-5}$$

where C is a constant, and subscript m stands for mean values. The last condition is nearly satisfied for gases with four or more atoms, to a certain approximation also for triatomic gases, and at extremely high or low temperatures for diatomic gases.

Similarity is lacking also in so far as non-reacting gases are free of heat sources or sinks. Prandtl overcame this difficulty by introducing a fourth condition:

$$q' = \pi r^2 q''' \tag{24-6}$$

where q' is the heat transferred from the surface per unit length, and r is the tube radius. Equation 24-6 involves the assumption that each volume element of the flow contributes the same amount to the heat given up to the surface and that no heat-flow resistance has to be considered for the transference of these amounts of heat through the fluid.

From Eqs. 24-3 to 6

$$q' = -\pi r^2 \frac{k}{\mu} \cdot \frac{\theta_m}{v_m} \cdot \frac{dp}{dx} \tag{24-7}$$

This equation, in its rather restricted range of validity, yields the heat transfer from the simple measurements of velocity, temperature, and pressure drop.

24-2. Original Form of the Combined Impulse and Boundary-layer Theories

The theory of similarity, as previously presented, did not make use of any assumptions about the mechanism of the process. A very general idea of the process was all that was needed, for instance, to know whether it was a case governed solely by conduction or also by convection. Now details of the mechanism of the process, though much simplified, will be considered.

The following procedure has been envisaged and used first by Prandtl [1910] and then, independently, by G. I. Taylor [1916].

It is assumed that in the bulk of a fluid, i.e., in the core of the flow, impulse as well as heat is exchanged by mixing movements only, and in a boundary layer which is in laminar motion along the wall impulse and heat exchange are solely due to the random movements of the molecules.

Considering again the flow through a tube of inner radius r, the balance of forces in axial direction becomes

$$2\pi r\mu \frac{dv}{dy} = -\pi r^2 \frac{dp}{dx} \tag{24–8}$$

where y is the radial distance of a point of the layer from the solid surface.

Supposing a linear velocity slope in this layer,

$$\frac{dv}{dy} = \frac{v_b}{b} \tag{24–9}$$

where b = the thickness of the boundary layer, and
v_b = the velocity at $y = b$, i.e., at the border between boundary layer and core of the flow.

From Eqs. 24–8 and 9

$$b = -2 \frac{\mu v_b}{r \cdot dp/dx} \tag{24–10}$$

Since all heat that is exchanged between surface and fluid must pass through the boundary layer,

$$q' = \frac{2\pi r k \theta_b}{b} \tag{24–11}$$

where q' = the rate of heat flow per unit length,
k = the thermal conductivity of the fluid, and
θ_b = the excess of the temperature t_b at $y = b$ over the surface temperature t_s.

Substituting b from Eq. 24–10 in Eq. 24–11,

$$q' = -\pi r^2 \frac{k}{\mu} \cdot \frac{\theta_b}{v_b} \cdot \frac{dp}{dx} \tag{24–12}$$

which is of the same type as, but not identical with, Eq. 24–7.

Considering now the core of the flow, Eqs. 24–1 and 2 are simplified by omitting the last term on the right side of each of them, since it is supposed that viscosity and thermal conductivity do not play a part.

Corresponding to Eq. 24–3,

$$\theta - \theta_b = C(v - v_b) \quad \text{and} \quad \theta_m - \theta_b = C(v_m - v_b) \tag{24–13}$$

From Eqs. 24-4, 6, and 13

$$q' = -\pi r^2 c_p \frac{\theta_m - \theta_b}{v_m - v_b} \cdot \frac{dp}{dx} \tag{24-14}$$

We have now derived two expressions for q', Eq. 24-12 holding for the boundary layer only, and Eq. 24-14, which is valid in the core of the flow. At their common interface, $y = b$, they must be identical. Hence

$$\theta_b = \frac{\dfrac{\mu c_p}{k}}{\dfrac{\mu c_p}{k} + \dfrac{v_m}{v_b} - 1} \theta_m \tag{24-15}$$

or, by substitution in Eq. 24-12,

$$q' = -\pi r^2 \frac{k}{\mu} \cdot \frac{\dfrac{\mu c_p}{k} \cdot \dfrac{1}{v_b}}{\dfrac{\mu c_p}{k} + \dfrac{v_m}{v_b} - 1} \theta_m \frac{dp}{dx} \tag{24-16}$$

For $N_{Pr} = \mu c_p/k = 1$ this equation simplifies to Eq. 24-7. Equation 24-16, however, should hold for any Prandtl numbers.

This comes from having eliminated Eq. 24-5, which contains N_{Pr}, by omitting the last term in Eqs. 24-1 and 2. But free from the restriction imposed by Eq. 24-5, N_{Pr} re-emerges when Eqs. 24-12 and 14 are identified at $y = b$. To pay for this, v_b had to be introduced; it has proved to be quite a liability in the balance.

Substituting $r = D/2$, $\mu c_p/k = N_{Pr}$, and $dp/dx = \Delta p/L$ (using Eq. 21-37) in Eq. 24-16,

$$q' = -\frac{\pi}{8} 0.3164 D \frac{k}{\mu} \cdot \frac{N_{Pr}}{v_m[1 + (v_b/v_m)(N_{Pr} - 1)]} \cdot \frac{\theta_m}{\sqrt[4]{N_{Re}}} \rho v_m^2$$

But $q' = h(\pi D)\theta_m$. Introducing this and the Nusselt number, as defined by Eq. 23-26,

$$N_{Nu} = 0.03955(N_{Re})^{3/4} \frac{N_{Pr}}{1 + (v_b/v_m)(N_{Pr} - 1)} \tag{24-17}$$

The weak point in this equation is the ratio v_b/v_m. There exists a whole literature about this ratio [see G. I. Taylor 1916; Prandtl 1928; Jakob 1933, p. 262; ten Bosch 1936; Hoffmann 1937, p. 162; McAdams 1942, pp. 112 and 119].

Taylor, starting from an equation on eddy formation by H. A. Lorentz, obtained $v_b/v_m = 0.38$ for flat surfaces, but 0.56 from Stanton and Pannell's experiments for the flow in tubes. Prandtl [1928] in his second paper on this subject represented the ratio by

$$\frac{v_b}{v_m} = B(N_{Re})^{-1/8} \tag{24-18}$$

with $B = 1.74$, but considered $B = 1.1$ to 1.2 as better. Hoffmann recommended the expression:

$$\frac{v_b}{v_m} = 1.5 N_{Re}{}^{-1/8} N_{Pr}{}^{-1/6} \tag{24-19}$$

and ten Bosch arrived at

$$\frac{v_b}{v_m} = E N_{Re}{}^{-0.1} N_{Pr}{}^{-0.185} \tag{24-20}$$

with $E = 1.4$ for heating, and $E = 1.12$ for cooling of the fluid.

It may be mentioned that Langmuir [1912] also recognized that the resistance against heat transfer centers in a thin layer close to the surface. He found that under free-convection conditions the heat output of a hot wire, placed in a gas, can be calculated as the rate of thermal conduction through a film of the gas, surrounding the wire and having a thickness which depends solely on the diameter of the wire and the kind of gas. The reason why this film concept did not influence the theory of heat transfer to such a degree as the Prandtl-Taylor theory seems to be its static character, compared to the dynamic elements in the combination of boundary-layer and impulse theories.

24-3. A Second Representation of the Combined Impulse and Boundary-layer Theories

Because of the importance of the Prandtl-Taylor theory another derivation, due to von Kármán [1939], may also be presented.

The shear stress S_s occurring in an arbitrary plane perpendicular to the y axis is equal to the total momentum transferred by molecular and turbulent exchange. It will be shown that it can be expressed by

$$S_s = \mu \frac{d\bar{v}_x}{dy} + \rho(\tilde{v}_x \tilde{v}_y)_m \tag{24-21}$$

where subscript m stands for mean, and $\tilde{v}_x$ and $\tilde{v}_y$ are fluctuating components, superimposed on mean values $\bar{v}_x$ and $\bar{v}_y$ so that the instantaneous components of velocity are

$$v_x = \bar{v}_x + \tilde{v}_x \tag{24–22}$$

$$v_y = \bar{v}_y + \tilde{v}_y \tag{24–23}$$

All means are taken over a time interval τ which may be any multiple of the period of fluctuation.

In the time $d\tau$ the mass $\rho v_y \cdot d\tau$ crosses the surface unit at the distance y from the wall, and $(\rho v_y\, d\tau)v_x$ is the momentum transported through

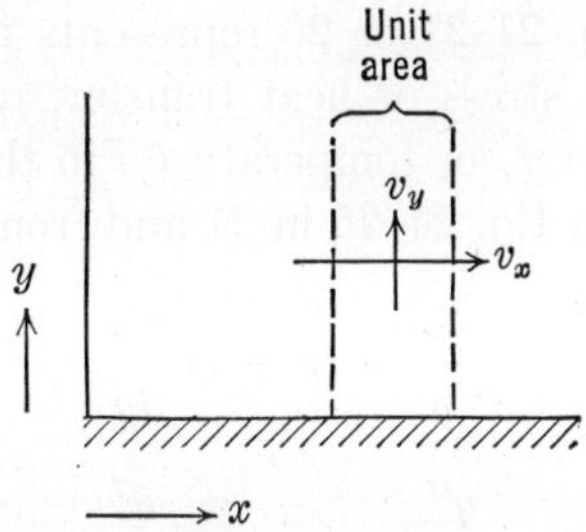

FIG. 24–1. Instantaneous velocity components in turbulent flow.

unit surface area (see Fig. 24–1). The time rate of momentum transportation is

$$\rho(v_x v_y)_m = \frac{1}{\tau}\int_0^\tau \rho v_x v_y \cdot d\tau \tag{24–24}$$

From Eqs. 24–22 and 23 by multiplication

$$v_x v_y = \bar{v}_x\bar{v}_y + \bar{v}_x\tilde{v}_y + \tilde{v}_x\bar{v}_y + \tilde{v}_x\tilde{v}_y$$

and by substitution in Eq. 24–24

$$\rho(v_x v_y)_m = \rho\bar{v}_x\bar{v}_y + \frac{\rho\bar{v}_x}{\tau}\int_0^\tau \tilde{v}_y\, d\tau + \frac{\rho\bar{v}_y}{\tau}\int_0^\tau \tilde{v}_x\, d\tau + \rho(\tilde{v}_x\tilde{v}_y)_m$$

Since $\bar{v}_y$ and both integrals are zero,

$$o(v_x v_y)_m = \rho(\tilde{v}_x\tilde{v}_y)_m = \rho\epsilon\frac{d\bar{v}_x}{dy} \tag{24–25}$$

This is the last term in Eq. 24–21, and ϵ is a coefficient of turbulent exchange, defined by this equation.

An equation for heat transfer from the fluid to the surface, corresponding to Eq. 24–21, is

$$q'' = k\,\frac{d\bar{t}}{dy} + \rho c_p(\tilde{t}\tilde{v}_y)_m \tag{24–26}$$

where $\bar{t}$ is the time mean, and $\tilde{t}$ is the fluctuating component of the temperature. The heat transportation $\rho c_p(\tilde{t}\tilde{v}_y)_m\,d\tau$ corresponds to the momentum transportation $\rho(\tilde{v}_x\tilde{v}_y)\,d\tau$, and in analogy * to Eq. 24–25

$$\rho c_p(tv_y)_m = \rho c_p(\tilde{t}\tilde{v}_y)_m = \rho c_p \epsilon\,\frac{d\bar{t}}{dy} \tag{24–27}$$

The last term of Eq. 24–21 or 26 represents the contribution of the eddy motions to shear stress or heat transfer, respectively, because of the exchange of velocity $\tilde{v}_x$ or temperature $\tilde{t}$ in the y direction.

By substitution from Eq. 24–25 in 21 and from Eq. 24–27 in 26:

$$\frac{S_s}{\rho} = (\nu + \epsilon)\,\frac{d\bar{v}_x}{dy} \tag{24–28}$$

$$\frac{q''}{\rho c_p} = (\alpha + \epsilon)\,\frac{d\bar{t}}{dy} \tag{24–29}$$

A concept similar to that used in these equations had previously been suggested by Murphree [1932]. He added a quantity which he called eddy viscosity to the dynamic viscosity μ, the sum being a quantity employed first by Boussinesq [1877] and known as exchange quantity.†

According to Eqs. 24–28 and 29, proportionality between shear stress and heat transfer can be expected:

a. If ν and α are negligible compared to ϵ, and

b. If ν and α are numerically equal.

Condition *a* is generally satisfied if the flow is turbulent, except for a laminar sublayer close to the wall. Consider two planes $y = y_1$ and $y = y_2$ in the turbulent region; then if S_s and q'' are independent of, or vary similarly ‡ with, y, by integration of Eqs. 24–28 and 29

$$\frac{q''}{c_p(t_2 - t_1)} = \frac{S_s}{v_{x,2} - v_{x,1}} \tag{24–30}$$

This is the analytical expression of Reynolds' analogy.

* Preliminary experiments with two-dimensional flow of air by Corcoran, Roudebush, and Sage [1947] seem to confirm that the same coefficient ϵ can be used in Eqs. 24–25 and 27.

† Translation of the term *Austauschgroesse* used by Prandtl.

‡ Deviations from similarity occur, but are not very significant in most cases.

Condition *b* is approximately satisfied by gases; for liquids, however, $\nu/\alpha > 1$, going up to 200 and more.

Von Kármán now introduces two dimensionless coefficients, the friction factor § f and a heat convection number C_h, defined by

$$S_s = f\frac{\rho v_m{}^2}{2} \tag{24-31}$$

and

$$q'' = C_h \rho c_p v_m \theta_{m,A} \tag{24-32}$$

where v_m is a suitably chosen reference velocity at some place of the flow, and $\theta_{m,A}$ is the temperature excess over surface temperature, taken at the same place. For instance, v_m may be the mean velocity in a tube, as defined by Eq. 21–2, and $\theta_{m,A}$ may be the excess of the mean temperature, the mean being taken over the cross-sectional area according to Eq. 21–6*a*.

The constant C_h can be considered the ratio of the rates of heat flow across and along the surface. Since the shear stress can be considered the rate of momentum transferred across the stream, and $\rho v_m{}^2$ as the momentum carried in the direction of the stream, the coefficient C_h is analogous to $f/2$.

Another conception, however, is commonly used in engineering practice. For molecular heat exchange per unit cross-sectional area:

$$q_k'' = k\frac{dt}{dy}$$

or, with $\theta_{m,A}$ as a fixed temperature difference for distance L,

$$q_k'' = k\frac{\theta_{m,\cdot}}{L}$$

Considering the more general case of molecular and turbulent heat transfer,

$$q'' = k_e\frac{\theta_{m,A}}{L} \tag{24-33}$$

where k_e would be an "equivalent" thermal conductivity. Comparison of Eqs. 24–29 and 33 shows that

$$k_e = \rho c_p(\alpha + \epsilon) \tag{24-34}$$

or, introducing an equivalent thermal diffusivity,

$$_e = \alpha + \epsilon \tag{24-35}$$

§ The quantity $f = \phi/4$, where ϕ is the coefficient of friction defined in Article *d* of Sect. 21–4.

Considering Eq. 24–28, equivalent viscosities ‖ can similarly be defined by

$$\mu_e = \rho(\nu + \epsilon) \tag{24-36}$$

and

$$\nu_e = \nu + \epsilon \tag{24-37}$$

The dimensionless numbers ν_e/ν and α_e/α are the ratios of total momentum and heat transfer to the respective quantities of molecular transference. The ratio k_e/k is identical with what generally is called the Nusselt number, as can be seen by comparison of Eq. 24–33 with the following relations:

$$q'' = h\theta_m = \frac{hL}{k} \cdot \frac{k}{L} \cdot \theta_{m,A} = N_{Nu} \frac{k}{L} \theta_{m,A} \tag{24-38}$$

where L is a characteristic length, for instance, the diameter of the tube.

Hence, the Nusselt number directly indicates what multiple of mere thermal conduction is transferred by convection.¶ Of course, this can be seen also from the usual definition of Nusselt's number:

$$N_{Nu} = \frac{hL}{k}$$

h indicating the heat transfer by convection, k/L that by conduction in the same fluid.

Another interesting relationship follows from Eqs. 24–32 and 38:

$$N_{Nu} = C_h N_{Re} N_{Pr} = C_h N_{Pe} \tag{24-39}$$

It is self-explanatory that these relations include the case $N_{Pr} = 1$. In this case

$$\frac{S_s}{\rho} = (\nu + \epsilon) \frac{dv_x}{dy} \tag{24-28}$$

and

$$\frac{q''}{\rho c_p} = (\nu + \epsilon) \frac{dt}{dy} \tag{24-40}$$

the latter one replacing Eq. 24–29. Integration yields

$$v_x - v_s = \frac{S_s}{\rho} \int_0^y \frac{dy}{\nu + \epsilon} \tag{24-41}$$

an

$$t - t_s = \frac{q''}{\rho c_p} \int_0^y \frac{dy}{\nu + \epsilon} \tag{24-42}$$

‖ The equivalent dynamic viscosity μ_e is identical with the exchange quantity as defined previously.

¶ The analogy with Martinelli's interpretation (see Eq. 23–44) is obvious.

where v_s and t_s are the velocity in the direction x and the temperature at the surface. Herewith the similarity of velocity and temperature distribution, as claimed by Reynolds and experimentally verified by Pannell [1916], is more exactly proved for the conditions a and b.

For $v_s = 0$, $v_x = v_m$, and $t - t_s = \theta_{m,A}$, Eqs. 24–41 and 42 yield

$$\frac{q''}{c_p\theta_{m,A}} = \frac{S_s}{v_m} \tag{24-43}$$

and by substitution from Eqs. 24–31 and 32

$$C_h = \frac{f}{2} \tag{24-44}$$

The following important conclusions for the case of $N_{Pr} = 1$ may be drawn:

1. If the shear stress S_s is proportional to the nth power of v_m, then the heat transfer q'' is proportional to the $(n - 1)$th power (from Eq. 24–43).
2. Roughness increases the friction and heat transfer at the same ratio (from Eq. 24–44).
3. Equation 24–44 is an approximation only because S_s and q'' vary differently with y.

Introducing now the boundary layer of thickness b and interface velocity v_b and proceeding in the same way as Prandtl and Taylor, the following simple equation is obtained:

$$C_h = \frac{f}{2}\cdot\frac{1}{1 + (N_{Pr} - 1)v_b/v_m} \tag{24-45}$$

This is identical with Eq. 24–16, as can be proved by substituting C_h from Eq. 24–32, $q'' = q'/(2\pi r)$, f from Eq. 24–31, and S_s from the balance of forces:

$$2\pi r\cdot S_s = \pi r^2 \cdot \frac{dp}{dx}$$

For $N_{Pr} = 1$, Eq. 24–45 simplifies to Eq. 24–44.

24–4. Improvements of the Combined Impulse and Boundary-layer Theories

The results of the Prandtl-Taylor surface-layer theory, with their original assumptions for v_b/v_m, are in agreement with experience up to $N_{Pr} \approx 2$. For higher values of the Prandtl number, as occurring with most liquids, the theory yields much too small values of h because it oversimplifies the problem by the assumption that there is only a boundary layer in which the flow is streamlined and a core with fully developed turbulence. By Eqs. 24–19 and 20 the range of application of the theory has been extended empirically up to $N_{Pr} \approx 300$.

Murphree [1932] seems to have been the first to improve the theory by defining the film in a different manner. He assumed that the eddy currents, while zero at the wall due to boundary conditions, increase continually from zero at the wall to a value which is constant for the main body of the fluid. He further assumed that in the region over which the eddy currents vary and which he calls "film," their strength is proportional to a power function of the distance from the surface. In his analysis Murphree showed that the eddy viscosity varies with no less power than the cube of the distance from the wall. He tried the third and fourth power and found the third preferable. Therefore he based his further derivation on the assumption that the eddy viscosity varies through the film as the cube of the distance from the surface, and is constant in the main body. His results are given in a table of $(12N_{Nu})$ as function of $(N_{Re})_m$ in the range of 3000 to 2,500,000 for different values of $\sqrt[3]{0.413N_{Pr}} = 0.5$, 1.0, and 2.0 and further in the range up to $(N_{Re})_m = 500{,}000$ for $\sqrt[3]{0.413N_{Pr}} = 4.0$ and 8.0; i.e., up to $N_{Pr} = 1240$.

Independently of this paper von Kármán [1939] improved the Taylor-Prandtl theory, using the assumption that in a buffer layer between the laminar and turbulent region neither ν nor ϵ can be considered negligible.*

For a *boundary layer*, which may include the buffer layer, according to Eqs. 24–28 and 29,

$$v_b = \frac{S_s}{\rho}\int_0^b \frac{dy}{\nu + \epsilon} \tag{24–46}$$

$$t_b - t_s = \frac{q''}{\rho c_p}\int_0^b \frac{dy}{\alpha + \epsilon} \tag{24–47}$$

Using the universal relation for the velocity distribution, given in Eq. 21–47 with the definitions of Eqs. 21–45 and 46, by differentiation

$$\Phi'(y^*) = \frac{dv^*}{dy^*} = \frac{dv^*}{dy}\cdot\frac{dy}{dy^*} = \frac{\nu\rho}{S}\cdot\frac{dv}{dy} \tag{24–48}$$

and by comparison with Eq. 24–28

$$\frac{S}{\rho\cdot dv/dy} = \frac{\nu}{dv^*/dy^*} = \nu + \epsilon \tag{24–49}$$

or

$$\frac{\epsilon}{\nu} = \frac{1}{\Phi'(y^*)} - 1 \tag{24–50}$$

This shows that ϵ/ν is a function of y^* alone.

* The concept of a buffer layer seems to go back to Eagle and Ferguson [1930, 1930(*a*)].

Substituting y from Eq. 21–46 and ϵ from Eq. 24–50 in Eqs. 24–46 and 47,

$$v_b = \left(\frac{S}{\rho}\right)^{1/2} \int_0^{b^*} \Phi'(y^*)\cdot dy^* = v_b^* \left(\frac{S}{\rho}\right)^{1/2} = K\left(\frac{S}{\rho}\right)^{1/2} \qquad (24\text{–}51)$$

and

$$t_b - t_s = \frac{q''}{\rho c_p \left(\frac{S}{\rho}\right)^{1/2}} \int_0^{b^*} \frac{dy^*}{\frac{\alpha}{\nu} - 1 + \frac{1}{\Phi'(y^*)}} = q'' \frac{M(N_{Pr})}{\rho c_p \left(\frac{S}{\rho}\right)^{1/2}} \qquad (24\text{–}52)$$

where K is a constant and M a function of N_{Pr} alone. From these two equations

$$\frac{\rho c_p(t_b - t_s)}{q''} - \frac{\rho v_b}{S} = \left(\frac{\rho}{S}\right)^{1/2} [M(N_{Pr}) - K]$$

For the *bulk* of the fluid, according to Eqs. 24–30 and 43,

$$\frac{\rho c_p(t_{m,A} - t_b)}{q''} - \frac{\rho}{S}(v_m - v_b) = 0$$

From these two equations by addition

$$\frac{\rho c_p(t_{m,A} - t_s)}{q''} - \frac{\rho v_m}{S} = \left(\frac{\rho}{S}\right)^{1/2} [M(N_{Pr}) - K] \qquad (24\text{–}53)$$

or, introducing again C_h and f,

$$C_h = \frac{f}{2} \cdot \frac{1}{1 + (f/2)^{1/2}[M(N_{Pr}) - K]} \qquad (24\text{–}54)$$

To determine the function $M(N_{Pr}) - K$, von Kármán used the representation of Nikuradse's experimental values as previously indicated in Sect. 21–6. The three sections of y^*, according to Eqs. 21–52, 54, and 48 are shown in Fig. 24–2.

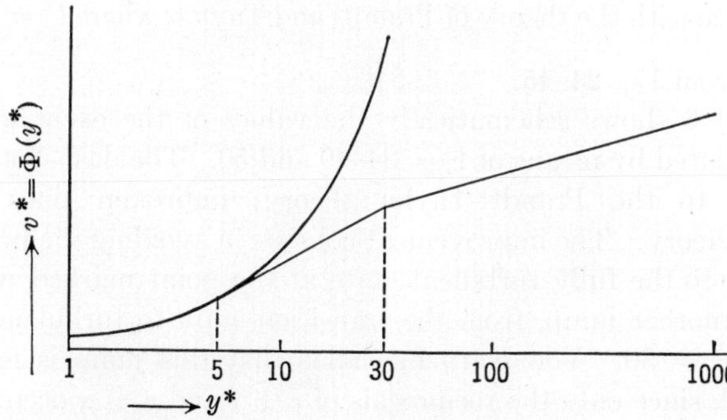

FIG. 24–2. Universal representation of the velocity distribution in the vicinity of a surface.

Now, from Eq. 24–51,

$$K = \int_0^{b^*} \Phi'(y^*) \cdot dy^*$$

and by substitution from Eqs. 21–51 and 53 for the ranges from $y^* = 0$ to 5 and from $y^* = 5$ to 30, respectively:

$$K = \int_0^5 dy^* + \int_5^{30} \frac{5}{y^*} dy^* = 5(1 + \ln 6) \tag{24–55}$$

Furthermore, from Eq. 24–52,

$$M(N_{Pr}) = \int_0^{b^*} \frac{dy^*}{\dfrac{\alpha}{\nu} - 1 + \dfrac{1}{\Phi'(y^*)}}$$

and by substitution, as above,

$$M(N_{Pr}) = \int_0^5 \frac{dy^*}{\alpha/\nu} + \int_5^{30} \frac{dy^*}{\dfrac{\alpha}{\nu} - 1 + \dfrac{y^*}{5}}$$

$$= 5[N_{Pr} + \ln(1 + 5N_{Pr})] \tag{24–56}$$

By substitution in Eq. 24–54

$$C_h = \frac{f/2}{1 + 5(f/2)^{1/2}\{N_{Pr} - 1 + \ln[1 + \frac{5}{6}(N_{Pr} - 1)]\}} \tag{24–57}$$

For small values of $N_{Pr} - 1$ by expansion of the logarithmic term:

$$C_h \approx \frac{f/2}{1 + C(f/2)^{1/2}(N_{Pr} - 1)} \tag{24–58}$$

in accordance with the theory of Prandtl and Taylor, where $C = \left(\frac{2}{f}\right)^{1/2} \frac{v_b}{v_m}$ as follows from Eq. 24–45.

Figure 24–3 shows schematically the values of the exchange coefficients calculated by means of Eqs. 24–49 and 50. The dash-dotted lines correspond to the Prandtl-Taylor theory; unbroken lines to von Kármán's theory. The improvement consists in avoiding the jump from the laminar to the fully turbulent state at the point marked with b_P^*. However, another jump, from the transition state to turbulence, takes place at $b_K^* = 30$. Von Kármán claims that this jump is much less detrimental, since only the reciprocals of $\nu + \epsilon$ and $\nu + \alpha$ occur in Eqs. 24–46 and 47. The accuracy of their integration depends chiefly on the

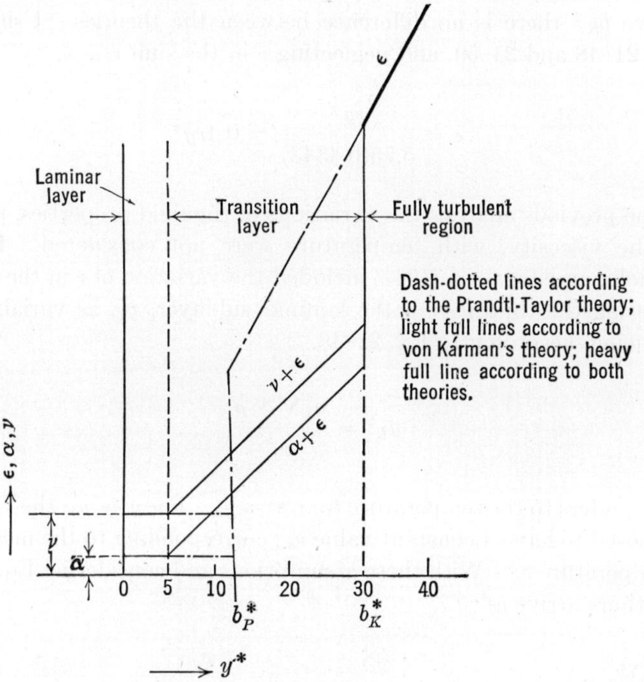

FIG. 24-3. Exchange coefficient ϵ, defined by Eqs. 24-36 and 37, in the vicinity of a surface.

correctness of the assumptions in a region where these quantities are small; hence an improvement of the assumptions in the neighborhood of $y^* = b_P^*$ favorably influences the calculation of S_s and q'' from these equations.

The point $y^* = b_P^*$ can be calculated from Eq. 24-10, which may be converted to

$$b = \frac{\mu v_b}{S}$$

From the definition of y^* (Eq. 21-46)

$$b_P^* = \frac{\mu v_b}{S\nu}\sqrt{\frac{S}{\rho}} = v_b\sqrt{\frac{\rho}{S}}$$

and from Eq. 21-45

$$b_P^* = v_b^* \tag{24-59}$$

According to Nikuradse's experiments (Curve A in Fig. 21-5), this seems to occur not far from $y^* = 10$, to be compared with the place of discontinuity $y^* = 30$ in von Kármán's set-up.

Above $b_K{}^*$ there is no difference between the theories. Using Eqs. 21–47, 21–48 and 24–50, and neglecting ν in the sum $\nu + \epsilon$,

$$\epsilon = \frac{\nu y^*}{5.75(0.4343)} \approx 0.4\nu y^* \tag{24–60}$$

In the previous analysis the variations of the fluid properties, particularly the viscosity, with temperature were not considered. Boelter, Martinelli, and Jonassen [1941] included the variation of ν in the theory, considering the thickness of the laminar sublayer, y_1, as variable, and introducing, according to Eq. 21–46,

$$y_1{}^* = \frac{y_1\sqrt{S/\rho}}{\nu_1} \tag{24–61}$$

where ν_1 refers to the temperature t_1 at $y = y_1$. For $y > y_1$, the viscosity is supposed to have a constant value ν_m, corresponding to the mean mixing temperature t_m. With these assumptions and considering Eq. 24–39, the authors arrive at

$$\frac{N_{Nu}}{(N_{Re})_m(N_{Pr})_m} = \frac{\sqrt{f/2}\cdot\theta_c/\theta_m}{y_1{}^*\left\{(N_{Pr})_1 + \ln\left[1 + (N_{Pr})_m\left(\frac{30}{y_1{}^*} - 1\right)\right] + \frac{2.5}{y_1{}^*}\ln\frac{(N_{Re})_m\sqrt{f/2}}{60}\right\}} \tag{24–62}$$

where the subscripts 1, m, and c refer to the temperatures t_1, t_m, and to the temperature in the center line t_c, respectively. As usual, $\theta_m = t_m - t_s$ and $\theta_c = t_c - t_s$.

Substituting experimental values of Morris and Whitman [1928] in Eq. 24–62, the values of Table 24–1 were obtained. It is seen that $y_1{}^*$ may vary from 2.3 to 5.0, whereas von Kármán had assumed $y_1{}^* = 5$.

In order to find t_1 the authors recommend to determine it first approximately from

$$\frac{t_s - t_1}{\theta_c} = \frac{(N_{Pr})_m}{(N_{Pr})_m + \ln\left[1 + 5(N_{Pr})_m\right] + 0.5\ln\frac{(N_{Re})_m\sqrt{f/2}}{60}} \tag{24–63}$$

TABLE 24-1

VALUES OF y_1^*, DEFINED BY EQ. 24-61 AND USED FOR CALCULATION OF LAMINAR SUBLAYER IN TURBULENT FLOW

μ_s/μ_1	$(N_{Re})_m$					
	3000	5000	10,000	20,000	40,000	100,000
0.2	2.30					
0.3	3.12	2.34				
0.4	3.70	3.07	2.40			
0.5	4.10	3.60	3.09	2.60	2.00	
0.6	4.43	4.03	3.62	3.34	2.90	2.50
0.7	4.68	4.40	4.10	3.90	3.62	3.40
0.8	4.82	4.70	4.50	4.40	4.25	4.10
0.9	4.95	4.90	4.80	4.75	4.70	4.65
1.0	5.00	5.00	5.00	5.00	5.00	5.00
1.5	4.95	4.80	4.60	4.35	4.00	3.75
2.0	4.89	4.52	4.20	3.75	3.00	2.50
3.0	4.75	4.18	3.60	2.70		
4.0	4.65	3.90	3.15			
5.0	4.57	3.70				
6.0	4.50					

Using μ_1 at this approximate value of t_1, an approximation of y_1^* can be obtained from Table 24-1, and t_1 will then be obtained more precisely from

$$\frac{t_s - t_1}{\theta_c} = \frac{(N_{Pr})_1}{(N_{Pr})_1 + \ln\left[1 + (N_{Pr})_m\left(\dfrac{30}{y_1^*} - 1\right)\right] + \dfrac{2.5}{y_1^*}\ln\dfrac{(N_{Re})_m\sqrt{f/2}}{60}} \tag{24-64}$$

which is a modification of Eq. 24-62.

It should be remarked that the authors' Eq. 24-62 contains a factor θ_c/θ_m, which occurred in neither the Prandtl-Taylor or von Kármán's analysis. These previous theories are all based on a mean temperature difference $\theta_{m,A}$ which may be converted into θ_m by multiplication with $\theta_m/\theta_{m,A}$. Boelter and coworkers represented θ_m/θ_c by a graph, from which the values of Table 24-2 have been taken.

TABLE 24-2

VALUES OF θ_m/θ_c FOR TURBULENT FLOW THROUGH A TUBE

N_{Pr}	N_{Re}		
	5000	10,000	100,000
0.6	0.795	0.805	0.82
0.8	0.815	0.825	0.84
1	0.835	0.84	0.86
5	0.925	0.925	0.93
10	0.95	0.95	0.955
50	0.985	0.985	0.985
100	0.99	0.99	0.99
500	1.00	1.00	1.00
1000	1.00	1.00	1.00

Boelter and coworkers also developed simple formulas showing how the mean viscosity in the sublayer, μ_f, and the interface velocity $v_x = v_1$ (at $y = y_1$) depend upon the ratio μ_1/μ_s.

By definition:

$$\mu_f = \frac{1}{y_1}\int_0^{y_1} \mu \cdot dy \tag{24-65}$$

An empirical expression of wide application is

$$\mu = \frac{a}{t^c} \tag{24-66}$$

where a and c are constants.

Furthermore, for the sublayer,

$$t = t_s + \frac{t_1 - t_s}{y_1} y \tag{24-67}$$

Substituting μ from Eq. 24-66 and y from Eq. 24-67 in Eq. 24-65, integrating, and resubstituting,

$$\mu_f = \mu_1 \frac{(\mu_1/\mu_s)^{\frac{1-c}{c}} - 1}{(1 - c)[(\mu_1/\mu_s)^{1/c} - 1]} \tag{24-68}$$

Further

$$S = \mu \frac{dv_x}{dy}$$

and by integration,

$$v_1 = \int_0^{y_1} \frac{S \cdot dy}{\mu}$$

Substituting μ and t as before, performing the integration, and introducing then the universal parameter $y_1{}^*$,

$$\frac{v_1}{\sqrt{S/\rho}} = y_1{}^* \frac{1 - (\mu_1/\mu_s)^{\frac{c+1}{c}}}{(1 + c)[1 - (\mu_1/\mu_s)^{1/c}]} \qquad (24\text{–}69)$$

The parameter $y_1{}^*$ can be taken from Table 24–1. Hence, if μ varies with the temperature, $v_1{}^* = v_1/\sqrt{S/\rho}$ loses its universal character and depends on μ_1/μ_s and the exponent c in Eq. 24–66.

Martinelli [1947] extended the analogy between heat and momentum transfer to cases of very low values of the Prandtl number, as occur with molten metals, particularly mercury ($N_{Pr} \approx 0.01$). This required considering the simultaneous transfer of heat by molecular conduction and eddy diffusion in the core of the flow. Calling their ratio G, it is necessary to multiply the third term in the denominator of Eq. 24–62 by a correction factor $F = 1/(G + 1)$. Table 24–3 contains some numerical values of this factor which are taken from a graph in Martinelli's paper. For details the original paper or its thorough discussion by McAdams [1948] should be consulted.

TABLE 24–3

CORRECTION FACTOR F FOR THE THIRD TERM IN THE DENOMINATOR OF EQ. 24–62

N_{Re}	N_{Pe}			
	100	1,000	10,000	100,000
2,000	0.25	0.77	1.00	1.00
10,000	0.175	0.67	0.92	1.00
100,000	0.095	0.45	0.82	1.00
1,000,000	0.050	0.28	0.65	0.95

24-5. Similarity between Heat Transfer and the Transport of Vorticity or Kinetic Energy

G. I. Taylor [1915] suggested regarding the dynamics of turbulent motion as an effect of diffusion of vorticity, rather than as a diffusion of momentum. As he explained more in detail in a later paper [G. I. Taylor 1932], the vorticity-transport theory can be employed more generally than the theory of momentum transfer. However, he admits that for the practically important cases of the flow along a plane plate and in a tube either theory can be used.

Bakhmeteff [1940] in a discussion of von Kármán's paper recommended interpreting the Reynolds analogy in terms of kinetic-energy transfer instead of momentum transfer. Concerning the advantages claimed for this procedure, reference is made to Bakhmeteff's paper.

24-6. Comparison of Temperature and Pressure Gradients in the Flow Direction

As early as 1874 Osborne Reynolds had conceived that the transmission of heat from a hot fluid to a surface was directly related to the fluid friction exerted on the surface, and in 1897 he claimed analogy between the "motion of heat" and the "motion of momentum." As a simple consequence of this postulate he started from the equation of the pressure drop of a fluid passing a cylindric tube and derived a similar expression for its temperature drop. These equations may be given in the form

$$\Delta p = C\left(\frac{vD}{\nu}\right)^{-n}\frac{L}{D}\rho v^2 \tag{24-70}$$

and

$$\Delta t = C\left(\frac{vD}{\nu}\right)^{-n}\frac{L}{D}\theta \tag{24-71}$$

where ρ = the density of the fluid,
ν = the kinematic viscosity of the fluid,
D = the inner diameter of the tube,
L = a length in axial direction,
v = a flow velocity (for instance, the true mean velocity) in this direction,
$\rho v^2/2$ = the excess of the kinetic energy, contained in unit volume of the fluid and due to its velocity v, above zero velocity at the surface,
θ = the temperature excess of the fluid above surface temperature, or

= the excess of the kinetic energy, contained in unit volume and due to the random velocities of the molecules, above the corresponding kinetic energy at surface temperature,

n = a positive number,

Δp = the pressure drop over the length L, and

Δt = the temperature drop over the length L.

From Eqs. 24–70 and 71

$$\frac{\Delta p}{\rho v^2} = \frac{\Delta t}{\theta} \tag{24–72}$$

On the other hand, for a channel of cross-sectional area A and inner surface S,

$$A \cdot \Delta p = f \cdot S \frac{\rho v^2}{2} \quad \text{or} \quad \frac{f}{2} = \frac{A}{S} \cdot \frac{\Delta p}{\rho v^2} \tag{24–73}$$

and

$$v \rho c_p \cdot A \cdot \Delta t = hS\theta \quad \text{or} \quad \frac{h}{c_p \dot{m}''} = \frac{A}{S} \cdot \frac{\Delta t}{\theta} \tag{24–74}$$

where c_p = the specific heat at constant pressure of the fluid,

$\dot{m}'' = \rho v$ = the mass velocity of the fluid,†

f = the friction factor, defined by Eq. 24–31, and

h = the film coefficient of heat transfer.

The dimensionless group $h/(c_p \dot{m}'')$ is called Stanton number ‡ (N_{St}). According to Eq. 24–74, it is the ratio of the heat transmissions perpendicular and parallel to the flow direction, each for unit temperature difference and area. Comparison with Eq. 24–32 shows that N_{St} is identical with von Kármán's "heat convection number" C_h.

From Eqs. 24–72, 73, and 74

$$\frac{h}{c_p \dot{m}''} = \frac{f}{2} \tag{24–75}$$

both sides being dimensionless. It will be seen later on (*cf.* Sect. 26–8) that Eq. 24–75 holds only if $N_{Pr} = 1$, which is the order of value of N_{Pr} occurring with gases.

† The American Standards Association [1943] recommends the symbol G for mass velocity. However, one should distinguish between $\dot{m}'' = \rho v$ and $G = \gamma v$ where γ is the specific weight.

‡ As suggested by a Round Table Conference at the Chicago meeting of the American Institute of Chemical Engineers in 1933.

Defining further the frictional resistance per unit area of inner surface by

$$R_f'' = \frac{A}{S}\,\Delta p \tag{24-76}$$

and using Eqs. 24–73, 75, and 76,

$$\frac{h}{c_p\dot{m}''} = \frac{R_f''}{\rho v^2} \quad \text{or} \quad h = \frac{c_p}{v}\,R_f'' \tag{24-77}$$

H. Lorenz [1930] seems to have been the first to plot values of the Stanton number versus Reynolds numbers. Colburn [1933(*a*)] improved

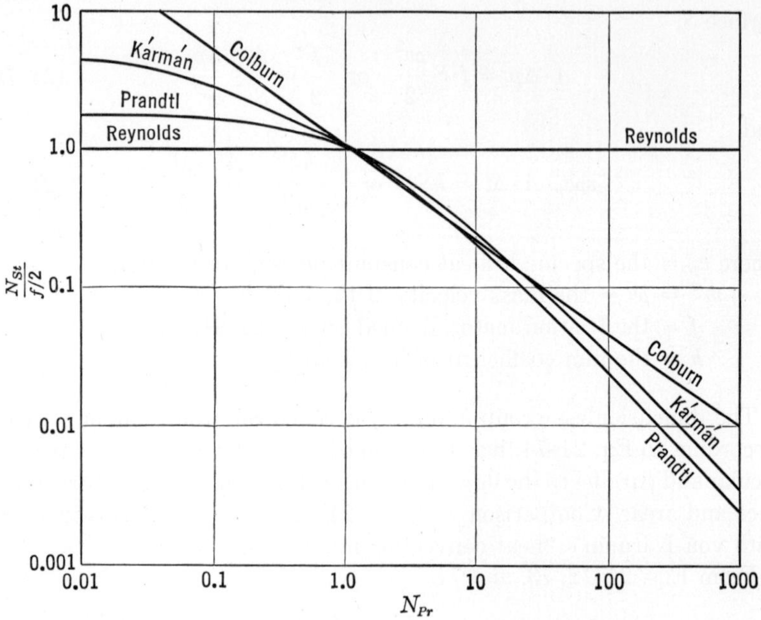

FIG. 24–4. Ratio of Stanton number to half of friction factor f versus Prandtl number.

From Colburn [1942].

this procedure by taking care of the influence of the Prandtl number also. Considering that the Nusselt number was found to be almost proportional to the third root of the Prandtl number, he defined a "heat-transfer factor" j by the following equation:

$$j = \frac{A}{S}\cdot\frac{\Delta t}{\theta}\left(\frac{\nu}{\alpha}\right)^{2/3} = \frac{h}{c_p\dot{m}''}\left(\frac{\nu}{\alpha}\right)^{2/3} = N_{St}(N_{Pr})^{2/3} \tag{24-78}$$

For $N_{Pr} = 1$

$$j = \frac{f}{2} \qquad (24\text{–}79)$$

Colburn checked Eq. 24–78 and Reynolds' original form, Eq. 24–75, by comparison with the more elaborate equations of Prandtl and von Kármán (see Fig. 24–4). In this figure $N_{St}/(f/2) = h/\frac{1}{2}fc_p\dot{m}''$ is used as ordinate, and N_{Pr} as abscissa. According to Reynolds (Eq. 24–75), the ordinate should be equal to 1, independent of N_{Pr}. However, it is seen that this ordinate occurs only when $N_{Pr} = 1$. Colburn's form (Eq. 24–78) is given by a straight line which in the range of air and water ($N_{Pr} = 0.73$ to 10) is close to the curves according to Prandtl's and von Kármán's equations, Eqs. 24–45 and 57, respectively. To draw Fig. 24–4, Colburn [1942, p. 27, Fig. 12] replaced the ratio v_b/v_m in Eq. 24–45 by 0.42, and $5(f/2)^{1/2}$ in Eq. 24–57 by 0.28.

CHAPTER 25

Experiments on Free Convection and Their Correlation

25–1. Free Convection on Horizontal Cylinders

Only a small selection of representative experiments on convection can be dealt with in this book. In order to correlate them in a proper way the principle of similarity will be applied.

Possibly the first correlation of heat-transfer data using this principle was made by Nusselt [1915(*a*)]. It referred to the natural cooling of horizontal cylinders in atmospheric air. The experiments performed by Kennelly, Wright, and Bylevelt; Wamsler; Langmuir; and Bylevelt covered a wide range of diameters (0.004 to 8.9 cm) and surface temperatures (49 to 1595°C).

The Grashof numbers (N_{Gr}) varied from 0.0002 to 5,000,000, i.e., as $1{:}25(10^9)$. Plotting log N_{Nu} versus log N_{Gr}, Nusselt obtained a unique line, with, however, some scattering of the points. The results of this correlation are shown in Table 25–1. They are supposed to be valid for any diatomic gases as well as for air.

Nusselt has recognized that the temperature difference between surface and gas has some influence upon the result and can be considered by introduction of a temperature ratio. Hermann [1932, 1936] suggested to use $(T_s - T_e)/T_e = \theta_s/T_e$ as characteristic temperature ratio, where temperatures are expressed in the absolute scale and s and e refer to surface and environment. Values of N_{Nu} for $\theta_s/T_e = 0$ and 0.65 which Hermann considered the most probable are given in the two last columns of Table 25–1. It seems that for small Grashof numbers Hermann's values are more nearly correct than Nusselt's original values, but that they are too low for very high Grashof numbers. However, his results allow the influence of the temperature difference to be considered.

TABLE 25-1

CORRELATION OF EXPERIMENTS ON NATURAL COOLING OF LONG HORIZONTAL CYLINDERS IN AIR

N_{Gr}	$\log N_{Gr}$	According to Nusselt		According to Hermann	
		$\log N_{Nu}$	N_{Nu}	N_{Nu} $(\theta_s/T_e = 0)$	N_{Nu} $(\theta_s/T_e = 0.65)$
10^{-5}	−5	0.650 − 1	0.447	......	......
10^{-4}	−4	0.670 − 1	0.468	0.484	0.415
10^{-3}	−3	0.710 − 1	0.512	0.520	0.446
10^{-2}	−2	0.767 − 1	0.585	0.612	0.524
10^{-1}	−1	0.855 − 1	0.716	0.809	0.688
1	0	0.957 − 1	0.905	1.10	0.966
10	1	0.08	1.20	1.50	1.36
10^2	2	0.23	1.70	2.18	1.95
10^3	3	0.42	2.63	2.99	2.87
10^4	4	0.64	4.36	4.47	4.42
10^5	5	0.90	7.95	7.08	7.08
10^6	6	1.16	14.5	12.4	12.4
10^7	7	1.42	26.3	22.1	22.1
10^8	8	....	....	39.3	39.3

In a later paper Nusselt [1929] generalized the results of his first correlation. In 1915 he had stated that the general equation

$$N_{Nu} = \Phi(N_{Gr}, N_{Pr}) \tag{25-1*}$$

takes the simpler form

$$N_{Nu} = \Psi(N_{Gr} \cdot N_{Pr}) \tag{25-2}$$

if the inertia terms † in Stokes' formulas are negligible compared to the viscosity terms. A. H. Davis [1922] found that his experiments on thin platinum wires in air, carbon tetrachloride, aniline, olive oil, and glycerine actually could be represented by Eq. 25-2. Assuming $N_{Pr} = 0.705$ for air, Nusselt multiplied the values of the first column of N_{Gr} in Table 25-1 by 0.705 and increased those of the second column by $\log 0.705 = -0.152$ in order to convert these columns to $(N_{Gr} \cdot N_{Pr})$ and $\log (N_{Gr} \cdot N_{Pr})$, respectively. ‡ In this way the table for cooling in

* Identical with Eq. 23-27.

† That is, the left-side terms of Eq. 3-14. In addition, the slowly streaming fluid is considered incompressible.

‡ Hermann's values in Table 25-1 are based on $N_{Pr} = 0.74$ for air.

air turns into a table for cooling in any gas or liquid. For simplification the first four columns of Table 25–2, proceeding in unit steps of $\log(N_{Gr} \cdot N_{Pr})$, have been derived from Table 25–1. McAdams [1942, p. 243] considered also experiments of different other observers. His correlation covered the following substances and ranges: hydrogen,

TABLE 25–2

CORRELATION OF EXPERIMENTS ON NATURAL COOLING OF LONG HORIZONTAL CYLINDERS IN GASES AND LIQUIDS

$N_{Gr} \cdot N_{Pr}$	$\log(N_{Gr} \cdot N_{Pr})$	Nusselt's Correlation		McAdams' Correlation	
		$\log N_{Nu}$	N_{Nu}	$\log N_{Nu}$	N_{Nu}
10^{-5}	−5	−0.349	0.448		
10^{-4}	−4	−0.325	0.473	−0.31	0.490
10^{-3}	−3	−0.283	0.521	−0.26	0.550
10^{-2}	−2	−0.220	0.603	−0.18	0.661
10^{-1}	−1	−0.134	0.735	−0.075	0.841
1	0	−0.025	0.944	0.035	1.08
10	1	0.100	1.26	0.18	1.51
10^{2}	2	0.258	1.81	0.325	2.11
10^{3}	3	0.455	2.85	0.50	3.16
10^{4}	4	0.68	4.79	0.73	5.37
10^{5}	5	0.94	8.71	0.97	9.33
10^{6}	6	1.20	15.9	1.21	16.2
10^{7}	7	1.46	28.8	1.46	28.8
10^{8}	8	1.72	52.5	1.71	51.3
10^{9}	9			1.97	93.3

nitrogen, air, carbon dioxide, water, aniline, carbon tetrachloride, toluene, glycerine; diameters from 0.003 to 16 cm; pressures of gases from 0.1 to 2000 atmospheres; surface temperatures from 18 to 1027°C. The values recommended by him are given in Columns 5 and 6 of Table 25–2. Nusselt's original values of N_{Nu} are 0 to 16½% smaller than McAdams' values, which can be considered more reliable. However, as Fig. 25–1 shows, the trend of the curves is the same and the differences are moderate, considering the enormous range of $(N_{Gr} \cdot N_{Pr})$ covered by the table and graph. §

§ If the influence of viscosity is negligible compared to that of acceleration, then, according to Boussinesq [1902], N_{Nu} should be a function of $N_{Gr}(N_{Pr})^2$. Fishenden and Saunders [1930] used this in the correlation of convection data for gases. See footnote on p. 530.

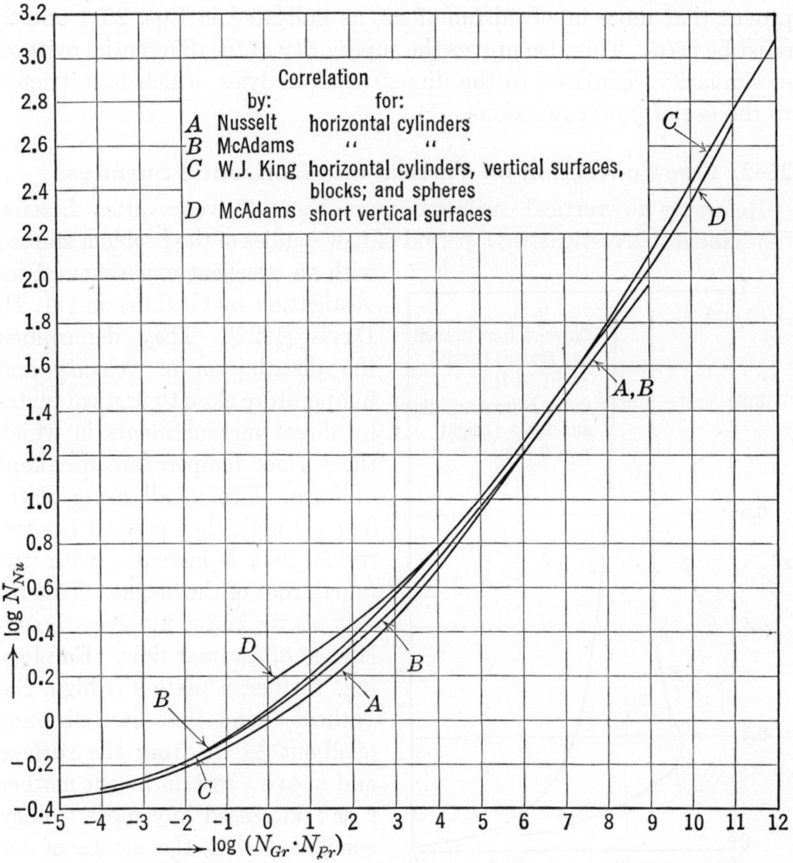

FIG. 25–1. Results of experiments on cooling of different surfaces by fluids in free convection.

For $(N_{Gr} \cdot N_{Pr}) < 10^{-5}$ the Nusselt number seems to approach a constant value of about 0.4, so that

$$h \approx 0.4 \frac{k}{L} \tag{25–3}$$

For $(N_{Gr} \cdot N_{Pr}) \gtreqless 10{,}000$ the lines A and B in Fig. 25–1 can be considered straight. With fair approximation they may be represented by the equation

$$N_{Nu} = 0.52\sqrt[4]{N_{Gr}N_{Pr}} \tag{25–4}$$

It should be kept in mind that the relationship between dimensionless groups cannot generally be represented by power functions as used in Eq. 25–4. The fact that the lines shown in Fig. 25–1 are not straight proves that more involved functions, as indicated in Eqs. 25–1 and 2, must be used. This also proves the superiority of the differential method of similarity compared to the dimensional analysis, which is restricted to the use of power functions.

25–2. Free Convection on Vertical and Horizontal Surfaces

Referring to vertical surfaces, more than 40 years after Lorenz' fundamental investigation, a period of new studies of the problem started with an excellent experimental investigation by Griffiths and A. H. Davis [1922]. They determined the distribution of velocity and temperature close to vertical plates by direct measurements in which the surface temperature was kept uniform. First of all, using plates 3 or 4 ft high, they verified Lorenz' result that h increases with the fourth root of the height. This result, as we know, involves the existence of laminar flow. Employing, further, a plate 9 ft high, the authors found that at a distance of about ⅕ in. from the surface and above a certain height neither v or t increased any more. They concluded that the strata of air nearest the plate were not themselves carrying away the heat from the upper part of the surface, but merely transmitted it to the outer layers which carried it away.

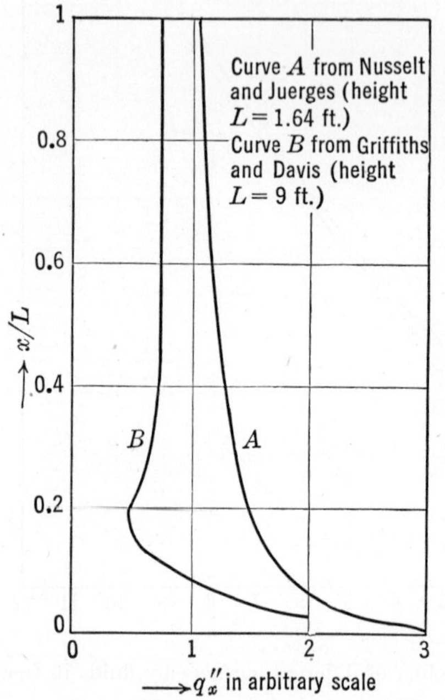

Fig. 25–2. Heat transfer from vertical surfaces on different heights to air by free convection.

Mere conduction, however, would not suffice for such transference of heat; thus they guessed that the motion of the air was turbulent. Referring to Fig. 25–2, W. J. King [1932] later compared the behavior of the heat transfer on low and higher vertical surfaces.

In this figure the local rate of heat flow per unit area q_x'' is plotted versus the ratio x/L, where x and L are the vertical distances of a point under consideration and the upper edge of the vertical plate, respectively, from the lower edge of the plate.

Curve A is typical for surfaces less than 2 ft high. Owing to the use of dimensionless ordinates, it is easily seen that the section below $x/L =$

0.20 contributes 1.5(20)/1(100) = 30% of the total heat transfer. On a 9-ft-high surface a distribution according to Curve B was found, and the following conclusions concerning the mechanism of the flow were drawn. At the lower edge cold air receives heat from the plate and commences its vertical flow in a streamline manner. The lowest surface elements lose more heat than the higher ones because these are swept by preheated air. The starting of turbulence is marked by the minimum of q_x'', located at $x/L \approx 0.17$. Above this point more heat is carried away by turbulent motion. Finally a sort of steady condition is reached, where in equal subsequent vertical sections the same amount of heat crosses the boundary layer and is transferred to the environment without any influence upon the boundary layer.

Representing the experimental results by a plot of $q''L/\theta$ versus $L^3\theta$, where q'' is the mean rate of heat flow per unit area, an inflection point occurred corresponding to the minimum of q_x'', and the upper extremity of the curve showed the following relationship:

$$\frac{q''L}{\theta} \sim (L^3\theta)^{1/3} \tag{25–5}$$

or

$$q'' = h\theta \sim \theta^{4/3} \tag{25–6}$$

A similar behavior had been found with electrically heated cylinders, hanging vertically in atmospheric air or placed on pillars of equal or smaller diameter. Using temperature differences from 20 to 180 F, the authors obtained the exponents given in Table 25–3.

TABLE 25–3

Exponents $(n + 1)$ in the Relation $q'' \sim \theta^{n+1}$

L [inches]	1.83	3.2	5.9	11.3	23	69	104
$n + 1$	1.23	1.17	1.18	1.25	1.25	1.30	1.34

These and similar observations indicate that with increasing Grashof number the mean heat transfer per unit area and therefore also the mean coefficient of heat transfer become independent of the height L. It is worth noting that an analogous phenomenon and Eq. 25–6 had been predicted by Nusselt [1915(a)] for air, cooling horizontal cylinders at $N_{Gr} > 10^7$. The occurrence of an exponent $n = 1/3$ in the equation

$$N_{Nu} = C(N_{Gr}N_{Pr})^n \tag{25–7}$$

was later verified by a correlation of W. J. King [1932] and by Jakob and Linke [1933] when they represented their own and Jakob and Fritz's [1931] experimental results on not too vehemently boiling water.

King correlated data of his own and other experiments with air, water, alcohol, and oil in contact with vertical cylinders and planes, horizontal wires, spheres, and blocks, by Curve *C* of Fig. 25–1, from which the values of Table 25–4 have been taken. It may be mentioned that in this

TABLE 25–4

Correlation of Experiments on Natural Cooling of Vertical and Other Surfaces in Gases and Liquids

$N_{Gr} \cdot N_{Pr}$	$\log (N_{Gr} N_{Pr})$	W. J. King's Correlation for Vertical Surfaces, Horizontal Cylinders, Blocks, and Spheres		McAdams' Correlation for Low Vertical Surfaces	
		$\log N_{Nu}$	N_{Nu}	$\log N_{Nu}$	N_{Nu}
10^{-4}	−4	−0.31	0.49		
10^{-3}	−3	−0.26	0.55		
10^{-2}	−2	−0.20	0.63		
10^{-1}	−1	−0.10	0.79		
1	0	0.03	1.07	0.16	1.45
10	1	0.19	1.55	0.28	1.91
10^2	2	0.36	2.29	0.42	2.63
10^3	3	0.56	3.63	0.59	3.89
10^4	4	0.78	6.03	0.78	6.03
10^5	5	1.00	10.0	1.02	10.5
10^6	6	1.24	17.4	1.27	18.6
10^7	7	1.52	33.1	1.52	33.1
10^8	8	1.80	63.1	1.77	58.9
10^9	9	2.10	126	2.02	105
10^{10}	10	2.41	257	2.33	214
10^{11}	11	2.78	603	2.70	501
10^{12}	12	3.06	1150		

correlation the experiments of Nusselt and Juerges [1928], and E. Schmidt and Beckmann [1930] mentioned previously, are included; they are in good agreement with Curve *C*. The table and figure include also, for comparison, the values by which McAdams [1942, p. 248] correlated the experimental data of Saunders [1936] for low plates in air. For up to 2- or 3-ft height King recommends the following simple formula derived from Curve C and expressed in British technical units:

$$h = 0.275 \sqrt[4]{\frac{\theta}{L}} \tag{25-8}$$

whereas for horizontal cylinders the following equation may be used which is more nearly in agreement with Curve B:

$$h = 0.24 \sqrt[4]{\frac{\theta}{D}} \tag{25-9}$$

These equations, however, are valid only in the range $N_{Gr} \cdot N_{Pr} = 10^4$ to 10^8 where the exponent $n = 1/4$ can be applied. In the range $N_{Gr} \cdot N_{Pr} = 10^8$ to 10^{12} the exponent $n = 1/3$ must be used, as can be seen from Fig. 25-1 or from a graph by which Jakob and Linke [1933] combined Nusselt's and King's correlations (Fig. 25-3). According to this figure

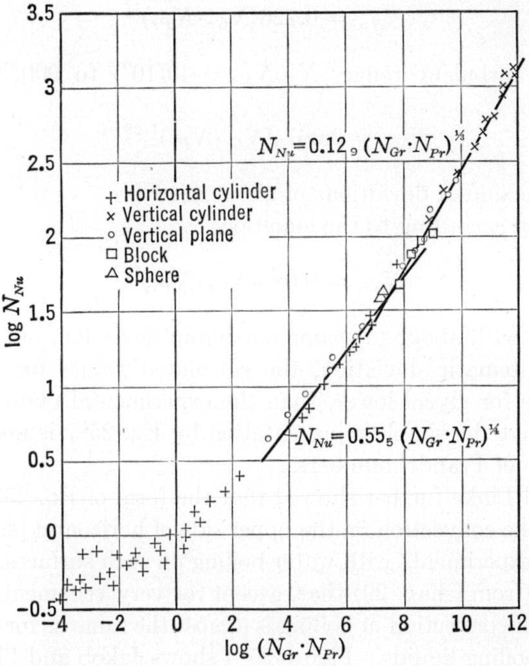

FIG. 25-3. Heat transfer from different surfaces to fluids by free convection.

the following equations may be used, which differ from King's equations only by a slight adjustment of the constant factors:

For the laminar range:

$$N_{Nu} = 0.555(N_{Gr} \cdot N_{Pr})^{1/4} \tag{25–10}$$

For the turbulent range:

$$N_{Nu} = 0.129(N_{Gr} \cdot N_{Pr})^{1/3} \tag{25–11}$$

More recently, Touloukian, Hawkins, and Jakob [1948] measured h on vertical cylinders in water and ethylene glycol. Since the cylinder diameter was 2.75 in., the influence of curvature was probably so small that the results will not differ appreciably from those for vertical plane surfaces. The cylinders were provided with unheated starting and trailing pieces of the same diameter. The temperatures were changed in such a range that N_{Pr} varied from 2.4 to 117.8 and N_{Gr} from $2.2(10^6)$ to $326(10^9)$.

In the laminar range, $N_{Gr} \cdot N_{Pr} = 0.2(10^9)$ to $40(10^9)$, the results could be represented by the expression

$$N_{Nu} = 0.726(N_{Gr} \cdot N_{Pr})^{1/4} \tag{25–10a}$$

and in the turbulent range, $N_{Gr}N_{Pr} = 40(10^9)$ to $900(10^9)$, by the expression ||

$$N_{Nu} = 0.0674[N_{Gr}(N_{Pr})^{1.29}]^{1/3} \tag{25–11a}$$

both with maximum deviations of about ±10%.

Correlation according to the equation

$$N_{Nu} = 0.086(N_{Gr} \cdot N_{Pr})^{1/3} \tag{25–11b}$$

was possible with about the same maximum deviation, but with a considerable systematic deviation, the calculated values for water being higher, those for glycol lower, than the experimental ones. It may be concluded that the simple representation by Eq. 25–7 is not possible in a wide range of Prandtl numbers.

Jakob and Linke further showed that the form of Eq. 25–11 can also be used for the convection on the upper side of horizontal plates, according to their experiments with water boiling on such surfaces. It may be anticipated (from Chap. 29) that, except for very vehement boiling, the mechanism of convection in boiling is almost the same as for free convection of not boiling liquids. Figure 25–4 shows Jakob and Linke's graph

|| Both viscosity and acceleration seem to be influential. See footnote on p. 524.

of log q'' versus log $(\zeta\theta^4)$, where $\zeta = g^2\beta\mu c_p k^2/\nu^2$. This magnitude comes from transformation of Eq. 25–5 to the dimensional form ¶

$$h = C(\zeta\theta)^{1/3} \tag{25–12}$$

or

$$q'' = C(\zeta\theta^4)^{1/3} \tag{25–13}$$

Since Fig. 25–4 is copied from the original paper, the representation with q'' in kcal hr^{-1} m^{-2} and ζ in kcal^3 hr^{-3} m^{-6} C^{-4} has been retained.

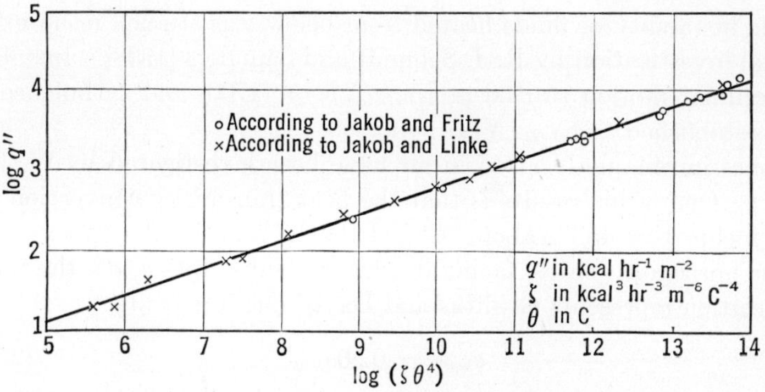

FIG. 25–4. Experimental verification of Eq. 25–13.

Using these, the factor C becomes dimensionless and general, corresponding to the definition in Eq. 25–7.

The figure shows that, in fact, n is exactly equal to $\frac{1}{3}$ and $C = 0.273$. The authors conclude that in the wide range of q'' used for the graph (7 to 5200 B hr^{-1} ft^{-2}) the heat transfer on a horizontal plate is independent of the place, as in vertical surfaces and horizontal cylinders in the turbulent range. In fact, this had to be expected, because it would not be understandable why the heat transfer on one place of a wide horizontal plate should be different from that at any other place except close to the edge. The mechanism of heat transfer, however, is different and much better than for homogeneous fluids on vertical surfaces, for which $C = 0.129$ had been found according to Eq. 25–11, i.e., less than half of the value for horizontal plates. A considerable part of this difference is probably due to the boiling of the liquid in the experiments. This is supposed, although not proved as yet, because the orientation of the surface has a much smaller influence for air. according to the experiments of Griffiths and A. H. Davis [1922].

¶ g^2 occurs because $N_{Pr} = g\mu c_p/k$ is employed by the authors with the physical properties expressed in metric technical units.

From their experiments one can derive the following equation for horizontal plates, facing upward in air:

$$q_{h,u}'' = 0.275\theta^{4/3} \tag{25-14}$$

in British technical units.

Comparing this with Griffiths and Davis' values of q_v'' for vertical plates 2 and 8⅔ ft high, we obtain

$$q_{h,u}'' \approx 1.28 q_v'' \tag{25-15}$$

Here the difference is 28%, instead of the 100% found for boiling water.

The instability of fluids heated from below was studied in an experimental investigation by R. J. Schmidt and Saunders [1938]. Instability and cellular motion started at $(N_{Gr} \cdot N_{Pr}) \approx 1700$, and turbulence was fully established at $(N_{Gr} \cdot N_{Pr}) > 45{,}000$.

Tubes in oblique location in air have been investigated by W. Koch [1927]. One of his results is that the heat transfer by convection on a tube inclined by 45° is about 92% of that of a horizontal tube.

For horizontal plates, facing downward (subscripts h, d), the following relation represents Griffiths and Davis' measurements:

$$q_{h,d}'' = 0.50 q_{h,u}'' \tag{25-16}$$

An even smaller factor than 0.50 might be expected, because of the fact that there is no reason for any convection when the hot air is above the colder one. Actually, some convection is caused by secondary influences, as temperature differences on the edge of the horizontal plate.

For heat transfer from air to colder horizontal plates, just the opposite behavior must be expected, i.e.,

$$q_{h,u}'' = 0.50 q_{h,d}'' \tag{25-16a}$$

It may be noted that

$$\frac{q_{h,u}'' + q_{h,d}''}{2} \approx \frac{1.28 + 0.64}{2} q_v'' = 0.96 q_v'' \approx q_v''$$

As has been mentioned previously, $n = 1/3$ holds only when $(N_{Gr} \cdot N_{Pr}) > 10^7$. This product depends on $L^3\theta$.

Columns 3 and 4 of Table 25-4 and Curve C of Fig. 25-1 cover also some tests with spheres and rectangular blocks. For such and similar bodies W. J. King [1932] used a characteristic length L defined by

$$\frac{1}{L} = \frac{1}{L_h} + \frac{1}{L_v} \tag{25-17}$$

where L_h and L_v are horizontal and vertical lengths occurring in such systems. Since for a sphere $L_h = L_v = D$, it follows that $L = D/2$ had to be used.

The heat transfer from vertical wires, 15.5 to 42.3 in. long, to air by natural convection has been investigated by Mueller [1942] in the range of temperature differences from 1 to 100 C. Inspection of the data showed no effect of wire length on the coefficient of heat transfer, as for long vertical plates, but did show that wire diameter D was a major variable. Using this in both the Nusselt and Grashof group, the relationship was surprisingly close to that for natural convection from horizontal wires. In the range of $(N_{Gr} \cdot N_{Pr}) = 10^{-7}$ to 10^{-2} and possibly up to $(N_{Gr} \cdot N_{Pr}) = 1$ it can be represented by the empirical equation

$$N_{Nu} = (N_{Gr} \cdot N_{Pr})^{0.1} \tag{25–18}$$

Some attempts to determine the temperature distribution on horizontal plates may just be quoted. R. Weise [1935] used square aluminum plates, about 16 and 24 cm in side length, 1.0 and 1.5 cm thick, respectively, which were hung up in a wide room and heated electrically. The results of a total of 70,000 readings were combined to patterns of isothermal lines, one of which is reproduced as Fig. 25–5. According to

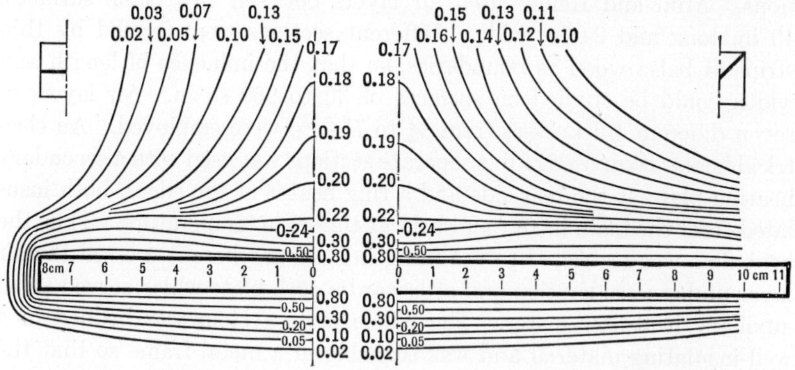

FIG. 25–5. Isothermals θ/θ_s around a horizontal square plate hanging in air. θ, θ_s = excess of air and surface temperature, respectively, above air temperature at far distance. Left side of figure: cross section parallel to plate edge; right side of figure: cross section through plate diagonal.

From R. Weise [1935].

these graphs, which vary slightly in different planes through the vertical axis, the boundary layer, in which the temperature drop is concentrated, is about 1.5 cm thick in the center below the plate and 1 cm close to the edge. Above the plate cold air is flowing from the edge to the center, causing the depression of the isothermals close to the plate at the center.

Weise calculated individual coefficients of heat transfer from the temperature drop, close to the plate, to check the mean coefficient as found by direct measurement.

25-3. Free Convection in Enclosed Plane Gas Layers

The heat transmission by free convection in horizontal and vertical plane air layers has been measured by different investigators; the most elaborate experiments seem to be those of Mull and Reiher [1930]. As a measure for the heat exchange in a gas layer they used an equivalent thermal conductivity, k_e. This includes the effect of convection and radiation and is defined by

$$k_e = k_c + k_r \tag{25-19}$$

where k_c is itself an equivalent thermal conductivity, including the effect of conduction and convection, and k_r is another equivalent conductivity superseding the effect of radiation.

Whereas k_r can be very simply calculated from Stefan-Boltzmann's law or the corresponding relations for gray or metallic radiators, the determination of k_c requires special experiments under varying conditions. Mull and Reiher used air layers between two plane surfaces, 40 in. long and 24 in. wide. Different sections were formed by thin strips of balsa wood as boundaries, so that the influence of length and width could be studied on surfaces of 30 to 960 sq in. Air layers of seven different thicknesses from ½ to 7¾ in. were employed. An electrical heater composed of five separate sections was used with a secondary heating plate at the rear side and a ring heater at the edges, both insulated from the main heater and held at the same temperature. Thus the heat flow was restricted to the front side, from where it crossed the air layer under consideration and afterward a cork plate which served as an auxiliary heat-flow meter. The whole plate system was bedded in a well-insulating material and was turnable in a metal frame so that the air layer could be brought into horizontal, vertical, or oblique position.

The rate of heat flow by convection between a warmer surface (subscript 1) and a cooler surface (subscript 2) can be expressed by

$$q_c = k_c \frac{A}{L}(t_1 - t_2) \quad \text{or} \quad q_c'' = \frac{k_c}{L}(t_1 - t_2) \tag{25-20}$$

where L = the thickness of the gas layer,
$A = H \cdot W$ = the area of this layer,
H = the height of the layer, when used in vertical position, and
W = the width of the layer.

Further, from the theory of similarity, for air and probably also for any other diatomic gases,

$$(N_{Nu})_L = \Phi\left[(N_{Gr})_L, \frac{H}{L}, \frac{W}{L}\right] \tag{25-21}$$

where slightly modified Nusselt and Grashof numbers are defined by

$$(N_{Nu})_L = \frac{q_c''}{t_1 - t_2} \cdot \frac{L}{k} = \frac{k_c}{k} \tag{25-22}$$

and

$$(N_{Gr})_L = \frac{\beta g}{\nu^2} L^3(t_1 - t_2) \tag{25-23}$$

From Eqs. 25–21 and 22

$$\frac{k_c}{k} = \Phi\left[(N_{Gr})_L, \frac{H}{L}, \frac{W}{L}\right] \tag{25-24}$$

Obviously

$$\frac{k_c}{k} \to 1 \quad \text{when} \quad (N_{Gr})_L \to 0 \tag{25-25}$$

Considering, now, first *horizontal layers* (subscript h), the experiments with heat flow upward showed that $k_{c,h}/k$ is independent of H and W. Then, for reasons of similitude, it will also be independent of H/L and W/L. Hence Eq. 25–24 simplifies to

$$\frac{k_{c,h}}{k} = \Phi(N_{Gr})_L \tag{25-26}$$

Mull and Reiher represented their experiments by plotting $k_{c,h}/k$ versus $\log N_{Gr}$ and obtained a smooth curve in the experimental range of $(N_{Gr})_L = 2100$ to 8,890,000. Jakob [1946], however, representing these results in bilogarithmic coordinates, has shown that there is at least one bend in the curve, in analogy to other cases of free convection. Figure 25–6 reveals that, in analogy to the cases of a heated surface facing upward and a vertical surface at high Grashof numbers,

$$\frac{k_{c,h}}{k} = 0.068(N_{Gr})_L^{1/3} \tag{25-27}$$

This is valid from $N_{Gr} \approx 400{,}000$ upward.

In the range from about 10,000 to 400,000, on the other hand, as for vertical surfaces in a medium range of Grashof numbers,

$$\frac{k_{c,h}}{k} = 0.195 N_{Gr}^{1/4} \tag{25-28}$$

For very small Grashof numbers log k_c/k approaches zero according to Eq. 25–25. This is indicated by the dotted line in Fig. 25–6.

The meaning of Eq. 25–27, obviously, is that, above a certain thickness L, the coefficient of heat transfer by convection does not change any more, so that there should be no difference between the heat transfer through the layer and that from a single horizontal plate facing upward.

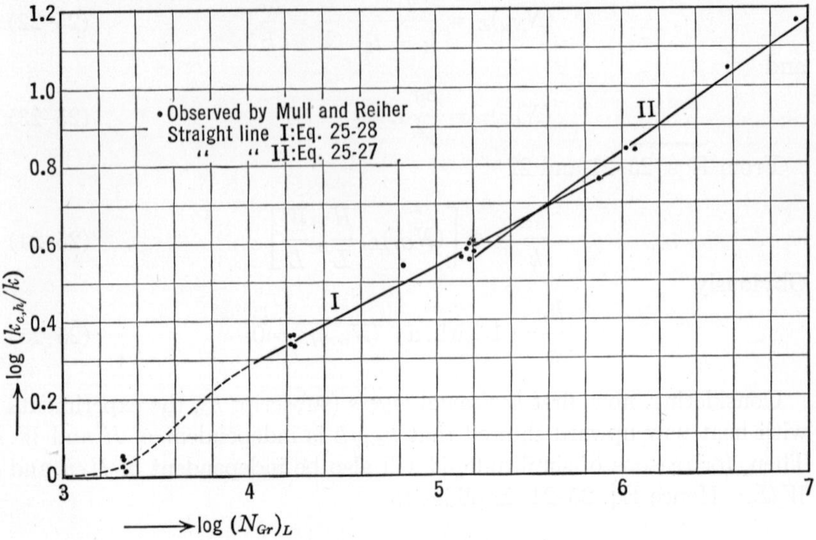

FIG. 25–6. Dimensionless representation of heat transfer through horizontal air layers by free convection.

This may be checked by calculating an example. According to Eqs. 25–22 and 27,

$$q_{c,h}'' = 0.068k\left(\frac{\beta g}{\nu^2}\right)^{1/3}(t_1 - t_2)^{4/3}$$

For $t_1 - t_2 = 50$ C and an atmospheric pressure of 730 mm Hg, Mull and Reiher used $\beta g/\nu^2 = 0.110(10^9)$ m^{-3} C^{-1} and $k = 23.0(10^{-3})$ kcal hr^{-1} m^{-1} C^{-1}. Employing these values and converting to British technical units,

$$\begin{aligned} q_{c,h}'' &= 0.068(23.0)10^{-3}(0.673)0.110^3(10^3)0.3048(1.8)(t_1 - t_2)^{4/3} \\ &= 0.277\theta^{4/3} \end{aligned}$$

in excellent agreement with Eq. 25–14.

For *vertical air layers* (subscript v) the conditions are less simple than for horizontal layers, because $k_{c,v}/k$ will not be independent of H and

W. Mull and Reiher, in one of their tests, kept H/L and N_{Gr} constant, but reduced W/L from 25.7 to 12.8, i.e., by 50%. However, the term $(k_{c,v}/k)\cdot(H/L)$ changed only by 1.3%. Hence, omitting W/L as an independent variable in Eq. 25–24, the corresponding equation for vertical air layers becomes

$$\frac{k_{c,v}}{k} = \Phi\left[(N_{Gr})_L, \frac{H}{L}\right] \tag{25–29}$$

As in the case of horizontal air layers, $k_{c,v}/k$ decreases with N_{Gr} and approaches $k_c/k = 1$ when N_{Gr} approaches zero. Hence, at this limit,

$$\frac{k_{c,v}}{k}\cdot\frac{H}{L} = \frac{H}{L} \tag{25–30}$$

For very large distances, the problem also simplifies because of the vanishing of the frictional resistance between two parallel vertical surfaces, very distant from each other. Obviously

$$t_1 - t_m = t_m - t_2$$

and

$$q_{c,v}'' = h(t_1 - t_m) = h(t_m - t_2)$$

or

$$q_{c,v}'' = \tfrac{1}{2}h(t_1 - t_2) \tag{25–31}$$

where h is the coefficient of heat transfer for a single vertical surface, which may be found from Eq. 25–10 or 11.

From Eqs. 25–20 and 31

$$k_{c,v} = \frac{hL}{2} \quad \text{or} \quad \frac{k_{c,v}}{k}\cdot\frac{H}{L} = \frac{h}{k}\cdot\frac{H}{2} = \frac{1}{2}\cdot\frac{h}{k}\cdot\frac{H}{L}L \tag{25–32}$$

Hence, $k_{c,v}$ becomes proportional to L for very thick air layers, and the resistance against thermal convection, $L/(k_{c,v}A)$ approaches a limit. The resistance against heat transfer by radiation is also constant when the surface temperatures and emissivities remain unvaried. Therefore, the insulating power of air layers cannot be increased infinitely by increasing the layer thickness, as would be the case with a solid body, but is limited. For this reason it is advisable to divide thick air layers by separating sheets or other solid bodies. The resistances against convection and radiation increase almost proportionally to the number of divisions. Under such conditions as exist in buildings the heat resistance has a maximum for vertical air layers of about 2 in. thickness, whereas horizontal layers of 2 to 8 in. thickness have about the same resistance.

For a general representation Mull and Reiher plotted $(k_{c,v}/k)\cdot(H/L)$, as measured, versus log N_{Gr} and built up 27 curves with H/L as parameter, making ample use of interpolation and extrapolation, since only 21 of their own and 4 other experimental points were available. Again, [Jakob 1946] bilogarithmic representation gives a much simpler and more reliable picture and leads to simple equations in the laminar and turbulent ranges. Figure 25–7 contains all the experimental points of Mull and Reiher and three points of other authors which they also used.

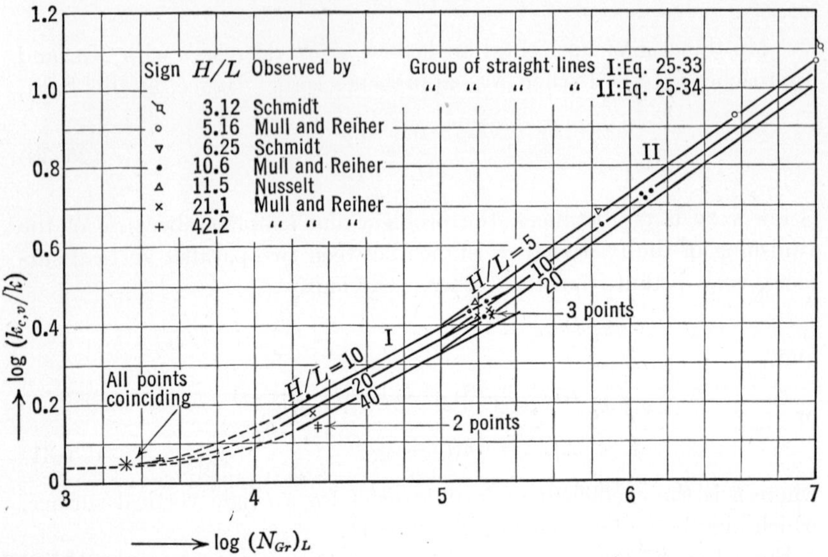

FIG. 25–7. Dimensionless representation of heat transfer through vertical air layers by free convection.

At Grashof numbers as small as 2000, Eq. 25–26 seems to hold, no influence of H/L upon $k_{c,v}/k$ existing in the range of the experiments ($H/L = 10.6$ to 42.2). For still smaller values of N_{Gr}, Eq. 25–25 seems to be verified.

In the range of $2000 \leqq (N_{Gr})_L \leqq 20{,}000$, the ratio $k_{c,v}/k$ increases with decreasing parameter H/L at $N_{Gr} =$ constant.

In the range of $20{,}000 \leqq (N_{Gr})_L \leqq 200{,}000$ the following equation represents the experimental data:

$$\frac{k_{c,v}}{k} = 0.18(N_{Gr})_L^{1/4}\left(\frac{H}{L}\right)^{-1/9} \tag{25–33}$$

and in the range of $200{,}000 \leqq (N_{Gr})_L \leqq 11{,}000{,}000$

$$\frac{k_{c,v}}{k} = 0.065 N_{Gr}^{1/3}\left(\frac{H}{L}\right)^{-1/9} \tag{25–34}$$

Again, obviously, the exponents ¼ and ⅓ are significant for laminar and turbulent flow.

The factor $(H/L)^{-1/9}$ has been found empirically. It certainly will not be valid down to $H/L = 0$, where it would yield an infinitely large heat transfer, whereas the behavior of single vertical surfaces should be approached (see Eq. 25–32). In the last-quoted paper it is demonstrated that Eq. 25–34 probably holds down to $H/L \approx 3$.

For *oblique layers* Mull and Reiher's paper contains only one measurement at 45° inclination, $H/L = 10.5$, and $(N_{Gr})_L = 1{,}050{,}000$, yielding $k_c/k = 6.13$. This may be compared with the mean of the values for horizontal and vertical position, calculated from Eqs. 25–27 and 34,

$$\frac{1}{2}\left(\frac{k_{c,h}}{k} + \frac{k_{c,v}}{k}\right) = \frac{1}{2}(6.92 + 5.09) = 6.005$$

It indicates that with good approximation the angle of inclination can be considered by linear interpolation between the formulas for horizontal and vertical layers.

Mull and Reiher's experiments were performed under rather ideal conditions, namely, uniformity of the surface temperatures and virtually non-conducting border strips. Griffiths and Davis [1922] and Wilkes and Peterson [1937] found 40 to 50% higher values of k_c, wherein, however, considerable amounts of heat conduction through wooden studs at the edges seem to be included [see Jakob 1946].

A transformation of Eqs. 25–27, 28, 33, and 34 to forms employable for other than diatomic gases and for liquids can be envisaged, since Eqs. 25–10 and 11 hold for gases and liquids. Considering that $N_{Pr} = 0.72$ was used for air, it is necessary only to multiply the right sides of Eqs. 25–33 and 34 by $(N_{Pr}/0.72)^n$ with the exponents $n = 1/4$ or $1/3$, respectively.

25–4. Free Convection in Cylindrical Gas Layers

The heat convection between coaxial cylindrical tubes, each 8 in. long, has been investigated by Beckmann [1931] in experiments with air, hydrogen, and carbon dioxide, and with diameter ratios of the annuli from 1.2 to 8.1. The range of Grashof numbers was from 640 to $15(10^6)$ in these experiments, and, including the results of some other observers, from 10^2 to 10^8.

Again, heat transfer by radiation will presently not be considered. As with plane layers, an equivalent conductivity k_c, including convection, may be defined by

$$q_c' = \frac{2\pi k_c}{\ln D_2/D_1}(t_1 - t_2) \tag{25–35}$$

where the prime sign indicates a rate per unit length, and subscripts 1 and 2 are symbols for the smaller and larger surface of the annulus, respectively. Figure 25-8 is a self-explanatory sketch indicating the flow in the annulus formed by coaxial horizontal tubes. It is obvious that convection will increase with $D_2 - D_1$; it becomes negligible when $D_2 - D_1 < \frac{1}{4}$ in., because of the effect of air viscosity.

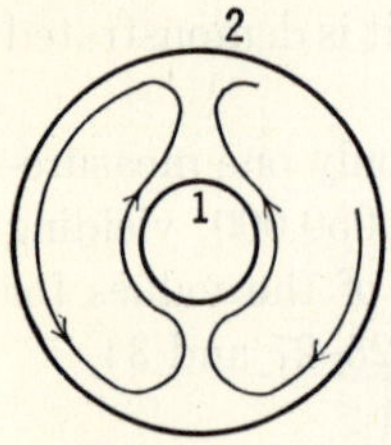

Fig. 25-8. Free convection in an annulus with horizontal axis.

Using again the principle of similarity, Beckmann started from

$$\frac{k_c}{k} = \Phi\left[(N_{Gr})_1, N_{Pr}, \frac{D_2}{D_1}\right] \tag{25-36}$$

where the subscript 1 of the Grashof number indicates that D_1 is taken as characteristic length. D_2, $D_2 - D_1$, or $(D_1 + D_2)/2$ could have been used as well. The physical properties, however, were taken as prescribed by Nusselt (see Eqs. 23-48 to 50, inclusive).

For gases with the same number of atoms in the molecule, Eq. 25-36 simplifies to

$$\frac{k_c}{k} = \Phi\left[(N_{Gr})_1, \frac{D_2}{D_1}\right] \tag{25-37}$$

Couples of tubes, giving different ratios D_2/D_1, had to be used in order to find the function Φ. The greater diameter, D_2, was changed. Instead of varying also the smaller diameter, D_1, which would have required changes of the electrical heater, Beckmann found it simpler and cheaper to vary the gas density, ρ, the square of which appears in $(N_{Gr})_1$. This is the reason why the kind of gas, as well as its pressure, was varied in the experiments.

The electrical heater was made by winding nichrome ribbon on an asbestos-covered steel tube. It was placed inside tube 1, whose outer diameter was $D_1 = 8.0$ cm. Seven different outer tubes were used having inner diameters D_2 from 9.5 to 64.8 cm, so that the thickness of the gas layer varied from 0.75 to 28.4 cm. The mean temperature was varied from 20 to 65°C, the difference $t_1 - t_2$ from 10 to 25 C, and the absolute pressure of the gases from $\frac{1}{6}$ to 1.9 atmospheres.

As for plane air layers, the ratio k_c/k approaches unity when the Grashof number approaches zero.

From the experimental results Beckmann constructed Fig. 25-9, the dotted lines being extrapolated. For $D_2/D_1 > 2$ the curves are based on measurements with air, and so are the upper parts of the curves for

$D_2/D_1 \gtreqless 2$. The lower parts of the latter are covered by measurements on hydrogen. The whole system is supposed to be valid for any diatomic gases.

Experiments with CO_2 were performed in the range from $D_2/D_1 =$ 1.2 to 2.1. The results can be brought to coincidence with the curves of Fig. 25–9 if $\log [N_{Gr} \cdot (N_{Pr})^2]$ instead of $\log N_{Gr}$ is used as abscissa.

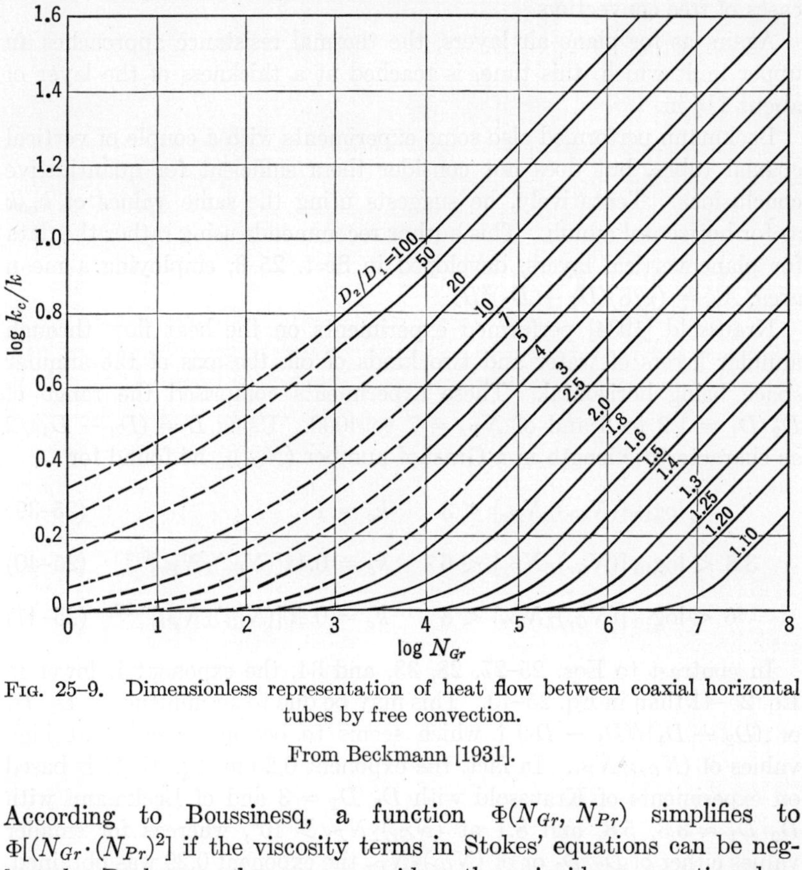

FIG. 25–9. Dimensionless representation of heat flow between coaxial horizontal tubes by free convection.

From Beckmann [1931].

According to Boussinesq, a function $\Phi(N_{Gr}, N_{Pr})$ simplifies to $\Phi[(N_{Gr} \cdot (N_{Pr})^2]$ if the viscosity terms in Stokes' equations can be neglected. Beckmann, however, considers the coincidence mentioned as incidental. In fact, a function $\Psi(N_{Gr} \cdot N_{Pr})$ is generally employed in other free convection problems, based on Nusselt's suggestion to neglect the acceleration terms and retain the viscosity terms. Using this procedure, however, Beckmann did not succeed in bringing his curves for carbon dioxide to coincidence with those for air and hydrogen.*

* A similar difficulty occurred in the experiments of Touloukian, Hawkins, and Jakob [1948]; see Sect. 25–2.

For the straight parts of the curves in Fig. 25–9 Beckmann used the relation

$$N_{Nu} = C(N_{Gr})^n \tag{25–38}$$

with $n = 0.258$, but he mentions that 0.25 could also have been used. The range of experiments is not sufficient to find out whether at high Grashof numbers the exponent turns into $n = 1/3$, as was found in other cases of free convection.

Again, as for plane air layers, the thermal resistance approaches an upper limit which, this time, is reached at a thickness of the layer of about 1½ in.

Beckmann performed also some experiments with a couple of vertical coaxial tubes, but does not consider them sufficient for quantitative conclusions. Tentatively, he suggests using the same values of k_c/k as for horizontal annuli. This author recommends using rather the data for plane vertical layers, developed in Sect. 25–3, employing a mean area, $A_m = (\pi/8)(D_1 + D_2)^2 H$.

Kraussold [1934] performed experiments on the heat flow through annular layers of water and two kinds of oil, the axis of the annular space being horizontal. These experiments comprised the range of $D_2/D_1 = 1.2$ to 3 and of $N_{Pr} = 7$ to 4000. Using $L = (D_2 - D_1)/2$ as characteristic length in a Grashof number $(N_{Gr})_L$, he found for

$$\log_{10} [(N_{Gr})_L N_{Pr}] < 3 \qquad k_c = 1 \tag{25–39}$$

$$3.8 < \log_{10} [(N_{Gr})_L N_{Pr}] < 6 \qquad k_c = 0.11[(N_{Gr})_L N_{Pr}]^{0.29} \tag{25–40}$$

$$6 < \log_{10} [(N_{Gr})_L N_{Pr}] < 8 \qquad k_c = 0.40[(N_{Gr})_L N_{Pr}]^{0.20} \tag{25–41}$$

In contrast to Eqs. 25–27, 28, 33, and 34, the exponent is lower in Eq. 25–41 than in Eq. 25–40. This may be due to an influence of D_2/D_1 or $(D_2 + D_1)/(D_2 - D_1)$ † which seems to become effective at high values of $(N_{Gr})_L N_{Pr}$. In fact, the exponent 0.20 in Eq. 25–41 is based on experiments of Kraussold with $D_2/D_1 = 3$ and of Beckmann with $D_2/D_1 \approx 3.3$, 5.8, and 8.1 at $(N_{Gr})_L N_{Pr} \gtreqless 10^6$, whereas for smaller values either of D_2/D_1 or of $(N_{Gr})_L N_{Pr}$ the exponent 0.29 was obtained, which comes close to 1/3 as found for plane gas layers.

† The influence of $(D_2 + D_1)/(D_2 - D_1)$ on convection may be compared with that of H/L for vertical plane layers (see Sect. 25–3).

CHAPTER 26

Experiments on Forced Convection and Their Correlation

26–1. Heat Transfer in the Laminar Flow of a Fluid through a Tube

Graetz's theoretical equations on the heat transfer in laminar flow in a tube are based on the assumption of constant properties of the fluid. Since the viscosity of a liquid is decreasing rapidly and that of a gas is increasing moderately with increasing temperature, the direct use of Graetz's equations is restricted to cases of small temperature differences. However, from his derivations the pertinent dimensionless group of properties, called Graetz number, emerged, which plays a role in the convection by laminar flow similar to the role of the Grashof number in free convection.

One of the first discussions of experimental data on the laminar-flow convection is that of Drew, Hogan, and McAdams [1931]. They started from Eqs. 22–49 and 50 and plotted the temperature function $\phi = 1 - \theta_m/\theta_0$ versus the Graetz number, which is defined by

$$N_{Gz} = \frac{c_p \dot{m}}{kL} \tag{26–1}$$

where $\dot{m}$ is the time rate of mass flow. For a cylindric tube

$$N_{Gz} = \frac{\pi s^2 v_m \rho c_p}{kL} \tag{26–2}$$

On the other hand, a quantity m_n was defined by Eq. 22–46 as follows:

$$m_n = \frac{\beta_n^{\,2}}{2} \cdot \frac{\alpha}{s^2 v_m}$$

By substitution

$$N_{Gz} = \frac{\pi \beta_n^{\,2}}{2 m_n L} \tag{26–3}$$

Equation 26–2 may also be written as

$$N_{Gz} = \frac{\pi}{4} \cdot \frac{v_m D}{\alpha} \cdot \frac{D}{L} = \frac{\pi}{4} N_{Pe} \frac{D}{L} \tag{26-4}$$

where $D = 2s$.

In the paper mentioned and its discussion the deviations of the experimental results of different investigators from those according to Graetz's formulas are treated in detail. In general the observed values of ϕ were above the theoretical values when heat was transmitted from the surface to the fluid. Jakob and Eck [1932] showed, from the experimental results by Holden [1923] and J. B. White [1927] as an example, that no cross-sectional velocity distribution whatsoever could have yielded these results when the flow was parallel to the axis of the horizontal tube. Probably radial flow components were superimposed on the axial flow.

Sieder and Tate [1936] correlated the convection data for viscous flow. The influence of temperature on viscosity and consequently on the flow pattern and the heat transfer was taken care of, in the following empirical equation:

$$\frac{h_a D}{k_a} = 1.86 \left(\frac{vD}{\nu_a}\right)^{1/3} \left(\frac{\nu_a}{\alpha_a}\right)^{1/3} \left(\frac{D}{L}\right)^{1/3} \left(\frac{\mu_a}{\mu_s}\right)^{0.14}$$

or

$$(N_{Nu})_a = 1.86 (N_{Re})_a^{1/3} (N_{Pr})_a^{1/3} \left(\frac{D}{L}\right)^{1/3} \left(\frac{\mu_a}{\mu_s}\right)^{0.14} \tag{26-5}$$

or

$$(N_{Nu})_a = 1.86 (N_{Pe})_a^{1/3} \left(\frac{D}{L}\right)^{1/3} \left(\frac{\mu_a}{\mu_s}\right)^{0.14}$$

where the subscripts s and a refer to the surface temperature and the arithmetic mean of the fluid temperature $t_a = (t_0 + t_L)/2$, respectively, t_0 and t_L being the entrance and exit temperatures.

For $t_a \gtrless t_s$ the ratio $\mu_a/\mu_s \lessgtr 1$, and it is claimed that introduction of the last term makes Eq. 26–5 suitable for both cases, i.e., for cooling or heating of the fluid.

Equation 26–5 can also be written in either of the following forms:

$$(N_{Nu})_a = 1.86 \sqrt[3]{\frac{4}{\pi}} (N_{Gz})^{1/3} \left(\frac{\mu_a}{\mu_s}\right)^{0.14} = 2.02 (N_{Gz})^{1/3} \left(\frac{\mu_a}{\mu_s}\right)^{0.14}$$

or

$$\psi = \frac{t_L - t_0}{t_s - t_a} = 1.86 \sqrt[3]{4\pi^2} (N_{Gz})^{-2/3} \left(\frac{\mu_a}{\mu_s}\right)^{0.14} = 6.33 (N_{Gz})^{-2/3} \left(\frac{\mu_a}{\mu_s}\right)^{0.14} \tag{26-6}$$

as can easily be verified, where ψ is a temperature ratio defined by this equation.

According to Eq. 22–50,

$$\phi = \frac{t_L - t_0}{t_s - t_0} \tag{26–7}$$

Hence

$$\frac{2\phi}{2 - \phi} = \frac{t_L - t_0}{t_s - (t_0 + t_L)/2} = \frac{t_L - t_0}{t_s - t_a} = \psi$$

or

$$\phi = \frac{2\psi}{2 + \psi} \tag{26–8}$$

Further, when $\mu_a = \mu_s$, from Eq. 26–6,

$$\psi = 6.33(N_{Gz})^{-2/3} \tag{26–9}$$

These two equations allow one to compare Graetz's and Lévêque's theoretical results with Sieder and Tate's empirical equation.

Table 26–1 gives values of ψ, according to the empirical equation 26–9, and of ϕ, calculated by Eq. 26–8. These values of ϕ are plotted in

TABLE 26–1

THE TEMPERATURE RATIOS ψ AND ϕ, CALCULATED FROM SIEDER AND TATE'S EMPIRICAL EQS. 26–9 AND 8 FOR FORCED CONVECTION IN LAMINAR FLOW THROUGH A TUBE

N_{Gz}	ψ	ϕ
3	3.043	1.208
5.63	2.000	1.000
10	1.364	0.812
30	0.656	0.494
100	0.294	0.257
300	0.141	0.132
1000	0.0633	0.0614
3000	0.0304	0.0300

Fig. 22–6 (Curve *d*). It is seen that Sieder and Tate's formula yields $\phi > 1$ at $N_{Gz} < 5.63$, which is impossible. For greater values of N_{Gz}, however, the empirical curve is in good agreement with the theoretical curves *a* and *b*, according to Graetz and Lévêque. Since, according to Table 26–1, ϕ is increasing with ψ, Eq. 26–6 yields higher values of ϕ

than are indicated in the table when $\mu_a > \mu_s$. Obviously, the velocity distribution over the cross section becomes closer to the uniform one which was assumed in setting up Curve c of Fig. 22–5.

In a more recent experimental investigation in the range $L/D = 10.5$ to 63 Cholette [1948] found values of h_aD/k which, up to $N_{Gz} \approx 10$, were in good agreement with the theoretical values according to Graetz.

A satisfactory agreement of the experimental data was obtained by the form

$$(N_{Nu})_a = C(N_{Gz})^n \tag{26–10}$$

with

$$C = 2/\pi \quad \text{and} \quad n = 1 \quad \text{for} \quad N_{Gz} < 5$$

and

$$C = 1.56 \quad \text{and} \quad n = 0.4 \quad \text{for} \quad 5 < N_{Gz} < 150$$

Free convection may cause an additional heat transfer as already mentioned in Sect. 22–2 (see footnote on p. 462).

For $N_{Gr} > 25{,}000$ Sieder and Tate recommend multiplying h_a as determined from Eq. 26–5 or 6 by a factor

$$\chi = 0.8[1 + 0.015(N_{Gr})_a^{1/3}] \tag{26–11}$$

which goes back to an earlier paper of Colburn [1933(*a*)].

Kern and Othmer [1943] performed tests with three kinds of oil under conditions where the induced free convection was from 100 to 380% of that which would be obtained from forced convection alone. The influence of free convection was found to be a function of the Reynolds number, as predicted by Martinelli, C. J. Southwell, Alves, Craig, Weinberg, Lansing, and Boelter [1942]. They suggest to replace Eq. 26–11 by the following equation:

$$\chi = \frac{2.25[1 + 0.010(N_{Gr})_a^{1/3}]}{\log N_{Re}}$$

This is said to hold within ±10 to 20% in the range of $N_{Re} = 20$ to 3000 and $(N_{Gr})_a = 300$ to 22,750,000. For high values of $(N_{Gr})_a$ it may be used down to $N_{Re} = 10$; for $(N_{Gr})_a < 5000$ only when $N_{Re} > 80$.

The authors conclude that, under conditions where free convection prevails, design differs radically from the general commercial practice, since reducing the velocity and increasing the tube diameter will improve the heat transfer.

Brinn, Friedman, Gluckert, and Pigford [1948] showed by experiments on two different sorts of sand falling inside vertical tubes, ⅝ and ⅞ in. diameter, 10 and 20 ft long, that this sort of flow can be considered like the axial motion of cylindric rods of the same diameters through the tubes. Therefore, Graetz's Eq. 22–60 could be used for describing the heat transfer in this case. It led to a satisfactory agreement with direct heat-transfer measurements for $N_{Gz} \geqq 4$. In addition, tests with white and colored sand confirmed the assumption of motion with the same velocity at any distance from the pipe axis.

26–2. Heat Transfer in the Turbulent Flow of a Fluid through a Tube

The following equation for turbulent flow similar to Eq. 26–5 and also based on data of Sieder and Tate [1936] was recommended by McAdams [1942, p. 167 ff.] for highly viscous fluids:

$$(N_{Nu})_a = 0.027(N_{Re})_a{}^{0.8}(N_{Pr})_a{}^{1/3}\left(\frac{\mu_a}{\mu_s}\right)^{0.14} \tag{26–12}$$

This equation is only the last of a series of similar equations, beginning with Eq. 23–37 from Nusselt's first paper on heat transfer in tubes, all having 0.8 as exponent of the Reynolds number, but exponents of the Prandtl number varying from 0.3 to 0.4. Most authors gave slightly different constant factors and exponents of the Prandtl number for cooling and heating, for instance, the well-known equations of Dittus and Boelter [1930]. Equation 23–37 with $n = 0.8$ should be used for gases and moderately viscous liquids, $m = 0.4$ being recommended when $t_s > t_a$, and $m = 0.3$ when $t_s < t_a$. The main contribution of Sieder and Tate is again the last term in Eq. 26–12, which makes the equation valid for heating and cooling.

Referring to the influence of the starting length, the coefficient of heat transfer will obviously start with an infinitely large value at $x = 0$ (see Sect. 22–5, immediately past Eq. 22–105) and decrease first rapidly and then less and less. From experiments of Burbach [1930] in the range $L/D = 10$ to 400 Nusselt [1931] concluded that h is proportional to $(D/L)^{1/18}$. He found this verified by experiments of Eagle and Ferguson in about the same range. Hence, Eq. 26–12 can be replaced by

$$(N_{Nu})_a = 0.036(N_{Re})_a{}^{0.8}(N_{Pr})_a{}^{1/3}\left(\frac{\mu_a}{\mu_s}\right)^{0.14}\left(\frac{D}{L}\right)^{1/18} \tag{26–13}$$

This yields 11% higher values than Eq. 26–12 at $L/D = 60$, 2% smaller values at $L/D = 240$, and 9% smaller values at $L/D = 1000$. However, the influence of L/D is much smaller than in the laminar range, as comparison between Eqs. 26–5 and 13 indicates. According to Cholette [1948], the influence of the starting length may be better expressed by $(D/L)^{0.1}$ than by the term used in Eq. 26–13.

Latzko's theoretical equations of the local coefficients in the starting length, for originally uniform velocity (Eqs. 22–109 and 110) or already fully developed velocity (Eq. 22–105), have recently been checked in a thorough experimental investigation by Boelter, Young, and Iverson [1948]. They found Eqs. 22–109 and 110 to agree almost exactly with their measurements in the range $x/D = 0.5$ to 5 and to yield values low by 2 to 8% from $x/D = 5$ to 17. The results of Eq. 22–105 are in less good agreement with the measured values, namely, low by 12% at $x/D = 0.5$, by 20% between $x/D = 1$ and 2, by 10% from $x/D = 10$ to 17. Designating by h_∞ the local coefficient at such distance where the temperature distribution is fully developed, the authors recommend in the range of $x/D > 5$ the following empirical equation for the average coefficient:

$$h_m = h_\infty \left(1 + \frac{KD}{x}\right)$$

where $K = 1.4$ (or, according to Latzko, $K = 0.067 N_{Re}{}^{0.25}$), if the velocity distribution is fully developed at the entrance, and $K = 0.7$ for initially uniform velocity distribution, the inlet temperature being uniform in both cases.

In the tests, covering a range of $N_{Re} \approx 25{,}000$ to 55,000, uniform velocity distribution was obtained by means of a bellmouth at the tube entrance, whereas a long starting piece of tube served for obtaining fully developed velocity distribution. The influence of various other inlet pieces (screens, orifices, tube bends) upon the development of the local coefficient was also investigated, and partly surprising results of practical importance were obtained which cannot be dealt with at this place.

In curved smooth tubes turbulence takes place at considerably higher Reynolds numbers than in straight tubes, as experiments of C. M. White [1929] have revealed. However, in the laminar range, the pressure drop increases with $N_{Re}\sqrt{D/D_c}$, where D is the inner diameter of the tube and $D_c/2$ is the radius of curvature of the tube axis. This is followed by a much stronger increase in the turbulent range. Correspondingly, also, an increase of the heat transfer by convection is to be expected. According to experiments of Jeschke [1925] with slightly rough tubes, the Nusselt number for straight tubes has to be multiplied by

$$\phi = 1 + 3.54 \frac{D}{D_c} \tag{26–14}$$

for curved tubes.

26–3. Heat Transfer in the Semi-turbulent Flow of a Fluid through a Tube

Heat transfer, as well as pressure drop in a tube, is governed by different laws in the laminar and turbulent regions, and there is no common state in between where the two laws would yield coinciding results. Calculating, for instance, $\frac{\Delta p}{\rho v_m^2/2} \cdot \frac{D}{L}$ from Eqs. 21–36 and 37, at $N_{Re} = 2300$, the best-known value for starting turbulence, the different values 0.0278 and 0.0457 are obtained. Hence, the pressure drop would jump to a 64% higher value if turbulence were fully developed at once. It is well known that this does not occur, but turbulence is only gradually developed in a transition region with increasing Reynolds number. The same holds for heat transfer, as could be checked by comparing Eqs. 26–5 and 13 for any suitable value of L/D. The heat transfer in the transition state, which may be called semi-turbulent flow, has been studied for heating and cooling. Colburn [1942] considers a correlation, made by Sieder and Tate [1936] and reproduced in Fig. 26–1, as the most

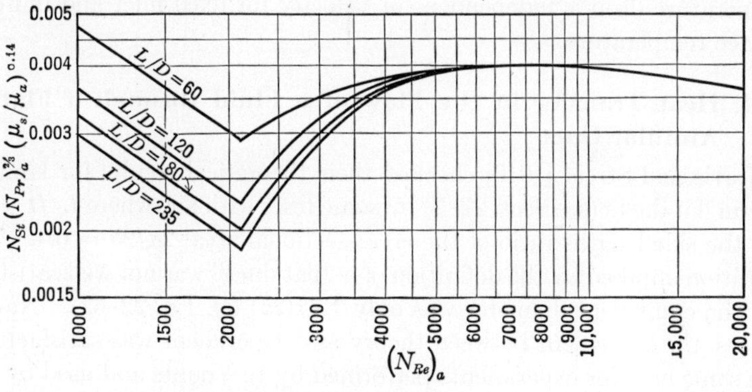

FIG. 26–1. Dimensionless representation of heat transfer in semi-turbulent flow. From Colburn [1942].

convenient practical solution. In this figure $N_{St}(N_{Pr})_a^{2/3}(\mu_s/\mu_a)^{0.14}$ is plotted versus $(N_{Re})_a$ where

$$N_{St} = \frac{h}{\rho c_p v_m} = \frac{h}{c_p \dot{m}''} \tag{26–15}$$

This dimensionless group is the Stanton number (see Sect. 24–6). Since the "mass velocity" $\dot{m}''$ is entirely, and c_p is nearly, independent of the temperature, the subscript a (used for $t_a = (t_1 + t_2)/2$) is omitted in N_{St}. Referring to the use of $N_{St}(N_{Pr})^{2/3}$, see Sects. 24–6 and 26–8.

More recently Norris and Sims [1942] investigated heat transfer by cooling liquids in vertical-downward flow in a smooth tube $(L/D = 234)$ in the range $(N_{Re})_a = 3500$ to 11,000, $(N_{Pr})_a = 35$ to 140, and $\mu_a/\mu_s = 0.32$ to 0.76. The results could be expressed with an accuracy of $\pm 5\%$ by any of the following three equations:

$$N_{St} = 0.0067(N_{Pr})_a{}^{-0.8}\left(\frac{\mu_a}{\mu_s}\right)^{0.14} \tag{26-16}$$

$$N_{Nu}(N_{Pr})_a{}^{-0.2} = 0.0067(N_{Re})_a\left(\frac{\mu_a}{\mu_s}\right)^{0.14} \tag{26-17}$$

$$\frac{t_1 - t_2}{\theta_m} = 0.0067(N_{Pr})_a{}^{-0.8}\,\frac{\pi DL}{\pi D^2/4}\left(\frac{\mu_a}{\mu_s}\right)^{0.14} \tag{26-18}$$

where the coefficient h in N_{St} and N_{Nu} is based on the log-mean temperature difference $\theta_{\ln}$.

According to Eq. 26-16, the coefficient h is independent of D (at least for the ratio $L/D = 234$, used in the experiments). According to Eq. 26-17, h is proportional to v_m. Finally, according to Eq. 26-18, the temperature drop is independent of velocity for fixed inlet and uniform surface temperature.

26-4. Heat Transfer in the Flow of a Fluid through a Flat or Annular Duct

Norris and Streid [1940] checked their theoretical results for laminar flow in flat ducts (see Sect. 22-2) by some tests with air, where $L/D = 11$ and the side-lengths ratio of the cross-sectional area, $2a/2b = 6.4$. The condition imposed by the definition of a "flat duct" was not well satisfied, and the equivalent diameter was only $1.74(2b)$ (*cf.* Eq. 22-62). Nevertheless, the agreement between theory and experiment was satisfactory. The same held for experiments performed by L. Younts and used by the authors, where $2a/2b = 12$ to 192 and $L/D = 8.7$ to 50. Also experimental results of Cope [see Bailey and Cope 1933] for heating and cooling water in the range of $2a/2b = 1$ to 8 and $L/D = 130$ to 320 are in reasonable agreement with the theoretical curves for flat ducts, given in Fig. 22-7.

Green and L. S. King [1946] investigated the heat transfer in tubes of originally 0.23 in. inside diameter, which were flattened to about half this size. Taking the equivalent diameter as characteristic length, it was found that the transition region for such tubes extended only from $N_{Re} = 2000$ to 3500, whereas the upper limit for cylindric tubes was about 8000. The Nusselt number was virtually the same for flat and round tubes in the laminar and turbulent region, but appreciably higher for the flat ones in the transition region. Dimpling the flat tubes by impressions from outside at distances of 0.5 in. increased the Nusselt number in the range of the tests ($N_{Re} \approx$ 1500 to 10,000). The authors assume that the dimples disturb the flow so that the experiments with dimples are to be considered as performed in the starting section, where, according to Nusselt, the heat transfer is increased.

Experimental data on the heat transfer in annuli for the laminar range are due to Chen, Hawkins, and Solberg [1946]. Coaxial brass tubes formed annuli with the diameter ratios $D_2/D_1 = 2.25/2$; $1.5/1.375$; $1/0.5$; $0.75/0.5$. The figures in these ratios are in inch units. Water flowing in the inner tube was used as heating and cooling fluid; water was also the medium streaming through the annulus. The film coefficient h, to be measured in the tests, was taken for the outer surface of the inner tube (diameter D_1) and for the temperature difference $\Delta t = t_1 - (t_i + t_0)/2$, where t_1 is the arithmetic mean temperature of the reference surface, and t_i and t_0 are the temperatures of the water at the inlet and outlet of the annulus. The tests led to the following empirical equation:

$$\frac{hD_e}{k} = 1.02\left(\frac{vD_e\rho}{\mu_a}\right)^{0.45}\left(\frac{\mu_a c_p}{k}\right)^{0.5}\left(\frac{\mu_a}{\mu_1}\right)^{0.14}\left(\frac{D_e}{L}\right)^{0.4}\left(\frac{D_2}{D_1}\right)^{0.8}\left(\frac{\rho^2\beta g\,\Delta t\,D_e{}^3}{\mu_a{}^2}\right)^{0.05} \tag{26–19}$$

where $D_e = D_2 - D_1 =$ the equivalent diameter of the annulus,*

$\mu_a =$ the dynamic viscosity of the water at the arithmetic-mean temperature, $t_a = \dfrac{t_1 + (t_i + t_0)/2}{2}$,

$\mu_1 =$ the dynamic viscosity of the water at the surface temperature t_1, and

$L =$ the heating or cooling length of the annular section.

The other symbols have the usual meaning. They can be expressed in any consistent units.

The equation is valid for heating and cooling alike in a range of Reynolds numbers $hD_e/k = 200$ to 2000; however, the experimental points deviate by ±6.6% from the calculated values. This is considered satisfactory in view of the variables encountered and the difficulties of thermometry.†

* The equivalent diameter is obtained from the definition $D_e = 4A/C$, where A is the cross-sectional area of the annular space and C its circumference.

† If a common exponent n had been found instead of the exponents 0.45, 0.5, and 0.4 in Eq. 26–19, this equation would have been simplified to

$$N_{Nu} = (N_{Gz})^n\left(\frac{D_2}{D_1}\right)^{0.8}(N_{Gr})^{0.05}\,.$$

The theory of heat transfer in annuli with uniformly heated surfaces (see Sect. 22–3) has been checked by Jakob and Rees [1941] by heating air, hydrogen, and ethylene in laminar flow through an annulus with $D_1 = 0.4$ and $D_2 = 0.8$ cm, the heat being supplied from the surface 2. The observed heat transfer was about 12% above the theoretical value as an average. The authors assume that this difference is due to limitations of their experimental accuracy rather than to a deficiency of their theory.

Monrad and Pelton [1942] performed experiments with water and air in turbulent flow through the annular space of coaxial tubes. The following ratios of the diameters of the surfaces of the annular space were used: $D_2/D_1 = 1.65$, 2.45, and 17. Introducing again an equivalent diameter $D_e = D_2 - D_1$ as characteristic length, the following empirical equation was obtained in the range of $N_{Re} = 12{,}000$ to 220,000:

$$(N_{Nu})_M = \frac{h_1 D_e}{k} = 0.020 \left(\frac{v_m D_e}{\nu}\right)^{0.8} \left(\frac{\nu}{\alpha}\right)^{1/3} \left(\frac{D_2}{D_1}\right)^{0.53} \qquad (26\text{–}20)$$

where h_1 is the coefficient of heat transfer at the inner surface of the annulus, and subscript M refers to the name of the senior author.

For the outer surface of the annulus, according to tests with one annulus ($D_2/D_1 = 1.85$), Eq. 26–12 may be used with D_e as characteristic length.

E. S. Davis [1943], on the other hand, represented the experimental results of different investigators in the following form:

$$(N_{Nu})_D = \frac{h_1 D_1}{k} = 0.038 \left(\frac{D_1 v_m}{\nu}\right)^{0.8} \left(\frac{\nu}{\alpha}\right)^{1/3} \left(\frac{\mu}{\mu_1}\right)^{0.14} \left(\frac{D_2}{D_1}\right)^{0.15} \qquad (26\text{–}21)$$

where subscript D refers to the name of the author, and μ_1 is taken at the temperature of surface 1, whereas the other properties seem to be taken at the arithmetic mean temperature t_a, as previously defined. When $\mu/\mu_1 = 1$, then Eq. 26–21 can be transformed into

$$(N_{Nu})_D = B(N_{Nu})_M \qquad (26\text{–}22)$$

where

$$B = \left(\frac{D_2}{D_1} - 1\right)^{-0.8} \left(\frac{D_2}{D_1}\right)^{-0.38}$$

But, according to the definition of $(N_{Nu})_D$ and $(N_{Nu})_M$,

$$(N_{Nu})_D = C(N_{Nu})_M$$

where

$$C = \left(\frac{D_2}{D_1} - 1\right)^{-1}$$

The calculation yields

For $\frac{D_2}{D_1} = 2$	10	20	100
$B = 1.46$	0.136	0.0576	0.00836
$C = 1$	0.111	0.0526	0.0101

This shows that for small values of D_2/D_1 Davis' formulation gives higher values of h_1 than Monrad and Pelton's formulation, and vice versa for great values of D_2/D_1. In the range of $D_2/D_1 = 10$ to 100 the two equations are in agreement within a tolerance of $\pm 20\%$.

In the range of $v_m D_1/\nu = 7$ to 120 Davis' Eq. 26–21 covers also Mueller's [1942] experiments with wires in tubes, ($D_2/D_1 = 605$ to 6800). Notwithstanding the small Reynolds numbers, these data belong to the turbulent state. For instance, a value of Mueller at $v_m D_1/\nu = 7.5$ belongs to $D_2/D_1 = 2590$. Hence

$$\frac{v_m D_e}{\nu} = \frac{v_m(D_2 - D_1)}{\nu} = \frac{D_2 - D_1}{D_1} \cdot \frac{v_m D_1}{\nu} = (2590 - 1)7.5 = 19{,}420$$

which obviously is in the turbulent state.

Davis also brought experimental values of Foust and Christian [1940] into line with his correlation, by some corrections which he found advisable.

In general Eq. 26–21 has been proved to be roughly valid for turbulent flow when

$$7 \gtreqless \frac{v_m D_1}{\nu} \gtreqless 180{,}000$$

and, at the same time,

$$1.2 \gtreqless \frac{D_2}{D_1} \gtreqless 6800$$

Carpenter, Colburn, Schoenborn, and Wurster [1946], using water flowing through an annulus with $D_1 = 5/8$ in. and $D_2 = 0.834$ in. ($D_2/D_1 = 1.334$), observed values between those of Monrad and Pelton, and of Davis, in a range from $N_{Re} = 2000$ to 16,000.

For heating from outside, Davis recommends the following formula:

$$\frac{h_2 D_e}{k} = 0.0225 \left(\frac{v_m D_e}{\nu}\right)^{0.8} \left(\frac{\nu}{\alpha}\right)^{1/3} \left(\frac{\mu}{\mu_2}\right)^{0.14} \left(\frac{D_2}{D_e}\right)^{-0.1} \qquad (26\text{–}23)$$

The last exponent is estimated only and may be some value between -0.05 and -0.15. The properties are taken at temperature t_a, except μ_2, which is for the temperature of annular surface 2.

For higher velocities Wiegand [1945] recommended somewhat different correlations for Reynolds numbers above 5000 and higher viscosities. Converted to a form similar to Eq. 26–23, they are, for the inner wall,

$$\frac{h_1 D_e}{k} = 0.023 \left(\frac{\dot{m}'' D_e}{\mu_f}\right)^{0.8} \left(\frac{\nu}{\alpha}\right)_f^{1/3} \left(\frac{D_2}{D_1}\right)^{0.45} \tag{26–24}$$

and for the outer wall,

$$\frac{h_2 D_e}{k} = 0.023 \left(\frac{\dot{m}'' D_e}{\mu_f}\right)^{0.8} \left(\frac{\nu}{\alpha}\right)_f^{1/3} \tag{26–25}$$

where the subscript f refers to film temperature.

26–5. Heat Transfer in the Flow of a Fluid Parallel to Plane or Moderately Curved Surfaces

Compared to the abundant experimental work on heat transfer in tubes and ducts, relatively little has been done concerning the flow of a fluid parallel to a heating or cooling surface when the fluid is not bounded by the walls of a tube or channel. Here, a Reynolds number for a point of the surface can be defined by

$$N_{Re} = \frac{v_m x}{\nu} \tag{26–26}$$

where x is the distance of that point, measured in the flow direction, from the place where the fluid has met the surface ($x = 0$).

For the critical Reynolds number, determining the place $x = x_{cr}$ where laminar flow ceases in the boundary layer, values between 100,000 and 500,000 have been measured, depending on the starting conditions, 300,000 being a fair average of the critical value according to Van der Hegge Zijnen [1924] and M. Hansen [1928]. With blunt leading edges, however, turbulent layers have been obtained straight from the start ($x = 0$) [see ten Bosch 1936, p. 139].

For the laminar range Pohlhausen's theoretical equation 22–91, employed for air with $N_{Pr} = 0.71$, becomes

$$N_{Nu} = 0.592\sqrt{N_{Re}} \tag{26–27}$$

This equation has been checked and proved by experiments of Jakob and Dow [1946]. They used air streaming parallel to an electrically heated smooth cylinder of 1.3 in. diameter with a spherical, ellipsoidal or conical nosepiece; unheated cylindric sections of different length were inserted between the spherical nose and the heating cylinder. The effect of these hydrodynamic-starting sections was corrected in a way that will be described later on. The authors further demonstrated that the influence of the surface curvature, if any, was within the range of the experimental accuracy, as the flow in the boundary layer was laminar.

Equation 26–27 was perfectly confirmed by these experiments except for a slight difference in the constant factor (0.590 instead of 0.592).

The major part of Jakob and Dow's investigation is devoted to the turbulent range. In this range the experimental results could be represented by an empirical equation of the form

$$N_{Nu} = C_0(N_{Re})^n \left[1 + C_1\left(\frac{L_{st}}{L_{tot}}\right)^m\right] = C(N_{Re})^n \tag{26–28}$$

and, particularly, for air,

$$N_{Nu} = 0.0280(N_{Re})^{0.8}\left[1 + 0.40\left(\frac{L_{st}}{L_{tot}}\right)^{2.75}\right] \tag{26–29}$$

where L_{tot} = the total length of the surface in the flow direction,
L_{st} = the length of the hydrodynamic (unheated) starting section of the plate,
$N_{Re} = v_0 L_{tot}/\nu_0$,
v_0 = the velocity of the arriving air,
ν_0 = its kinematic viscosity at the temperature t_0,
$N_{Nu} = hL_{tot}/k_f$,
k_f = the thermal conductivity of the air at the film temperature $t_f = (t_0 + t_s)/2$ where t_s is the uniform surface temperature of the heated section, and
h = the mean coefficient of heat transfer by convection, taken over the heated section and based on the temperature difference $t_s - t_0$.

From a theoretical viewpoint Eq. 26–29 should be valid also for other diatomic gases, as far as the Prandtl number is the same.

How closely the experimental results in the range of $L_{st}/L_{tot} = 0.1$ to 0.6 fit with this equation is seen from Fig. 26–2, in which

$$\frac{N_{Nu}}{[1 + 0.40(L_{st}/L_{tot})^{2.75}]}$$

is plotted versus N_{Re} in logarithmic coordinates. The transition from laminar to turbulent flow is also represented in the graph. It depends on the shape of the nose of the cylindrical heater rather than on the length of cylindrical unheated pieces interposed between nose and heating section. Turbulence started at $N_{Re} \approx 50{,}000$ with the spherical nose

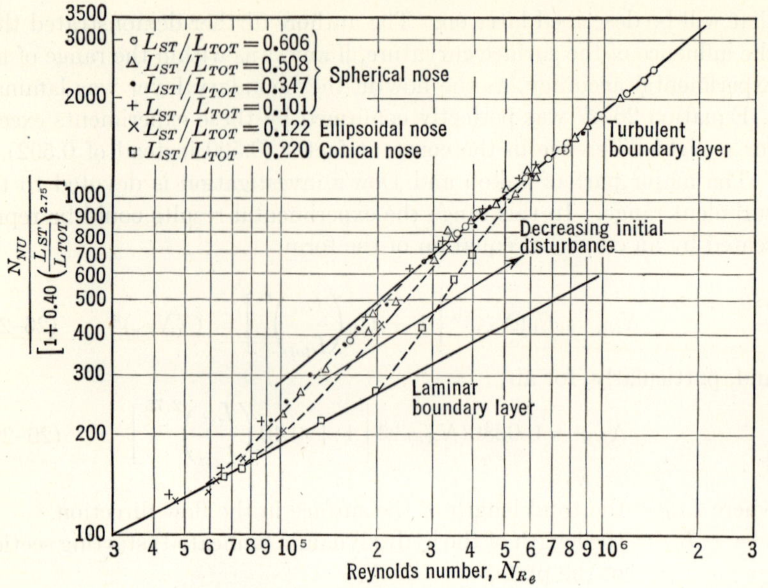

FIG. 26–2. Heat transfer from a cylinder to air in parallel flow, considering the influence of unheated starting sections.

and $N_{Re} \approx 100{,}000$ with the long conical nose and was fully established at $N_{Re} \approx 300{,}000$ and $N_{Re} \approx 600{,}000$, respectively.

In Fig. 26–3, the experimental results are compared with those of Juerges, obtained on a flat plate with $L_{st}/L_{tot} = 0.38$, and of Éliás on flat plates with $L_{st}/L_{tot} = 0.2$ to 0.5. The leading edge of Juerges'

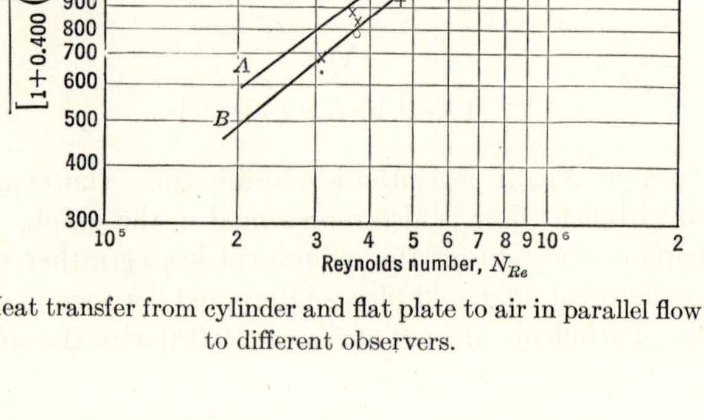

FIG. 26–3. Heat transfer from cylinder and flat plate to air in parallel flow according to different observers.

plate was not smoothed; hence the heat transfer found in his experiments was higher than it would otherwise have been. In addition, the heat loss by mixing on the side edges will have increased the heat transfer measured. An explanation of this effect is given below in the discussion of some experiments of Slegel and Hawkins.

Éliás derived his values from flow and temperature measurements in the boundary layer which are difficult and not too accurate. Since the exponent n of Eq. 26–28 is about 0.9, according to his experiments, it can be assumed that they refer to transition between streamline and turbulent flow. Juerges' results can be expressed by using Eq. 26–29 if $C_0 = 0.0322$ instead of 0.0280 is used.

Jakob and Dow, however, come to the conclusion that a curved surface should rather give higher values of h than too small ones.

Using an equation of von Kármán, they express the thickness of the boundary layer by

$$b = 0.366x'(N_{Re}')_{x'}^{-0.2} \tag{26–30}$$

where $x' = x - x_0$,

$(N_{Re}')_{x'} = v_a x'/\nu$,

$x = x_0 =$ that distance from the leading edge at which fully established turbulence ought to start in order to bring the boundary layer to the same thickness b at $x = x_{cr}$ which it actually attains in the laminar range.

Equating, therefore, b according to Eqs. 22–93 and 26–30, for $x = x_{cr}$, the authors were able to calculate the fictitious length x_0. The calculation for two extreme cases, $N_{Re} = 50{,}000$ and $1{,}500{,}000$, both with $(N_{Re})_{cr}$, yielded 7.5 and 19.6% larger coefficients for heat transfer than for plane surfaces, as an average a surplus of 13.5%.

The experiments on the cylindric specimen mentioned, on the other hand, led to

$$N_{Nu} = 0.0280(N_{Re})^{0.80} \tag{26–31}$$

for $L_{st} = 0$, i.e., about 10% more than according to Latzko's equation 22–117 when applied to air, taking $N_{Pr} = 0.71$.

This is a satisfactory agreement between theory and experiment. However, considering also Seibert's theoretical reasoning [1938], which is at variance with their theory, Jakob and Dow recommend as a compromise to use their values, determined on a cylinder, for plane plates also until more experimental evidence is available concerning the influence of curvature.

It must be mentioned that Slegel and Hawkins [1946] in experiments on a vertical plane surface obtained about 45% higher values than Jakob and Dow. After considerable speculation about possible reasons for such differences [see Jakob 1946(*a*)] the author finally came to the conclusion that the high values observed by different investigators are mainly due to flow mixing on the side edges of their plane plates. The assumption of strictly unidirectional flow on the edges, made in the calculation of h, is at variance with the concept of turbulence, which is based on mixing movements in all directions. Such mixing of a fraction of the fluid at practically undisturbed initial temperature with the flow on the plate under investigation increases the rate of heat needed for a fixed temperature difference between surface and fluid, which generally is measured somewhere close to the middle of the plate, as compared to the heat which would be needed for the same temperature difference if no mixing on the edges occurred. As emphasized already in Jakob and Dow's paper, a cylinder is free of such secondary effect.

Referring to fins as subject to the equations of heat transfer on plane surfaces, the most favorable distance is sought. If the size of the surface were the only controlling factor, it would be best to use very small distances. Then, however, the fin surfaces would be in contact with air which was already heated by the neighbor fin, and the coefficient of heat transfer would be small because the velocity between the fins would have to be small in order to keep the pressure drop within reasonable limits. Therefore neighboring fins should be so far distant that the fluid boundary layers on their surface will not appreciably influence each other. Referring to Eq. 22–93, fins can be arranged at smaller distances if the length, $x = L$, in the direction of flow is smaller and if the velocity v_0 of the fluid is higher. This holds in a range up to $N_{Re} = v_0 L/\nu = 100{,}000$ to 400,000, the higher figure being valid if the arriving air is free of eddies. With usual lengths of cooling fins the range of laminar boundary layer is seldom transgressed. Should this occur, then Eq. 26–30 may be used, but only for $x \gtreqqless x_{cr}$, because it is unlikely ‡ that b becomes smaller with increasing length. A similar procedure was previously suggested by Burgers and van der Hegge Zijnen [1924].§ However, some caution must be recommended in the application of values of h, as found under unbounded flow conditions, to the heat convection in the space between fins because of the mentioned discrepancy between results for unbounded flow and flow in a channel.

26-6. Heat Transfer in the Flow of a Fluid Parallel to a Wire

As mentioned in Sect. 26–5, the heat transfer from a cylinder to a fluid which flows parallel to its axis will not be much different from that

‡ It is unlikely, but not impossible, since starting of turbulence might tear off some of the laminar boundary layer.

§ See also van der Hegge Zijnen [1924].

between a plane plate and a fluid in parallel flow, provided that the diameter of the cylinder is large. If it is small, however, then the eddies occurring in turbulent flow may be great compared with the cylinder diameter, and the flow will become more similar to that perpendicular to the axis.

This has actually been found by Mueller [1942] in experiments on fine wires, extended in axial direction in vertical tubes. The result of the experiments is presented in Fig. 26–4. The abscissae are Reynolds

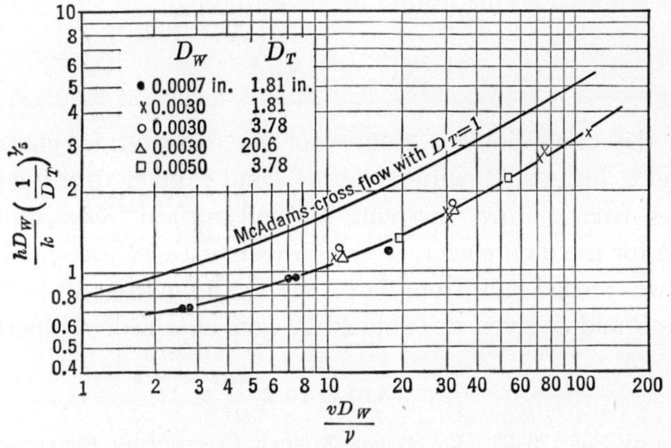

FIG. 26–4. Heat transfer to air in parallel flow to heating wires. Courtesy Dr. A. C. Mueller [1942].

numbers with the wire diameter D_W as characteristic length. The ordinates are not dimensionless, D_T being the tube diameter in foot-units. Mueller compared his results with those for cross flow, assuming $D_T = 1$ in that case. The curve for parallel flow has almost the same shape, though 40% smaller ordinates than the line for cross flow. It may be that the size of eddies inside the tube, compared to the tube diameter or to the distance of the two parallel wires used in the experiments, plays a role. For the time being, the empirical result represented in the figure must be used.

26–7. Heat Transfer in the Flow of a Fluid Perpendicular to Cylinders and Spheres

Exact measurements of the heat transfer between heated wires or circular tubes and air streams, in a direction perpendicular to the axis, have been performed by Hilpert [1933]. The diameter of the wires was varied from 0.002 to 0.1 cm, the diameter of the cross section of the tubes from 0.3 to 15 cm. Employing a nozzle of rectangular cross section,

Hilpert produced and used air streams of uniform velocity over the jet cross section. When log N_{Nu} was plotted versus N_{Re}, the test points fell almost exactly in a sequence of straight lines with bends at log N_{Re} = 1.6; 3.6; and 4.6; a fourth bend at log $N_{Re} = 0.6$ was less pronounced, but had also been observed by L. V. King [1914] and Kennely, Wright, and van Bylevelt [1909] in similar experiments. The bends obviously indicate changes in the character of the flow such as have also been found in mere flow experiments.

Hilpert represented his results in the form of

$$N_{Nu} = BN_{Re}{}^{n} \tag{26-32}$$

referring the dimensionless groups to the diameter as characteristic length and to mean arithmetic values of ν and k, taken over the temperature range from t_0 to t_s (arriving air and surface, respectively), and defining h for the difference $t_s - t_o$. In tests with 12 wires and tubes in which t_s was varied only from 93 to 109°C and t_0 from 19 to 24°C, the constants B and n, given in Table 26-2, were obtained. Hilpert's coefficients of heat transfer seem to be very reliable; however, they are smaller than those of some other investigators who had less homogeneous air streams.

TABLE 26-2

CONSTANTS OF EQS. 26-32 AND 33 FOR FORCED CONVECTION PERPENDICULAR TO CYLINDERS

(According to Hilpert)

N_{Re}		n	B	C
From	To			
1	4	0.330	0.891	0.872
4	40	0.385	0.821	0.802
40	4,000	0.466	0.615	0.600
4,000	40,000	0.618	0.174	
40,000	250,000	0.805	0.0239	

With one wire (0.003 cm diameter) t_s was varied from 61 to 1046°C. The coefficient h was about 6% larger at 1000°C than at 100°C. Supplementing these experiments by experiments of L. V. King [1914], the following relationship was found:

$$N_{Nu} = C\left[N_{Re}\left(\frac{T_s}{T_0}\right)^{1/4}\right]^n \tag{26–33}$$

The ranges of these experiments and the constant C are also given in Table 26–2, the exponent n being unchanged.

The values of Table 26–2 may be compared with results of Reiher [1925], Ulsamer [1932], and W. J. King [1932(*a*)] given in Table 26–3.

TABLE 26–3

CONSTANTS OF EQ. 26–32 FOR FORCED CONVECTION PERPENDICULAR TO CYLINDERS

(According to Different Authors)

Author	N_{Re}		n	B
	From	To		
Ulsamer	0.1	4	0.305	0.875
"	4	50	0.41	0.764
King		300	0.41	0.764
Ulsamer	50	10,000	0.50	0.536
Reiher	400	6,000	0.560	0.350
King	300	40,000	0.585	0.282

Reiher's data were obtained with air streams which had passed honeycomb and sieve so that they were turbulent, but without large and irregular eddies. He performed also some experiments in an air stream which had passed a two-row tube bank, producing irregular eddies. In this case $n = 0.648$ and $B = 0.264$ were found, which indicates a heat transfer more than 50% better, because of the disturbance of the flow. Comparing Tables 26–2 and 3 and in similar cases, the following should be kept in mind. Small differences in n, which occur in measuring the slope of the straight line which in bilogarithmic coordinates represents Eq. 26–32, generally cause considerable changes in the constant B.

Reiher also measured the temperature distribution over the perimeter of a steel tube, 2.8 cm outer diameter and 0.2 cm wall thickness. In experiments with different velocities and temperatures of the flowing air he found nearly sinusoidal temperature distribution on the wall, with a maximum on the front side and a minimum on the back side of the cross section. For instance, with 200°C air temperature and about 10 ft/sec velocity, these extreme values were 63 and 26°C. In general, however, it was found that an average between these extremes gives almost exactly the mean surface temperature.

The relationship between N_{Nu} and N_{Re} according to Eq. 26–32 holds also for tubes of other than circular cross section when the diameter of a circular tube of equal exposed surface is used as characteristic length. The constants for such profiles are given in Table 26–4. They are based

TABLE 26–4

CONSTANTS OF EQ. 26–32 FOR FORCED CONVECTION PERPENDICULAR TO NON-CIRCULAR TUBES

Flow Direction and Profile	N_{Re}		n	B	Observer
	From	To			
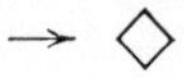	5,000	100,000	0.588	0.222	Hilpert
	2,500	15,000	0.612	0.224	Reiher
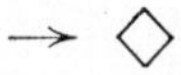	2,500	7,500	0.624	0.261	"
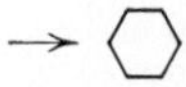	5,000	100,000	0.638	0.138	Hilpert
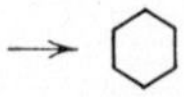	5,000	19,500	0.638	0.144	"
	5,000	100,000	0.675	0.092	"
	2,500	8,000	0.699	0.160	Reiher
	4,000	15,000	0.731	0.205	"
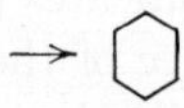	19,500	100,000	0.782	0.035	Hilpert
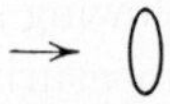	3,000	15,000	0.804	0.085	Reiher

on experiments of Reiher [see Knoblauch and Reiher 1925, p. 313, footnote 1] and Hilpert. The table is arranged according to increasing values of n, that is, for increasing steepness of lines in a plot of N_{Nu} versus N_{Re}. The bend at $N_{Re} = 19{,}500$ for the hexagonal profile is remarkable. It can be assumed that for other profiles also bends may occur below or beyond the range of the experiments. In general the non-circular profiles offer better heat transfer than the circular ones.

The influence of mechanical disturbances of the flow at the surface may be seen from Table 26–5, also based on experiments of Reiher

TABLE 26–5

CONSTANTS OF EQ. 26–32 FOR CYLINDERS WITH FINS OR GROOVES

(According to Reiher)

Surface Disturbance	N_{Re} From	N_{Re} To	n	B
None	400	6000	0.560	0.350
Longitudinal fin, 0.1 D thick, on front of tube	1000	4000	0.603	0.248
12 longitudinal grooves, 0.07 D wide	3500	7000	0.747	0.082
Same, with burs	3000	6000	0.86	0.0368

[1925]. The disturbance indicated in the last line of the table causes an increase of the heat transfer by 40% at $N_{Re} = 6000$, compared to the undisturbed flow.

The values given for n, B, and C may safely be used for other diatomic gases than air. For liquids, however, N_{Pr} enters the equation

$$N_{Nu} = E(N_{Re})^n(N_{Pr})^m \tag{26–34}$$

Values of n, m, E, as given by Ulsamer [1932], are presented in Table 26–6.

TABLE 26–6

CONSTANTS OF EQ. 26–34 FOR FORCED CONVECTION PERPENDICULAR TO CYLINDERS

(According to Ulsamer)

N_{Re} From	N_{Re} To	n	m	E
0.1	50	0.385	0.31	0.91
50	10,000	0.50	0.31	0.60

The behavior of tube banks will be discussed in Vol. II.

For the heat transfer on a sphere in a unidirectional air stream McAdams [1942, p. 237] shows data correlated by G. C. Williams. In the range of $N_{Re} = 20$ to 150,000 they can be represented by Eq. 26–32 with $n = 0.6$ and $B = 0.33$. Instead of these values, $n = 0.52$ and $B = 0.70$ can be used as well, according to Nottage and Boelter [1940, Fig. 16], but only in the range of $10^3 < N_{Re} < 10^5$.

26-8. Comparison of Heat Transfer and Fluid Friction

Colburn [1933(*a*)] is not in favor of the most popular method of representing forced heat-convection data, i.e., to plot $N_{Nu}(N_{Pr})^{-1/3}$ versus N_{Re}. Instead of this he suggested to plot the heat-transfer factor $j = N_{St}(N_{Pr})^{2/3}$ versus N_{Re}, which is the same in principle, but has some practical advantages which will now be explained. It can easily be shown that

$$N_{Nu}(N_{Pr})^{-1/3} = [N_{St}(N_{Pr})^{2/3}]N_{Re} = jN_{Re} = \left[\frac{A}{S} \cdot \frac{\Delta t}{\theta} N_{Pr}{}^{2/3}\right] N_{Re} \qquad (26\text{-}35)$$

using the same nomenclature as in Sect. 24–6. Colburn points out that the term in brackets contains the data which are generally found in experiments and used in design of heat exchangers. This term does not vary widely as compared with variations in N_{Re}. Thus plotting jN_{Re} versus N_{Re} almost means plotting a function against itself, whereas plotting j versus N_{Re} shows the characteristic variations in a clearer form. In particular, j is directly proportional to the temperature change Δt in a heat exchanger, so that the effect which by variation of the fluid flow is exerted upon the exit temperature is indicated at once by this form of plotting. It further allows a comparison with the friction factor || $f/2$, which may or may not be equal to the heat-transfer factor j.

Considering first the viscous (laminar) flow in tubes, the friction factor $f/2$ is decreasing faster with increasing N_{Re} than the heat-transfer factor j.

In turbulent flow, however, the following equation of Drew, Koo, and McAdams may be used for comparison:

$$\tfrac{1}{2}f = 0.0007 + 0.0625(N_{Re})^{-0.32} \qquad (26\text{-}36)$$

Experiments by different observers with water led to slightly higher values of j than $f/2$, according to this formula. Experiments of Nusselt with air and oxygen gave values of j about 10% smaller as an average. The latter difference, however, is in agreement with a theoretical equa-

|| The friction factor f is defined by Eq. 24–31.

tion of Boelter, Martinelli, and Jonassen [1941], according to which j actually is 10% smaller than $f/2$ in this region. This was also verified on hydrogen-nitrogen mixtures with $N_{Pr} \approx 0.5$.

In a more recent thorough investigation McAdams, Nicolai, and Keenan [1945, 1946] represented the heat transfer in tubes by the following equation:

$$(N_{St})_e = \frac{h_e}{c_p \dot{m}''} = 0.033(N_{Re})_m{}^{-0.23} \tag{26-37}$$

where h_e is based on the temperature difference $t_s - t_{th}$ and is called effective coefficient of heat transfer,
t_s is the temperature of the surface in contact with the fluid, and
t_{th} is the temperature which the surface would assume with the same bulk temperature of the fluid when the surface was impervious for heat ("adiabatic").¶

This equation is based on experiments of different research workers in the range of 0.28 to 0.99 in. diameter, 0.2 to 3 atmospheres absolute pressure, 10 to 400°F temperature, Mach numbers $N_{Ma} = 0$ to 1, and $N_{Re} = 10{,}000$ to 400,000. Individual test points of the authors and other observers deviate by ±7% as a maximum.

The coefficient h_e, as defined in this paper, has the advantage of being independent of the temperature difference in contrast to definitions of h based on other temperature differences.

Coming now to the flow parallel to a plane plate, Pohlhausen's theoretical equation is

$$N_{Nu} = (N_{Re})^{1/2}\psi(N_{Pr}) \tag{26-38}$$

with the distance L from the leading edge as characteristic length in N_{Nu} and N_{Re} and

$$\psi(N_{Pr}) \approx 0.664(N_{Pr})^{1/3} \tag{26-39}$$

In Colburn's form this becomes

$$N_{St}(N_{Pr})^{2/3} = j = 0.664(N_{Re})^{-1/2} \tag{26-40}$$

Values of j, according to this equation, are in perfect agreement with the experimental data of Jakob and Dow [1946] for heat transfer and with the theoretical values of $f/2$ according to Hopf [1927, pp. 156 and 157] when the frictional resistance per unit area (*cf.* Eq. 24–76) is defined by

$$R_f'' = \tfrac{1}{2} f \rho v^2 \tag{26-41}$$

¶ The subscript *th* means "thermometric" and indicates that the surface of a thermometer will assume the temperature t_{th}. This will be dealt with in detail in Vol. II (Chapter 33).

In the turbulent region the semi-empirical equation

$$N_{Nu} = 0.0356 N_{Re}{}^{0.8} N_{Pr}{}^{1/3} \tag{26-42}$$

can be converted into

$$N_{St}(N_{Pr})^{2/3} = j = 0.0356(N_{Re})^{-0.2} \tag{26-43}$$

This is also in good agreement with $f/2$ according to Hopf's graph, which shows that the theoretical line is exactly verified by experiments of C. Wieselsberger with air and Fr. Gebers with water in the range of $N_{Re} \approx 250{,}000$ to $50{,}000{,}000$.

For the flow across cylinders or tubes, however, $f/2$ is greater than j, especially in the turbulent region. This is attributed to the additional resistance to a flow around an obstacle.

Application to tube banks will be dealt with in Vol. II.

CHAPTER 27

Optical Methods of Determining Heat Transfer by Convection

27-1. General Features of the Optical Methods

As previously demonstrated, the distributions of velocity and temperature in a thin boundary layer on the surface are decisive for heat transfer by convection. In order to measure these distributions, some instruments such as thermocouples may be placed inside the boundary film, but will disturb the distributions more or less. Optical methods are free from this handicap. The two methods used so far are both based on the fact that the density of the fluid changes with temperature, so that decreasing of temperature across the film corresponds to an increase of density, and light travelling along the surface is deflected because of the existing differences in density. According to E. Schmidt's [1932] "schlieren" method,* pictures of caustic lines and shadows are thrown on photographic plates and show a surprisingly simple relationship to the rate and film coefficient of heat transfer. In Kennard's [1932] method, which was published in the same year as Schmidt's method, fringes, produced by an interferometer, are photographed. The pattern of fringes represents the temperature field across the boundary layer in all its details, whereas Schmidt's photos yield only the limits of that temperature field and the distribution of the heat transfer along the heating surface. Both methods are restricted to plane and single curved surfaces, i.e., surfaces on which parallel straight lines can be drawn; they are unfit for double-curved surfaces such as spheres and ellipsoids. Transportation of heat by the motion of the fluid in the boundary layer is neglected in both methods.

* The German word *Schlieren* was originally used for streaks in glass, which were produced by differences in the density and made it unfit for optical purposes. The British translation "striae" does not seem to be specific enough. Therefore the word schlieren has been adopted in the American scientific literature. See, for instance, G. M. Barnes [1945].

27-2. Method of Light Deflection by Schlieren

a. Theory of the Schlieren Method

Schmidt's schlieren method does not require any lenses, but plainly operates with shadow pictures which show the space limits, in which the temperature of the medium is influenced by the heating surface; the distances of caustic lines, to be seen in these photos, from the heating surface are proportionate to the heat flow at the surface.

Neglecting some anomalies discovered by Gans [1915], geometrical optics reveals the following behavior of a light wave in an inhomogeneous medium whose index of refraction varies in one direction only. Referring to Fig. 27-1, let line 1 be the front of a light wave, at the time τ,

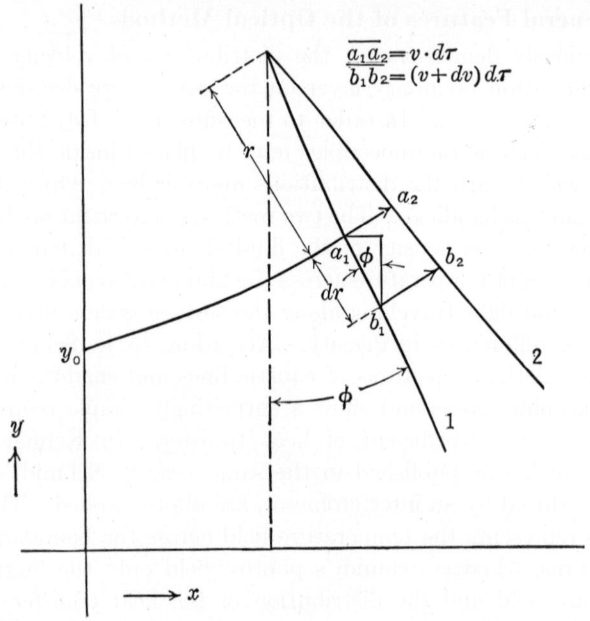

FIG. 27-1. Curved path of a light beam in an inhomogeneous medium.

which moves in a medium, for instance, air in which the temperature t varies in space. Since the index of refraction n changes with temperature, it is a function of the place, and the same holds for the velocity of light v, because

$$v = \frac{c}{n} \tag{27-1}$$

where c is the light velocity in an absolute vacuum. Let v and $v + dv$

be the light velocities at the points a_1 and b_1. Then, after a time interval $d\tau$, the wave front is at the line 2, which is inclined a little with respect to line 1. The light beam, which at the point a_1 appears in a direction perpendicular to the wave front, describes a curve with the radius of curvature r. According to the figure,

$$\frac{1}{r} = \frac{1}{v} \cdot \frac{dv}{dr} \tag{27-2}$$

Substituting v from Eq. 27–1,

$$\frac{1}{r} = -\frac{1}{n} \cdot \frac{dn}{dt} \cdot \frac{dt}{dr} = -\frac{d(\ln n)}{dt} \cdot \frac{dt}{dr} \tag{27-3}$$

Introducing rectangular coordinates and representing by $y = f(x, y_0)$ the path of that beam which meets the axis of ordinates ($x = 0$) at $y = y_0$,

$$\frac{dt}{dr} = \frac{\partial t}{\partial x} \cdot \frac{\partial x}{\partial r} + \frac{\partial t}{\partial y} \cdot \frac{\partial y}{\partial r} = \frac{\partial t}{\partial x} \sin \phi - \frac{\partial t}{\partial y} \cos \phi$$

where $\tan \phi = dy/dx = y'$. As usual, $d^2y/dx^2 = y''$ will also be employed. Expressing $\sin \phi$ and $\cos \phi$ in terms of y' and considering the geometrical relation between y', y'', and r,

$$\frac{1}{r} = -\frac{1}{n} \cdot \frac{dn}{dt} \left(\frac{y'}{\sqrt{1 + y'^2}} \cdot \frac{\partial t}{\partial x} - \frac{1}{\sqrt{1 + y'^2}} \cdot \frac{\partial t}{\partial y} \right) = \frac{y''}{(1 + y'^2)^{3/2}} \tag{27-4}$$

or

$$\frac{y''}{1 + y'^2} = -\frac{1}{n} \cdot \frac{dn}{dt} \left(y' \frac{\partial t}{\partial x} - \frac{\partial t}{\partial y} \right) \tag{27-5}$$

This is the differential equation which coordinates the trace of a light beam to the temperatures t. The term $\frac{1}{n} \cdot \frac{dn}{dt}$ depends upon the kind of medium in which the beam is observed.

Equation 27–5 simplifies considerably if only small deviations of the beam are considered. Let $y = 0$ be the surface of a heating source, and consider a bundle of rays arriving in the positive direction of x with grazing incidence to the surface; then, according to the increase of air density in the direction perpendicular to the surface, the wave front will be deflected. In Schmidt's experiments $y' \approx 0.01$ was the greatest tangent of deflection. Therefore y'^2 on the left side of Eq. 27–5 could be neglected. It may further be assumed that the temperature decreases in the y direction, but is constant in the x direction, except for the two ends of the heating surface where $\partial t/\partial x$ be of the same order of magnitude as $\partial t/\partial y$. But in Eq. 27-5, $\partial t/\partial x$ occurs only with the factor y', which is supposed to be small. Therefore, $y' \cdot \partial t/\partial x$ can be neglected against $\partial t/\partial y$, and the equation becomes

$$y'' = \frac{1}{n} \cdot \frac{dn}{dt} \cdot \frac{\partial t}{\partial y} = \frac{d(\ln n)}{dt} \cdot \frac{\partial t}{\partial y} \tag{27-6}$$

In a range of the temperature field where changes of n and $\partial t/\partial y$ with x are negligibly small, integration of Eq. 27-6 yields

$$y' = \frac{1}{n} \cdot \frac{dn}{dt} \cdot \frac{\partial t}{\partial y} x \tag{27-7}$$

and

$$y = \frac{1}{n} \cdot \frac{dn}{dt} \cdot \frac{\partial t}{\partial y} \cdot \frac{x^2}{2} = f(x) \tag{27-8}$$

Denoting the length of the heater surface by L_h, a beam, arriving in the direction x and travelling the length L_h, will eventually have the slope

$$y_h' = \frac{1}{n} \cdot \frac{dn}{dt} \cdot \frac{\partial t}{\partial y} L_h \tag{27-9}$$

The tangent, drawn at $x = L_h$ on the curve $y = f(x)$ (see Fig. 27-2), intersects the axis of abscissae at $x = L_h/2$.

Hence a beam of direction x which arrives at $x = 0$, $y = 0$ will hit a screen, located at $x = (L_h/2) + L$, at the ordinate

$$y_{scr} = \int_{x=\frac{L_h}{2}}^{x=\frac{L_h}{2}+L} y_h' \, dx = \frac{1}{n} \cdot \frac{dn}{dt} \cdot \frac{\partial t}{\partial y} L_h L \tag{27-10}$$

Measuring the distance y_{scr}, the equation yields $\partial t/\partial y$. Hence the rate of heat given up by a heating plate of length L_h and width W becomes

$$q = -k \frac{W y_{scr}}{L} n \frac{dt}{dn} \tag{27-11}$$

where k = the thermal conductivity of the air.

Schmidt in his experiments operated with $L \approx 50L_h$ and $y_{scr} \approx 5$ cm, corresponding to $y_h \approx 0.05$ cm. At this distance from the heating surface $\partial t/\partial y$ is still sensibly constant, according to direct measurements of Schmidt and Beckmann (see Sect. 22–1). Therefore the conditions are satisfied under which Eqs. 27–7 and 8 were derived.

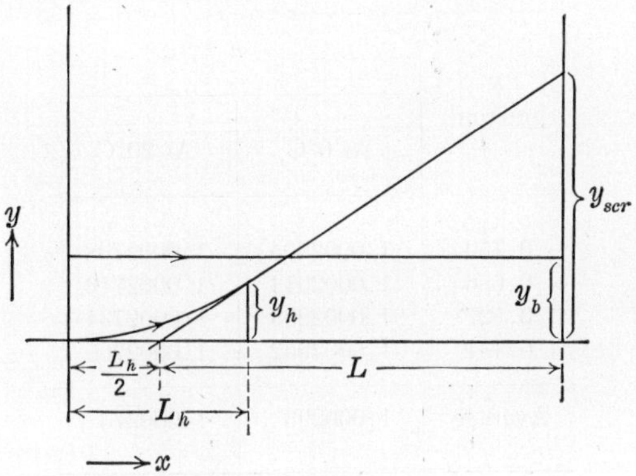

L_h = Length of plate
$\frac{L_h}{2} + L$ = Distance of screen from leading edge of plate.
y_h = Distance of beam from surface at $x = L_h$
y_{scr} = Distance of beam from surface when hitting the screen.
y_b = Thickness of boundary layer in which air above L_h is sensibly heated.
Only beams arriving at $x = 0$ between $y = 0$ and $y = y_b$ are deflected.

FIG. 27–2. Path of light beam meeting the surface of a hot plate with grazing incidence.

The essential property of the medium which occurs in Eq. 27–10 is the relative change of the refractive index with temperature, $\frac{1}{n} \cdot \frac{dn}{dt}$. Considering a gas like air, let

n_0 = refractive index at 0°C = 273°K and 760 mm Hg,

n = refractive index at t°C = T°K and b mm Hg.

Then, according to an old empirical formula of Dale and Gladstone [1863],

$$\frac{n - 1}{n_0 - 1} = \frac{273}{T} \cdot \frac{b}{760} \tag{27–12}$$

Further, n depends on the wavelength, but much less than on temperature, as may be seen from Table 27–1.

TABLE 27-1

DEPENDENCE OF REFRACTIVE INDEX OF AIR AT 760 MM HG ON WAVELENGTH λ OF THE LIGHT

λ [micron]	n At 0°C	n At 20°C
0.759	1.0002905	1.0002708
0.656	1.0002914	1.0002716
0.527	1.0002933	1.0002734
0.431	1.0002962	1.0002760
Average	1.000293	1.000273

Taking an average $n_0 = 1.000293$, Eq. 27–12 yields

$$n = 1 + \frac{0.000293(273)}{T} \cdot \frac{b}{760} = 1 + \frac{0.0800}{T} \cdot \frac{b}{760} = 1 + u \quad (27\text{–}13)$$

where u is defined by this equation. Furthermore,

$$\frac{1}{n} \cdot \frac{dn}{dt} = \frac{d(\ln n)}{dt} = \frac{d}{dt}\left(\frac{u}{1} - \frac{u^2}{2} + \frac{u^3}{3} - \cdots\right)$$

Because 0.08 is small compared to T, it is sufficient to use only the first term of the series, so that

$$\frac{1}{n} \cdot \frac{dn}{dt} = -\frac{0.0800}{T^2} \cdot \frac{b}{760}$$

For the beam which grazes the heating surface, Eq. 27–10 yields

$$y_{scr} = -\frac{0.0800}{T_s^{\,2}} \cdot \frac{b}{760}\left(\frac{\partial t}{\partial y}\right)_s L_h L \quad (27\text{–}14)$$

where the subscript s refers to the heating surface. But

$$h(T_s - T_e) = -k_s\left(\frac{\partial t}{\partial y}\right)_s \quad (27\text{–}15)$$

where T_e is the absolute environment temperature, i.e., the temperature of the air at great distance from the wall, and
k_s is the thermal conductivity of air at the temperature T_s.

From Eqs. 27–14 and 15

$$h = \frac{k_s}{T_s - T_e} \cdot \frac{T_s^2}{0.0800} \cdot \frac{760}{b} \cdot \frac{1}{L_h L} y_{scr} \tag{27–16}$$

For a horizontal tube of diameter D

$$N_{Nu} = \frac{hD}{k_s} = 12.5 \frac{D}{L_h L} \cdot \frac{760}{b} \cdot \frac{T_s^2}{T_s - T_e} y_{scr} \tag{27–17}$$

Regarding the units in this equation it must be considered that the factor 12.5 is the reciprocal value of 0.0800 in Eq. 27–13 and therefore has the dimension of C^{-1}.

For air $d(\ln n)/dt \approx -0.58(10^{-6})\ C^{-1}$ at $t_s = 100°C$. Thus for $L_h = 20$ cm, $L = 1000$ cm, $y_h = 0.05$ cm, and a gradient $\partial t/\partial y = -50$ C/cm, Eq. 27–10 yields $y_{scr} = 5.8$ cm.

b. Application of the Schlieren Method

The practical application of the schlieren method will be shown with a cylindric heating surface which receives light at grazing incidence. The abscissae axis of Fig. 27–2 may be supposed to be on the surface of a cylindric tube and parallel to the cylinder axis which would be at $y = -D/2$.

Figure 27–3 shows in a schematic manner what happens. The arrows between a_0 and b_0 represent the velocities of parallel rays which arrive

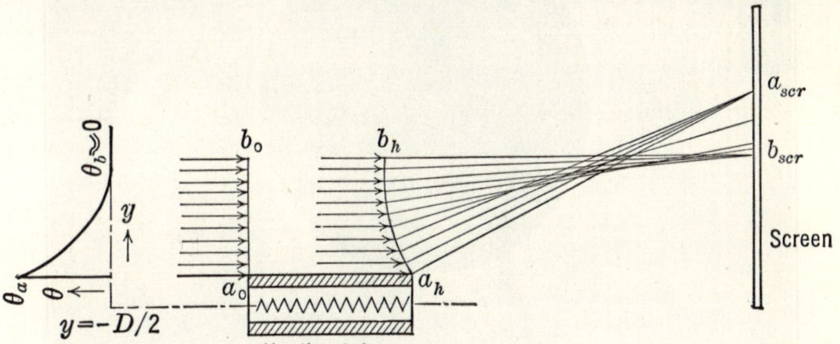

FIG. 27–3. Schematic representation of deflection of light in air due to an inhomogeneous temperature field.

at the leading edge of the heating tube and have a plane wave front $\overline{a_0 b_0}$. At the left the temperature distribution across the heated air layer is indicated by the curve $\overline{\theta_a \theta_b}$ with $\theta = t - t_e$ as abscissae, where t_e is the environment temperature (bulk air temperature). From theory

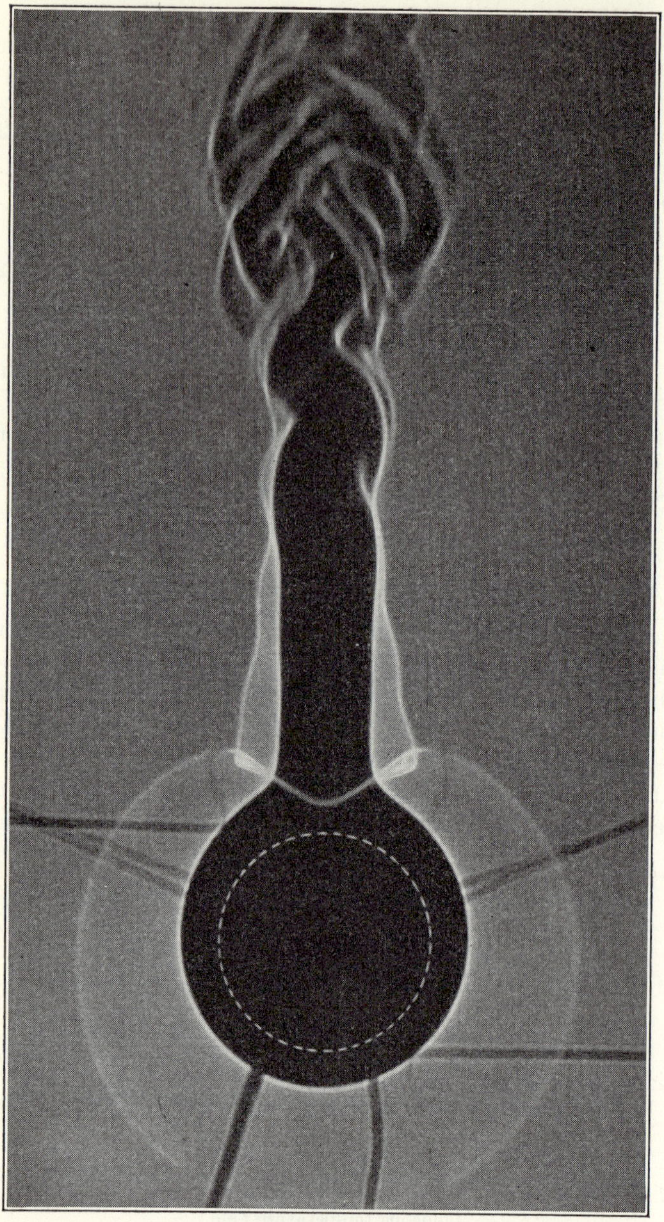

FIG. 27–4. Schlieren photo of a hot cylinder, 2 in. diam., 11½ in. long, placed horizontally in air at rest.
White dotted line is border of shade which would be cast by the cylinder when unheated.
Courtesy Dr. E. Schmidt [1932].

and experiment it is known that this line is almost straight on its two ends. Corresponding to the temperature drop and density increase in the direction y, the velocities are changing in the indicated manner, and close to the right end of the cylinder the wave front has taken the curved shape $\overline{a_h b_h}$. Now, according to a basic law of geometrical optics, light propagation is always perpendicular to the wave front. Hence light is thrown to the screen as indicated by the straight lines between the envelopes $\overline{a_h a_{scr}}$ and $\overline{b_h b_{scr}}$. Because of the small curvature of the ends of line $\theta_a \theta_b$ the ends of $a_h b_h$ are also nearly straight, and therefore light is concentrated in caustic bundles close to $\overline{a_h a_{scr}}$ and $\overline{b_h b_{scr}}$. They would form two bright circles on the screen if the temperature and density field had circular symmetry. In fact such symmetry does not exist as can be seen from Fig. 27–4, which is taken from Schmidt's paper. Since no light falls below b_{scr} of Fig. 27–3, a deep shade is formed inside the inner bright contour line of the photo, which obviously is the border inside of which the originally uniform temperature field is disturbed by the heater. This limit appears in natural scale because θ_b, b_0, b_h, and b_{scr} have all the same ordinate. The bright dotted circle of diameter D indicates the limit of the shade which would be thrown by the cylinder when unheated. The form of the shade indicates that the warmed air flows upward around the tube in an increasingly thicker layer and rises above the top of the cylinder, first still free of turbulence and then in eddies.

Above a_{scr} no deflected light falls upon the screen. The heart-shaped bright line is formed by all rays grazing the cylinder surface. The radial distance between this line and the dotted circle equals y_{scr} which, according to Eq. 27–11, is proportional to the rate of heat flow and, according to Eq. 27–16, is proportional to the film coefficient h, if the surface temperature t_s is uniform.

Hence, the heart-shaped curve represents the distribution of h around the cylinder. The figure shows that h has a maximum at the bottom, decreases slowly around the cylinder, and reaches a minimum at the top. Regarding the actual dimensions, it may be mentioned that Fig. 27–4 was obtained with a heated tube of 2 in. outer diameter and 11½ in. length.

The heart-shaped curve can be separately obtained by placing a circular diaphragm of slightly larger opening than the tube diameter concentrically before the tube.

Figure 27–5 was obtained in this way. It shows the results of Schmidt's experiments performed with a tube of 6 in. outer diameter and 10 in. length. The inner circle was taken with unheated tube, the three heart-shaped lines with different tube temperatures.

Schmidt's paper contains photographs of a great number of heating bodies, having different profiles (vertical and horizontal plates, cylinders of rectangular, hexagonal, and egg-shaped profile). One of them is the

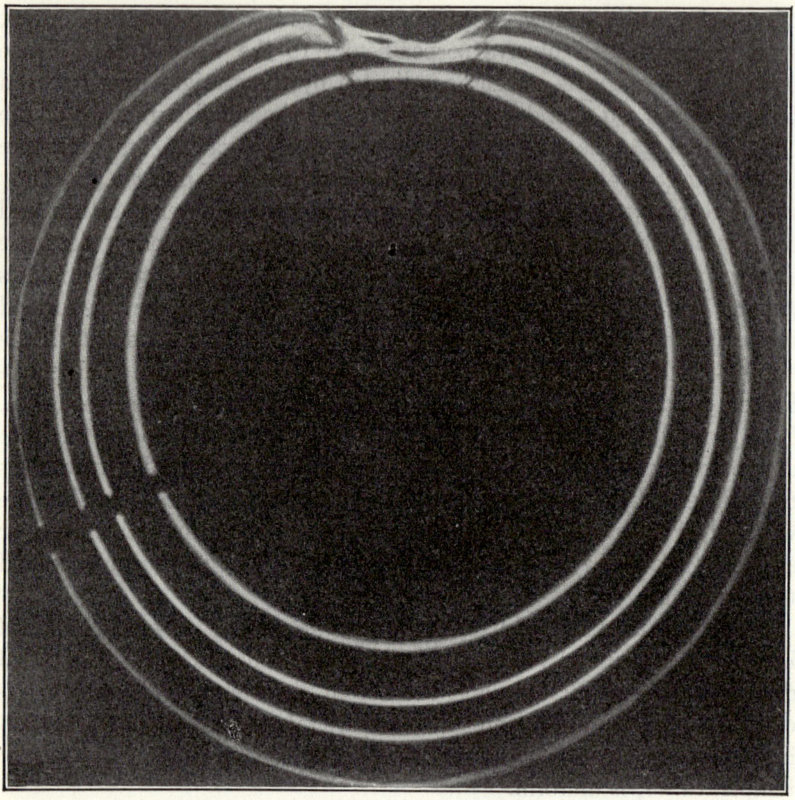

FIG. 27–5. Schlieren photos of a hot horizontal cylinder, 6 in. diam., 10 in. long, at four different temperatures.

A disc with circular opening, slightly wider than the diameter of the heating cylinder, has been used to blind off all light except the beams just grazing the cylinder surface.

From E. Schmidt [1932].

photo for a horizontal plate, reproduced as Fig. 27–6. It demonstrates the stable conditions on the bottom and the fluttering which occurs at the upper surface, due to the unstability of fluid and heat flow.

Finally, just a few words may be said concerning the experimental arrangement, which is very simple. In Schmidt's experiments an arc lamp with shutter, the electric-heated test sample, and a screen which could be replaced by photographic plate were all the apparatus needed.

The distance between lamp and test sample (heater) was 100 to 165 ft, and that between heater and screen 30 to 50 ft. These long paths had to be free of schlieren; therefore the experiments were performed in long basement corridors having homogeneous temperature. Where such lengths of optically homogeneous air are not available, lenses or mirrors will have to be inserted in the optical system in an appropriate manner.

Hermann [1936], in his previously mentioned paper (see Sect. 25–1), used Schmidt's schlieren method to find the critical Reynolds number at

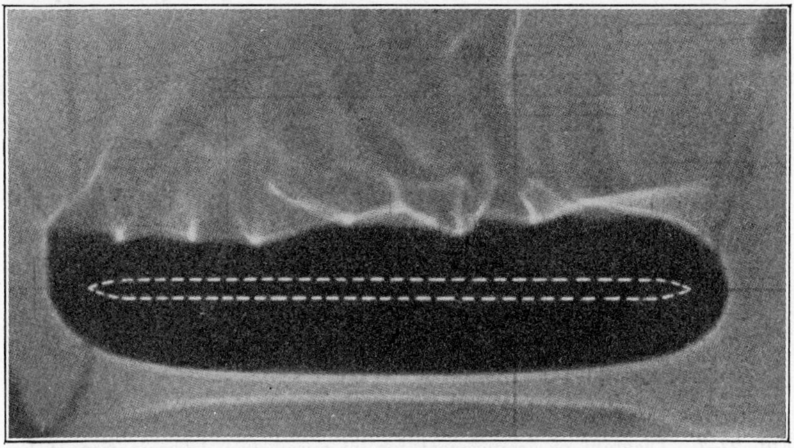

FIG. 27–6. Schlieren photo of a hot horizontal plate in air at rest. From E. Schmidt [1932].

which in vertical walls and horizontal cylinders turbulence starts. This occurred for the plate at $N_{Gr} \approx 1.0(10^9)$ and for the cylinder at $N_{Gr} \approx 3.5(10^8)$ with height of plate and diameter of cylinder as characteristic lengths. The critical Reynolds number, however, was at $N_{Re} \approx 300$ in both cases, where $N_{Re} = vs/\nu$, and v stands for the mean velocity in the boundary layer of thickness s.

27–3. Interferometric Method

The differences of density and light velocity close to a heating surface can also be used to produce interference fringes. This seems to have been first suggested by G. Hansen [1931], but Kennard's [1931, 1932] work, done independently, goes far beyond Hansen's attempt.

Figure 27–7 is a diagram of the experimental arrangement, consisting of light source, interferometer with heating body, and camera. The light from a mercury lamp L passed through a 4-in. aperture A and was brought to a slightly convergent beam by two condensing lenses, C_1

and C_2, the first of them shiftable, so that the focal length of the combination and the degree of convergence could be varied. The lamp housing was separated from the interferometer I by a tube T, 24 in. long, with a shutter on the open end. It served to protect the interferometer from stray air currents produced by the heat of the lamp. Before entering the interferometer, the light passed through a Wratten 77 filter F, which transmits only the mercury green light. A Zehnder interferometer was used, having four mirrors M_1, M_2, M_3, and M_4, the mirrors M_2 and M_3 being adjustable. The heating body H was placed

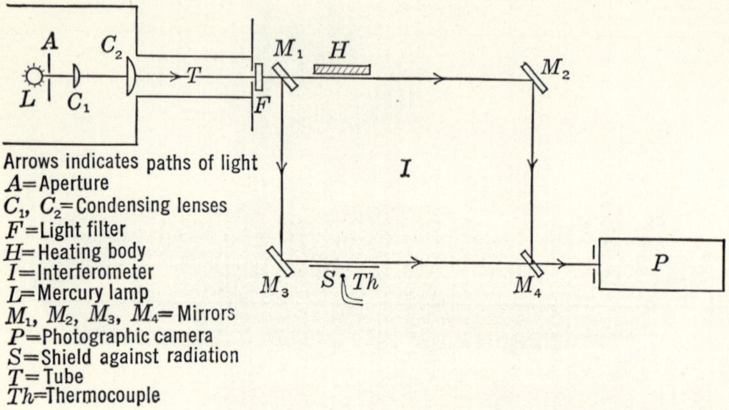

FIG. 27-7. Diagram of arrangement for interferometric determination of heat transfer in free convection.

in one arm of the interferometer. The interferometer had a frame of 2-in. steel pipe through which cooling water was flowing to avoid any distortion of the instrument by radiation from H. It was mounted so that a free flow of air about the heating surface without any chimney effects was secured. The bulk-air temperature was measured by a thermocouple Th, which was protected from radiation by a shield S. The photographic camera P gave a magnification of 2.4 diameters, and enlarged positives were made from the negatives, the total magnification being 7 diameters. A double exposure was made on the same plate to show also a millimeter scale in the object field. The experiments were performed in a basement room with 4-in. cork lining of walls and ceiling, so that disturbances by vibrations or stray convection currents were avoided.

The main adjustment consisted in bringing the axis of the beam of light to coincidence with that of the interferometer and to grazing incidence at the hot surface. The fringes were generally brought to a focus at the end of the heating body and in one case (hot wire) at the middle of the heating length. Figure 27–8 is the schematic sketch of a photograph of fringes. If the temperature were uniform throughout, the center line of fringe A instead of that of B would pass through the point P. The shift of one fringe at this point indicates that there is one less wave of light in the path through P than there would be if the temperature at P were the same as at O.

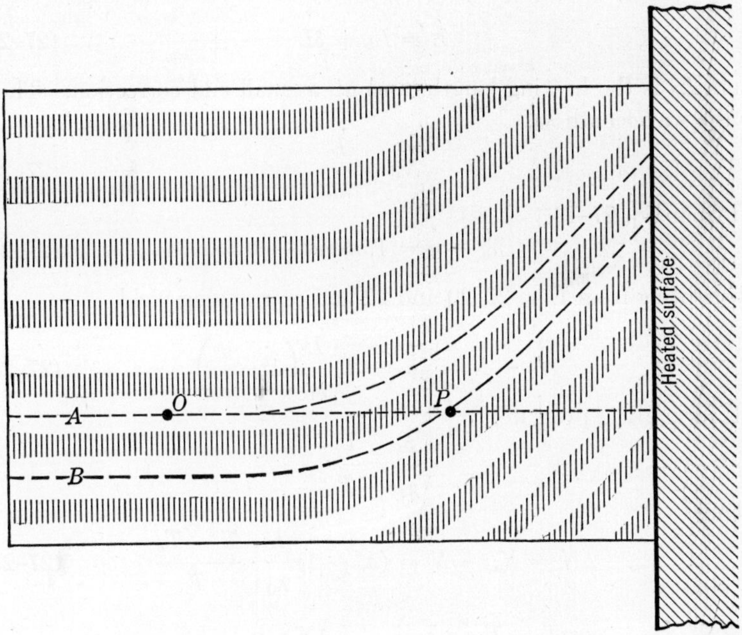

FIG. 27–8. Schematic sketch of a photo of fringes.

If the temperature were uniform, fringe centerline AO would go through P instead of turning upward, whereas for a heated surface the fringe at B proceeds to P.

The relation between refractive index n and air density ρ is given by the so-called Lorenz-Lorentz equation

$$\frac{n^2 - 1}{n^2 + 2} = K\rho \tag{27–18}$$

wherein K is a constant.

According to Table 27–1, the index n is of the order of 1.0003. Hence Eq. 27–18 reduces very closely to

$$n - 1 = 1.5K\rho \tag{27–19}$$

which is a special case of Eq. 27–12. Denoting by N the number of light waves over a given path in a medium having the refractive index n and by the subscripts 0 and e vacuum and another set of conditions:

$$n_e = \frac{N_e}{N_0}; \quad n = \frac{N}{N_0} \tag{27-20}$$

The path length under consideration may be

$$L = L_h + \delta L \tag{27-21}$$

where L_h is the heating length, and δL a small end correction. Then, for the wavelength λ_0,

$$N_0 = \frac{L}{\lambda_0} \tag{27-22}$$

and from Eq. 27–19,

$$n_e - n = 1.5K(\rho_e - \rho) \tag{27-23}$$

Substituting from Eqs. 27–20 and 22 and rearranging yield

$$N_e - N = (n_e - 1)\frac{L}{\lambda_0}\left(1 - \frac{\rho}{\rho_e}\right) \tag{27-24}$$

But, for a given pressure,

$$\frac{\rho}{\rho_e} = \frac{T_e}{T}$$

whence

$$\Delta N = N_e - N = (n_e - 1)\frac{L}{\lambda_0} \cdot \frac{T - T_e}{T} \tag{27-25}$$

or

$$\frac{T - T_e}{T_e} = \frac{\Delta N}{(n_e - 1)(L/\lambda_0) - \Delta N} \tag{27-26}$$

Since in the interferometer the change in number of light waves for a given path is equal to the fringe displacement, ΔN, Eq. 27–25 gives ΔN for a temperature change from T_e to T. If the subscript e refers to environmental (bulk of air) conditions, then Eq. 27–26 yields the temperature distribution in the heated air layer from the fringe displacement number ΔN.

The end correction δL takes into account the fact that the layer of heated air extends beyond the ends of the hot surface. In order to find the value of δL for † $y = 0$ from Eq. 27–26, Kennard measured the surface temperature $T = T_s$ by means of a thermocouple and compared the values of ΔN, as calculated from the equation, with those counted from Fig. 27–8. The difference was of the order of 5% when a heating plate

† y = distance from the surface as in Sect. 27–2.

of 4 in. length had been used. For $y > 0$, δL is not much different from the value for $y = 0$, as is shown in Kennard's paper. In Schmidt's method a similar correction should occur.

Some fringe photos, taken by Kennard, and temperature distributions drawn according to them are reproduced in Figs. 27–9 to 15.

Figure 27–9 gives the fringes in front of the half height of a heated plate, 21 cm high and 103 cm long, for four cases *a*, *b*, *c*, and *d* with the

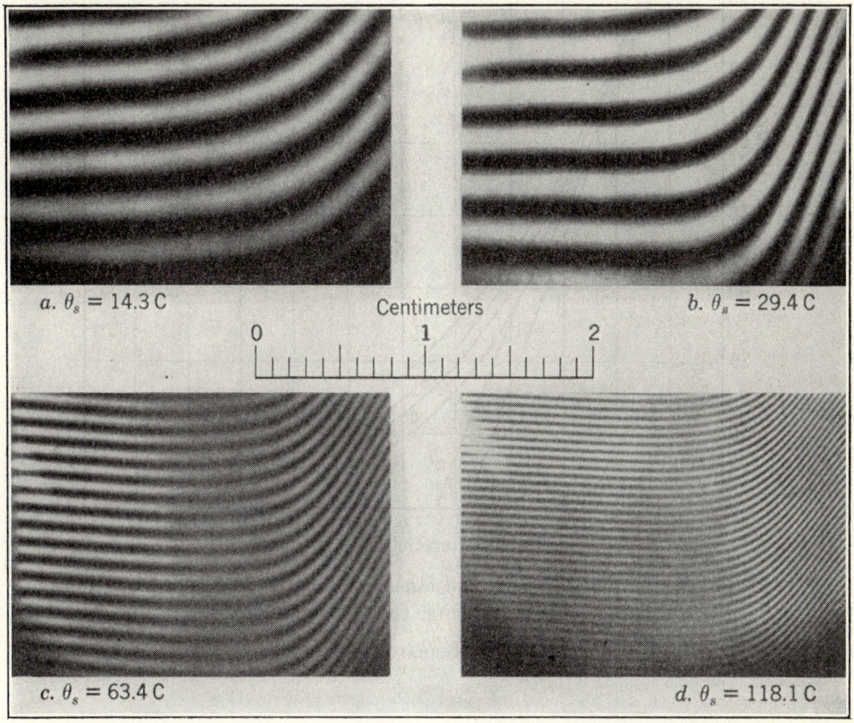

FIG. 27–9. Fringes in front of the upper half of a heated plate.

Courtesy Dr. R. B. Kennard [1932] and *Bureau of Standards Journal of Research.*

temperature excesses‡ $\theta_s = 14.3$, 29.4, 63.4, and 118.1 C and with t_e varying from 21.7 to 25.2°C. Figure 27–10 shows the ratio $100\theta/\theta_s$ plotted versus the distance y from the plate and includes the slope of the two extreme curves at $y = 0$.

A comparison between Fig. 27–10 and E. Schmidt and Beckmann's [1930] graph of directly measured temperature distribution (taken for $z = 10.5$ cm and $\theta_s = 53$ C) leads to the values of Table 27–2.

‡ As usual, the temperature excess is defined as $\theta = t - t_e$.

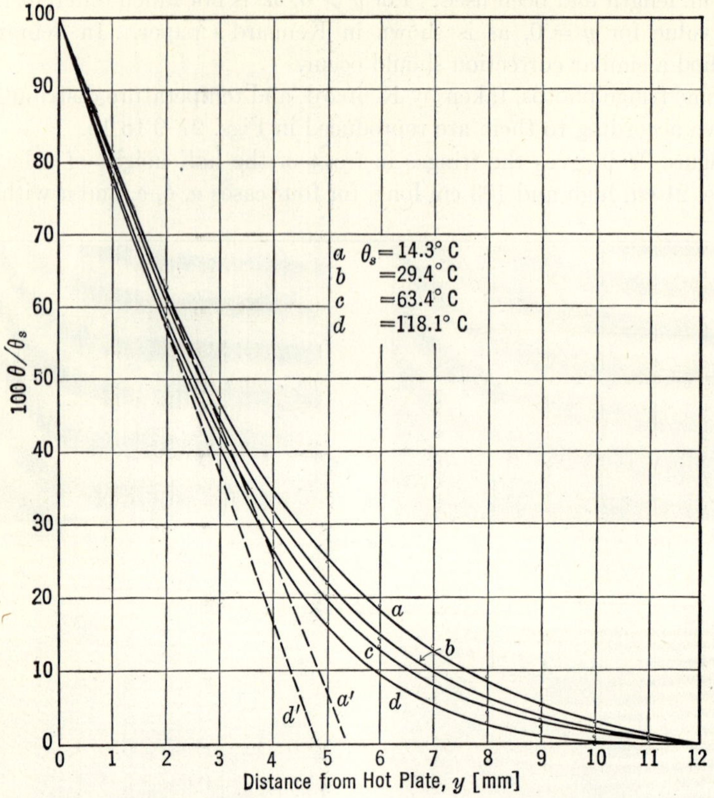

FIG. 27-10. Ratio 100 θ/θ_s versus distance y from the surface calculated from fringes in Fig. 27-9. (Free convection on a surface.)

From Kennard [1932].

TABLE 27-2

$100\theta/\theta_s$ IN THE BOUNDARY LAYER ON A HEATED VERTICAL PLATE IN AIR AT REST

y [centimeters]	From Direct Measurements	From Interferometric Measurements
0.25	60	52
0.5	26	21
1.0	2	2

Considering the difficulties of either method of measurement, the agreement is satisfactory.

Figure 27–11 is composed from four photos and refers to a horizontal

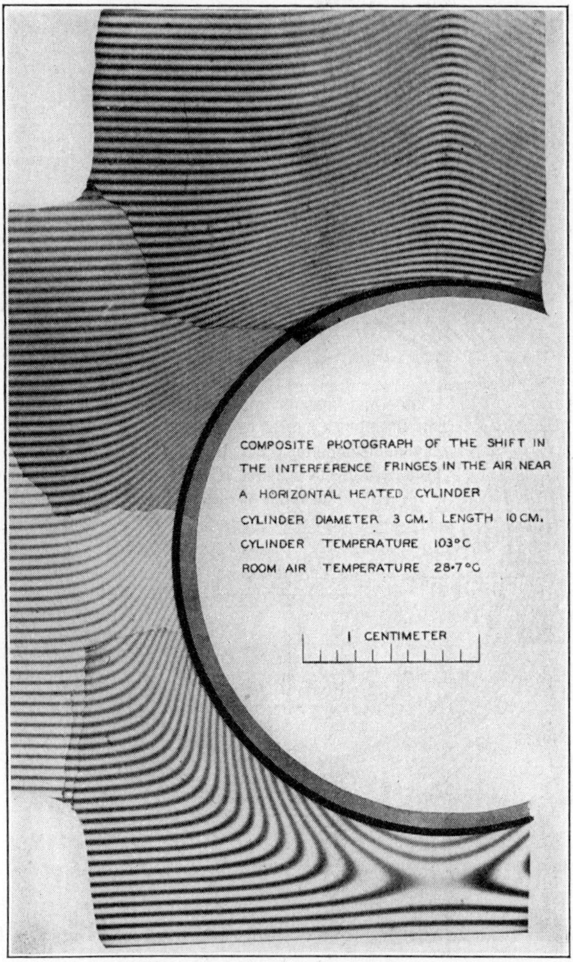

FIG. 27–11. Fringes on a horizontal heated cylinder, 3 cm diam., composed from four photos.

Courtesy Dr. R. B. Kennard [1932] and *Bureau of Standards Journal of Research.*

cylinder, $D = 3$ cm, $L = 10$ cm, at $t_s = 103°C$ and $t_e = 28.7°C$. Isothermals taken from this picture are given in Fig. 27–12.

Kennard's paper also includes photos for $D = 1.27$ and 0.5 cm. Finally, Figs. 27–13 and 14 were obtained with a wire, $D = 0.04$ cm.

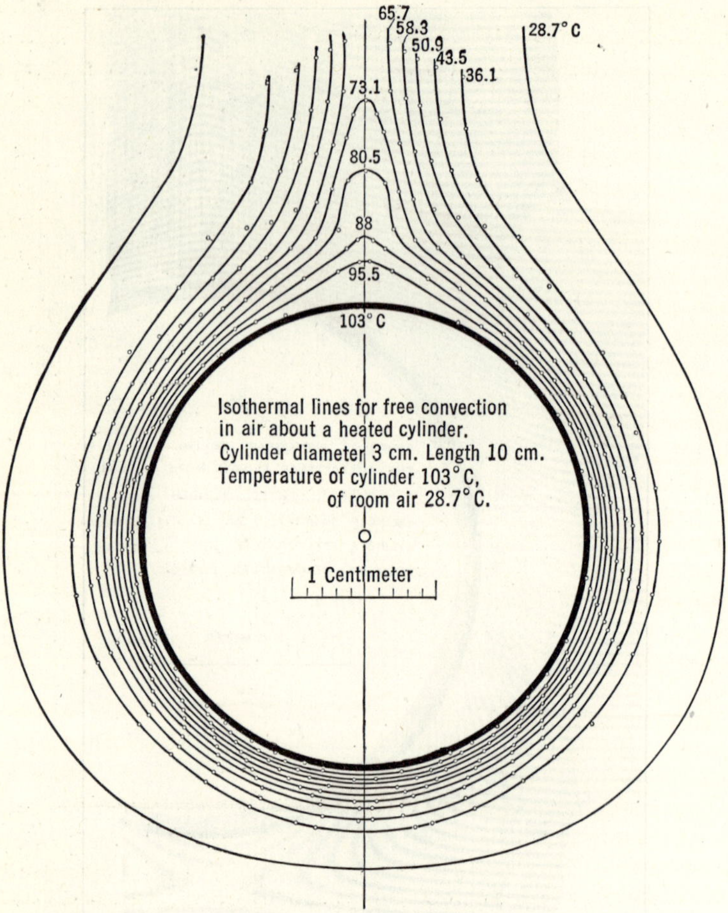

FIG. 27-12. Isothermals according to the fringes of Fig. 27-11.

From Kennard [1932].

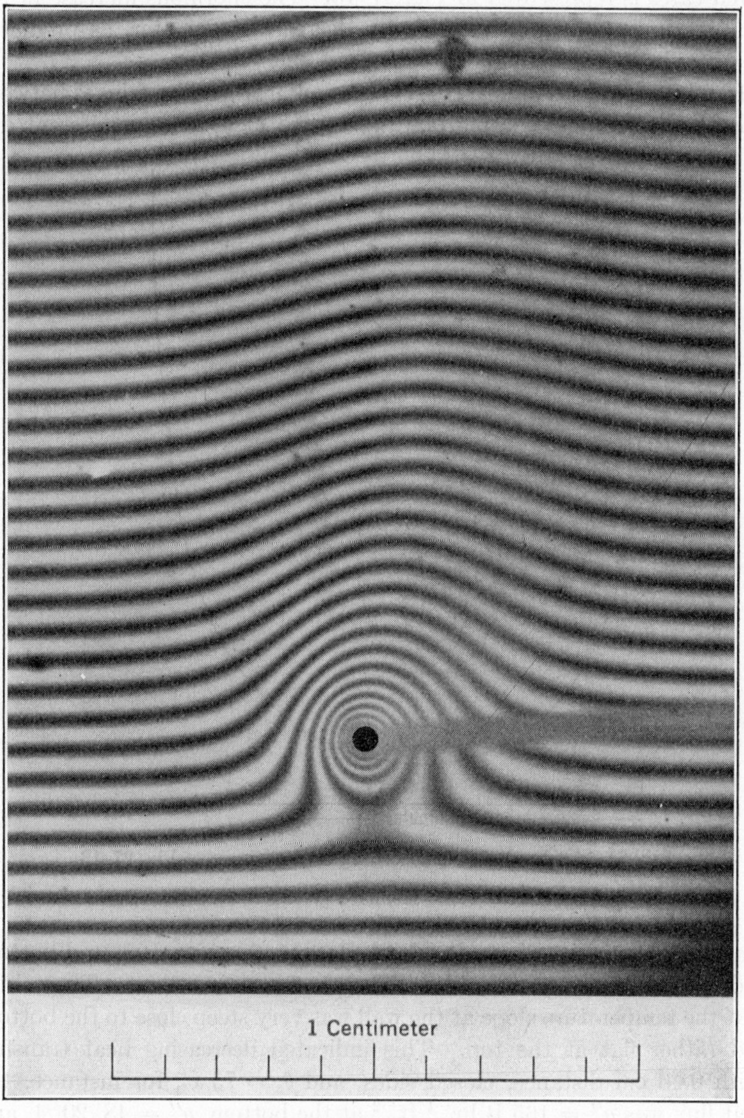

FIG. 27-13. Fringes on a horizontal heated wire, 0.04 cm diam. Courtesy Dr. R. B. Kennard and *Bureau of Standards Journal of Research.*

For a horizontal radius the temperature distribution in the four investigated cases is represented in Fig. 27-15. The enormous increase in the temperature gradient at $y = 0$ with decreasing diameter, as predicted by Nusselt, is remarkable.

In a later paper Kennard [1941] also took interference fringes between two parallel vertical heating plates, 21.4 cm high, at four different dis-

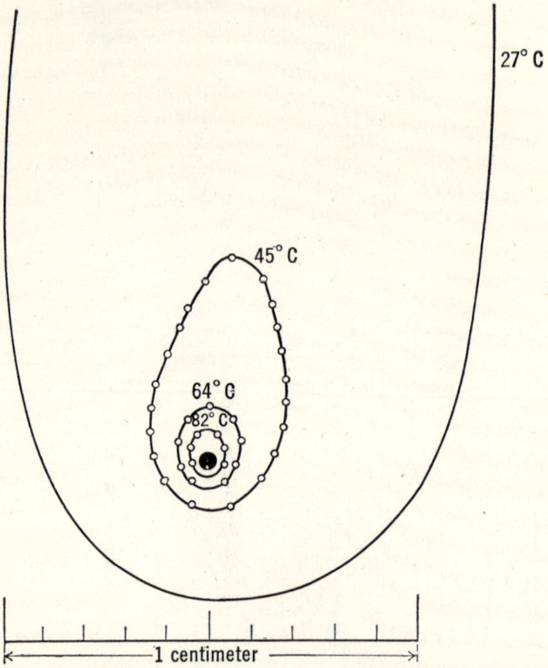

FIG. 27-14. Isothermals according to the fringes of Fig. 27-13. From Kennard [1932].

tances. The air layer between these plates was open at top and bottom and open or closed on the sides. The photos are excellent. They show that the temperature slope at the wall was very steep close to the bottom and rather flat at the top. This indicated decreasing heat transfer. With 0.46 cm distance, closed sides, and $\theta_s \approx 75$ C, for instance, the heat flux was $q'' = 165$ B hr^{-1} ft^{-2} at the bottom, $q'' = 48$, 20, 4, and 4 B hr^{-1} ft^{-2}, respectively, at ½, 1, 4, and 8 in. above the bottom. Unfortunately, the back sides of the plates were not insulated; hence the heat loss from these and the edges was a multiple of the heat transfer to the air layer. Therefore, a comparison between the interferometric and calorimetric measurements led to inconclusive results.

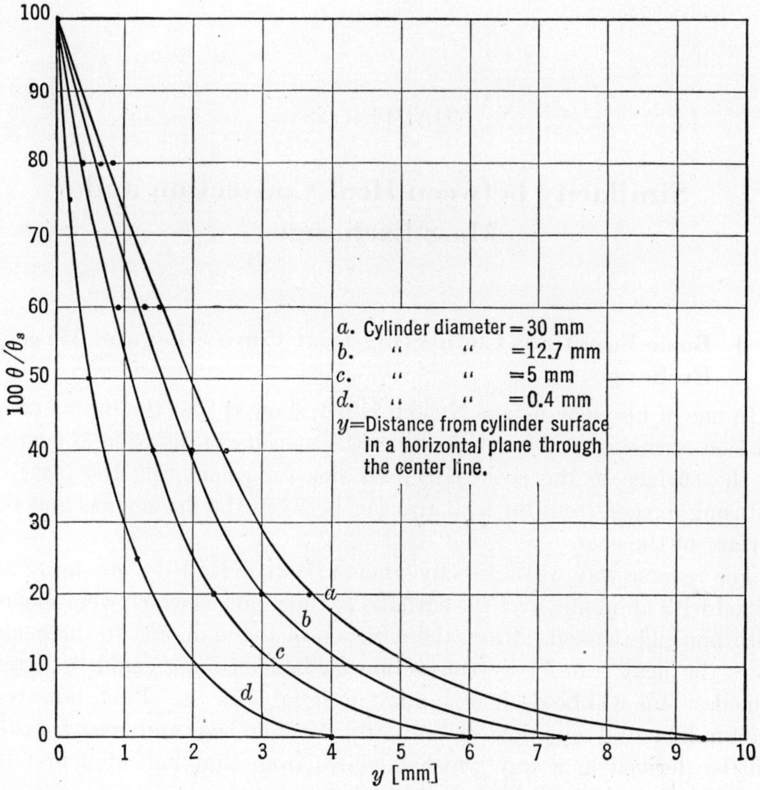

Fig. 27–15. Temperature distribution close to the surfaces of horizontal heated cylinders.

From Kennard [1932].

PART E. HEAT CONVECTION INCLUDING CHANGES OF PHASE

CHAPTER 28

Similarity between Heat Convection and Mass Exchange

28-1. Basic Equations Connecting Heat Convection and Mass Exchange

In one of his early papers Nusselt [1916] showed that the rate of combustion of coal in a stoker depends on the velocity of diffusion of oxygen to the surface of the coal. He used this for calculating the mass of diffusing oxygen from the heat transfer between the flowing gas and the surface of the coal.

The reverse way was taken by Thoma [1921]. He led a stream of air mixed with ammonia over the surfaces of tube models which were soaked with phosphoric acid. From the diffusion of the ammonia to these surfaces the heat transfer by convection on boiler tubing could be determined. This will be dealt with more in detail later on. First, however, the fundamental equations of similarity between heat and mass transfer will be derived in a way which deviates from that indicated first by E. Schmidt [1929] and Nusselt [1930] in one essential point.

Although Nusselt [1916(*a*) and 1930] mentioned that for gases and vapors the partial-pressure difference is the driving force in diffusional processes, both he and E. Schmidt developed similarity relations under the assumption that concentration differences are similar to temperature differences. This is valid only when these differences do not occur simultaneously in the same field, whereas the similarity between partial-pressure differences and temperature differences used in the following deductions is free of this restriction.*

The relationship between thermal conduction and mechanical diffu-

* Ackermann [1937] has already clearly expressed the fact that concentration differences cannot be used in the similarity treatment if the temperature and partial-pressure differences are large.

sion is based on the similarity of the respective differential equations for moving fluids:

$$\frac{\partial T}{\partial \tau} + v_x \frac{\partial T}{\partial x} + v_y \frac{\partial T}{\partial y} + v_z \frac{\partial T}{\partial z} = \alpha \left(\frac{\partial^2 T}{\partial x^2} + \frac{\partial^2 T}{\partial y^2} + \frac{\partial^2 T}{\partial z^2} \right) \quad (28\text{–}1)$$

$$\frac{\partial p}{\partial \tau} + v_x \frac{\partial p}{\partial x} + v_y \frac{\partial p}{\partial y} + v_z \frac{\partial p}{\partial z} = \delta \left(\frac{\partial^2 p}{\partial x^2} + \frac{\partial^2 p}{\partial y^2} + \frac{\partial^2 p}{\partial z^2} \right) \quad (28\text{–}2)$$

The first equation is a special case of Eq. 3–24. Absolute temperatures, T, are used for convenience because later on thermodynamical terms will occur which are based on absolute temperature. The second equation defines the mechanical diffusivity, δ, of one substance into a second one from the behavior of the field of partial pressure, p, in the same way as the thermal diffusivity, α, of a homogeneous substance was obtained by studying the space-time relations in a temperature field.† It is seen that both α and δ are kinematic quantities. Having the same physical dimensions (ft^2/hr in the British system) they deserve to have equal names.

Extending the treatment from conduction and diffusion to convection and the mixing motion of different fluids, the following two pairs of equations can be set up:

$$q'' = -k \frac{dT}{dn} = -\alpha \frac{de}{dn} \quad (28\text{–}3)$$

$$\dot{m}'' = -\delta \frac{1}{RT} \cdot \frac{dp}{dn} = -\delta \frac{dc}{dn} \quad (28\text{–}4)$$

† The mutual diffusivity δ is defined for a field of uniform temperature. If temperature differences occur, the so-called thermal diffusion will be superimposed over the mechanical diffusion and will complicate the situation as was first envisaged by Enskog [1911] and more precisely stated by Chapman [1916] and by Chapman and Cowling [1939]. Whereas in a field of uniform temperature under equilibrium conditions the concentration of each constituent is the same over all the volume, a temperature gradient causes heavier or larger gas molecules to diffuse in the direction of diminishing temperature, and vice versa for the lighter or smaller molecules. Therefore, the composition of the gas mixture becomes non-uniform. This anomaly, however, will be appreciable only when the atomic weights of the diffusing gases or their molecular diameters differ greatly and at the same time large temperature differences exist. According to Chapman and Dootson [1917], the anomaly is greatest when the gases are mixed in nearly equal proportions by volume. Here it will be assumed that the effect of thermal diffusion is taken care of in the value of the diffusivity δ, which in this way has the character of an apparent or equivalent diffusivity rather than of the true diffusivity as defined for uniform temperature.

and

$$q'' = h(T_s - T_0) = a(e_s - e_0) \tag{28-5}$$

$$\dot{m}'' = b\frac{1}{RT_m}(p_s - p_0) = b\frac{c_sT_s - c_0T_0}{T_m} \tag{28-6}$$

where q'' = the rate of heat flow per unit area,
$\dot{m}''$ = the rate of mass flow per unit area (mass velocity),
n = the direction of heat and mass flow,
k = the thermal conductivity of the medium,
$e = \rho c_p T$ = an energy quantity per unit volume,
h = the coefficient of heat transfer by convection,
$a = h/\rho c_p$,
b = a coefficient of mass transfer,
$T_m = (T_s + T_0)/2$ = a reference temperature,
R = the individual gas constant of the diffusing gas,
0 = a subscript for environment conditions, and
s = a subscript for surface conditions.

The quantities e and a have been introduced for convenience in order to make the similarity between heat and mass transfer as conspicuous as possible. For $T_s = T_0$ (isothermal diffusion) Eq. 28–6 simplifies to

$$\dot{m}'' = b(c_s - c_0) \tag{28-7}$$

which may also be used for small temperature differences. The same holds wherever the term $(c_sT_s - c_0T_0)/T_m$ occurs in this chapter.

From Eqs. 28–3 to 6 the following ones are readily derived:

$$\left.\begin{aligned} h &= -k\frac{1}{T_s - T_0}\cdot\frac{dT}{dn} \\ &\text{or} \\ a &= -\alpha\frac{1}{T_s - T_0}\cdot\frac{dT}{dn} \end{aligned}\right\} \tag{28-8}$$

and

$$b = -\delta\frac{1}{p_s - p_0}\cdot\frac{dp}{dn} = -\delta\frac{T_m}{c_sT_s - c_0T_0}\cdot\frac{dc}{dn} \tag{28-9}$$

It should be noted that a and b are also kinematic properties, each of them having the dimension of a velocity.

Supposing now as a special case,

$$\frac{1}{T_s - T_0}\cdot\frac{dT}{dn} = \frac{1}{p_s - p_0}\cdot\frac{dp}{dn} = \frac{T_m}{c_sT_s - c_0T_0}\cdot\frac{dc}{dn} \tag{28-10}$$

then

$$a = \frac{\alpha}{\delta}b \tag{28-11}$$

If, further, $\alpha = \delta$, then

$$a = \frac{h}{\rho c_p} = b \tag{28-12}$$

This relation was first derived by W. K. Lewis [1922]. It is based on the condition $\alpha = \delta$, as was first noticed by E. Schmidt [1929]. This condition is analogous to the condition $\alpha = \nu$ or $N_{Pr} = 1$, which occurred in Prandtl's first theory of similarity between heat transfer and impulse transfer. Neither of these conditions is generally satisfied. From Table 28-1, which contains some values of α/δ, it is seen that

TABLE 28-1

RATIO α/δ FOR DIFFERENT FLUIDS AT 0°C AND 760 MM HG

Diffusion of	Into Hydrogen ($\alpha_2 = 5.0$ ft²/hr)			Into Air ($\alpha_2 = 0.70$ ft²/hr)			Into Carbon Dioxide ($\alpha_2 = 0.335$ ft²/hr)		
	$\frac{M_1}{M_2}$	δ	$\frac{\alpha_2}{\delta}$	$\frac{M_1}{M_2}$	δ	$\frac{\alpha_2}{\delta}$	$\frac{M_1}{M_2}$	δ	$\frac{\alpha_2}{\delta}$
Hydrogen (H_2)	1.00		1.00	0.069		0.25	0.045	2.085	0.161
Water vapor (H_2O)	9.0	2.65	1.88	0.622	0.765	0.917	0.41	0.51	0.657
Air	14.5		1.95	1.00		1.00	0.66		0.61
Methyl alcohol (CH_4O)	16	1.935	2.58	1.10	0.511	1.37	0.73	0.340	0.984
Carbon dioxide (CO_2)	22	2.105	2.37	1.52	0.549	1.28	1.00		1.00
Ethyl alcohol (C_2H_6O)	23	1.46	3.42	1.59	0.393	1.79	1.05	0.2655	1.26
Isobutylisobutyrate ($C_8H_{16}O_2$)	72	0.73	6.85	4.97	0.1816	3.87	3.28	0.142	2.36
Isobutylvalerate ($C_{10}H_{12}O$)	74	0.666	7.50	5.28	0.1644	4.78	3.36	0.1182	2.84

α/δ increases with the molecular weight M_1 of the diffusing substance for diffusion into any gas of a fixed molecular weight M_2. For diffusion of water vapor into air, α/δ is close to 1 so that the condition of similarity is almost satisfied and Lewis' rule (Eq. 28-12) can be used. The same holds for some other pairs of substances; the extreme values of α/δ, however, in the range of the table are 7.50 and 0.16. A table for 33 diffusing substances including values of δ has been published by Hilpert

[1932]. If the partial pressure of the diffusing substance is small compared to the total pressure of the mixture, values of α, referring to the second substance, may be used. Where $\alpha \neq \delta$ the more general similarity relation between heat and mass transfers (Eq. 28–11) must be used.

In the case of *free convection* it is practical to replace the factor $\beta\theta$ in the Grashof number by $(T_s - T_0)/T_0$, thus restricting the treatment to ideal gases; the ideal-gas state, however, can mostly be assumed in diffusional and gas-mixing processes. With L as characteristic length of an arrangement:

$$N_{Nu} = \frac{hL}{k} = \frac{hL}{\alpha \rho c_p}$$

Now, according to Eq. 23–27,

$$\frac{h}{\rho c_p} = \frac{\alpha}{L}\,\Phi(N_{Gr}, N_{Pr})$$

or

$$a = \frac{\alpha}{L}\,\Phi\left[\frac{gL^3}{\nu^2}\left(\frac{T_s}{T_0} - 1\right), \frac{\nu}{\alpha}\right] \tag{28–13}$$

and for a similar diffusional process under isothermal conditions by analogy

$$b = \frac{\delta}{L}\,\Phi\left[\frac{gL^3}{\nu^2}\left(\frac{M_0}{M_s} - 1\right), \frac{\nu}{\delta}\right] \tag{28–14}$$

where M_0 is the molecular weight of the medium before being penetrated by the diffusing substance, i.e., at a large distance from the surface,

M_s is the molecular weight of the medium at the surface, and

$\nu/\delta = N_{Sc}$, called Schmidt number.‡

It is obvious that a smaller molecular weight has the same effect on buoyancy as a correspondingly higher temperature, and that, owing to Eqs. 28–1 and 2, the mechanical diffusivity, δ, supersedes the thermal diffusivity, α.

If in corresponding arrangements (subscripts 1 and 2)

$$\frac{gL_1^3}{\nu_1^2}\left(\frac{T_{s,1}}{T_{0,1}} - 1\right) = \frac{gL_2^3}{\nu_2^2}\left(\frac{M_{0,2}}{M_{s,2}} - 1\right) \tag{28–15}$$

and

$$\frac{\nu_1}{\alpha_1} = \frac{\nu_2}{\delta_2} \tag{28–16}$$

then Eq. 28–11 is satisfied.

‡ Proposed at a Round Table Conference of the Chicago meeting of the American Institute of Chemical Engineers in 1933.

So far, a process of heat transfer without mass exchange was compared with a process of mass exchange without heat flow. If both processes occur simultaneously and in the same medium, a modified Grashof number will govern the combined process, which can be defined by

$$(N_{Gr})_{\alpha,\delta} = \frac{gL^3}{\nu^2}\left(\frac{M_0 T_s}{M_s T_0} - 1\right) \tag{28–17}$$

where the pair of subscripts α, δ refers to the combined process. It is an achievement of the theory of similarity that relations like Eq. 28–17 can be set up just by reasoning without much of a proof. In this way the following set of similar equations was obtained:

$$a_{\alpha,\delta} = \frac{h_{\alpha,\delta}}{\rho c_p} = \frac{\alpha}{L} \cdot \Phi\left[\frac{gL^3}{\nu^2}\left(\frac{M_0 T_s}{M_s T_0} - 1\right), \frac{\nu}{\alpha}, \frac{\nu}{\delta}\right] \tag{28–18}$$

$$b_{\alpha,\delta} = \frac{\delta}{L} \cdot \Phi\left[\frac{gL^3}{\nu^2}\left(\frac{M_0 T_s}{M_s T_0} - 1\right), \frac{\nu}{\delta}, \frac{\nu}{\alpha}\right] \tag{28–19}$$

The function Φ has the same form in both equations; however, the quantities ν/α and ν/δ change their places. If they enter the first equation with different exponents, then their exponents will be exchanged in the second equation. If $\alpha = \delta$, then obviously Lewis' relation (Eq. 28–12) holds also for simultaneous heat and mass transfer.

For *forced convection* the following equations may be used if heat transfer and mixing are similar, but do not occur in the same field:

$$a = \frac{h}{\rho c_p} = \frac{\alpha}{L} \cdot \Psi\left(\frac{vL}{\nu}, \frac{\nu}{\alpha}\right) \tag{28–20}$$

$$b = \frac{\delta}{L} \cdot \Psi\left(\frac{vL}{\nu}, \frac{\nu}{\delta}\right) \tag{28–21}$$

Here Eq. 28–11 is satisfied if

$$\frac{v_1 L_1}{\nu_1} = \frac{v_2 L_2}{\nu_2} \tag{28–22}$$

and

$$\frac{\nu_1}{\alpha_1} = \frac{\nu_2}{\delta_2} \tag{28–16}$$

For heat and mass transfer in the same field

$$a_{\alpha,\delta} = \frac{\alpha}{L} \cdot \Psi\left(\frac{vL}{\nu}, \frac{\nu}{\alpha}, \frac{\nu}{\delta}\right) \tag{28–23}$$

$$b_{\alpha,\delta} = \frac{\delta}{L} \cdot \Psi\left(\frac{vL}{\nu}, \frac{\nu}{\delta}, \frac{\nu}{\alpha}\right) \tag{28–24}$$

It is easily understood how cases of combined free and forced convection can be handled.

The range of temperature and concentration differences in which the above equations can be used is somewhat restricted according to the following lack of similarity. In the case of diffusion, for instance, of ammonia to a surface where it is absorbed, a component of velocity perpendicular to the surface will exist at the very surface, in contrast to the assumption of zero velocity at the surface on which the equations of pure heat transfer are based. Also, if no change of phase occurs, there may be heat transfer, but no mass transfer at the surface. Hence, a process of mere mass transfer which involves a change of phase can be assumed to be similar to a process of mere heat transfer only in case of small temperature and concentration differences. If both kinds of transfer occur simultaneously, the lack of similarity mentioned does not exist. However, another difficulty arises which will be dealt with in Sect. 28–4.

28–2. Sweat-water Formation on a Cold Vertical Wall

The equations of the last section will be first illustrated by the case of sweat water forming on a vertical cold wall. Following Schmidt's derivations, the difference in the influence of ν/α and ν/δ in Eqs. 28–23 and 24 will be neglected, and Lorenz' equation of heat transfer by convection will be used in the simplified form

$$h = N\sqrt[4]{T_0 - T_s} = N\sqrt[4]{T_0}\cdot\sqrt[4]{1 - (T_s/T_0)} = K\sqrt[4]{1 - (T_s/T_0)} \tag{28–25}$$

where T_0 and T_s are the absolute temperatures of the air and surface, respectively, N is a constant, and $K = N\sqrt[4]{T_0}$ is also a constant if only cases at a fixed air temperature are considered.

If diffusion takes place at the same time, the theory and Eq. 28–25 lead to

$$h_{\alpha,\delta} = K\sqrt[4]{1 - \frac{M_0}{M_s}\cdot\frac{T_s}{T_0}} = h\sqrt[4]{\left(1 - \frac{M_0}{M_s}\cdot\frac{T_s}{T_0}\right)\Big/\left(1 - \frac{T_s}{T_0}\right)} \tag{28–26}$$

where M_0 and M_s are the apparent molecular weights for air in the states far off, and at the surface, respectively. Since Φ symbolizes identical mathematical operations in Eqs. 28–18 and 19,

$$b_{\alpha,\delta} = \frac{\delta}{\alpha}\cdot\frac{h_{\alpha,\delta}}{\rho c_p} = \frac{\delta}{\alpha}\cdot\frac{K}{\rho c_p}\sqrt[4]{1 - \frac{M_0}{M_s}\cdot\frac{T_s}{T_0}} \tag{28–27}$$

Now, according to Eq. 28–6, the quantity of water condensed in unit time at unit area will be

$$\dot{m}'' = b_{\alpha,\delta}\frac{c_0T_0 - c_sT_s}{T_m} = b_{\alpha,\delta}\frac{1}{RT_m}(p_0 - p_s) \tag{28–28}$$

The sign is changed, compared to Eq. 28–6, because the water-vapor concentration of the air far from the wall will be greater than at the surface, where it will be close to zero because of condensation.

If λ is the enthalpy of condensation, then an amount, $q_\lambda'' = \dot{m}''\lambda$, is transferred to unit area of the surface in unit time. This heat energy had to be taken from the environment and carried to the wall, as well as that part of energy which was brought to the wall by the humid air at the rate of

$$q_{\alpha,\delta}'' = h_{\alpha,\delta}(T_0 - T_s) \tag{28–29}$$

An apparent coefficient of heat transfer, h_λ, for the heat of condensation may be defined by

$$q_\lambda'' = \dot{m}''\lambda = h_\lambda(T_0 - T_s) \tag{28–30}$$

Substituting $\dot{m}''$ from Eq. 28–28 and $b_{\alpha,\delta}$ from Eq. 28–27,

$$h_\lambda = b_{\alpha,\delta}\lambda\frac{1}{T_m}\cdot\frac{c_0T_0 - c_sT_s}{T_0 - T_s} = \frac{\delta}{\alpha}\cdot\frac{h_{\alpha,\delta}}{\rho c_p}\lambda\frac{1}{T_m}\cdot\frac{c_0T_0 - c_sT_s}{T_0 - T_s} \tag{28–31}$$

The total rate of heat energy transferred to unit area of the surface becomes

$$\begin{aligned} q'' &= q_{\alpha,\delta}'' + q_\lambda'' = (h_{\alpha,\delta} + h_\lambda)(T_0 - T_s) \\ &= h_{\alpha,\delta}\left(1 + \frac{\delta}{\alpha}\cdot\frac{\lambda}{\rho c_p}\cdot\frac{1}{T_m}\cdot\frac{c_0T_0 - c_sT_s}{T_0 - T_s}\right)(T_0 - T_s) \\ &= h\sqrt[4]{\left(1 - \frac{M_0}{M_s}\cdot\frac{T_s}{T_0}\right)\Big/\left(1 - \frac{T_s}{T_0}\right)} \\ &\quad \times\left(1 + \frac{\delta}{\alpha}\cdot\frac{\lambda}{\rho c_p}\cdot\frac{1}{T_m}\cdot\frac{c_0T_0 - c_sT_s}{T_0 - T_s}\right)(T_0 - T_s) \end{aligned} \tag{28–32}$$

Herewith the coefficient of heat transfer in humid air, including the formation of sweat water, can be calculated from the known value of h for dry air (Eq. 28–25). Equation 28–32 is simple to handle. The temperatures are given; the density of humid air and the concentration of the water vapor can be taken from tables; in particular, $\alpha/\delta = 0.92$ from Table 28-1 and $N = 0.39$ B hr^{-1} ft^{-2} F$^{-5/4}$, from experience.

The rate of mass flow per unit area is also determined by Eqs. 28-26, 30, and 31 and can be expressed by

$$\dot{m}'' = h \sqrt[4]{\left(1 - \frac{M_0}{M_s} \cdot \frac{T_s}{T_0}\right) \Big/ \left(1 - \frac{T_s}{T_0}\right)} \cdot \frac{\delta}{\alpha \rho c_p} \cdot \frac{c_0 T_0 - c_s T_s}{T_m} \tag{28-33}$$

Hence the amount of drop-water formation on a vertical wall can also be calculated from the known value of h for dry air.

Concerning numerical calculations, it may be considered that on the wet wall the water vapor is in the saturated state for surface temperature t_s (neglecting the temperature drop in the water layer §) and that $c_s = 1/\rho_{g,s} = v_{g,s}$ where $\rho_{g,s}$ and $v_{g,s}$ are the density and specific volume, respectively, of saturated steam which can be taken from steam tables. The apparent molecular weight M_s of the mixture at temperature t_s can be determined from the universal gas equation, considering the additivity of partial pressures or densities according to Dalton's law. The necessary data for the humid air at large distance from the surface can be obtained in a similar way if the relative humidity is given.

28-3. Evaporation from a Liquid Surface into a Gas Stream at the Same Temperature

The heat transfer between a plane solid surface and air flowing parallel to it has been investigated by different observers (see Sect. 26-5). Their results may be used to determine the evaporation from a liquid surface into a gas stream parallel to the surface when both have the same temperature.

From Eqs. 28-20 and 21

$$\frac{aL}{\alpha} = \Psi\left(\frac{v_0 L}{\nu}, \frac{\nu}{\alpha}\right) \tag{28-34}$$

and

$$\frac{bL}{\delta} = \Psi\left(\frac{v_0 L}{\nu}, \frac{\nu}{\delta}\right) \tag{28-35}$$

where L is a characteristic length, for instance, the length of the path of the wind along the surface. If an air stream of uniform velocity v_0, temperature t_0, and $\nu/\alpha = 0.71$ strikes a solid plane surface at tempera-

§ In Chapter 30 it will be shown how the temperature drop in the water film can be calculated. Herewith a closer approximation to the actual state may be obtained.

ture t_s in parallel flow, then, according to Jakob and Dow [1946, Table 4] ||

$$\frac{hL}{k} = \frac{aL}{\alpha} = C_0 \left(\frac{v_0 L}{\nu}\right)^{0.8} = 0.028 \left(\frac{v_0 L}{\nu}\right)^{0.8} \tag{28-36}$$

or for any gases of arbitrary value of ν/α

$$\frac{aL}{\alpha} = C_0 0.71^{-0.8} \left(\frac{v_0 L}{\alpha}\right)^{0.8} = 0.037 \left(\frac{v_0 L}{\alpha}\right)^{0.8} \tag{28-37}$$

Since the function Ψ holds for both heat transfer and diffusion,

$$\frac{bL}{\delta} = 0.037 \left(\frac{v_0 L}{\delta}\right)^{0.8} \tag{28-38}$$

where, according to Mache [1910],

$$\delta = \frac{160}{p} \left(\frac{T}{273}\right)^{1.89} \tag{28-39}$$

in cm^2/sec, p being the atmospheric pressure in millimeters of mercury and T the absolute temperature of the air in degrees Kelvin. Of course, consistent units have to be used in these equations.

By this method the coefficient of mass transfer b and, using Eq. 28-6, the mass velocity $\dot{m}''$ of evaporated liquid can be determined from the heat transfer on a solid surface if the concentrations c_s and c_0 are given. For water evaporating into wind c_0 is known from the humidity of the bulk of the flowing air, and c_s can be assumed to be equal to the density of water vapor at the temperature t_s [see Hilpert 1932, p. 10].

28-4. Automatic Compensation of a Diffusional Flow by Convection

At the end of Sect. 28-1 it has been mentioned that similarity between pure heat and mass transfer is not perfect if mass is absorbed or given up by a surface, and therefore a finite velocity perpendicular to the surface occurs which is missing in pure heat transfer. This restricts the methods for determining mass transfer from pure heat-transfer measurements and vice versa to cases where the indicated influence of velocity is small, i.e., to small temperature and concentration differences.

Heat and mass transfer in the same field, as dealt with in Sect. 28-2, are free from this inconvenience, since both are governed by the convec-

|| The value $C_0 = 0.028$ observed by the authors on a cylinder of 1.3 in. diameter is assumed to hold for plane plates, since it lies between Latzko's theoretical values $C_0 = 0.253$ and Juerges' somewhat questionable value $C_0 = 0.0320$.

tion streams. However, there occurs another difficulty from a secondary convection stream which was discovered and first used in experiments by Stefan [1874] and explained more theoretically by Meyer [1877]. Nusselt [1930] again analyzed the process in a refined way and showed how similarity can be assured by taking care of this effect, which plays a role in Thoma's ammonia-absorption method as well as in the evaporation of a liquid in contact with a flowing gas (cooling towers).

First the one-dimensional isothermal diffusion of ammonia and air in a tube will be considered. By experimental means the total pressure p_0 and absolute temperature T_0 in the tube are kept constant, and the concentration of ammonia is held different, but unvariable, at the two ends. Then ammonia and air diffuse in opposite directions with mass velocities $\dot{m}''$ and $\dot{m}_2''$, respectively.¶ Let y be the abscissae, taken positive in the direction of the air diffusion, y_0 the tube length, c_1 and c_2 the concentrations of ammonia and air, and δ the diffusivity of the two substances. Then

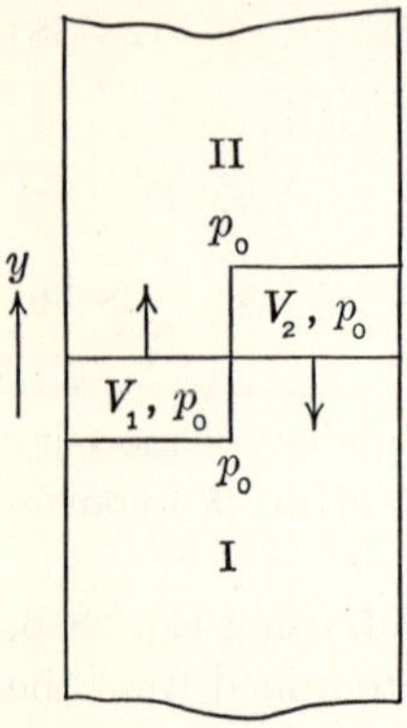

FIG. 28-1. Schematic representation of volume exchange by diffusion.

$$\dot{m}_1'' = \delta \cdot \frac{dc_1}{dy} \quad \text{and} \quad \dot{m}_2'' = -\delta \cdot \frac{dc_2}{dy} \tag{28-40}$$

Furthermore

$$p_0 = p_1 + p_2 = \text{constant} \tag{28-41}$$

where p_1 and p_2 are the partial pressures of ammonia and air. Figure 28-1 indicates in a schematic way the volume exchange by diffusion between two adjacent sections I and II of the tube. In unit time the ammonia volume V_1 is exchanged against the air volume V_2, both at the total pressure p_0. These volumes must be equal; otherwise the pressures in I and II would be changed. Denoting the time rates of diffusing volumes and masses by $\dot{V}_1$, $\dot{V}_2$ and $\dot{m}_1$, $\dot{m}_2$, respectively, and the respective gas constants by R_1 and R_2,

$$p_0\dot{V}_1 = \dot{m}_1 R_1 T_0 \quad \text{and} \quad p_0\dot{V}_2 = \dot{m}_2 R_2 T_0 \tag{28-42}$$

$$\dot{V}_1 = \dot{V}_2 = \dot{V} \tag{28-43}$$

Hence

$$\dot{m}_1 R_1 T_0 = \dot{m}_2 R_2 T_0 = p_0 \dot{V} \tag{28-44}$$

¶ Only the numerical (absolute) values of the mass velocities will be considered; hence it would be more exact to write always $|\dot{m}_1''|$ and $|\dot{m}_2''|$.

Furthermore, from the definition of concentration,

$$c_1 = \frac{p_1}{R_1 T_0} \quad \text{and} \quad c_2 = \frac{p_2}{R_2 T_0} \tag{28–45}$$

From Eqs. 28–40, 43, 44, and 45

$$\frac{p_0 \dot{V}}{A R_1 T_0} = \frac{\delta}{R_1 T_0} \cdot \frac{dp_1}{dy} \quad \text{and} \quad \frac{p_0 \dot{V}}{A R_2 T_0} = -\frac{\delta}{R_2 T_0} \cdot \frac{dp_2}{dy}$$

where A is the cross-sectional area of the tube, or

$$p_0 \frac{\dot{V}}{A} = \delta \cdot \frac{dp_1}{dy} \quad \text{and} \quad p_0 \frac{\dot{V}}{A} = -\delta \cdot \frac{dp_2}{dy} \tag{28–46}$$

By integration

$$p_0 \frac{\dot{V}}{A} y = \delta \cdot p_1 + C_1 \quad \text{and} \quad p_0 \frac{\dot{V}}{A} y = -\delta \cdot p_2 + C_2 \tag{28–47}$$

The boundary conditions (see Fig. 28–2) are

$$p_1 = p_{1,s} \quad \text{and} \quad p_2 = p_{2,s} \quad \text{at} \quad y = 0$$

$$p_1 = p_{1,0} \quad \text{and} \quad p_2 = p_{2,0} \quad \text{at} \quad y = y_0$$

Substituting in Eq. 28–47,

$$\frac{p_0 \dot{V}}{A} = \frac{\delta}{y_0} \cdot (p_{1,0} - p_{1,s}) = \frac{\delta}{y_0} \cdot (p_{2,s} - p_{2,0}) \tag{28–48}$$

From Eqs. 28–42, 45, and 48

$$\dot{m}_1'' = \frac{\delta}{y_0} \cdot (c_{1,0} - c_{1,s}) \quad \text{and} \quad \dot{m}_2'' = \frac{\delta}{y_0} \cdot (c_{2,s} - c_{2,0}) \tag{28–49}$$

Equations 28–47 and 45 show that the partial pressures and concentrations change linearly with y (see also Fig. 28–2).

It is different with Thoma's [1921] diffusion arrangement, which was mentioned in Sect. 28–1, or with a cooling tower. These would have to be compared with a tube closed on one end, for instance, at $y = 0$. If this end is covered with plotting paper, soaked with phosphoric acid, one has the scheme of Thoma's experiment; if it is closed by a water layer, one has an element of a cooling tower. It will be shown that here the partial pressures do not change linearly with y.

Again air and ammonia (or water vapor) diffuse in opposite directions. But the end at $y = 0$ being impervious to air, no air transportation can actually occur in the tube. Hence the diffusion of air, a molecular transportation in the direction y, will automatically be compensated

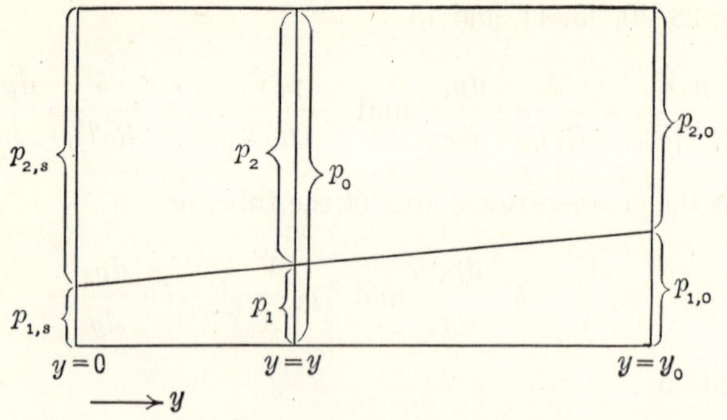

FIG. 28-2. Pressure distribution in the mutual diffusion of two gases through a tube.

for by convection of air in the opposite direction, as had been recognized by Meyer [1877].

This means a sensible air flow in the direction $(-y)$, and ammonia, which is mixed with the air, is carried in the same direction and with the same velocity, the absolute value of which may be called u. It follows that

$$\dot{m}_2'' = 0 = -uc_2 - \delta \cdot \frac{dc_2}{dy} \quad \text{or} \quad 0 = up_2 + \delta \cdot \frac{dp_2}{dy} \tag{28-50}$$

Correspondingly, for the flow of ammonia,

$$\dot{m}_1'' = uc_1 + \delta \cdot \frac{dc_1}{dy} \quad \text{or} \quad \dot{m}_1'' R_1 T_0 = up_1 + \delta \cdot \frac{dp_1}{dy} \tag{28-51}$$

From Eqs. 28-50, 51, and 41

$$\dot{m}_1'' R_1 T_0 = u(p_1 + p_2) + \delta \cdot \frac{d(p_1 + p_2)}{dy} = up_0 + \delta \cdot \frac{dp_0}{dy} = up_0 \tag{28-52}$$

From Eqs. 28-52 and 44

$$u = \frac{\dot{m}_1'' R_1 T_0}{p_0} = \frac{\dot{V}}{A} \tag{28-53}$$

This is a remarkable equation, showing that the velocity of sensible flow (convectional velocity) is numerically equal to the diffusional average velocity and is constant, as is $\dot{m}_1''$ over all the tube length.

Substituting this in Eq. 28-51,

$$\dot{m}_1''R_1T_0 = \delta \cdot \frac{p_0}{p_0 - p_1} \cdot \frac{dp_1}{dy} \qquad (28\text{-}54)^*$$

By integration within the limits $p_{1,s}$ (for $y = 0$) and p_1

$$\frac{p_0 - p_{1,s}}{p_0 - p_1} = e^{\frac{\dot{m}_1''R_1T_0}{\delta \cdot p_0}y} \quad \text{or} \quad p_1 = p_0 - (p_0 - p_{1,s})e^{\frac{\dot{m}_1''R_1T_0}{\delta \cdot p_0}y} \qquad (28\text{-}55)$$

It is seen that the partial pressures do not change linearly with y, but according to an e function.

Substituting $p_1 = p_{1,0}$ for $y = y_0$,

$$\dot{m}_1'' = \frac{\delta \cdot p_0}{R_1T_0y_0} \ln \frac{p_0 - p_{1,s}}{p_0 - p_{1,0}} \qquad (28\text{-}56)$$

These equations were derived first by Stefan [1874] and formed the basis of the experimental determination of δ by Stefan and Winkelmann. They show that the diffusing quantity is not proportional to dp_1/dx. However, if $p_1 \ll p_0$, Eq. 28-56 simplifies † to

$$\dot{m}_1'' = \frac{\delta}{R_1T_0} \cdot \frac{p_{1,0} - p_{1,s}}{y_0} = \delta \cdot \frac{c_{1,0} - c_{1,s}}{y_0} \qquad (28\text{-}57)$$

Since for pure heat transfer the velocity at $x = L$ is zero, whereas it is u for pure diffusion, the model experiments of Thoma are correct only when $\dot{m}_1''$ and therefore p_1 is very small so that u may be taken as equal to zero. The same holds for Nusselt's own derivation in Sect. 28-3. In these cases there was no way but to neglect u. It would have been possible to consider it in Schmidt's derivation of the sweat-water formation (Sect. 28-2). This, however, will not be followed up, but another case will be dealt with in the next section showing how Nusselt used u in a practical case.

28-5. Evaporation from a Liquid Surface into a Gas Stream of Different Temperature

Considering simultaneous heat transfer and evaporation from a liquid surface to a gas flowing parallel to it, Eqs. 28-34 and 35 may be used for small partial pressures of the vapor. However, at 123°F the partial

* The reader of Nusselt's original derivation will find that the signs are confused; this may be caused by errors in printing; otherwise Nusselt could not have arrived at Eq. 28-54.

† Using only the first term in the series $\ln z = 2\left(\frac{z-1}{z+1} + \cdots\right)$.

pressure of water vapor is already 1.84 lb/sq in., that is ⅛ of the standard atmospheric pressure. Hence a modification of these equations is needed. Assuming that diffusion proceeds perpendicularly to a liquid surface, it would be equal to the diffusion in a closed tube as dealt with in Sect. 28–4. This assumption is valid in the range of the laminar boundary layer or sublayer.

Referring to a characteristic length L of the surface in the direction of the flowing gas and arbitrary width W, let

$$N_{Re} = \frac{v_0 L}{\nu} \quad \text{(Reynolds number)}$$

$$N_{Nu} = \frac{hL}{k} \quad \text{(Nusselt number)}$$

$$N_{Pr} = \frac{\nu}{\alpha} \quad \text{(Prandtl number)}$$

$$N_{Sc} = \frac{\nu}{\delta} \quad \text{(Schmidt number)}$$

where again subscript 0 refers to the environment at considerable distance from the surface. Then, according to the theory of similarity,

$$N_{Nu} = \frac{hL}{k} = \Psi(N_{Re}, N_{Pr}) \tag{28-58}$$

Further, introducing y for the direction of evaporation normal to the surface: ‡

$$-\left(\frac{dt}{dy}\right)_s = \frac{h}{k}(t_s - t_0) \tag{28-59}$$

From these equations and by analogy

$$-\left(\frac{dt}{dy}\right)_s = \frac{t_s - t_0}{L}\Psi(N_{Re}, N_{Pr}) \tag{28-60}$$

$$-\left(\frac{dc}{dy}\right)_s = \frac{c_s T_s - c_0 T_0}{L T_m}\Psi(N_{Re}, N_{Sc}) \tag{28-61}$$

Now, as explained for the ammonia diffusion in Sect. 28–4, at the water surface air cannot flow perpendicularly to it, but is flowing by diffusion in the direction $(-y)$. Hence, by convection air must flow in the direc-

‡ The direction y is the same as defined in Sect. 28–4.

tion y, and water vapor is carried along with the same velocity. Thus, corresponding to Eq. 28-51, the mass velocity of the vapor is

$$\dot{m}_1'' = uc_{1,s} - \delta \cdot \left(\frac{dc_1}{dy}\right)_s \quad \text{or} \quad \dot{m}_1''R_1T_s = up_{1,s} - \delta \cdot \left(\frac{dp_1}{dy}\right)_s \tag{28-62}$$

where subscript s refers to the liquid surface and u, in analogy to Eq. 28-53, becomes

$$u = \frac{\dot{m}_1''R_1T_s}{p_0} = \frac{\dot{V}}{A} \tag{28-63}$$

But

$$R_1T_s = \frac{p_{1,s}}{c_{1,s}} \tag{28-64}$$

Hence

$$uc_{1,s} = \frac{\dot{m}_1''p_{1,s}}{p_0} \tag{28-65}$$

By substitution in Eq. 28-62 the velocity u will be eliminated:

$$-\left(\frac{dc_1}{dy}\right)_s = \frac{\dot{m}_1''}{\delta}\left(1 - \frac{p_{1,s}}{p_0}\right) \tag{28-66}$$

For the heat transportation, in analogy to Eq. 28-62,

$$\frac{q''}{\rho c_p} = u(t_s - t_0) - \alpha\left(\frac{dt}{dy}\right)_s \tag{28-67}$$

where ρ and c_p are the density and specific heat of the mixture.

Using Eqs. 28-64 and 65 for elimination of u,

$$-\left(\frac{dt}{dy}\right)_s = \frac{1}{k}\left[q'' - \dot{m}_1''\frac{R_1T_s}{p_0}\rho c_p(t_s - t_0)\right] \tag{28-68}$$

Introducing the molecular weights M_0 and M_1,

$$-\left(\frac{dt}{dy}\right)_s = \frac{1}{k}\left[q'' - \dot{m}_1''\frac{M_0}{M_1}c_p(t_s - t_0)\right] \tag{28-69}$$

From Eqs. 28-60 and 69

$$k\frac{t_s - t_0}{L}\Psi(N_{Re}, N_{Pr}) = q'' - \dot{m}_1''\frac{M_0}{M_1}c_p(t_s - t_0) \tag{28-70}$$

From Eqs. 28-61 and 66

$$\delta \cdot \frac{c_sT_s - c_0T_0}{LT_m}\Psi(N_{Re}, N_{Sc}) = \dot{m}_1''\left(1 - \frac{p_{1,s}}{p_0}\right) \tag{28-71}$$

From the last two equations

$$\frac{q''}{\dot{m}_1''} = \left(1 - \frac{p_{1,s}}{p_0}\right) \frac{k(t_s - t_0)T_m}{\delta \cdot (c_s T_s - c_0 T_0)} \cdot \frac{\Psi(N_{Re}, N_{Pr})}{\Psi(N_{Re}, N_{Sc})} + \frac{M_0}{M_1} c_p(t_s - t_0) \tag{28-72}$$

But

$$N_{Nu} = \Psi(N_{Re}, N_{Pr}) = C\left(\frac{v_0 L}{\nu}\right)^m \left(\frac{\nu}{\alpha}\right)^n \tag{28-73}$$

and

$$\Psi(N_{Re}, N_{Sc}) = C\left(\frac{v_0 L}{\nu}\right)^m \left(\frac{\nu}{\delta}\right)^n \tag{28-74}$$

By substitution in Eq. 28–72

$$\frac{q''}{\dot{m}_1''} = \left(1 - \frac{p_{1,s}}{p_0}\right) \frac{k}{\delta} \cdot \frac{(t_s - t_0)T_m}{c_s T_s - c_0 T_0} \left(\frac{\delta}{\alpha}\right)^n + \frac{M_0}{M_1} c_p(t_s - t_0) \tag{28-75}$$

From Eqs. 28–71 and 74

$$\dot{m}_1'' = \delta \cdot \frac{c_s T_s - c_0 T_0}{L T_m} \cdot \frac{1}{1 - p_{1,s}/p_0} C\left(\frac{v_0 L}{\nu}\right)^m \left(\frac{\nu}{\delta}\right)^n \tag{28-76}$$

Using this equation, the mass transfer of one gas into and through another one can be determined by measurement of the partial-pressure gradient and application of the heat-transfer relation of Eq. 28–73. This is a tremendous achievement of the theory of similarity.

By substituting $\dot{m}_1''$ from Eq. 28–76 in Eq. 28–75

$$q'' = \frac{C(t_s - t_0)}{L}\left[k(N_{Pr})^n + \frac{M_0}{M_1} \cdot \frac{c_p(c_s T_s - c_0 T_0)}{(1 - p_{1,s}/p_0)T_m} (N_{Sc})^n\right] N_{Re}{}^m \tag{28-77}$$

This is a case where representation by one product of power functions, as usual in dimensional analysis, would fail. The equation allows one to calculate the heat transfer in evaporation from measurements of the temperature and concentration gradients and formula Eq. 28–73, which is valid for heat transfer without changes of phase and can easily be used for any shapes of surfaces for which C, m, and n are known.

If $c_s T_s = c_0 T_0$, then no evaporation takes place, and Eq. 28–77 turns into the ordinary form

$$q'' = C \frac{t_s - t_0}{L} k(N_{Re})^m (N_{Pr})^n$$

or

$$\frac{hL}{k} = C(N_{Re})^m (N_{Pr})^n$$

From a quantitative viewpoint, however, it should not be forgotten that it is not known as yet whether heat transfer on a liquid surface (without phase change) follows the same law as heat transfer on a solid surface. Even so, wave formation may make it difficult to use the analogy at higher velocities.

However, this difficulty can be overcome by using wetted porous plates or ice or other sublimating substances (see Sect. 28–6).

28–6. Experiments on Heat and Mass Exchange

Only a few applications of the methods described in this chapter will be mentioned.

The first to deal with the similarity of forced convection and mass exchange was Thoma [1921], as mentioned in Sect. 28–1. He made models of tube banks out of filter paper which was soaked with a weighed quantity of phosphoric acid. When an air-ammonia mixture was driven through the tube bank, the ammonia was absorbed so eagerly that its concentration at the wall became negligibly small. The rest of the acid remaining in the filter paper was then determined by titration with a weak ammonia solution, using methyl orange as indicator. From the neutralized quantity of acid, on the other hand, the quantity of ammonia absorbed was determined. By this method Thoma found a change of the coefficient of heat transfer in tube banks with the 0.6th power of the Reynolds number in reasonable, although not exact, agreement with direct measurements which were performed later on (to be dealt with in Vol. II). Thoma further found that by staggering the model tubes the heat transfer was increased by 20% for a bank with about 2-diameter transverse pitch of the tubes and 1¼-diameter pitch in the flow direction, which is of the order of magnitude directly measured later on. These are quite remarkable results, considering the fact that Thoma was not yet aware of some of the theoretical requirements dealt with above, in this chapter.

Similar experiments, but refined and on larger scale, were performed by Lohrisch [1929]. He concentrated his efforts on the determination of the heat transfer by diffusion measurements on tube banks with staggered and in-line arrangement in a set-up represented in Fig. 28–3, which is self-explanatory.

He further did some experiments on single tubes, using ammonia as well as water vapor as diffusing gas. Sodium hydroxide rods, 1 cm in diameter, served as tube models and absorbing material. Because NaOH is very hygroscopic the rods were always covered with a wet film consisting of sodium lye. This made some corrections necessary, but the results were quite satisfactory and just between the results of direct measurements of the heat transfer by Reiher [1925] and Hughes [1916].

Very remarkable is an arrangement to use the ammonia-absorption method for measuring the heat-transfer distribution over the perimeter

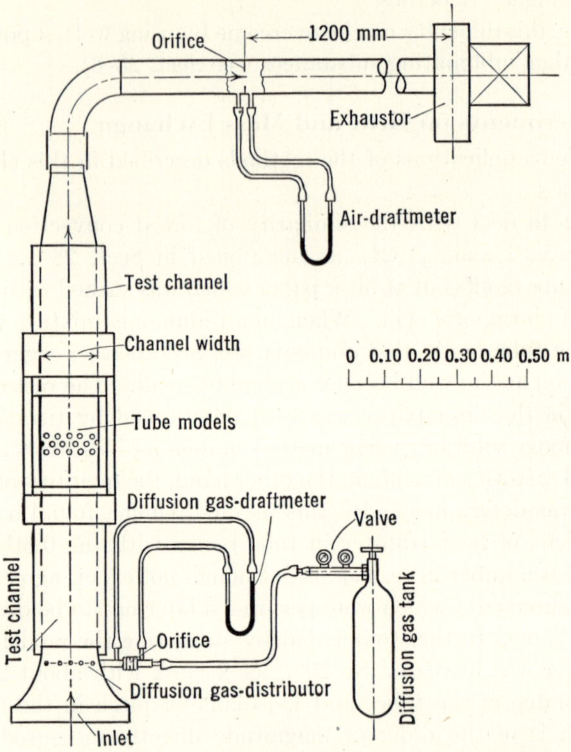

FIG. 28–3. Arrangement for determination of heat-transfer coefficients by diffusion measurements.

From Lohrisch [1929].

of a tube. This was covered with plotting paper, and longitudinal strips of the paper only were soaked with phosphoric acid. The results are represented in Fig. 28–4. A by-product of these experiments were flow pictures, obtained by taking photos of the ammonia chloride clouds which were formed when the model surfaces were soaked with chloric acid. Figures 28–5, 6, and 7, taken from Lohrisch's paper, show several interesting details. In Fig. 28–5, for instance, it is seen that the backstream side of the tube is covered by eddies which correspond to a steep temperature drop in Fig. 28–4. Figures 28–6 and 7 are helpful for understanding the behavior of tube banks.

Hilpert [1932] investigated evaporation and heat transfer on vertical plates in still air. Porous ceramic plates of square shape (side lengths 2 to 5 in.), able to absorb liquid up to about ⅓ of their own volume § were soaked with water or a hydrocarbon (benzene, toluene, chlorobenzene, bromobenzene, carbon tetrachloride, and ethylene bromide).

A compartment with air at controlled humidity was provided in which the plates were vertically suspended on one arm of a balance. The weight loss of the wet plates was measured. The behavior of plates soaked with cold or hot water was investigated. In addition to this, heat-transfer experiments were performed on dry aluminum plates.

In the cooling of wet plates it was found that the air originally flows upward, but then is reversed to the opposite flow direction. In the moment of reversal the evaporation has a minimum. A similar minimum was noticed by Powell [1940] in an investigation of the distributions of temperature and vapor pressure around a horizontal wet cylinder.

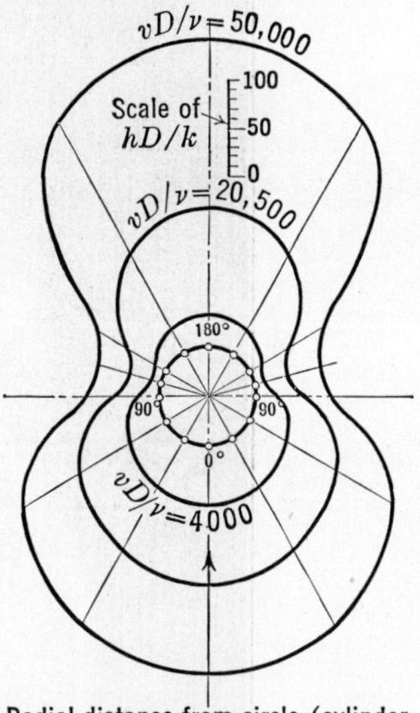

FIG. 28–4. Distribution of Nusselt number around a cylinder exposed to air flowing perpendicularly to axis.

From Lohrisch [1929].

Both the mass and heat-transfer coefficients (b and h) were found to be proportional to the fourth root of a modified Grashof number. The ratio h/b was found to be $0.92\rho c_p$ and about $2\rho c_p$, respectively, for the evaporation of water and hydrocarbons into air. From experiments on spheres Powell [1940(a)] obtained $h/b = 0.90\rho c_p$ in good agreement with Hilpert's result. Both Hilpert's and Powell's investigations with still air are rich in interesting details which cannot be dealt with here.

The last part of this section will rather be devoted to a discussion of the experiments on evaporation under forced-convection conditions, which were performed in the British National Physical Laboratory. The investigators, Powell and Griffiths [1935], used plates and cylinders covered with tightly stretched linen sheets or with filter paper, which were soaked and covered with water and could be held at uniform temperature by means of a series of separately controlled electric heaters.

§ The material used is called "Osmose Ton" (osmose clay).

The apparatus shown in Fig. 28-8, for instance, was used for measurement of the distribution of evaporation along the surface of a plane plate, which was placed in a wind tunnel. The sketch is self-explanatory.

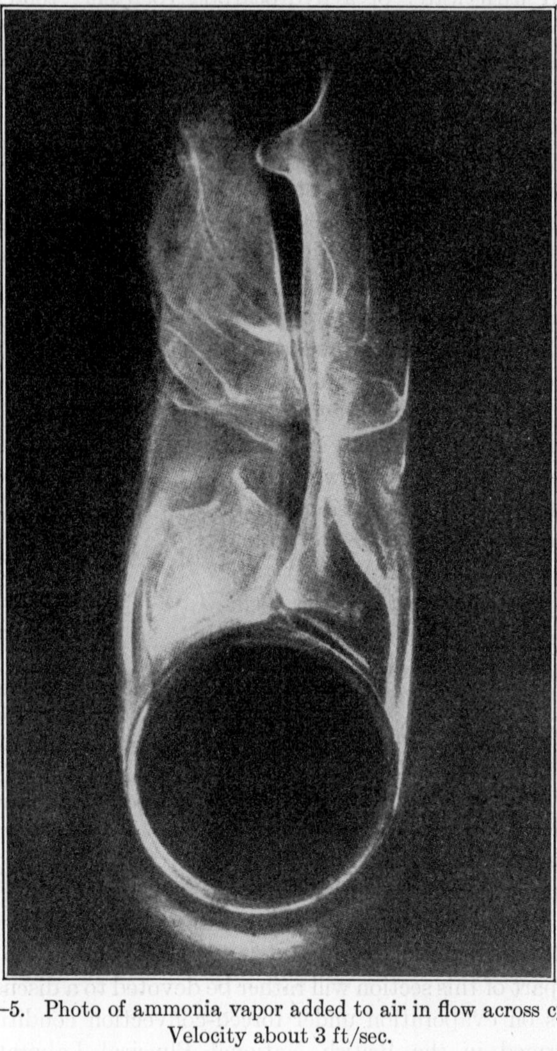

FIG. 28-5. Photo of ammonia vapor added to air in flow across cylinder. Velocity about 3 ft/sec.
From Lohrisch [1929].

The rate of evaporation was determined in two ways: directly, by measuring the quantity of water evaporated, and, indirectly, from the difference between the electric energies needed to keep the surface temperature uniform and constant at wet or dry surface.

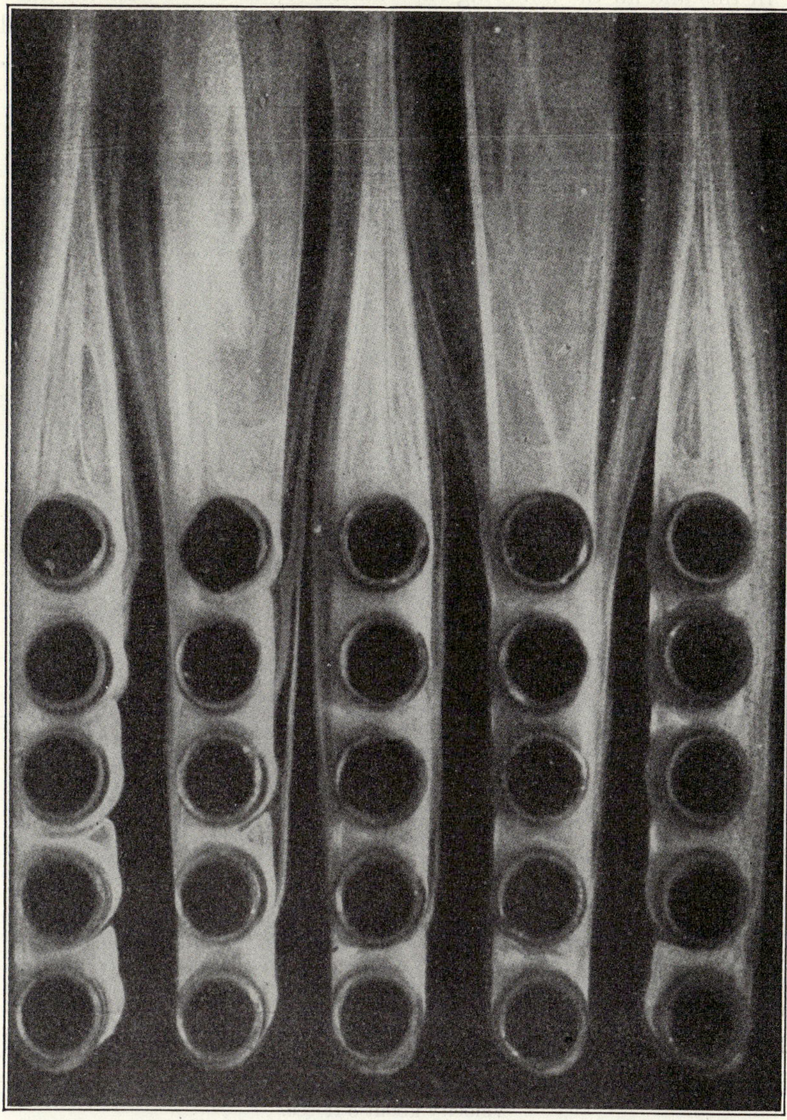

FIG. 28-6. Photo of ammonia vapor added to air in flow across tube bank (in-line arrangement).

Velocity about 5 ft/sec.

From Lohrisch [1929].

In this and all the following experiments it was found that the mass flow per unit area, $\dot{m}''$, was proportional to the difference Δp of the partial pressures of water vapor at the surface and in the core of the flowing air.

FIG. 28-7. Photo of ammonia vapor added to air in flow across tube bank (staggered arrangement).

Velocity about 5 ft/sec.

From Lohrisch [1929].

One special result of this part of the investigation will be indicated somewhat later.

In measurements with wind in cross flow to vertical cylinders nine vertical partitions were used, and distributions of the evaporation around the cylinder similar to those of heat transfer (see Fig. 28–4) were observed with maxima in the front and rear and a minimum at the side of the circular cross section. The ratio of evaporation from the front half to that from the rear half of a cylinder was found in satisfactory agreement with the ratio of diffusion obtained by Lohrisch [1929].

In a continuation of the evaporation measurements Powell [1940(*a*)] first of all observed about the same evaporation into air flowing parallel to the axis of a wet cylinder, 4.36 cm in diameter, as he and Griffiths

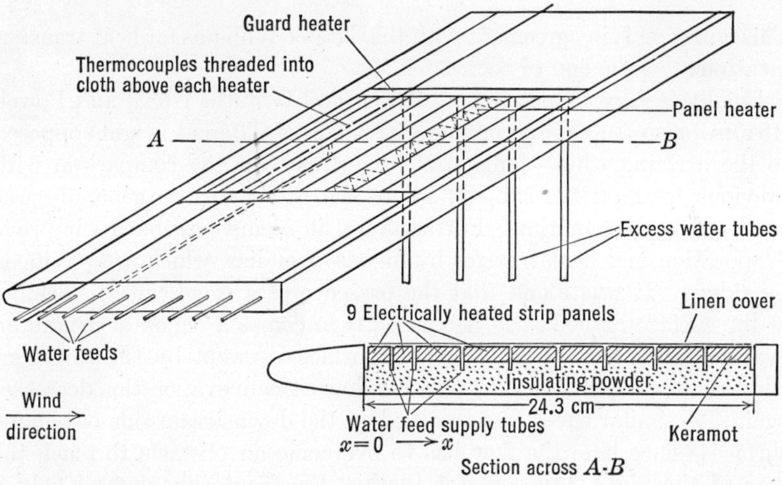

FIG. 28–8. Device for measurement of local rates of evaporation from a horizontal plate into an air stream.

had found with the plane plate, both cylinder and plate being provided with streamlined wooden nose pieces. Furthermore, $\dot{m}''L \sim (N_{Re})^{0.8}$ was obtained in the range of $N_{Re} = 5000$ to 360,000, in agreement with Jakob and Dow's [1946] measurements on heat transfer from a cylinder of similar size to an airstream.

Placing cylinders transverse to the airstream, Powell found $\dot{m}''L \sim N_{Re}{}^{0.6}$ in the range of $N_{Re} = 260$ to 13,000. The exponent 0.6 seems to be somewhat high, compared with the corresponding exponent for heat transfer according to Tables 26–2 and 3.

For circular discs of diameter D facing wind, the exponent 0.56 was observed, i.e., nearly the same as the exponent 0.57 measured by Jakob and Kezios for the heat transfer in perpendicular impact [see Jakob 1947]. Another interesting result of these tests was that the evaporation was only about 20% smaller when the surface faced downstream than when facing the wind, and that the evaporation from surfaces parallel to the wind was only slightly larger and sometimes even smaller than from surfaces facing downstream.

For spherical surfaces the exponent 0.59 was obtained. In another paper Powell [1939–1940] found a common curve for evaporation of water and sublimation of ice (at −8 to −12°C), which, with the denotations of Eq. 28–6, can be expressed in the dimensionless form ‖

$$\frac{bD}{\delta} = \frac{RT_m}{\Delta p} \cdot \frac{\dot{m}''D}{\delta} = 28.3(10^{-9}) \left(\frac{vD}{\nu}\right)^{0.62} \tag{28-78}$$

This equation is in agreement with the data of Williams for heat transfer, mentioned at the end of Sect. 26–7.

Finally it may be noted that Powell and Griffiths [1935] and Powell [1940(*a*)] also studied the influence of ridges of different heights opposed to the arriving wind. This is of importance for the comparison with previous tests on the subject, where the water surface generally was bordered by a pan rim. Furthermore, it seems possible to improve evaporation and heat transfer by means of eddies, which are produced by ridges. It was found that the presence of a ridge causes shielding of the surface immediately behind it; then comes a region of maximum evaporation, where presumably the surface is swept by the turbulent eddy of air set up by the ridge; further downstream evaporation decreases again. A similar effect was observed on the downstream side of oblique surfaces, since here the flow has to overcome an obstacle to reach the rear of the plate. In contrast to this, the front side showed only a moderate increase of evaporation when the plate was transferred from impact to parallel direction, which is again in agreement with the results of Jakob and Kezios' heat-transfer measurements.

The papers of Powell and Griffiths contain also a thorough discussion of the pertinent literature, partly taken from remote sources.

The heat-mass transfer analogy was also used by R. L. Perry [1944] in experiments on the dehydration of prunes in a tunnel through which hot air was flowing. In reversing the purpose of such experiments it may be possible to determine the vapor concentration on the surface of fruits or vegetables during the drying process from thermal-convection measurements on dry surfaces of similar shape.

Since the result of Powell's sublimation experiments on ice spheres is encouraging, similar experiments with discs of naphthalene and other compounds, exposed to an air stream, are being performed by H. Sogin in cooperation with the present author in Illinois Institute of Technology. Various reports about this work have been submitted to the

‖ The left side of Powell's original equation does not seem to be dimensionless.

United States Air Corps, the first of them in August 1947. The mass and heat transfer in tube banks has been studied by Winding and Cheney, Jr. [1948], who also used naphthalene models. The total sublimation from the discs or tube models is obtained by weighing them before and after exposure to the air stream; distribution of the sublimation over the exposed surface is determined by measuring the change of thickness at different places of the surface. Because of the small vapor pressure, these measurements must be very precise. Furthermore, exact values of the diffusivity δ are needed for the determination of heat-transfer data from diffusion measurements.

CHAPTER 29

Heat Transfer in Evaporation

29-1. The Different Kinds of Evaporation

Evaporation is the conversion of a solid or liquid into vapor. It requires heat energy, and therefore some heat transfer is necessarily involved.

The direct conversion of a solid to a vapor is called sublimation. It presents a simpler problem of heat transfer than the conversion of a liquid to vapor. Only vaporization of liquids will be dealt with in this chapter.

Evaporation is a very general process. Whenever a liquid surface is in contact with a gaseous environment, molecules leave the liquid surface and mix with the gas. This may or may not have another constitution than the evaporating liquid and can be at any pressure. Evaporation from inside a liquid body under the formation of vapor bubbles is called ebullition or boiling.

In Chap. 28 the evaporation without boiling from a liquid into a gas of different constitution has been dealt with. In the present chapter the evaporation of a liquid into a vapor of the same constitution, and particularly the process of boiling, will be studied.

Referring to the ordinary boiling of water, it is generally known that the vapor formed on the free surface is first invisible, but becomes visible when the heating is increased. With further increase of the heat input boiling starts somewhere below the water surface; steam bubbles are formed, rise to the surface, and break through it to join the vapor above the water surface. Since this sort of boiling is bound to the presence of some solid particles or other sorts of nuclei on which bubbles are formed, it is called nucleate boiling. We may also simply call it ordinary boiling.

Another sort of boiling, called film boiling, occurs when water is spilled on a hot stove. Then the so-called Leidenfrost [1756] phenomenon occurs; i.e., water globules dance on the hot surface for quite some time without sensible evaporation. In fact, a film of superheated steam exists below each water globule, serves as a thermal insulation between the heating surface and the water, and thus prevents the bulk of the water from fast evaporating. This sort of retarded evaporation, however, is only a special case of nucleate boiling, as will be understood, considering the first stage of quenching a glowing steel bar by throwing it into water or another liquid. Again, on the metal surface an insulating vapor layer is formed through which only a little heat penetrates to the liquid. In this case the vapor formed cannot escape sidewards as under the Leidenfrost droplets but increases the thickness of the vapor film which surrounds the surface of the steel bar. When the vapor layer has grown so much that its buoyancy can overcome the capillary forces which keep it in place and together, it blows up, and more or less of the vapor is released into the bulk of the liquid, where it immediately takes the form of a vapor bubble different only by its larger original size from the bubbles formed in nucleate boiling. Like those, it rises, and eventually pierces the free surface of the liquid. After the blowing up of the vapor film on the steel bar the film grows again, and the play repeats until the steel bar is cooled down to a state where the Leidenfrost phenomenon will be stopped and ordinary nucleate boiling may occur.

Heat transfer in nucleate boiling will be the main subject of this chapter; however, evaporation without boiling will also be briefly dealt with, and film boiling will be explained more in detail than has been done above.

29–2. Superheating of Liquid and Evaporation

Evaporation, like various other processes, has been studied for a long time from a merely thermodynamical viewpoint, i.e., the latent heat of vapors has been thoroughly measured and taken into consideration in the heat balances which can be made up for the change of phase which is called evaporation. The driving forces for, and the actual resistances against, such changes were of no concern to the thermodynamists.

A small temperature excess of boiling water over the leaving steam was first noticed by Henning [1909] in measurements of the latent heat between 100 and 180°C. When Jakob [1928(*a*)] observed the same effect at 180°C and 210°C, he began to study whether this result was due to errors of observation or to actual physical reasons. This was done in two series of investigations, one connected with precision measurements of the latent heat of water, the other devoted to the process of evaporation.

In the experiments on the latent heat, first of all, by continued improvement of arrangement and technique of measurement, all conceivable sources of error were excluded, for instance, impurities of the water, formation of drops on the walls, influence of hydraulic pressure, and losses by thermal conduction [Jakob and Fritz 1929, 1930].

In the latter paper, it was verified that the temperature of boiling water was higher than that of the steam, and exact pressure measurements showed that the steam temperature was identical with the equilibrium temperature of saturation so that the water must have been superheated. Finally Jakob and Fritz [1935] derived the influence of pressure on the temperature excess from about 34 measurements in different arrangements at pressures between 14.7 and 2873 lb/sq in., corresponding to saturation temperatures of 100 to 365°C (Fig. 29–1). The temperature difference tends toward zero if the critical point (3200 lb/sq in., 374.1°C) is approached, where the surface and all differences between liquid and vapor disappear. [See also Jakob 1936.*]

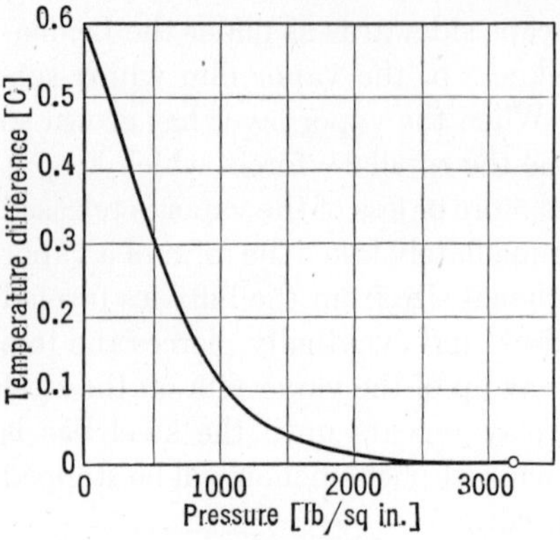

FIG. 29–1. Temperature difference between boiling water and steam.

The tiny temperature difference shown in the figure causes a heat flow in the water toward any interface between the liquid and the vapor, and evaporation of water on that interface, whether this be the free water surface of a boiler vessel or the interface between water and steam in a steam bubble. It really is the driving force in the process of evaporation, and its observation was the incentive for much of the research on the process of boiling dealt with in this chapter.

To understand the interrelation between superheating of the liquid and evaporation the behavior under equilibrium conditions may first be examined.

When the liquid and gaseous phases of a substance exist together, one of the conditions of equilibrium is that of masses; i.e., the same number of molecules must always pass in both directions through an area separating the two phases. Therefore, the process of evaporation—i.e., the continual one-way migration of molecules into the space of the gaseous phase—is not strictly in accordance with the conditions of thermodynamic equilibrium.

It may be asked how a vapor bubble is formed in the liquid. It cannot spontaneously originate in a saturated liquid. This would require latent

* Much of this paper is used in the present chapter.

heat of evaporation to be taken from the liquid, which would become cooler, an obvious violation of the second law of thermodynamics. If the vapor is able to exist as a separate phase, a surface of separation is necessary which in the case of a vapor bubble will be curved; but the liquid offers a resistance to the formation of such a curved surface which manifests itself in the so-called surface tension. To overcome this, work must be done. In the formation of a bubble, such work consists in separating the liquid (work of separation). W. Thomson (Lord Kelvin) has shown that the vapor pressure, $p_{g,r}$, on a concave liquid surface of radius of curvature r is smaller than that on a plane surface, p_g, at the same temperature, according to the equation

$$p_{g,r} = p_g - \frac{2\sigma}{r} \cdot \frac{\rho_g}{\rho_f - \rho_g} \qquad (29\text{–}1)$$

where σ = the surface tension of the liquid against the vapor,
ρ_g = the density of the saturated vapor, and
ρ_f = the density of the saturated liquid.

This equation can be understood easily from the mechanical equilibrium in a capillary tube (Fig. 29–2). The surface tension σ is acting on the perimeter $2\pi r$ with the force $\sigma \cdot 2\pi r$, producing a stress $(\sigma \cdot 2\pi r)/\pi r^2 = 2\sigma/r$ used for lifting the liquid level to the height H The hydrostatic pressure is decreasing by this amount if one passes the curved surface in downward direction, i.e., the pressure on the liquid side of the interface becomes

$$p_{f,r} = p_{g,r} - \frac{2\sigma}{r} \qquad (29\text{–}2)$$

Further, obviously,

$$H\rho_f g = \frac{2\sigma}{r} \quad \text{or} \quad H = \frac{2\sigma}{r} \cdot \frac{1}{\rho_f g} \qquad (29\text{–}3)$$

and

$$p_{g,r} = p_g - H\rho_g g = p_g - \frac{2\sigma}{r} \cdot \frac{\rho_g}{\rho_f} \qquad (29\text{–}4)$$

Since $\rho_g \ll \rho_f$, except for very high pressure, Eq. 29–4 can be considered identical with Eq. 29–1.

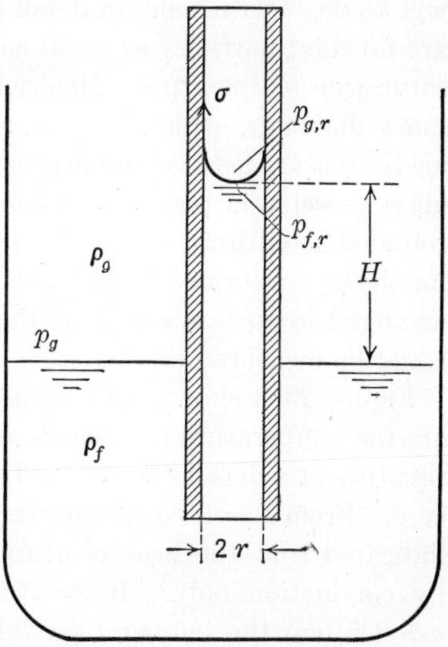

FIG. 29–2. Rise of liquid level in a capillary tube due to surface tension.

Now, according to the observations mentioned, the vapor of a boiling liquid is at the saturation temperature. But $p_{g,r} < p_g$, according to Eq. 29–1. Hence, if the liquid temperature around a vapor bubble is equal to or higher than the saturation temperature which corresponds to the pressure p_g, then under equilibrium conditions the liquid in the vicinity of a vapor bubble must be superheated with respect to the vapor inside the bubble, which, for simplicity, may be considered as being spherical and subject, in each greatest circle of its surface, to the same surface tension as indicated in Fig. 29–2. [See also Jakob 1941, Fig. 2.†]

According to Eq. 29–1, the temperature excess above saturation will be the larger, the smaller the radius of the bubble. From this follows the strange result that in an absolutely pure liquid no vapor bubbles could originate at all if the bubbles had to start from an infinitely small diameter. Indeed, pure water can be highly superheated without ebullition, thus existing in an unstable state of equilibrium. When boiling conditions suddenly take place at any spot in the liquid, then a great portion of vapor may instantaneously be formed in an explosive manner, as is well known.

29–3. Evaporation on a Free Liquid Surface without Ebullition

Heidrich [1931] carefully investigated the temperature distribution and heat flow under conditions which were almost ideal for the evaporation of water without ebullition. As has previously been mentioned and will be dealt with more in detail later on, boiling does not occur if there are no solid particles or small gas bubbles in a liquid held at or above saturation temperature. Heidrich used a cylindric glass vessel of 8.3 cm inner diameter, which was cleaned by means of concentrated sulfuric acid. Water, distilled and degasified under vacuum, was filled in, and the glass vessel was placed in a hot chlorine calcium solution which surrounded it entirely except for some capillary tubes serving as outlets for the steam formed and for the leads of a fine thermocouple. This was arranged in such a way that the water temperature below the surface could be measured exactly.

Figure 29–3 shows some results of such measurements. The curves are for temperature differences, $t_{W,z} - t_{St}$, where $t_{W,z}$ is the water temperature at a depth z below the free surface and t_{St} is the steam temperature. From $z = 0$ to 0.4 mm the temperature increased linearly, which indicated that the heat of evaporation was carried through the water by conduction only. In greater depth convection occurred, and for $z \gtreqless 1.5$ mm the temperature did not change any more. The amounts of water evaporated at the surface were 4.39, 5.58, and 15 kg hr^{-1} m^{-2},

† Parts of this paper are used in the following presentation of the boiling problem.

respectively.‡ These figures should be proportional to the heat energy carried to the free surface per unit time and unit area, and used there for evaporation. However, the calculation of the left straight part of the curves, assuming pure conduction, led to a steeper increase than the experiment gave. Heidrich attributes this difference to temperature differences in the water which occurred in horizontal planes. He further plotted the mass velocity of evaporation, $\dot{m}''$, versus the temperature

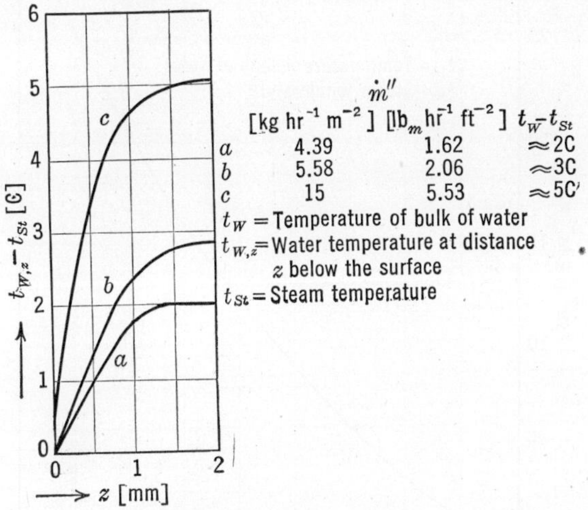

FIG. 29-3. Temperature distribution below a surface of not boiling superheated water.

From Heidrich [1931].

difference, $\Delta t = t_W - t_{St}$, where t_W is the mean temperature of the bulk of the water. This plot gave rather exactly the curve represented in Fig. 29-4. It shows the time rate of bubble-free evaporation per unit area up to a rather considerable temperature excess of water above the saturation point.

It is remarkable that $\dot{m}''$ increased much more than proportionally to Δt, for instance, from 5 to 30 kg hr^{-1} m^{-2} (ratio 1:6) when Δt increased from 2.65 to 8.85 C (ratio 1:3⅓). Comparison of Figs. 29-3 and 4 shows that this was caused by a considerable reduction in thick-

‡ Here and in some following cases the original graphs and values are presented, in which metric units are used. It may be recalled that

$$1 \text{ kcal hr}^{-1} \text{ m}^{-2} = 0.368 \text{ B hr}^{-1} \text{ ft}^{-2}$$

$$1 \text{ kcal hr}^{-1} \text{ m}^{-2} \text{ C}^{-1} = 0.205 \text{ B hr}^{-1} \text{ ft}^{-2} \text{ F}^{-1}$$

$$1 \text{ kcal hr}^{-1} \text{ m}^{-1} \text{ C}^{-1} = 0.672 \text{ B hr}^{-1} \text{ ft}^{-1} \text{ F}^{-1}$$

ness of the layer of linear temperature change with increasing $\dot{m}''$. In other words, the influence of convection became stronger, and it extended upward to a region which was closer to the free water surface. According to Heidrich, this came from an increase of the temperature gradient in radial direction due to the fact that heat was carried to the glass vessel

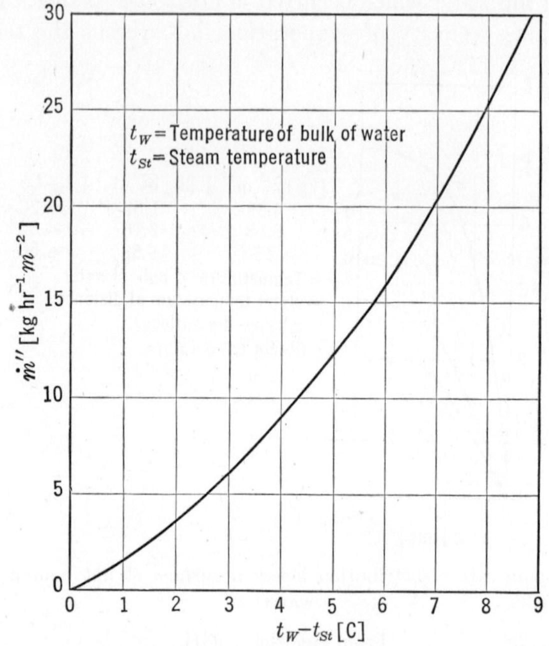

FIG. 29-4. Mass velocity of evaporation from a surface of not boiling superheated water.

From Heidrich [1931].

in all directions from outside. However, even without this, a considerable convection may always be expected if evaporation occurs on a free liquid surface because the liquid will be warmer and therefore lighter in greater depth than just below the surface, so that the mechanical equilibrium of the liquid is necessarily unstable.

Prueger [1940, 1941] performed refined experiments with water and carbon tetrachloride, similar to those of Heidrich and with similar results, and contributed essentially to the theoretical understanding of the process.

29-4. The Temperature Distribution in Boiling Liquids

As mentioned above, a bubble cannot originate from $r = 0$ and boiling will not start unless finite curvatures are present, such as are formed by wall roughnesses or by gas bubbles and other impurities in the liquid. Then tiny vapor bubbles spring up at a temperature of the liquid exceeding the saturation temperature, which corresponds to the pressure inside the bubbles, whereas the vapor produced is at the saturation temperature. The temperature between liquid and vapor tends to equalize, but cannot completely because of the quick rising of the bubbles. Therefore, the superheating of the liquid may last, even when the radius of the bubble has increased so much that $p_{g,r} - p_{f,r}$ becomes negligibly small.

Indeed, all our experiments showed that the bulk of the liquid is at a higher temperature than the saturation temperature, so that a steep drop occurs on the free surface between the liquid and the vapor.

Figure 29-5 gives two characteristic examples of this behavior obtained by Jakob and Fritz [1931; see also Jakob 1936 and 1941] with

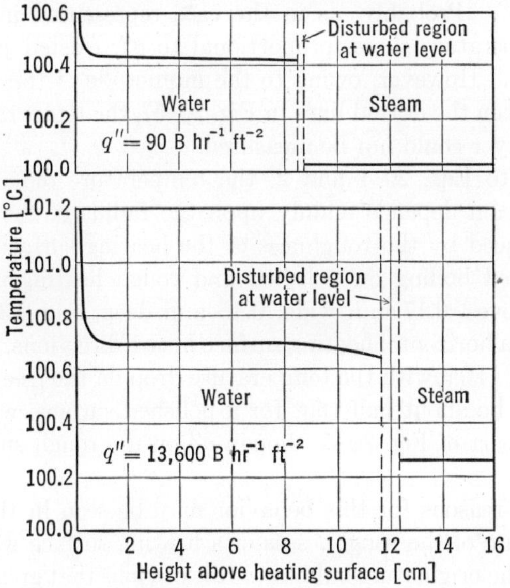

FIG. 29-5. Temperature distribution in nucleate boiling from a horizontal smooth surface.

water which was boiled upon a horizontal polished heating plate. Beginning with the temperature of the heating surface, the water temperature decreases greatly in the close proximity of the surface. This indicates that a thin layer exists which is not mixed by convection. The temperature drop in this layer (not to be seen completely in the figure) depends upon the heat flow per unit time and unit area, q''. This was

90 B hr^{-1} ft^{-2} in the upper part of the graph, and 13,600 B hr^{-1} ft^{-2} in the lower one, the corresponding temperature drops being 0.4 and 10.1 C, respectively.

Proceeding in the direction toward the water level, a negligible decrease of temperature was observed in the bulk of the water, whereas, owing to the change of hydraulic pressure, a decrease of 0.2 and 0.3 C, respectively, would have to be expected. This means that even with a heating of only 90 B hr^{-1} ft^{-2} sufficient mixing by convection occurred to compensate the effect of hydraulic pressure.

The most interesting result, however, was the discovery of how little the heat flow q'' influenced the temperature difference between the bulk of water and steam. Whereas the lower part of the graph belongs to a heating 150 times larger than the upper part, the temperature drop at the liquid level was almost the same (about 0.4 C) in both cases, and also in other experiments of the authors the temperature drop did not change much with q''. Probably, as in the case represented in Fig. 29-3, a uniform temperature slope, proportional to q'', existed just below the water surface. However, owing to the inquietude of the water level in boiling (between the dotted lines in Fig. 29-5), the temperature distribution in this layer could not be measured.

According to Eqs. 29-1 and 2, the temperature difference between water and steam depends mainly upon the radius of the bubbles, and this is influenced by the roughness of the heating surface. Jakob and Fritz compared boiling on polished and rough heating surfaces. For instance, grooves, 0.17 mm wide, 0.15 mm deep, and 0.3 mm distant, were cut into a horizontal heating surface in two directions, perpendicular to each other. Herewith the temperature drop on the free water surface was found to be about half that for a polished surface, which is shown in the lower part of Fig. 29-5, though q'' on the rough surface was 6% larger.

One of the reasons for this behavior may be seen in the fact that a statistical mean of the roughnesses of a heating surface will control the mean size of the originating bubbles. It is obvious that greater roughness facilitates evaporation on the solid surface, the elements of roughness having larger mean radii of curvature than occur on smooth surfaces. However, according to Larson [1945], this effect can be expected only if the rough surface is a heating surface. In radiating heat from outside to a ⅛-in. bronze ball in water, he could not find any difference between the behavior of the ball whether it was smooth or roughened.

Returning to the experiments of Jakob and Fritz, it was surprising that, despite the increase of the saturation temperature with the pressure, the temperature of boiling water in greater depth was not higher than just below the water level. Fritz and Homann [1936] investigated this more exactly in an arrangement designed by the author [see also Jakob 1941]. In these experiments the depth of boiling water was varied in a cylindrical container having a horizontal heater on the bottom and being perfectly insulated against heat losses. The width of the boiling vessel was 6 in., and tubes, 2 or 4 in. wide, open on both ends, could be placed in it coaxially in order to study the influence of width.

The investigation, which was conducted in the range of 1840 to 25,800 B hr^{-1} ft^{-2}, confirmed the fact that variations of q'' do not influence the temperature distribution at all. Some other results, however, were rather unexpected.

Figure 29–6 shows temperature differences, Δt, between the water and the steam at different heights, z, above the heating surface for four total water depths z_0. The curves are averages of all tests. No measurements were made at distances $z < 1$ cm. The temperature excess of the heating surface above water temperature was 2 to 11 C, depending upon the heat flow, and one must imagine that the curves would bend up sharply from their left end to these values or, as an average, to 6.5 C, at $z/z_0 = 0$.

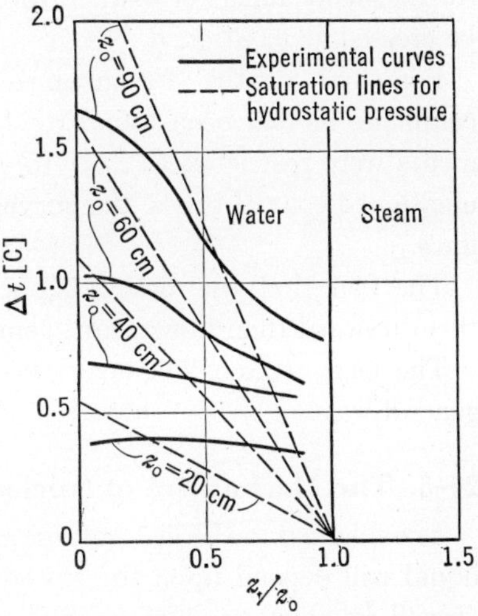

FIG. 29–6. Influence of height of water level above horizontal heating surface upon the temperature distribution in nucleate boiling.

From Fritz and Homann [1936].

The dotted straight lines in the graph show the temperature differences for equilibrium conditions due to the hydraulic pressure. Actually the water temperature does not follow these lines at all, but a very odd temperature distribution exists. Starting from about 6.5 C and sharply decreasing with increasing z/z_0, the temperature curve, when drawn from $z/z_0 = 0$ on, would twice intersect the equilibrium lines, indicating that close to the water level ($z/z_0 = 1$) the water is superheated in any case, namely, about 0.3 C for the water depth $z_0 = 20$ cm and 0.75 C for $z_0 = 90$ cm.

This strange temperature distribution is caused by the convection, which drives superheated water from the heating surfaces up to the free water surface; although it is cooled in this way, it arrives at still higher than equilibrium temperature, which is less at the water level than at the heating surface. The cooler upper water, on the other hand, is moved toward the heating surface; though it is heated in this way, it arrives at still lower than equilibrium temperature in the vicinity of the heating surface. Finally, in a thin layer above the heating surface the convection is negligible, and the small conductivity of water is responsible for the steep change of temperature in which the equilibrium lines would be crossed as mentioned.

For small depths (for instance $z_0 = 20$ cm) the mixture is almost complete, as had been found by Jakob and Fritz [1931], whose experiments were restricted to $z_0 \gtreqless 15$ cm. Damping the convection by the mentioned coaxial tubes, the curves of Δt come closer to the equilibrium lines.

The fact that the circulating water carries steam bubbles, more of them upward than downward, complicates the process still more.

The next section will deal with the role of the vapor bubbles in a general way.

29–5. The Mechanism of Nucleate Boiling

As explained in the last section, the degree of superheating of a boiling liquid will depend upon the statistical mean of the radius of curvature of wall roughnesses or of bubbles of gas adhering to the walls on which the liquid can evaporate. According to this idea, a rapidly boiling liquid could not be greatly superheated, since the bubbles arising on such surface curvatures would disturb the state of high surface tension which is necessary for high superheating, and the convection of the rising bubbles would equalize the temperatures. On the other hand, heat transmission from the heating surfaces through the liquid to the bubbles is necessarily bound to differences of temperature in the liquid.

From such considerations the slight superheating observed in our experiments could be understood qualitatively. A quantitative treatment of the heat transfer to a bubble from the liquid surrounding it is owed to Bošnjaković [1930]. He considered the fact that for a time element $d\tau$ the product of the mass increase dm of a growing bubble and of the latent heat λ of vaporization must be equal to the heat energy dQ conveyed to the bubble. The same heat energy will pass the surface area A_b of the bubble under the action of the difference Δt between liquid outside and vapor inside the bubble, and a coefficient of heat transfer h_b. Hence

$$dQ = \lambda \cdot dm = h_b A_b \cdot \Delta t \cdot d\tau$$

and, introducing the time rate of heat flow, q,

$$\Delta t = \frac{q}{h_b A_b} = \frac{\rho_g \lambda}{h_b A_b} \cdot \frac{dV_b}{d\tau} \tag{29-5}$$

where ρ_g = the density of the saturated vapor, and
V_b = the volume of the bubble.

By means of this equation the coefficient h_b can be calculated from the temperature excess Δt, the surface area A_b of the bubble at any time τ, and the volume increase per unit time.

Introducing a length element dr, defined by

$$dV_b = A_b \cdot dr$$

and denoting by z the height of the gravitational center of the bubble above a horizontal heating surface, Eq. 29-5 converts to

$$\Delta t = \frac{\rho_g \lambda}{h_b} \cdot \frac{dr}{d\tau} = \frac{\rho_g \lambda}{h_b} \cdot \frac{dr}{dz} \cdot \frac{dz}{d\tau} = \frac{\rho_g \lambda}{h_b} v_b \frac{dr}{dz} \tag{29-6}$$

where v_b = the vertical component of the velocity of the bubble. If the bubble is spherical, r is the radius of the sphere.

The author and coworkers [Jakob 1932, 1936; Fritz and Ende 1936], using a high-speed camera, performed experiments by which A_b and $dV_b/d\tau$ and therefore also v_b and dr/dz for the actual shapes of the bubbles were obtained. First, however, Jakob and Fritz [1931] studied the influence of the heating surface upon the formation of steam bubbles. For this purpose smooth surfaces, chromium plated and polished, were compared with surfaces roughened by means of a sand blast or fitted with square cavities by cutting grooves in the surface, as described above; these may be called screen surfaces. Three typical shapes of steam bubbles on the heating surface were obtained (see Fig. 29-7).

Case *a* occurs on unwetted surfaces. Such a surface was produced by covering the screen surface with a thin layer of oil. Case *b* (partial wetting) is obtained by evaporation from a smooth surface. Case *c* (total wetting) is produced on the clean screen surface, since the roughness counteracts the effect of surface tension.

Bubbles of the first type spread themselves out on the unwetted surface. The edge of the bubble is drawn into a fine wedge between liquid and heating surface and thus is held fast against the action of buoyancy. Therefore, the bubble gets rather large (up to about ⅓ in.) before being detached, and is replaced at once by a new one. In this way mercury will boil if no wetting ingredient is used. Because of the relatively large interface between surface and vapor, more of the heating energy will be transmitted immediately to the bubble than in the other cases, and the vapor will be slightly superheated. The contact area between heating surface and liquid, on the other hand, is too small for a strong heat transfer, and the temperature drop on the heater surface will be relatively large.

The boiling on smooth surfaces (type *b*) is entirely different from the former type. The bubbles originate at only a few small spots of the plate which do not seem to be distinguished from the rest of the surface and are irregularly distributed over it. One has the impression that columns of steam are issuing from these spots and become larger as they rise.

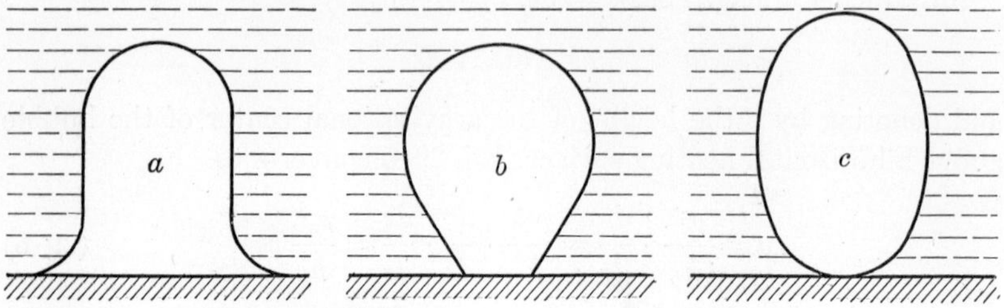

FIG. 29-7. Typical shapes of steam bubbles.

Photographs show that these columns actually consist of a great number of steam bubbles, one rising after the other. The number of these spots increases with the load; for example, on a horizontal plate having a diameter of 4 in., 5 such spots were counted (all of them at the edge of the disc), the heat flow being $q'' = 7500$ kcal hr^{-1} m^{-2}; 8 places were counted (5 of them at the edge) for $q'' = 15{,}000$ kcal hr^{-1} m^{-2}, and 24 places (again 5 at the edge) for $q'' = 30{,}000$ kcal hr^{-1} m^{-2}.

This has been explained [Jakob 1936] by considering the influence of the thin layer of liquid near the heating surface, which is more highly superheated than the main mass of the liquid and whose temperature rises with the rate of heating energy as has been shown in Sect. 29-4. Since the bubble originates in this layer, it is, in the first stage of its development, exposed to an environment which becomes hotter when the heating energy is increased. Therefore more heat energy will be transmitted to its surface, which is at first extremely small and is in contact with the heating surface. Thus the bubble will increase and be torn off more quickly, so that the element of roughness in question will be more often available for the formation of a new bubble. On the other hand, when the liquid close to the heating surface is superheated to a higher degree, vapor bubbles can originate at smaller curvatures of the surface elements, which are formed by roughnesses or traces of gases, adsorbed on the surface or suspended in the liquid and getting free in the form of minute bubbles. Therefore such places of smaller curvature are coming into action, and more and more columns of vapor bubbles start when the rate of heating increases.

As an average, and in the range up to about 18,400 B hr^{-1} ft^{-2}, it could be concluded [Jakob 1941] that for every increase of the heat flow

FIG. 29–8. Boiling of water on a horizontal surface covered with oil.

Time of exposure: $\frac{1}{200}$ sec.

FIG. 29–9. Boiling of water on a clean horizontal surface.

Time of exposure: $\frac{1}{200}$ sec.

by 700 B hr^{-1} ft^{-2} one more steam column was formed on a polished horizontal heating surface. From this it is easily calculated that each column of steam bubbles received about 59 B/hr, and that 0.061 lb steam of atmospheric pressure was formed per hour in each column.

The effect of forced convection produced by the rising bubbles will be dealt with in Sect. 29–6.

In the case of bubbles of type *c* (see Fig. 29–7) the liquid tends to push away and to shear off the bubble, which rests at only one point and becomes globular or oval and leaves the surface while still very small.

The photographs reproduced in Figs. 29–8 and 9 were taken with the same rate of heat flow on the screen surface mentioned, the first when it was covered with oil, the second when it was clean. In both cases the time of exposure was 1/200 sec. Figure 29–8 shows a comparatively small number of big bubbles (type *a*) so that the fluid appears dark. In the

Fig. 29–10. Boiling of water on a smooth vertical surface.

Time of exposure: 1/100,000 sec.

Courtesy Dampney Company of America.

Fig. 29–11. Boiling of water on a vertical surface painted with a graphited mass.

Time of exposure: 1/100,000 sec.

Courtesy Dampney Company of America.

second case, on the contrary, a great number of bubbles of type *c* was formed, which caused the visual field to appear bright.

Figures 29–10 and 11 are reproductions of photographs taken in 1/100,000 sec by the light of a 16,000-volt spark.§ Each of them shows

§ The author is indebted to the Dampney Company of America, Boston, Massachusetts, for having secured him these photographs.

a short piece of $1\frac{1}{2}$ in. outer-diameter steel boiler tube, which contained a 300-watt electric cartridge-type heater element fitting snugly into the tube and closed by a brass disc at the bottom. The tube in Fig. 29–10 was smoothed and that of Fig. 29–11 was uniformly painted with a graphited mass called Apexior No. 1. On the first one evaporation occurred at only a few places, the columns of steam being clearly resolved into single bubbles, following each other. The second photo shows that the rough painted surface gave rise to the development of bubbles at many places, probably owing to many elements of roughness formed by the graphite and the gas adhering to it.

The first picture also shows the increase and change in the shape of the rising steam bubbles if these are compared with succeeding ones in the steam column. This can be still better recognized in an illustration (Fig. 29–12) obtained by means of the high-speed-camera experiments

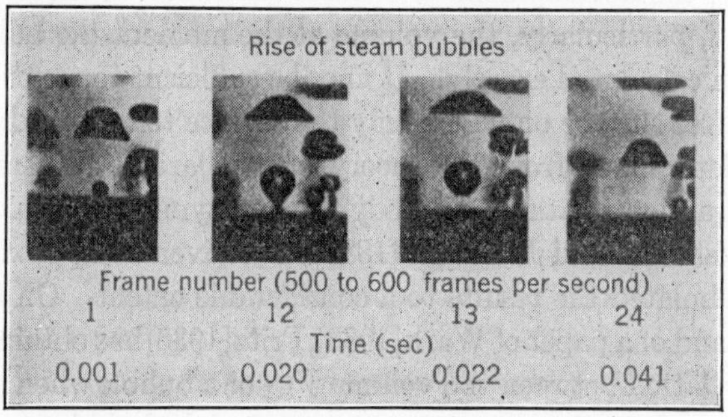

FIG. 29–12. Formation and rise of steam bubbles.

mentioned. The film reel, moving continually, was exposed by means of four converging lenses which revolved with the film so that a picture stood still against the moving film reel during the exposure. The frequency of pictures was 500 to 600 per second.

The steam bubbles were produced by a horizontal electrical heater inside a glass vessel having plane parallel walls. They were illuminated from behind by an arc lamp, and a screen of oilpaper was placed between the lamp and the vessel, in order to obtain diffused light. Thus the bubbles gave distinct black images on a bright background.

The illustrations shown in Fig. 29–12 were taken with a heating flux ($q'' = 19{,}300$ kcal hr^{-1} m^{-2}) which produced eight columns on a horizontal smooth disc of 4 in. diameter. Four out of 24 frames taken in 0.04 sec are reproduced to show some characteristic phases of two bubbles. In Frame No. 1 a bubble is forming at the heating surface; 0.02 sec later (Frame No. 12) it is about to leave the base; 0.002 sec still later

(Frame No. 13) it is free. The tail of the bubble, seen in No. 12, has jerked upward and indents the lower portion of the bubble surface. The surface tension, together with the flow of the rising bubble, forces the bubble into the form of a steel helm (Frame No. 24). In Frame No. 1 the preceding bubble will also be seen, looking like a hat. In the two following frames, this hat is flattening. In Frame No. 24 it has passed out of the field of view.

To understand the mechanism of nucleate boiling one has to distinguish between what happens as long as a vapor bubble adheres to the heating surface and how it grows after breaking off that surface. These two phases differ from each other by the acting differences of temperature as well as by the manner of the hydrodynamic motion.

The growth of the bubble at rest could be observed with sufficient exactness only on rather large, and not with small, bubbles. But, fortunately, the development of the bubble at rest is theoretically well known; and, particularly, the volume at the moment the bubble breaks off has been calculated exactly. It may be sufficient to mention that the shape of a free surface on which only the surface tension and gravity are acting is determined from the theory of capillarity. Whereas the differential equation relating to a body of axial symmetry could be solved exactly, Bashforth and J. Adams [1883] have given an approximate solution and tabulated the results to a considerable extent. On the basis of their work and of a paper of Wark[1933], Fritz[1935] has obtained a simple empirical relation between the volume $V_{b,1}$ of a bubble when breaking off the heating surface, the outer contact angle β of the bubble at this instant, and Laplace's constant, which is defined by

$$a = \sqrt{\frac{2\sigma}{g(\rho_f - \rho_g)}} \tag{29–7}$$

In the range of $\beta = 0$ to 140° this relationship can be expressed by the empirical equation

$$\sqrt[3]{V_{b,1}}/a = 0.0119\beta \tag{29–8}$$

with β in angular degrees. The quantity $\sqrt[3]{V_{b,1}}/a$ is dimensionless as can easily be seen.

Herewith Fritz has calculated values of the maximum volume $V_{b,1}$ for different fluids and angles. For instance, when $\beta = 100°$, it would be

95 cu mm for an air bubble in water at 20°C.

75 cu mm for a steam bubble at atmospheric pressure.

25 cu mm for a steam bubble at 50 atmospheres or a bubble of mercury vapor at atmospheric pressure.

10 cu mm for a steam bubble at 100 atmospheres or a bubble of carbon tetrachloride vapor at atmospheric pressure.

Figure 29–13 shows a straight line according to Eq. 29–8 and a number of experimental points. The small circles refer to angle measurements on photographs of steam bubbles made by the author's former coworker, H. Kipcke. The numerous crosses refer to hydrogen bubbles, formed in greatly diluted solutions at 20°C, according to measurements by Kabanow and Frumkin [1933]. The deviation between the theoretical curve

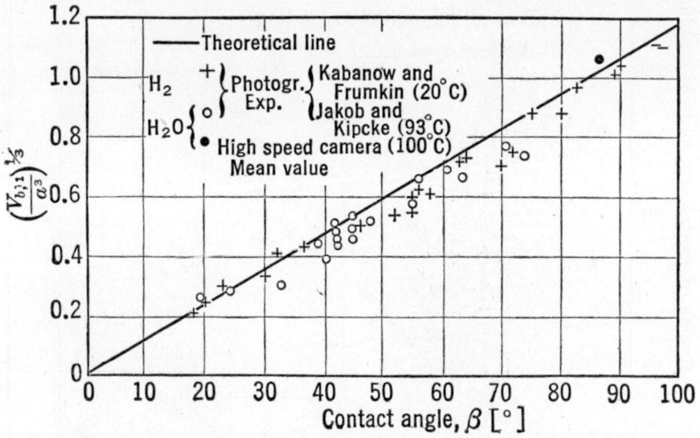

FIG. 29–13. Volume of vapor and gas bubbles when breaking off the heating surface.

and the experimental points may partly be caused by disturbances of the equilibrium in the experiments; considerably larger disturbances will occur in vigorous ebullition.

Coming now to the evaluation of the moving pictures of bubbles, it may be mentioned that pictures of three hundred steam bubbles were enlarged and analyzed. The paths of the bubbles and the times required were calculated. An example [Jakob 1932] is given in Fig. 29–14. In the time $\tau_0 \approx 1/40$ sec, during which the bubble adheres to the heating surface, its center of gravity rises with almost the same velocity with which the bubble later rises when it leaves the surface. The mean speed of rise was $v_{b,m} = 23$ cm/sec in this case. The average velocity of six bubbles, having diameters between 1 and 8 mm, was found to be $v_{b,m} =$ 26 cm/sec.

According to Fig. 29–14, new bubbles originated at intervals of about $1/20$ sec, i.e., with the frequency of 20 per second. After each bubble broke off (marked by a vertical dotted line), a pause of about $1/40$ sec, i.e., half of a period, took place, until a new bubble started to be developed. This delay occurs because the spot where the bubbles originate is cooled slightly by water when a bubble has left and must be heated again to the temperature needed for formation of a new bubble.

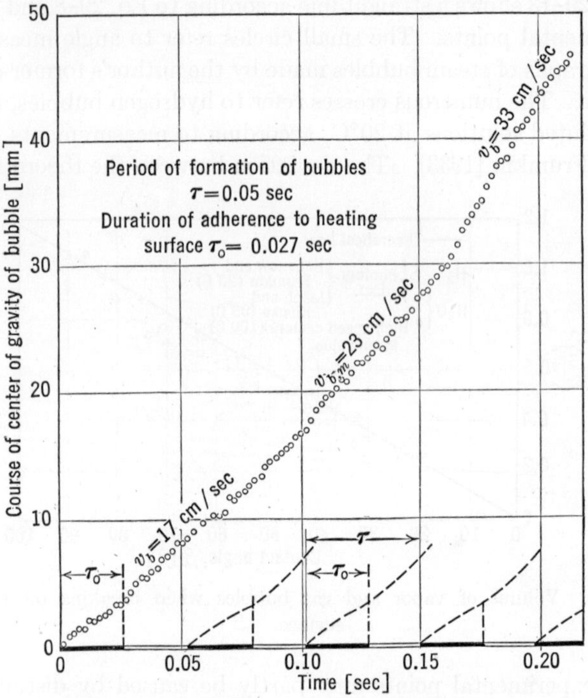

Fig. 29–14. Rise of steam bubbles versus time.

The frequency f of bubble formation depends on the size of the bubbles at the moment of breaking off. In Fig. 29–15 the frequency is plotted versus the diameter of a sphere of the same volume as the bubble had when leaving the heating surface (Curve a). The circles belong to observations on steam bubbles of 0.5 to 55 cu mm volume at the time of breaking off. The crosses are statistical means of frequency, according to stroboscopic observations on water and carbon tetrachloride. A hyperbola, roughly representing the relation between diameter and frequency, is drawn as Curve b.

The surface area A_b and the volume V_b of the steam bubbles were obtained by graphical integration of enlarged pictures taken with the high-speed camera. The volume development of two bubbles is represented in Fig. 29–16. The upper curve refers to a large bubble which grew to 52 cu mm on the heating surface and then increased to about 200 cu mm as it rose in the water. When resting on its base the bubble grew much faster than when rising. This effect is partly due to the high temperature of the water layer bordering immediately upon the heating surface, and partly to the large ratio of surface area to volume of the still small bubble. Curve *b* refers to a bubble of mean size, all points being taken in the rising stage.

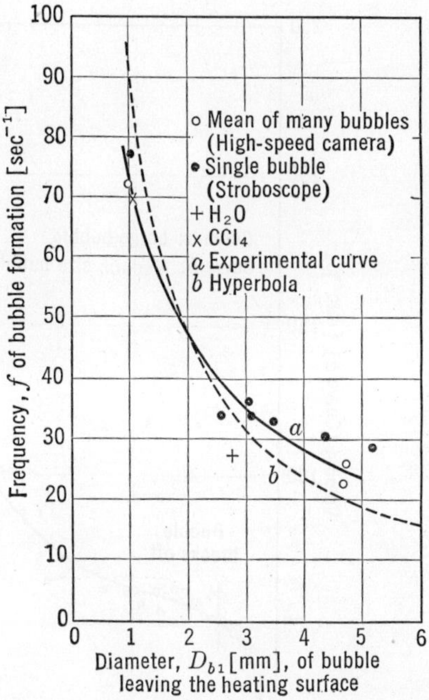

FIG. 29–15. Frequency of vapor bubble formation.

Since the temperature of the steam above the water-level surface and that of the water from the heating surface up to the water-level surface were exactly measured, all data that enter Eq. 29–5 were known, and it was possible to calculate the course of the coefficient of heat transfer h_b on the surface of the growing and rising bubble. When the bubble originates, the coefficient h_b is amazingly high. Values of about 40,000 B hr^{-1} ft^{-2} F^{-1} have been observed. Within $\frac{1}{100}$ sec, however, h_b decreased to about 4000 B hr^{-1} ft^{-2} F^{-1}, and then remained at about 3000 for the main part of the $\frac{1}{5}$ sec of observation in which the free water level, 2 in. above the heating surface, was reached. This behavior has been explained [Jakob 1936] as follows.

Generally, a vapor bubble will be surrounded by a very thin boundary layer of liquid across which the temperature decreases from the higher value existing in the superheated liquid to the saturation value. Then h_b may be expressed approximately by

$$h_b = \frac{k}{b} \tag{29–9}$$

where k is the thermal conductivity of the liquid, and b is the thickness of the boundary layer. But at the moment when the bubble originates, such a layer does not exist yet. By the explosion on a microscopic scale which produces the bubble, the temperature of its surface decreases immediately from the higher water temperature to the lower saturation temperature so that b will be extremely small and h_b enormously large. Immediately, a cold wave in radial outward direction sets in, and the temperature gradient at the inner surface of the bubble decreases quickly.

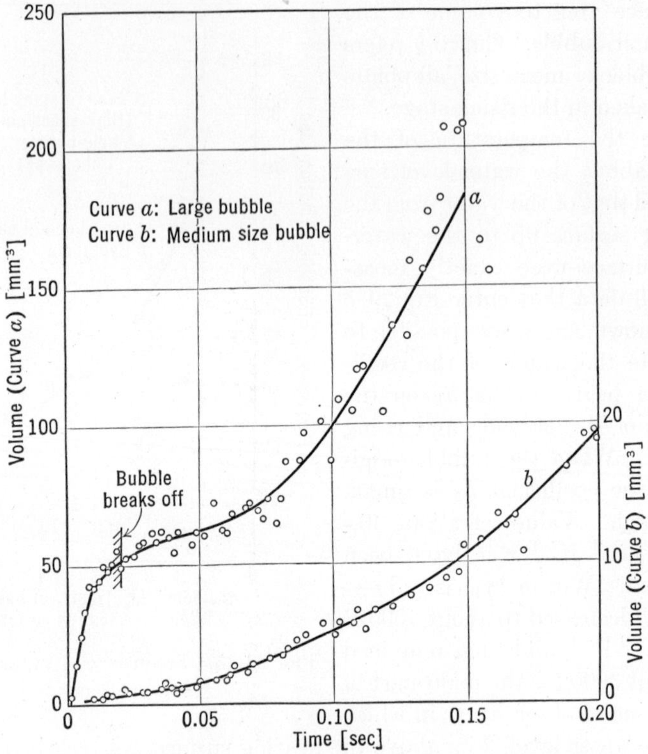

Fig. 29–16. Volume development of two steam bubbles.

Numerical calculation [Fritz and Ende 1936] led to a qualitative agreement between the values of h_b calculated according to this theory, and those obtained from the high-speed camera measurements.

Summarizing our knowledge of the mechanism of nucleate boiling on a clean, smooth or rough, heating surface, it can be said that only a very small part of the heat produced in a heater is directly transferred to the interior of bubbles adhering on the surface. The main part of the energy makes a detour through the liquid. The prerequisite of boiling is formation of a thin, considerably superheated layer of liquid on the heating surface and slight superheating of the bulk of the liquid. Vapor bubbles originate on roughnesses or from gas bubbles which exist at the heating surface. The bubble develops first by evaporation because of the considerable temperature excess of the liquid layer mentioned over the saturation temperature. The bubble breaks off when its volume has grown so much that the buoyancy exceeds the capillary forces which bind it to the heating surface. In the ensuing rise of the bubble through the bulk of the liquid the excess of liquid temperature and the coefficient of heat transfer on the bubble surface are smaller. However, this surface and the time available for heat transfer are so much greater that the vapor formation is largest during the free rise of the bubble (see Fig. 29–16). Vapor bubbles subsequently originating on the same spot form a sort of swaying column; with increasing heating energy and temperature excess spots of smaller roughness satisfy the capillary conditions of bubble formation and start as new sources of vapor columns. The influence of forced convection exerted by these vapor columns is dealt with in the next section which, in general, will be devoted to the results of direct measurements of the total heat transfer from heating surfaces to boiling liquids.

29–6. Heat Transfer at Atmospheric Pressure in Nucleate Boiling from a Submerged Heater

The analysis of the mechanism of nucleate boiling revealed that most of the heat is first carried to the liquid and only then to the vapor bubbles. This fact can be expressed by the following double heat balance:

$$q = hA(t_s - t_f) = \Sigma[h_b A_b(t_f - t_g)] \tag{29–10}$$

where q = the rate of heat given up by the heating surface,
h = the mean film coefficient of heat transfer on the heating surface,
t_s = the temperature of the heating surface,
t_f = the temperature of the bulk of the liquid,
Σ = a sum sign to be extended over all bubbles existing simultaneously in boiling under steady-state heating conditions,
h_b = the film coefficient of heat transfer on the surface of one bubble,
A_b = the surface area of that bubble, and
t_g = the temperature of saturated steam.

However, in the second term of Eq. 29–10, that part of heat which is transferred to bubbles existing in the thin warm layer on the heating surface is not separately considered, but included in the definition of h. On the other hand, the high temperatures in this layer instead of t_f should be used in the third term of the equation for the bubbles present in that layer. In other words, Eq. 29–10 has the character of a practical approximation rather than that of an entirely precise relation.

The first part of this equation can also be written as

$$q'' = h\theta_s \tag{29–11}$$

where $q'' = q/A$ and $\theta_s = t_s - t_f$. Most of the practical experiments and correlations are based on Eq. 29–11.

The influence of roughnesses and gases adsorbed on the heating surface is illustrated by Figs. 29–17 and 18 [Jakob and Fritz 1931; Jakob 1936] which are two different plots of the same experimental results.|| The conditions under which the curves of these graphs were obtained are explained in Table 29–1.

TABLE 29–1

Influence of Surface Conditions upon Heat Transfer in Boiling

(Curves 1 to 7 in Figs. 29–17 and 18)

Curve No.	State of Heating Surface
1	Recently roughened by a sand blast
2	Same, oxidized by longer use
3	Same, roughened once more by sand blast
4	Same, after long use
5	Recently fitted with a roughness screen and cleaned
6	Same, after 4 hours of boiling, followed by 24 hours of lying in water
7	Same, after 8 more hours of boiling, followed by another 24 hours of lying in water

In Curve 5 the coefficient h nearly reached 14,000 kcal hr^{-1} m^{-2} C^{-1}; this may partly be due to the 80 per cent increase of area as a result of the grooves which form the screen.¶ Curves 6 and 7 in Fig. 29–17 are above Curve 5, probably because the number of air particles, adsorbed

|| In his book on heat transmission McAdams [1942] demonstrates that of the three possibilities of plotting (h versus q''; θ_s versus q''; and h versus θ_s) the last one shows existing discrepancies most distinctly, the first one reveals them least. This comes from the fact that the first means plotting q''/θ_s versus q'', the last plotting q''/θ_s versus θ_s. However, the scattering in almost all known tests and correlations is so great that the more usual first two sorts of representation are sufficient in most cases; they make correlations easier because deviations do not show up so much, which sometimes would be more confusing than revealing. The present author generally uses the first two sorts of plotting, his preference being the second one, in which none of the pertinent quantities, q'', θ_s, and h, appears simultaneously as a factor or divisor in both abscissa and ordinate values.

¶ This increase was not considered in A when h was calculated by means of Eq. 29–10.

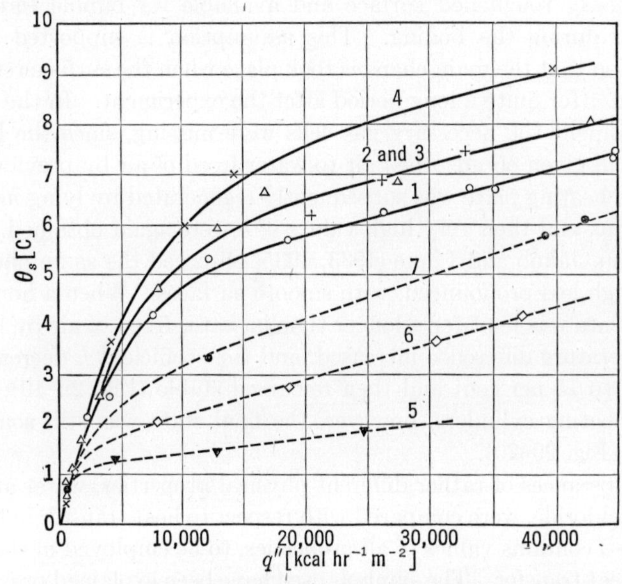

FIG. 29–17. Influence of roughness upon temperature difference between horizontal heating surface and boiling water.

For explanation of surface conditions, see Table 29–1.

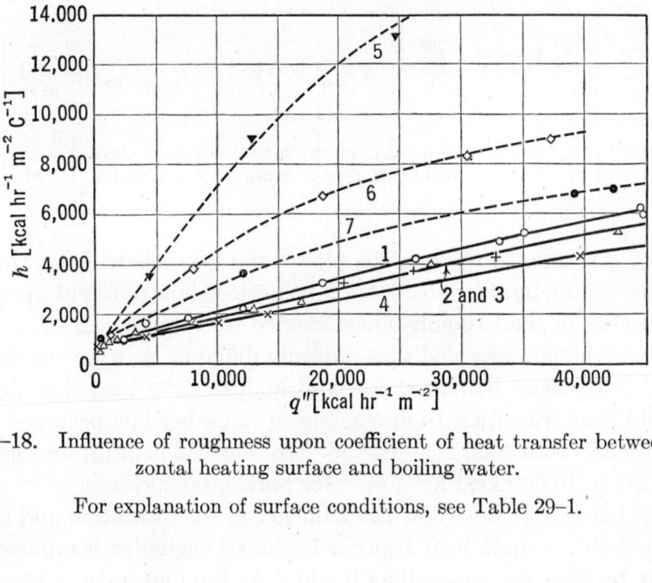

FIG. 29–18. Influence of roughness upon coefficient of heat transfer between horizontal heating surface and boiling water.

For explanation of surface conditions, see Table 29–1.

by the newly roughened surface and available for bubble formation, decreased during the boiling. This assumption is supported by the observation that the main changes took place when the surface remained under water for quite a long period after the experiment. In the following experiment the necessary gas cells were missing, since the heating surface had given off adsorbed air to water freed of air by previous boiling. The heating plate was subsequently regenerated by lying in air for a long time, and then very high values of h were again obtained.

Later on, Jakob and Linke [1933, 1935] observed the same phenomenon, though less pronounced, with smooth surfaces. When a horizontal heating plate was kept for a longer time in water freed of air by boiling, the temperature difference increased, and the coefficient h decreased by about 10 to 15 per cent and then remained stable (Fig. 29–19). With vertical heating cylinders, however, the final state was still somewhat unstable (Fig. 29–20).

Two substances of rather different physical properties, water and carbon tetrachloride, were compared with respect to heat transfer in boiling. Table 29–2 contains values of all properties, to be employed in the equations of heat transfer. The symbols used have been explained previously;

TABLE 29–2

PROPERTIES OF WATER AND CARBON TETRACHLORIDE AT STANDARD SATURATION TEMPERATURE

Substance	t_g [°C]	ρ [kg m^{-3}]	β [C^{-1}]	$\mu \cdot 10^9$ [kg hr m^{-2}]	c_p [kcal kg^{-1} C^{-1}]	k [kcal hr^{-1} m^{-1} C^{-1}]	N_{Pr}	v_g [m^3 kg^{-1}]	λ_g [kcal kg^{-1}]	$\sigma \cdot 10^3$ [kgm^{-1}]
H_2O	100	959	750	8.0	1.011	0.588	1.75	1.675	539	6.00
CCl_4	76.7	1483	1224	13.8	0.209	0.078	4.2	0.182	46	2.06

subscript g relates to the gaseous phase; the symbols without this subscript are for the liquid state. It will be noticed how different are almost all properties of the two substances elected for comparison.

The experiments revealed that radically different laws govern the heat transfer in nucleate boiling at low and high rates of heat flow per unit area, and that transition from the one to the other law occurred in the range from 2000 to 4000 kcal hr^{-1} m^{-2} for vertical heating surfaces, and from 5000 to 10,000 kcal hr^{-1} m^{-2} for horizontal surfaces.

At the low rate of heat flow the influence of the formation and motion of vapor bubbles upon heat transfer is almost negligible compared with the heat transfer for non-boiling liquid. At the high rate of heat flow,

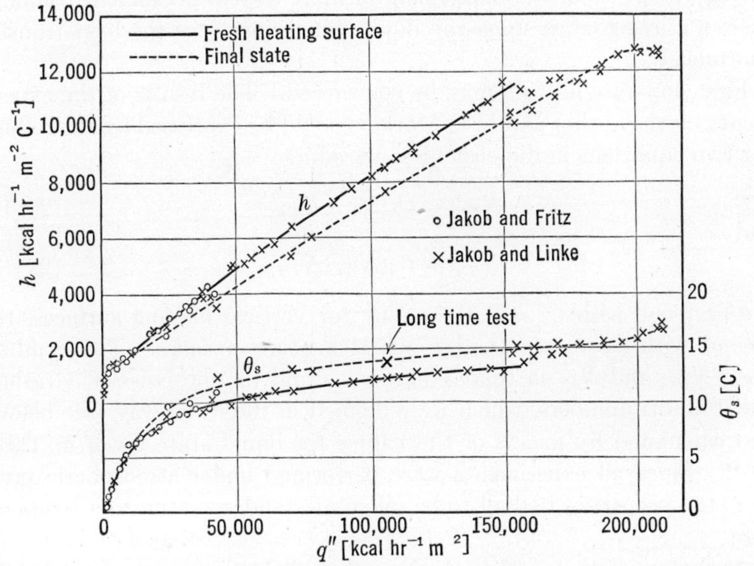

FIG. 29-19. Coefficient of heat transfer h and temperature drop θ_s on a smooth horizontal heating surface in boiling water.

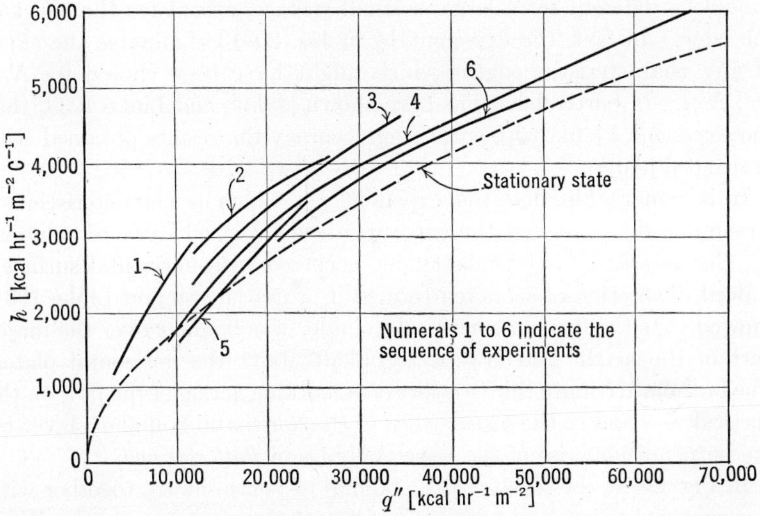

FIG. 29-20. Coefficient of heat transfer h on a smooth vertical heating surface in boiling water.

however, the vapor bubbles, rising in more or less permanent columns, exert a stirring effect upon the liquid which increases the heat transfer enormously.

First, low-rate heating may be considered. The results of the experiments by the author and his coworkers could be expressed by the following two equations in dimensionless variables:

$$N_{Nu} = C_1(N_{Gr} \cdot N_{Pr})^{1/4} \tag{29-12}$$

and

$$N_{Nu} = C_2(N_{Gr} \cdot N_{Pr})^{1/3} \tag{29-13}$$

the first one with $C_1 = 0.61$ holding for vertical heating surfaces, the second with $C_2 = 0.16$ for horizontal heating surfaces. The symbols N_{Nu}, N_{Gr}, and N_{Pr} in these equations stand for the Nusselt, Grashof, and Prandtl numbers, which were defined in the usual way (see below) and calculated by means of the values for liquid state given in Table 29–2. Since all experiments were performed under atmospheric pressure, the properties had all to be taken at standard saturation temperature.

As characteristic length in the Nusselt and Grashof numbers the height of the vertical surface was used. For horizontal surface, however, the same reasoning as previously applied leads to the conclusion that no length influence can exist, since it would not be understandable why q'' should be different for a large or small surface, except for the effect of the edge. In fact, the exponent $\frac{1}{3}$ in Eq. 29–13 eliminates the effect of any characteristic lengths which might have been chosen for N_{Nu} and N_{Gr}. In particular it has been shown [Jakob and Linke 1933] that the exponent $\frac{1}{3}$ fits with surprising accuracy the results obtained with horizontal plates.

It is remarkable that the exponent $\frac{1}{4}$, which is characteristic for streamline flow, covered the experiments on vertical surfaces, whereas $\frac{1}{3}$, characteristic for turbulent flow, occurred with horizontal surfaces. Indeed, inspection of schlieren formed in water [Jakob and Linke 1935] showed that the flow on a vertical cylinder was laminar over the major part of the height and always turbulent above the horizontal plates. The authors attribute this to a sort of smoothing action of the flow on the vertical wall and to the disturbance of the horizontal boundary layer by the perpendicular rise of the vapor bubbles in *statu nascendi*.

In Fig. 29–21 the results of the boiling tests are shown, together with experimental points and two straight lines, taken from a paper by W. J. King [1933]. The boiling experiments are all in the range of $N_{Gr} \cdot N_{Pr} = 10^8$ to 10^{11}.

Starting with the left ends of the four boiling curves of the graph, it is seen that the slope of those for a vertical surface is virtually ¼, and that for a horizontal surface is ⅓, like the slope of the two straight lines of King. However, they are a little higher, the constant factors of Eqs. 29–12 and 13 being $C_1 = 0.61$ and $C_2 = 0.16$, to be compared with the corresponding constants of King, $C_1 = 0.56$ and $C_2 = 0.13$.

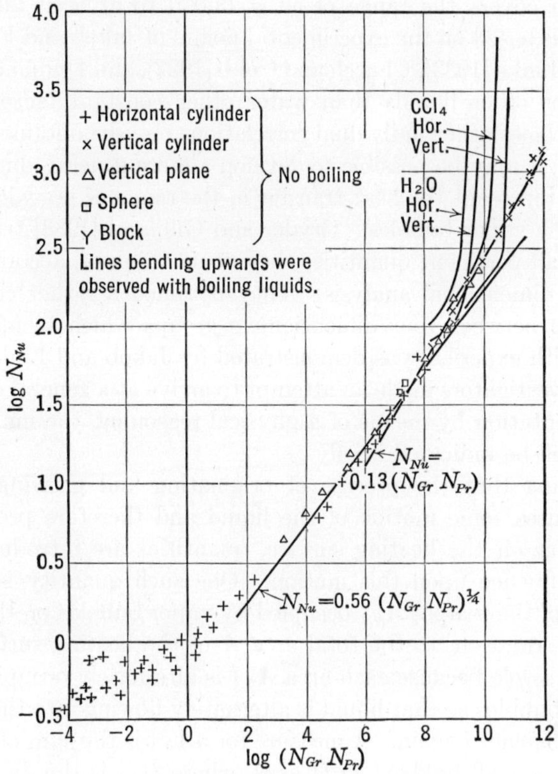

FIG. 29–21. Dimensionless representation of heat transfer (suitable within range of free convection).

This is probably due to a slight stirring effect of the vapor bubbles which adds to the main free convection of the liquid.

It would be possible to replace the left parts of the two curves for a vertical surface by a unique straight line and those for a horizontal surface by another straight line, which means that the dimensionless equations 29–12 and 13 are independent of the substance.

At higher heating load, the boiling curves suddenly split individually and rise very steeply, indicating that the two equations do not hold any more.

It is difficult to determine these slopes exactly. For water boiling over a horizontal surface, Fritz [1937] recommends the equation

$$h = 1.48(q'')^{0.75} = 4.80\theta_s^{\,3} \tag{29–14}$$

with θ_s in C, q'' in kcal hr^{-1} m^{-2}, and h in kcal hr^{-1} m^{-2} C^{-1}. This equation, which covers the range of $q'' = 13{,}000$ to at least 200,000 kcal hr^{-1} m^{-2}, is based on the experimental points of Jakob and Fritz [1931], Jakob and Linke [1933], Church and Cobb [1922], and Dunn and Vincent [1931]. For other liquids than water other constant factors and for vertical surfaces other individual correlations may be obtained. On the other hand, it may be possible to develop a more general, dimensionless relation to represent the heat transfer in the range of prevalent stirring action of the vapor bubbles. Cryder and Gilliland [1933] tried this by combining all pertinent quantities they could think of, according to the method of dimensional analysis. They obtained a rather complicated relationship between four dimensionless groups which seems to be at variance with experience as demonstrated by Jakob and Linke [1935].

These investigators made an attempt to arrive at a general dimensionless representation by means of a physical reasoning, the main features of which will be indicated briefly.

Considering that the process of originating and growing of vapor bubbles causes some motion of the liquid and therefore promotes the heat transfer on the heating surface, quantities are introduced which have an influence upon this motion. One such quantity is the ratio $(nA_{b,1}/A)$ of the n areas $A_{b,1}$ occupied by vapor bubbles on the n places of bubble formation, to the total area A of the heating surface. This ratio plays a role because each area $A_{b,1}$ is alternately occupied by and free of the bubbles so that liquid is alternately flowing into the liberated space and displaced again. A measure for $nA_{b,1}$ is the sum of the cross-section areas of all bubbles, $n\pi D_{b,1}{}^2/4$, where $D_{b,1}$ is the diameter of a sphere which has the same volume as a bubble of average size when breaking off. A second influential quantity is the ratio of the volumes of the bubble when breaking off the heating surface and when breaking through the free liquid surface, $V_{b,1}/V_{b,2}$, because this ratio for a fixed value of $nA_{b,1}/A_1$ is a measure of the continual displacement of liquid which produces a circulation of water along each vapor column. Hence, the following relationship was set up:

$$N_{Nu} = \frac{hD_{b,1}}{k} = \Psi\left(\frac{nA_{b,1}}{A} \cdot \frac{V_{b,2}}{V_{b,1}}\right) \tag{29–15}$$

where h is the coefficient of heat transfer between heating surface and liquid, and $D_{b,1}$ is taken as a characteristic length in the process of vehement boiling.

Introducing the diameter $D_{b,2}$ of the bubble at the free surface of the liquid and the frequency f of bubble formation in each vapor column,

$$\frac{nA_{b,1}}{A} \cdot \frac{V_{b,2}}{V_{b,1}} = \frac{3}{2} \cdot \frac{n}{A} \cdot \frac{(\pi/6)D_{b,2}{}^3 f}{D_{b,1} f}$$

But the total volume of vapor formed in unit time due to the heat energy taken from unit area of heating surface is

$$\frac{n}{A} \cdot \frac{\pi}{6} D_{b,2}{}^3 f = \frac{q''}{\rho_g \lambda}$$

where ρ_g and λ are the density of saturated vapor and the latent heat of vaporization, respectively. Hence

$$\frac{hD_{b,1}}{k} = \Psi\left(\frac{q''}{\rho_g \lambda} \cdot \frac{1}{D_{b,1} f}\right) \tag{29–16}$$

Jakob and Linke took photographs of air and steam bubbles in water on a smooth horizontal surface and found as an average $\beta \approx 50°$, for which, according to Fritz, $V_{b,1}/a^3 = 0.205$. Substituting for a from Eq. 29–7 and neglecting ρ_g compared to ρ_f,

$$D_{b,1} = 1.034 \sqrt{\frac{\sigma}{\rho_f}} \tag{29–17}$$

Using this equation and data given in Table 29–2, the values of Table 29–3 were obtained. Again the data for water and carbon tetrachloride

TABLE 29–3

QUANTITIES PERTINENT TO THE EVALUATION OF EQ. 29–16 FOR THE HEAT TRANSFER IN BOILING

Substance	$D_{b,1}10^3$ (Calculated by Eq. 29–17) [m]	$D_{b,1}10^3$ (From Photographs) [m]	f (From Stroboscopic Measurements) [1/sec]	$D_{b,1} \cdot f \cdot 10^3$ (From Columns 3 and 4) [m/sec]
H_2O	2.6	2.8	28	78.4
CCl_4	1.2	1.1	70	77.0

are very different, except for $D_{b,1} \cdot f$, which was virtually the same and therefore was assumed to be a constant which may be called w_b and amounts to 280 m/hr in the metric system of units.

Substituting this value and $D_{b,1}$ from Eq. 29–17 and including the factor 1.034 in the function Ψ without changing this symbol, Eq. 29–16 converts to

$$N_{Nu} = \frac{h}{k}\sqrt{\frac{\sigma}{\rho_f}} = \Psi\left(\frac{q''}{\rho_g \lambda w_b}\right) \qquad (29\text{–}18)$$

Herewith a relation is obtained which, as usual, presents h as a function of q'', all other quantities being properties of the boiling substance, except for the empirical constant w_b. This equation seems to have been the first attempt to express the heat transfer in boiling in terms of surface tension and other thermodynamical quantities, i.e., to introduce the capillary theory of vapor bubbles in the heat balance of boiling.

Using the measured values of h and q'', values of N_{Nu} were plotted versus $q''/(\rho_g \cdot \lambda \cdot w_b)$ in bilogarithmic coordinates (Fig. 29–22). It is

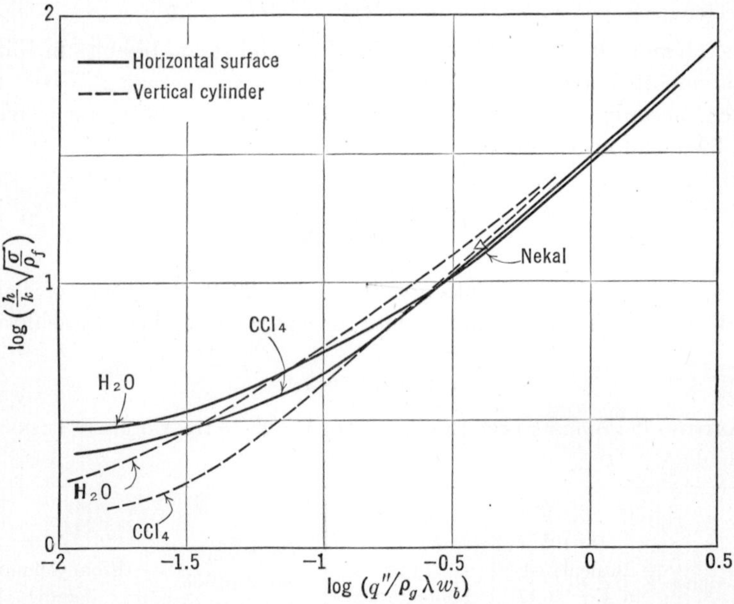

Fig. 29–22. Dimensionless representation of heat transfer in boiling (suitable in range of stirring effect of vapor bubbles).

seen that in the range of prevailing stirring action of the vapor bubbles, i.e., for $q''/(\rho_g \cdot \lambda \cdot w_b) > 10$ the lines for water and carbon tetrachloride, for both horizontal and vertical heating surfaces, approach each other, whereas they were split up in four quite distinct lines in Fig. 29–20. A common empirical equation * for the four lines is

$$N_{Nu} = \frac{h}{k}\sqrt{\frac{\sigma}{\rho_f}} = 30\left(\frac{q''}{\rho_g \lambda w_b}\right)^{0.8} \tag{29–19}$$

a special form of Eq. 29–18.

In Fig. 29–22 one point is shown which has been taken after adding less than 1% of an organic "wetting" solution (called Nekal BX) to water, which reduced the surface tension by about 50% without sensibly affecting any other pertinent property of the water. This led to a considerably smaller value of h for the same abscissa, but to no change of the ordinate, thus confirming the validity of Eq. 29–19.

It may also be mentioned that experiments of Kaiser [1937], carried out on a brewing pan of 15-ft diameter and covering the range of $q'' =$ 9000 to 50,000 B $\text{hr}^{-1}\,\text{ft}^{-2}$, are in excellent agreement with this equation, based on experiments with a heating surface only $\frac{1}{2000}$ as large. This proved in particular that the dimensions of a horizontal heating surface do not influence the heat transfer, exactly as has been concluded in the discussion of Eq. 29–13 for moderately strong boiling.

However, Eq. 29–19, and the underlying theoretical idea, should be considered a start rather than a final solution of the problem. In particular, the assumption of an average value of 50° for the contact angle β is a rough approximation only [see Jakob 1938]. Variation of β may change very considerably the constant factor, which has been assumed to be 30 in Eq. 29–19. Whenever superheat is observed, the angle β will be smaller than 90°, as Larson [1945] has demonstrated theoretically. This seems to be an upper limit for practical application, except for certain metals (for instance, aluminum, zinc, and magnesium) on which appreciable undercooling instead of superheating has been observed by Larson; however, in his experiments the metal surface was not a heating surface. At all events, a more thorough study of contact angles in boiling liquids will be necessary for further development of the theory of nucleate boiling.

A careful experimental investigation with water, carbon tetrachloride, isopropanol, and a 40% sucrose solution was performed by Insinger and Bliss [1940]. They used a vertical heating cylinder, having almost the same length (6 in.), diameter (about $1\frac{1}{4}$ in.), and hydraulic starting and

* Jakob and Linke's original derivation contains some inaccuracies which have been corrected in this presentation. However, the final result is the same as in the original paper.

trailing sections as Jakob and Linke's [1935] heater. The result of these experiments was expressed by the empirical equation

$$\log Y = 0.363 + 0.923 \log X - 0.047(\log X)^2 \qquad (29\text{–}20)$$

where X and Y are dimensionless groups defined by

$$X = \frac{q''}{\gamma_f\sqrt{Jg\lambda^3}}10^{10}$$

$$Y = \frac{h\sqrt{\gamma_g\sigma}}{(\gamma_f{}^3c_p\cdot k\sqrt{J^3g\lambda^3})^{0.5}}10^{10}$$

with † J = mechanical equivalent of heat = 778 ft lb/B,
g = acceleration due to gravity = 417 (10^6) ft/hr^2,
γ_f = specific weight of the liquid in lb/ft^3,
γ_g = specific weight of the saturated vapor in lb/ft^3,
σ = surface tension in lb/ft,
c_p = specific heat of the liquid in B/lb F, and
λ = latent heat of evaporation in B/lb,
q'' and h being expressed in British technical units as usual.

This equation satisfies not only the authors' experimental results, but also those of Jakob and Linke [1933, 1935] on water and carbon tetrachloride, Linden and Montillon [1930] and Dunn and Vincent [1931] on water, and Akin and McAdams [1939] on water, isopropanol, isobutanol, and n-butanol. Furthermore, as was done by Jakob and Linke in some experiments, a wetting agent (called Triton W-30) was added to water, 0.207% by volume of which reduced the surface tension by 33% and increased h by 27% (for q'' = 12,800), but without sensible influence upon the dimensionless representation.

Plotting $\log Y$ versus $\log X$, a slightly curved line was obtained. Without much loss of accuracy this could be replaced by a straight line, for instance, one according to the equation

$$Y = 4X^{0.68} \qquad (29\text{–}21)$$

In this form the relationship of Insinger and Bliss's equation with equation 29–19 becomes obvious, since their equation could have been written in the form:

$$\frac{h}{\sqrt{k}}\cdot\frac{\sqrt{\sigma}}{\gamma_f{}^{0.82}} = (4J^{0.41}g^{-0.09}10^{-3.2})\frac{(q'')^{0.68}}{\gamma_g{}^{0.5}\lambda^{0.27}}c_p{}^{0.5} \qquad (29\text{–}22)$$

† It may be noticed that here lb stands for lb force.

Since γ_f and γ_g are numerically equal to ρ_f and ρ_g, respectively, and the term in brackets is a constant, it is seen that Eq. 29–21 differs from Eq. 29–19 mainly in the exponents used and by the addition of the factor $c_p{}^{0.5}$. It will particularly be noticed that the factor $\sqrt{\sigma}$, which was introduced in Eq. 29–19 from reasons of the capillary theory, has been retained in the authors' equation.

In Table 29–4 values for different substances, as calculated by the authors, are represented.

TABLE 29–4

VALUES OF X/q'' AND Y/h ACCORDING TO EQ. 29–20 FOR THE HEAT TRANSFER IN BOILING

Substance	X/q'' [B^{-1} hr ft^2]	Y/h [B^{-1} hr ft^2 F]
Water with 0.207% by volume Triton W-30	0.0097	0.092
Water	0.0097	0.112
40% sucrose	0.0083	0.124
Isopropanol	0.079	0.997
n-Butanol	0.0945	1.13
Isobutanol	0.0993	1.18
Toluene	0.186	2.56
Carbon tetrachloride	0.249	3.85

29–7. Influence of Pressure upon Heat Transfer in Nucleate Boiling from a Submerged Heater

Both Jakob and Linke's and Insinger and Bliss's equations have the disadvantage of not expressing the influence of pressure on heat transfer. For this reason Jakob [1938] modified his theory and obtained the following equation:

$$N_{Nu} = \frac{h}{k}\sqrt{\frac{\sigma}{\rho_f - \rho_g}} = 31.6\,\frac{\nu_a}{\nu}\left[\left(\frac{D_{b,1}}{D_{b,a}}\right)^2 \frac{q''}{\rho_{g,a}\cdot\lambda_a\cdot w_{b,a}}\right]^{0.8}$$

$$= 31.6\,\frac{\nu_a}{\nu}\left(\frac{\rho_{f,a}}{\rho_f}\cdot\frac{\sigma}{\sigma_a}\cdot\frac{q''}{\rho_{g,a}\cdot\lambda_a\cdot w_{b,a}}\right)^{0.8} \qquad (29\text{–}23)$$

where the subscript a relates to atmospheric pressure,
ν is the kinematic viscosity of the liquid, and
ν_a is the kinematic viscosity of saturated water at 212°F.

In the derivation of this equation it was considered that both $D_{b,1}$ and f, and their product, the characteristic velocity, w_b, depend upon the pressure. Furthermore, one had to restore ρ_g from Eq. 29–7 because it cannot be neglected at high pressure. Finally an empirical ratio ν_a/ν has been added, which was of minor influence when the vapors of water and carbon tetrachloride at atmospheric pressure were compared, but increases appreciably with the saturation temperature. The constant factor of Eq. 29–19 had to be adjusted slightly.

Bonilla and Ch. W. Perry [1941] modified Eq. 29–23 so that it would correlate the results of thorough experiments which they had performed with water, ethanol, butanol, acetone, and numerous binary mixtures of these substances boiling on horizontal plates. The modification consisted in changing the exponent from 0.8 to 0.73, adding the square root of the Prandtl number $N_{Pr} = \mu c_p/k$ as a factor, and adjusting the constant factor of the equation. In this way they arrived at

$$N_{Nu} = \frac{h}{k}\sqrt{\frac{\sigma}{\rho_f}} = 16.6\,\frac{\nu_a}{\nu}\left(\frac{\rho_{f,a}}{\rho_f}\cdot\frac{\sigma}{\sigma_a}\cdot\frac{q''}{\rho_{g,a}\lambda_a w_{b,a}}\right)^{0.73}\left(\frac{\mu c_p}{k}\right)^{0.5} \qquad (29\text{–}24)$$

Comparison with Eq. 29–22 shows that the authors considered the influence of k and c_p in the same manner as Insinger and Bliss. Addition of the factor $\mu^{0.5}$, however, weakens the influence of ν, introduced in Eq. 29–23, which should rather be reinforced with regard to the influence of pressure. Anyway, this modification works at atmospheric pressure as the authors show for $q'' = 50{,}000$ B hr^{-1} ft^{-2}, where the extreme deviations of their experimental data from those calculated by means of Eq. 29–24 were reduced to $\pm 30\%$ in the range of $N_{Pr} = 1$ to 15 for all pure substances and mixtures under consideration.

Equation 29–23, on the other hand, gave rather satisfactory though somewhat too small values for steam at high pressure and about 18% smaller values than measured by Cichelli and Bonilla [1945, 1946] with benzene, *n*-pentane, *n*-heptane, and water at atmospheric pressure and $q'' = 70{,}000$ B hr^{-1} ft^{-2}. However, it gave much too small values for organic substances at high pressure. This could be improved by introducing a higher than the first power for ν_a/ν in Eq. 29–23. The mentioned addition of $\mu^{0.5}$ by Bonilla and Perry acted in the opposite direction so that their formula yields extremely small values at high pressure.

Insinger and Bliss found a better approximation to the experimental values by multiplying the values of h, obtained from Eq. 29–20, with $(p/p_a)^{0.4}$, where p and p_a are the existing and atmospheric pressure, respectively.

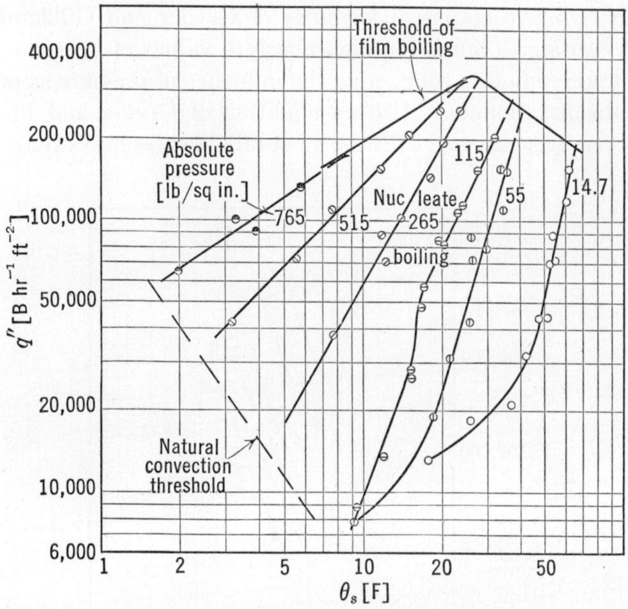

FIG. 29-23. Boiling of pure ethyl alcohol from a clean horizontal surface. From Cichelli and Bonilla [1945].

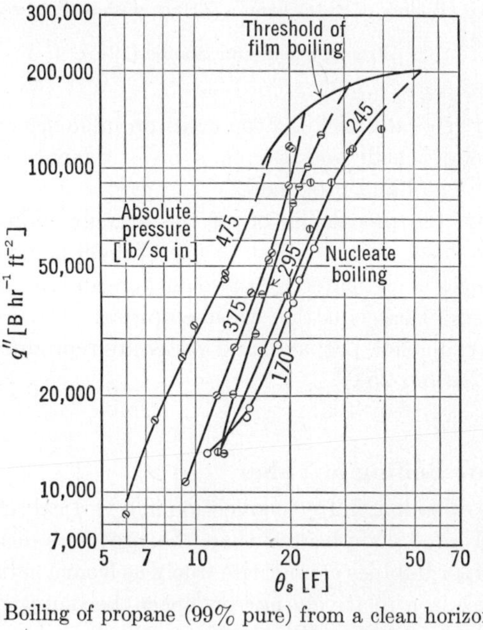

FIG. 29-24. Boiling of propane (99% pure) from a clean horizontal surface. From Cichelli and Bonilla [1945].

The first of two empirical formulas of Cryder and Gilliland [1932], although leading to much too high absolute values of h, seems to yield the best approximation concerning the influence of pressure as measured by Cichelli and Bonilla. Also an equation of Cryder and Finalborgo [1937], though at variance with the most reliable absolute values of other

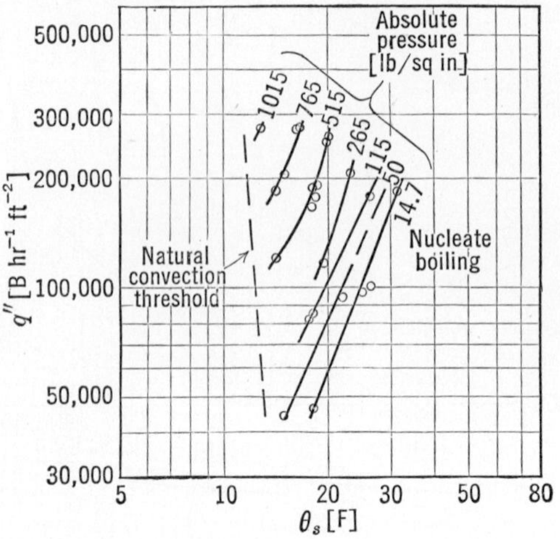

Fig. 29-25. Boiling of water from a surface with a slight scale deposit.

From Cichelli and Bonilla [1945].

observers, and generally yielding too great an influence of pressure, is in agreement with Cichelli and Bonilla's measurements of the pressure influence for ethanol and propane.

Summarizing, the precalculation of the pressure influence is still so uncertain that it seems to be best to use Cichelli and Bonilla's graphs for the substances investigated by these authors and for other substances the Insinger and Bliss equation with empirical pressure factor. The graphs for ethyl alcohol, propane, and water are reproduced as examples (Figs. 29-23, 24, and 25).

29-8. Nucleate Boiling in Tubes

The process of boiling in free convection in a vertical tube heated from outside is still more complicated than the processes dealt with above, because the vapor bubbles cannot rise freely as from a submerged heater, but have to rise in a relatively narrow space, becoming more and more crowded and displacing more liquid. Hence the fluid changes in composition, density, and velocity. No wonder, therefore, that our knowledge of the process of boiling in tubes is rather poor.

Data of different investigators collected by Fritz [1937], which refer to water boiling at atmospheric pressure in tubes of 1½ to 6½ ft length, can be approximately expressed by the empirical formula

$$h = 0.83(q'')^{0.75} = 0.475\theta_s^{\,3} \tag{29–25}$$

in the usual British units. This includes data of Linden and Montillon [1930] for a tube inclined by 45°. Comparison with Eq. 29–14 shows that, for the same flux q'', the coefficient is 28% higher. This experience was used by Jakob [1938] when he multiplied the right side of Eq. 29–23 by 1.25 in order to make it applicable to the boiling inside tubes. In fact a satisfactory agreement between the adjusted equation and experimental results with water in the range of 1.8 to 1420 lb/sq in. was obtained.

The forced convection in a vertical tube was investigated by Boarts, Badger, and Meisenburg [1937 and 1937(*a*)]. They used a tube, 0.76 in. wide, 12 ft long, in which water flowing upward was partly evaporated by means of steam condensing in a jacket. The boiling temperature of the water was varied between 140 and 212°F. In a range of Reynolds numbers $N_{Re} = 65{,}000$ to 307,000 the following approximate equation was obtained:

$$N_{Nu} = C(N_{Re})^{0.8}(N_{Pr})^{0.4} \tag{29–26}$$

where $C = 0.028$, not much larger than the value $C = 0.0243$ or 0.024 for heating a fluid flowing through a tube without boiling, according to Dittus and Boelter [1930] or Sherwood and Petrie [1932], respectively. Obviously the velocity of the water was so large (2.4 to 15 ft/sec at the inlet) and the fraction of evaporation so small (0.25 to 4%), that the formation and motion of vapor bubbles did not appreciably disturb the liquid convection. The temperature excess θ_s of the heating surface varied from 2 to 24 F. For $N_{Re} < 65{,}000$, i.e., lower velocity and larger evaporation, an attempt was made to separate a starting section in which $C = 0.029$ was found from the section of vehement boiling where a value for h was obtained almost double that of the starting section, as an average.

Oliver [1939], according to a report of McAdams [1942, Fig. 170], ran tests of boiling in a vertical nickel tube, 0.495 in. wide and only 1.72 ft long, in a range of heat flow $q'' = 38{,}000$ to 260,000 B hr^{-1} ft^{-2}. An immediate comparison with the results of Boarts, Badger, and Meisenburg is difficult because the temperature drop is defined in a different way; however it appears that for $N_{Re} > 65{,}000$ and increasing temperature difference, the constant C of Eq. 29–26 increased to about 0.040, and for $N_{Re} < 65{,}000$ it increased up to about 0.10, compared with $C = 0.023$, as assumed by McAdams for non-boiling liquid.

Concerning forced-convection boiling in horizontal or slightly inclined tubes, only overall coefficients seem to have been measured. This subject will be dealt with in Vol. II.

29–9. Film Boiling

Film boiling is an instance of the so-called spheroidal state, as described first in a scientific form by Leidenfrost [1756] (see Sect. 29–1). Some remarks about this state may be quoted from an excellent paper of Drew and Mueller [1937] dealing with the subject:

"From their books it appears that modern physicists have relegated the spheroidal state to their limbo for exceedingly curious but not particularly important phenomena. This is unfortunate because this phenomenon is technically important. . . . Briefly, the spheroidal state is a non-equilibrium condition assumed by two or more bodies of solid or liquid when an attempt is made to bring them together while their temperatures differ by more than a determinate amount; it is characterized by the subsistence of a layer of vapor between the bodies, which resists and prevents their being brought into direct contact."

Film boiling occurs in the mentioned phenomenon of droplets dancing on a hot plate, in quenching of steel, and in boiling of a liquid on the surface of a tube in which a vapor is condensed whose saturation temperature considerably exceeds that of the boiling liquid, or on the surface of electrical heaters under high load.

Figure 29–26 shows q'' and h plotted versus θ_s in bilogarithmic coordinates as measured by Farber and Scorah [1948] with water boiling on a Chromel C wire under atmospheric pressure.

It is seen that q'' and h both increase with the temperature difference θ_s up to a maximum, which is $q'' = 1{,}000{,}000$ B hr^{-1} ft^{-2} at $\theta_s = 100$ F, and $h = 12{,}100$ B hr^{-1} ft^{-2} F^{-1} at $\theta_s = 72.6$ F. In the range $h = 100$ to 5000 the increase can be represented by either of the equations

$$q'' = 0.055(\theta_s)^{4.26} \quad \text{or} \quad h = 0.055(\theta_s)^{3.26} \quad \text{or} \quad h = 0.5057(q'')^{0.765}$$

This is in the range of nucleate boiling and yields values of the same order as Eq. 29–23 for a plane plate.

When with increasing wattage of the heating wire the maximum of q'' is surpassed, then obviously the Joulean heat cannot any more be taken up entirely by the boiling liquid, and the wire will burn through unless the wattage is immediately reduced to a value corresponding to the branch of the curve, which descends with increasing θ_s. Also Jakob and Linke [1935], in their experiments with water boiling on a horizontal electrical heating plate, observed a maximum, but at the relatively low flux $q'' \approx 74{,}000$ B hr^{-1} ft^{-2} (see Fig. 29–19). Jakob [1936] described this phenomenon as follows: "In this region the number of steam bubbles formed on the plate becomes so large that the bubbles disturb one another as to their stirring effects. We even observed a drop in h when

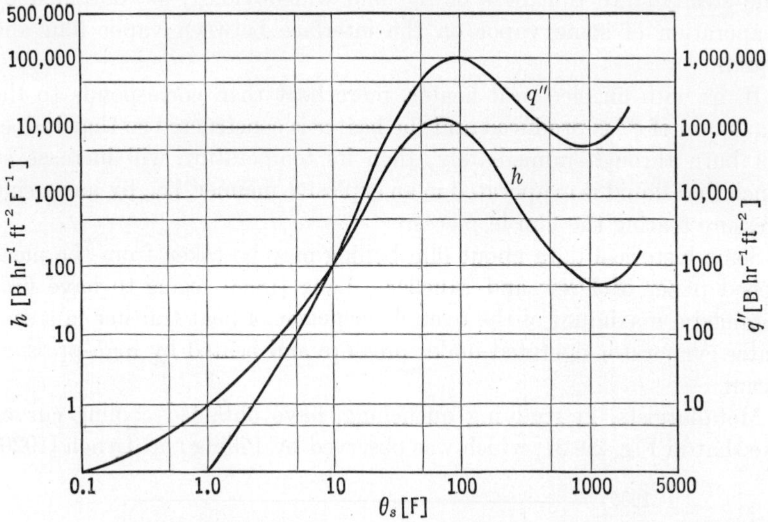

FIG. 29–26. Density of heat flow q'' and coefficient of heat transfer h for water boiling on a Chromel C wire under atmospheric pressure.

From Farber and Scorah [1947].

increasing the heating energy still more. This, of course, was connected with a rapid rise of temperature, soon leading to a burning through of the heating resistance."

Whether the spheroidal state is always initiated by coalescence of steam bubbles or may spontaneously occur, will be difficult to decide. Evidence is also missing whether the maximum generally occurs at a much smaller load with a horizontal heating surface than with a wire.

When a vapor condensing in a tube is used as heat source for the boiling of a liquid of lower saturation temperature on the outer surface, the process becomes more simple. Using, for instance, steam at elevated pressure inside the tube and increasing its pressure, the maximum of q'' can be surpassed without any dangerous effect, because then automatically less vapor condenses inside the tube, and a stationary state of heat transfer is obtained corresponding to a point of the descending branch of the q'' curve. If the temperature difference θ_s increases so much that radiation through the film becomes considerable, then eventually a minimum of q'' (57,500 B hr^{-1} ft^{-2} at $\theta_s = 1000$ F in Fig. 29–26) will be obtained, and q'' will start again to rise.

In the range where radiation does not play an essential role, the decrease of q'' with increasing θ_s must be due to an increase of the thickness of the vapor film which is relatively larger than the increase of θ_s. The steady-state thickness of the film will obviously be obtained by evaporation of some vapor on the interface between vapor film and liquid.

If, as with an electrical heater, more heat than corresponds to the maximum of q'' is produced and the heater is constructed so that it does not burn through immediately, then its temperature will increase so much that liquid is evaporated in an explosive manner, i.e., by increasing pressure tearing the film in pieces.

Some historical data about film boiling may be taken from the mentioned paper of Drew and Mueller. Lang [1888] seems to have first obtained a maximum of the overall coefficient of heat transfer in a seawater evaporator operated under pressure and heated by high-pressure steam.

Metallurgists, in studying quenching, have obtained cooling curves like that in Fig. 29–27, which was observed by Pilling and Lynch [1920]

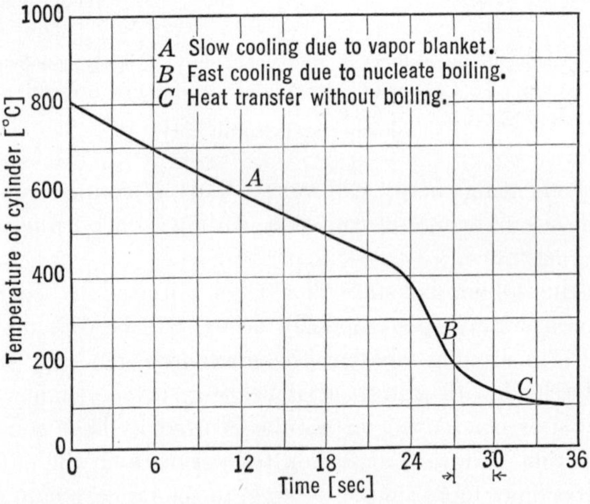

FIG. 29–27. Cooling of a silicon-nickel rod quenched in water.

Diameter of rod ¼ in. Initial temperature of rod 800°C. Initial temperature of water 98°C.

From Drew and Mueller [1937].

when a short rod of silicon-nickel alloy, initially at about 800°C, was quenched in water at 98°C. In this curve the section A is due to slow cooling by radiation and conduction across a vapor blanket (corresponding to the right ends of the curves in Fig. 29–26); the inflection point of section B would correspond to the maximum of q''; and the right end of the curve (zone C) to heat transfer without boiling. Nukiyama [1934], using a platinum wire, 0.14 mm in diameter, seems to have been the first to succeed in obtaining a substantially complete boiling curve with a maximum boiling rate at $\theta_s \approx 45$ C. Drew and Mueller themselves measured maxima of h for different chemical compounds, using a loop of ¼-in. copper tubing in which steam under pressure was condensed, as heater.

Very interesting data were obtained by Mosciki and Broder [1926] by merging a vertical platinum wire, 0.0039 in. diam., in water. They are reproduced in Table 29–5, taken from a recent publication of

TABLE 29–5

CRITICAL (BURN-OUT) DATA FOR A PLATINUM WIRE, 0.0039 IN. DIAM., IN WATER

t_c [°F]	t_b [°F]	$\theta_c = t_c - t_b$ [F]	q_c'' [B hr^{-1} ft^{-2}]
267	68	199	1,200,000
266	104	162	1,030,000
266	140	126	758,000
266	211	55	178,000

McAdams [1948]. Subscripts b and c refer to the bulk-of-water temperature and to the critical temperature, respectively, at which the wire burnt out. According to this table the critical wire temperature t_c is very low, showing that local overheating causes the burning out, and is independent of the water temperature, whereas low water temperature allows a much higher heat load before the wire burns through. Considering these results, McAdams emphasizes that unexpectedly high values of h, observed in that part of tubular evaporators in which there was no net generation of vapor, may have been due to local boiling.

Farber and Scorah [1947] in their investigation measured temperatures at only one point of the heating wire by use of a welded-on thermocouple. McAdams, Addoms, Rinaldo, and Day [1948] used horizontal platinum wires of different diameters (D = 0.004 to 0.024 in.) which served as heaters and resistance thermometers at the same time. End effects were eliminated by employing fine platinum potential leads spot-welded to the heating wire, and electrolytic corrosion was prevented by a special circuit. Therefore these experiments have a more quantitative character than the earlier ones. Only a few results, based on Rinaldo's measurements, can be indicated.

In both zones of free convection ($\theta_s \approx$ 5 to 15 F and 1500 to 2500 F) the thinner wires carried much more heat per unit surface than the thicker ones; for instance, at θ_s = 10 F, $q'' \approx$ 16,000 and 5000 B hr^{-1} ft^{-2} for D = 0.004 and 0.024 in., respectively; at θ_s = 2000 F, $q'' \approx$ 450,000 and 250,000 for D = 0.008 and 0.024, respectively. It may be noticed that $h = q''/\theta_s$ is much smaller in the high-temperature zone. In the range of nucleate boiling the effect of wire thickness was smaller, and finally at θ_s = 40 F reversed to $q'' \approx$ 180,000 and 375,000 for D = 0.004 and 0.016 or 0.024, respectively. The largest flux density was obtained with the wire of 0.008-in. diameter and amounted to about 600,000 B hr^{-1} ft^{-2} at θ_s = 2500 F, i.e., about double the peak density (called q_c'' in Table 29–5) where film boiling started. About 25,000 B hr^{-1} ft^{-2} of this amount was due to radiation through the film.

Data for water above atmospheric pressure have been obtained by Farber and Scorah [1947]. They indicate a considerable change of both q_{max}'' and the corresponding θ_s. With the same heating wire with which the curves of Fig. 29–26 were obtained at atmospheric pressure (q_{max}'' = 1,000,000 at θ_s = 100 F), the authors found $q_{max}'' \approx$ 1,500,000 at $\theta_s \approx$ 15 F with 114.7 lb/sq in. (338°F saturation temperature). For h the maxima were 12,100 and 100,000 at 14.7 and 114.7 lb/sq in., respectively, and θ_s = 72.6 and 13.1 F.

In a much larger pressure range Day observed an almost linear increase of q_{max}'' with the pressure, using a horizontal platinum wire of 0.012-in. diameter [see McAdams, Addoms, Rinaldo, and Day 1948]. For instance, q_{max}'' = 400,000; 1,400,000; and 2,100,000 were obtained at atmospheric pressure, 600, and 1200 lb/sq in., respectively.

Farber and Scorah also found an influence of the heating-wire material upon the maxima. With Chromel C, the maximum of h was 11,900 B hr^{-1} ft^{-2} F^{-1} at θ_s = 71.7 F; with Chromel A, 28,000 at θ_s = 20.2 F; with nickel, 24,500 at θ_s = 14.7 F. This may be due to the difference in the interfacial energy occurring at the contact of a given fluid with different metals [see Larson 1945]. This had previously been made evident by Rhodes and Bridges [1939], who boiled water on the surface of a tube through which hot mercury was flowing. They definitely found that film boiling is promoted by making the surface less wettable, for instance, by covering it with a thin film of wax or by adding a minute quantity of oleic acid or mineral oil to the water. Addition of some sodium carbonate or sodium chloride restored nucleate boiling. This effect is easily understood from the shape of vapor bubbles indicated in

Fig. 29–7. Those formed at non-wettable surfaces are fit to blanket the surface.

A tendency of chromium-plated surfaces to cause film boiling was also observed, and two possible reasons for this behavior were indicated. It may be due either to a true lack of wettability of chromium metal by water or to the presence of a very thin film of chromium oxide that is not readily wetted. The action of the sodium carbonate may be attributed either to the effect of this salt in increasing the interfacial tension or to its effect in removing the film of oxide. The maximum observed by Jakob and Linke at relatively low heating load of their chromium-plated heater, as mentioned above, may have been caused by this peculiarity of chromium surfaces.

Data for the maximum overall temperature difference between condensing steam and boiling organic liquids have been published by Sauer, Cooper, Akin, and McAdams [1938], based on unpublished theses of the first three authors. Measurements of θ_s are due to Kaulakis and Sherman [1938], Braunlich [1941], and Akin and McAdams [1939, 1939(*a*)].

Some reliable average values of the temperature difference θ_s at which the maximum of q'' occurs in boiling on outer surfaces of tubes are 45 F for water, 65 F for *n*-butanol, 80 F for isobutanol, and 85 F for isopropanol. According to the investigations just mentioned, the difference θ_s for a maximum of q'' does not seem to be influenced considerably by the pressure in the vacuum range; the maximum of q'', however, increases with the saturation temperature t_f. When boiling isopropanol, for instance, the flux maximum increases from 70,000 to 110,000 B hr^{-1} ft^{-2} if the saturation temperature is raised from 130 to 175°F. Similarly, for *n*-butanol, $q_{max}'' = 80{,}000$ and 105,000 at $t_f = 175$°F and 241°F, respectively; and for water, $q_{max}'' = 160{,}000$; 250,000; 360,000 at $t_f = 130$; 170; and 212°F, respectively.

Finally, the practical importance of the transition from nucleate to film boiling may be emphasized. The quenching process is entirely dependent on this transition, and for heat exchangers, particularly those employed in chemical engineering, it is desirable to come close to the maximum of h, but never to surpass the maximum of q'' where a steep drop of heat transfer and unstable conditions will occur. With electrical heaters, surpassing the maximum means destruction of the apparatus.

CHAPTER 30

Heat Transfer in Condensation

30–1. The Different Kinds of Condensation

Corresponding to nucleate and film boiling, two modes of condensing are known, called dropwise condensation and film condensation. However, while bubblewise boiling is the most important kind of boiling, film formation being an extreme case which has to be avoided (except in quenching), the reverse holds for condensing: dropwise condensation is the more seldom case, though it would be more desirable, and film condensation is the usual, but less effective, form. In the following analysis, condensation of steam is preferably dealt with, and a former presentation of the subject by Jakob [1936(*a*)] has extensively been used.

Film condensation occurs on a wettable cooling surface. A coherent film covers and passes the wall, driven by mechanical forces, but is immediately renewed by the condensing liquid, since the cooling surface is never free of liquid.

The process of wetting is essential only at the start of the condensation process. Once the surface is wet, only the binary liquid-vapor system need be considered, as the radius of curvature of the usual condensing surfaces is so large that the effect of capillary forces disappears.

On non-wettable cooling surfaces the vapor is condensed in drop form. The drops are extremely small when they originate, in accordance with the laws of capillarity. As they grow by further condensation, they roll over the surface, under the influence of gravity and friction exerted by adjacent moving parts of the vapor. In rolling along, the droplets unite with other ones, forming larger drops.

The conditions under which either kind of condensation occurs were first studied by experiments on steam by E. Schmidt, Schurig, and Sellschopp [1930], then by Spoelstra [1931], by Roecke and by Jakob and Erk [Jakob 1932], and, particularly, by Nagle and Drew [1933–1934], Nagle, Bays, Blenderman, and Drew [1935], and Drew, Nagle, and W. Q. Smith [1935].

On the basis of these experiments the last mentioned authors have ascertained the following points:

"Clean steam, whether or not it contains non-condensable gas, always condenses in a film on clean surfaces, rough or polished.

"Dropwise condensation of steam does not occur unless the cooling surface is in some way contaminated.

"Although numerous substances, while actually on the surface, will make it non-wettable, only those that are strongly absorbed or otherwise firmly held are significant as drop promoters in a condenser. Some contaminants seem to depend for their activity as promoters on the amount of non-condensable gas present. Some contaminants are specifically effective on certain metals (e.g., mercaptans * on copper alloys); others are quite generally effective (e.g., fatty acids)."

It is understood that film condensation may occur on some parts of a cooling surface, and dropwise condensation on others. This may be called mixed condensation.

Drew and his associates obtained the three kinds of condensation in a single experiment, condensing steam on the exteriors of three tubes which they arranged in a glass vessel, Fig. 30–1.† The tube on the left was of copper, that in the middle of nickel, and that on the right of an alloy of nickel, copper, and zinc. When these tubes were cooled by means of water flowing inside, and pure steam was used outside, all of them exhibited film condensation. With plant steam, on the other hand, the copper surface, adsorbing a contaminant contained in the steam, showed dropwise condensation with many fine-grained drops; the non-adsorbing nickel tube, film condensation; and the alloy, mixed condensation with a few coarse-grained drops.

30–2. Some Particular Features of Film and Drop Condensation

As mentioned already, dropwise condensation is much more effective than film condensation, yielding in some cases up to 15 or 20 times more condensate. To understand this the main features of both kinds of condensation must be considered.

Film condensation has been theoretically analyzed by Nusselt [1916(*a*)]. If a vapor is condensing in a film on a vertical cooling surface, the film flows down, under the influence of gravity, but is retarded by shearing forces because of the viscosity of the liquid. The latent heat freed by condensation passes through the film from the condensing vapor

* Mercaptans contain the radical SH. For instance, the chemical formula of benzyl mercaptan is $C_6H_5 \cdot CH_2 \cdot SH$. (*Note by the author.*)

† I am indebted to Professor W. H. McAdams for placing this illustration at my disposal.

to the wall. The film resists this heat flow because of the low conductivity of liquids, and, therefore, an appreciable temperature drop across the film occurs. Where the film is in contact with the cooling surface its temperature is equal to the surface temperature. Nusselt further

FIG. 30–1. The different kinds of condensation demonstrated in a single experiment. Courtesy Professor W. H. McAdams.

assumed that the interface between film and vapor is at the saturation temperature. If this were exactly correct no condensation could occur. It is often remarked that the film opposes a strong resistance to the heat, freed by condensation, which is to be conducted through the film, and that the film becomes thicker as condensation increases, thus limiting the rate of condensation. This argument holds only if the interface between vapor and film is below the temperature of the bulk of vapor. Otherwise it would be uncomprehensible how the vapor could know the thickness of the liquid layer on which the rate of its condensation is said to depend. Obviously the temperature of the liquid surface in contact with condensing vapor must be slightly below saturation temperature and will decrease with decreasing wall temperature. This modification of Nusselt's assumption is indispensable, since otherwise the increased condensation, caused by lowering the wall temperature, cannot be understood. However, the temperature difference between vapor and interface can be, and obviously is, very small, just sufficient to yield the pressure difference necessary for carrying vapor to the surface. In any case, condensation depends upon the thickness of the liquid film and the conductivity of the liquid and therefore will be comparatively small.

In dropwise condensation, on the other hand, the cooling surface is in direct contact with the vapor. Thus the effective temperature difference is much larger, and formation of droplets is much more rapid than the corresponding formation of liquid on the film surface. It may be objected that formation of the droplets is only a part of the condensation process and growth of drops is governed by the same law as film condensation. This is true only as far as the growth is not due to coalescence with other droplets, which probably is the major factor,‡ and it must be added (at least for vertical and oblique surfaces) that the liquid is carried away more quickly than in the case of film condensation, as the drops roll down and take other droplets with them.

Notwithstanding all the advantages of drop condensation, film condensation is of paramount importance, being established more easily and remaining much more stable in operation. It will be dealt with first and more in detail in this chapter than the other modification of condensing.

30–3. The Flow of a Liquid Film on a Solid Surface

Nusselt [1916(a)] in his theory of film condensation has combined the laws of laminar flow of a liquid film and of heat conduction through it. First the equations for the flow of the film had to be developed. A liquid film is supposed to form and flow down a surface which may be inclined by the angle ϕ against a horizontal plane. Let v be the downward velocity parallel to the surface at a distance y from the surface. Then a shear stress occurs in the liquid which can be expressed by the well-known equation

$$S = \mu \frac{dv}{dy} \tag{30–1}$$

‡ Jakob [1936(a)] mentioned that the ratio of surface to volume is much larger for drops than for a film and that the coefficient of heat transfer is greater than that of a plane surface, the small curvature being an essential factor. However, it must be considered that the heat freed by condensation cannot be easily removed from the droplets and that this is retarding their growing.

where μ is the dynamic viscosity of the liquid. If, further, γ denotes the specific weight of the liquid, the equation of mechanical equilibrium will be

$$dS = -\gamma \cdot \sin \phi \cdot dy \quad \text{or} \quad \frac{dS}{dy} = -\gamma \cdot \sin \phi \tag{30–2}$$

From Eqs. 30–1 and 2

$$\frac{d^2v}{dy^2} = -\frac{\gamma}{\mu} \sin \phi \tag{30–3}$$

and by integration

$$v = -\frac{\gamma \cdot \sin \phi}{2\mu} y^2 + C_1 y + C_2 \tag{30–4}$$

Since the liquid adheres to the wall,

$$v = 0 \quad \text{at} \quad y = 0$$

and therefore

$$C_2 = 0$$

Herewith

$$v = -\frac{\gamma \cdot \sin \phi}{2\mu} y^2 + C_1 y \tag{30–5}$$

The value of C_1 depends upon the flow of the condensing vapor, for, if the vapor streams along the film with the velocity v_v, it exerts a shearing force upon the liquid.

For vapor which is virtually at rest no sensible friction occurs at its interface with the film. Let $y = Y$, the thickness of the film at any place; then, according to Eq. 30–1,

$$\left(\frac{dv}{dy}\right)_{y=Y} = 0 \tag{30–6}$$

Herewith, from Eq. 30–5,

$$C_1 = \frac{\gamma \cdot \sin \phi}{\mu} Y \tag{30–7}$$

so that for vapor at rest

$$v = -\frac{\gamma \cdot \sin \phi}{2\mu} (y^2 - 2Yy) \tag{30–8}$$

If the condensing vapor of density ρ_v flows along the surface with the velocity v_v, for instance, inside a pipe of width D, then the shearing stress occurring on the film surface can be calculated from the equation of pressure drop in tubes (Eq. 21–38), which may be written as

$$\frac{dp}{dx} = -\frac{4f}{D} \cdot \frac{\rho_v v_v^2}{2} \tag{30–9}$$

where f is a function of the Reynolds number,§ and the coordinate x is in the direction of main flow of the condensate film.

On the other hand, equilibrium on the inner surface of the tube ‖ requires

$$S_{y=Y}\pi D = -\frac{\pi D^2}{4} \cdot \frac{dp}{dx} \tag{30–10}$$

From Eqs. 30–9 and 10

$$S_{y=Y} = f\frac{\rho_v v_v^2}{2} \tag{30–11}$$

Differentiating Eq. 30–5 and substituting $y = Y$, using also Eqs. 30–1 and 11,

$$S_{y=Y} = \pm f\frac{\rho_v v_v^2}{2} = \mu\left(\frac{dv}{dy}\right)_{y=Y} = -\gamma Y \cdot \sin\phi + C_1\mu$$

or

$$C_1 = \pm\frac{f}{\mu} \cdot \frac{\rho_v v_v^2}{2} + \frac{\gamma}{\mu} Y \cdot \sin\phi \tag{30–12}$$

The positive sign of the first term on the right holds for vapor flow downward, the negative sign for flow upward.

From Eqs. 30–5 and 12

$$v = \left(\pm\frac{f}{\mu} \cdot \frac{\rho_v v_v^2}{2} + \frac{\gamma}{\mu} Y \cdot \sin\phi\right) y - \frac{\gamma \cdot \sin\phi}{2\mu} y^2 \tag{30–13}$$

By integration one obtains the mean velocity of the condensate film:

$$v_m = \frac{1}{Y}\int_0^Y v \cdot dy = \frac{\gamma \sin\phi}{3\mu} Y^2 \pm \frac{f}{2\mu} \cdot \frac{\rho_v v_v^2}{2} Y \tag{30–14}$$

30–4. Theory of Film Condensation of Vapor at Rest on a Plane Surface

The expression vapor at rest should not be considered verbally; vapor, having a velocity of a few feet per second, will not sensibly affect the flow of a thin condensate film and is therefore included in the concept of

§ As mentioned in Sect. 24–3, the friction coefficient $4f$ is identical with ϕ, used in Eq. 21–38. Nusselt assumed f to be a constant.

‖ The film is usually so thin that its free surface can be identified with the inner tube surface.

"vapor at rest." In this sense, vapor at rest exists in most condensers in which vapor is condensed on the outside of tubes through which a cooling liquid is streaming. This, therefore, is a very common and practically important case.

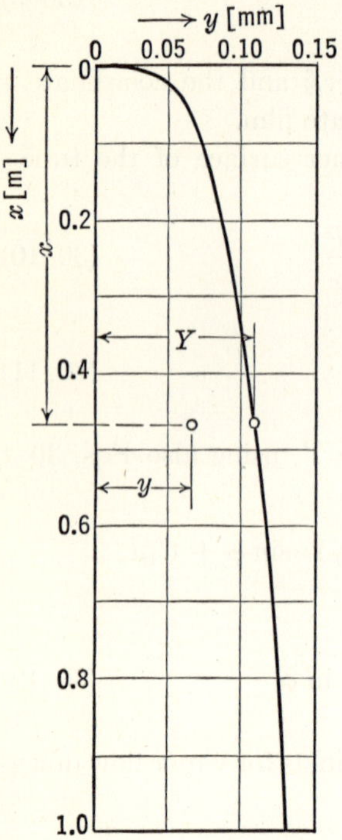

FIG. 30-2. Growing of a water film on a vertical condensing surface.

A vapor may be condensed on a vertical surface of height H and arbitrary width. Unit width will be considered in the following treatment of the problem. Figure 30-2 represents a vertical section through the film, x being the distance in downward direction from the place where condensation starts.

Considering that in Eq. 30-14 the vapor velocity $v_v = 0$ and $\sin \phi = 1$, the time rate of weight ¶ of the film flowing downward will be

$$\dot{W}' = \gamma Y v_m = \frac{\gamma^2}{3\mu} Y^3 \qquad (30\text{-}15)$$

At the ordinate $x + dx$ the rate of flow is larger by

$$d\dot{W}' = \gamma \cdot d(Y v_m) = \frac{\gamma^2}{\mu} Y^2 \, dY \qquad (30\text{-}16)$$

This increase comes from the condensation on the differential height dx.

The quantity $d\dot{W}'$ can also be expressed in terms of the time rate of heat dq' traversing the film in the area $1 \cdot dx$. Denoting the latent heat by λ (expressed in heat units per unit weight),

$$dq' = \lambda \cdot d\dot{W}' \qquad (30\text{-}17)$$

Since laminar flow of the film has been assumed, the usual law of heat conduction can be applied:

$$dq' = k \frac{t_f - t_s}{Y} dx \qquad (30\text{-}18)$$

where k = the thermal conductivity of the liquid,
t_f = the saturation temperature, and
t_s = the temperature of the cooling surface.

It is assumed that the film surface in contact with the vapor is at satura-

¶ It should be noticed that weight is considered as a force.

tion temperature, neglecting the small existing difference mentioned previously.

From Eqs. 30–17 and 18

$$d\dot{W}' = \frac{k}{\lambda} \cdot \frac{t_f - t_s}{Y} dx \tag{30–19}$$

Thus the condensing rate $d\dot{W}'$ is represented hydrodynamically by Eq. 30–16 and in thermal quantities by Eq. 30–19. Equating these expressions,

$$\frac{k}{\lambda} \cdot \frac{t_f - t_s}{Y} dx = \frac{\gamma^2}{\mu} Y^2 \, dY \tag{30–20}$$

and by integration

$$x = \frac{\gamma^2 \lambda Y^4}{4\mu k(t_f - t_s)} + C \tag{30–21}$$

But $C = 0$, since $Y = 0$ at $x = 0$. Hence

$$Y = \sqrt[4]{\frac{4\mu k(t_f - t_s)x}{\gamma^2 \lambda}} \tag{30–22}$$

showing that the thickness of the film increases only with the fourth root of x.

Now from Eqs. 30–18 and 22

$$q'' = \frac{dq'}{dx} = \sqrt[4]{\frac{\gamma^2 \lambda k^3 (t_f - t_s)^3}{4\mu x}} \tag{30–23}$$

and by integration over the entire surface of height H and unit width,

$$q' = \frac{4}{3} \sqrt[4]{\frac{\gamma^2 \lambda k^3 H^3 (t_f - t_s)^3}{4\mu}} \tag{30–24}$$

Introducing a coefficient of heat transfer h, defined by

$$q'' = h(t_f - t_s) \tag{30–25}$$

Eq. 30–23 yields

$$h = \sqrt[4]{\frac{\gamma^2 \lambda k^3}{4\mu x(t_f - t_s)}} \tag{30–26}$$

This coefficient applies to the position x. A mean coefficient of heat transfer h_m may be defined by

$$q' = h_m H(t_f - t_s) \tag{30–27}$$

Substituting this in Eq. 30–24,

$$h_m = \frac{4}{3}\sqrt[4]{\frac{\gamma^2 \lambda k^3}{4\mu H(t_f - t_s)}} \qquad (30\text{–}28)\ *$$

Herewith the calculation of the heat transfer in condensation is reduced to geometrical, mechanical, and thermal quantities. The most important result is that q' and q'' are not proportional to $(t_f - t_s)$ and H, but to $(t_f - t_s)^{3/4}$ and $H^{3/4}$. Doubling, for instance, the height H yields only 69% more condensate. Referring to the influence of the properties of the liquid, γ, μ, λ, and k, it may easily be seen that, on the same surface and with the same temperature drop across the film, h_m for water is about four times larger than for ethyl alcohol.

In order to give an idea of the thickness of condensate films, it may be mentioned that Fig. 30–2 is for $H = 1$ m and for steam at atmospheric pressure. With $t_f = 100°\text{C}$ and $t_s = 90°\text{C}$, the film would attain a thickness of only 0.14 mm on the lower end of the plate.

If the surface is inclined by the angle ϕ against a horizontal plane, then, using the equations of Sect. 30–3, instead of Eqs. 30–22 and 28, the following formulas are obtained:

$$Y = \sqrt[4]{\frac{4\mu k(t_f - t_s)x}{\gamma^2 \lambda \cdot \sin \phi}} \qquad (30\text{–}29)$$

$$h_m = \frac{4}{3}\sqrt[4]{\frac{\gamma^2 \lambda k^3 \sin \phi}{4\mu H(t_f - t_s)}} \qquad (30\text{–}30)$$

For instance, with $\phi = 45°$ and $30°$, respectively, 8% and 16% less condensate will be formed than for the vertical surface.

If the vapor which is to be condensed is superheated, having the temperature $t_v > t_f$, then the heat balance becomes

$$h_v(t_v - t_f)\, dx + \frac{\gamma^2 \lambda}{\mu} Y^2\, dY = \frac{k}{Y}(t_f - t_s)\, dx \qquad (30\text{–}31)$$

or

$$\frac{dY}{dx} = \frac{\mu k(t_f - t_s)}{\gamma^2 \lambda Y^3}(1 - mY) \qquad (30\text{–}32)$$

where

$$m = \frac{h_v(t_v - t_f)}{k(t_f - t_s)}$$

* Nusselt has also derived an equation which considers the slight horizontal component of the flow in the liquid film.

The first term of Eq. 30–31 represents the heat transfer between vapor and film before condensation, governed by a coefficient of heat transfer h_v, which will not sensibly differ from that between a vapor and a solid surface at temperature t_f. The second term represents the heat freed by condensation (according to Eqs. 30–16 and 17), and the term on the right side gives the heat conduction through the film.

Integration yields

$$x = \frac{\gamma^2\lambda}{\mu k(t_f - t_s)}\left(\frac{1}{m^4}\ln\frac{1}{1 - mY} - \frac{Y^3}{3m} - \frac{Y^2}{2m^2} - \frac{Y^3}{m^3}\right) \tag{30–33}$$

Since necessarily $dY/dx > 0$, Eq. 30–32 requires $mY < 1$. Hence, the logarithm in Eq. 30–33 can be expanded in a series, and the equation turns into

$$x = \frac{\gamma^2\lambda}{\mu k(t_f - t_s)}\left(\frac{Y^4}{4} + \frac{mY^5}{5} + \frac{m^2Y^6}{6} + \cdots\right) \tag{30–34 †}$$

For $t_v = t_f$, it converts to Eq. 30–22.

For a given vertical length $x = H$, Eq. 30–34 yields

$$H = \frac{\gamma^2\lambda}{\mu k(t_f - t_s)}\left(\frac{{Y_H}^4}{4} + \frac{m{Y_H}^5}{5} + \frac{m^2{Y_H}^6}{6} + \cdots\right) \tag{30–35}$$

from which the thickness of the film at the lower end, $y = Y_H$, can be obtained by trial and error. Substituting this in Eq. 30–15 yields the rate $\dot{W}'$ of condensation of superheated vapor.

30–5. Theory of Film Condensation of Vapor at Rest on Cylindric Surfaces

The equations for vertical plane surfaces hold also for vertical cylinders, since the film is usually so thin that the curvature can be neglected. For the condensation of a vapor on the outside of a horizontal tube, cooled from the inside, Nusselt performed the following derivation.

Let $D = 2r$ be the outer diameter, and in a cross section perpendicular to the tube axis, ϕ the angle which any radius r forms with a vertical line. Then, according to Eq. 30–14, the mean velocity of the film at the end of that radius will be

$$v_m = \frac{\gamma}{3\mu} Y^2 \sin\phi \tag{30–36}$$

† The series can also be written as follows:

$$Y^4\left(\frac{1}{4} + \frac{mY}{5} + \frac{m^2Y^2}{6} + \cdots\right)$$

Since $mY < 1$, the geometric series $1 + mY + m^2Y^2 + \cdots$ converges; hence also the above series converges.

In a tube section of unit length and arc element $r \cdot d\phi$ the following amount of heat will be exchanged in unit time:‡

$$d\left(\frac{q'}{2}\right) = \frac{kr(t_f - t_s)\, d\phi}{Y} = \gamma\lambda \cdot d(v_m Y) \tag{30–37}$$

the last term being the amount of latent heat liberated by condensation.

From Eqs. 30–36 and 37

$$d\left(\frac{q'}{2}\right) = \frac{\gamma^2\lambda}{3\mu}\, d(Y^3 \sin\phi) \tag{30–38}$$

From Eqs. 30–37 and 38

$$B \cdot d\phi = Y \cdot d(Y^3 \sin\phi) = \tfrac{3}{4} \sin\phi \cdot dY^4 + Y^4 \cos\phi \cdot d\phi \tag{30–39}$$

where

$$B = \frac{3\mu kr(t_f - t_s)}{\gamma^2\lambda}$$

Introducing

$$z = \psi^4 = \frac{Y^4}{B} \tag{30–40}$$

it follows that

$$\frac{3}{4} \sin\phi \cdot \frac{dz}{d\phi} + z \cdot \cos\phi - 1 = 0 \tag{30–41}$$

The solution of this linear differential equation is

$$z = \frac{1}{\sin^{4/3}\phi}\left(\frac{4}{3}\int \sin^{1/3}\phi \cdot d\phi + C\right) \tag{30–42}$$

If no liquid is dropping on the top of the tube, then for $\phi = 0$ the film thickness Y and therefore z will be finite. This requires

$$C = 0$$

and therefore

$$z = \frac{4}{3\sin^{4/3}\phi}\int_0^{\phi} \sqrt[3]{\sin\phi}\, d\phi \tag{30–43}$$

By graphical integration the values of Table 30–1 have been obtained.

‡ The differential flux at any angle ϕ is called $d(q'/2)$. The same flux will exist at the angle $-\phi$.

TABLE 30-1

RESULTS OF NUMERICAL INTEGRATION OF EQ. 30-43 FOR THE HEAT TRANSFER IN CONDENSATION ON A HORIZONTAL TUBE

ϕ [angular degrees]	ψ	$\frac{1}{\psi}$	$\psi^4 = \frac{Y^4}{B}$	$\frac{4}{3}\int_0^\phi \sqrt[3]{\sin\phi}\, d\phi$
0	1.000	1.000	1.000	0
5	0.991	1.009	0.963	0.0372
10	0.998	1.002	0.991	0.0960
20	1.003	0.997	1.013	0.2423
30	1.012	0.988	1.049	0.4163
40	1.023	0.977	1.097	0.6085
50	1.037	0.964	1.156	0.8104
60	1.057	0.946	1.247	1.029
70	1.080	0.926	1.359	1.251
80	1.109	0.902	1.513	1.482
90	1.144	0.874	1.714	1.714
100	1.187	0.843	1.985	1.946
110	1.240	0.806	2.365	2.177
120	1.306	0.766	2.905	2.399
130	1.390	0.719	3.733	2.617
140	1.501	0.666	5.081	2.819
150	1.660	0.603	7.588	3.012
160	1.910	0.524	13.317	3.186
170	2.422	0.413	34.385	3.332
175	3.061	0.327	87.733	3.391
180	∞	0	∞	3.428

From Eqs. 30-40 and 43

$$Y = \psi \cdot \sqrt[4]{B} = \psi \cdot \sqrt[4]{\frac{3\mu k r(t_f - t_s)}{\gamma^2 \lambda}} \tag{30-44}$$

It is seen that the film thickness is proportional to ψ. Hence, according to Table 30-1, it does not start with zero, but with finite thickness on the top of the tube, then decreases slightly, then increases moderately to jump to infinity at $\phi = 180°$. This is the mathematical expression of the fact that water drops or flows down from the bottom of the tube. Figure 30-3, taken from Nusselt's paper, refers to a tube of 0.026 m ($\approx$1 inch) outer diameter and 90°C surface temperature, on which steam of atmospheric pressure condenses. At $\phi = 170°$ the condensate film is still only 0.1 mm thick.

The local coefficient of heat transfer at the angle ϕ becomes

$$h = \frac{k}{Y} = \frac{1}{\psi}\sqrt[4]{\frac{\gamma^2\lambda k^3}{3\mu r(t_f - t_s)}} = \frac{k}{\psi\cdot\sqrt[4]{B}} \tag{30-45}$$

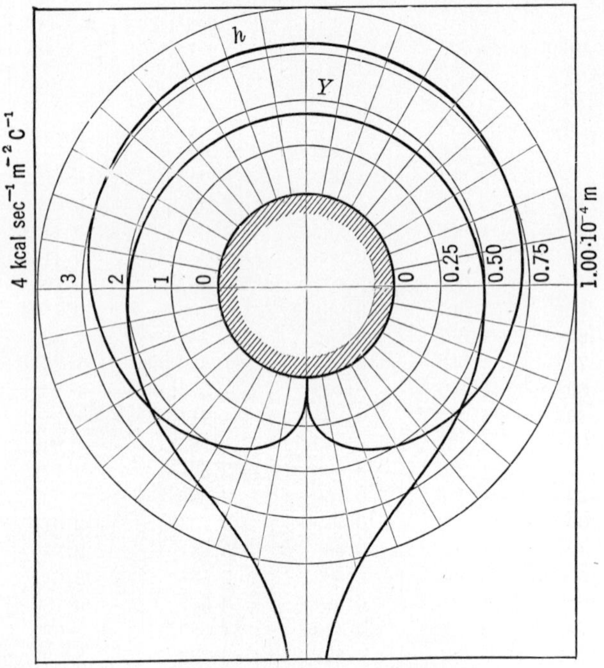

FIG. 30-3. Coefficient of heat transfer h and water film thickness Y on the surface of a horizontal condensing tube.

From Nusselt [1916(a)].

By integration over the range from ϕ_1 to ϕ_2 the following mean coefficient is obtained:

$$[h_m]_{\phi_1}^{\phi_2} = \frac{k}{\sqrt[4]{B}(\phi_2 - \phi_1)}\int_{\phi_1}^{\phi_2}\frac{d\phi}{\psi} \tag{30-46}$$

Graphical integration of this expression showed that, under the assumed conditions, 59.4% of the total condensate was produced in the upper half of the tube and 40.6% in the lower one.

Finally it may be mentioned that the equations derived above are also valid for outside cooling of a tube and inside condensing of vapor if r is used for the inside radius.

Comparison of Eqs. 30-28 and 46, taking $\phi_2 - \phi_1 = 180°$, reveals that the following relation holds between the mean film coefficient h_{hor}

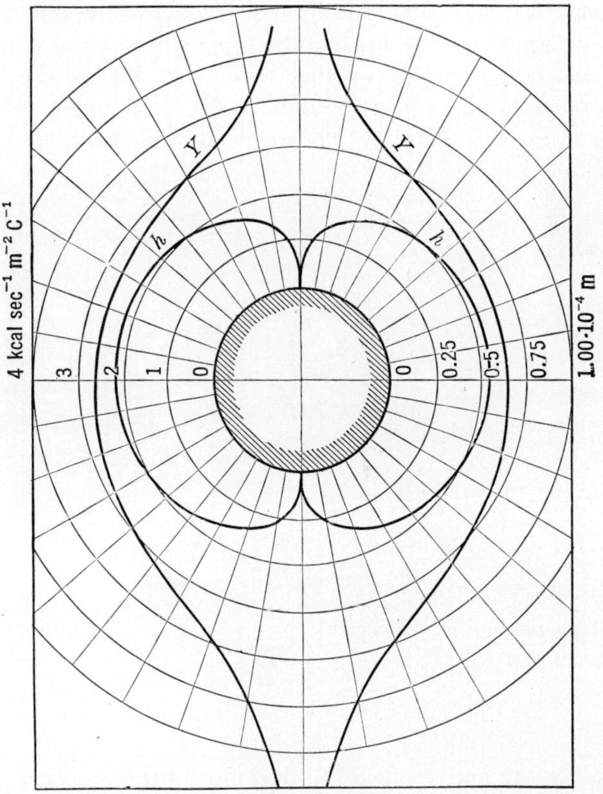

Fig. 30–4. Coefficient of heat transfer h and water film thickness Y on the surface of a horizontal condensing tube placed below the one dealt with in Fig. 30–3. From Nusselt [1916(*a*)].

for a horizontal tube of diameter D and h_{vert} for a vertical tube or other surface of height H:

$$\frac{h_{hor}}{h_{vert}} = 0.770 \sqrt[4]{\frac{H}{D}} \tag{30–47}$$

Taking, for instance, a tube of ¾ in. outer diameter and 37½ in. length, then in horizontal position 2.05 times the amount of vapor in vertical position will be condensed.

Substituting the value of h_m from Eq. 30–28 for h_{vert} in Eq. 30–47,

$$h_{hor} = 0.725 \sqrt[4]{\frac{\gamma^2 \lambda k^3}{\mu D(t_f - t_s)}} \tag{30–48}$$

If several horizontal tubes are located one above the other, then the condensate formed on any upper tube drops upon the next lower one, reducing the heat exchange on that tube. Let $\dot{W}_1'$ be the weight of liquid dropping in unit time from unit length of the upper tube. Now, according to Eq. 30–14, the rate of weight flowing down on one side of the lower tube will be

$$\gamma Y v_m = \frac{\gamma^2}{3\mu} Y^3 \sin\phi$$

Hence

$$\frac{\dot{W}_1'}{2} = \frac{\gamma^2}{3\mu} (Y^3 \sin\phi)_{\phi=0}$$

and

$$Y_{\phi=0}{}^3 = \left(\frac{3\mu \dot{W}_1'}{2\gamma^2 \sin\phi}\right)_{\phi=0} = \infty \tag{30–49}$$

But, according to Eq. 30–42,

$$\frac{Y_{\phi=0}{}^4}{B} = 1 + \left(\frac{C}{\sin^{4/3}\phi}\right)_{\phi=0} \tag{30–50}$$

Since 1 is negligible compared to $Y_{\phi=0}{}^4/B = \infty$, eliminating $\sin\phi$ from Eqs. 49 and 50,

$$C = \frac{1}{B}\left(\frac{3\mu \dot{W}_1'}{2\gamma^2}\right)^{4/3} \tag{30–51}$$

Hence Eqs. 30–42 and 51 yield the thickness of the condensate film for any place on a tube the top of which receives the flow $\dot{W}_1'$ from a tube above.

Since, according to Table 30–1, $\frac{4}{3}\int_0^\phi \sqrt[3]{\sin\phi}\, d\phi = 3.428$, Eq. 30–42 leads to a value of $z = z_1 = 3.428/\sin^{4/3}\phi$ for the bottom of the first tube, the top of which may not receive any liquid from above. Continuing with this amount of z at the second tube, starting again with $\phi = 0$, it is seen that $z_1 = C/\sin^{4/3}\phi$ for the second tube, where C would be due to the dropping of liquid from the first tube. At the bottom of the second tube, z again has increased by $z_1 = 3.428/\sin^{4/3}\phi$, so that for the third tube $2z_1 = C/\sin^{4/3}\phi$, where C would be due to the dropping from the second tube, etc. Hence C is increasing by the amount 3.428 for each tube of the row, and the following recursion formula can be used to find the thickness of the film at any angle ϕ of the nth tube:

$$\frac{Y^4}{B} = \frac{1}{\sin^{4/3}\phi}\left[(n-1)3.428 + \frac{4}{3}\int_0^\phi \sqrt[3]{\sin\phi}\, d\phi\right] \tag{30–52}$$

Graphical integration of h_m for the second tube of a row (under the same conditions as used above for the first tube) shows that 35.5% and 32.5% of the amount condensed on the first tube will be condensed on the upper and lower half, respectively, of the second tube. As a whole the second tube produces 68% of the condensate of the first tube.

Figure 30–4 gives Y and h for the second tube. At $\phi = 170°$ the film thickness is 0.125 mm, compared to about 0.105 mm at the same angle of the first tube.

Equations 30–40, 46, and 52 can be used to find an average h_m for n tubes located below one another, namely:

$$h_m = \frac{k}{\sqrt[4]{Bn\pi}} \int_n \frac{\sin^{1/3} \phi \cdot d\phi}{\sqrt[4]{\frac{4}{3} \int_n \sin^{1/3} \phi \cdot d\phi}}$$

where the limit sign n of the integral indicates that the integration has to be performed n times in the limits $\phi = 0$ and $\phi = \pi$. Then

$$h_m = \frac{k}{\sqrt[4]{Bn\pi}} \cdot \frac{n \int_0^{\pi} \sin^{1/3} \phi \cdot d\phi}{\sqrt[4]{n \cdot 3.428}} \sim \frac{1}{\sqrt[4]{n}}$$

Hence, Eq. 30–48 generalizes to

$$h_m = 0.725 \sqrt[4]{\frac{\gamma^2 \lambda k^3}{n\mu D(t_f - t_s)}} \tag{30–53}$$

30–6. Dimensionless Relations for Film Condensation of Vapor at Rest

The equations for h_m can be expressed in dimensionless form, as has been demonstrated by Kirkbride [1934]. Considering a *vertical surface*, the local coefficient of heat transfer, h, may be expressed in different ways:

$$h = \frac{k}{Y} = \frac{dq'}{(t_f - t_s)\, dx} = \frac{\lambda \cdot d\dot{W}'}{(t_f - t_s)\, dx} \tag{30–54}$$

On the other hand, according to Eq. 30–27,

$$h_m = \frac{\lambda \dot{W}_H'}{H(t_f - t_s)} \tag{30–55}$$

where the subscript H refers to the point $x = H$.

From Eqs. 30–54 and 55

$$Y = \frac{k\dot{W}_H'\,dx}{h_m H\cdot d\dot{W}'} \tag{30-56}$$

and from Eq. 30–15

$$Y = \sqrt[3]{\frac{3\mu\dot{W}'}{\gamma^2}} \tag{30-57}$$

From Eqs. 30–56 and 57

$$k\sqrt[3]{\frac{\gamma^2}{3\mu}}\frac{dx}{H} = h_m\frac{\sqrt[3]{\dot{W}'}\,d\dot{W}'}{\dot{W}_H'} \tag{30-58}$$

Integrating over the height H,

$$h_m = 0.925k\sqrt[3]{\frac{\gamma^2}{\mu\dot{W}_H'}} \tag{30-59}$$

Now a Reynolds number for the film, N_{Re}, may be defined, considering that $D_e = 4Y$ is the equivalent diameter of a fluid layer on one side of which friction is acting: §

$$(N_{Re})_H = \frac{(v_m)_H 4Y_H}{\nu} = \frac{4\dot{W}_H'}{g\mu} \tag{30-60} \|$$

Substituting in Eq. 30–59 and rearranging,

$$\left(\frac{g\mu^2 h_m{}^3}{\gamma^2 k^3}\right)^{1/3} = 1.47\left(\frac{4\dot{W}_H'}{g\mu}\right)^{-1/3} = 1.47(N_{Re})_H{}^{-1/3} \tag{30-61}$$

For a *horizontal tube* the procedure is analogous:

$$h = \frac{k}{Y} = \frac{dq'}{2(t_f - t_s)r\cdot d\phi} = \frac{\lambda\cdot d\dot{W}'}{2(t_f - t_s)r\cdot d\phi} \tag{30-62}$$

where $\dot{W}'/2$ and $q'/2$ are the weight flux and heat flux, respectively, for any point of the surface.

In analogy to Eq. 30–55

$$h_m = \frac{\lambda W_D'}{2\pi r(t_f - t_s)} \tag{30-63}$$

where the subscript D refers to the point $\phi = \pi$.

§ $D_e = 4A/C$, where A is the area of a cross section through the film, and C the part of its circumference on which friction is acting.

‖ Here and in some other places the gravitational acceleration g occurs in the dimensionless groups when all dimensionless quantities are expressed in units of the British gravitational system (see Appendix B, Note 5).

From Eqs. 30–62 and 63

$$Y = \frac{k\dot{W}_D' r \cdot d\phi}{h_m(\pi r)\, d\dot{W}'} \tag{30–64}$$

and from Eq. 30–14

$$Y = \sqrt[3]{\frac{3\mu \dot{W}'}{2\gamma^2 \sin\phi}} \tag{30–65}$$

From Eqs. 30–64 and 65

$$k \cdot \sqrt[3]{\frac{\gamma^2 \sin\phi}{3\mu}} \cdot \frac{r \cdot d\phi}{\pi r} = h_m \frac{\sqrt[3]{\dot{W}'/2}}{\dot{W}_D'}\, d\dot{W}' \tag{30–66}$$

Integrating within the limits $\phi = 0$ and $\phi = \pi$,

$$\frac{k}{\pi} \sqrt[3]{\frac{\gamma^2}{3\mu}} \cdot \frac{4}{3} \int_0^\pi \sqrt[3]{\sin\phi}\, d\phi = h_m \sqrt[3]{\frac{\dot{W}_D'}{2}} \tag{30–67}$$

According to Table 30–1,

$$\tfrac{4}{3}\int_0^\pi \sqrt[3]{\sin\phi}\, d\phi = 3.428$$

Hence

$$h_m = 0.952k \sqrt[3]{\frac{\gamma^2}{\mu \dot{W}_D'}} \tag{30–68}$$

Introducing

$$(N_{Re})_D = (v_m)_D \cdot \frac{4Y_D}{\nu} = \frac{4(\dot{W}_D'/2)}{g\mu} \tag{30–69}$$

which holds ¶ for $\phi = \pi$:

$$\left(\frac{g\mu^2 h_m{}^3}{\gamma^2 k^3}\right)^{1/3} = 1.51 \left(\frac{2\dot{W}_D'}{g\mu}\right)^{-1/3} = 1.2(N_{Re})_D{}^{-1/3} \tag{30–70}$$

30–7. Theory of Film Condensation of Vapor Flowing through a Vertical Tube

Nusselt [1916(*a*)] also analyzed the case of saturated vapor flowing downward or upward on a vertical cooling surface. It may be assumed that the cooling surface is a vertical tube with the vapor inside (diameter D).

¶ A second stream of liquid of the same strength comes down to $\phi = -\pi$.

If v_v is the vapor velocity for *downward flow*, then, according to Eq. 30–14,

$$v_m = \frac{\gamma}{3\mu} Y^2 + \frac{f}{8\mu} \cdot \frac{\rho_v v_v^2}{2} Y = \frac{\gamma}{\mu}\left(\frac{Y^2}{3} + \frac{3}{8} bY\right) \tag{30–71}$$

where

$$b = \frac{f}{3\gamma} \cdot \frac{\rho_v v_v^2}{2} \tag{30–72}$$

having the physical dimension of a length.

The weight of liquid passing in unit time a horizontal plane at distance x from the top of the vertical tube will be

$$\dot{W} = \pi D \gamma v_m Y \tag{30–73}$$

Its increase for a differential dx, using Eq. 30–71, will be

$$d\dot{W} = \pi D\gamma \cdot d(v_m Y) = \frac{\pi D \gamma^2}{\mu}\left(Y^2 + \frac{3}{4} bY\right) dY \tag{30–74}$$

On the other hand, from a heat balance

$$d\dot{W} = \frac{k(\pi D)(t_f - t_s)}{\lambda Y} dx \tag{30–75}$$

From Eqs. 30–74 and 75

$$dx = a(4Y^3 + 3bY^2)\, dY \tag{30–76}$$

where

$$a = \frac{\gamma^2 \lambda}{4\mu k(t_f - t_s)} \tag{30–77}$$

The solution of this equation is

$$\frac{x}{a} = Y^4 + bY^3 + C \tag{30–78}$$

where the constant of integration C is again determined by the amount of liquid which may come from a place $x < 0$, i.e., from above the top of the vertical tube. Since $Y = 0$ at $x = 0$ has been assumed, $C = 0$ and Eq. 30–78 simplifies to

$$\frac{x}{a} = Y^4 + bY^3 \tag{30–79}$$

For given values of a and b the equation yields the film thickness Y as a function of x. The exact solution of the equation, as derived by Klaften, is given and discussed in Nusselt's paper. However, it is very clumsy, and therefore Nusselt has determined the only positive and real solution of the equation by approximation. The exact treatment revealed the following dimensionless relationship:

$$\frac{h_m}{k}\sqrt[4]{\frac{H}{a}} = \Phi\left(b\sqrt[4]{\frac{a}{H}}\right) \qquad (30\text{–}80)$$

where h_m is the mean coefficient taken over the height H, and Φ is a function sign. Values of this function are given in Table 30–2.

TABLE 30–2

NUMERICAL SOLUTION OF EQ. 30–80 FOR THE CONDENSATION OF A VAPOR FLOWING THROUGH A VERTICAL TUBE

$b\sqrt[4]{\frac{a}{H}}$	$\frac{h_m}{k}\cdot\sqrt[4]{\frac{H}{a}}$
0	1.333
0.144	1.412
0.577	1.588
1.290	1.831
2.308	2.114
3.61	2.376
8.11	3.06
14.43	3.67
22.50	4.25
32.47	4.79

For large values of $b\sqrt[4]{a/H}$, that is, in particular, at high vapor velocity, Eq. 30–80 can be replaced by the simple form

$$\frac{h_m}{k}\sqrt[4]{\frac{H}{a}} = \frac{3}{2}\sqrt[3]{b\sqrt[4]{\frac{a}{H}}}$$

or *

$$h_m = \frac{3k}{2}\sqrt[3]{\frac{ab}{H}} = \frac{3}{2}\sqrt[3]{\frac{f\gamma\lambda k^2}{3\mu H(t_f - t_s)}\cdot\frac{\rho_v v_v^{\,2}}{2}} \qquad (30\text{–}81)$$

Since for turbulent flow f is proportional to $v_v^{-1/4}$, the coefficient h_m increases with the 0.58th power of the vapor velocity.† Comparison

* Equation 30–81 differs slightly from Nusselt's original equation because of the difference in f (see Eq. 30–9). In addition to this, not less than four typographical errors had to be corrected in Nusselt's equation and definitions belonging to it. Students of the original paper of Nusselt are warned not to overlook any of the numerous misprintings in that paper.

† Nusselt's original equation leads to the ⅔rd power.

with Eq. 30–28 for vapor at rest shows that with flowing vapor the influence of vapor pressure, temperature difference, and wall height is greater. Since the first of these influences is somewhat complex, a numerical comparison may be given.

Whereas for $H = 1$ m and $(t_f - t_s) = 10$ C, h_m is about 800, 1000, and 1100 B hr^{-1} ft^{-2} F^{-1} for steam at rest and under 0.1, 1, and 5 atmospheres absolute pressure, respectively, the corresponding values for steam flowing at 500 ft/sec are about 1500, 3600, and 6800 B hr^{-1} ft^{-2} F^{-1}, respectively.

For *upward flow* of the vapor, the sign of the second term at the right side of Eq. 30–71 must be changed. Then

$$v_m = \frac{\gamma}{\mu}\left(\frac{Y^2}{3} - \frac{3}{8}bY\right) \tag{30–82}$$

and

$$\frac{x}{a} = Y^4 - bY^3 + C \tag{30–83}$$

If the vertical tube is extended on both ends, the end pieces being well insulated, then at high vapor velocity all condensate will be driven upward, so that

$$Y = 0 \quad \text{for} \quad x = H$$

Herewith, from Eq. 30–83,

$$C = \frac{H}{a} \tag{30–84}$$

and

$$\frac{x}{a} - \frac{H}{a} = Y^4 - bY^3 \tag{30–85}$$

The discussion of this equation becomes rather involved. If the vapor velocity is below a certain critical value, then a part of the liquid flows upward, another part downward, and an unsteadiness in the film thickness occurs at the dividing point of the two flows. Referring to the details, see Nusselt's [1916(*a*)] paper.

So far, it has been assumed that the vapor velocity v_v does not change along the cooling surface. This restricts the use of the theory to the condensation of only a small part of the vapor in a tube, for the velocity of vapor flowing through a tube necessarily decreases while condensation proceeds.

Jakob, Erk, and Eck [1935] extended the theory to the case of decreasing v_v and therefore to the condensation of a large fraction of the flowing vapor. Only the case of downward flowing vapor has been dealt with, and it was assumed that only the velocity v_v, not the density ρ_v, of the vapor is changing. For simplification Nusselt's assumption that f in Eqs. 30–9, 13, and 14 is a constant has also been retained.

Denoting the entrance velocity and weight rate of the vapor by $v_{v,0}$ and $\dot{W}_{v,0}$, respectively,

$$\dot{W}_{v,0} = \rho_v \frac{\pi}{4} D^2 v_{v,0} \tag{30–86}$$

The weight rate of condensate at a distance x from the entrance, on the other hand, is

$$\dot{W} = \pi D \gamma v_m Y \tag{30–73}$$

From these equations, obviously,

$$v_v = \frac{\dot{W}_{v,0} - \dot{W}}{\rho_v(\pi/4)D^2} = \frac{\dot{W}_{v,0} - \gamma\pi D Y v_m}{\rho_v(\pi/4)D^2} \tag{30–87}$$

Substituting this in Eq. 30–13 and integrating,

$$v_m = \frac{N}{MY^3}[1 + MY^2 - \Phi(Y)] \tag{30–88}$$

where

$$M = \frac{f\gamma v_{v,0}}{\mu D}$$

$$N = \frac{\rho_v D v_{v,0}}{4\gamma}$$

and

$$\Phi(Y) = \sqrt{1 + 2MY^2 - \frac{2}{3}\cdot\frac{M}{N}\cdot\frac{\gamma}{\mu}\cdot Y^5} \tag{30–89}$$

In a differential section dx the rate of condensation increases by

$$d\dot{W} = \pi D\gamma d(v_m Y) = \frac{2N\pi\gamma D}{Y}\left[\frac{PY^3 - 1}{\Phi(Y)} + \frac{\Phi(Y) - 1}{MY^2}\right] dY \tag{30–90}$$

where

$$P = \frac{5}{6N}\cdot\frac{\gamma}{\mu}$$

For superheated steam a heat balance, similar to the one of Eq. 30–31, can be set up:

$$\lambda\cdot d\dot{W} + h_v\pi D(t_v - t_f)\,dx = k\frac{\pi D}{Y}(t_f - t_s)\,dx \tag{30–91}$$

From Eqs. 30-90 and 91

$$\frac{dx}{dY} = \phi(Y) \tag{30-92}$$

where

$$\phi(Y) = 2N\gamma\lambda \frac{\dfrac{PY^3 - 1}{\Phi(Y)} + \dfrac{\Phi(Y) - 1}{MY^2}}{k(t_f - t_s) - h_v Y(t_v - t_f)}$$

This equation also includes the case of saturated vapor ($t_v - t_f = 0$).

The authors solved the equation numerically by a method of finite differences. The procedure will briefly be indicated.

Corresponding to Eq. 30-92,

$$\Delta x = \phi(Y) \cdot \Delta Y \tag{30-93}$$

Assuming an appropriate value of ΔY, the values Δx_1, Δx_2, Δx_3, $\cdots$ can be calculated which belong to $Y = \Delta Y$, $Y = 2 \cdot \Delta Y$, $Y = 3 \cdot \Delta Y$, $\cdots$ respectively. Then in first approximation the film thickness up to $Y = n \cdot \Delta Y$ will be

$$x_n \approx \Delta x_1 + \Delta x_2 + \cdots \Delta x_{n-1} + \tfrac{1}{2}\,\Delta x_n \tag{30-94}$$

A more nearly exact equation is also given in the authors' paper; however the results of the first approximation were found to be close to those of the second approximation in some numerical examples.

For instance, considering the condensation of superheated steam at atmospheric pressure ($t_v = 300°C$) and entrance velocity $v_{v,0} = 80$ m/sec in a tube of inner diameter $D = 0.04$ m, where $t_f - t_s = 25$ C, it was asked at what length x the film would reach a thickness $Y = 0.15$ mm.

Dividing $Y = 0.15$ into 15 parts (i.e., $\Delta Y = 0.01$ mm), $x = 1.600$ m was found; assuming 5 parts ($\Delta Y = 0.03$ mm), $x = 1.610$ m, and assuming 3 parts ($\Delta Y = 0.05$ mm), $x = 1.635$ m were obtained. Division into a few parts was obviously sufficient.

30-8. Experiments on Film Condensation of Vapors

Much information has been published about the film condensation of vapors. Because of the sound basis and mathematical perfection of Nusselt's theory, the results of experiments used to be compared with the predictions of the theory.

A few main reasons for deviations between experimental and theoretical results will be indicated before entering into details.

First of all, mixed condensation may occur on some places of the cooling surface. If dropwise condensation exists on some small spots only, then it may be difficult to discover it. The effect, of course, will be a better performance of condensation than occurs according to the film theory.

Turbulence of the film has a similar effect. Nusselt's theory is based upon the assumption of streamline flow of the condensate. However, turbulence may occur when the film thickness and velocity exceed critical values, which will particularly occur on the end section of a cooling surface. In order to check Nusselt's assumption of laminar flow Cooper, Drew, and McAdams [1934] represented experimental data of various investigators in dimensionless form. These measurements dealt with the isothermal flow of liquid layers on smooth vertical and inclined walls in the range of $N_{Re} = 1.5$ to 150,000, where the Reynolds number is defined as in Eq. 30–60. According to the theory, the coefficient of friction for a plane layer in laminar flow should be

$$f = \frac{24}{N_{Re}} = \frac{96 v_m Y}{\nu} \tag{30–95}$$

From the tests dealt with in the paper mentioned and other ones reviewed by McAdams [1940] it is seen that this equation can be used up to $N_{Re} \approx 2000$. For $N_{Re} > 2000$, however, turbulence and a much better heat transfer have been observed.

In the same year an investigation of Shea and Krase [1940] was published. They had measured heat transfer for different heights, using a vertical surface 23 in. high and five separate cooling-water sections, each of 4.6 in. height. The average film coefficients obtained in this way were about 10% larger than according to the theory for very short plates and about 30% larger for 23-in. height. The authors assumed that these deviations are partly due to ripples and mostly to turbulence in the film, which is more likely to occur on the lower sections because of thicker condensate layers.

Condensation measured on the outside of single horizontal tubes generally deviates not more than $\pm 10\%$ from the theoretical value, obviously, because the height of the condensation surface is small. Values smaller than the theoretical ones were measured by Trapp [1941] with ammonia and by Othmer and R. E. White [1941] on some alcohols.

A third kind of disturbance of the film, mentioned just before, consists in the formation of liquid ripples. Very considerable deviations from the mean thickness Y have been observed [Friedman and Miller 1941], and it is easy to understand that these deviations cause a smaller thermal resistance of the film as an average. Such a waviness of the film is particularly to be expected when liquid is dropping from one tube to another as described in Sect. 30–5.

In fact, most experiments performed on vertical tubes gave higher values than Nusselt's theory. For $N_{Re} < 1300$, McAdams [1940] recommends an addition of 20% to the theoretical value of h_m. Above this Reynolds number, the influence of turbulence of the film prevails, and for this range an empirical dimensionless formula may be used which has been given by Colburn [1933–1934, 1934]:

$$\frac{hY}{c_p\dot{W}'}\left(\frac{\mu c_p}{k}\right)^{2/3} = 0.027\left(\frac{4\dot{W}'}{g\mu}\right)^{-0.2} \tag{30–96}$$

Here local values of h and W' are employed. Mean values h_m have been derived from this equation, which at $N_{Pr} = 5$ are in good agreement with experimental data of Kirkbride [1933–1934, 1934].

For horizontal tubes Eqs. 30–48 and 70 have been found to yield reasonable agreement with experimental values.

Regarding the vapors investigated and tube dimensions used by various workers, the excellent review of McAdams [1940] may be consulted.

An interesting experimental investigation on condensation of dichlorodifluoromethane ("Freon-12") in a bank of horizontal tubes was carried out by Young and Wohlenberg [1942]. Their contribution is particularly valuable because "Freon-12" is one of the poorest conducting liquids, whereas water, most frequently used in condensing experiments, is the best conducting non-metallic liquid. Five copper tubes, $\frac{3}{4}$ in. in outer diameter, 41 in. long, were placed with axes in a vertical plane and cooled from inside by water. The values of h_m for the four lower tubes were found to be about 20% larger than according to Nusselt's theory. About 3% of this amount was explained by Jakob [1942] as due to the change of $t_f - t_s$ from tube to tube in the tests. Assuming an accuracy of the experiments of about ±5%, an excess of 10% of measured above theoretical values seems to be real. This may be caused by some turbulence at the top of each tube due to liquid flowing down from the tube above, and to ripples. It is recommended to add 10% to h for all tubes, except the first one, in order to compensate for the difference mentioned of actual conditions from those assumed in the theory.

Referring to condensation of saturated and superheated steam of atmospheric pressure, flowing through cooled tubes, the author with some coworkers has checked Nusselt's theory in different investigations. First [Jakob and Erk 1928] the condensation in a vertical tube, 17 mm wide, cooled over 460 mm, was measured at entrance velocities $v_{v,0}$ of 10 to 110 m/sec, entrance temperatures $t_{v,0}$ of 101 and 325°C, and temperature differences, $t_f - t_s = 1$ to 25 C. Then [Jakob, Erk, and

Eck 1929, Jakob 1936(*a*)] the same tube was investigated in horizontal position. Much higher rates of heat transfer were observed in this position. The temperature of the tube wall was measured by thermocouples movable in the axial direction in the wall. By rotating the tube around the axis, the temperature at different angles against the horizontal axis could be determined. In these tests the temperature at the bottom was found to be much smaller than at the top, because of the increase of the film thickness from top to bottom. The flow of condensate in axial direction caused an increase of this difference in the flow direction. In this way the mean temperature difference $t_f - t_s$ was obtained as a function of the location on the perimeter, and from the theory the distribution of the film thickness at the same places was calculated, as illustrated by Fig. 30–5.

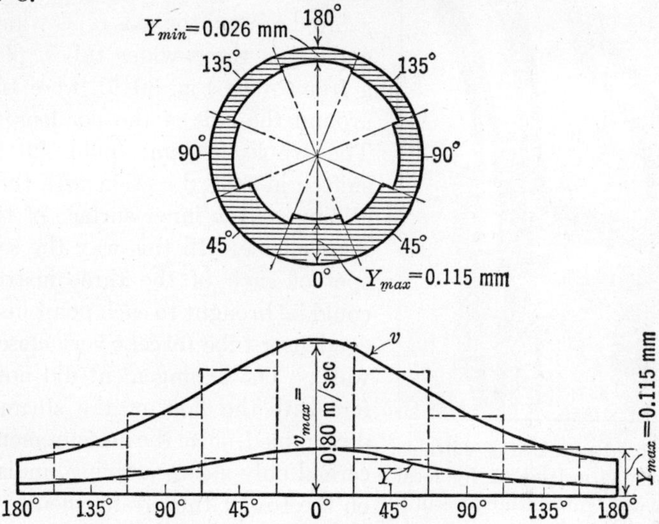

FIG. 30–5. Thickness and velocity of water film inside a horizontal tube in which flowing saturated steam is condensed.

The illustration refers to superheated steam at $t_{v,0} = 617°F$, entering with $v_{v,0} = 95$ m/sec. In the upper part of the figure the thickness of the water film is recorded, the scale being 30 times as great as the scale of the tube diameter. The thickness of the film on the top is only 0.026 mm, but on the bottom it is 0.115 mm. In the lower diagram, the lower curve shows the variation of the film thickness as a function of the location on the perimeter, expressed in terms of the angular positions indicated in the upper diagram. The upper curve represents the velocity of the water film with a maximum of 0.80 m/sec at the bottom. From these distribution curves the weight rate of the condensate was calculated as $\dot{W} = 14.3$ lb/hr, whereas $\dot{W} = 13.8$ lb/hr was found by weighing the condensate formed. The agreement between theory and experiment is excellent.

The final experiments [Jakob, Erk, and Eck 1932, 1935; Jakob 1936(*a*)] were performed with a vertical condenser tube, 40 mm inner diameter and 1210 mm cooled length. Three fine tubes (see Fig. 30–6), issuing like fingers from somewhat wider tubes, were placed inside the condensing tube. One of these fine tubes served to measure the static pressure, another the total pressure, and the third, containing a thermocouple, the temperature. Each fine tube could be rotated around the axis of its wider tube, and these three wider tubes, placed in a frame (see Fig. 30–6), were turnable around the axis of the condenser tube. The whole system could further be shifted in axial direction with the frame gliding on the inner surface of the condenser tube. In this way the sensitive spot of each of the three instruments could be brought to each point inside the condenser tube (except very close to the wall). The arrangement did not interfere with the flow of the steam to be determined, since the measurements concerned only steam arriving undisturbed on the top of the arrangement.

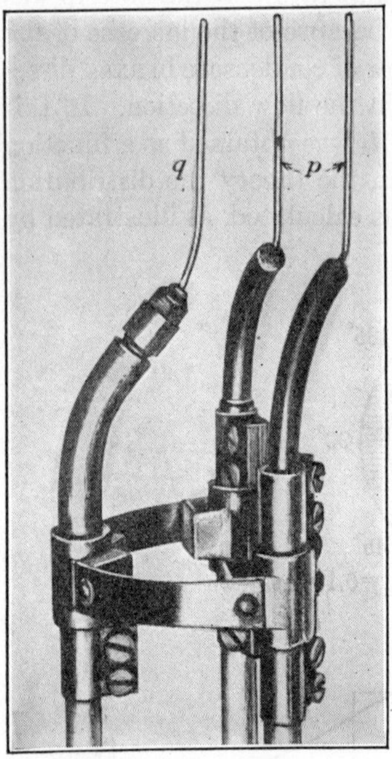

Fig. 30–6. Upper end of shiftable and turnable instrument for local velocity and temperature measurements.
p = Impact and static pressure tubes.
q = Tube containing thermocouple.

Herewith, first, the following anomaly was studied and quantitatively cleared, which had previously been observed and explained qualitatively by Jakob, Erk, and Eck [1930]. If superheated steam always entered the condenser tube with the same velocity and temperature, then the temperature, measured in the center of the lower end of the condenser tube, rose if the temperature of the tube wall was decreased by stronger cooling from the outside. For instance, when the steam entered at 325°C, it left at 280°C if the tube wall was kept at 95°C, and it left at 315°C when the wall temperature was reduced to 75°C. This phenomenon was explained in the following manner. Cooling of the core of steam, apart from expansion, is caused by "hot" (i.e., fast-

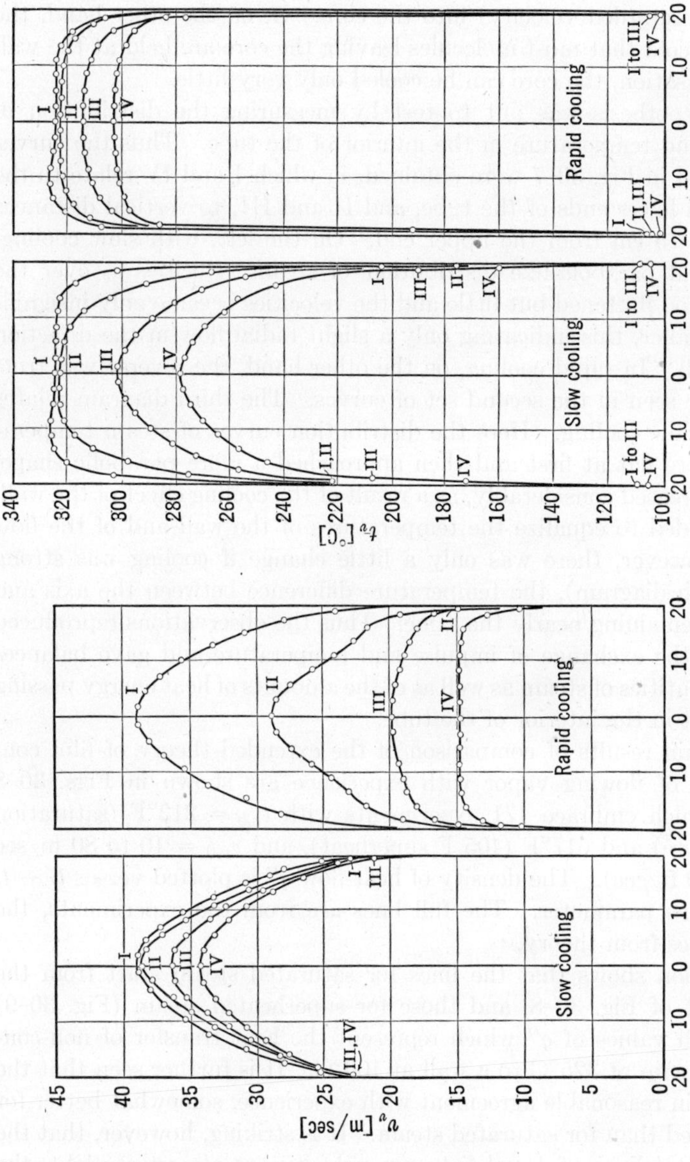

FIG. 30–7. Distribution of velocity and temperature of superheated steam condensing in a vertical tube. I at upper end of tube, IV at lower end of tube. Steam entrance at upper end.

moving) molecules coming out of the core, then touching the "colder" (i.e., slower) molecules close to the surface, and finally returning "cooled" (i.e., with retarded velocity) into the core. If, on the other hand, the wall is so cold that most molecules leaving the core are held at the wall by condensation, the core can be cooled only very little.

This hypothesis was put to test by measuring the distribution of velocity and temperature in the interior of the tube. Thus the curves reproduced in Fig. 30-7 were obtained, in which I and IV relate to the upper and lower ends of the tube, and II and III, to vertical distances of 25 and 75 cm from the upper end. On the left, with slow cooling, the original parabola-like distribution of steam velocities v_v over the cross section flattened but little and the velocities became only insignificantly smaller, this indicating only a slight radial flow in the direction to the wall. In rapid cooling, on the other hand, the reverse was true, as may be seen in the second set of curves. The third diagram relates again to slow cooling. Here the distribution curves of steam temperature t_v were flat at first and then approached a more parabolic shape, and t_v decreased considerably, as a result of the cooling effect of the wall, which tended to equalize the temperatures of the wall and of the flow core. However, there was only a little change if cooling was strong (see fourth diagram), the temperature difference between the axis and the wall remaining nearly the same. Thus the observations reproduced perfectly the exchange of impulse and temperature and gave balances of the quantities of steam as well as of the amounts of heat energy passing each point in the interior of the tube.

The main results of comparison of the extended theory of film condensation of flowing vapor with experience are shown in Figs. 30-8 and 9, which embrace 171 experiments with $t_{v,0} = 212°F$ (saturation temperature) and 617°F (405 F superheat), and $v_{v,0} = 10$ to 80 m/sec (33 to 260 ft/sec). The density of heat flow q'' is plotted versus $t_f - t_s$ with $v_{v,0}$ as parameter. The full lines are from the experiments, the dotted lines from theory.

Inspection shows that the lines for saturated steam start from the zero point of Fig. 30-8, and those for superheated steam (Fig. 30-9) from small values of q'' which represent the heat transfer of non-condensing steam at 325°C to a wall at 100°C. It is further seen that the theory is in reasonable agreement with experience, somewhat better for superheated than for saturated steam. It is striking, however, that the experimental lines of equal entrance velocity are almost straight, the theoretical lines more nearly parabolic.

In particular, the comparison shows the following two differences:

1. For large values of $t_f - t_s$, i.e., in strong cooling, the heat transfer observed was always better than that calculated by the theory. This

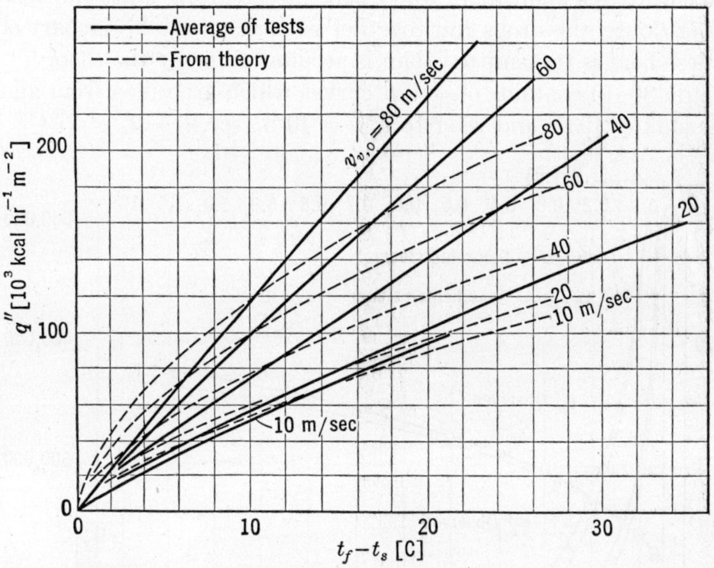

FIG. 30–8. Heat transfer by condensation of saturated steam at atmospheric pressure.

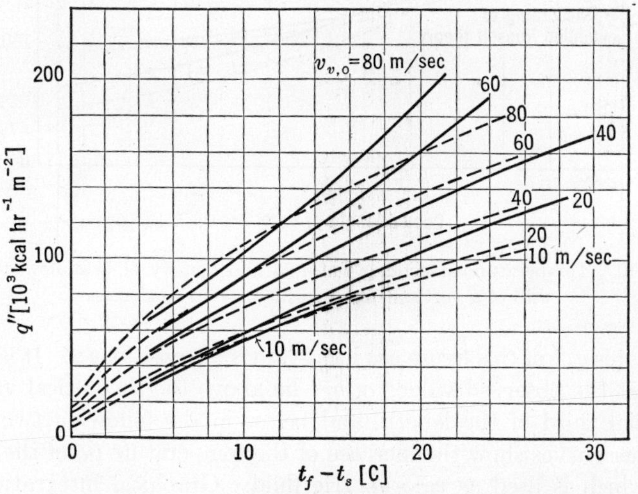

FIG. 30–9. Heat transfer by condensation of superheated steam at atmospheric pressure and 325°C.

is partially due to dropwise condensation on the upper end of the cooling surface. At this point more water is formed than by film condensation. Farther down, the drops run together so that on the lower part of the tube less heat is transmitted than is accounted for by the theory.

Figure 30–10 contains observed curves which are taken from another paper [Jakob 1934] and refer to $v_{v,0} = 40$ m/sec, $t_f - t_s \approx 27$ C. The

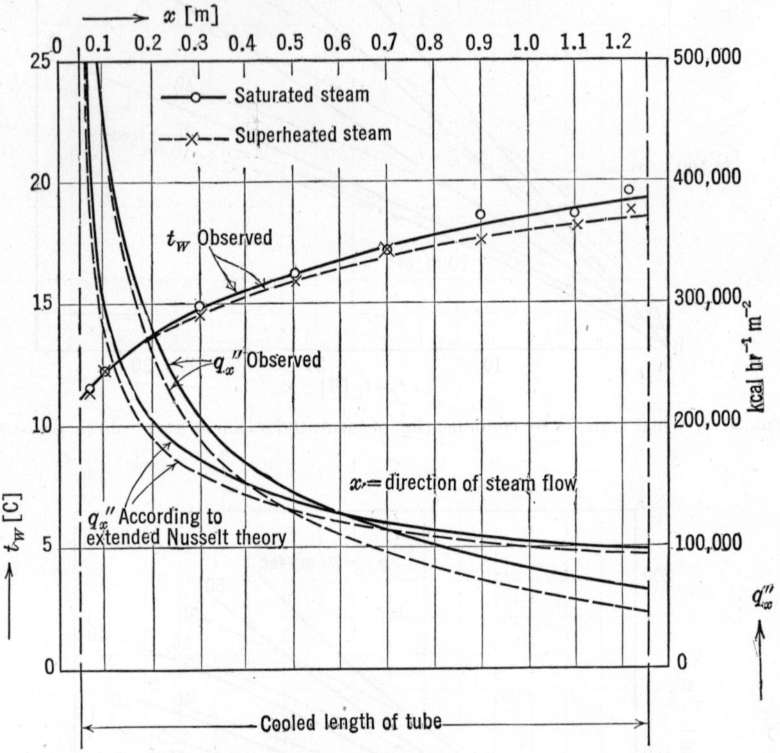

FIG. 30–10. Temperature of cooling water t_w and density of heat flow across the wall of a vertical tube in which steam condenses.

coordinates q_x'' of this figure are local values for positions x. It is clearly seen that the observed values of q_x'' lie above the theoretical values in the initial third of the length, and below in the following two thirds. The other curves show the increase of the temperature t_W of the cooling water which is used as calorimetric fluid. Graphical integration over the whole cooled length gave

$$\left(\int q_x'' \cdot dx\right)_{obs} - \left(\int q_x'' \cdot dx\right)_{calc} > 0 \qquad (30\text{–}97)$$

the difference being larger for superheated than for saturated steam, in accordance with the respective average values of Figs. 30–8 and 9.

Since in the experiments the Reynolds number of the water film, defined by Eq. 30–60, reached values up to 2300 and the critical value may be as low as 1500 or 1600, q'' will be raised by turbulence at the highest values of $v_{v,0}$. This partly explains the steeper increase of $\dot{q}''$ at the high values of $t_f - t_s$ and $v_{v,0}$ in Figs. 30–8 and 9.

2. For small values of $t_f - t_s$ the experiments with saturated steam yielded lower values of q'' than the theory, particularly at the high entrance velocity of 80 m/sec. The steam may have been undercooled in this case, a possibility which Nusselt [1928, p. 23] has already considered in order to explain the observation of dry starting lengths by Holmboe. According to the curves for $v_{v,0} = 80$ m/sec in Fig. 30–8, assuming $t_f - t_s = 5$ C, $q_{calc}'' = 78{,}000$, and $q_{obs}'' = 59{,}000$ kcal hr^{-1} m^{-2}. Film formation on two thirds of the cooled length would be sufficient to set free $q'' = 59{,}000$ as an average. Hence, the first third of the tube might remain entirely dry, and in this case much less heat would be transferred to the wall than undercooled steam can release before being cooled down to the Wilson line, which is the limit at which condensation must occur. It may be mentioned that Claassen [1933] had already presumed that the first section of a tube would remain dry; this, however, refers to superheated steam, where Fig. 30–9 does not show much of such an effect.

As a simple empirical representation of the experimental results of Figs. 30–8 and 9 the authors recommend:

For saturated steam:

$$q'' = (3400 + 100v_{v,0})(t_f - t_s)\left(\frac{1.21}{H}\right)^{1/3} \tag{30–98}$$

For superheated steam:

$$q'' = (3500 + 51v_{v,0})(t_f - t_s)\left(\frac{1.21}{H}\right)^{1/3} \tag{30–99}$$

where H is the cooled length in m,

$v_{v,0}$ is the entrance velocity in m/sec,

t_f is the saturation temperature in °C,

t_s is the wall temperature in °C, and

q'' is the heat flux in kcal hr^{-1} m^{-2}.

30–9. Condensation of Mixtures of Vapors

Since there is no difference in principle between vapors and gases, mixtures of vapors may include gases which are non-condensable under the thermodynamical conditions considered. Such mixtures, however, will be dealt with in the next section, and the term mixtures of vapors may be restricted to condensable vapors.

Binary mixtures with components of complete mutual solubility of the liquids will behave exactly as simple substances. Therefore, the film theory as presented in this chapter and empirical formulas like Eqs. 30–98 and 99 can be used. However, it is necessary to consider that instead of the saturation temperature of vapors the dew-point temperature must be used and that the composition of the condensate formed is different from the composition of the vapor mixture at the same temperature and total pressure. It will be recalled that the dew point is that temperature at which liquid first forms from a vapor mixture cooled at constant pressure. Whereas the temperature of simple vapors remains the same until everything is condensed, the point of complete condensation at constant pressure is at a lower temperature, called the bubble point (i.e., the point where bubbling starts in heating up the liquid mixture). This, of course, complicates the precalculation of condensation. Obviously, in order to apply the film theory, the saturation temperature t_f of the free film surface has to be replaced by the dew-point temperature of the mixture at the existing total pressure, and the mole fraction of the condensed liquid has to be taken, for the same pressure and temperature, from a composition diagram of the constituents.

If the temperature of the vapor mixture to be condensed is above the dew point, then the procedure described for superheated vapors (see, for instance, Eq. 30–31) has to be used.

Experimental data on the condensation of ethanol and steam obtained by Wallace and Davison [1938] and Hagenbuch [1941] and correlated by McAdams [1942] scatter considerably and deviate from the theoretical values.

In condensing binary mixtures of organic vapors and steam which give immiscible condensates, drops of water in the organic liquid film were observed, first by Cogan [1934] and then by E. M. Baker and Mueller [1937, 1937(*a*)] and by E. M. Baker and Tsao [1940, 1940(*a*)]. Further experimentation will be needed to elucidate more clearly these complex processes.

30–10. Influence of Non-condensable Gas upon the Condensation of a Vapor

For a long time it has been observed that in the condensation of steam much larger cooling surfaces are required when the steam contains some air. In this case the temperature of the free surface of the film t_f will be equal to the saturation temperature of steam at a pressure equal to its partial pressure on that surface. Since the saturation temperature decreases considerably with the pressure, the difference $t_f - t_s$ will become much smaller when air is present; hence much less heat will be transferred, which means correspondingly less condensation. The temperature t_f can be found from a heat balance, considering that the heat conducted through the liquid film must be equal to the sum of the heat transferred by convection from the bulk of the steam-air mixture at temperature t_v to the free film surface and the heat liberated by condensation. To determine this, the amount of steam must be known which diffuses to the liquid surface. The difference between the partial pressure of the water vapor in the bulk of the vapor mixture and its partial pressure on the free liquid surface is the driving force for this diffusion. Using the procedures dealt with in Chapter 28, the temperature t_f may be found by trial and error. However, its value will change along the liquid film, as will be explained later on.

Langen [1931] has performed experiments on the condensation of steam at rest on the outside of a horizontal tube of 30 mm outer diameter which was cooled from inside over a 900-mm length. Platinum wires were stretched through holes in the tube wall and served as integrating resistance thermometers in angle intervals of $360/8 = 45°$ around the axis. The total pressure was varied from 0.15 to 0.65 kg/cm^2 absolute. Air contents between 0.2 and 20% of the steam weight were covered. Nusselt's [1916(*a*)] theory of condensation and equation for the diffusing steam were used; however, they were eventually replaced by empirical equations. In this way Figs. 30–11 and 12 were obtained, in which curves for air-free steam were inserted by Erk [1933]. These charts are for 0.2 kg/cm^2 condenser pressure and vertical surfaces of 1 m height.

A more thorough treatment of the problem would have to consider that, because of the greater thickness of the liquid film in the lower part of the surface, less heat would be transferred there if t_f were the same as on the upper part. If this were true, the diffusion of steam from the bulk of the vapor mixture to the free liquid surface would be the same. Hence, equal amounts of steam would be carried to equal upper and lower portions of the surface. Assuming that all steam arriving on the upper portion would be condensed, not all steam arriving on the lower part could be condensed; uncondensed steam would collect there which, of course, cannot occur under steady-state conditions. Therefore, t_f must become larger on the lower parts of the film than on the upper ones. Hence on the lower parts the difference of partial pressures $p_v - p_f$ and the diffusion of steam due to that difference will be reduced, and, on the other hand, the heat flow through the film will be increased so much that heat- and mass-flow equilibria will be retained. This, of course, means a considerable complication of the process, which Nusselt

[1916(*a*)] does not mention, though the procedure recommended by him will encounter this difficulty. The same holds for a trial-and-error

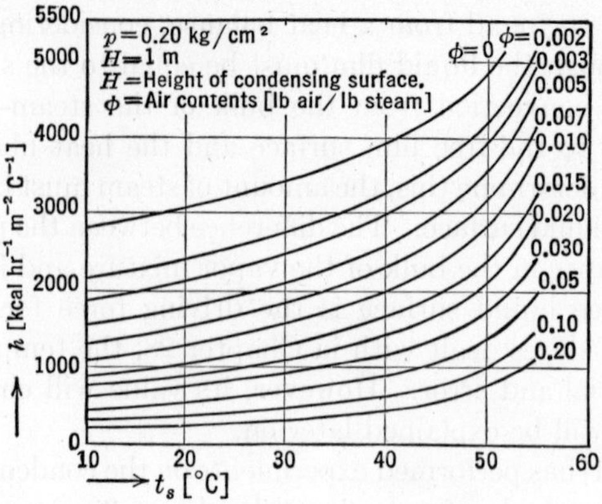

FIG. 30–11. Coefficient of heat transfer h in the condensation of steam containing air versus surface temperature t_s.

method which has been suggested by Colburn and Hougen [1934] and which proceeds in the same way.

Another complication occurs under vigorous cooling conditions, where a thin layer of fog may be formed close to the water surface. The

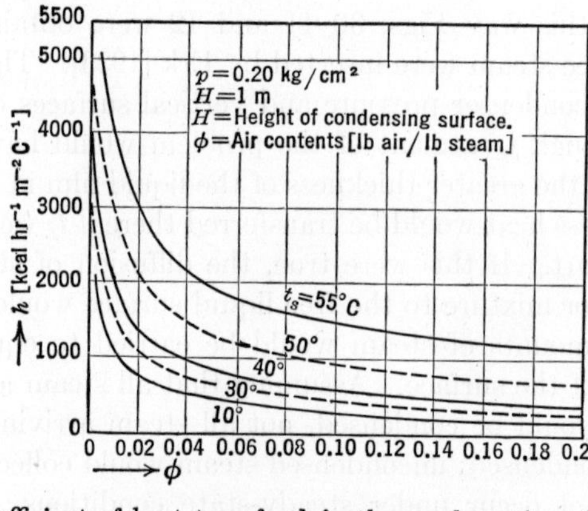

FIG. 30–12. Coefficient of heat transfer h in the condensation of steam containing air versus air contents ϕ.

assumptions made in the film theory of condensation do not hold any more in this case.

The analysis given in this section is not restricted to steam-air mixtures, but is also valid for other vapor-gas mixtures.

30–11. Theories of Dropwise Condensation

The occurrence of film or dropwise condensation is due to the forces acting between solid surface, liquid, and vapor (see Fig. 30–13). When the outer contact angle β is made equal to zero, then the spreading tension σ_{spr} (generally given in dynes per centimeter) can be expressed as the sum of the interfacial tensions:

$$\sigma_{spr} = \sigma_{s,g} - \sigma_{s,f} - \sigma_{f,g} \quad (30\text{–}100)$$

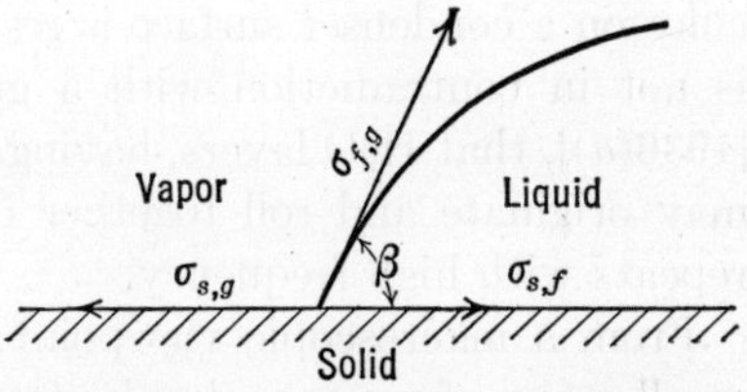

FIG. 30–13. Forces acting on the interfaces between solid, liquid, and vapor.

where the subscripts s, f, and g refer to the solid, liquid, and gaseous phase, respectively. For $\sigma_{spr} > 0$ the liquid will spread over the surface. If $\sigma_{spr} < 0$, the liquid will draw off the solid surface until equilibrium is obtained, i.e.,

$$\sigma_{s,g} = \sigma_{s,f} + \sigma_{f,g} \cos \beta \quad (30\text{–}101)$$

From Eqs. 30–100 and 101

$$\sigma_{spr} = \sigma_{f,g}(\cos \beta - 1) \quad (30\text{–}102)$$

By measuring the surface tension $\sigma_{f,g}$ and β, the spreading tension can be determined.

Theories of dropwise condensation have been proposed by Eucken [1937] and Emmons [1939].

Emmons remarks that $\sigma_{spr} < 0$ is a necessary, but not sufficient, condition for dropwise condensation. The drops of liquid which form on the cooling surface grow, touch, and merge rapidly. If, after touching, two drops merge so slowly that other drops combine with them on all sides, the surface will be covered with a film, formed by the joining of drops. Therefore Emmons indicates four conditions favoring dropwise condensation:

1. Low rate of condensation.
2. Low viscosity of the condensate.
3. High surface tension and large negative values of σ_{spr}, these being the forces which produce the drop shape.
4. Smoothness of the cooling surface.

So far no material which has such a combination of mechanical, chemical, and thermal properties that always $\sigma_{spr} < 0$ has been found for the construction of condenser tubes. It is therefore necessary to cover the cooling surface with a very thin layer of some substance which has an extremely low affinity for the vapor, but a high affinity for the solid used, so that the layer will not be washed off too readily by the condensate drops. The most effective drop promoters have been mentioned in Sect. 30–1. Emmons has claimed from the theory and verified by experiments that one and only one complete layer of promoter molecules on a condenser surface is responsible for drop condensation. This is not in contradiction with a guess, previously published by Jakob [1936(*a*)], that H_2O layers, having a thickness of the order of $1/1000$ mm, may originate and roll together to a droplet, and that this procedure repeats with high frequency.

From a macroscopic viewpoint one might think that any arbitrarily small group of vapor molecules touching the solid surface would immediately be converted to liquid. This, however, is not the case. Eucken, in the paper mentioned above, used the kinetic theory of gases to calculate what heat transfer would occur in the condensation of steam at atmospheric pressure if all molecules which hit the cooling surface were immediately condensed. He arrived at $q'' = 72(10^6)$ B hr^{-1} ft^{-2}, that is, 260 times more than has ever been observed with dropwise condensation. According to this result, as a maximum, about 0.4% of the steam molecules hitting a cooling surface will be immediately converted into the liquid state. This is in good agreement with results of evaporation experiments performed by Alty and Nicoll [1931] at 12°C, in which 1.5% was found. The authors mention that this figure will decrease by 0.01 for 1 C temperature increase, so that at 100°C about 0.6% of the hitting molecules would be condensed.

The reflection of more than 99% of the molecules is in agreement with other experiences with interfaces. According to Langmuir [1918], the first step in the condensation of a vapor consists in the catching of single molecules by the colder surface and the formation of a very thin layer of supersaturated (undercooled) vapor. To explain this, Emmons mentions that vapor molecules which strike the bare cooling surface evaporate at the temperature of that surface and are therefore at a lower temperature than saturated vapor. Hence a blanket of supersaturated vapor will cover the bare cooling surface between the droplets.

Eucken compares the formation of the extremely small droplets with the formation of nuclei in a supersaturated solution. In the close neighborhood of the droplets the monomolecular layer is not supersaturated any more, since supersaturation vanishes when condensation occurs. The density ("surface concentration") of the layer close to the droplets is therefore somewhat smaller than at some distance. As in solutions, a concentration gradient causes mass transportation toward the droplets, which in three-dimensional bodies is known as diffusion and in monomolecular interfacial layers is called surface diffusion. In this way the droplets are fed from the supersaturated adsorbed surface layer. According to Eucken's guess, the field in which the gradient of surface concentration is sufficiently large for this process may be only $\frac{1}{1000}$ or $\frac{1}{100}$ mm wide. It is cooled by the solid surface, which itself is cooled by the adsorption of liquid. The practical consequence would be to increase the number of droplets so that a relatively large part of the cooling surface would be covered by very small droplets. This may be obtained by the use of suitable material, e.g., very finely grained two-phase alloys.

Emmons assumes mechanical influences. He thinks that by the rapid condensation a local reduction of pressure will occur, which in turn would set up violent local eddy currents in the vapor between the drops (see Fig. 30–14). Thus, his concept deviates from Eucken's theory by considering turbulent flow instead of molecular diffusion as the driving force.

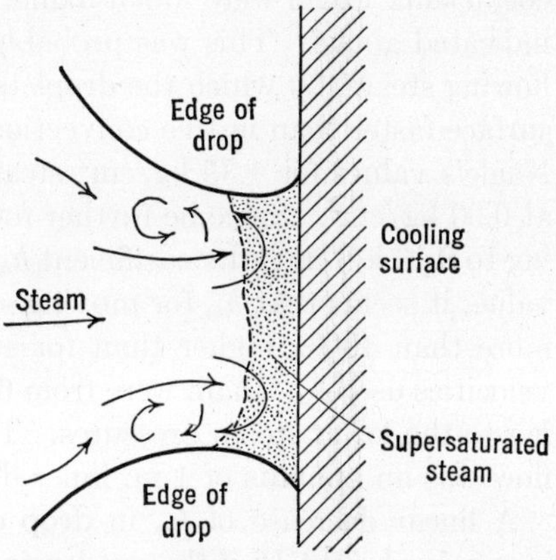

FIG. 30–14. Eddy currents in the vapor between drops on a condensing surface.

The present author is more in favor of Eucken's explanation of the mechanism; it may be that, as in many heat-transfer processes, molecular and macromechanical exchanges concur.

30–12. Experimental Results on Dropwise Condensation

E. Schmidt, Schurig, and Sellschopp [1930], who probably were the first to study systematically dropwise condensation, used a polished or chromium-plated copper plate, the rear side of which was vigorously cooled by a jet of water. They claim to have obtained the most stable drop condensation with clean polished surfaces; treatment of the surface with oil or fat had no effect at all, and treatment with kerosene a temporary effect only. The mean coefficients h_m of heat transfer which they found were between 6000 and 8000 B hr^{-1} ft^{-2} F^{-1}.

Nagle, Bays, Blenderman, and Drew [1935], using oleic acid on a chromium-plated copper tube, found values of $h_m \approx 14{,}000$ B hr^{-1} ft^{-2} F^{-1} with deviations of only $\pm 10\%$, whereas q'' was changed from 77,000 to 170,000 B hr^{-1} ft^{-2}. These results were obtained with filtered steam, whereas steam containing rust gave much smaller values. Fitzpatrick, Baum, and McAdams [1939], who used benzyl mercaptan as promoter on a vertical copper tube, obtained values of $h_m = 6800$ to 16,400 B hr^{-1}

ft^{-2} F^{-1}. These were apparently independent of the heat flux, which ranged from 68,000 to 252,000 B hr^{-1} ft^{-2}. With rust in the steam $h_m = 4200$ B hr^{-1} ft^{-2} F^{-1} was found.

Also Gnam [1937], in experiments with steam driven downward on the surface of a vertical tube, which was cooled from inside, found that h_m was substantially independent of the heat flux. However, he obtained coefficients which were much higher than those of Nagle and coworkers indicated above. This was probably caused by the wiping effect of the flowing steam, by which the droplets were removed from the condensing surface faster than in free convection. A value of h_m, 100% larger than Nagle's values for 1.35 kg/cm² steam pressure, was obtained by Gnam at 0.90 kg/cm². Since he further found that, with the pressure decreasing to 0.16 kg/cm², the coefficient h_m decreased somewhat below Nagle's value, it seems that h_m for moving steam at 1.35 kg/cm² would be even more than 100% higher than for steam at rest. The steam entrance velocities used by Gnam were from 6 to 30 ft/sec, the higher ones belonging to the lower steam pressures. The cross-sectional area of the steam flow was an annulus of 1 in. inner diameter and 1¼ in. outer diameter.

A linear decrease of h_m in drop condensation of steam at rest with increasing height H of the condensing surface was observed by Shea and Krase [1940]. The values $h_m = 17{,}800$ B hr^{-1} ft^{-2} F^{-1} for $H = 4$ in. and $h_m = 15{,}300$ for $H = 24$ in. are taken from a graph in their paper.

In general, our knowledge of the process of dropwise condensation is still rather limited. Further intensive theoretical and experimental studies will be needed to establish the same standard as with respect to film condensation.

PART F. HEAT RADIATION IN SPACES OF SIMPLE CONFIGURATION
and
PART G. SELECTED FIELDS OF APPLICATION
will form the main subjects of Volume II.

APPENDIX A

Problems

NOTE: Each problem is identified by two numbers, the first indicating the chapter to which the problem mainly belongs, and the second being its own number in the sequence for that chapter.

2–1. Derive the differential equation for the temperature t as a function of the coordinates x, y, z, and time τ in a body having uniform and constant density and specific heat, but different conductivities k_x, k_y, and k_z, in the directions x, y, and z, provided that these conductivities are linear functions of t. No heat energy is generated or absorbed in the substance of the body under consideration, but storage of heat energy is supposed to occur.

2–2. Derive the differential equation for the temperature t in a cylinder, considering the radial distance from its axis r and the time τ as independent variables, and assuming perfect symmetry with respect to the axis of the cylinder. Any heat flow in axial direction may be prevented by perfect insulation of the ends of the cylinder. Density, specific heat, and thermal conductivity of the substance are uniform and constant, and no heat energy is generated or absorbed in that substance, but storage of heat energy is supposed to occur.

2–3. Derive the differential equation for the temperature t in a sphere, considering the radial distance from its center r and the time τ as independent variables, and assuming perfect symmetry with respect to the center of the sphere. Density, specific heat, and thermal conductivity of the substance are uniform and constant, and no heat energy is generated or absorbed in that substance, but storage of heat energy is supposed to occur.

2–4. The temperature in an infinitely extended homogeneous body at a certain instant of time may be:

$$t = C(x^4 + y^4 - z^4)$$

where x, y, z are the coordinates of a cartesian system, and C is a constant of positive value.

Where is the temperature t increasing with time and where is it decreasing, provided the physical properties of the substance are independent of temperature?

3–1. Derive the differential equation of continuity of flow for the case of the radial flow of a fluid between two coaxial cylinders, using the radial distance r and the time τ as independent variables.

3–2. Derive the differential equation for the temperature t in the space between two coaxial cylinders 1 and 2, having the radii r_1 and r_2, respectively, when a fluid is flowing at a constant time rate in radial direction from cylinder 1 to cylinder 2. The two cylinders may, for instance, be perforated tubes and the radial velocity in the ring space may be v_1 and v_2 at $r = r_1$ and r_2, respectively. The temperature is supposed to be a function of the radial distance from the cylinder axis r and of the time τ only. Density, specific heat, and thermal conductivity of the fluid are uniform and constant, and no heat energy is generated or absorbed in the fluid, but storage of heat energy is supposed to occur.

4-1. The monochromatic areal radiant intensity of the radiation of a black body should approach zero when the wavelength approaches zero. Prove that Planck's equation of radiation satisfies this requirement.

4-2. The electric resistance R_e of an aluminum wire (cross-sectional area $A = 0.01$ cm^2, length $L = 100$ cm) is 0.0249 ohm at 0°C, 0.0366 ohm at 100°C, and varies linearly with the temperature t within these limits. The wire is placed in a perfect vacuum and radiates against a very cold environment, the re-radiation of which is negligible. The temperature differences existing in the wire can also be neglected.

a. Find an equation representing the time rate of radiation q of the cylindric surface S of the wire as a function of its temperature t, using the theory of radiation of pure metals.

b. Calculate q in watts for $t = 0$, 50, and 100°C.

c. What would the surface temperature be for the same q as found under (*b*) for 100°C if the surface of the wire were perfectly black?

6-1. The thermal conductivity k of platinum has the following values at the indicated temperatures t:

t [°C] =	−200	0	600
k [B hr^{-1} ft^{-1} F^{-1}] =	45.0	40.6	47.1

Find the temperature and value of the minimum of k.

7-1. An ice plate, 1 sq ft in area, receives radiating heat from a moderately warm ordinary radiator, an amount of 32 B/hr hitting the plate, the temperature of which is just below 32°F.

How much ice will melt per hour (*a*) if the plate is directly exposed to the radiation? (*b*) if a very thin oxidized steel sheet is covering the exposed surface? (*c*) if a very thin polished copper sheet is covering it?

Neglect thermal resistance or heat storage by these sheets.

8-1. A cylindric shell is composed of materials having different thermal conductivities, namely k_{12} between the radii r_1 and r_2, k_{23} between the radii r_2 and r_3, etc., up to $k_{n-1,n}$ between the radii r_{n-1} and r_n. If the temperatures at the radii r_1 and r_n are held constant and equal to t_1 and t_n, respectively, and the length of the cylinder is L, what is the thermal resistance of the body in radial direction between these limits, provided that the cylinder ends are perfectly heat insulated?

8-2. The inner surface of a plane furnace wall of thickness L is held at the temperature t_0, the outer surface at t_L. The thermal conductivity of the wall material at any temperature t is given by $k = k_0 + \beta(t - t_0)$, where k_0 and β are constants. The distance of any point of the wall from the inner surface may be called x.

a. Derive an equation for the temperature distribution across the wall and find location x and amount Δt of the maximum deviation of the temperature from linear distribution.

b. Assuming $L = \frac{5}{6}$ ft, $k_0 = 1.399$ B hr^{-1} ft^{-1} F^{-1}, and $\beta = 0.000373$ B hr^{-1} ft^{-1} F^{-2}, $t_0 = 1980°$F, $t_L = 180°$F, represent the temperature distribution by a graph.

c. Explain the curvature of the temperature line.

8–3. Two coaxial metal tubes form an annulus whose surfaces have the radii r_1 and r_3, where $r_3 > r_1$, and the corresponding temperatures t_1 and t_3. The annulus is to be filled with two concentric layers of insulating material. The thermal conductivities of these materials are k_{12} and k_{23}, respectively, where $k_{12} > k_{23}$. The latter material can stand no higher temperature than t_2, where $t_2 < t_1$. Considering this limitation, what must the thicknesses $r_2 - r_1$ and $r_3 - r_2$ of the two insulating layers be in order that the heat exchange between the two tubes will be a minimum?

Calculate also a numerical example of the problem, assuming $r_1 = 1$ in., $r_3 = 2$ in., $t_1 = 350°$F, $t_2 = 250°$F, $t_3 = 110°$F, $k_{12} = 0.7$, and $k_{23} = 0.1$ B hr^{-1} ft^{-1} F^{-1}.

8–4. A rectangular plate of a homogeneous material is placed on the plane $z = 0$ of a cartesian system of coordinates with one corner at $x = 0$, $y = 0$, the second at $x = a$, $y = 0$, the third at $x = a$, $y = b$, and the fourth at $x = 0$, $y = b$. The temperature at one edge of the plate ($y = 0$) may be given as function $t_{y=0} = f(x)$; the other three edges are held at $t = 0°$C. The two main faces of the plate (at $z = 0$ and $z = c$) are perfectly insulated so that heat is conducted perpendicularly only to the z axis.

a. Derive an equation for the temperature distribution in a x, y plane inside the plate.

b. Assuming $a = b = 4$ ft and $f(x) =$ constant $= 100°$C, calculate the temperature at $y = 1$, 2, and 3 ft, each for $x = 1$ and 2 ft.

It is recommended to start from Laplace's differential equation and to assume $t = X \cdot Y$, where X and Y are functions of x and y, respectively. Considering the given boundary conditions t can be expressed by an infinite series. Finally, $f(x)$ may be expanded into a Fourier series.

9–1. Equation 9–2 can be written in the form

$$\frac{d^2\theta}{dx^2} = c\theta$$

where $\theta = t - t_i$. The general solution of this differential equation is

$$\theta = Me^{-\sqrt{c}\,x} + Ne^{\sqrt{c}\,x}$$

where M and N are arbitrary constants.

Derive Eq. 9–14, employing this form of solution instead of the one used in Sect. 9–4.

10–1. The two ends of a cylindric metal wire of length $2L$, which is perfectly insulated against radial heat flow, are held at the same temperature t_0. The two ends of the wire may be called the left and the right end. The left end and the middle between both ends are connected with a source of electricity so that a current flows through the left half of the wire and produces Joulean heat at the uniform rate of q''' per unit volume of the wire. No heat is generated in the right half of the wire. The thermal conductivity k of the wire metal may be independent of the temperature. Assuming steady-state conditions,

a. Express the temperature as a function of the distance from the left end. Two different equations will be found for the left and right half-lengths of the wire.

b. Will the temperature have a maximum somewhere between the two ends and, if so, at what distance from the left end?

c. Express the maximum temperature, if any, in mathematical form.

10–2. The two ends of a metallic cylindric rod which is perfectly insulated against radial heat flow are held at the same temperature, t_0. The ends and the middle of the rod are connected with sources of electricity and resistors in such a way that a current, i_1, passes through one (for instance, the left) half of the rod, and a current, i_2, smaller than i_1, passes through the other (for instance, the right) half of the rod. The electric and thermal conductivities of the rod material may both be considered independent of the temperature.

The following quantities are given: the dimensions of the rod; the Joulean heat energies, q_1''' and q_2''', developed in the two halves of the rod per unit time and volume; the thermal conductivity k of the rod material; and the temperature t_0.

a. Find for steady-state conditions: (1) the equations of temperature distribution over the length of the rod; (2) the temperature at the middle of the rod length; (3) the distance from the left end at which the temperature has its maximum; (4) the maximum temperature.

b. Solve the above questions for the following numerical conditions: A nichrome wire 6 ft 6¾ in. long, 1⁄25 in. thick, $k = 7.8$ B hr^{-1} ft^{-1} F^{-1}, having an electric resistance of 2.2 ohms. Currents $i_1 = 0.12$ amp and $i_2 = 0.09$ amp. Temperature $t_0 = 212$°F.

c. Draw the curve of temperature distribution.

d. State whether you find something surprising in the numerical results and, if so, try to explain it.

10–3. On a cylindric electric coil, the outer and inner surface temperatures, t_2 and t_1, respectively, have been measured under steady-state conditions, and it has been found that $t_2 < t_1$. Heat losses through the two end surfaces of the coil may be negligible so that the heat flow in the coil can be considered strictly radial. The thickness of the coil is so small compared to its diameters that the curvature of the coil can be neglected. Then a cross section through the coil, formed by a plane through the center line, can be considered as the cross section of a plane plate in which Joulean heat is produced at the amount of q''' in unit volume and unit time.

Take x as the distance of a point of the plate from that surface at which the temperature t_1 is held and $x = s$ as the thickness of the plate, and introduce $\theta = t - t_2$ for convenience. The generation of heat in the plate may be expressed by

$$\frac{q'''}{k} = m + n\theta = m(1 + \epsilon\theta)$$

where the constant k is the apparent conductivity of the coil material, and m, n, and ϵ are constants, defined by

$$m = \frac{q_2'''}{k} > 0 \quad \text{and} \quad n = \frac{\epsilon q_2'''}{k} > 0$$

The subscript 2 is used for the place where the temperature $t = t_2$.

a. Derive an equation for the distribution of the temperature excess θ in the coil, i.e., within the range $x = 0$ and $x = s$.

b. Derive from this equation the condition under which θ has a maximum in this range.

c. Find an equation for the distance x at which the maximum occurs.

d. Calculate this distance x and the maximum temperature from the following numerical data:

$$s = 2.5 \text{ cm}$$
$$q_2''' = 0.16 \text{ watt/cm}^3$$
$$m = 30 \text{ C/cm}^2$$
$$n = 0.1 \text{ cm}^{-2}$$
$$t_1 = 100°\text{C}$$
$$t_2 = 76°\text{C}$$

It is recommended to perform the calculation without changing the system of units.

10–4. At the outer and inner radius r_2 and r_1 of a cylindric electric coil, the temperatures t_2 and t_1, respectively, have been measured under steady-state conditions, and it has been found that $t_2 < t_1$. The heat flow in the coil may be considered as strictly radial.

Take r as the distance of a point of the coil from the center line and introduce $\theta = t - t_2$ for convenience. Again the three equations, given in Problem 10–3, may be valid.

a. Derive an equation for the distribution of the temperature excess θ in the coil, i.e., within the range from $r = r_1$ to $r = r_2$, where $(r_2 - r_1)/r_2$ is not negligibly small.

b. Find the condition for a maximum of θ in this range.

c. Find an equation for the radius r at which the maximum occurs.

d. Calculate this radius and the maximum temperature for $r_1 = 0.5$ cm, $r_2 = 3.0$ cm, using the data given under *d* in Problem 10–3.

e. Compare the results with those of Problem 10–3.

11–1. A steel tube is placed in rather quiet air. The tube is heated at one end at a given constant rate of heat flow and perfectly insulated at the other end. What will be the temperatures at either end of the tube, assuming that no appreciable heat flow occurs in the air inside the tube and that radial temperature differences in the steel are negligibly small? Assume, further, that the thermal conductivity of the steel is independent of the temperature and that the coefficient of heat transfer is the same over all the outer tube surface.

The result should be given in mathematical formulas and for the following special case:

Tube length = 1 ft
Outer diameter of tube = 1 in.
Inner diameter of tube = ¾ in.
Thermal conductivity of steel = 29.4 B hr^{-1} ft^{-1} F^{-1}
Temperature of air outside the tube = 75°F
Coefficient of heat transfer on the outside = 1.4 B hr^{-1} ft^{-2} F^{-1}
Rate of heat flow = 10 B/hr

11–2. A circular ring of 5 in. mean diameter and rectangular cross section (1 in. high, 0.02 in. thick) is located in air at 60°F and is held at 212°F in one cross-sectional area. What will be the temperature in the coldest cross-sectional area (diametrically opposite to the place of heating), when a steady state of heat exchange has been reached, (*a*) if the ring consists of glass (thermal conductivity = 0.58 B hr^{-1} ft^{-1} F^{-1})? (*b*) if the ring consists of copper (thermal conductivity = 200 B hr^{-1} ft^{-1} F^{-1})?

Assume that any influence of the curvature of the ring and temperature differences in any cross-sectional area can be neglected. The coefficient of heat transfer between the ring and the air may be taken as 2 B hr^{-1} ft^{-2} F^{-1} at each point of the surface.

11-3. Two vessels are kept at 270°F and 120°F, respectively, in a room at 70°F air temperature. The vessels are connected by 4 non-insulated nickel-steel bolts, each ½ in. in diameter, 1 ft long. The thermal conductivity of the nickel steel is 24 B hr^{-1} ft^{-1} F^{-1}, and the surface coefficient of heat transfer between the rods and the air is 1.0 B hr^{-1} ft^{-2} F^{-1}.

How much heat energy per hour is conducted away from the warmer vessel and how much is supplied to the colder vessel by the connecting rods?

11-4. A steel rod, 1 in. in diameter, 2 ft 7 in. long, connects the walls of two vessels which contain hot liquids and are insulated perfectly throughout with the exception of the connecting rod. The wall temperature of each vessel is 380°F; the room temperature is 80°F.

The thermal conductivity of the steel is 26 B ft^{-1} hr^{-1} F^{-1}, and the coefficient of heat transfer from the rod to the air is 1.3 B ft^{-2} hr^{-1} F^{-1}, including the effect of radiation.

a. What will be the total heat loss per hour from the rod, and the temperature of the rod at half distance from the vessels?

b. (1) What will be the heat loss if, instead of the connecting rod, a rod of the same material and cross section, but half the length, protrudes from each vessel, and the coefficient of heat transfer from the free ends of these two rods to the air is 1.3, like that from the other parts of the surface?

(2) What temperature will occur at the free ends?

c. What mistake in the calculation of the heat loss will be made if the heat loss by the free ends is supposed to be equal to zero? What temperature at the free ends will be calculated under this condition?

d. (1) What mistake in the calculation of the heat loss will be made if the loss by the free ends is not entirely neglected, but the rods are considered as infinitely long? This means that the free ends are assumed to have that temperature which would occur at the same distance from the heat source in an infinitely long rod.

(2) What would be the heat loss from the free ends in this case?

e. (1) What would be the temperature at the free ends according to the latter calculation?

(2) But what temperature would be necessary if the amount of heat loss through the free ends as asked for in item *d*(2) should really be given up with the supposed coefficient of heat transfer and air temperature?

(3) Explain the surprising results of part *e*(2).

11-5. Two vessels, the one at 212°F, the other one at 104°F, are perfectly insulated against heat losses, with the exception of a non-insulated metallic connection between the vessels, consisting of an iron rod, 1 ft long, of rectangular cross section, 1 in. by ⅛ in.

How much heat energy will be given up by the first vessel, and how much will be received by the second vessel, if the air environing the rod is at 77°F, the thermal conductivity, k, of the iron is 40 B hr^{-1} ft^{-1} F^{-1}, and the coefficient of heat transfer, h, on the rod surface is 1.25 B hr^{-1} ft^{-2} F^{-1}, the latter including also the heat transfer by radiation?

11-6. Two cylindric rods, each of diameter D and length L, are welded together at one end and held at constant temperature t_0 at the other end. The thermal conductivities of the rods are k_1 and k_2, respectively; the coefficient of heat transfer is the same all over their surfaces and equal to h. The system is placed in air at temperature t_e.

Find a formula for the temperature of the welded cross section.

11-7. Prove the following theorem: For any potential flow a channel of uniform cross section has the smallest resistance of all channels having the same length and volume.

11-8. Simulate the distributions of temperature and heat flow in a straight fin of rectangular profile and in a straight fin having a profile which requires the least material by electric analogues.

11-9. Compare the heat losses of a straight, rectangular fin, as calculated from Eq. 11-30 or from Eq. 11-33 for the whole range $h/km = 0$ to ∞. Is there a value of h/km at which both equations give the same result?

11-10. 387.4 B/hr are to be dissipated by a circular fin of uniform thickness on a cylinder of 4 in. diameter. The temperature of the fin should not exceed 180°F at its root when the temperature of the environment is 80°F. Assume $k = 120$ B $\text{hr}^{-1}\ \text{ft}^{-1}\ \text{F}^{-1}$ and $h = 3.7$ B $\text{hr}^{-1}\ \text{ft}^{-2}\ \text{F}^{-1}$.

Calculate the width and thickness of the fin so that a minimum of material will be required.

11-11. For the conditions indicated in Problem 11-10, calculate the ratio of tip diameter to root diameter of a fin of that profile for which the minimum of material would be required. Compare the quantities of material needed for such a fin and for the fin of uniform thickness, dealt with in Problem 11-10.

11-12. A surface is to be cooled by extensions having the form of conical pins, base diameter $2r_0$, tip diameter $2r_L$, length L. The rate of heat to be carried away by each pin is q_0; the surface coefficient of heat transfer is h; the thermal conductivity of the pin material is k, all these quantities having constant values.

Denoting by r the radius of the pin cross section at an axial distance x from the pin base and by θ the temperature excess at this distance above environment, find a differential equation for θ, with x as independent variable.

Show also how the heat transfer from the tip surface to the environment can be considered, assuming that the coefficient h is the same as for the conical part of the pin surface.

12-1. A fluid is streaming perpendicularly to the axis of a metal wire whose diameter and length may be denoted by D and $2L$, respectively. The thermal conductivity of the wire, k, can be considered constant, and temperature differences in any cross section negligible. The fluid stream and the two ends of the wire are kept at the same constant temperature t_0. Introducing, further, t as the temperature of the wire and $\theta = t - t_0$ as its temperature excess, the electric resistance of unit length of the wire may be $R_e' = a + b\theta$, where a and b are positive constants. The coefficient of heat transfer between wire and fluid is given as h = constant. An electric current i passes permanently through the wire.

a. Derive a differential equation of θ as function of the axial distance x from the middle of the wire axis.

b. Find the solutions of the differential equation for the two cases, $h \gtrless bi^2/\pi D$, considering the data given above.

c. Discuss the results in comparison with related cases dealt with in a previous chapter.

12-2. To determine the thermal conductivity of a mixture of asbestos and diatomaceous earth a device according to Fig. 12-2 was used, for which the following data were given: $2L = 2.96$ m; $2r_i = 0.0598$ m; $2r_e = 0.181$; $(kA)_a = 0.053$ kcal hr^{-1} m C^{-1}.

The load of the heating element was 403 watts, and the following temperatures were measured:

$$t_e = 43.8°\text{C}$$
$$t_i = 293.5°\text{C} \quad \text{at} \quad x = 0$$
$$t_i = 210°\text{C} \quad \text{at} \quad x = L$$
$$t_i = 192°\text{C} \quad \text{at} \quad x = -L$$

Assume $t_i = 201°\text{C}$ for $x = L$ as well as for $x = -L$, and calculate the thermal conductivity of the material, considering the axial heat loss in the cylindric heating element.

13-1. A steel sphere, 7¾ in. in diameter, which is heated to 536°F, is thrown into an oil bath at 86°F. The density of the steel is 7.7 gram/cm^3, its specific heat is 0.13 cal gram^{-1} C^{-1}, and its thermal conductivity is 33.6 B hr^{-1} ft^{-1} F^{-1}. The coefficient of heat transfer between the sphere and the oil may be assumed to be 102.4 B hr^{-1} ft^{-2} F^{-1}.

a. Find the temperature in the center and at the surface of the sphere, 36 sec, 3 min, and 12 min after its plunging into the oil. How much heat energy will have been given up by the sphere at these instants?

b. For the instant 3 min after plunging, find the temperature distribution along a radius of the sphere by the approximative graphical construction indicated in Sect. 13-8, Article *d*.

13-2. A plain brick wall, 20 in. thick, separates a room from the free atmosphere. Initially a steady state of heat flow existed with 32°F room temperature and −4°F outdoor temperature. Starting at an instant, called $\tau = 0$, the wall was heated from the inside at an invariable rate of heat of 40 B hr^{-1} ft^{-2}, whereas the outdoor temperature remained unchanged.

Assuming coefficients of heat transfer $h_1 = 4$ and $h_2 = 6$ B hr^{-1} ft^{-2} F^{-1}, inside and outside, respectively, and $\rho = 100$ lb_m/cu ft, $c_p = 0.2$ B lb_m^{-1} F^{-1}, and $k = 0.4$ B hr^{-1} ft^{-1} F^{-1}, calculate and show in graph form the temperature distribution at the instant $\tau = 8$ hr. Also calculate the room temperature for that instant.

Procedure suggested: Determine first the initial temperature distribution in the wall and then extrapolate this distribution beyond the outside surface as though the wall were infinitely thick. Do not mind that this leads to temperatures which are impossibly low from a physical viewpoint. The result will justify the assumption made.

13-3. In a furnace, inside temperature 1652°F, a long iron cylinder, 15¾ in. in diameter, is to be heated from an initial homogeneous temperature of 212°F. Assume radial heating only, and neglect the influence of the ends of the rod. The thermal conductivity of the iron is 22.8 B hr^{-1} ft^{-1} F^{-1}; its thermal diffusivity is 0.189 sq ft/hr. The coefficient of heat transfer on the surface of the cylinder is 24.6 B hr^{-1} ft^{-2} F^{-1}, which includes convection and radiation.

The cylinder is taken out of the furnace when the temperature in the center line is 180 F lower than on the surface. After what time of heating will this occur, and what will then be the temperature in the center line?

The difference 180 F will occur twice, first shortly after the start of the heating, and a second time later on. The temperature at the later instant of time is asked for.

Tables or graphs may be used for the calculation.

13–4. A cylindric copper rod, ¼ in. in diameter, 2½ in. long, is perfectly insulated against radial heat losses. Initially its one end is held at 70°F, its other end at 32°F. Starting from a certain instant $\tau = 0$, the warm end is not heated any more from the outside, but is perfectly heat insulated, whereas the cold end is permanently maintained at 32°F.

Calculate analytically and represent graphically the change of the temperature distribution in the rod with time, assuming $\alpha = 4.26$ ft²/hr as the thermal diffusivity of copper.

13–5. The bulk of water in a big reservoir virtually retains its temperature $t_w > 0$°C when ice formation occurs in a surface layer because of low temperature of the atmospheric air. Assume that the air was also at the temperature t_w originally, but suddenly changed to the temperature $t_a < 0$°C at the time $\tau = 0$ and then remained at this temperature. Let x be the thickness of the ice layer at the time instant $\tau = \tau$, ρ and k the density and thermal conductivity, respectively, of the ice, h_w and h_a the coefficients of heat transfer on the water and air side, respectively, of the ice, λ the melting heat (enthalpy increase in melting of unit mass of ice), all these quantities being considered constant.

Neglecting the heat energy liberated by the cooling of the ice below 0°C, find an equation with the time τ and the thickness x of the ice, formed in that time, as variables.

In deriving the differential equation the solution of which is required, the heat liberated by freezing must be considered the only heat source in the process. Furthermore, heat is supposed to flow in the x direction only.

13–6. A long tube of inner radius s is originally filled with water at 0°C. Its inner surface is suddenly brought to a temperature $t_s < 0°$ at the time $\tau = 0$ and then is held at this temperature. Freezing is supposed to proceed in radial direction inwards only.

Assuming the same physical properties of the ice as in Problem 13–5 and a constant coefficient of heat transfer h_w between ice and water,

a. Find an equation with the time τ and the radius r as variable, where $s - r$ is the thickness of the cylindric ice shell formed in that time.

b. Find the time in which all water in the tube will be frozen. Again, as in Problem 13–5, neglect the heat energy liberated by cooling the ice below 0°C and consider the heat liberated by freezing as the only heat source in the process.

14–1. The wall of a regenerator consists of a refractory material the thermal diffusivity of which is 0.0256 ft²/hr. The regenerator is switched every hour so that the time of a full cycle is 2 hr.

a. Supposing that the switching of the regenerator is equivalent to a sinusoidal change of surface temperature, find in what depth of the wall the temperature amplitude of the oscillation has been reduced to 1% of the amplitude at the inner surface of the wall.

b. In what depth will the amplitude of the third harmonic of the surface-temperature fluctuation have been reduced to 1% of its value?

14–2. Both sides of a large plane plate undergo the same sinusoidal change of temperature, the full cycle of change taking 4 hours. What will be that thickness of the plate for which most heat energy will be stored per unit area if the thermal diffusivity of the wall material is 0.0256 ft²/hr?

14-3. The average daily change of temperature of the earth's surface at a certain place may be considered as corresponding to a cosine function with ±5.4 C as amplitude. Assume the thermal diffusivity of the earth to be 873 cm^2/day.

In what depth will the amplitude of temperature oscillation be reduced to 1.99 C?

14-4. In a 2-stroke Diesel engine, running 100 revolutions per minute, the change of gas temperature during 1 period (= 2 strokes = 1 revolution) has been observed by means of fine thermocouples. By a Fourier analysis the following equation was found:

$$\begin{aligned} t_g = 1040 &- 1152 \cos\left(12{,}000\pi\tau + \frac{2\pi}{360}120\right) \\ &- 324 \cos\left(24{,}000\pi\tau + \frac{2\pi}{360}69\right) \\ &- 194.4 \cos\left(36{,}000\pi\tau + \frac{2\pi}{360}75\right) \\ &- 221.4 \cos\left(48{,}000\pi\tau + \frac{2\pi}{360}37\right) \\ &- 147.6 \cos\left(60{,}000\pi\tau + \frac{2\pi}{360}10\right) \end{aligned}$$

where t_g = the gas temperature, in degrees Fahrenheit, and
τ = the time, in hours.

The density, specific heat, and thermal conductivity of the wall material (cast iron) were

$$\rho = 450 \text{ lb}_m/\text{cu ft}$$

$$c_p = 0.13 \text{ B lb}_m{}^{-1} \text{ F}^{-1}$$

$$k = 33.6 \text{ B hr}^{-1} \text{ ft}^{-1} \text{ F}^{-1}$$

The coefficient of heat transfer was $h = 102.5$ B hr^{-1} ft^{-2} F^{-1}.

a. Write an equation corresponding to that for t_g, which expresses the change of the temperature t_s of the inner surface of the cylinder wall, and find the maximum temperature change of this surface which takes place in the course of 1 period. The average value of the inner surface temperature was found to be 464°F in the experiments.

b. The thickness of the wall was 1⅜ in. Show that a wall of this thickness can be considered infinitely thick for the required calculation.

14-5. The temperature at and below one surface of a thick plane wall may change according to the following equation:

$$\theta = \theta_{1,a}e^{-m_1x} \cos(\omega_1\tau - m_1x) - \theta_{3,a}e^{-m_3x} \cos(\omega_3\tau - m_3x)$$

where τ = the time, in hours,
x = the distance, in feet, of any point of the wall from the mentioned surface,
θ = the temperature excess, in Fahrenheit degrees, of a point of the wall at depth x and time τ above a mean temperature level,
$\theta_{1,a}$ and $\theta_{3,a}$ are the amplitudes of the first and third harmonics, respectively, of the temperature oscillation around the temperature level mentioned,

m_1, m_3, ω_1, and ω_3 are constants which can be calculated by means of the following data: The full periods (cycles) of the first and third harmonics are 120 and 40 min, respectively; $\theta_{1,a} = 1200$ F, $\theta_{3,a} = 600$ F, and the thermal diffusivity of the wall material is 0.025 ft^2/hr. These are figures which may occur in the operation of an industrial regenerator of refractory material.

a. What will be the temperature excess θ above the temperature level mentioned at the time $\tau = \frac{1}{2}$ hr in the depth of 1 in. from the surface mentioned?

b. Calculate the amplitudes of the temperature excess of the first and third harmonics in that depth.

c. Find by a graphical plot or by trial and error the maximum temperature excess occurring in that depth.

17–1. A homogeneous medium may exist between two infinitely wide, plane surfaces, which are perpendicular to the z axis of a cartesian system of coordinates and have the distance L from each other. Initially the medium may be at uniform temperature t_e. The temperature excess $\theta = t - t_e$ at any point of the medium will vary in time.

a. Derive an equation for the temperature distribution in the medium, due to an instantaneous line source of heat, corresponding to Eq. 17–55 for a point source. It is recommended to start with a procedure similar to the one which has lead to Eq. 17–69. The volume dV, introduced in Eq. 17–54, may be replaced by $L \cdot dA$, where dA is an area differential in any x,y plane in the medium.

b. Derive the equation for a continuous line source, corresponding to Eq. 17–60 assuming that the total capacity of that source at $x = 0$, $y = 0$ can be expressed by Eq. 17–56.

c. Prove that the temperature excess above environment at any point between the two parallel surfaces and at any time τ can be expressed by the equation

$$\theta = \frac{q_i}{2\pi k L} Z$$

where

$$Z = \int_{\zeta}^{\zeta=\infty} \frac{e^{-\zeta^2}}{\zeta} d\zeta,$$

$$\zeta = \frac{r}{(2\sqrt{\alpha\tau})},$$

$$r = \sqrt{(x')^2 + (y')^2}, \quad \text{and}$$

$$q_i = \phi(\tau')$$

in analogy to the denotations and assumptions in Sect. 17–4, Articles *b* and *c*.

For numerical calculations L. R. Ingersoll, Zobel, and A. C. Ingersoll [1948] have tabulated the integral Z. They also indicate that for $x < 0.2$

$$Z = \ln\frac{1}{\zeta} + \frac{\zeta^2}{2} - \frac{\zeta^4}{8} - 0.2886$$

17–2. A pipe, 1.2 in. outer diameter, belonging to a heat-pump system, is buried in rather humid soil in such depth that the influence of the free surface of the ground can be neglected. The soil (ρ = 100 lb_m/ft^3, c_p = 0.5 B lb_m^{-1} F^{-1}, k = 1.0 B hr^{-1} ft^{-1} F^{-1}) is originally at the uniform temperature of 60°F, and so is the pipe. From the start of operation of the heat pump the pipe permanently takes 20 B hr^{-1} ft^{-1} from the soil.

Imagining the pipe to be replaced by a linear heat sink at the pipe axis, answer the following questions:

a. What will be the temperature at the pipe surface 3 hr 20 min, 50 hr, and 1250 hr after start of operation?

b. What will be the temperature at 12 in. radial distance from the center line 1250 hr after start of operation?

It is recommended to use the last equation indicated in Problem 17–1.

17–3. Two electric cables are located parallel to each other in soil which has the physical properties given in Problem 17–2. The center lines of the cables are distant by 12 in.; the energy loss of 1 ft length of each cable is 5.86 watt. The soil and the cables are originally at 60°F.

What will be the temperature in the middle between the cables 1250 hr after the current was switched on?

Neglect any influences from the free surface of the ground or from the material of the cables. Consider these rather as parallel line sources at 12 in. distance which concur in heating up an infinitely extended medium.

Again use the last equation given in Problem 17–1.

17–4. A cable is buried in soil, originally at 60°F, which has the physical properties given in Problem 17–2. The center line of the cable is only 8 in. below the ground surface, so that the influence of this surface, which is permanently held at 60°F, cannot be neglected. The energy loss of 1 ft length of the cable is 5.86 watt.

What will be the temperatures 8 in. below the center line of the cable 1250 hr after the current was switched on?

To solve this problem, assume that a fictitious line sink, absorbing 5.86 watts per foot length, is located 16 in. above the center line of the cable and that the soil is infinitely extended upwards.

Again use the last equation given in Problem 17–1.

17–5. A steel wire, ¼ in. in diameter, is to be heated from an initial temperature of 60°F to 1540°F by being passed with constant velocity v through an induction coil which is so short that the coil field can be assumed to act as a plane source of heat. Considering the efficiency of the coil, it may deliver 0.97 kw to the wire. Assuming 486 lb_m/cu ft, 0.116 B lb_m^{-1} F^{-1}, and 40 B hr^{-1} ft^{-1} F^{-1}, respectively, as density, specific heat, and thermal conductivity of the steel, and neglecting surface losses of the wire, answer the following questions:

a. How many feet of wire can be processed per hour?

b. At what distance before the wire enters the heating coil will its temperature be raised by 14.8 F (= 1/100 of the final temperature excess)?

17–6. In a room at 100°F copper is being drawn through a thin 1-in. die at the rate of 40 ft/min. The heat of friction may be assumed to be given up to the bar at a plane perpendicular to the die axis. The wire arrives at this plane with a temperature of 100°F and leaves it at 225°F. Assuming 550 lb_m/cu ft, 0.100 B lb_m^{-1} F^{-1}, and 220 B hr^{-1} ft^{-1} F^{-1}, respectively, as density, specific heat, and thermal conductivity of copper, and a coefficient of heat transfer of 3 B hr^{-1} ft^{-2} F^{-1} between the bar surface and room temperature, including radiation, what will the temperature of the drawn copper be 100 ft away from the die?

17-7. A straight crack on the upper horizontal surface of a large and thick steel slab is to be repaired by electrowelding. The electrode, which consists of the same steel as the slab, moves along the surface crack at the rate of 4.50 in./min, and 2.30 in. of the electrode are consumed per minute. Its arc can be considered as a point source having a strength of 88,000 B/hr. The dimensions of the slab are such that it may be considered a semi-infinite body. Take $k = 44$ B hr^{-1} ft^{-1} F^{-1} and $\alpha = 0.80$ ft^2/hr for the steel and 90°F as initial temperature of the slab.

a. Find the temperatures at three points, each 1 in. below the slab surface. The first point may be located immediately below the point source, the second point $\sqrt{3}$ in. behind, and the third point $\sqrt{3}$ in. ahead of the point source.

b. If the diameter of the electrode is $\frac{5}{16}$ in., its fusion temperature 2500°F, and its initial temperature 90°F, how much heat energy enters the electrode in 1 hr? Assume $\rho c_p = 51.3$ B ft^{-3} F^{-1}.

17-8. Referring to the illustrative problem in Sect. 17-6, Article h, find q_0 if the rod is not insulated but gives up heat to an environment which is permanently at $t = 0$, the coefficient of heat transfer being h.

17-9. Referring to the problem in Sect. 17-6, Article h, find t, q_0, and Q_0 if the temperature t_0 is not constant but is given by $t_0 = a + b\tau + c\tau^2$.

17-10. Referring to the problem in Sect. 17-6, Article h, assume that the body initially at temperature t_0 is also a rod, extending from $x = 0$ to $x = -\infty$. Show that the temperature at the junction of the two rods is independent of time and is given by $\dfrac{t_0}{1 + \dfrac{k_2\sqrt{\alpha_1}}{k_1\sqrt{\alpha_2}}}$, and find expressions for the temperatures t_1 and t_2 as functions of x and τ, subscripts 1 and 2 referring to $x < 0$ and $x > 0$, respectively.

17-11. Between two tubes of elliptical confocal cross section and length L, a free space exists. A cross section through this space, perpendicular to the axis of the tubes, is bounded by two ellipses which, in cartesian coordinates u and v, have the equation

$$\frac{u^2}{(r + 1/r)^2} + \frac{v^2}{(r - 1/r)^2} = \frac{1}{4}$$

where r is a constant (r_1 for the inner, r_2 for the outer, ellipse).

The space between the tubes is filled with a material of constant thermal conductivity k. The surfaces of the two cylinders of elliptic cross section bounding this material are kept at the temperatures t_1 and t_2, and heat flow in axial direction is prevented.

Find the rate of heat flow between the two tubes by means of conformal mapping, using the following equation of transformation:

$$w = u + iv = \frac{z + 1/z}{2}$$

which defines a z plane in terms of the coordinates of a w plane.

18-1. An exothermic reaction occurs in a homogeneous pulverized substance. In order to find the approximate rate of the heat energy developed, 7 thermocouples are bedded in the substance in such a manner that, considering one of them (No. 0) to be at the zero point ($x = 0$, $y = 0$, $z = 0$) of a rectangular system of coordinates, the other 6 thermocouples are at the distances x, y, z, from the zero point, which are given in the following table:

Thermocouple	x [inches]	y [inches]	z [inches]	$t - t_0$ [F]
No. 1	$-\frac{1}{2}$	0	0	+8
No. 2	$+\frac{1}{2}$	0	0	−2
No. 3	0	−1	0	+13
No. 4	0	+1	0	−7
No. 5	0	0	$-\frac{3}{4}$	+8
No. 6	0	0	$+\frac{3}{4}$	−7

The substance is heated in a furnace until the reaction works, and then the heating is turned off. At a certain instant the temperatures t, indicated by thermocouples No. 1 to 6, are in excess of the temperature t_0, indicated by thermocouple No. 0, by amounts $t - t_0$, which likewise are given in the table.

Furthermore, it is observed that the temperature t_0 increases by 12 F in the time interval from ½ min before, to ½ min after, the simultaneous readings of the temperatures of the 6 thermocouples.

The apparent specific weight of the reacting pulverized substance is 15 lb/cu ft; its specific heat is 0.35 B lb_m^{-1} F^{-1}; its thermal conductivity is 0.03 B hr^{-1} ft^{-1} F^{-1}.

How much heat, approximately, is developed per hour in 1 lb of the substance?

18–2. Solve part *b* of Problem 8–4, using the method of relaxation.

18–3. A straight fin of rectangular profile area (0.05 ft thick, 0.5 ft long) is held at 100°F at one end (root) and perfectly heat insulated at the other end (tip). The thermal conductivity of the fin material is 50 B hr^{-1} ft^{-1} F^{-1}. The fin is exposed to air at 0°F, the surface coefficient of heat transfer being h = 10 B hr^{-1} ft^{-2} F^{-1}. The heat flow in the direction perpendicular to the fin-profile area may be considered negligible, and the temperature across the fin may be uniform at any distance from the root in the length direction of the profile.

a. Find the temperature distribution in the length direction and the heat loss of a piece of the fin, extending 1 ft in the direction perpendicular to the profile, using the method of relaxation. Explain the procedure chosen for the calculation.

b. Check the result of the numerical solution by analytical solution.

18–4. Referring to Fig. 18–4, discuss the possible influence of the corner of the furnace upon the result of the numerical calculation. How would you try to consider this influence? Show numerically how your suggestion would work out in the present case.

18–5. Solve Problem 13–4 by means of the method of numerical approximation.

18–6. A large cast-iron slab, 1 in. thick, lies on a bed of insulating material. Slab and surroundings are at 100°F. A large copper slab, 2 in. thick, having been heated uniformly to 400°F, is placed on the iron slab and sprayed with water at 100°F.

Edge effects may be neglected so that the heat flow can be considered one-dimensional. Surface resistance between the slabs and conduction from the iron to the insulating bed may also be negligible.

Assume for copper: $\rho = 556$; $c_p = 0.094$; $k = 218$;
" " iron: $\rho = 442$; $c_p = 0.118$; $k = 27$;
and $h = 500$; 700; and 900 at the sprinkled surface for surface temperatures 400; 350; and 300°F, respectively.

Cooling curves at ½-in. intervals in the two slabs down to a surface temperature of 300°F are to be determined by the numerical method dealt with in Sect. 18–3.

The following procedure is recommended: Calculate $M_{\text{Cu}}/M_{\text{Fe}}$ and N for the largest value of h occurring; herefrom M_{Cu} and $\Delta\tau$. Further determine the initial temperature of the surfaces of copper and iron when brought into contact, considering their values of ρc_p. After these preliminary calculations start with the numerical procedure.

19–1. Solve Problem 8–4 by drawing curvilinear squares.

19–2. Solve Problem 13–2 by the method of finite differences.

21–1. In a cylindric pipe through which air is flowing the velocity distribution and the temperature distribution over a cross section have been measured, and it was found that these distributions can be represented by the formulas

$$v = 100\left(\frac{s-r}{s}\right)^{1/7}$$

and

$$t = t_s + 50\left(\frac{s-r}{s}\right)^{1/5}$$

where v is the velocity, in feet per second, at the distance r from the center line,
t is the temperature, in degrees Fahrenheit, at the same distance,
s is the inside radius of the pipe, and
$t_s = 77°\text{F}$ = the temperature of the inside surface of the pipe.

What is (*a*) the true mean velocity of the flowing air? (*b*) its true mean temperature ("cup temperature")? For calculating (*b*), the specific heat of the air can be considered independent of the temperature.

22–1. A hot gas streaming parallel to the upper surface of a horizontal porous plate is cooled by a liquid which is forced through the plate. The liquid approaches the plate in laminar flow upward at the constant velocity v. At a distance s below the plate the liquid temperature is t_0, at the lower surface t_s.

Derive an equation for the heat loss in steady state from the plate to the arriving liquid per unit area and unit time, assuming that no heat will be exchanged in horizontal direction, and that the density ρ, the specific heat c_p, and the thermal conductivity k of the liquid are given.

23–1. It may be known that the surface coefficient of heat transfer between the inner surface of a cylindric tube and a fluid in streamline flow through the tube depends upon the following variables: diameter, ratio of diameter to length, velocity, thermal conductivity, and thermal diffusivity of the fluid.

a. Employing dimensional analysis, derive a dimensionless equation for the heat transfer under the given condition.

b. Using this equation, what would be the smallest number of experiments needed to express h numerically in terms of the quantities given above?

23–2. It is known that the volume V to which a vapor bubble grows on a horizontal heating surface before breaking off depends upon the gravitational acceleration g, the surface tension σ between bubble and surface, and the difference between the densities of the saturated liquid and vapor $(\rho_f - \rho_g)$. This maximum volume is also proportional to β^3, where β is the outer contact angle between bubble and heating surface.

Find by means of dimensional analysis an equation which connects V, g, $\rho_f - \rho_g$, σ, and β. A constant factor in this equation remains unknown and would have to be determined by experiments.

23-3. A hot gas at constant temperature $t_{g,1}$ enters at the bottom of a wide bed of particles which all have the same geometrical form and size (as, for instance, balls of the same diameter D) and the same initial temperature $t_{b,0}$. At the height L above the bottom the gas leaves at a temperature $t_{g,2}$. Introducing $\theta_{g,1} = t_{g,1} - t_{b,0}$ and $\theta_{g,2} = t_{g,2} - t_{b,0}$ and considering beds of equal ratio L/D, the temperature-excess ratio $\theta_{g,2}/\theta_{g,1}$ depends upon D, τ, C_b, $\dot{m}c_p$, k_b, and h, where D = a characteristic length of the particles, τ = the time counted from $\tau = 0$ at which the bed was at the temperature $t_{b,0}$, C_b = the specific heat of the bed of particles per unit volume, $\dot{m}$ = the rate of mass flow of the gas, c_p = the specific heat at constant pressure of the gas, k_b = the thermal conductivity of the bed material, and h = the surface coefficient of heat transfer between gas and particles.

Using dimensional analysis, express θ_{g2}/θ_{g1} as a function of dimensionless moduli composed of the quantities mentioned.

23-4. The mass m of a drop of liquid falling from a vertical capillary tube is found to depend upon the radius of the tube r, the acceleration by gravity g, the density of the liquid ρ, and the surface tension σ.

By dimensional analysis find an expression for m.

24-1. Two coaxial tubes, 100 cm long, form a ring space with the diameters 1.2 cm and 1.0 cm. Ethylene at about atmospheric pressure and 100°C is flowing through with the mean velocity 2000 cm/sec. Calculate the pressure drop in this annulus and then, from the pressure drop, the heat energy picked up by the gas per hour if the temperature of the tubes is 10 C above the mean temperature of the gas at any point of the tube length. The thermal resistance of the tube walls and the change of the physical properties of the gas inside the ring space may be negligible.

Use the following data for ethylene:

Density = $0.924(10^{-3})$ gram/cm^3
Dynamic viscosity = $124(10^{-6})$ poise
Specific heat at constant pressure = 0.44 cal gram^{-1} C^{-1}
Thermal conductivity = $64(10^{-6})$ cal sec^{-1} cm^{-1} C^{-1}

24-2. 3000 lb/hr of a hydrocarbon oil are flowing through a smooth tube of ¾ in. inner diameter, 20 ft long, the wall of which is held at 250°F by steam condensing at the outside. To what temperature will the oil be heated in the tube if it enters at 100°F?

The specific gravity of the oil relative to water of 60°F can be expressed by

$$G = 0.82 - 0.0004t$$

and its dynamic viscosity by

$$\mu = 0.055\left(\frac{60}{t}\right)^{1.25}$$

with t in degrees Fahrenheit and μ in poise.

The specific heat of the oil is 0.5 B lb_m^{-1} F^{-1}, and its thermal conductivity is 0.1 B hr^{-1} ft^{-1} F^{-1}. Both may be assumed to be constant in the range of temperature under consideration.

Compare the results obtained from (*a*) von Kármán's theory; (*b*) Boelter, Martinelli, and Jonassen's theory; (*c*) Dittus and Boelter's empirical equation.

25–1. The heat given up by convection from a vertical surface, 3 ft high and 2 ft wide, to atmospheric air at rest, which is 40 F colder than the surface, may be known. How great a percentage more or less of heat would be given up by a surface, 2 ft high, 3 ft wide, and 36.9 F warmer than the air, all other conditions being the same? Neglect the heat transfer by radiation and any particular influence of the edges of the surface upon convection.

25–2. An enclosed horizontal square air layer, 9 sq ft in area, 5 in. thick, is heated from below at a rate of 50 B/hr while its upper surface is held at 77°F by means of a cooling device.

What will be the temperature at the bottom of the layer if the thermal conductivity of the air is 0.01545 B hr^{-1} ft^{-1} F^{-1} and its kinematic viscosity is 0.17 cm^2/sec?

Radiation and heat losses at the edges of the layer may be neglected.

25–3. An enclosed vertical square air layer, 3 ft wide, 3 ft high, 5 in. thick, is heated from one of its large surfaces at a rate of 50 B/hr while the opposite surface is held at 77°F.

Calculate the temperature of the warmer surface with the values of air properties indicated in Problem 25–2.

25–4. It is required that the surface temperature of an electrically heated wire, 0.025 in. in diameter, should not exceed 300°F so that its hard-rubber insulation may not be damaged.

How thick should this insulation be made in order that the wire can carry a maximum of watts per foot length when horizontally placed in quiet air at 70°F? What will be the maximum wattage?

Assume that the insulated wire is perfectly protected against radiation losses by means of a radiation shield located at such distance that the convection losses are not influenced by it. The thermal conductivity of rubber can be taken as constant and equal to 0.09 B hr^{-1} ft^{-1} F^{-1}.

a. Assume, first, the film coefficient h at the outer surface of the insulation to be constant.

b. Find a more exact result by assuming that hD_0/k depends on D_0 only, D_0 being the diameter of the insulated wire.

26–1. The inside surface of a condenser tube, 0.94 in. in inner diameter, is held at 124°F by water of 95°F, flowing through the tube with an average velocity of 3 ft/sec.

From experiments on the amount of heat given up to a liquid in turbulent flow in a tube, it is known that log ($N_{Nu}N_{Pr}^{-0.4}$) is a linear function of log N_{Re}. In particular, $N_{Nu}N_{Pr}^{-0.4} = 40$ and 255, at $N_{Re} = 10{,}000$ and 100,000, respectively.

What is the surface coefficient of heat transfer from the inside surface of the tube to the water flowing in that tube?

Use the following values of the properties of water:

61.9 lb_m/cu ft for the density
1.0 B lb_m^{-1} F^{-1} for the specific heat
0.000016 slug ft^{-1} sec^{-1} for the viscosity
0.37 B hr^{-1} ft^{-1} F^{-1} for the thermal conductivity

26–2. In a dairy-products factory cream cheese is forced through a 6-in. tube, 8 ft long, at the rate of 4000 lb/hr, entering at 59°F. The tube is submerged in a tank of hot water and in this way held at 176°F. Calculate the mean exit temperature of the cheese and the heat picked up by it in flowing through the tube.

Density, dynamic viscosity, and specific heat of the cheese are, respectively, 1.08 gram/cm^3, 500 poise, and 0.63 B lb_m^{-1} F^{-1}.

The value of the thermal conductivity of the cheese was not directly available. However, the following experiment was performed, from which it may be calculated. A cylinder of cheese, 12¾ in. in diameter, 12¾ in. long, at an initial uniform temperature of 52°F was suddenly brought into contact with air which was held at 72°F. One junction of a thermocouple was located in the center of the cylinder, the second junction in melting ice. The temperature at the center rose 18 F in 50 hours. The coefficient of heat transfer between air and cheese was about 1 B hr^{-1} ft^{-2} F^{-1} in this experiment.

It is recommended to determine first the mean exit temperature by means of Graetz's method, assuming parabolic velocity distribution (see Sect. 22–2). Herewith the heat transmitted to the cheese can directly be calculated. Check the result by calculating the coefficient of heat transfer by one of the empirical formulas available. Discuss briefly possible reasons of differences, if any, between the two results.

26–3. How would measurements of the thermal conductivity of a liquid be best arranged when the experimentation is to be based upon the flow of the fluid through a cylindric tube and the heating or cooling of the fluid from the tube wall?

Explain the choice of the experimental conditions and of the equation which you would use for the evaluation of the experimental data.

26–4. Using one of the solutions of Problem 12–1, assume that a platinum wire, 0.010 in. in diameter, 17 in. long, is exposed to a stream of air at standard atmospheric pressure and 150°F. The wire carries a current of 3.0 amp. Its resistance is 0.638 ohm/ft at 20°C; its temperature coefficient $b/a = 0.0039\ \text{C}^{-1}$. The ends of the wire are connected to a copper strip which is held at 150°F.

a. Find the distribution of the temperature over the length of the wire for the air velocity of 100 ft/sec.

b. Find the temperature in the middle of the wire length at air velocities of 50 and 150 ft/sec.

c. What arrangement would be needed if it were desired to use this wire as a hot-wire anemometer?

28–1. Prove that the partial-pressure difference, and not the concentration difference, is the driving force for diffusion. The proof shall be based upon the behavior of a plane layer of a binary mixture of ideal gases with end faces at constant temperatures t_1 and t_2, respectively, under steady-state conditions, if the same total pressure exists all over the layer. Show that the concentration of each gas varies across the layer without diffusion occurring. Neglect in the proof the anomaly indicated in the footnote on page 589.

28–2. The film coefficient of heat transfer, h, between a plane plate and dry air flowing parallel to the plate with the velocity, v, can be expressed by:

$$h = 0.057 \left(\frac{k}{L}\right)^{0.22} (\rho c_p v)^{0.78}$$

where L is the length of the plate in the flow direction, and

ρ, c_p, and k are the density, specific heat at constant pressure, and thermal conductivity of the air, respectively.

This equation has been found in experiments with $L = 50$ cm and a fluid-dynamic starting length of 30 cm.

a. Prove that the above formula holds for any consistent units.

b. Find, by means of the similarity between diffusion and heat transfer, how much water of 30°C will evaporate in 1 hr from the surface of a square water basin, 50 cm in side length, if dry air at 30°C is blown parallel to the water surface in the direction of one side of the basin at an average velocity of 500 cm/sec.

It is supposed that the air flows over a plane plate, 30 cm long, before meeting the water surface, which is at the same plane. The edge of the basin is also supposed to be flush with this plane.

The specific volume of saturated steam at 30°C is 32,930 cu cm/gram. The coefficient of diffusion ("diffusivity") of water vapor into air at 30°C is 0.25 cm^2/sec.

28–3. The rate of evaporation from a horizontal surface (0.5 sq ft) of carbon tetrachloride (CCl_4) into dry air at rest is to be calculated, using a dimensionless equation for the heat transfer by convection from a horizontal surface to a gas at rest above the surface. The liquid surface temperature is 100°F; the air temperature is 70°F. Assume 0.15 ft^2/hr as the mechanical diffusivity of CCl_4 and other physical properties according to Table 29–2.

30–1. A condenser unit is formed by five horizontal copper tubes, each 41 in. long, ¾ in. in outer diameter, with axes in a vertical plane. Saturated "Freon-12" (dichlorodifluoromethane) at 104.8°F is the vapor to be condensed on the outer tube surfaces. These are held at 89.8°F by water flowing inside the tubes.

Calculate how much "Freon" will be condensed in 1 hr, assuming the vapor to be virtually at rest and adding 10% for some deficiencies of Nusselt's theory.

Assume the following values for the properties of "Freon-12" at the average of the two temperatures given above:

Density of liquid = 79.8 lb_m/cu ft
Dynamic viscosity of liquid = $1.417(10^{-9})$ lb_f hr ft^{-2}
Latent heat of evaporation = 57.6 B/lb_m
Thermal conductivity of liquid = 0.0483 B hr^{-1} ft^{-1} F^{-1}

30–2. Water vapor at 1.095 lb/sq in. absolute pressure has the same saturation temperature (104.8°F) as assumed in Problem 30–1 for "Freon-12."

How much water having this temperature would be condensed per hour in the same condenser unit and under the same conditions as indicated in Problem 30–1?

APPENDIX B

Nomenclature

Explanatory Notes

NOTE 1. Most of the symbols used are approved by the American Standards Association [1943]; some minor deviations are explained in a footnote * on page 726.

NOTE 2. In the list on pages 717 to 726, the symbols are arranged in the following sequence:

Group *a*: Simple Latin letter symbols
" *b*: Simple Greek letter symbols
" *c*: Latin letter subscripts
" *d*: Greek letter subscripts
" *e*: Numeral subscripts
" *f*: Special signs
" *g*: Selected composite symbols

NOTE 3. This nomenclature is not complete; however, it contains most of the symbols employed. Not listed are symbols of a merely mathematical character, as the coordinates x,y,z,r,ξ,η,ζ; symbols for functions and running numbers; constants, factors, and the usual mathematical signs. An exception is made with listing M and N, preferably used as constants of integration instead of C,C', etc., because the symbol C is reserved for circumference (perimeter), which frequently occurs in the equations.

NOTE 4. Most symbols had to be used for more than one quantity. Then the quantities occurring most often are listed first, those occurring less often in the second line, and those seldom used last and in parentheses.

NOTE 5. The physical dimension is added to each symbol of Groups *a*, *b*, and *g*, generally in the British gravitational system with the following basic units: foot (ft), hour (hr), pound force (lb), degree Fahrenheit (F), and British thermal unit (B). Occasionally the time unit second (sec) is used instead of the hour (hr). The mass unit in this system is the slug, defined as that mass which is accelerated at the rate of 1 ft/sec^2 by 1 lb force. Since the use of the slug is not very popular, the British system with the pound mass (lb_m) as basic unit is occasionally employed when no forces are involved. Some symbols are expressed in the units of the physical centimeter-gram-second system, centimeter (cm), second (sec), gram mass (gram), degree centigrade (C), calorie (cal), or in electrophysical units, as volt, ampere, ohm, farad. The practical metric system, including the kilocalorie (kcal), is used at various places in the book, but not in the present nomenclature.

NOTE 6. In the literature the expression "degree Fahrenheit" is used in two different senses, either for a certain point of the Fahrenheit temperature scale or as a unit difference in this scale. In this book the symbol °F is used with the first, the symbol F with the second, of these meanings [see Jakob and Dow, 1946]. This has different advantages. One is that formal discrepancies like the following are avoided:

41 degrees Fahrenheit is equal to 5 degrees centigrade: 41 F = 5 C
1 degree Fahrenheit is equal to $\frac{5}{9}$ degree centigrade: 1 F = $\frac{5}{9}$ C

Instead of this, according to the author's suggestion,

$$41°\text{F} = 5°\text{C}$$

$$1\ \text{F} = \tfrac{5}{9}\ \text{C}$$

where the ° sign indicates that the point in a scale is meant, whereas F and C alone designate just units.

Another example:

280 degrees Kelvin is equal to 7 degrees centigrade: 280 K = 7 C
1 degree Kelvin is equal to 1 degree centigrade: 1 K = 1 C

Instead of this, the author recommends:

$$280°\text{K} = 7°\text{C}$$

$$1\ \text{K} = 1\ \text{C}$$

It is seen that, as *units*, C ≡ K, and F ≡ R, where R stands for degree Rankine (absolute Fahrenheit scale). Therefore, the units K and R are not needed, and the physical dimension of Stefan-Boltzmann's constant σ, for instance, can be written as B $\text{hr}^{-1}\ \text{ft}^{-2}\ \text{F}^{-4}$ or cal $\text{sec}^{-1}\ \text{cm}^{-2}\ \text{C}^{-4}$.

It might be objected that the unit symbols F or C cannot be used for the absolute temperature T, since T is counted from the absolute zero point and not from 0°F or 0°C. However, in using the unit symbol ft for a length x or hr for a time τ, nobody cares from what zero point of coordinates or initial instant of time the quantities x or τ are counted.

a. Simple Latin Letter Symbols

SYMBOL	QUANTITY	UNIT
A	Area, particularly cross-sectional area	ft^2
a	Half-length of the side of a square or of one side of a rectangle (sometimes: full length)	ft
$a = h/\rho c_p$	(A heat-transfer quantity, defined as ratio of surface coefficient of heat transfer to product of density and specific heat at constant pressure)	ft hr^{-1}
a	(Area)	ft^2
b	Half-length of one side of a rectangle (sometimes: full length)	ft
b	Thickness of boundary layer	ft
$b = h/k$	Ratio of surface coefficient of heat transfer to thermal conductivity	ft^{-1}

Symbol	Quantity	Unit
b	Mass transfer coefficient	ft hr^{-1}
b	(Barometer level)	mm Hg
C	Circumference or perimeter	ft
C	Coefficient of total radiation, defined by Eq. 7–1	B hr^{-1} ft^{-2}
c	Concentration	lb_m ft^{-3}
c	(Light velocity)	cm sec^{-1}
c	(Sound velocity)	cm sec^{-1}
D	Diameter	ft
D	(Mean distance of neighboring molecule centers)	cm
d	Characteristic length of roughness element	ft
e	Voltage	volt
e	(Electric potential)	$cm^{1/2}$ $gram^{1/2}$ sec^{-1} (electrostatic units)
$e = \rho c_p T$	(An energy quantity per unit volume, defined as product of density, specific heat at constant pressure, and absolute temperature)	B ft^{-3}
$f = \phi/4$	Friction factor, defined by Eq. 24–31. See also Coefficient of friction ϕ.	....
f	(Frequency)	sec^{-1}
G	Inertia force, as gravity, per unit volume	lb ft^{-3}
G	(Specific gravity relative to water)	....
H	Height	ft
H	Enthalpy; stored heat energy	B
H	Heaviside's unit function (see Eq. 17–127)	....
h	Surface coefficient of heat transfer	B hr^{-1} ft^{-2} F^{-1}
h	(Planck's constant)	erg sec
i	Electric current	amp
J	Mechanical equivalent of heat unit	ft lb B^{-1}
j	Heat-transfer factor, defined by Eq. 24–78	....
k	Thermal conductivity	B hr^{-1} ft^{-1} F^{-1}
k	(Boltzmann's constant)	erg C^{-1}
L	Length; distance	ft
L	Laplace transform, defined by Eq. 17–152	....
L	(Lorenz number)	$volt^2$ C^{-2}
M	Constant of integration	
M	Molecular weight	....
m	Mass	slug
$m = \sqrt{hC/kA}$	A quantity composed from surface coefficient of heat transfer h, thermal conductivity k, circumference (perimeter) C, and cross-sectional area A	ft^{-1}
$m = q_s'''/k$	Ratio of rate of heat generated in unit volume at surface temperature to thermal conductivity	ft^{-2} F
m	Logarithmic decrement of radiation	ft^{-1}
$m = \sqrt{\omega/2\alpha}$	A quantity composed from angular velocity ω and thermal diffusivity α	ft^{-1}
m	(Momentum of doublet defined by Eq. 17–42)	B hr^{-1} ft

Symbol	Quantity	Unit
N	Constant of integration	
N	Areal radiant intensity = radiant energy per unit of time, unit of projected area, and unit of solid angle	B hr^{-1} ft^{-2}
N	Avogadro's number = number of molecules in one mole	....
N	(Number of fringes = number of light-wave interferences)	....
n	Direction normal to surface	
n	Coordinate normal to surface	ft
n	Index of refraction	....
$n = \epsilon q_s'''/k$	Ratio of the product of temperature coefficient of electric resistance and rate of heat generation in unit volume of an electric coil to its apparent thermal conductivity k	ft^{-2}
P	Porosity	% by volume
P	(Dynamic force per unit volume, as the pressure drop in a fluid per unit length)	lb ft^{-3}
P	(Power)	ft lb hr^{-1}
p	Pressure; partial pressure	lb ft^{-2}
p	Heaviside's operator, defined by Eq. 17–125	hr^{-1}
Q	Heat energy *	B
q	Rate of heat flow	B hr^{-1}
R	Thermal resistance †	B^{-1} hr F
R	Individual gas constant ‡	ft^2 sec^{-2} F^{-1}
R	(Radius)	ft
r	Radial distance; (radius)	ft
S	Surface area, in contrast to cross-sectional area	ft^2
S	Shearing stress	lb ft^{-2}
S	(Entropy)	B F^{-1}
s	Distance of surface of plane plate, cylinder, or sphere from median plane, center line, or center, respectively	ft
s	(Plate thickness; lubricating-film thickness)	ft
T	Temperature in the absolute scale	°R
t	Temperature in the international scale	°F
U	Overall coefficient of heat transfer	B hr^{-1} ft^{-2} F^{-1}
U	(Internal energy)	B
u	Space density of radiation	B ft^{-3}
u	(Velocity)	ft hr^{-1}
V	Volume	ft^3
V	Half volume of a fin	ft^3
V	(Frictional force per unit volume, as caused by viscosity)	lb ft^{-3}

* Except with subscript e; see Q_e in Group g.

† Except with subscripts e or f; see R_e and R_f in Group g.

‡ Units are obtained from the gas equation $p/\rho = RT$ with p in lb/sq ft, ρ in slug/cu ft, T in Fahrenheit units.

Symbol	Quantity	Unit
v	Velocity	ft hr^{-1}
W	Radiant flux density	B hr^{-1} ft^{-2}
W	Weight	lb
W	Width	ft
w	Velocity; in particular, molecular random velocity	cm sec^{-1}
w	(Water contents)	% by volume
Y	Thickness of condensate film	ft
z	Height	ft
z	Depth	ft

b. Simple Greek Letter Symbols

Symbol	Quantity	Unit
α	Thermal diffusivity	ft^2 hr^{-1}
α	Radiation absorptivity	....
β	Coefficient of cubic thermal expansion	F^{-1}
β	Temperature coefficient of thermal conductivity of solids	F^{-1}
β	Outer contact angle between vapor bubble and solid surface	....
γ	Specific weight	lb ft^{-3}
δ	Mechanical diffusivity	ft^2 hr^{-1}
δ	(Grain thickness)	ft
ϵ	Radiation emissivity	....
ϵ	Temperature coefficient of electric resistivity	F^{-1}
ϵ	Coefficient of turbulent exchange	....
ϵ	(Elementary charge of electron)	coulomb
ζ	Coefficient of velocity discontinuity of a gas on a surface, defined by Eq. 6–25	cm
H	Basic unit of heat energy	B
η	Effectiveness	....
η	Reduction factor of temperature amplitude	....
$\eta = y/s$	(Ratio of the distance of a point in a lubricating film from the bearing surface to the distance between bearing and journal surfaces)	....
Θ	Basic unit of temperature	F
Θ	(Temperature discontinuity of a gas on a solid surface)	C
θ	Temperature excess over a reference temperature	F
κ	Dielectric constant	Dimensionless in electrostatic system of units
$\kappa = c_p/c_v$	Ratio of specific heats at constant pressure and volume	....
Λ	Basic unit of length	ft
λ	Wavelength of radiation	cm
λ	Mean free path length of molecules	cm

Symbol	Quantity	Unit
λ	Temperature wavelength	ft
λ	Latent heat of evaporation	B lb^{-1}
M	Basic unit of mass	slug
μ	Dynamic viscosity	lb hr ft^{-2} or poise = gram $cm^{-1}sec^{-1}$
ν	Kinematic viscosity	ft^2 hr^{-1}
$\xi = x/s$	Ratio of distance of a point to surface distance from median plane of a plane plate	....
$\xi = r/s$	Ratio of radial distance of a point to surface distance from center line or center of cylinder or sphere, respectively	....
ξ	(Coefficient of temperature discontinuity of a gas on a surface, defined by Eq. 6–24)	cm
ρ	Density	slug ft^{-3}
ρ	Radiation reflectivity	....
ρ	Ratio of radii of circular fin, defined by Eq. 11–86	....
σ	Stefan-Boltzmann's constant	B hr^{-1} ft^{-2} F^{-4}
σ	Surface tension	lb ft^{-1}
$\sigma = s\sqrt{n}$	A dimensionless quantity composed from distance s of surface from median plane, center line, or center of plane plate, cylinder, or sphere, respectively, and $n = \epsilon q_s'''/k$. See also n'.	....
T	Basic unit of time	hr
τ	Time	hr
τ	(Radiation transmissivity)	....
Φ	Radiant flux	B hr^{-1}
Φ	(Velocity discontinuity of a gas on a surface)	cm sec^{-1}
ϕ	Angle of incidence = angular deviation from normal direction	....
ϕ	Spheric pole distance	....
ϕ	Shape factor, defined by Eq. 8–12	....
$\phi = \theta_m/\theta_0$	Ratio of mean to maximum temperature excess	....
ϕ	Temperature ratio, defined by Eq. 20–3	....
$\phi = 4f$	Coefficient of friction, defined by Eq. 21–38. See also Friction factor f.	
ϕ	(Fibril angle = angle of timber fibers against longitudinal direction)	
X	Shape factor, defined by Eq. 22–69	....
ψ	Stream function	ft^2 hr^{-1}
ψ	Spherical longitude	....
ω	Solid angle	....
ω	Angular velocity	hr^{-1}

c. Latin Letter Subscripts

Symbol	Quantity
A	Area A
A	Air
a	Atmosphere; standard atmosphere; (air)
a	Apparent
a	Arithmetic mean
a	One of the half side lengths of rectangle
a	(Amplitude)
ad	Adiabatic
b	Black body
b	One of the half side lengths of rectangle
b	Boundary; distance of free surface of boundary layer from solid or liquid surface
b	Vapor bubble
b	(Burning temperature of a ceramic)
b	(Bearing surface)
c	Convection; conduction + convection
c	Circumferential, circular
c	Corrected
c	Cycle
c	Center line
c	Characteristic
c	(Coil; cable)
c	(Curvature)
cr	Critical
d	Downward
e	Environment
e	Equivalent; effective
e	Electric
e	External; outer
f	Surface film
f	Saturated liquid
f	Frictional
f	Fictitious
f	Most favorable
f	Fully established turbulence
g	Saturated vapor; gas
g	Glass
gr	Gray
H	Stored heat energy
h	Convective
h	Heated
h	Horizontal
i	Incoming; inlet
i	Initial
i	Electric insulator
i	Internal, inner
j	Journal surface of a bearing
k	Conductive
L	Distance L
ln	Logarithmic mean
M	Mole
M	Material
m	Mean; cup-mixing mean; mixed
m	Metal
m	(Middle)
max	Maximum
min	Minimum
n	Normal direction; normal polarized
O	Center point; zero point
o	Outgoing
o	Observed
o	Oscillation
opt	Optimum
p	Constant pressure
p	Parallel polarized; in parallel direction
p	Partly established turbulence
p	Particle
r	Radial distance r
r	Radiation
r	Rectangular
r	Reference
r	(Rotation)
r	(Required)
s	Surface; distance of surface from a zero place
s	Solid
scr	Screen
spr	Spreading
St	Steam
st	Starting length
st	Hydrostatic
t	Triangular
t	Translation

Symbol	Quantity	Symbol	Quantity
t	Transverse direction	v	Constant volume
th	Thermometer	W	Water
tot	Total length	W	(Winding)
u	Upward	w	Wedge-shaped
V	Volume V	w	Wetted
v	Vapor	w	Wall
v	Vertical		

d. Greek Letter Subscripts

Symbol	Quantity	Symbol	Quantity
α	Thermal diffusion	λ	Latent heat λ
δ	Mechanical diffusion	ϕ	Angle ϕ
λ	Wavelength λ		

e. Numeral Subscripts

Symbol	Quantity
0	For a quantity = 0; at a zero point of coordinates; at a standard state; arriving undisturbed flow of fluid; ice-melting point; at environment condition; (at distance y_0)
273	At $T = 273°K$
∞	For infinite thickness; at infinite distance; at infinite time

f. Special Signs

Sign	Significance	Example of Application
$\cdot$	Per unit time	$\dot{m}$ = mass flowing in unit time
$'$	Per unit length	q' = Radial heat conduction in cylinder of unit length
$''$	Per unit area	q'' = Heat conduction through unit area
$'''$	Per unit volume	q''' = Heat generation in unit volume
—	To designate a line	$\overline{ab}$ = Line from a to b (in ab as subscript the bar is omitted)
—	Mean value	$\bar{v}$ = Mean of fluctuating velocity
~	Fluctuating component	$\tilde{v}$ = Fluctuating velocity component
*	Belonging to a maximum	λ^*T^* = Product of wavelength and absolute temperature at maximum of intensity of radiation according to Wien's displacement law
*	Dimensionless, generalized quantity	v^* = Ratio of velocity to "friction velocity," defined by Eq. 21–45

g. Selected Composite Symbols

Symbol	Quantity	Unit
a_L	Semi-profile area, defined by Eq. 11–36	ft^2
C_e	Electric capacity	farad
$C_h = f/2$	Heat-convection number, defined by Eq. 24–32; see also Friction factor f	

Symbol	Quantity	Unit
$C_p = Mc_p$	Mole heat at constant pressure = product of molecular weight and specific heat at constant pressure	cal gram^{-1} C^{-1}
$C_p = \rho C_p$	Specific heat of a mass occupying unit volume, at constant pressure	B ft^{-3} F^{-1}
$C_v = Mc_v$	Mole heat at constant volume = product of molecular weight and specific heat at constant volume	cal gram^{-1} C^{-1}
c_1	First constant of Planck's law of radiation	erg cm^2 sec^{-1}
c_2	Second constant of Planck's law of radiation	cm C
c_p	Specific heat at constant pressure §	B lb_m^{-1} F^{-1}
c_v	Specific heat at constant volume	B lb_m^{-1} F^{-1}
$D_e = 4A/C$	Equivalent diameter, where A = cross-sectional area and C = circumference or perimeter	ft
H_i	Strength of instantaneous heat source	B
h_e	Effective coefficient of heat transfer	B hr^{-1} ft^{-2} F^{-1}
h_λ	Coefficient of latent-heat transfer	B hr^{-1} ft^{-2} F^{-1}
i_C	Reactive current	amp
i_R	Active current	amp
k_a	Apparent thermal conductivity, including conduction, convection, and radiation	B hr^{-1} ft^{-1} F^{-1}
k_c	Equivalent thermal conductivity, including conduction and convection	B hr^{-1} ft^{-1} F^{-1}
k_e	Electric conductivity	ohm^{-1} cm^{-1}
k_r	Equivalent thermal conductivity, superseding heat radiation	B hr^{-1} ft^{-1} F^{-1}
$\dot{m}$	Rate of mass flow	slug hr^{-1}
$\dot{m}''$	Density of rate of mass flow	slug hr^{-1} ft^{-2}
$N_{Fo} = \alpha\tau/L^2$	Fourier number, where α = thermal diffusivity, τ = time, L = a characteristic length	....
$N_G = gD^3/\nu^2$	Gravity number, where g = gravitational acceleration, D = a characteristic length, ν = kinematic viscosity	....
$N_{Gr} = gD^3\beta\theta/\nu^2$	Grashof number with symbols as in N_G and, further, β = cubic coefficient of thermal expansion, θ = temperature excess	....
$N_{Gz} = \dot{m}c_p/kL$	Graetz number, where k = thermal conductivity, L = a characteristic length. See also $\dot{m}$ and c_p.	....
$N_{Ma} = v/c$	Mach number = ratio of fluid velocity to sound velocity	....

§ Sometimes slug or lb_f is used instead of lb_m.

Symbol	Quantity	Unit
$N_{Pe} = vD/\alpha$	Péclet number, where v = velocity, D = a characteristic length, α = thermal diffusivity	....
$N_{Pr} = \nu/\alpha$	Prandtl number = ratio of kinematic viscosity to thermal diffusivity	....
$N_{Re} = vD/\nu$	Reynolds number, where v = velocity, D = a characteristic length, ν = kinematic viscosity	....
$N_{Sc} = \nu/\delta$	Schmidt number = ratio of kinematic viscosity to mechanical diffusivity	....
$N_{St} = h/c_p\dot{m}'' = f/2$	Stanton number, where h = surface coefficient of heat transfer, f = friction factor. See also c_p and $\dot{m}''$.	
$N_T = T_{s,1}/T_{0,1} = T_{s,2}/T_{0,2}$	Temperature number, where subscripts s and 0 stand for surface and environment, and 1 and 2 for two similar systems	....
$N_\tau = \nu\tau/L^2$	A modified Fourier number, where ν = kinematic viscosity, τ = time, L = a characteristic length	....
q', q'', q'''	See under Examples of Group f.	
R_e	Electric resistance	ohm
R_f	Frictional resistance	lb
Q_e	Electrical charge	$cm^{3/2}\ gram^{1/2}\ sec^{-1}$ (in electrostatic units)
S_s	Shearing stress (S is also used)	lb ft^{-2}
$t_f = (t_s + t_e)/2$	Film temperature = mean between surface and environment temperatures	°F
t_m	Cup-mixing temperature	°F
$\dot{V}$	Rate of volume flow	$ft^3\ hr^{-1}$
X_e	Capacitive electric resistance	ohm
y_1	Thickness of laminar sublayer	ft
α_e	Equivalent thermal diffusivity, defined by Eq. 24–35	ft^2/hr
β_ν	νth zero place of Bessel function of zero order and first kind (J_0)	....
Δp	Pressure drop	lb ft^{-2}
Δt	Temperature drop	F
Δ_x	Finite difference sign for varying length x	....
Δ_τ	Finite difference sign for varying time τ	....
θ_a	Arithmetic mean temperature difference, defined by Eq. 21–14	F
θ_{ln}	Logarithmic-mean temperature difference, defined by Eq. 21–16	F
λ_c	Characteristic length in a model defined by Eq. 20–6	ft
λ_s	Average distance of effect of thermoelastic wave (see Eq. 6–36)	cm

Symbol	Quantity	Unit
μ_e	Equivalent dynamic viscosity = exchange quantity, defined by Eq. 24–36	lb hr ft^{-2} or poise = gram $cm^{-1} sec^{-1}$
ν_e	Equivalent kinematic viscosity, defined by Eq. 24–37	ft^2/hr
σ_{fg}	Surface tension (also σ)	dyne cm^{-1}
σ_{spr}	Spreading tension	dyne cm^{-1}

* The present nomenclature deviates in a few items only from the symbols of the American Standards Association [1943].

The symbol C is employed for circumference and perimeter of tubes and ducts instead of L_p as recommended in the standards in order that the symbol may not be confused with L used for the axial length. For a similar reason the subscript ln is introduced for logarithmic mean instead of the recommended subscript L.

Reserving the letter D without or with subscript for diameter and distance, the author has chosen the notation δ for mechanical diffusivity in analogy to the approved symbol α for thermal diffusivity.

Following a mathematical usage, the time rate of mass flow $dm/d\tau$ is denoted by $\dot{m}$. Since double prime is used with the meaning "per unit area" (see Article *f*), it follows that $\dot{m}''$ stands for mass velocity. The letter G might designate weight velocity. The standards recommend G as symbol for both weight and mass velocity. It is the author's opinion that a difference should be made between symbols for weight (force) and mass, as the unit lb_f is distinguished from lb_m (see Explanatory Note 5).

APPENDIX C

Conversion Factors

Explanatory Notes

NOTE 1. The conversion factors are arranged in the following three groups:
a. Conversion within the British system of units.
b. Conversion between physical, metric, and British systems of units.
c. Conversion between electrophysical and British systems of units.
The distinction made between physical and electrophysical systems concerns only the use of the calorie or watt-second as the unit of heat energy.

NOTE 2. Conversion tables *b* and *c* are prepared for ready conversion of numerical values of a quantity given in the physical, metric, or electrophysical system into the corresponding values of the British gravitational system, which is preferably used in this book. Therefore the tables indicate in the first line which values in the British system correspond to unit values in the other systems. In some cases the values in the other systems for unit values in the British system are also listed.

NOTE 3. As an example of the use of the tables assume that Stefan-Boltzmann's constant σ be given as $1.377(10^{-12})$ cal sec^{-1} cm^{-2} C^{-4}. Then, according to the last line of Table *b*, it follows that

$$\sigma = (1265)1.377(10^{-12}) = 0.1742(10^{-8})\text{B hr}^{-1}\text{ ft}^{-2}\text{ F}^{-4}$$

Table a. Conversion within the British System of Units

Symbol of Quantity to Be Converted	Corresponding Values in Different Units		
v	1 ft sec^{-1}	= 3600 ft hr^{-1}	= 0.6818 miles hr^{-1}
	1.467 ft sec^{-1}	= 5280 ft hr^{-1}	= 1 mile hr^{-1}
g	1 ft sec^{-2}	= $12.96(10^6)$ ft hr^{-2}	
m	1 slug	= 32.17 lb$_m$	
μ	1 lb ft^{-2} hr	= $12.96(10^6)$ slug ft^{-1} hr^{-1}	
	1 lb$_m$ ft^{-1} hr^{-1}	= 0.03108 slug ft^{-1} hr^{-1}	
Q	1 B	= 778.3 ft lb	

Table b. Conversion between Physical, Metric, and British Systems of Units

Symbol of Quantity to Be Converted	Corresponding Values in Units of the Physical System	Metric Gravitational System	British Gravitational System
L	2.540 cm	= 0.02540 m	= 1 inch
	30.48 cm	= 0.3048 m	= 1 ft
	100 cm	= 1 m	= 3.281 ft
A	1 cm^2	= 0.0001 m^2	= 0.1550 sq in.
	10,000 cm^2	= 1 m^2	= 10.765 sq ft
V	1 cm^3	= 10^{-6} m^3	= 0.06102 cu in.
	10^6 cm^3	= 1 m^3	= 35.32 cu ft
g	980.7 cm sec^{-2}	= 9.807 m sec^{-2}	= 32.17 ft sec^{-2}
$\nu = \mu/\rho$ or $\alpha = k/\rho c_p$	1 cm^2 sec^{-1}	= 0.3600 m^2 hr^{-1}	= 3.874 ft^2 hr^{-1}
		1 m^2 hr^{-1}	= 10.765 ft^2 hr^{-1}
F	10^6 dyne	= 1.020 kg	= 2.249 lb
		1 kg	= 2.205 lb
		0.4536 kg	= 1 lb
$p = F/A$		1 kg cm^{-2} (= 1 technical atmosphere)	= 14.23 lb/sq in.
	1 atmosphere	= 1.033 kg cm^{-2}	= 14.70 lb/sq in.
		1 kg m^{-2}	= 0.2048 lb/sq ft
$\gamma = F/A$		1 kg m^{-3}	= 0.06243 lb/cu ft
m	1000 gram	= 1 kg_m	= 0.06854 slug
		1 kg_m	= 2.205 lb_m
$\rho = m/V$	1 gram cm^{-3}	= 1000 kg_m m^{-3}	= 1.941 slug/cu ft
μ	1 gram cm^{-1} sec^{-1} (= 1 poise)	= 2.8325(10^{-6}) kg m^{-2} hr	= 0.5802(10^{-6}) lb ft^{-2} hr
		= 360.0 kg_m m^{-1} hr^{-1}	= 7.519 slug ft^{-1} hr^{-1}
			= 241.9 lb_m ft^{-1} hr^{-1}
t	n°C	= n°C	= (1.8n + 32)°F
	5/9(n − 32)°C	= 5/9(n − 32)°C	= n°F
T	n°K	= n°K	= 1.8n°R
	n°K	= (n + 273.2)°C	= (1.8n + 459.7)°F
$\Delta t = \Delta T$	1 C	= 1 C	= 1.8 F
Q	1000 cal	= 1 kcal	= 3.968 B
	252.0 cal	= 0.2520 kcal	= 1 B
q	0.2778 cal sec^{-1}	= 1 kcal hr^{-1}	= 3.968 B hr^{-1}
	1 cal sec^{-1}	= 3.600 kcal hr^{-1}	= 14.285 B hr^{-1}
q''	27.78(10^{-6}) cal sec^{-1} cm^{-2}	= 1 kcal hr^{-1} m^{-2}	= 0.3686 B hr^{-1} ft^{-2}
	1 cal sec^{-1} cm^{-2}	= 36,000 kcal hr^{-1} m^{-2}	= 13.27(10^3) B hr^{-1} ft^{-2}
q'''	0.2778(10^{-6}) cal sec^{-1} cm^{-3}	= 1 kcal hr^{-1} m^{-3}	= 0.1123 B hr^{-1} ft^{-3}
k	2.778(10^{-3}) cal sec^{-1} cm^{-1} C^{-1}	= 1 kcal hr^{-1} m^{-1} C^{-1}	= 0.6719 B hr^{-1} ft^{-1} F^{-1}

Symbol of Quantity to Be Converted	Physical System	Metric Gravitational System	British Gravitational System
k	$4.134(10^{-3})$ cal $\sec^{-1}$ cm^{-1} C^{-1}	= 1.488 kcal hr^{-1} m^{-1} C^{-1}	= 1 B hr^{-1} ft^{-1} F^{-1}
h	$27.78(10^{-6})$ cal $\sec^{-1}$ cm^{-2} C^{-1}	= 1 kcal hr^{-1} m^{-2} C^{-1}	= 0.2048 B hr^{-1} ft^{-2} F^{-1}
		4.883 kcal hr^{-1} m^{-2} C^{-1}	= 1 B hr^{-1} ft^{-2} F^{-1}
	1 cal $\sec^{-1}$ cm^{-2} C^{-1}	= 36,000 kcal hr^{-1} m^{-2} C^{-1}	= 7373 B hr^{-1} ft^{-2} F^{-1}
	$27.78(10^{-6})$ cal $\sec^{-1}$ cm^{-2} C^{-4}	= 1 kcal hr^{-1} m^{-2} C^{-4}	= 0.03512 B hr^{-1} ft^{-2} F^{-4}
σ	1 cal $\sec^{-1}$ cm^{-2} C^{-4}	= 36,000 kcal hr^{-1} m^{-2} C^{-4}	= 1264 B hr^{-1} ft^{-2} F^{-4}

Table c. Conversion between Electrophysical and British Systems of Units

Symbol of Quantity to Be Converted	Electrophysical System	British Gravitational System
q	0.2930 watt	= 1 B hr^{-1}
	1 watt	= 3.413 B hr^{-1}
q''	1 watt cm^{-2}	= 3170 B hr^{-1} ft^{-2}
q'''	1 watt cm^{-3}	= 96630 B hr^{-1} ft^{-3}
k	1 watt cm^{-1} C^{-1}	= 57.79 B hr^{-1} ft^{-1} F^{-1}
h	1 watt cm^{-2} C^{-1}	= 1761 B hr^{-1} ft^{-2} F^{-1}
σ	1 watt cm^{-2} C^{-4}	= 302.0 B hr^{-1} ft^{-2} F^{-4}

APPENDIX D

Bibliography and Author Index

Explanatory Notes

NOTE 1. In the text the name of the author and the year in brackets (with (*a*), (*b*), etc., added for more than one publication in the same year) refer to items in the bibliography.

NOTE 2. The bibliography is arranged in alphabetical order by the names of the authors and, where the same author has more than one publication, in the chronological order of the publications, with (*a*), (*b*), etc., added to the year for the second and all following publications by the same author in the same year.

NOTE 3. The page or pages of the text on which any publication is mentioned are added to each entry of the bibliography so that it also forms an index of author names. To make it complete the names of coauthors not only are given in the entries of the publications in which they participated, but are inserted also in the alphabetical sequence with a reference to the senior author.

NOTE 4. The names of magazines are printed in italics and followed by the volume (in boldface), the page number, and the year.

REFERENCE	PAGE
ACKERMANN, G., *VDI-Forschungsheft* No. 382; 1937.	588
ACTON, F. S., *see under* WILHELM.	
ADAMS, E. A., *see under* LANGMUIR.	
ADAMS, J., *see under* BASHFORTH.	
ADAMS, L. H., *see under* WILLIAMSON.	
ADDOMS, J. N., *see under* MCADAMS.	
AKIN, G. A., and MCADAMS, W. H., *Trans. Am. Inst. Chem. Engrs.* **35**, 137; 1939.	646, 657
AKIN, G. A., and MCADAMS, W. H., *Ind. Engg. Chem.* **31**, 481; 1939(*a*).	657
AKIN, G. A., *see also under* SAUER.	
ALTY, T., and NICOLL, F. H., *Canadian Journ. Res.* **4**, 547; 1931.	694
ALVES, G., *see under* MARTINELLI.	
AMERICAN SOCIETY FOR TESTING MATERIALS, C177–45: *Standard method of test for thermal conductivity of materials by means of the guarded hot plate*, Philadelphia; 1945.	154

REFERENCE PAGE

AMERICAN STANDARDS ASSOCIATION, *American standard letter symbols for heat and thermodynamics including heat flow* (Standard Z10.4–1943, published by American Society of Mechanical Engineers); June 1943. 487, 519, 716, 726

ANGELL, M. F., *Phys. Rev.* **33**, 421; 1911. 173

ÅNGSTROEM, A. J., *Annalen d. Physik u. Chemie* **114**, 513; 1861. 313

ÅNGSTROEM, A. J., *Annalen d. Physik u. Chemie* **118**, 423; 1863. 313

ÅNGSTROEM, A. J., *Annalen d. Physik u. Chemie* **123**, 628; 1864. 313

ASCHKINASS, E., *Annalen d. Physik* **17**, 960; 1905. 45

AVRAMI, M., and LITTLE, J. B., *Journ. Appl. Physics* **13**, 255; 1942. 224, 414

AVRAMI, M., and PASCHKIS, V., *Trans. Am. Inst. Chem. Engrs.* **38**, 631; 1942. 414

AWBERY, J. H., *Phil. Mag.* (7) **4**, 629; 1927. 305, 321

AWBERY, J. H., and SCHOFIELD, F. H., Effect of shape on heat loss through insulation, *Proc. Fifth Internat. Congr. on Refrigeration* **3**, 591; 1929. 404

BADGER, W. L., *see under* BOARTS.

BAILEY, A., and COPE, W. F., *Aeronaut. Research Comm.* (*Great Britain*) *Techn. Rep.* 43, p. 199; 1933. 550

BAKER, E. M., and MUELLER, A. C., *Trans. Am. Inst. Chem. Engrs.* **33**, 531; 1937. 690

BAKER, E. M., and MUELLER, A. C., *Ind. Engg. Chem.* **29**, 1067; 1937(*a*). 690

BAKER, E. M., and TSAO, U., *Trans. Am. Inst. Chem. Engrs.* **36**, 517, 783; 1940. 690

BAKER, E. M., and TSAO, U., *Ind. Engg. Chem.* **32**, 1115; 1940(*a*). 690

BAKER, H. D., *see under* PASCHKIS.

BAKHMETEFF, B. A., *The mechanics of turbulent flow*, Princeton University Press; 1936. 438

BAKHMETEFF, B. A., *Mech. Engg.* **58**, 625; 1936(*a*). 430, 438

BAKHMETEFF, B. A., *Trans. Am. Soc. Mech. Engrs.* **62**, 551; 1940. 518

BARNES, G. M., *Mech. Engg.* **67**, 827; 1945. 567

BASHFORTH, FR., and ADAMS, J., *Capillary action*, Cambridge, England; 1883. 630

BAUM, S., *see under* FITZPATRICK.

BAYS, G. S., *see under* NAGLE.

BECKMANN, W., *Forschung a. d. Geb. d. Ingenieurwes.* **2**, 165, 213, 407; 1931. 539, 541

BECKMANN, W., *see also under* SCHMIDT, E.

BERG, E. J., *Heaviside's operational calculus*, McGraw-Hill Book Company, New York; 1929. 352

BERGET, A., *Comptes rendus* **105**, 224; 1887. 152

BERGET, A., *Journ. de physique* (2) **7**, 2; 1888. 152

BERGGREN, W. P., *see under* PERRY, R. L.

BEUKEN, CL. L., Dissertation, Saechs. Bergakademie Freiberg, Triltsch and Huther, Berlin; 1936. 399

BEUKEN, CL. L., *Compte-rendu du Congrès internationale des applications électrocalorifiques et électrochimiques*, Scheveningue, p. 204; 1936(*a*). 399

BEUKEN, CL. L., *Economisch Technisch Tijdschrift* **19**, 43; 1939. 399, 411

BEUKEN, CL. L., and PASCHKIS, V., *Economisch Technisch Tijdschrift* **17**, 163; 1937. 270, 411

BEUKEN, CL. L., *see also under* PASCHKIS.

BIERMASZ, TH., *see under* HAAS.

BINDER, L., Dissertation, Techn. Hochschule Muenchen, W. Knapp, Halle a. S.; 1911. 380

REFERENCE	PAGE
Biot, J. B., *Bibliothéque Britannique* **27**, 310; 1804.	2, 208
Biot, J. B., *Traité de physique*, Vol. 4, p. 669, Paris; 1816.	2
Blasius, H., *Zeitschr. f. Math. u. Physik* **56**, 1; 1908.	471
Blenderman, L. M., *see under* Nagle.	
Bliss, H., *see under* Insinger.	
Boarts, R. M., Badger, W. L., and Meisenburg, S. J., *Trans. Am. Inst. Chem. Engrs.* **33**, 363; 1937.	651
Boarts, R. M., Badger, W. L., and Meisenburg, S. J., *Ind. Engg. Chem.* **29**, 912; 1937(*a*).	651
Boelter, L. M. K., Cherry, V. H., and Johnson, H. A., *Heat transfer*, University of California Press, Berkeley and Los Angeles; 1942.	402
Boelter, L. M. K., Martinelli, R. C., and Jonassen, F., *Trans. Am. Soc. Mech. Engrs.* **63**, 447; 1941.	440, 514, 565
Boelter, L. M. K., and Tribus, M., Numerical solutions for thermal systems, Chapter 6 of L. E. Grinter, *Numerical methods of analysis in engineering*, The Macmillan Company, New York; 1949.	379
Boelter, L. M. K., Young, G., and Iversen, H. W., *Nat. Advis. Comm. Aeronaut. Techn. Note* 1451; 1948.	548
Boelter, L. M. K., *see also under* Dittus; *under* Lee; *under* Martinelli; *under* Nottage.	
Boltzmann, L., *Wiedemanns Annalen* **22**, 291; 1884.	3, 32
Bonilla, Ch. F., and Perry, Ch. W., *Trans. Am. Inst. Chem. Engrs.* **37**, 685; 1941.	648
Bonilla, Ch. F., *see also under* Cichelli.	
Bornefeld, H., *Techn. Zentralblatt f. praktische Metallbearbeitung* **43**, 14; 1933.	343
Bosch, M. ten, *Die Waermeuebertragung*, 3rd Ed., Julius Springer, Berlin; 1936.	503, 554
Bošnjaković, F., *Techn. Mechanik u. Thermodynamik* **1**, 358; 1930.	624
Boussinesq, J., *Journ. de math. pure et appliquée* (2) **13**, 377; 1868.	442
Boussinesq, J., *Essai sur la théorie des eaux courantes*, Mémoires présentés par divers savants à l'Académie des Sciences de l'Institut de France, Vol. 23; 1877.	506
Boussinesq, J., *Journ. de physique*, (4) **1**, 65; 1902.	524
Bragg, W. L., *Journ. Institution Electr. Engrs. London* **69**, 1239; 1931.	99
Brainerd, J. G., *see under* Emmons.	
Braunlich, R. H., Thesis, Massachusetts Institute of Technology; 1941.	657
Bridges, C. H., *see under* Rhodes.	
Bridgman, P. W., *Proc. Nat. Acad. Sciences* **9**, 341; 1923.	80
Brinn, M. S., Friedman, S. J., Gluckert, F. A., and Pigford, R. L., *Ind. Engg. Chem.* **40**, 1050; 1948.	546
Broder, J., *see under* Mosciki.	
Brown, A. J., and Marco, S. M., *Introduction to heat transfer*, McGraw-Hill Book Company, New York; 1942.	492
Brown, P. W., *see under* Harper.	
Bueche, W., and Schau, Fr., *Archiv f. Waermewirtsch. u. Dampfkesselwes.* **17**, 67; 1936.	238–240
Burbach, Th., Waermeuebergang in Rohren, in P. Hermann and Th. Burbach, *Stroemungswiderstand und Waermeuebergang in Rohren*, Leipzig; 1930.	547

REFERENCE PAGE

BURGERS, J. M., and HEGGE ZIJNEN, B. G. VAN DER, *Verhandel. d. k. Akad. van Wetenschappen, Amsterdam*, 1st Sect., Part XIII, No. 3; 1924 (also *Mededeeling* No. 5, *Lab. voor Aerodynamica en Hydrodynamica*, Techn. Hoogeschool, Delft). 558

BUTLER, F. H., *see under* SHANKLIN.

BYLEVELT, J. S. VAN, *see under* KENNELY.

CAMERON, R. H., *see under* ROSENTHAL.

CARPENTER, F. G., COLBURN, A. P., SCHOENBORN, E. M., and WURSTER, A., *Trans. Am. Inst. Chem. Engrs.* **42**, 165; 1946. 553

CARSLAW, H. S., *Introduction to the mathematical theory of the conduction of heat in solids*, 2nd Ed., Macmillan and Company, London; 1921 (also Dover Publications, New York; 1945). 252, 332, 337

CARSLAW, H. S., and JAEGER, J. C., *Proc. London Math. Soc.* (2) **46**, 361; 1940. 265, 268

CARSLAW, H. S., and JAEGER, J. C., *Operational methods in applied mathematics*, Clarendon Press, Oxford; 1941. 352, 364

CARSLAW, H. S., and JAEGER, J. C., *Conduction of heat in solids*, Oxford Press; 1947. 252

CHAPMAN, S., *Phil. Trans.* (A) **211**, 459; 1912. 65, 73

CHAPMAN, S., *Proc. Roy. Soc. London* (A) **93**, 1; 1916. 589

CHAPMAN, S., and COWLING, T. G., *The Mathematical theory of non-uniform gases*, University Press, Cambridge, England; 1939. 589

CHAPMAN, S., and DOOTSON, F. W., *Phil. Mag.* (6) **33**, 248; 1917. 589

CHEN, C. Y., HAWKINS, G. A., and SOLBERG, H. L., *Trans. Am. Soc. Mech. Engrs.* **68**, 99; 1946. 551

CHENEY, A. J., JR., *see under* WINDING.

CHERRY, V. H., *see under* BOELTER; *under* LEE.

CHOLETTE, A., *Chem. Engg. Progress* **44**, 81; 1948. 546, 547

CHRISTIAN, G. A., *see under* FOUST.

CHRISTOPHERSON, D. G., and SOUTHWELL, R. V., *Proc. Roy. Soc.* (A) **168**, 317; 1938. 365

CHURCH, J. W., and COBB, H. B., Thesis, Massachusetts Institute of Technology; 1922. 642

CHURCHILL, R. V., *Modern operational mathematics in engineering*, McGraw-Hill Book Company, New York; 1944. 352, 364

CICHELLI, M. T., and BONILLA, CH. F., *Trans. Am. Inst. Chem. Engrs.* **41**, 755; 1945. 648

CICHELLI, M. T., and BONILLA, CH. F., *Trans. Am. Inst. Chem. Engrs.* **42**, 411; 1946. 648–650

CLAASSEN, H., *Zeitschr. d. Ver. deutsch. Ing.* **77**, 1141; 1933. 689

CLARKE, MARTHA, *see under* JAEGER, J. C.

CLEMENT, J. K., and EGY, W. L., *Phys. Rev.* **28**, 71; 1909. 149

COBB, H. B., *see under* CHURCH.

COGAN, C. A., Thesis, Massachusetts Institute of Technology; 1934 (according to McAdams [1942]). 690

COLBURN, A. P., *Ind. Engg. Chem.* **25**, 873; 1933. 428

COLBURN, A. P., *Trans. Am. Inst. Chem. Engrs.* **29**, 174; 1933(*a*). 520, 546, 564

COLBURN, A. P., *Trans. Am. Inst. Chem. Engrs.* **30**, 187; 1933–1934. 682

REFERENCE PAGE

COLBURN, A. P., *Ind. Engg. Chem.* **26,** 432; 1934. 682

COLBURN, A. P., *Engg. Bull. Purdue University,* Vol. 26, No. 1 (Engineering Experiment Station, Res. Series No. 84); January 1942. 520, 521, 549

COLBURN, A. P., and HOUGEN, O. A., *Ind. Engg. Chem.* **26,** 1178; 1934. 692

COLBURN, A. P., *see also under* CARPENTER.

COLLIER, D. W., *see under* WILHELM.

COOPER, C. M., DREW, TH. B., and MCADAMS, W. H., *Ind. Engg. Chem.* **26,** 428; 1934. 681

COOPER, H. B. H., *see under* SAUER.

COPE, W. F., *see under* BAILEY.

CORCORAN, W. H., ROUDEBUSH, B., and SAGE, B. H., *Chem. Engg. Progress* **43,** 135; 1947. 506

COWLING, T. G., *see under* CHAPMAN.

CRAIG, H. L., *see under* MARTINELLI.

CROSS, H., *Trans. Am. Soc. Civil Engrs.* **96,** 1; 1932. 365

CRYDER, D. S., and FINALBORGO, A. C., *Trans. Am. Inst. Chem. Engrs.* **33,** 346; 1937. 649

CRYDER, D. S., and GILLILAND, E. R., *Ind. Engg. Chem.* **24,** 1382; 1932. 649

CRYDER, D. S., and GILLILAND, E. R., *Refrigerating Engg.* **25,** 78; 1933. 642

DALE, T. P., and GLADSTONE, I. H., *Phil. Trans.* **153,** 317; 1863. 571

DANILOFF, M., *Journ. Am. Chem. Soc.* **54,** 1328; 1932. 78

DAVIES, S. J., and WHITE, C. M., *Engg.* **128,** 69; 1929. 442

DAVIS, A. H., *Phil. Mag.* (6) **44,** 920; 1922. 523

DAVIS, A. H., *see also under* GRIFFITHS.

DAVIS, E. S., *Trans. Am. Soc. Mech. Engrs.* **65,** 755; 1943. 552

DAVISON, A. W., *see under* WALLACE.

DAVISSON, C., and WEEKS, J. R., *Journ. Opt. Soc. Am.* **8,** 581; 1924. 46

DAY, R. S., *see under* MCADAMS.

DEBYE, P., *Vortraege ueber die kinetische Theorie der Materie und Elektrizitaet,* Leipzig and Berlin; 1914. 99

DESPRETZ, C. M., *Annales de chim. et physique* **19,** 97; 1822. 208

DESPRETZ, C. M., *Annales de chim. et physique* **36,** 422; 1828. 208

DESPRETZ, C. M., *Annalen d. Physik u. Chemie* **12,** 281; 1828(*a*). 208

DIESSELHORST, H., *Zeitschr. f. Instrumentenkunde* **22,** 115; 1902. 172

DIESSELHORST, H., Waermeleitung, Physikalischer Teil, in *Encyklopaedie der mathematischen Wissenschaften,* Vol. 5, Part 1, Article 4, II (article finished 1904), B. G. Teubner, Leipzig and Berlin; 1903–1921. 317

DIESSELHORST, H., *see also under* JAEGER, W.

DITTUS, F. W., and BOELTER, L. M. K., *University of California Publications in Engineering,* **2,** 443; 1930. 547, 651

DOOTSON, F. W., *see under* CHAPMAN.

DOW, W. M., *see under* JAKOB.

DREW, TH. B., Mathematical attacks on forced convection problems: a review. *Publications from the Massachusetts Institute of Technology,* Vol. 67, No. 59, Publ. Ser. No. 869 (Dept. Chem. Engg. Serial No. 282); 1931. 451, 459, 460, 473

DREW, TH. B., HOGAN, J. J., and MCADAMS, W. H., *Publications from the Massachusetts Institute of Technology,* Vol. 67, No. 61, Publ. Ser. No. 871 (Dept. Chem. Engg. Serial No. 284); 1931. 543

REFERENCE PAGE

DREW, TH. B., and MUELLER, A. C., *Trans. Am. Inst. Chem. Engrs.* **33**, 449; 1937. 652, 654

DREW, TH. B., NAGLE, W. M., and SMITH, W. Q., *Trans. Am. Inst. Chem. Engrs.* **31**, 605; 1935. 658

DREW, TH. B., *see also under* COOPER; *under* NAGLE.

DRUDE, P., *Annalen der Physik* **1**, 566; 1900. 112

DULONG and PETIT, *Annales de chim. et physique* **1**, 225 and 337; 1817. 444

DUNN, P. S., and VINCENT, A. D., Thesis, Massachusetts Institute of Technology; 1931. 642, 646

DUSINBERRE, G. M., *Trans. Am. Soc. Mech. Engrs.* **67**, 703; 1945. 365, 373

EAGLE, A., and FERGUSON, R. M., *Proc. Roy. Soc. London* (A) **127**, 540; 1930. 510

EAGLE, A., and FERGUSON, R. M., *Proc. Institution Mech. Engrs.* (London) **2**, 985; 1930(*a*). 510

ECK, H., *see under* JAKOB.

ECKERT, E., *Forschung a. d. Geb. d. Ingenieurwes.* **9**, 251; 1938. 26

ECKERT, E., and WEISE, W., *Forschung a. d. Geb. d. Ingenieurwes.* **12**, 40; 1941. 475

ECKERT, E., *see also under* SCHMIDT, E.

EDWARDS, J. D., *see under* TAYLOR, C. S.

EGY, W. L., *see under* CLEMENT.

EMMONS, H. W., *Trans. Am. Inst. Chem. Engrs.* **35**, 109; 1939. 693

EMMONS, H. W., *Trans. Am. Soc. Mech. Engrs.* **65**, 607; 1943. 365

EMMONS, H. W., *Quarterly Appl. Math.* **2**, 173; 1944. 365

EMMONS, H. W., and BRAINERD, J. G., *Journ. Appl. Mech.* **8**, A–105; 1941. 476

ENDE, W., *see under* FRITZ.

ENSKOG, D., *Physik. Zeitschr.* **12**, 533; 1911. 589

ERK, S., Den Waermeaustausch hindernde Massnahmen, Chapter VIII of Vol. I, Part 1 of A. EUCKEN and M. JAKOB, *Der Chemie-Ingenieur*, Akademische Verlagsgesellschaft, Leipzig; 1933. 691

ERK, S., Den Waermeaustausch foerdernde Massnahmen, Chapter VII of Vol. I, Part 1 of A. EUCKEN and M. JAKOB, *Der Chemie-Ingenieur*, Akademische Verlagsgesellschaft, Leipzig; 1933(*a*). 91

ERK, S., *see also under* GROEBER; *under* JAKOB.

EUCKEN, A., *Physik. Zeitschr.* **12**, 1005; 1911. 95

EUCKEN, A., *Annalen d. Physik* (4) **34**, 185; 1911(*a*). 98, 150, 163

EUCKEN, A., *Physik. Zeitschr.* **14**, 324; 1913. 73

EUCKEN, A., *VDI-Forschungsheft* No. 353; 1932. 103

EUCKEN, A., *Naturwissenschaften* **25**, 209; 1937. 693

EUCKEN, A., *Forschung a. d. Geb. d. Ingenieurwes.* **11**, 6; 1940. 74, 76, 85, 101, 117

EULER, L., *Hist. de l'Acad. Berlin* **10**, 227; 1754 (publ. 1756), German translation by E. A. BRAUER and M. WINKELMANN, *Ostwald's Klassiker der exakten Wissenschaften* No. 182, Leipzig; 1911. 19

FARBER, E. A., and SCORAH, R. L., *Trans. Am. Soc. Mech. Engrs.* **70**, 369; 1948. 652, 653, 655, 656

FERGUSON, R. M., *see under* EAGLE.

FINALBORGO, A. C., *see under* CRYDER.

FISHENDEN, MARGARET, and SAUNDERS, O. A., *Engg.* **130**, 177, 193; 1930. 524

REFERENCE PAGE

FITZPATRICK, J. P., BAUM, S., and MCADAMS, W. H., *Trans. Am. Inst. Chem. Engrs.* **35,** 97; 1939. 695

FOOTE, P. D., *Bull. Bur. Standards* **11,** 607; 1915. 49

FORBES, J. D., *Trans. Roy. Soc. Edinburgh* **23,** 133; 1864. 314

FORBES, J. D., *Trans. Roy. Soc. Edinburgh* **24,** 73; 1867. 314

FOURIER, J. B. J., *Théorie analytique de la chaleur*, Paris; 1822 (comprising some of his papers since 1807), English translation by Freeman, Cambridge; 1878. German by Weinstein, Berlin; 1884. 2, 252

FOUST, A. S., and CHRISTIAN, G. A., *Trans. Am. Inst. Chem. Engrs.* **36,** 541; 1940. 553

FRANZ, R., *see under* WIEDEMANN.

FRIEDMAN, S. J., and MILLER, C. O., *Ind. Engg. Chem.* **33,** 885; 1941. 681

FRIEDMAN, S. J., *see also under* BRINN.

FRITZ, W., *Physik. Zeitschr.* **36,** 379; 1935. 630

FRITZ, W., *Zeitschr. d. Ver. deutsch. Ing., Beiheft "Verfahrenstechnik,"* No. 5, p. 149; 1937. 642, 651

FRITZ, W., *Forschung a. d. Geb. d. Ingenieurwes.* **14,** 1; 1943. 97

FRITZ, W., and ENDE, W., *Physik. Zeitschr.* **37,** 391; 1936. 625, 634

FRITZ, W., and HOMANN, F., *Physik. Zeitschr.* **37,** 873; 1936. 623

FRITZ, W., and MOSER, H., *Feuerungstechnik* **28,** 97; 1940. 97

FRITZ, W., *see also under* JAKOB; *under* KOCH, B.

FROCHT, M. M., *Journ. Appl. Phys.* **17,** 730; 1946. 365

FRUMKIN, A., *see under* KABANOW.

FURTHMANN, E., *see under* SCHMIDT, H.

GANS, R., *Annalen d. Physik* **47,** 708; 1915. 568

GARDNER, K. A., *Trans. Am. Soc. Mech. Engrs.* **67,** 621; 1945. 241

GELISSEN, H. C. J. H., *Ind. Heating* **15,** 402; 1948. 411

GIEBE, E., *Verhandl. d. deutsch. phys. Ges.* **5,** 60, 1903. 316

GILLILAND, E. R., *see under* CRYDER.

GLADSTONE, I. H., *see under* DALE.

GLUCKERT, F. A., *see under* BRINN.

GNAM, E., *VDI-Forschungsheft* No. 382, Berlin, 1937. 696

GOENS, E., *see under* GRUENEISEN.

GRAETZ, L., *Annalen d. Physik* (N.F.) **18,** 79; 1883. 459

GRAETZ, L., *Annalen d. Physik* (N.F.) **25,** 337; 1885. 451

GREEN, F. H., and KING, L. S., *Trans. Am. Soc. Mech. Engrs.* **68,** 115; 1946. 550

GRIFFITHS, E., and DAVIS, A. H., *The transmission of heat by radiation and convection*, Department of Scientific and Industrial Research, Food Investigation Board, Special Rep. No. 9, His Majesty's Stationery Office, London, 1922. 526, 531, 539

GRIFFITHS, E., and KAYE, G. W. C., *Proc. Roy. Soc. London* (A) **104,** 71; 1923. 164

GRIFFITHS, E., *see also under* POWELL.

GROEBER, H., Dissertation, Techn. Hochschule Muenchen, R. Oldenbourg, Muenchen; 1908. 148

GROEBER, H., *Zeitschr. d. Ver. deutsch. Ing.* **54,** 1319; 1910. 148

GROEBER, H., *Die Grundgesetze der Waermeleitung und des Waermeueberganges*, Julius Springer, Berlin; 1921. 252, 270, 299, 301, 342, 451, 487

GROEBER, H., *Zeitschr. d. Ver. deutsch. Ing.* **69,** 705; 1925. 270, 274, 275, 278, 281

REFERENCE PAGE

GROEBER, H., *Forschungsarb. a. d. Geb. d. Ingenieurwes.* No. 300, p. 3; 1928. 293

GROEBER, H., and ERK, S., *Die Grundgesetze der Waermeuebertragung*, Julius Springer, Berlin; 1933. 252, 303

GRUENEISEN, E., *Annalen d. Physik* (4) **3,** 43; 1900. 316

GRUENEISEN, E., and GOENS, E., *Zeitschr. f. Physik* **44,** 615; 1927. 111, 116

GRUESS, H., Technische Gasanalyse durch Messung des Waermeleitvermoegens, in A. EUCKEN and M. JAKOB, *Der Chemie-Ingenieur*, Vol. 2, Part 4, pp. 75 ff., Leipzig; 1933. 70

GRUESS, H., and SCHMICK, H., *Veroeffentl. a. d. Siemens-Konzern* **7,** 202; 1928. 70

GURNEY, H. P., and LURIE, J., *Ind. Engg. Chem.* **15,** 1170; 1923. 270

HAAS, W. J. DE, and BIERMASZ, TH., *Physica* **2,** 673; 1935. 102

HAGENBUCH, W. H., Thesis, Massachusetts Institute of Technology; 1941 (according to McAdams [1942]). 690

HAGG, A. G., *Journ. Appl. Mech.* **11,** A–72; 1944. 205

HAGG, A. G., *Journ. Appl. Mech.* **12,** A–126; 1945. 205

HANSEMANN, G., *see under* KIRCHHOFF.

HANSEN, G., *Zeitschr. f. techn. Physik* **12,** 436; 1931. 577

HANSEN, M., *Zeitschr. f. angew. Math. u. Mech.* **8,** 185; 1928. 554

HARPER, D. R., and BROWN, P. W., *Nat. Advis. Comm. Aeronaut. Techn. Rep.* 158, p. 677; 1922. 216, 218

HAWKINS, G. A., SIBBITT, W. L., and SOLBERG, H. L., *Trans. Am. Soc. Mech. Engrs.* **70,** 19; 1948. 66

HAWKINS, G. A., SOLBERG, H. L., and POTTER, A. A., *Trans. Am. Soc. Mech. Engrs.* **62,** 677; 1940. 66

HAWKINS, G. A., SOLBERG, H. L., and SIBBITT, W. L., *Power Plant Engg.*; November 1941. 19

HAWKINS, G. A., *see also under* CHEN; *under* JAKOB; *under* SLEGEL; *under* TOULOUKIAN.

HEAVISIDE, O., *Proc. Roy. Soc. London* (A) **52,** 504; 1893. 352

HEGGE ZIJNEN, B. G. VAN DER, Thesis, Delft, 1924 (also *Mededeeling* No. 6, *Lab. voor Aerodynamica en Hydrodynamica*, Techn. Hoogeschool, Delft). 554, 558

HEGGE ZIJNEN, B. G. VAN DER, *see also under* BURGERS.

HEIDRICH, A., Dissertation, Techn. Hochschule, Aachen; 1931. 618–620

HEILMAN, R. H., *Trans. Am. Inst. Chem. Engrs.* **31,** 165; 1934. 122

HEISLER, M. P., *Trans. Am. Soc. Mech. Engrs.* **69,** 227; 1947. 270, 284–291, 414–420

HELD, E. F. M. VAN DER, *Gesundheits-Ingenieur* **59,** 736; 1936. 192

HENNING, FR., *Annalen d. Physik* (4) **29,** 454; 1909. 615

HERMANN, R., *Physik. Zeitschr.* **33,** 425; 1932. 522

HERMANN, R., *VDI-Forschungsheft* No. 379; Berlin, 1936. 450, 522, 577

HIGGINS, TH. J., *Trans. Am. Soc. Mech. Engrs.* **66,** 665; 1944 (reprinted in *Res. Publ. Illinois Inst. Technology* **6,** 136; 1948). 181, 192, 194

HIGGINS, TH. J., *Trans. Am. Inst. Electr. Engrs.* **64,** 190; 1945 (reprinted in *Res. Publ. Illinois Inst. Technology* **6,** 142; 1948). 196

HIGGINS, W. F., *see under* KAYE.

HILD, K., *Mitt. a. d. Kaiser-Wilhelm-Institut f. Eisenforschung* **14,** 59; 1932. 123

HILPERT, R., *VDI-Forschungsheft* No. 355; 1932. 591, 592, 597, 607

HILPERT, R., *Forschung a. d. Geb. d. Ingenieurwes.* **4,** 215; 1933. 559

REFERENCE PAGE

HOBSON, E. W., Waermeleitung, Mathematischer Teil, in *Encyklopaedie der mathematischen Wissenschaften*, Vol. 5, Part 1, Article 4, I (article finished 1904), B. G. Teubner, Leipzig and Berlin; 1903–1921. 252, 298, 332

HOFFMANN, E., *Zeitschr. f. d. ges. Kaelte-Ind.* **44**, 99; 1937. 503

HOGAN, J. J., *see under* DREW.

HOHENEMSER, K., and PRAGER, W., *Dynamik der Stabwerke. Eine Schwingungslehre fuer Bauingenieure*, Julius Springer, Berlin; 1933. 300

HOLBORN, L., *Zeitschr. f. Physik*, **8**, 58; 1921. 174

HOLDEN, P. P., Thesis, Massachusetts Institute of Technology; 1923. 544

HOLST, G., *see under* ONNES.

HOMANN, F., *see under* FRITZ.

HONDA, K., and SIMIDU, T., *Science Rep. Tôhoku University* (1), **6**, 219; 1917. 115, 154

HOPF, L., *Zaehe Fluessigkeiten*, in H. GEIGER and K. SCHEEL, *Handbuch der Physik*, Vol. VII, Chapter 2, Berlin; 1927. 565

HOUGEN, O. A., *see under* COLBURN.

HUGHES, I. A., *Phil. Mag.* **31**, 118; 1916. 606

HUMBURG, K., *Elektrotechn. u. Maschinenbau* **27**, 677 and 696; 1909. 176

INGENHOUSS, I., *Journ. de physique* **34**, 68 and 380; 1789. 164, 207

INGERSOLL, A. C., *see under* INGERSOLL, L. R.

INGERSOLL, L. R., ZOBEL, O. J., and INGERSOLL, A. C., *Heat conduction with engineering and geological applications*, McGraw-Hill Book Company, New York; 1948. 251, 707

INSINGER, TH. H., JR., and BLISS, H., *Trans. Am. Inst. Chem. Engrs.* **36**, 491; 1940. 645

IVERSEN, H. W., *see under* BOELTER.

JAEGER, J. C., and CLARKE, MARTHA, *Proc. Roy. Soc. Edinburgh* (A) **61**, 223; 1942. 265

JAEGER, J. C., *see also under* CARSLAW.

JAEGER, W., and DIESSELHORST, H., *Wissenschaftl. Abhandl. d. Physik.-Techn. Reichsanstalt* **3**, 269; 1900. 170, 172

JAKOB, M., *Archiv f. Elektrotechnik* **8**, 117; 1919. 175, 177

JAKOB, M., *Zeitschr. d. Ver. deutsch. Ing.* **63**, 69; 1919(*a*). 149, 154

JAKOB, M., *Annalen d. Physik* (4) **63**, 537; 1920. 155

JAKOB, M., *Zeitschr. d. Ver. deutsch. Ing.* **66**, 688; 1922. 161

JAKOB, M., *Zeitschr. f. d. ges. Kaelte-Ind.* **29**, 83; 1922(*a*). 399

JAKOB, M., *Zeitschr. d. Ver. deutsch. Ing.* **67**, 126; 1923. 103

JAKOB, M., *Zeitschr. f. Instrumentenkunde* **44**, 108; 1924. 161

JAKOB, M., *Zeitschr. f. d. ges. Kaelte-Ind.*, p. 21; 1926. 134

JAKOB, M., Waermeleitung, Chapter 2 of Vol. XI in H. GEIGER and K. SCHEEL, *Handbuch der Physik*, Berlin; 1926(*a*). 164–166

JAKOB, M., *Zeitschr. f. techn. Physik* **9**, 21; 1928. 19, 20

JAKOB, M., *Forschungsarb. a. d. Geb. d. Ingenieurwes.* No. 310, p. 9; 1928(*a*). 615

JAKOB, M., *Zeitschr. d. Ver. deutsch. Ing.* **76**, 1161; 1932. 625, 631, 658

JAKOB, M., Allgemeine Grundlagen der Waermeuebertragung, Chapter 5 of Vol. I, Part 1, in A. EUCKEN and M. JAKOB, *Der Chemie-Ingenieur*, Leipzig; 1933. 200, 269, 451, 457, 503

REFERENCE PAGE

JAKOB, M., Allgemeine hydrodynamische Grundlagen der Stroemung in Rohrleitungen, Chapter 1 of Vol. I, Part 1, in A. EUCKEN and M. JAKOB, *Der Chemie-Ingenieur*, Leipzig; 1933(*a*). 434

JAKOB, M., *Centralblatt f. d. Zuckerind.* **42**, 474; 1934. 688

JAKOB, M., *Mech. Engg.* **58**, 643; 1936. 616, 621, 625, 626, 633, 636, 653

JAKOB, M., *Mech. Engg.* **58**, 729; 1936(*a*). 658, 661, 683, 684, 694

JAKOB, M., *Proc. 5th Internat. Congr. Appl. Mech.*, p. 561; 1938 (reprinted in *Techn. Bull. Armour Inst. Technology*, Vol. 2, No. 1; July 1939). 645, 647, 651

JAKOB, M., *Phil. Mag.* (7) **28**, 571; 1939. 216

JAKOB, M., *Trans. Am. Inst. Chem. Engrs.* **35**, 563; 1939(*a*). 464, 467

JAKOB, M., *Trans. Am. Inst. Chem. Engrs.* **36**, 657, 659; 1940. 435

JAKOB, M., *Trans. Am. Soc. Mech. Engrs.* **36**, 533; 1940(*a*). 464

JAKOB, M., Local temperature differences as occurring in evaporation, condensation, and catalytic reaction, in *Temperature, its measurement and control in science and industry*, Reinhold Publishing Corporation, New York, p. 834; 1941 (reprinted in *Res. Publ. Illinois Inst. Technology* **2**, 159; 1942). 618, 621, 623, 627

JAKOB, M., *Trans. Am. Soc. Mech. Engrs.* **64**, 792; 1942. 682

JAKOB, M., *Trans. Am. Soc. Mech. Engrs.* **65**, 593; 1943 (reprinted in *Res. Publ. Illinois Inst. Technology* **6**, 123; 1948). 181, 184, 194, 196, 199

JAKOB, M., *Trans. Am. Soc. Mech. Engrs.* **65**, 613; 1943(*a*). 380

JAKOB, M., *Trans. Am. Soc. Mech. Engrs.* **67**, 629; 1945. 243

JAKOB, M., *Trans. Am. Inst. Electr. Engrs.* **64**, 493; 1945(*a*) (reprinted in *Res. Publ. Illinois Inst. Technology* **6**, 147; 1948). 192

JAKOB, M., *Trans. Am. Soc. Mech. Engrs.* **68**, 189; 1946. 535, 538, 539

JAKOB, M., *Engg. Bull. Purdue University*, Vol. 30, No. 3, p. 16 (Engineering Experiment Station, Res. Series No. 97); May 1946(*a*). 558

JAKOB, M., *Proc. Phys. Soc.* (London) **59**, 726; 1947. 55, 197, 199, 611

JAKOB, M., *Trans. Am. Soc. Mech. Engrs.* **70**, 25; 1948 (reprinted in *Res. Publ. Illinois Inst. Technology* **6**, 164; 1948). 192, 197, 199

JAKOB, M., and DOW, W. M., *Trans. Am. Soc. Mech. Engrs.* **68**, 123; 1946. 473, 481, 554, 565, 597, 611, 717

JAKOB, M., and ECK, H., *Forschung a. d. Geb. d. Ingenieurwes.* **3**, 121; 1932. 544

JAKOB, M., and ERK, S., *Zeitschr. f. Physik* **35**, 670; 1926. 214

JAKOB, M., and ERK, S., *Forschungsarb. a. d. Geb. d. Ingenieurwes.* No. 310, p. 1; 1928. 682

JAKOB, M., and ERK, S., *Zeitschr. f. techn. Physik* **10**, 623; 1929; and *Zeitschr. f. d. ges. Kaelte-Ind.* **36**, 229; 1929. 96, 161

JAKOB, M., ERK, S., and ECK, H., *Zeitschr. d. Ver. deutsch. Ing.* **73**, 1517; 1929. 682, 683

JAKOB, M., ERK, S., and ECK, H., *Techn. Mech. u. Thermodynamik* **1**, 46; 1930. 684

JAKOB, M., ERK, S., and ECK, H., *Forschung a. d. Geb. d. Ingenieurwes.* **3**, 161; 1932. 684

JAKOB, M., ERK, S., and ECK, H., *Physik. Zeitschr.* **36**, 73; 1935. 678, 684

JAKOB, M., and FRITZ, W., *Zeitschr. d. Ver. deutsch. Ing.* **73**, 629; 1929; and *Festschrift, A. Stodola zum Geburtstag*, Orell Fuessli, Zuerich and Leipzig, p. 289; 1929. 616

JAKOB, M., and FRITZ, W., *Techn. Mech. u. Thermodynamik* **1**, 173; 1930. 616

REFERENCE PAGE

JAKOB, M., and FRITZ, W., *Forschung a. d. Geb. d. Ingenieurwes.* **2,** 435; 1931. 528, 621, 624, 625, 636, 642

JAKOB, M., and FRITZ, W., *Physik. Zeitschr.* **36,** 651; 1935. 616

JAKOB, M., and HAWKINS, G. A., *Elements of heat transfer and insulation*, John Wiley and Sons, New York, 1942. 176, 488, 489

JAKOB, M., and LINKE, W., *Forschung a. d. Geb. d. Ingenieurwes.* **4,** 75; 1933. 528, 529, 638, 640, 642, 646

JAKOB, M., and LINKE, W., *Physik. Zeitschr.* **36,** 267; 1935. 638, 640, 642, 646, 653

JAKOB, M., and REES, K. A., *Trans. Am. Inst. Chem. Engrs.* **37,** 619; 1941. 464, 467, 552

JAKOB, M., *see also under* KNOBLAUCH; *under* TOULOUKIAN.

JEANS, J. H., *Phil. Mag.* **10,** 91; 1905. 38

JESCHKE, H., in *Technische Mechanik* [Ergaenzungsheft (supplement number) to Vol. 69 of *Zeitschr. d. Ver. deutsch. Ing.*], p. 24; 1925. 548

JOHNSON, H. A., *see under* BOELTER.

JOHNSON, W. C., *see under* WILHELM.

JONASSEN, F., *see under* BOELTER.

JUERGES, W., *see under* NUSSELT.

KABANOW, W., and FRUMKIN, A., *Zeitschr. f. physik. Chemie* (A) **165,** 316 and 433; 1933. 631

KAISER, F., *Zeitschr. d. bayer. Revisionsver.* **40,** 25, 37, 45; 1937. 645

KARDOS, A., *Zeitschr. f. d. ges. Kaelte-Ind.* **41,** No. 1/2; 1934. 78, 79

KARDOS, A., *Forschung a. d. Geb. d. Ingenieurwes.* **5,** 14; 1934(*a*). 82

KÁRMÁN, TH. VON, *Zeitschr. f. angew. Math. u. Mech.* **1,** 233; 1921. 481

KÁRMÁN, TH. VON, *Trans. Am. Soc. Mech. Engrs.* **61,** 705; 1939. 439, 504, 510

KAULAKIS, A. F., and SHERMAN, L. M., Thesis, Massachusetts Institute of Technology; 1938. 657

KAYE, G. W. C., and HIGGINS, W. F., *Proc. Roy. Soc. London* (A) **117,** 459; 1928. 163

KAYE, G. W. C., and HIGGINS, W. F., *Proc. Roy. Soc. London* (A) **122,** 633; 1929. 163

KAYE, G. W. C., *see also under* GRIFFITHS.

KEENAN, J. H., *see under* MCADAMS.

KELVIN, LORD, *see under* THOMSON.

KENNARD, R. B., *Phys. Rev.* **38,** 1921; 1931. 577

KENNARD, R. B., *Bur. Standards Journ. Res.* **8,** 787; 1932. 567, 577, 581–587

KENNARD, R. B., Temperature distribution and heat flux in and by interferometry, in *Temperature, its measurement and control in science and industry*, Reinhold Publishing Corporation, New York, p. 685; 1941. 586

KENNEDY, R. N., *see under* MCINTIRE.

KENNELY, A. E., WRIGHT, C. A., and BYLEVELT, J. S. VAN, *Trans. Am. Inst. Electr. Engrs.* **28,** 363; 1909. 560

KERN, D. Q., and OTHMER, D. F., *Trans. Am. Inst. Chem. Engrs.* **39,** 517; 1943. 546

KIMBALL, W. S., and KING, W. J., *Phil. Mag.* (7) **13,** 888; 1932. 450

KING, L. S., *see under* GREEN.

KING, L. V., *Phil. Trans. Roy. Soc. London* (A) **214,** 373; 1914. 560, 561

KING, R. W., *Phys. Rev.* (2) **6,** 437; 1915. 318

KING, W. J., *Mech. Engg.* **54,** 347; 1932. 526, 528, 532

REFERENCE	PAGE
KING, W. J., *Mech. Engg.* **54**, 410; 1932(*a*).	560
KING, W. J., *Refrigerating Engg.* **25**, 83; 1933.	640
KING, W. J., *see also under* KIMBALL.	
KINSBURY, A., *Mech. Engg.* **55**, 685; 1933.	202
KIRCHHOFF, G., *Monatsber. d. preuss. Akad. d. Wiss.*, p. 783; 1859.	24
KIRCHHOFF, G., *Poggendorffs Annalen* **109**, 275; 1860.	24, 25
KIRCHHOFF, G., *Vorlesungen ueber die Theorie der Waerme*; 1894.	192
KIRCHHOFF, G., and HANSEMANN, G., *Wiedemanns Annalen* **13**, 406; 1881.	316
KIRKBRIDE, C. G., *Trans. Am. Inst. Chem. Engrs.* **30**, 170; 1933–1934.	682
KIRKBRIDE, C. G., *Ind. Engg. Chem.* **26**, 425; 1934.	673, 682
KNOBLAUCH, O., and JAKOB, M., *Sitzungsber. d. math.-physik. Klassed. bayer. Akad. d. Wiss.* **35**, 441; 1906.	70
KNOBLAUCH, O., and JAKOB, M., *Forschungsarb. a. d. Geb. d. Ingenieurwes.*, Nos. 35 and 36, p. 109; 1906(*a*).	70
KNOBLAUCH, OSC., and REIHER, H., Waermeuebertragung, in W. WIEN and F. HARMS, *Handbuch d. Experimentalphysik*, Vol. 9, Part 1, p. 189, Leipzig; 1925.	562
KOCH, B., and FRITZ, W., *Waerme-u. Kaeltetechnik* **42**, 113 (No. 8/9); 1940.	117
KOCH, W., *Gesundh.-Ing. Beihefte*, Reihe 1, No. 22; 1927.	532
KOENIGSBERGER, J., *Annalen d. Physik* **12**, 342; 1903.	55
KOHLRAUSCH, FR., *Annalen d. Physik* (4) **1**, 200; 1872.	170
KOHLRAUSCH, FR., *Zeitschr. f. Instrumentenkunde* **18**, 139; 1898.	170
KRASE, N. W., *see under* SHEA.	
KRAUSSOLD, H., *Forschung a. d. Geb. d. Ingenieurwes.* **5**, 186; 1934.	542
KUNDT, A., and WARBURG, E., *Poggendorffs Annalen* **155**, 337 and 525; 1875.	89
KURLBAUM, E., *see under* LUMMER.	
LAMBERT, J. H., *Photometria*, Augsburg; 1760.	27
LANDOLT-BOERNSTEIN, *Physikalisch-Chemische Tabellen*, 5th Ed., Julius Springer, Berlin; 1923 (reprinted by Edwards Brothers, Ann Arbor; 1943).	111
LANDOLT-BOERNSTEIN, *Physikalisch-Chemische Tabellen*, 5th Ed., Ergaenzungsband IIIa, Julius Springer, Berlin; 1935 (reprinted by Edwards Brothers, Ann Arbor; 1943).	76
LANG, C., *Trans. Inst. Engrs. Shipbuilders Scot.* **32**, 279; 1888.	654
LANGBERG, CHR., *Annalen d. Physik u. Chemie* **66**, 1; 1845.	216
LANGEN, E., *Forschung a. d. Geb. d. Ingenieurwes.* **2**, 359; 1931.	691
LANGMUIR, I., *Phys. Rev.* **34**, 401; 1912.	504
LANGMUIR, I., *Phys. Rev.* **7**, 151; 1916.	172
LANGMUIR, I., *Journ. Am. Chem. Soc.* **40**, 1930; 1918.	694
LANGMUIR, I., ADAMS, E. Q., and MEIKLE, G. S., *Trans. Am. Electrochem. Soc.* **24**, 53; 1913.	399
LANSING, N. E., *see under* MARTINELLI.	
LARSON, R. F., *Ind. Engg. Chem.* **37**, 1004; 1945.	622, 645, 656
LATZKO, H., *Zeitschr. f. angew. Math. u. Mech.* **1**, 268; 1921.	476, 480
LEE, A., NELSON, W. O., CHERRY, V. H., and BOELTER, L. M. K., *Proc. 5th. Internat. Congress Appl. Mech.*, p. 571; 1938.	453, 455
LEES, CH. H., *Phil. Trans.* **191**, 399; 1898.	146
LEHMANN, TH., *Elektrotechn. Zeitschr.* **30**, 995 and 1019; 1909.	380

REFERENCE PAGE

LEIDENFROST, J. G., *De aquae communis nonnullis qualitatibus tractatus*, Duisburg; 1756 (also included in collected works: *Opuscula physicochemica et medica*, Vol. 3, pp. 1–175 in *Biblipolio Meyeriano*, Lengoviae 1797; available in British Museum); quotation taken from Drew and Mueller [1937]. 652

LÉVÊQUE, J., *Annales des mines* (12) **13**, 201, 305 and 381; 1928. 460

LEWIS, W. K., *Mech. Engg.* **44**, 445; 1922. 591

LINDEN, C. M., and MONTILLON, G. H., *Ind. Engg. Chem.* **22**, 708; 1930. 646, 651

LINKE, W., *see under* JAKOB.

LITTLE, J. B., *see under* AVRAMI.

LODGE, O. J., *Phil. Mag.* (5) **5**, 110; 1878. 164

LOHRISCH, W., *Forschungsarb. a. d. Geb. d. Ingenieurwes.* No. 322, p. 46; 1929. 605–610

LORENZ, H., in *Vortraege a. d. Geb. d. Aerodynamik u. verwandter Gebiete* (Aachen; 1929), edited by A. GILLES, L. HOPF, and TH. VON KÁRMÁN, Berlin, p. 183; 1930. 520

LORENZ, L., *Poggendorffs Annalen* **147**, 429; 1872. 112

LORENZ, L., *Wiedemanns Annalen* **13**, 581; 1881. 317, 443

LUMMER, O., and KURLBAUM, E., *Verhandl. d. phys. Ges. zu Berlin* **17**, 106; 1898. 46

LUMMER, O., and PRINGSHEIM, E., *Verhandl. d. deutsch. phys. Ges.* **1**, 215; 1899. 46

LURIE, J., *see under* GURNEY.

LYNCH, T. D., *see under* PILLING.

MACHE, H., *Sitzungsber. d. Kais. Akad. d. Wiss. Wien, math.-naturwiss. Klasse* **119**, 1417; 1910. 597

MACLEAN, J. D., *Heating, Piping, Air Conditioning* **13**, 380; 1941 (abstract in *Mech. Eng.* **63**, **734**; 1941). 93

MARCO, S. M., *see under* BROWN, A. J.

MARTINELLI, R. C., *Trans. Am. Soc. Mech. Engrs.* **69**, 947; 1947. 517

MARTINELLI, R. C., and BOELTER, L. M. K., *University of California Publications in Engineering* **5**, 23; 1942. 462

MARTINELLI, R. C., SOUTHWELL, C. J., ALVES, G., CRAIG, H. L., WEINBERG, E. B., LANSING, N. E., and BOELTER, L. M. K., *Trans. Am. Inst. Chem. Engrs.* **38**, 943; 1942. 546

MARTINELLI, R. C., *see also under* BOELTER.

MAXWELL, CL., *Phil. Mag.* (4) **19**, 20; 1860. 63

MCADAMS, W. H., *Publications from the Massachusetts Institute of Technology*, Vol. 67, No. 61, Publ. Ser. No. 871, p. 32 (Dept. Chem. Engg. Ser. No. 284); October 1931. 462

MCADAMS, W. H., *Trans. Am. Inst. Chem. Engrs.* **36**, 1; 1940. 681, 682

MCADAMS, W. H., *Heat transmission*, 2nd Ed., McGraw-Hill Book Company, New York; 1942. 503, 524, 528, 547, 564, 636, 651, 690

MCADAMS, W. H., *Engg. Bull. Purdue University*, Vol. 32, No. 2 (Engineering Experiment Station, Res. Series No. 104); March 1948. 517, 655

MCADAMS, W. H., ADDOMS, J. N., RINALDO, P. M., and DAY, R. S., *Chem. Engg. Progress* **44**, 639; 1948. 655, 656

MCADAMS, W. H., NICOLAI, LL. A., and KEENAN, J. H., *Nat. Advis. Comm. Aeronaut. Techn. Notes* 985, Washington; 1945. 565

REFERENCE PAGE

McAdams, W. H., Nicolai, Ll. A., and Keenan, J. H., *Trans. Am. Inst. Chem. Engrs.* **42**, 907; 1946. 565

McAdams, W. H., *see also under* Akin; *under* Cooper; *under* Drew; *under* Fitzpatrick; *under* Sauer.

McIntire, O. R., and Kennedy, R. N., *Chem. Engg. Progress* **44**, 727; 1948. 88

Meikle, G. S., *see under* Langmuir.

Meisenburg, S. J., *see under* Boarts.

Meissner, A., *Elektrotechn. u. Maschinenbau* **53**, No. 25; 1935. 100, 107

Meissner, W., *Annalen d. Physik* (4) **47**, 1001; 1915. 172

Meyer, O. E., *Kinetische Theorie der Gase*, 1st Ed., Breslau; 1877. 598, 600

Miller, C. O., *see under* Friedman.

Miller, K. W., and Wollaston, F. O., *Trans. Am. Inst. Electr. Engrs.* **52**, Part 1, p. 98; March 1933. 312

Monrad, C. C., and Pelton, J. F., *Trans. Am. Inst. Chem. Engrs.* **38**, 593; 1942. 552

Montillon, G. H., *see under* Linden.

Moore, A. D., *Fundamentals of electrical design*, McGraw-Hill Book Company, New York; 1927. 196

Morgan, F., *see under* Muskat.

Morris, F. H., and Whitman, W. G., *Ind. Engg. Chem.* **20**, 234; 1928. 514

Mosciki, I., and Broder, J., *Rocznicki Chemje* **6**, 319; 1926 (English translation on file at Engineering Research Laboratory, Experimental Station, E. I. duPont de Nemours & Co., Wilmington, Del.). 655

Moser, H., *see under* Fritz.

Mueller, A. C., *Trans. Am. Inst. Chem. Engrs.* **38**, 613; 1942. 533, 553, 559

Mueller, A. C., *see also under* Baker; *under* Drew.

Mull, W., and Reiher, H., *Gesundh.-Ing. Beihefte*, Reihe 1, No. 28; 1930. 534

Murphree, E. V., *Ind. Engg. Chem.* **24**, 726; 1932. 506, 510

Muskat, M., and Morgan, F., *Journ. Appl. Mech.* **10**, p. A–131; 1943. 202

Nagle, W. M., Bays, G. S., Blenderman, L. M., and Drew, Th. B., *Trans. Am. Inst. Chem. Engrs.* **31**, 593; 1935. 658, 695

Nagle, W. M., and Drew, Th. B., *Trans. Am. Inst. Chem. Engrs.* **30**, 217; 1933–1934. 658

Nagle, W. M., *see also under* Drew.

Navier, *Mém. de l'Académie* **6**, 389; presented 1823 (printed 1827). 19

Nelson, W. O., *see under* Lee.

Nessi, A., and Nissolle, L., *Méthodes graphiques pour l'étude des installations de chauffage et de réfrigération en régime discontinu*, Dunod, Paris; 1929. 380

Neumann, F., *Annales de chim. et physique* (3) **66**, 183; 1862. 313

Newton, I., *Phil. Trans. Roy. Soc. London* **22**, 824; 1701. 2

Nicolai, Ll. A., *see under* McAdams.

Nicoll, F. H., *see under* Alty.

Nikuradse, J., *Ingenieur-Archiv* **1**, 306; 1930. 442

Nikuradse, J., *VDI-Forschungsheft* No. 356; 1932. 438

Nikuradse, J., *VDI-Forschungsheft* No. 361; 1933. 435

Nissole, L., *see under* Nessi.

Niven, C., *Proc. Roy. Soc. London* (A) **76**, 34; 1905. 149

Norris, R. H., and Sims, M. W., *Trans. Am. Inst. Chem. Engrs.* **38**, 469; 1942. 550

REFERENCE PAGE

NORRIS, R. H., and STREID, D. D., *Trans. Am. Soc. Mech. Engrs.* **36**, 525; 1940. 462, 550

NORTHRUP, E. F., *Trans. Am. Electrochem. Soc.* **24**, 85; 1913. 408

NOTTAGE, H. B., and BOELTER, L. M. K., *Trans. Am. Soc. Heating, Ventilating Engrs.* **46**, 41; 1940. 564

NUKIYAMA, S., *Journ. Soc. Mech. Engrs. Japan* **37**, 367; (English abstr. on p. S–53); 1934. 655

NUKIYAMA, S., and YOSIKATA, H. J., *Journ. Soc. Mech. Engrs. Japan* **33**, No. 1; March 1930. 402

NUSSELT, W., *Forschungsarb. a. d. Geb. d. Ingenieurwes.* Nos. 63 and 64; 1909. 141, 146, 444

NUSSELT, W., *Zeitschr. d. Ver. deutsch. Ing.* **53**, 1750; 1909(*a*). 483

NUSSELT, W., *Zeitschr. d. Ver. deutsch. Ing.* **54**, 1154; 1910. 451, 460

NUSSELT, W., *Forschungsarb. a. d. Geb. d. Ingenieurwes.* No. 89, Berlin; 1910(*a*). 483, 490

NUSSELT, W., *Zeitschr. d. bayer. Revisionsver.* Nos. 13 and 14; 1913. 85

NUSSELT, W., *Zeitschr. d. Ver. deutsch. Ing.* **58**, 361; 1914. 499

NUSSELT, W., *Zeitschr. f. d. ges. Kaelte-Ind.* Nos. 1 and 2; 1915. 399

NUSSELT, W., *Gesundheits-Ing.* **38**, 477 and 490; 1915(*a*). 483, 496, 498, 522, 527

NUSSELT, W., *Zeitschr. d. Ver. deutsch. Ing.* **60**, 102; 1916. 588

NUSSELT, W., *Zeitschr. d. Ver. deutsch. Ing.* **60**, 541 and 569; 1916(*a*). 588, 659, 661, 670, 671, 675, 678, 691, 692

NUSSELT, W., *Forschungsarb. a. d. Geb. d. Ingenieurwes.* No. 264; 1923. 55

NUSSELT, W., *Zeitschr. d. Ver. deutsch. Ing.* **67**, 206; 1923(*a*). 462

NUSSELT, W., *Zeitschr. d. Ver. deutsch. Ing.* **71**, 85; 1927. 398

NUSSELT, W., *Forschungsarb. a. d. Geb. d. Ingenieurwes.* No. 300, p. 15; 1928. 689

NUSSELT, W., *Zeitschr. d. Ver. deutsch. Ing.* **73**, 1475; 1929. 523

NUSSELT, W., *Zeitschr. f. angew. Math. u. Mech.* **10**, 105; 1930. 588, 598

NUSSELT, W., *Forschung a. d. Geb. d. Ingenieurwes.* **2**, 309; 1931. 547

NUSSELT, W., and JUERGES, W., *Zeitschr. d. Ver. deutsch. Ing.* **72**, 597; 1928. 444, 528

OLIVER, E., Thesis, Massachusetts Institute of Technology; 1939. 651

OLLENDORFF, F., *see under* ROTHE.

ONNES, H. K., and HOLST, G., *Commun. Univ. Leiden* No. 142c; 1914. 113

OTHMER, D. F., and WHITE, R. E., *Trans. Am. Inst. Chem. Engrs.* **37**, 135; 1941. 681

OTHMER, D. F., *see also under* KERN.

PALMER, W., *see under* SMITH, C. S. A.

PANNELL, J. R., *Brit. Aeronautical Research Committee, Memor.* 243, London; 1916. 509

PAPAPETRU, A., *Physik. Zeitschr.* **35**, 527; 1934. 102

PASCHKI, *Journ. Russian Chem. Soc.*, p. 276; 1915. 80

PASCHKIS, V., *Heating, Piping, Air Conditioning* **14**, 133; 1942. 414

PASCHKIS, V., and BAKER, H. D., *Trans. Am. Soc. Mech. Engrs.* **64**, 105; 1942. 411, 414

PASCHKIS, V., and BEUKEN, CL. L., *Elektrotechn. u. Maschinenbau* **56**, 98; 1938. 411

PASCHKIS, V., *see also under* AVRAMI; *under* BEUKEM.

PATERSON, S., *Phil. Mag.* **32**, 384; 1941. 309

PÉCLET, E., *Traité de la Chaleur*, Vol. I, Paris; 1860. 146

REFERENCE PAGE

PEIERLS, R., *Annalen d. Physik* (5) **3**, 1055; 1929. 102

PELTON, J. F., *see under* MONRAD.

PERRY, CH. W., *see under* BONILLA.

PERRY, R. L., *Trans. Am. Soc. Mech. Engrs.* **66**, 447; 1944. 612

PERRY, R. L., and BERGGREN, W. P., *University of California Publications in Engineering* **5**, 59; 1944. 253, 265, 268, 387, 393

PETERSON, C. M. F., *see under* WILKES.

PETIT, *see under* DULONG.

PETRIE, J. M., *see under* SHERWOOD.

PIGFORD, R. L., *see under* BRINN.

PILLING, N. B., and LYNCH, T. D., *Trans. Am. Inst. Mining Met. Engrs.* **62**, 665; 1920. 654

PIRANI, M., *Journ. Scientific Instr.* **16**, 372; 1939. 122

PLANCK, M., *Verhandl. d. deutsch. physik. Ges.* **2**, 202 and 237; 1900. 38

PLANCK, M., *Annalen d. Physik* **4**, 553; 1901. 38

POENSGEN, R., *Forschungsarb. a. d. Geb. d. Ingenieurwes.* No. 130; 1912. 152

POHLHAUSEN, E., *Zeitschr. f. angew Math. u. Mech.* **1**, 115; 1921. 469, 474

POHLHAUSEN, E., *Zeitschr. f. angew. Math. u. Mech.* **1**, 252; 1921(*a*). 473

POHLHAUSEN, K., *see under* ROTHE.

POISSON, S. D., *Journ. de l'école polytechn.*, Vol. 13, cah. 20, p. 1; 1831. 19

POTTER, A. A., *see under* HAWKINS.

POWELL, R. W., *Phil. Mag.* (7) **27**, 677; 1939. 151

POWELL, R. W., *Proc. Brit. Assoc. Refrigeration*, Vol. 36; 1939–1940. 612

POWELL, R. W., *Phil. Mag.* (7) **29**, 274; 1940. 607

POWELL, R. W., *Trans. Institution Chem. Engrs.* (Brit.) **18**, 36; 1940(*a*). 607, 611, 612

POWELL, R. W., and GRIFFITHS, E., *Trans. Institution Chem. Engrs.* (Brit.) **13**, 175; 1935. 607, 612

POWELL, R. W., and GRIFFITHS, E., *Proc. Roy. Soc. London* (A) **163**, 189; 1937. 96, 117

PRAGER, W., *see under* HOHENEMSER.

PRANDTL, L., *Verhandl. d. 3. Internat. Mathematiker-Vereinigung in Heidelberg*, B. G. Teubner, Leipzig; 1905. 436

PRANDTL, L., *Physik. Zeitschr.* **11**, 1072; 1910. 436, 500, 501

PRANDTL, L., *Ergebnisse der Aerodynam. Versuchsanstalt zu Goettingen* No. I, p. 136; 1921. 438

PRANDTL, L., *Physik. Zeitschr.* **29**, 487; 1928. 503, 504

PRINGSHEIM, E., *see under* LUMMER.

PRUEGER, W., *Zeitschr. f. Physik* **115**, 202; 1940. 620

PRUEGER, W., *Forschung a. d. Geb. d. Ingenieurwes.* **12**, 258; 1941. 620

RAYLEIGH, LORD, *Phil. Mag.* (5) **49**, 539; 1900. 38

RAYLEIGH, LORD, *Phil. Mag.* (6) **22**, 381; 1911. 332

REED, CH. E., *see under* SHERWOOD.

REES, K. A., *see under* JAKOB.

REIHER, H., *Forschungsarb. a. d. Geb. d. Ingenieurwes.* No. 269; 1925. 561, 563, 606

REIHER, H., *see also under* KNOBLAUCH; *under* MULL.

REYNOLDS, O., *Proc. Lit. Phil. Soc. of Manchester* Vol. 14; 1875 (Oct. 6, 1874) (also *Papers on mechanical and physical subjects*, Vol. 1, Paper No. 14, p. 81, University Press, Cambridge, England; 1900). 500

REFERENCE PAGE

RHODES, F. H., and BRIDGES, C. H., *Trans. Am. Inst. Chem. Engrs.* **35, 73**; 1939. 656

RICHARDS, J. W., *Trans. Am. Electrochem. Soc.* **24,** 106; 1913. 410

RICHTER, R., *Elektrische Maschinen,* Vol. 1, Julius Springer, Berlin; 1924. 196

RINALDO, P. M., *see under* MCADAMS.

RINSUM, W. VAN, *Forschungsarb. a. d. Geb. d. Ingenieurwes.* No. 228; 1920. 149, 248

RODGER, J. W., *see under* THORPE.

ROGOWSKI, W., *Archiv f. Elektrotechnik* **7,** 41; 1918. 175

ROSENTHAL, D., *Trans. Am. Soc. Mech. Engrs.* **68,** 849; 1946. 343

ROSENTHAL, D., and CAMERON, R. H., *Trans. Am. Soc. Mech. Engrs.* **69,** 961; 1947. 352

ROTHE, R., OLLENDORFF, F., and POHLHAUSEN, K., *Theory of functions as applied to engineering problems* (translation by A. Herzenberg), Technology Press, Massachusetts Institute of Technology, Cambridge, 1933. 331, 335, 336, 342

ROUDEBUSH, B., *see under* CORCORAN.

ROUSE, H., *Fluid mechanics for hydraulic engineers,* 1st Ed., 2nd impr., McGraw-Hill Book Company, New York; 1938. 19

ROWLEY, F. B., *Heating, Piping, Air Conditioning* **5,** 313; 1933. 91

SAGE, B. H., *see under* CORCORAN.

SAUER, E. T., COOPER, H. B. H., AKIN, G. A., and MCADAMS, W. H., *Mech. Engg.* **60,** 669; 1938. 657

SAUNDERS, O. A., *Proc. Roy. Soc. London* (A) **157,** 278; 1936. 529

SAUNDERS, O. A., *see also under* FISHENDEN; *under* SCHMIDT, R. J.

SCHACK, A., *Zeitschr. f. techn. Physik* **5,** 267; 1924. 52

SCHAU, FR., *see under* BUECHE.

SCHLEIERMACHER, A., *Wiedemanns Annalen* **34,** 623; 1888. 150

SCHMICK, H., *Physik. Zeitschr.* **29,** 633; 1928. 70

SCHMICK, H., *see also under* GRUESS.

SCHMIDT, E., On the application of the calculus of finite differences to technical heating and cooling problems (title translated), in *Festschrift zum siebzigsten Geburtstag August Foeppls,* Julius Springer, Berlin, p. 179; 1924. 380

SCHMIDT, E., *Zeitschr. d. Ver. deutsch. Ing.* **70,** 885 and 947; 1926. 218

SCHMIDT, E., *Gesundh.-Ing. Beihefte,* Ser. 1, No. 20; 1927. 41, 44, 46, 52

SCHMIDT, E., *Gesundh.-Ing.* **52,** 525; 1929. 588, 591

SCHMIDT, E., *Forschung a. d. Geb. d. Ingenieurwes.* **3,** 181; 1932. 567, 574, 576, 577

SCHMIDT, E., *Forschung a. d. Geb. d. Ingenieurwes.* **5,** 1; 1934. 41, 44

SCHMIDT, E., and BECKMANN, W., *Techn. Mech. u. Thermodynamik* **1,** 341 and 391; 1930. 444, 445, 449, 528, 581

SCHMIDT, E., and ECKERT, E., *Forschung a. d. Geb. d. Ingenieurwes.* **6,** 175; 1935. 41, 44, 49, 52

SCHMIDT, E., SCHURIG, W., and SELLSCHOPP, W., *Techn. Mech. u. Thermodynamik* **1,** 53; 1930. 658, 695

SCHMIDT, E., and SELLSCHOPP, W., *Forschung a. d. Geb. d. Ingenieurwes.* **3,** 277; 1932. 151, 331

SCHMIDT, H., and FURTHMANN, E., *Mitt. a. d. Kaiser-Wilhelm-Institut f. Eisenforschung* **10,** 225; 1928. 49, 126

REFERENCE PAGE

SCHMIDT, R. J., and SAUNDERS, O. A., *Proc. Roy. Soc. London* (A) **165**, 216; 1938. 532

SCHOENBORN, E. M., *see under* CARPENTER.

SCHOFIELD, F. H., *Proc. Roy. Soc. London* (A) **107**, 206; 1925. 155

SCHOFIELD, F. H., *Phil. Mag.* (7) **10**, 480; 1930. 404

SCHOFIELD, F. H., *Phil. Mag.* (7) **12**, 329; 1931. 404

SCHOFIELD, F. H., *see also under* AWBERY.

SCHULZE, F. A., *Wiedemanns Annalen* **66**, 207; 1898. 316

SCHURIG, W., *see under* SCHMIDT.

SCORAH, R. L., *see under* FARBER.

SEIBERT, O., *Jahrbuch 1938 der deutsch. Luftfahrtforschung*, pp. II 245–II 256 (translated in *Nat. Advis. Comm. for Aeronautics, Techn. Memor.* 1044, Washington; 1943). 557

SELLSCHOPP, W., *Forschung a. d. Geb. d. Ingenieurwes.* **5**, 162; 1934. 79, 151

SELLSCHOPP, W., *see also under* SCHMIDT, E.

SÉNARMONT, H. DE, *Comptes rendus* **25**, 459 and 707; 1847. 164

SÉNARMONT, H. DE, *Annales de chim. et physique* **21**, 457; 1847(*a*). 164

SHANKLIN, G. B., and BULLER, F. H., *Gen. Electr. Review*, p. 523; September 1931. 312

SHEA, F. L., and KRASE, N. W., *Trans. Am. Inst. Chem. Engrs.* **36**, 463; 1940. 681, 696

SHERMAN, L. M., *see under* KAULAKIS.

SHERWOOD, TH. K., and PETRIE, J. M., *Ind. Engg. Chem.* **24**, 736; 1932. 651

SHERWOOD, TH. K., and REED, CH. E., *Applied mathematics in chemical engineering*, McGraw-Hill Book Company, New York; 1939. 241

SIBBITT, W. L., *see under* HAWKINS.

SIEDER, E. N., and TATE, G. E., *Ind. Engg. Chem.* **28**, 1429; 1936. 544, 547, 549

SIMIDU, T., *Science Rep. Tôhoku University* (1) **6**, 111; 1917. 114, 172

SIMIDU, T., *see also under* HONDA.

SIMS, M. W., *see under* NORRIS.

SLEGEL, L., and HAWKINS, G. A., *Engg. Bull. Purdue University*, Vol. 30, No. 3 (Engineering Experiment Station, Res. Series No. 97); May 1946. 557

SMITH, C. ST. A., and PALMER, E. W., *Metals Technology*, Vol. 2, No. 6 (*Techn. Publ.* 648); September 1935 116

SMITH, J. F. D., *Trans. Am. Soc. Mech. Engrs.* **58**, 719; 1936. 78, 83

SMITH, W. Q., *see under* DREW.

SMOLUCHOWSKI, M. v., *Wiener Berichte* (2a) **107**, 304; 1898. 89

SMOLUCHOWSKI, M. v., *Krakauer Anzeiger* (A), p. 129; 1910; and *Report about the 2nd International Congress on Heat*, Vol. 2, p. 166, Wien; 1910. 90

SOLBERG, H. L., *see under* CHEN; *under* HAWKINS.

SOUTHWELL, C. J., *see under* MARTINELLI.

SOUTHWELL, R. V., *see under* CHRISTOPHERSON.

SPOELSTRA, H. J., *Arch. Suikerind. in Nederl.-Indië*, Part 3, No. 23, p. 905; 1931. 658

STEFAN, J., *Sitzungsber. d. Kais. Akad. d. Wiss. Wien, math.-naturwiss. Klasse* **68**, 385; 1874. 598, 601

STEFAN, J., *Sitzungsber. d. Kais. Akad. d. Wiss. Wien, math.-naturwiss. Klasse* **79**, 391; 1879. 3

REFERENCE PAGE

Stokes, G. G., *Trans. Cambridge Phil. Soc.* **8,** 287; 1845 (reprinted in Stokes' *Mathematical and Physical Papers*, Vol. 1, p. 75, University Press, Cambridge, England; 1905). 18

Streid, D. D., *see under* Norris.

Sutherland, W., *Phil. Mag.* (5) **36,** 507; 1893. 64

Tanasawa, Y., *Trans. Soc. Mech. Engrs. Japan* **1,** 217; (English abstr. on p. S–68); 1935. 320

Tate, G. E., *see under* Sieder.

Taylor, C. S., and Edwards, J. D., *Heating, Piping, Air Conditioning* **11,** 59; 1939; also *Trans. Am. Soc. Heating, Ventilating Engrs.* **45,** 179; 1939 (Paper 1112). 127

Taylor, G. I., *Phil. Trans. Roy. Soc. London* (A) **215,** 1; 1915. 518

Taylor, G. I., *Brit. Advis. Comm. Aeronaut., Rep. and Mem.* 272; May 1916. 501, 503

Taylor, G. I., *Proc. Roy. Soc. London* (A) **135,** 685; 1932. 518

Taylor, J. P., *Trans. Am. Soc. Mech. Engrs.* **65,** 201; 1943. 200

Thoma, H., *Hochleistungskessel*, Julius Springer, Berlin; 1921. 588, 599, 605

Thomas, H. A., *Mech. Engg.* **65,** 110; 1943. 486

Thomson, Sir William (Lord Kelvin), Heat, in *Encyclopaedia Britannica*, 9th Ed., Vol. 11, § 82; 1880. 10, 332

Thomson, Sir William (Lord Kelvin), *Mathematical and Physical Papers*, Vol. II, Article 72, p. 41, University Press, Cambridge, England; 1884. 10, 332

Thorpe, T. E., and Rodger, J. W., *Phil. Trans. Roy. Soc. London* **185,** 397; 1894. 60

Thorpe, T. E., and Rodger, J. W., *Phil. Trans. Roy. Soc. London* **189,** 90; 1897. 60

Timrot, D. L., *Journ. Physics U.S.S.R.* **2,** 419; 1940. 66

Timrot, D. L., and Vargaftik, N. B., *Journ. Physics U.S.S.R.* **2,** 101; 1940. 70, 117

Touloukian, Y. S., Hawkins, G. A., and Jakob, M., *Trans. Am. Soc. Mech. Engrs.* **70,** 13; 1948. 530, 541

Trapp, A., *Zeitschr. d. Ver. deutsch. Ing.* **85,** 959; 1941. 681

Tribus, M., *see under* Boelter.

Tsao, U., *see under* Baker.

Ulsamer, J., *Forschung a. d. Geb. d. Ingenieurwes.* **3,** 94; 1932. 561, 563

Vargaftik, N. B., *see under* Timrot.

Vidmar, M., *Elecktrotechnik u. Maschinenbau* **36,** 65; 1918. 175

Vincent, A. D., *see under* Dunn.

Voigt, W., *Wiedemanns Annalen* **60,** 350; 1897. 165

Voigt, W., *Wiedemanns Annalen* **64,** 95; 1898. 165

Wallace, J. L., and Davison, A. W., *Ind. Engg. Chem.* **30,** 948; 1938. 690

Wang, Chi-Teh, *Journ. Appl. Mech.* **13,** p. A–77; 1946. 440

Wangaard, F. F., *Trans. Am. Soc. Mech. Engrs.* **65,** 127; 1943. 92

Warburg, E., *Ueber Waermeleitung und andere ausgleichende Vorgaenge*, Julius Springer, Berlin; 1924. 252

REFERENCE PAGE

WARBURG, E., *see also under* KUNDT.
WARK, J. W., *Journ. Phys. Chem.* **37,** 623; 1933. 630
WASSILJEWA, A., *Physik. Zeitschr.* **54,** 481; 1917. 70
WEEKS, J. R., *see under* DAVISSON.
WEINBERG, E. B., *see under* MARTINELLI.
WEISE, R., *Forschung a. d. Geb. d. Ingenieurwes.* **6,** 281; 1935. 450, 533
WEISE, W., *see under* ECKERT.
WETZLER, K., Dissertation, Techn. Hochschule, Darmstadt; 1927. 122
WHITE, C. M., *Proc. Roy. Soc. London* (A) **123,** 645; 1929. 548
WHITE, C. M., *see also under* DAVIES.
WHITE, J. B., Thesis, Massachusetts Institute of Technology, 1927. 544
WHITE, R. E., *see under* OTHMER.
WHITMAN, W. G., *see under* MORRIS.
WIEDEMANN, G., and FRANZ, R., *Annalen d. Physik u. Chemie* **89,** 530; 1853. 112
WIEGAND, J. H., *Trans. Am. Inst. Chem. Engrs.* **41,** 147; 1945. 554
WIEN, W., *Berichte d. preuss. Akad. d. Wiss.*, p. 55; 1893. 34
WIEN, W., *Annalen d. Physik* **58,** 662; 1896. 37
WILHELM, R. H., JOHNSON, W. C., and ACTON, F. S., *Ind. Engg. Chem.* (Industr. Ed.) **35,** 562; 1943. 199
WILHELM, R. H., JOHNSON, W. C., WYNKOOP, R., and COLLIER, D. W., *Chem. Engg. Progress* **44,** 105; 1948. 405
WILKES, G. B., and PETERSON, C. M. F., *Heating, Piping, Air Conditioning* **9,** 505; 1937. 539
WILLIAMSON, E. D., and ADAMS, L. H., *Phys. Rev.* **14,** 99; 1919. 264
WINDING, C. C., and CHENEY, A. J., JR., *Ind. Engg. Chem.* **40,** 1087; 1948. 613
WOHLENBERG, W. J., *see under* YOUNG, F. L.
WOLLASTON, F. O., *see under* MILLER, K. W.
WOOD, W. A., *Proc. Roy. Soc. London* (A) **163,** 199; 1937. 97
WRIGHT, C. A., *see under* KENNELY.
WURSTER, A., *see under* CARPENTER.
WYNKOOP, R., *see under* WILHELM.

YOSIKATA, H. J., *see under* NUKIYAMA.
YOUNG, F. L., and WOHLENBERG, W. J., *Trans. Am. Soc. Mech. Engrs.* **64,** 787; 1942. 682
YOUNG, G., *see under* BOELTER.

ZAHM, A. F., *Nat. Advis. Comm. Aeronaut. Rep.* 287, p. 185; 1928. 428, 429
ZOBEL, O. J., *see under* INGERSOLL, L. R.

APPENDIX E

Subject Index

NOTE: This subject index will serve only for locating special items. All major items are indicated and should be looked up in the very detailed lists of Contents, Tables, and Figures at the front of the book.

C

D

E

P

Q

R

S

T